Series in Contemporary Mathematics

Volume 3

Series in Contemporary Mathematics (SCM), featuring high-quality mathematical monographs, is to presents original and systematic findings from the fields of pure mathematics, applied mathematics and math-related interdisciplinary subjects. It has a history of over fifty years since the first title was published by Shanghai Scientific & Technical Publishers in 1963. Professor HUA Luogeng (Lo-Keng Hua) served as Editor-in-Chief of the first editorial board, while Professor SU Buqing acted as Honorary Editor-in-Chief and Professor GU Chaohao as Editor-in-Chief of the second editorial board since 1992. Now the third editorial board is established and Professor LI Tatsien assumes the position of Editor-in-Chief. The series has already published twenty-six monographs in Chinese, and among the authors are many distinguished Chinese mathematicians, including the following members of the Chinese Academy of Sciences: SU Buqing, GU Chaohao, LU Qikeng, ZHANG Gongqing, CHEN Hanfu, CHEN Xiru, YUAN Yaxiang, CHEN Shuxing etc. The monographs have systematically introduced a number of important research findings which not only play a vital role in China, but also exert huge influence all over the world. Eight of them have been translated into English and published abroad. The new editorial board will inherit and carry forward the former traditions and strengths of the series, and plan to further reform and innovation in terms of internalization so as to improve and ensure the quality of the series, extend its global influence, and strive to forge it into an internationally significant series of mathematical monographs.

More information about this series at http://www.springer.com/series/13634

Doina Cioranescu · Alain Damlamian
Georges Griso

The Periodic Unfolding Method

Theory and Applications to Partial Differential Problems

Springer

Doina Cioranescu
Laboratoire Jacques-Louis Lions
University Pierre et Marie Curie
Paris, France

Georges Griso
Laboratoire Jacques-Louis Lions
University Pierre et Marie Curie
Paris, France

Alain Damlamian
Laboratoire d'Analyse et Mathématiques
Appliquées
Université Paris-Est Créteil Val de Marne
Creteil, Cedex, France

ISSN 2364-009X ISSN 2364-0103 (electronic)
Series in Contemporary Mathematics
ISBN 978-981-13-3031-5 ISBN 978-981-13-3032-2 (eBook)
https://doi.org/10.1007/978-981-13-3032-2

Library of Congress Control Number: 2018962117

This Springer imprint is published by the registered company Springer Nature Singapore Pte Ltd.
The registered company address is: 152 Beach Road, #21-01/04 Gateway East, Singapore 189721, Singapore

To Ion, Jean-Michel and Joseph;
to my teachers and to my students

To Jeanne and our daughters

To Anne-Marie, our children and grandchildren

Preface

In the late 1960's and early 1970's, the theory of homogenization became in its own right a new part of the branch of mathematics concerning partial differential equations and their numerical approximations. It had its origin in several works from physics and mechanics, where equivalent (spatially) homogeneous macroscopic structures were obtained for microscopic heterogeneous media (see for example Bruggeman [40](1935), Hashin and Shtrikman [133](1963)). The interest of applied mathematicians and numerical analysts started in earnest in the late 1960's. Seminal works appear in De Giorgi and Spagnolo [96](1973), Babuska [19](1975), Bensoussan, Lions and Papanicolaou [21](1978), Sanchez-Palencia [179](1980), among others.

Various mathematical methods were introduced to understand the limit of PDE's with highly oscillating but bounded coefficients, from the formal asymptotic expansions ([21, 179]) to G-and Gamma-convergences ([96]) and the method of oscillating test functions (Murat and Tartar [164]), thus generating a great range of results in different settings and a wide bibliography (e.g., the books of Sanchez-Palencia [179](1980), Attouch [17](1984), Bakhvalov and Panasenko [20](1989), Dal Maso [82](1993), Jikov, Kozlov and Oleinik [137](1994), Braides and Defranceschi [34](1998), Cioranescu and Donato [70](1999), Carbone and De Arcangelis [45](2001)). In particular, various kinds of problems concerning both scalar-valued and vector-valued configurations were considered also for convex and non convex energies.

In 1987, the notion of "two-scale convergence" was originally introduced as an abstract theoretical tool by Nguetseng [167](published in 1989). It was developed and applied to periodic homogenization first by Allaire [4](1989), then by many others (see Lukkassen, Nguetseng and Wall [154](2002) and references therein). It was generalized to some multi-scale periodic problems by Ene and Saint Jean Paulin [106](1995), Allaire and Briane [6](1996), Lions, Lukkassen, Persson and Wall [150](2001) (see also [35, 153, 154]).

The method of rescaling had long been used successfully to understand the passage from a thin three dimensional problem to a two or one dimensional problem (e.g. dimensional reductions for the study of plate or beams; see Ciarlet-Destuynder [58](1979), Caillerie [43](1980)). In some way, this method is similar to the well-known "blow-up" method for the resolution of singularities in differential geometry as well as in algebraic geometry. Many thought that a similar method could be applied for periodic homogenization. But it required some more time.

Already in [14](1990) Arbogast, Douglas and Hornung used a "dilation" operator (similar to the unfolding operator) to study homogenization for a periodic medium with double porosity in a domain which is always a union of small cells (but they did not envisage a general method for homogenization from that point, nor did any other reader of their paper). Similar dilations were used again, independently and under different names in many papers: see Bourgeat, Luckhaus and Mikelic [33](1996) (where it is called "periodic modulation"), Lenczner et al [144–147](1997-2004) (specifically for discrete electrical networks), Allaire and C. Conca [7](1998), Allaire, Conca and Vanninathan [8](1999) (for Bloch decomposition), Casado-Díaz et al [47–50](2000-2003) (for reticulated structures). Here again, the authors did not realize there was an underlying general method.

Independently in the mid 1990's, Griso, in [116], used some dilation for the study of truss-like structures (with two small parameters) but he also introduced the difference of the solution with its natural Q_1-interpolate as a tool to study the limit problem (without using two-scale convergence). This led to the periodic unfolding method as a general tool to study periodic homogenization, presented first in [60], then developed in [59, 62, 63]). It can be seen as the proper extension of the blow-up method in the case of periodicity.(1)

This book is dedicated to a detailed presentation of the periodic unfolding method. It combines the dilation technique with ideas from finite element approximations. It replaces the two-scale convergence simply by the weak convergence of the unfolded sequence in an appropriate fixed space (for the periodic homogenization of quadratic energies, the appropriate spaces are Hilbert spaces), and by introducing the adjoint of the unfolding operator, it allows for very simple and general but precise corrector results.

It is a **method**, and as such, its application procedure depends significantly upon the problem at hand.

But its **philosophy** is straightforward:

• apply an appropriate unfolding operator, obtain uniform estimates for the unfolded functions and pass to the weak limit at least for a subsequence in a fixed space;

  • establish an unfolded limit problem;
  • extract an homogenized problem from the unfolded limit problem;(2)

(1) The French name for the method, "éclatement périodique", follows the "blow-up" terminology. But, due to the close proximity with the events of 9/11/2001, the name "periodic unfolding method", coined by Christopher Schwab, was adopted for the English version.

(2) If the microstructure carries too much information this last step is not always possible (for an example of such a situation see Section 8.2).

• obtain a direct corrector result (possibly making use of the adjoint of the unfolding operator).

In this book, we present the method in various settings and give several corresponding examples where it applies. This shows the efficiency and the resilience of the method. In most cases, the domains are not unions of ε-periodicity cells and this fact generates some serious complications which are localized in a vanishing layer near their boundary. These difficulties are completely handled in the method.

Part I presents unfolding in fixed domains with one or several small periodic scales. A detailed study of the error estimates is also given.

Part II presents unfolding in perforated domains with one or several small periodic scales. Here it has the definite advantage of setting the unfolded problems in a fixed domain.

The brief Part III is concerned with partial unfolding and rapidly oscillating boundaries. Here also the problem, when simple, is transferred to a fixed domain. A more general framework of varying domain is also introduced in this setting.

Part IV deals with unfolding in the case of small holes in the bulk or along hyperplanes. In these situations, it is known that "strange" terms can appear. Again, in some sense, the problem is transferred to a fixed domain.

Part V is dedicated to an example in linear elasticity where both unfolding and dimension reduction are combined.

Finally, Part VI gives a detailed presentation of sharp error estimates and correctors for the model problem.

Each part presents several detailed examples in a very accessible form.

The prerequisites for this book are basic undergraduate functional analysis, measure theory and integration, Sobolev spaces, and elliptic partial differential equations. Some notions of the theory of homogenization can be useful but not absolutely necessary (for this, we refer to any of the following books: [21, 34, 70, 78, 156, 158, 160, 168, 175, 186]).
This volume can be used as a textbook for a graduate course introducing homogenization theory. The more technical sections are indicated with a star ($*$) and can be omitted in a first reading.

Since it was first presented in [60], the method has been used in many research papers (too many to list here).(3)

It is the hope of the authors that this book will inspire many more extended and successful applications of the method.

Doina Cioranescu, Alain Damlamian, Georges Griso (July 2018)

(3) It is perhaps a measure of the success of the unfolding method that several (unaware) authors use it for the very *definition* of "two-scale convergence" ... !(see Proposition 1.19).

Contents

Part I

Unfolding in Fixed Domains

The first part of this book presents the basics of the periodic unfolding method and a sample of its possible applications in fixed domains. In [60], a general and elementary approach was presented for classical or multi-scale periodic homogenization (for the classical case, the single small scale is given by ε, a small parameter which will tend to 0). Originally restricted to the case of domains consisting of a union of ε-cells defined from a given "periodic cell" Y (a parallelotope), it was extended to general domains (see [86] and [62]).

The periodic unfolding method starts with the cell Y which is defined from the set of macroscopic periods attached to the considered problem.

In Chapter 1, the three operators intimately connected to the εY-tiling of the domain, are defined. The first is the unfolding operator $\mathcal{T}_\varepsilon$, similar to the dilation operator. The second is the local average operator $\mathcal{M}_\varepsilon$. The third is the adjoint of $\mathcal{T}_\varepsilon$ is introduced as the averaging operator $\mathcal{U}_\varepsilon$. A variant is the operator $\mathcal{Q}_\varepsilon$, which is the $\mathbf{Q}_1$-approximation operator from the theory of Finite Elements. The properties of each of these operators are presented, most importantly when applied to sequences in the spaces of type L^p and $W^{1,p}$.

Extending the results of Chapter 1, Chapter 2 considers various generalizations concerning possible choices for the basic period, unfolding in the case of partial information on derivatives (such as divergence, curl or even a single derivative), as well as unfolding for L^1 functions and measures. Some of its technical sections (indicated with a $*$) can be omitted in a first reading.

Chapter 3 is devoted to examples of the use of the unfolding method for some partial differential equations with highly oscillating coefficients. It presents the case of linear second order elliptic diffusion equations with various conditions on the boundary. A first result for correctors is given using the unfolding operators. An example of multiscale linear diffusion is also presented. The chapter ends with two examples of the homogenization of non-linear problems. [(4)]

[(4)] The last chapter of this book (Chapter 14) covers a detailed and quite technical proof for precise optimal error estimates in the homogenization of the standard second order Dirichlet problem.

Chapter 1

Unfolding operators in fixed domains

The periodic unfolding method starts with the cell Y which is defined from the set of macroscopic periods attached to the considered problem. Next, three operators intimately connected to the εY-tiling of the domain, are defined for measurable functions and functions in a Lebesgue space.

The first is the unfolding operator $\mathcal{T}_\varepsilon$, similar to the dilation operator (see Section 1.1). By definition, the operator $\mathcal{T}_\varepsilon$ associates to any function w in $L^p(\Omega)$, a function $\mathcal{T}_\varepsilon(w)$ in $L^p(\Omega \times Y)$ via Definition 1.2. An immediate property of $\mathcal{T}_\varepsilon$ is that it enables to transform any integral over Ω in an integral over $\Omega \times Y$. Indeed, by Proposition 1.8 below

$$\int_\Omega w(x)\, dx \sim \frac{1}{|Y|} \int_{\Omega \times Y} \mathcal{T}_\varepsilon(w)(x, y)\, dx\, dy, \quad \forall w \in L^1(\Omega),$$

equality holding when Ω is the exact union of ε-cells. The unfolded function $\mathcal{T}_\varepsilon(w)$ is constant with respect to the variable x on each of the ε-cells. Therefore, a useful intermediary to connect w with its unfolding is the local average $\mathcal{M}_\varepsilon(w)$ of w which is the piece-wise constant function of x alone taking on each ε-cell the average value of w on that cell. The operator $\mathcal{M}_\varepsilon$ is introduced in Section 1.2. In Section 1.3 the adjoint of $\mathcal{T}_\varepsilon$ is defined as the averaging operator $\mathcal{U}_\varepsilon$. Most importantly, the averaging operator is almost a left inverse of $\mathcal{T}_\varepsilon$ and will play an essential role to obtain corrector results in applications. At this point, Proposition 1.19 shows that the notion of "two-scale convergence" of a sequence is exactly equivalent to the weak convergence of the corresponding unfolded sequence in the appropriate Lebesgue space (the latter is not only simpler to check, but also gives more precise information on the limit).

The second ingredient of the periodic unfolding method concerns functions which are in a Sobolev space. It consists in separating the characteristic scales

D. Cioranescu et al., *The Periodic Unfolding Method*, Series in Contemporary Mathematics 3, https://doi.org/10.1007/978-981-13-3032-2_1

by decomposing every function w belonging to $W^{1,p}(\Omega)$ in two parts, one corresponding to an average behavior of w and the other collecting the oscillations of scale ε of w. In Section 1.4 this scale splitting is achieved by using the local average $\mathcal{M}_\varepsilon(w)$ and the local deviation $w - \mathcal{M}_\varepsilon(w)$. In Section 1.6, the original proof of this scale-splitting, inspired by the Finite Element Method, is presented. The confrontation of the two methods of Section 1.4 and 1.6 is interesting in itself (Theorem 1.41 and Proposition 1.78).

As an interesting variation, in Section 1.5, a theory of unfolding with parameters and iterated unfolding (unfolding for multiscales) is briefly presented.

1.1 The unfolding operator

In $\mathbb{R}^N$, let Ω be an open set and $\mathbf{G}$ an additive subgroup generated by a basis $\mathbf{B} \doteq (\mathbf{b}_1, \dots, \mathbf{b}_N)$ of $\mathbb{R}^N$ (the group of reference periods). Let Y be the open parallelotope associated with this basis. For example, $\mathbf{G} = \mathbb{Z}^N$ and $Y = (0,1)^N$, which is classical in standard homogenization.

By analogy with the notation in the one-dimensional case, for almost every (a.e.) $z \in \mathbb{R}^N$, $[z]_Y$ denotes the unique integer combination $\sum_{j=1}^N k_j \mathbf{b}_j$ in $\mathbf{G}$ such that $z - [z]_Y$ belongs to Y, and set

$$\{z\}_Y = z - [z]_Y \in Y. \tag{1.1}$$

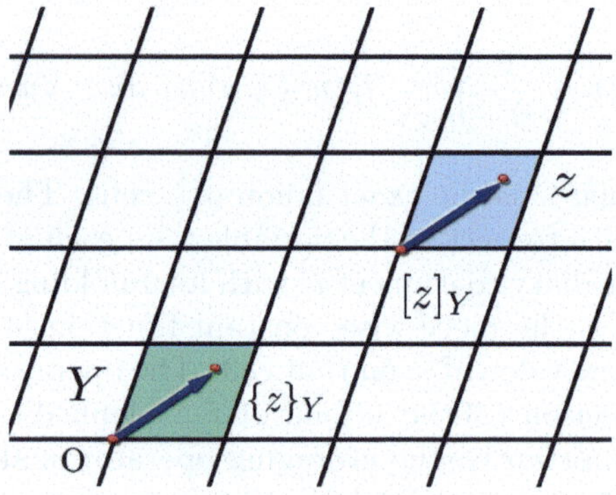

Figure 1.1: Definition of $[z]_Y$ and $\{z\}_Y$

Figure 1.1 gives a paving on $\mathbb{R}^N$. We scale this paving by ε, a strictly positive number in a sequence going to 0^+. Then for a.e. $x \in \mathbb{R}^N$ and $\varepsilon > 0$,

$$x = \varepsilon\Big(\Big[\frac{x}{\varepsilon}\Big]_Y + \Big\{\frac{x}{\varepsilon}\Big\}_Y\Big). \tag{1.2}$$

For Ω a domain in $\mathbb{R}^N$, consider a covering using the notations

$$\begin{aligned} &\Xi_\varepsilon = \{\xi \in \mathbf{G},\, \varepsilon(\xi + Y) \subset \Omega\}, \\ &\widehat{\Omega}_\varepsilon = \text{interior}\{\bigcup_{\xi \in \Xi_\varepsilon} \varepsilon(\xi + \overline{Y})\}, \qquad \Lambda_\varepsilon = \Omega \backslash \widehat{\Omega}_\varepsilon. \end{aligned} \tag{1.3}$$

The set $\widehat{\Omega}_\varepsilon$ is the largest union of $\varepsilon(\xi + \overline{Y})$ cells ($\xi \in \mathbf{G}$) included in Ω, while Λ_ε is the subset of Ω containing the parts from $\varepsilon(\xi + \overline{Y})$ cells intersecting the boundary $\partial\Omega$ (See Figure 1.2).

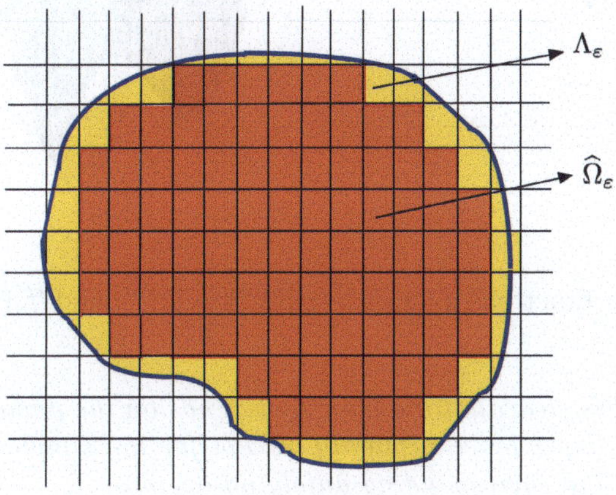

Figure 1.2: The domains $\widehat{\Omega}_\varepsilon$ and Λ_ε

Remark 1.1. *Note that, if $\Omega = \mathbb{R}^N$, the corresponding set Λ_ε is empty, and so it is if Ω is the union of ε-cells (i.e. a union of $\varepsilon(\xi + Y), \xi \in \mathbf{G}$ up to a Lebesgue null set). More generally,*

$$\Lambda_\varepsilon \subset \{x \in \Omega \mid \; dist(x, \partial\Omega) < \varepsilon \, diameter(Y)\}.$$

Hence its characteristic function $\mathbf{1}_{\Lambda_\varepsilon}$ converges to 0 a.e. in Ω. Consequently, if $\partial\Omega$ is bounded,

$$|\Lambda_\varepsilon| \doteq measure(\Lambda_\varepsilon) \to 0.$$

Definition 1.2 (The unfolding operator $\mathcal{T}_\varepsilon$). *For ϕ Lebesgue-measurable on $\widehat{\Omega}_\varepsilon$, the unfolding operator $\mathcal{T}_\varepsilon$ is defined as follows:*

$$\mathcal{T}_\varepsilon(\phi)(x, y) = \begin{cases} \phi\Big(\varepsilon\Big[\dfrac{x}{\varepsilon}\Big]_Y + \varepsilon y\Big) & \text{for a.e.} \quad (x, y) \in \widehat{\Omega}_\varepsilon \times Y, \\ 0 & \text{for a.e.} \quad (x, y) \in \Lambda_\varepsilon \times Y. \end{cases}$$

More precisely, $\mathcal{T}_\varepsilon$ acts on functions defined on sets larger than $\widehat{\Omega}_\varepsilon$ (such as Ω) by operating on their restriction to $\widehat{\Omega}_\varepsilon$.

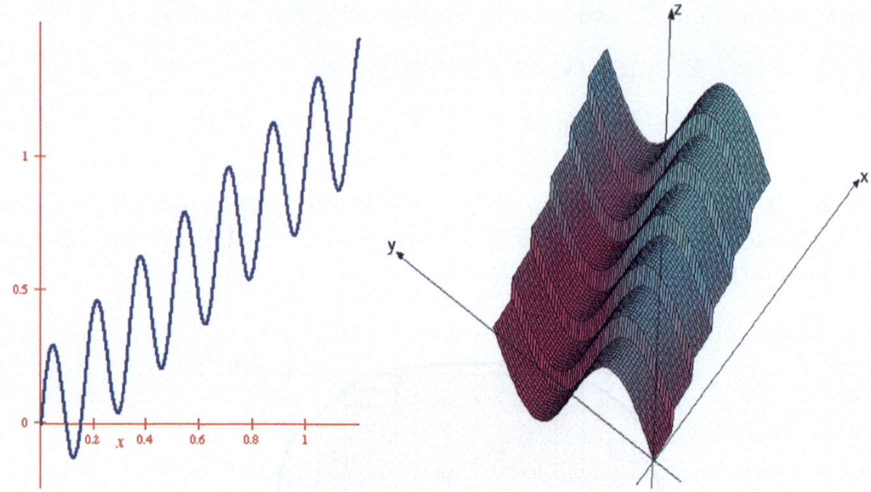

Figure 1.3: The functions $f_\varepsilon(x) = \dfrac{1}{4}\sin\left(2\pi\dfrac{x}{\varepsilon}\right) + x$ and $\mathcal{T}_\varepsilon(f_\varepsilon)$ for $\varepsilon = \dfrac{1}{6}$

Remark 1.3. *It is clear that the map $\mathcal{T}_\varepsilon$ depends on the domain Ω. We shall not indicate this dependence explicitly, except for the simplest case $\Omega \equiv \mathbb{R}^N$ where we denote the corresponding unfolding operator $\widetilde{\mathcal{T}}_\varepsilon$.*

Remark 1.4. *The introduction of the boundary layer Λ_ε is necessary to deal with the unavoidable complications arising from the fact that Ω is not a union of cells. Recently, some papers proposed a variant for the definition of $\mathcal{T}_\varepsilon$ by setting it to be the identity on Λ_ε, so that it is an isometry. This does not improve on any of the results but makes matters more intricate when dealing with derivatives (with their definition formula* (1.34) *is false on Λ_ε!).*

Observe that the function $\mathcal{T}_\varepsilon(\phi)$ is Lebesgue-measurable on $\Omega \times Y$ and vanishes for x outside of the set $\widehat{\Omega}_\varepsilon$.

As in classical periodic homogenization, two different scales appear in the definition of $\mathcal{T}_\varepsilon$: the "macroscopic" scale x gives the position of a point in the domain Ω, while the "microscopic" scale y $(= x/\varepsilon)$ gives the position of a point in the cell Y. The unfolding operator doubles the dimension of the space and put all the oscillations in the second variable, in this way separating the two scales (see Figures 1.3 and 1.4 as well as Figures 1.5 and 1.6 for other examples).

The following properties of $\mathcal{T}_\varepsilon$ are simple consequences of Definition 1.2 for v and w Lebesgue-measurable, and for every continuous function H on $\mathbb{R}$ (e.g. $r \mapsto |r|$ or $r \mapsto |r|^p$),

$$\begin{aligned} \mathcal{T}_\varepsilon(vw) &= \mathcal{T}_\varepsilon(v)\, \mathcal{T}_\varepsilon(w), \\ \mathcal{T}_\varepsilon(H(v)) &= H(\mathcal{T}_\varepsilon(v)). \end{aligned} \tag{1.4}$$

Another easy consequence of Definition 1.2, is the following result concerning highly oscillating functions:

Proposition 1.5. *For f measurable on Y, extended by Y−periodicity to the whole of $\mathbb{R}^N$, define the sequence $\{f_\varepsilon\}_\varepsilon$ by*

$$f_\varepsilon(x) = f\Big(\frac{x}{\varepsilon}\Big) \quad \textit{for a.e. } x \in \mathbb{R}^N. \tag{1.5}$$

Then

$$\mathcal{T}_\varepsilon(f_{\varepsilon|\Omega})(x,y) = \begin{cases} f(y) & \textit{for a.e. } \ (x,y) \in \widehat{\Omega}_\varepsilon \times Y, \\ 0 & \textit{for a.e. } \ (x,y) \in \Lambda_\varepsilon \times Y. \end{cases} \tag{1.6}$$

If f belongs to $L^p(Y)$, $p \in [1,+\infty[$, and if Ω is bounded,

$$\mathcal{T}_\varepsilon(f_{\varepsilon|\Omega}) \to f \quad \textit{strongly in } L^p(\Omega \times Y). \tag{1.7}$$

Proof. Formula (1.6) is straightforward from Definition 1.2. Convergence (1.7) follows from the a.e. convergence to 0 of $\mathbf{1}_{\Lambda_\varepsilon}$ (cf. Remark 1.1). □

Remark 1.6. *An equivalent way to define f_ε in (1.5) is to take simply*

$$f_\varepsilon(x) \doteq f\Big(\Big\{\frac{x}{\varepsilon}\Big\}_Y\Big).$$

For example, with

$$f(y) = \begin{cases} 1 & \textit{for } y \in \Big(0, \dfrac{1}{2}\Big), \\ 2 & \textit{for } y \in \Big(\dfrac{1}{2}, 1\Big), \end{cases}$$

the function f_ε is the highly oscillating periodic function with period ε from Figure 1.4.

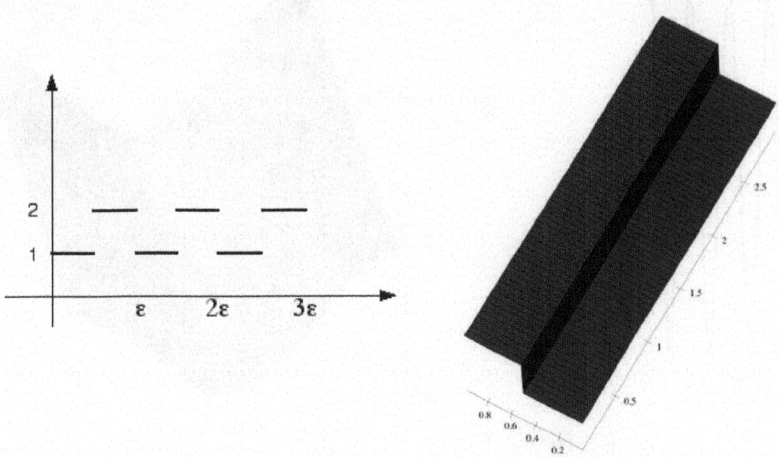

Figure 1.4: The functions f_ε and $\mathcal{T}_\varepsilon(f_\varepsilon)$

Remark 1.7. *Suppose $p \in (1, +\infty)$. Let f in $L^p(Y)$ and f_ε be defined by (1.5). It is well-known that $\{f_{\varepsilon|\Omega}\}$ converges weakly in $L^p(\Omega)$ to the mean value of f on Y, and not strongly, unless f is a constant (see Proposition 1.14 below for a proof of both statements).*

For a smooth function "with no oscillations", the unfolding exhibits small jumps in x and is smooth in y, see Figure 1.5 below.

Figure 1.5: The function $f(x) = \dfrac{1}{2}x(2-x)$ and $\mathcal{T}_\varepsilon(f)$ for ε small.

For a highly oscillating function, the unfolding can be smooth. In the case of f below, the unfolding is simply $\cos(2\pi y)$ (note that the ε is the same in the function and in the operator).

Figure 1.6: The functions $f(x) = \cos\dfrac{2\pi x}{\varepsilon}$ and $\mathcal{T}_\varepsilon(f)$ for ε small.

The next two results, essential in the study of the properties of the unfolding operator, are also straightforward from Definition 1.2.

Proposition 1.8. *Suppose $p \in [1, +\infty]$. The operator $\mathcal{T}_\varepsilon$ is linear and continuous from $L^p(\widehat{\Omega}_\varepsilon)$ to $L^p(\Omega \times Y)$. For every ϕ in $L^1(\widehat{\Omega}_\varepsilon)$, ψ in $L^1(\Omega)$ and w in $L^p(\Omega)$,*

$$(i) \quad \frac{1}{|Y|}\int_{\Omega\times Y} \mathcal{T}_\varepsilon(\phi)(x,y)\,dxdy = \int_{\widehat{\Omega}_\varepsilon} \phi(x)\,dx,$$

$$(ii) \quad \left|\int_\Omega \psi\,dx - \frac{1}{|Y|}\int_{\Omega\times Y} \mathcal{T}_\varepsilon(\psi)\,dxdy\right| \leqslant \int_{\Lambda_\varepsilon} |\psi|\,dx,$$

$$(iii) \quad \|\mathcal{T}_\varepsilon(w)\|_{L^p(\Omega\times Y)} \leqslant |\,Y\,|^{1/p}\,\|w\|_{L^p(\Omega)}.$$

Proof. Recalling Definition 2.2 of $\widehat{\Omega}_\varepsilon$, one has

$$\begin{aligned}\frac{1}{|Y|}\int_{\Omega\times Y} \mathcal{T}_\varepsilon(\phi)(x,y)\,dxdy &= \frac{1}{|Y|}\int_{\widehat{\Omega}_\varepsilon\times Y} \mathcal{T}_\varepsilon(\phi)(x,y)\,dxdy \\ &= \frac{1}{|Y|}\sum_{\xi\in\Xi_\varepsilon}\int_{(\varepsilon\xi+\varepsilon Y)\times Y} \mathcal{T}_\varepsilon(\phi)(x,y)\,dxdy.\end{aligned} \tag{1.8}$$

On each $(\varepsilon\xi+\varepsilon Y)\times Y$, by definition, $\mathcal{T}_\varepsilon(\phi)(x,y) = \phi(\varepsilon\xi+\varepsilon y)$ is constant in x. Hence each integral in the sum on the right-hand side successively equals

$$\begin{aligned}\int_{(\varepsilon\xi+\varepsilon Y)\times Y} \mathcal{T}_\varepsilon(\phi)(x,y)\,dxdy &= |\varepsilon\xi+\varepsilon Y|\int_Y \phi(\varepsilon\xi+\varepsilon y)\,dy \\ &= \varepsilon^N|Y|\int_Y \phi(\varepsilon\xi+\varepsilon y)\,dy \\ &= |Y|\int_{(\varepsilon\xi+\varepsilon Y)} \phi(x)\,dx.\end{aligned}$$

Summing over Ξ_ε, property (i) follows from (1.8) and from the equality

$$\sum_{\xi\in\Xi_\varepsilon}\int_{(\varepsilon\xi+\varepsilon Y)} \phi(x)\,dx = \int_{\widehat{\Omega}_\varepsilon} \phi(x)\,dx.$$

Inequality (ii) is immediate from (i). By (1.4) for $H(r) = |r|^p$, $r \in \mathbb{R}$, (i) implies

$$\|\mathcal{T}_\varepsilon(w)\|_{L^p(\Omega\times Y)} = |\,Y\,|^{1/p}\,\|w\|_{L^p(\widehat{\Omega}_\varepsilon)}, \tag{1.9}$$

which gives (iii) for p finite. The case $p = +\infty$ is straightforward. □

Property (ii) in Proposition 1.8 above shows that any integral of a function on Ω, is "almost equivalent" to the integral of its unfolded on $\Omega \times Y$, the "integration defect" arises only from the cells intersecting the boundary $\partial\Omega$ and is controlled by its integral over Λ_ε.

Property (iii) means that $\mathcal{T}_\varepsilon$ is bounded uniformly with respect to ε from $L^p(\Omega)$ to $L^p(\Omega \times Y)$. The connection of $\mathcal{T}_\varepsilon$ with strong convergences in L^p-spaces is presented in the next proposition, and that with weak convergences in Proposition 1.12.

Proposition 1.9. *Suppose $p \in [1, +\infty)$.*
(i) For $w \in L^p(\Omega)$,

$$\mathcal{T}_\varepsilon(w) \to w \quad \text{strongly in } \ L^p(\Omega \times Y).$$

(ii) Let $\{w_\varepsilon\}_\varepsilon$ be a sequence in $L^p(\Omega)$ such that

$$w_\varepsilon \to w \quad \text{strongly in } \ L^p(\Omega).$$

Then

$$\mathcal{T}_\varepsilon(w_\varepsilon) \to w \quad \text{strongly in } \ L^p(\Omega \times Y).$$

Proof. *(i)* For w in $C^0_c(\Omega)$ (the space of continuous function with compact support in Ω) one has

$$\|\mathcal{T}_\varepsilon(w) - w\|_{L^\infty(\widehat{\Omega}_\varepsilon \times Y)} \leqslant \varpi\big(\varepsilon\, diameter(Y)\big),$$

where ϖ is the modulus of uniform continuity of w. Because of the compact support of w, for ε small enough, $\mathcal{T}_\varepsilon(w)$ and w vanish in $\Lambda_\varepsilon \times Y$. This implies the convergence of $\mathcal{T}_\varepsilon(w)$ to w in all $L^p(\Omega \times Y)$ for every $p \in [1, +\infty)$.

If $w \in L^p(\Omega)$ with $p \in [1, +\infty)$, let $\phi \in C^0_c(\Omega)$. Then by using (iii) from Proposition 1.8,

$$\begin{aligned}\|\mathcal{T}_\varepsilon(w) - w\|_{L^p(\Omega \times Y)} &= \|\mathcal{T}_\varepsilon(w - \phi) + \big(\mathcal{T}_\varepsilon(\phi) - \phi\big) + (\phi - w)\|_{L^p(\Omega \times Y)}\\ &\leqslant 2|Y|^{1/p}\|w - \phi\|_{L^p(\Omega)} + \|\mathcal{T}_\varepsilon(\phi) - \phi\|_{L^p(\Omega \times Y)},\end{aligned}$$

hence,

$$\limsup_{\varepsilon \to 0} \|\mathcal{T}_\varepsilon(w) - w\|_{L^p(\Omega \times Y)} \leqslant 2|Y|^{1/p}\|w - \phi\|_{L^p(\Omega)},$$

from which statement (i) follows by density.
(ii) The following estimate is a consequence of Proposition 1.8(iii):

$$\|\mathcal{T}_\varepsilon(w_\varepsilon) - \mathcal{T}_\varepsilon(w)\|_{L^p(\Omega \times Y)} \leqslant |\,Y\,|^{\frac{1}{p}}\, \|w_\varepsilon - w\|_{L^p(\Omega)}, \quad \forall w \in L^p(\Omega).$$

This, together with (i), gives the result. □

To investigate the relationship of $\mathcal{T}_\varepsilon$ with weak convergences in the spaces L^p when ε goes to 0, requires to introduce the mean value operator $\mathcal{M}_Y$.

Definition 1.10 (The mean value operator $\mathcal{M}_Y$). *For $p \in [1, +\infty]$, the mean value operator*

$$\mathcal{M}_Y : L^p(\Omega \times Y) \longrightarrow L^p(\Omega),$$

is defined for Φ in $L^p(\Omega \times Y)$, as follows:

$$\mathcal{M}_Y(\Phi)(x) = \frac{1}{|Y|}\int_Y \Phi(x, y)\, dy \quad \text{for a.e.} \quad x \in \Omega. \tag{1.10}$$

A consequence of this definition is the estimate

Proposition 1.11. *Suppose* $p \in [1, +\infty]$. *For every* $\Phi \in L^p(\Omega \times Y)$,

$$\|\mathcal{M}_Y(\Phi)\|_{L^p(\Omega)} \leqslant |Y|^{-1/p} \|\Phi\|_{L^p(\Omega \times Y)}. \tag{1.11}$$

Proof. The case $p = +\infty$ is straightforward. For $p \in [1, +\infty)$, Hölder's inequality implies for a.e. $x \in \Omega$,

$$|\mathcal{M}_Y(\Phi)(x)|^p = |Y|^{-p} \Big| \int_Y \Phi(x, y)\, dy \Big|^p \leqslant |Y|^{-p} |Y|^{p/p'} \int_Y |\Phi(x, y)|^p\, dy,$$

where p' is the conjugate of p. Then inequality (1.11) is obtained by integration over Ω. □

Proposition 1.12. *Let* $p \in (1, +\infty)$. *Suppose that a sequence* $\{w_\varepsilon\}_\varepsilon$ *is bounded in* $L^p(\Omega)$, *then so is the sequence* $\{\mathcal{T}_\varepsilon(w_\varepsilon)\}_\varepsilon$ *in* $L^p(\Omega \times Y)$. *Furthermore the following implications hold.*
(i) If

$$\mathcal{T}_\varepsilon(w_\varepsilon) \rightharpoonup \overline{w} \quad \textit{weakly in } L^p(\Omega \times Y),$$

then

$$w_\varepsilon \rightharpoonup \mathcal{M}_Y(\overline{w}) \quad \textit{weakly in } L^p(\Omega).$$

(ii) If $\{w_\varepsilon\}_\varepsilon$ *converges to some* w *in* $L^p(\Omega)$, *then, up to a subsequence, there exists* $\widehat{w}$ *in* $L^p(\Omega \times Y)$ *with* $\mathcal{M}_Y(\widehat{w}) = 0$ *such that*

$$\mathcal{T}_\varepsilon(w_\varepsilon) \rightharpoonup w + \widehat{w} \quad \textit{weakly in } L^p(\Omega \times Y),$$

and

$$\|w + \widehat{w}\|_{L^p(\Omega \times Y)} \leqslant |Y|^{1/p} \liminf_{\varepsilon' \to 0} \|w_{\varepsilon'}\|_{L^p(\Omega)}. \tag{1.12}$$

The same results hold for $p = +\infty$ *with weak topology replaced by weak-∗ topology in the corresponding spaces*[(1)].

Remark 1.13. *Under the first hypotheses of Proposition 1.12, let* $\mathcal{W}$ *be the (non-empty) set of weak limit points of* $\{w_\varepsilon\}_\varepsilon$ *in* $L^p(\Omega)$ *and* $\widehat{\mathcal{W}}$ *be the (non-empty) set of weak limit points of* $\{\mathcal{T}_\varepsilon(w_\varepsilon)\}_\varepsilon$ *in* $L^p(\Omega \times Y)$. *Then Proposition 1.12 implies that*

$$\mathcal{W} \equiv \mathcal{M}_Y(\widehat{\mathcal{W}}) \doteq \{\mathcal{M}_Y(\psi) \mid \psi \in \widehat{\mathcal{W}}\}.$$

In particular, a necessary and sufficient condition for $\{w_\varepsilon\}_\varepsilon$ *to converge weakly in* $L^p(\Omega)$ *is that* $\mathcal{M}_Y(\widehat{\mathcal{W}})$ *has a single element. Then, that element is the weak limit of* $\{w_\varepsilon\}_\varepsilon$.

(1) See Proposition 2.33 for the case $p = 1$.

Proof of Proposition 1.12. (*i*) From (1.4) and Proposition 1.8(ii) applied to the function $w_\varepsilon \psi$ (with ψ arbitrary in $L^{p'}(\Omega)$), it follows that

$$\int_\Omega w_\varepsilon(x)\,\psi(x)\,dx - \frac{1}{|Y|}\int_{\Omega\times Y} \mathcal{T}_\varepsilon(w_\varepsilon)(x,y)\,\mathcal{T}_\varepsilon(\psi)(x,y)\,dxdy \to 0. \qquad (1.13)$$

Note that by Proposition 1.9(i), $\{\mathcal{T}_\varepsilon(\psi)\}_\varepsilon$ converges to ψ strongly in $L^{p'}(\Omega)$. Therefore, (1.13) implies

$$\int_\Omega w_\varepsilon(x)\,\psi(x)\,dx \to \frac{1}{|Y|}\int_{\Omega\times Y} \widetilde{w}(x,y)\,\psi)(x)\,dxdy,$$

from which (i) follows.
(*ii*) Let $\widehat{w}$ be a weak limit-point for the bounded sequence $\{\mathcal{T}_\varepsilon(w_\varepsilon) - w\}_\varepsilon$ in $L^p(\Omega \times Y)$, associated with the subsequence $\{\varepsilon'\}$. Then (i) applied to the sequence $\{\varepsilon'\}$ implies that $\mathcal{M}_Y(\widehat{w}) = 0$. Inequality (1.12) follows from Proposition 1.8(iii). □

Proposition 1.14. *Suppose* $p \in (1, +\infty)$. *Let* f *be in* $L^p(Y)$ *and* $\{f_\varepsilon\}_\varepsilon$ *be the sequence defined by* (1.5). *Let* Ω *be any bounded subset of* $\mathbb{R}^N$. *Then*

$$f_{\varepsilon|\Omega} \rightharpoonup \mathcal{M}_Y(f) \quad \textit{weakly in } L^p_{loc}(\Omega) \quad (L^p(\Omega) \textit{ when } \Omega \textit{ is bounded}). \qquad (1.14)$$

Furthermore, convergence (1.14) *is strong if and only if* f *is constant.*

Proof. Proposition 1.5 gives the strong (hence weak) convergence of the sequence $\{\mathcal{T}_\varepsilon(f_{\varepsilon|\Omega})\}_\varepsilon$ to f in $L^p(\Omega\times Y)$. Convergence (1.14) follows from Proposition 1.12(i).

Assume that $f_{\varepsilon|\Omega}$ converges strongly in $L^p(\Omega)$ to the constant function $\mathcal{M}_Y(f)$. Then Proposition 1.9(ii) implies $\mathcal{T}_\varepsilon(f_{\varepsilon|\Omega}) \to \mathcal{M}_Y(f)$ strongly in $L^p(\Omega \times Y)$. But by (1.7), $\mathcal{T}_\varepsilon(f_{\varepsilon|\Omega}) \to f$ strongly in $L^p(\Omega \times Y)$. Hence $f = \mathcal{M}_Y(f)$ a.e. in Y. □

Remark 1.15. *This proof of Proposition 1.14, using the unfolding operator, is strikingly simpler and shorter than the classical one (see for example [70], Theorem 2.6).*

Remark 1.16. *Note that if* $\{w_\varepsilon \mathbf{1}_{\Lambda_\varepsilon}\}_\varepsilon$ *is bounded in* $L^p(\Omega)$, *and if* $\{\mathcal{T}_\varepsilon(w_\varepsilon)\}_\varepsilon$ *converges weakly in* $L^p(\Omega \times Y)$, *so does* $\{w_\varepsilon\}_\varepsilon$ *by Proposition 1.12(iii) since the sequence* $\{w_\varepsilon\}_\varepsilon$ *is bounded in* $L^p(\Omega)$.

In general, if for every ε, Λ_ε *is not a null set, there is no relation between the behaviors of* $\{w_\varepsilon \mathbf{1}_{\Lambda_\varepsilon}\}_\varepsilon$ *and* $\{\mathcal{T}_\varepsilon(w_\varepsilon)\}_\varepsilon$. *Even if* $\{w_\varepsilon \mathbf{1}_{\Lambda_\varepsilon}\}_\varepsilon$ *converges strongly to 0 in* $L^p(\Omega)$ *and* $\{\mathcal{T}_\varepsilon(w_\varepsilon)\}_\varepsilon$ *also converges strongly, this does not imply the strong convergence of* $\{w_\varepsilon\}_\varepsilon$, *as it is shown by the sequence* $\{f_{\varepsilon|\Omega}\}_\varepsilon$ *in Proposition 1.14, unless* f *is a constant on* Y.

Recall an elementary lemma of integration theory.

Lemma 1.17. *Suppose* $p \in (1, +\infty)$. *Let* Ω *be an open subset of* $\mathbb{R}^N$ *and* $\{u_\varepsilon\}_\varepsilon$ *a weakly convergent sequence in* $L^p(\Omega)$ *with limit* u, *and* $\{\phi_\varepsilon\}_\varepsilon$ *a bounded sequence in* $L^\infty(\Omega)$ *with limit* ϕ *a.e. in* Ω(2). *Then*

$$u_\varepsilon\, \phi_\varepsilon \rightharpoonup u\, \phi \quad \textit{weakly in} \;\; L^p(\Omega).$$

Proof. Let w be an arbitrary element of $L^{p'}(\Omega)$. By the Lebesgue dominated convergence theorem, the sequence $\{\phi_\varepsilon w\}_\varepsilon$ converges to $\phi\, w$ strongly in $L^{p'}(\Omega)$. Therefore, by weak/strong convergence,

$$\int_\Omega u_\varepsilon\, \phi_\varepsilon w\, dx \to \int_\Omega u\, \phi\, w\, dx.$$

This implies that $u\,\phi$ is the weak limit of the sequence $\{u_\varepsilon \phi_e\}_\varepsilon$. □

A consequence of this result, with important applications in homogenization theory, is the following corollary which allows to pass to the limit in some products of (only) weakly convergent sequences under appropriate assumptions.

Corollary 1.18. *Suppose* $p \in (1, +\infty]$. *Let* $\{\phi_\varepsilon\}_\varepsilon$, $\{u_\varepsilon\}_\varepsilon$ *and* $\{v_\varepsilon\}_\varepsilon$ *be bounded sequences in* $L^\infty(\Omega)$, $L^p(\Omega)$ *and* $L^{p'}(\Omega)$, *respectively, such that* (3)

$$\begin{aligned} &\mathcal{T}_\varepsilon(\phi_\varepsilon) \to \Phi && \textit{a.e. in } \Omega \times Y, \\ &\mathcal{T}_\varepsilon(u_\varepsilon) \rightharpoonup U && \textit{weakly in } L^p(\Omega \times Y) \;\; (\textit{weakly-}* \textit{ if } p = +\infty)\,, \\ &\mathcal{T}_\varepsilon(v_\varepsilon) \to V && \textit{strongly in } L^{p'}(\Omega \times Y). \end{aligned}$$

Then

$$\int_{\widehat{\Omega}_\varepsilon} u_\varepsilon(x)\, v_\varepsilon(x)\, \phi_\varepsilon(x) dx \to \frac{1}{|Y|} \int_{\Omega \times Y} U(x,y)\, V(x,y)\, \Phi(x,y) dx dy. \tag{1.15}$$

Moreover, if

$$\int_{\Lambda_\varepsilon} |u_\varepsilon|^p\, dx \to 0 \quad \textit{or} \quad \int_{\Lambda_\varepsilon} |v_\varepsilon|^{p'}\, dx \to 0, \tag{1.16}$$

then

$$\int_\Omega u_\varepsilon(x)\, v_\varepsilon(x)\, \phi_\varepsilon(x) dx \to \frac{1}{|Y|} \int_{\Omega \times Y} U(x,y)\, V(x,y)\, \Phi(x,y) dx dy. \tag{1.17}$$

(2) In view of the relationship between convergence in local measure and convergence almost everywhere, this hypothesis can be replaced by the more general condition that $\{\phi_\varepsilon\}_\varepsilon$ converge locally in measure to ϕ.

(3) As in the previous lemma, the first condition below can be replaced by the more general condition of convergence in local measure in $\Omega \times Y$.

Proof. First observe that due to (1.4) $\mathcal{T}_\varepsilon(u_\varepsilon\phi_\varepsilon) = \mathcal{T}_\varepsilon(u_\varepsilon)\mathcal{T}_\varepsilon(\phi_\varepsilon)$, and by Lemma 1.17, the sequence $\{\mathcal{T}_\varepsilon(u_\varepsilon\phi_\varepsilon)\}_\varepsilon$ converges weakly to $U\,\Phi$ in $L^p(\Omega\times Y)$.

By Proposition 1.8,

$$\int_{\widehat{\Omega}_\varepsilon} u_\varepsilon(x)\, v_\varepsilon(x)\, \phi_\varepsilon(x) dx = \frac{1}{|Y|}\int_{\Omega\times Y} \mathcal{T}_\varepsilon(u_\varepsilon\phi_\varepsilon)(x,y)\, \mathcal{T}_\varepsilon(v_\varepsilon)(x,y) dx dy,$$

so that (1.15) follows by duality.

If (1.16) holds, by the Hölder inequality,

$$\int_{\Lambda_\varepsilon} u_\varepsilon(x)\, v_\varepsilon(x)\, \phi_\varepsilon(x)\, dx \to 0,$$

so that the conclusion (1.17) is a consequence of (1.15). □

Recall the definition of two-scale convergence from Nguetseng [167] and Allaire [5]:

Two-scale convergence. *Suppose $p \in (1, +\infty)$. A bounded sequence $\{w_\varepsilon\}_\varepsilon$ in $L^p(\Omega)$, two-scale converges to some w belonging to $L^p(\Omega \times Y)$ whenever, for every smooth function φ on $\Omega\times\mathbb{R}^N$ with bounded support in Ω with respect to the first variable and Y-periodic in the second,*

$$\int_\Omega w_\varepsilon(x)\varphi\Big(x, \frac{x}{\varepsilon}\Big)\, dx \to \frac{1}{|Y|}\int_{\Omega\times Y} w(x,y)\, \varphi(x,y)\, dx dy.$$

The next result reduces two-scale convergence of a sequence, to a mere weak $L^p(\Omega \times Y)$-convergence of the unfolded sequence.

Proposition 1.19. *Suppose $p \in (1, +\infty)$. Let $\{w_\varepsilon\}_\varepsilon$ be a bounded sequence in $L^p(\Omega)$. Then the following assertions are equivalent :*

(i) $\{\mathcal{T}_\varepsilon(w_\varepsilon)\}_\varepsilon$ *converges weakly to w in $L^p(\Omega \times Y)$,*

(ii) $\{w_\varepsilon\}_\varepsilon$ *two-scale converges to w.*

Proof. By Proposition 1.8(ii) and in view of the fact that, for every test-function admissible for two-scale convergence φ (for instance, φ uniformly continuous on $\Omega \times Y$), $\|\varphi(\cdot, \frac{\cdot}{\varepsilon})\mathbf{1}_{\Lambda_\varepsilon}\|_{L^{p'}(\Omega)}$ goes to 0 with ε, it is enough to check that $\big\{\mathcal{T}_\varepsilon\big(\varphi(\cdot, \frac{\cdot}{\varepsilon})\big)\big\}_\varepsilon$ converges strongly to φ in $L^{p'}(\Omega \times Y)$. From the definition of $\mathcal{T}_\varepsilon$, indeed,

$$\mathcal{T}_\varepsilon\Big(\varphi\Big(\cdot, \frac{\cdot}{\varepsilon}\Big)\Big)(x,y) = \varphi\Big(\varepsilon\Big[\frac{x}{\varepsilon}\Big]_Y + \varepsilon y, y\Big),$$

converges uniformly to φ. □

Remark 1.20. *Proposition 1.19 shows that the two-scale convergence of a sequence in $L^p(\Omega)$, is equivalent to the weak-$L^p(\Omega \times Y)$ convergence of the unfolded sequence. The fundamental (and not easy) result of two scale-convergence is that from a bounded sequence, one can extract a subsequence* *which two-scale converges. This now reduces to the weak compactness of bounded sets in $L^p(\Omega \times Y)$. Also, one difficulty in two-scale convergence is to properly define a set of test functions which is sufficiently large. The unfolding presentation does away with this difficulty.*

1.2 The local average operator

In this subsection, we consider the classical average operator associated to the partition of Ω by ε-cells Y (setting it to be zero on the cells intersecting the boundary $\partial\Omega$).

Definition 1.21 (Local average operator). *For $p \in [1, +\infty]$, the local average operator*

$$\mathcal{M}_\varepsilon : L^p(\Omega) \longrightarrow L^p(\Omega),$$

is defined by

$$\mathcal{M}_\varepsilon(\phi)(x) = \begin{cases} \dfrac{1}{|Y|}\displaystyle\int_Y \phi\Big(\varepsilon\Big[\frac{x}{\varepsilon}\Big]_Y + \varepsilon z\Big)dz & \text{if } x \in \widehat{\Omega}_\varepsilon, \\ 0 & \text{if } x \in \Lambda_\varepsilon. \end{cases}$$

This operator will be denoted $\widetilde{\mathcal{M}}_\varepsilon$ when Ω is the whole space $\mathbb{R}^N$.

The local average $\mathcal{M}_\varepsilon(\phi)$ of a function ϕ is piecewise constant on each ε-cell (see Figure 1.7).

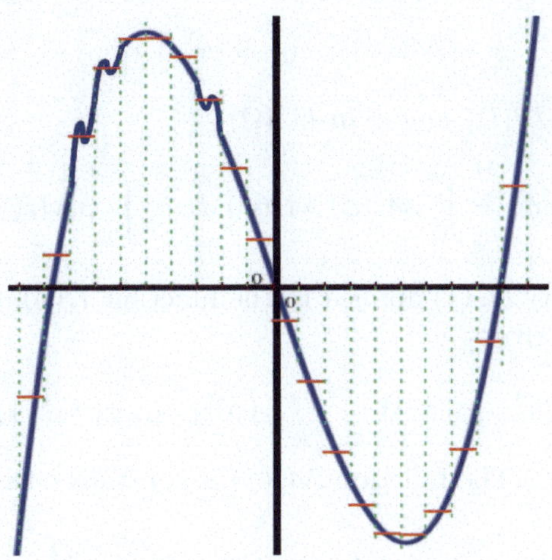

Figure 1.7: A function and its piece-wise constant local average

Remark 1.22. By convention the value of $\mathcal{M}_\varepsilon(\phi)$ on the cell $\varepsilon(\xi + Y)$ is simply denoted $\mathcal{M}_\varepsilon(\phi)(\varepsilon\xi)$.

Note that the local average $\mathcal{M}_\varepsilon$ is connected to the unfolding operator $\mathcal{T}_\varepsilon$ via the average over Y, namely

$$\mathcal{M}_\varepsilon(\phi) = \frac{1}{|Y|}\int_Y \mathcal{T}_\varepsilon(\phi)(\,\cdot\,, y)\,dy = \mathcal{M}_Y \circ \mathcal{T}_\varepsilon(\phi). \tag{1.18}$$

It is easily seen that

$$\mathcal{M}_\varepsilon \circ \mathcal{M}_\varepsilon = \mathcal{M}_\varepsilon \quad \text{and} \quad \mathcal{T}_\varepsilon \circ \mathcal{M}_\varepsilon = \mathcal{M}_\varepsilon. \tag{1.19}$$

Remark 1.23. *The local average operator is different but somewhat related to the well-known "Steklov smoothing" associated with Y defined as*

$$\mathfrak{S}_\varepsilon(f)(x) \doteq \frac{1}{|Y|}\int_Y f(x-\varepsilon y)\,dy \quad \textit{for } f \textit{ locally integrable,}$$

(see [193] where the Steklov smoothing is used in the context of homogenization).

One could conceive of an unfolding of the form $f(x+\varepsilon y)$, for which an integral equality similar to Proposition 1.8(i) does hold when Ω is the whole space. For general Ω, using this alternative definition would lead to some difficulties due to the presence of the boundary. In the case of perforated domains as in Part II, these difficulties seem to preclude its use.

Proposition 1.24. *Suppose $p \in [1,+\infty]$.*
(i) For any ϕ in $L^p(\Omega)$,

$$\|\mathcal{M}_\varepsilon(\phi)\|_{L^p(\Omega)} \leqslant \|\phi\|_{L^p(\Omega)}.$$

(ii) For any ϕ in $L^p(\Omega)$ and ψ in $L^{p'}(\Omega)$,

$$\int_\Omega \mathcal{M}_\varepsilon(\phi)\,\psi\,dx = \int_\Omega \mathcal{M}_\varepsilon(\phi)\,\mathcal{M}_\varepsilon(\psi)\,dx = \int_\Omega \phi\,\mathcal{M}_\varepsilon(\psi)\,dx. \tag{1.20}$$

Proof. (*i*) Inequality (1.11) applied to the function $\mathcal{T}_\varepsilon(\phi)$, together with (iii) of Proposition 1.8 gives

$$\|\mathcal{M}_\varepsilon(\phi)\|_{L^p(\Omega)} \leqslant |Y|^{-1/p}\|\mathcal{T}_\varepsilon(\phi)\|_{L^p(\Omega\times Y)} \leqslant \|\phi\|_{L^p(\Omega)}.$$

(*ii*) Let ξ be in Ξ_ε. The function $\mathcal{M}_\varepsilon(\phi)$ is constant on $\varepsilon(\xi+Y)$, hence

$$\begin{aligned}\int_{\varepsilon(\xi+Y)} \mathcal{M}_\varepsilon(\phi)_{|\varepsilon(\xi+Y)}\,\psi(x)\,dx &= \varepsilon^N \mathcal{M}_\varepsilon(\phi)_{|\varepsilon(\xi+Y)}\int_Y \psi(\varepsilon\xi+\varepsilon y)\,dy \\ &= \varepsilon^N |Y| \mathcal{M}_\varepsilon(\phi)_{|\varepsilon(\xi+Y)}\,\mathcal{M}_\varepsilon(\psi)_{|\varepsilon(\xi+Y)}.\end{aligned}$$

Summing over Ξ_ε gives the first equality. The second inequality follows by symmetry. ∎

Proposition 1.25.
(i) Suppose $p \in [1, +\infty)$*. Let* $\{v_\varepsilon\}_\varepsilon$ *be a sequence such that*

$$v_\varepsilon \to v \quad \textit{strongly in } L^p(\Omega).$$

Then

$$\mathcal{M}_\varepsilon(v_\varepsilon) \to v \quad \textit{strongly in } L^p(\Omega).$$

In particular, for every $\phi \in L^p(\Omega)$,

$$\mathcal{M}_\varepsilon(\phi) \to \phi \quad \textit{strongly in } L^p(\Omega). \tag{1.21}$$

(ii) Suppose $p \in (1, +\infty)$*. If* $\{v_\varepsilon\}_\varepsilon$ *is a bounded sequence in* $L^p(\Omega)$*, the following convergences are equivalent:*

$$\begin{aligned} &(a) \quad v_\varepsilon \rightharpoonup v \quad \textit{weakly in } L^p(\Omega), \\ &(b) \quad \mathcal{M}_\varepsilon(v_\varepsilon) \rightharpoonup v \quad \textit{weakly in } L^p(\Omega). \end{aligned}$$

The same holds true with weak-∗ convergences in $L^\infty(\Omega)$[4]*.*

Proof. (i) Note that

$$\mathcal{M}_Y(\mathcal{T}_\varepsilon(v_\varepsilon) - v\mathbf{1}_{\widehat{\Omega}_\varepsilon}) \equiv \mathcal{M}_\varepsilon(v) - v\mathbf{1}_{\widehat{\Omega}_\varepsilon}.$$

Then, applying inequality (1.11) to $\mathcal{T}_\varepsilon(v_\varepsilon) - v\mathbf{1}_{\widehat{\Omega}_\varepsilon}$ gives

$$\|\mathcal{M}_\varepsilon(v_\varepsilon) - v\mathbf{1}_{\widehat{\Omega}_\varepsilon}\|^p_{L^p(\widehat{\Omega}_\varepsilon)} \leqslant \frac{1}{|Y|} \|\mathcal{T}_\varepsilon(v_\varepsilon) - v\mathbf{1}_{\widehat{\Omega}_\varepsilon}\|^p_{L^p(\widehat{\Omega}_\varepsilon \times Y)}.$$

Adding to both sides of this inequality the term

$$\|v\mathbf{1}_{\Lambda_\varepsilon}\|^p_{L^p(\Omega)} = \frac{1}{|Y|} \|v\mathbf{1}_{\Lambda_\varepsilon}\|^p_{L^p(\Omega \times Y)},$$

one obtains

$$\|\mathcal{M}_\varepsilon(v_\varepsilon) - v\|^p_{L^p(\Omega)} \leqslant \frac{1}{|Y|} \|\mathcal{T}_\varepsilon(v_\varepsilon) - v\|^p_{L^p(\Omega \times Y)}.$$

To finish, one uses (ii) of Proposition 1.9.

(ii) $(a) \Rightarrow (b)$. Apply (1.20) for ϕ in $L^{p'}(\Omega)$,

$$\int_\Omega \mathcal{M}_\varepsilon(\phi)\, \mathcal{M}_\varepsilon(v_\varepsilon)\, dx = \int_\Omega \phi\, \mathcal{M}_\varepsilon(v_\varepsilon)\, dx = \int_\Omega \mathcal{M}_\varepsilon(\phi)\, v_\varepsilon\, dx.$$

One concludes with (i).

$(b) \Rightarrow (a)$. The previous result implies that $v \doteq \text{weak} \lim_{\varepsilon \to 0} \mathcal{M}_\varepsilon(v_\varepsilon)$ is the only weak limit point for the sequence $\{v_\varepsilon\}_\varepsilon$.

The proof for $p = +\infty$ uses the fact that a bounded sequence in $L^\infty(\Omega)$ which converges weakly in every $L^p_{loc}(\Omega)$ for p finite, converges weakly-∗ in $L^\infty(\Omega)$. □

[4] See Proposition 2.34 for the case $p = 1$.

1.3 The averaging operator

In this section, we introduce the averaging operator $\mathcal{U}_\varepsilon$ which is the adjoint of $\mathcal{T}_\varepsilon$ (up the normalizing constant $1/|Y|$). To do so, for p in $[1,+\infty]$, let v be in $L^p(\Omega\times Y)$ and u be in $L^{p'}(\Omega)$. Successively,

$$\begin{aligned}
&\frac{1}{|Y|}\int_{\Omega\times Y}\mathcal{T}_\varepsilon(u)(x,y)\,v(x,y)\,dxdy\\
&\qquad=\frac{1}{|Y|}\int_{\widehat{\Omega}_\varepsilon\times Y}\mathcal{T}_\varepsilon(u)(x,y)\,v(x,y)\,dxdy\\
&\qquad=\frac{1}{|Y|}\sum_{\xi\in\Xi_\varepsilon}\int_{\varepsilon(\xi+Y)\times Y}u(\varepsilon\xi+\varepsilon y)\,v(x,y)\,dxdy\\
&\qquad=\frac{1}{|Y|}\sum_{\xi\in\Xi_\varepsilon}\int_{Y\times Y}u(\varepsilon\xi+\varepsilon y)\,v(\varepsilon\xi+\varepsilon z,y)\varepsilon^N\,dzdy\\
&\qquad=\frac{1}{|Y|}\sum_{\xi\in\Xi_\varepsilon}\int_Y dz\int_{\varepsilon(\xi+Y)}u(x)\,v\Big(\varepsilon\Big[\frac{x}{\varepsilon}\Big]_Y+\varepsilon z,\Big\{\frac{x}{\varepsilon}\Big\}_Y\Big)\,dx\\
&\qquad=\int_{\widehat{\Omega}_\varepsilon}u(x)\Big(\frac{1}{|Y|}\int_Y v\Big(\varepsilon\Big[\frac{x}{\varepsilon}\Big]_Y+\varepsilon z,\Big\{\frac{x}{\varepsilon}\Big\}_Y\Big)\,dz\Big)dx.
\end{aligned}\tag{1.22}$$

This suggests the following definition:

Definition 1.26. *For $p\in[1,+\infty]$, the averaging operator*

$$\mathcal{U}_\varepsilon : L^p(\Omega\times Y)\to L^p(\Omega),$$

is defined as follows:

$$\mathcal{U}_\varepsilon(\Phi)(x)=\begin{cases}\dfrac{1}{|Y|}\displaystyle\int_Y\Phi\Big(\varepsilon\Big[\frac{x}{\varepsilon}\Big]_Y+\varepsilon z,\Big\{\frac{x}{\varepsilon}\Big\}_Y\Big)\,dz & \text{for a.e. } x\in\widehat{\Omega}_\varepsilon,\\ 0 & \text{for a.e. } x\in\Lambda_\varepsilon.\end{cases}$$

As for the unfolding operator and the local average operator, the averaging operator corresponding to the case $\Omega\equiv\mathbb{R}^N$ is denoted $\widetilde{\mathcal{U}}_\varepsilon$.

Observe that if Φ does not depend upon y (so it depends only on x), then

$$\mathcal{U}_\varepsilon(\Phi)=\mathcal{M}_\varepsilon(\Phi),\quad\forall\Phi\in L^p(\Omega),\tag{1.23}$$

and therefore, recalling Proposition 1.25,

$$\mathcal{U}_\varepsilon(\Phi)\to\Phi\quad\text{strongly in } L^p(\Omega).\tag{1.24}$$

The computation (1.22), shows that for ψ in $L^p(\Omega)$ and Φ in $L^{p'}(\Omega\times Y)$,

$$\int_\Omega\mathcal{U}_\varepsilon(\Phi)(x)\,\psi(x)\,dx=\frac{1}{|Y|}\int_{\Omega\times Y}\Phi(x,y)\,\mathcal{T}_\varepsilon(\psi)(x,y)\,dxdy.\tag{1.25}$$

From Definition 1.26 and the decomposition (1.2), it follows that $\mathcal{U}_\varepsilon$ is almost a left-inverse of $\mathcal{T}_\varepsilon$ since,

$$\mathcal{U}_\varepsilon \circ \mathcal{T}_\varepsilon(\phi) = \phi\, \mathbf{1}_{\widehat{\Omega}_\varepsilon}, \quad \forall \Phi \in L^p(\Omega \times Y). \tag{1.26}$$

Also, for every Φ in $L^p(\Omega \times Y)$,

$$\mathcal{T}_\varepsilon \circ \mathcal{U}_\varepsilon(\Phi)(x,y) = \begin{cases} \dfrac{1}{|Y|}\displaystyle\int_Y \Phi\Big(\varepsilon\Big[\frac{x}{\varepsilon}\Big]_Y + \varepsilon z, y\Big)\, dz & \\ & \text{for a.e. } (x,y) \in \widehat{\Omega}_\varepsilon \times Y, \\ 0 & \text{for a.e. } (x,y) \in \Lambda_\varepsilon \times Y. \end{cases} \tag{1.27}$$

As consequence of Hölder's inequality and of Proposition 1.8(iii), the following result is straightforward:

Proposition 1.27. *Suppose $p \in [1, +\infty]$. The averaging operator is linear and continuous from $L^p(\Omega \times Y)$ to $L^p(\Omega)$ and*

$$\|\mathcal{U}_\varepsilon(\Phi)\|_{L^p(\Omega)} \leqslant |Y|^{-1/p}\, \|\Phi\|_{L^p(\Omega \times Y)}. \tag{1.28}$$

The operator $\mathcal{U}_\varepsilon$ maps the space $L^p(\Omega \times Y)$ into $L^p(\Omega)$. It allows to replace the function $x \mapsto \Phi(\frac{\cdot}{\varepsilon})$ which is meaningless in general, by a function which always makes sense.

Remark 1.28. *This and (1.25) imply that the largest set of test sequences adapted to two-scale convergence is actually the set $\{\mathcal{U}_\varepsilon(\Phi) \mid \Phi \in L^{p'}(\Omega \times Y)\}$.*

Proposition 1.29. *Suppose $p \in [1, +\infty)$.*
(i) Let $\{\Phi_\varepsilon\}_\varepsilon$ be a sequence which converges strongly in $L^p(\Omega \times Y)$ to Φ. Then

$$\mathcal{T}_\varepsilon \circ \mathcal{U}_\varepsilon(\Phi_\varepsilon) \to \Phi \quad \text{strongly in } L^p(\Omega \times Y),$$
$$\mathcal{M}_\varepsilon \circ \mathcal{U}_\varepsilon(\Phi_\varepsilon) \to \mathcal{M}_Y(\Phi) \quad \text{strongly in } L^p(\Omega \times Y).$$

In particular, for every Ψ in $L^p(\Omega \times Y)$,

$$\mathcal{T}_\varepsilon \circ \mathcal{U}_\varepsilon(\Psi) \to \Psi \quad \text{strongly in } L^p(\Omega \times Y).$$

(ii) Let $\{w_\varepsilon\}_\varepsilon$ be a sequence in $L^p(\Omega)$ and W be in $L^p(\Omega \times Y)$. Then the following assertions are equivalent:

$$(a) \quad \mathcal{T}_\varepsilon(w_\varepsilon) \to W \quad \text{strongly in } L^p(\Omega \times Y) \quad \text{and} \quad \int_{\Lambda_\varepsilon} |w_\varepsilon|^p\, dx \to 0,$$

$$(b) \quad w_\varepsilon - \mathcal{U}_\varepsilon(W) \to 0 \quad \text{strongly in } L^p(\Omega).$$

(iii) In particular, if in (ii) W is **independent** *of y, then*

$$(c) \quad \mathcal{T}_\varepsilon(w_\varepsilon) \to W \quad \text{strongly in } L^p(\Omega \times Y) \quad \text{and} \quad \int_{\Lambda_\varepsilon} |w_\varepsilon|^p\, dx \to 0,$$

is equivalent to

$$(d) \quad w_\varepsilon \to W \quad \text{strongly in } L^p(\Omega).$$

Proof. (i) The proof of the first assertion follows exactly the same lines as that of (ii) of Proposition 1.9. The second follows from the first by the strong continuity of $\mathcal{M}_Y$.
(ii) Suppose (a) holds. Inequality (1.28) applied to $\Phi_\varepsilon = \mathcal{T}_\varepsilon(w_\varepsilon) - \widehat{w}$, together with (1.26), implies

$$w_\varepsilon \mathbf{1}_{\widehat{\Omega}_\varepsilon} - \mathcal{U}_\varepsilon(\widehat{w}) \to 0 \quad \text{strongly in } L^p(\Omega).$$

Then (b) holds because $\displaystyle\int_{\Lambda_\varepsilon} |w_\varepsilon|^p \, dx \to 0.$

Conversely, suppose that (b) is true. Then, by Definition 1.26 of $\mathcal{U}_\varepsilon$,

$$\int_{\Lambda_\varepsilon} |w_\varepsilon|^p \, dx \to 0.$$

Proposition 1.9(ii) applied to $w_\varepsilon - \mathcal{U}_\varepsilon(\widehat{w})$ gives

$$\mathcal{T}_\varepsilon(w_\varepsilon - \mathcal{U}_\varepsilon(\widehat{w})) \to 0 \quad \text{strongly in } L^p(\Omega \times Y).$$

One concludes since, by assertion (i), $\mathcal{T}_\varepsilon \circ \mathcal{U}_\varepsilon(\widehat{w})$ converges strongly to $\widehat{w}$ in $L^p(\Omega \times Y)$.
(iii) If (c) holds, by (ii) $w_\varepsilon - \mathcal{U}_\varepsilon(W) \to 0$ strongly in $L^p(\Omega)$. Since W is independent of y, by (1.23) $\mathcal{U}_\varepsilon(W) = \mathcal{M}_\varepsilon(W)$ which converges strongly to W by (1.21), hence (d). The converse is just Proposition 1.9(ii). □

Remark 1.30. *1. Assertion (i) implies that for every $\Phi \in L^p(\Omega \times Y)$,*

$$\mathcal{U}_\varepsilon(\Phi) \rightharpoonup \mathcal{M}_Y(\Phi) \quad \textit{weakly in } L^p(\Omega).$$

2. Assertion (ii)(b) is a corrector-type result.
3. Some authors define the notion of "strong two-scale convergence" via a convoluted duality argument (from the weak two-scale convergence). Assertion (ii)(a) of Proposition 1.29 can be seen as an equivalent direct and simpler definition of such a notion.

Proposition 1.31. *Suppose $p \in (1, +\infty)$.*
(i) Let $\{\Phi_\varepsilon\}_\varepsilon$ be a sequence which converges weakly in $L^p(\Omega \times Y)$ to Φ. Then

$$\mathcal{U}_\varepsilon(\Phi_\varepsilon) \rightharpoonup \mathcal{M}_Y(\Phi) \quad \textit{weakly in } L^p(\Omega).$$

(ii) Let $\{\Phi_\varepsilon\}_\varepsilon$ be a sequence such that $\Phi_\varepsilon \rightharpoonup \Phi$ weakly in $L^p(\Omega \times Y)$. Then

$$\begin{aligned} &\mathcal{T}_\varepsilon \circ \mathcal{U}_\varepsilon(\Phi_\varepsilon) \rightharpoonup \Phi \quad &&\textit{weakly in } L^p(\Omega \times Y),\\ &\mathcal{M}_\varepsilon \circ \mathcal{U}_\varepsilon(\Phi_\varepsilon) \rightharpoonup \mathcal{M}_Y(\Phi) \quad &&\textit{weakly in } L^p(\Omega). \end{aligned}$$

For $p = +\infty$, the same convergences hold for the weak-$*$ topologies[5].

[5] See Proposition 2.35 for the case $p = 1$.

Proof. *(i)* The statement is obtained from Proposition 1.9(i) by duality as $1 < p < +\infty$ (cf. (1.25)).

For $p = +\infty$, the result follows since $\mathcal{U}_\varepsilon(\Phi_\varepsilon)$ is bounded in $L^\infty(\Omega)$ as soon as Φ_ε is bounded in $L^\infty(\Omega \times Y)$, and because by the previous argument, it converges weakly in every $L^p_{loc}(\Omega)$.

(ii) Observe that $\mathcal{T}_\varepsilon \circ \mathcal{U}_\varepsilon$ is its own adjoint. Therefore for every Ψ in $L^{p'}(\Omega)$,

$$\frac{1}{|Y|}\int_{\Omega\times Y} \mathcal{T}_\varepsilon \circ \mathcal{U}_\varepsilon(\Phi_\varepsilon)(x,y)\,\Psi(x,y)\,dxdy$$
$$= \frac{1}{|Y|}\int_{\Omega\times Y} \Phi_\varepsilon(x,y)\mathcal{T}_\varepsilon \circ \mathcal{U}_\varepsilon(\Psi)(x,y)\,dxdy,$$

which gives the first convergence by Proposition 1.29(i). The second follows from the weak continuity of $\mathcal{M}_Y$.

For $p = +\infty$ the same argument as in (i) applies. □

Remark 1.32. *The statement of Proposition 1.29(ii) does not hold for weak convergences instead of strong ones. Indeed, weak convergence in (ii)(a) implies weak convergence in (ii)(b) by Proposition 1.12(iv) and Proposition 1.31(i). The converse is not true as can be seen by the following counterexample.*

Consider a function $\widehat{w} \not\equiv 0$ *in* $L^p(Y)$ *with* $\mathcal{M}_Y(\widehat{w}) = 0$. *By Proposition 1.31(i),*

$$\mathcal{U}_\varepsilon(\widehat{w}) \rightharpoonup \mathcal{M}_Y(\widehat{w}) = 0 \quad \textit{weakly in } L^p(\Omega).$$

This implies

$$w_\varepsilon - \mathcal{U}_\varepsilon(\widehat{w}) \rightharpoonup 0 \quad \textit{weakly in } L^p(\Omega)$$

for $\{w_\varepsilon\}_\varepsilon$ *the null sequence. Hence* $\mathcal{T}_\varepsilon(w_\varepsilon) \equiv 0$ *converges to* 0 *weakly (and strongly) in* $L^p(\Omega \times Y)$. *Since* $\widehat{w} \neq 0$, *this shows that (b) does not imply (a) in the case of weak convergences.*

Lemma 1.33. *Suppose* $p \in (1, +\infty)$.
(i) Let $\Psi \in L^p(\Omega \times Y)$, *such that*

$$\mathcal{U}_\varepsilon(\Psi) \to \mathcal{M}_Y(\Psi) \quad \textit{strongly in } L^p(\Omega). \tag{1.29}$$

Then Ψ *is independent of* y.
(ii) Let $\{\Phi_\varepsilon\}_\varepsilon$ *be a sequence in* $L^p(\Omega \times Y)$ *such that* $\mathcal{U}_\varepsilon(\Phi_\varepsilon)$ *lies in a strongly compact set of* $L^p(\Omega)$ *and*

$$\Phi_\varepsilon \rightharpoonup \Phi \quad \textit{weakly in } L^p(\Omega \times Y). \tag{1.30}$$

Then Φ *is independent of* y.

Proof. *(i)* Under hypothesis (1.29) by Proposition 1.9(ii),

$$\mathcal{T}_\varepsilon \circ \mathcal{U}_\varepsilon(\Phi) \to \mathcal{M}_Y(\Phi) \quad \text{strongly in } L^p(\Omega \times Y).$$

On the other hand, Proposition 1.29(i) implies that

$$\mathcal{T}_\varepsilon \circ \mathcal{U}_\varepsilon(\Phi) \to \Phi \quad \text{strongly in } L^p(\Omega \times Y),$$

therefore $\Phi \equiv \mathcal{M}_Y(\Phi)$, hence Φ is independent of y.
(ii) Hypothesis (1.30) and assertion (i) of Proposition 1.31, immediately give

$$\mathcal{U}_\varepsilon(\Phi_\varepsilon) \rightharpoonup \mathcal{M}_Y(\Phi) \quad \text{weakly in } L^p(\Omega),$$

hence strongly in the same space by hypothesis. Consequently, by 1.9(ii),

$$\mathcal{T}_\varepsilon \circ \mathcal{U}_\varepsilon(\Phi_\varepsilon) \to \mathcal{M}_Y(\Phi) \quad \text{strongly in } L^p(\Omega \times Y).$$

On the other hand, by (ii) from Proposition 1.31, $\{\mathcal{T}_\varepsilon \circ \mathcal{U}_\varepsilon(\Phi_\varepsilon)\}_\varepsilon$ converges to Φ weakly in $L^p(\Omega \times Y)$, so that $\Phi = \mathcal{M}_Y(\Phi)$. □

We complete this section with a somewhat unusual convergence property involving the averaging operator $\mathcal{U}_\varepsilon$ and which is applied in Theorem 5.10.

Proposition 1.34. *For $p \in [1, +\infty)$, suppose that α is in $L^p(\Omega)$ and β in $L^\infty(\Omega; L^p(Y))$. Then, the product $\mathcal{M}_\varepsilon(\alpha)\mathcal{U}_\varepsilon(\beta)$ belongs to $L^p(\Omega)$ and*

$$\mathcal{U}_\varepsilon(\alpha\beta) - \mathcal{M}_\varepsilon(\alpha)\mathcal{U}_\varepsilon(\beta) \to 0 \quad \textit{strongly in } L^p(\Omega). \tag{1.31}$$

Proof. It suffices to prove (1.31) and to do so, consider first the function $\mathcal{M}_\varepsilon(\alpha)\mathcal{U}_\varepsilon(\beta)$ in the set $\varepsilon(\xi + Y) \times Y$, $\xi \in \Xi_\varepsilon$. Using the fact that on this set $\mathcal{M}_\varepsilon(\alpha)$ is constant on the cell $\varepsilon(\xi + Y)$, one has $\mathcal{M}_\varepsilon(\alpha)\mathcal{U}_\varepsilon(\beta) = \mathcal{U}_\varepsilon\big(\mathcal{M}_\varepsilon(\alpha)\beta\big)$. Therefore,

$$\int_{\varepsilon(\xi+Y)} |\mathcal{U}_\varepsilon(\alpha\beta) - \mathcal{M}_\varepsilon(\alpha)\mathcal{U}_\varepsilon(\beta)|^p\, dx = \int_{\varepsilon(\xi+Y)} \big|\mathcal{U}_\varepsilon\big(\beta(\alpha - \mathcal{M}_\varepsilon(\alpha))\big)\big|^p\, dx.$$

Summing over $\xi \in \Xi_\varepsilon$ gives

$$\begin{aligned}\|\mathcal{U}_\varepsilon(\alpha\beta) - \mathcal{M}_\varepsilon(\alpha)\mathcal{U}_\varepsilon(\beta)\|^p_{L^p(\Omega)} &= \int_\Omega \big|\mathcal{U}_\varepsilon\big(\beta(\alpha - \mathcal{M}_\varepsilon(\alpha))\big)\big|^p\, dx \\ &\leqslant \frac{1}{|Y|}\, \|(\alpha - \mathcal{M}_\varepsilon(\alpha))\beta\|^p_{L^p(\Omega\times Y)} \\ &\leqslant \frac{1}{|Y|}\, \|\alpha - \mathcal{M}_\varepsilon(\alpha)\|^p_{L^p(\Omega)}\, \|\beta\|^p_{L^\infty(\Omega; L^p(Y))}.\end{aligned}$$

The last expression goes to 0 by (1.21). □

1.4 Unfolding and gradients

Our aim is to study the properties of the restriction of the unfolding operator to the space $W^{1,p}(\Omega)$. Recall the definitions of the Sobolev spaces on a domain Ω in $\mathbb{R}^N$ where $p \in [1, +\infty]$,

$$\begin{aligned} W^{1,p}(\Omega) &= \big\{\varphi \in L^p(\Omega) \,|\, \nabla\varphi \in L^p(\Omega)^N \text{ in the sense of distributions in } \Omega\big\}, \\ W_0^{1,p}(\Omega) &= \text{the closure of the space } \mathcal{C}_c^\infty(\Omega) \text{ in } W^{1,p}(\Omega).\end{aligned}$$

For $p \neq +\infty$, the dual space of $W_0^{1,p}(\Omega)$ is denoted $W_0^{-1,p'}(\Omega)$; it is a subspace of the space of distributions $\mathcal{D}'(\Omega) \equiv (\mathcal{C}_c^\infty(\Omega))'$. For $p = 2$, the corresponding spaces are usually denoted $H^1(\Omega), H_0^1(\Omega)$ and $H^{-1}(\Omega)$, respectively. They are Hilbert spaces.

Also introduce the notations

$$\begin{aligned} W_{per}^{1,p}(Y) &\doteq \{\varphi \in W_{loc}^{1,p}(\mathbb{R}^N) \mid \varphi \text{ is } \mathbf{G}\text{- periodic}\}, \\ W_{per,0}^{1,p}(Y) &\doteq \{\varphi \in W_{per}^{1,p}(Y) \mid \mathcal{M}_Y(\varphi) = 0\}. \end{aligned} \tag{1.32}$$

Both of these spaces are isometric to closed subspaces of $W^{1,p}(Y)$ when endowed with the corresponding norm. We shall consider them as subspaces of $W_{loc}^{1,p}(\mathbb{R}^N)$ as well as of $W^{1,p}(Y)$ (in the second case, its elements are extended to $\mathbb{R}^N$ by Y-periodicity). The second space plays a central role in the unfolding method for gradients. For $p = 2$, these spaces will be denoted $H_{per}^1(Y)$ and $H_{per,0}^1(Y)$, respectively.

Recall the **Poincaré-Wirtinger inequality** in Y (see [37]; we refer to Subsection 2.3.1 for more detailed consequences of this deep inequality). It is stated as follows (the constant depends only on Y and can be chosen independent of p):

$$\exists C \in \mathbb{R}^+, \quad \forall \psi \in W^{1,p}(Y), \quad \|\psi - \mathcal{M}_Y(\psi)\|_{L^p(Y)} \leqslant C\|\nabla\psi\|_{L^p(Y)}. \tag{1.33}$$

The Poincaré-Wirtinger inequality implies that $\|\nabla_y\varphi\|_{L^p(Y)}$ is a norm on $W_{per,0}^{1,p}(Y)$, equivalent to $\|\varphi\|_{W^{1,p}(Y)}$.

The next proposition states the relationship between $\mathcal{T}_\varepsilon$ and gradients.

Proposition 1.35. *Suppose $p \in [1, +\infty]$. The operator $\mathcal{T}_\varepsilon$ maps $W^{1,p}(\Omega)$ into $L^p(\Omega; W^{1,p}(Y))$ and*

$$\forall w \in W^{1,p}(\Omega), \quad \nabla_y(\mathcal{T}_\varepsilon(w)) = \varepsilon\mathcal{T}_\varepsilon(\nabla w) \quad \textit{a.e. in } \Omega \times Y. \tag{1.34}$$

Proof. Assume first p finite. Let w be any element of $W^{1,p}(\Omega)$ and let $\{w_n\}_n$ be a sequence in $C^1(\Omega) \cap W^{1,p}(\Omega)$ which converges to w in $W^{1,p}(\Omega)$. Formula (1.34) is straightforward for each w_n. Furthermore, by Proposition 1.8(iii),

$$\mathcal{T}_\varepsilon(w_n) \to \mathcal{T}_\varepsilon(w) \quad \text{strongly in } L^p(\Omega \times Y),$$

$$\nabla_y(\mathcal{T}_\varepsilon(w_n)) = \varepsilon\mathcal{T}_\varepsilon(\nabla w_n) \to \varepsilon\mathcal{T}_\varepsilon(\nabla w) \quad \text{strongly in } L^p(\Omega \times Y)^N.$$

From these convergences, it follows that $\nabla_y(\mathcal{T}_\varepsilon(w)) = \varepsilon\mathcal{T}_\varepsilon(\nabla w)$ in $L^p(\Omega \times Y)$, which concludes the proof for p finite.

For $p = +\infty$, one uses the previous argument and the fact that $W^{1,\infty}(\Omega)$ is included in every $W^{1,p}(\Omega)$. This implies that (1.34) is true also for $p = +\infty$. From this identity follows that $\mathcal{T}_\varepsilon$ maps $W^{1,\infty}(\Omega)$ into $L^\infty(\Omega; W^{1,\infty}(Y))$. □

In the next subsections we first consider the case of sequences in the space $W^{1,p}(\Omega)$ which are bounded in $L^p(\Omega)$, while their gradients are of order ε^{-1} in $L^p(\Omega)^N$. Next, we consider the case of bounded sequences in $W^{1,p}(\Omega)$. Finally, the intermediate case is investigated.

In Chapter 2, we shall consider the more general case of sequences which are bounded in $L^p(\Omega)$ while only some of their derivatives are of order ε^{-1} or bounded in $L^p(\Omega)$.

1.4.1 First case: $\|w_\varepsilon\|_{L^p(\Omega)} + \varepsilon\|\nabla w_\varepsilon\|_{L^p(\Omega)}$ bounded

Theorem 1.36. *Suppose $p \in (1, +\infty)$. Let $\{w_\varepsilon\}_\varepsilon$ be in $W^{1,p}(\Omega)$ satisfying*

$$\|w_\varepsilon\|_{L^p(\Omega)} + \varepsilon\|\nabla w_\varepsilon\|_{L^p(\Omega)} \leqslant C. \tag{1.35}$$

Then there exist a subsequence (still denoted ε), w in $L^p(\Omega)$ and $\widehat{w}$ in $L^p(\Omega; W^{1,p}_{per,0}(Y))$, such that

$$\begin{aligned} &w_\varepsilon \rightharpoonup w \quad \text{weakly in } L^p(\Omega), \\ &\varepsilon\nabla w_\varepsilon \rightharpoonup 0 \quad \text{weakly in } L^p(\Omega)^N, \\ &\mathcal{T}_\varepsilon(w_\varepsilon) \rightharpoonup w + \widehat{w} \quad \text{weakly in } L^p(\Omega; W^{1,p}(Y)), \\ &\mathcal{T}_\varepsilon(\varepsilon\nabla w_\varepsilon) = \nabla_y(\mathcal{T}_\varepsilon(w_\varepsilon)) \rightharpoonup \nabla_y\widehat{w} \quad \text{weakly in } L^p(\Omega \times Y)^N. \end{aligned} \tag{1.36}$$

Similar results hold for $p = +\infty$ replacing the weak topology by the weak- topology on the corresponding spaces.*

Proof. Up to a subsequence, convergences (1.36) are a simple consequence of estimate (1.35) together with Propositions 1.12 and 1.35.

It remains to prove the periodicity of $\widehat{w}$. Without loss of generality, we show periodicity in the $\mathbf{b}_N$-direction.

Let Y' be the $N-1$ open parallelotope associated to $(\mathbf{b}_1, \dots, \mathbf{b}_{N-1})$. Set $y = y' + y_N\mathbf{b}_N$, with y' in Y' and $y_N \in (0,1)$. The sequence $\{\mathcal{T}_\varepsilon(w_\varepsilon)\}_\varepsilon$ is bounded in

$$L^p(\Omega \times Y'; W^{1,p}(0,1)) = W^{1,p}((0,1); L^p(\Omega \times Y')),$$

so that, by the standard trace theorem, the function $\{s \mapsto \mathcal{T}_\varepsilon(w_\varepsilon)|_{\{y_N=s\}}\}$ converges to the function $\{s \mapsto \widehat{w}|_{\{y_N=s\}}\}$ in $C([0,1]; L^p(\Omega \times Y'))$.

The periodicity in the $\mathbf{b}_N$-direction results from the following computation for arbitrary ψ in $C_c^\infty(\Omega \times Y')$, using Definition 1.2 of $\mathcal{T}_\varepsilon$, performing an obvious change of variables and using Definition 1.2 again:

$$\begin{aligned}
&\int_{\Omega\times Y'} \big[\mathcal{T}_\varepsilon(w_\varepsilon)(x,y'+\mathbf{b}_N) - \mathcal{T}_\varepsilon(w_\varepsilon)(x,y')\big]\,\psi(x,y')\,dxdy' \\
&= \int_{\Omega\times Y'} \Big\{ w_\varepsilon\Big(\varepsilon\Big[\frac{x}{\varepsilon}\Big]_Y + \varepsilon(y'+\mathbf{b}_N)\Big) - w_\varepsilon\Big(\varepsilon\Big[\frac{x}{\varepsilon}\Big]_Y + \varepsilon\, y'\Big)\Big\}\psi(x,y')\,dxdy' \\
&= \int_{\Omega\times Y'} w_\varepsilon\Big(\varepsilon\Big[\frac{x}{\varepsilon}\Big]_Y + \varepsilon\, y'\Big)\big[\psi(x-\varepsilon\,\mathbf{b}_N,y') - \psi(x,y')\big]\,dxdy', \\
&= \int_{\Omega\times Y'} \mathcal{T}_\varepsilon(w_\varepsilon)(x,y')\big[\psi(x-\varepsilon\,\mathbf{b}_N,y') - \psi(x,y')\big]\,dxdy'.
\end{aligned}$$

Take now the limit as $\varepsilon \to 0$ to obtain, since the last term above goes to zero,

$$\begin{aligned}
\lim_{\varepsilon\to 0} &\int_{\Omega\times Y'} \big[\mathcal{T}_\varepsilon(w_\varepsilon)(x,y'+\mathbf{b}_N) - \mathcal{T}_\varepsilon(w_\varepsilon)(x,y')\big]\,\psi(x,y')\,dxdy' \\
&= \int_{\Omega\times Y'} \big[\widehat{w}(x,y'+\mathbf{b}_N) - \widehat{w}(x,y')\big]\,\psi(x,y')\,dxdy = 0,
\end{aligned}$$

from which the equality of the two traces follows, proving so the periodicity of $\widehat{w}$ in the $\mathbf{b}_N$-direction. The proof is similar for the other basis directions. ☐

1.4.2 Second case: $\{w_\varepsilon\}_\varepsilon$ bounded in $W^{1,p}(\Omega)$

In this case, we start with a corollary of Theorem 1.36.

Corollary 1.37. *Suppose $p \in (1,+\infty)$. Let $\{w_\varepsilon\}_\varepsilon$ be a sequence in $W^{1,p}(\Omega)$ such that*

$$w_\varepsilon \rightharpoonup w \quad \textit{weakly in } W^{1,p}(\Omega).$$

Then

$$\mathcal{T}_\varepsilon(w_\varepsilon) \rightharpoonup w \quad \textit{weakly in } L^p(\Omega; W^{1,p}(Y)).$$

Furthermore, if

$$w_\varepsilon \to w \quad \textit{strongly in } L^p(\Omega)^{(6)},$$

then

$$\mathcal{T}_\varepsilon(w_\varepsilon) \to w \quad \textit{strongly in } L^p(\Omega; W^{1,p}(Y)).$$

Similar results hold for $p = +\infty$ replacing the first weak convergence by the weak-$*$ convergence and the second strong convergence by the strong convergence in every $L^p_{loc}(\Omega; W^{1,p}(Y))$ for p finite.

Proof. By Theorem 1.36, there exist a subsequence (denoted ε) and $\widehat{w}$ in $L^p(\Omega; W^{1,p}(Y))$ with

$$\begin{aligned}
&\mathcal{T}_\varepsilon(w_\varepsilon) \rightharpoonup \widehat{w} \quad \text{weakly in } L^p(\Omega; W^{1,p}(Y)), \\
&\nabla_y(\mathcal{T}_\varepsilon(w_\varepsilon)) \rightharpoonup \nabla_y \widehat{w} \quad \text{weakly in } L^p(\Omega\times Y)^N.
\end{aligned} \tag{1.37}$$

(6) Due to the Sobolev embedding theorem (see [2]), this is the case when Ω is bounded with Lipschitz boundary.

Formula (1.34) implies the estimates

$$\|\nabla_y(\mathcal{T}_\varepsilon(w_\varepsilon))\|_{L^p(\Omega\times Y)} \leqslant \varepsilon C,$$

so that

$$\nabla_y(\mathcal{T}_\varepsilon(w_\varepsilon)) \to \nabla_y \widehat{w} = 0 \text{ strongly in } L^p(\Omega \times Y)^N. \tag{1.38}$$

Consequently, $\widehat{w}$ does not depend on y, and Proposition 1.12(i-ii) yields

$$w = \mathcal{M}_Y(\widehat{w}) = \widehat{w}.$$

Moreover, convergence (1.37) holds for the whole sequence $\{\varepsilon\}$.

Finally, if $\{w_\varepsilon\}_\varepsilon$ converges strongly to w in $L^p(\Omega)$, so does $\{\mathcal{T}_\varepsilon(w_\varepsilon)\}_\varepsilon$, thanks to Proposition 1.9(ii) and to the strong convergence (1.38). ☐

In the sequel, the following results on the local average operator $\mathcal{M}_\varepsilon$ will be applied to functions in $W^{1,p}(\Omega)$:

Proposition 1.38. *Suppose $p \in [1, +\infty]$. Then, for every φ in $W^{1,p}(\Omega)$,*

$$\begin{aligned} \|\varphi - \mathcal{M}_\varepsilon(\varphi)\|_{L^p(\widehat{\Omega}_\varepsilon)} &\leqslant C\varepsilon\|\nabla\varphi\|_{L^p(\Omega)}, \\ \|\varphi - \mathcal{T}_\varepsilon(\varphi)\|_{L^p(\widehat{\Omega}_\varepsilon\times Y)} &\leqslant C\varepsilon\|\nabla\varphi\|_{L^p(\Omega)}, \end{aligned} \tag{1.39}$$

with the constant C depending only on Y. Moreover,

$$\frac{1}{\varepsilon}(\varphi \mathbf{1}_{\widehat{\Omega}_\varepsilon} - \mathcal{M}_\varepsilon(\varphi)) \rightharpoonup 0 \quad \textit{weakly in } L^p(\Omega). \tag{1.40}$$

The last convergence holds in $L^\infty(\Omega)$ weakly - for $p = +\infty$.*

Proof. The rescaled Poincaré-Wirtinger inequality (1.33) becomes for every ξ in Ξ_ε,

$$\|\varphi - \mathcal{M}_\varepsilon(\varphi)\|^p_{L^p(\varepsilon(\xi+Y))} \leqslant (C\varepsilon)^p \, \|\nabla\varphi\|^p_{L^p(\varepsilon(\xi+Y))},$$

from which $(1.39)_1$ follows by summation over Ξ_ε.

Since $\mathcal{T}_\varepsilon \circ \mathcal{M}_\varepsilon \equiv \mathcal{M}_\varepsilon$, applying unfolding to $(1.39)_1$ gives

$$\|\mathcal{T}_\varepsilon(\varphi) - \mathcal{M}_\varepsilon(\varphi)\|_{L^p(\widehat{\Omega}_\varepsilon\times Y)} \leqslant C\varepsilon\|\nabla\varphi\|_{L^p(\Omega)},$$

which combined with $(1.39)_1$ implies $(1.39)_2$.

Convergence (1.40) can be proved directly from (1.39) with a density argument. In Corollary 1.43, a generalization of (1.40) will be proved independently. ☐

Proposition 1.39. *Suppose $p \in [1, +\infty)$. Let $\{w_\varepsilon\}_\varepsilon$ be a sequence in the space $W^{1,p}(\Omega)$ such that*

$$w_\varepsilon \to w \quad \textit{strongly in } W^{1,p}(\Omega).$$

Then

(i) $\mathcal{T}_\varepsilon(\nabla w_\varepsilon) \to \nabla w$ *strongly in* $L^p(\Omega \times Y)$,

(ii) $\dfrac{1}{\varepsilon}\Big(\mathcal{T}_\varepsilon(w_\varepsilon) - \mathcal{M}_\varepsilon(w_\varepsilon)\Big) \to \sum_{j=1}^{N} y_j^c \dfrac{\partial w}{\partial x_j}$ *strongly in* $L^p(\Omega; W^{1,p}(Y))$,

where

$$y^c = y - \mathcal{M}_Y(y). \tag{1.41}$$

Proof. *(i)* The first assertion follows directly from Proposition 1.9.
(ii) Set now

$$Z_\varepsilon = \frac{1}{\varepsilon}\Big(\mathcal{T}_\varepsilon(w_\varepsilon) - \mathcal{M}_\varepsilon(w_\varepsilon)\Big) - \sum_{j=1}^{N} y_j^c \frac{\partial w}{\partial x_j}. \tag{1.42}$$

Observe that due to (1.18) and to definition (1.41), the function Z_ε satisfies

$$\mathcal{M}_Y(Z_\varepsilon) = 0. \tag{1.43}$$

Then, applying the Poincaré-Wirtinger inequality (1.33) gives

$$\|Z_\varepsilon\|_{L^p(\Omega\times Y)} \leqslant C\|\nabla_y Z_\varepsilon\|_{L^p(\Omega\times Y)}. \tag{1.44}$$

On the other hand,

$$\nabla_y Z_\varepsilon = \frac{1}{\varepsilon}\nabla_y\big(\mathcal{T}_\varepsilon(w_\varepsilon)\big) - \nabla w = \mathcal{T}_\varepsilon(\nabla w_\varepsilon) - \nabla w. \tag{1.45}$$

Therefore by assertion (i), $\nabla_y Z_\varepsilon$ converges strongly to 0 in $L^p(\Omega \times Y)$. So (1.44) entails

$$Z_\varepsilon \to 0 \quad \text{strongly in } L^p(\Omega; W^{1,p}(Y)),$$

whence convergence (ii). □

Remark 1.40. *For the case* $Y = (0,1)^N$

$$y^c = \Big(y_1 - \frac{1}{2}, \dots, y_N - \frac{1}{2}\Big) \quad \textit{and} \quad \sum_{j=1}^{N} y_j^c \frac{\partial w}{\partial x_j} = y^c \cdot \nabla w.$$

In the general case, the basic periods are not orthonormal and the last quantity is not the scalar product in the ambient space $\mathbb{R}^N$.
However, for simplicity, in the remainder of this book, the tensor condensed notation $y^c \cdot \nabla w$ ***will be used in the general case, even though it is not connected to the scalar product of*** $\mathbb{R}^N$.

In Proposition 1.39, the sequence $\{w_\varepsilon\}_\varepsilon$ is strongly convergent in $W^{1,p}(\Omega)$. One may ask whether with $\{w_\varepsilon\}_\varepsilon$ converging only weakly, there is still some information about the sequence $\{\mathcal{T}_\varepsilon(w_\varepsilon)\}_\varepsilon$ of its unfoldings. The answer is given in the next result.

Theorem 1.41. *Suppose $p \in (1, +\infty)$. Let $\{w_\varepsilon\}_\varepsilon$ be a sequence in $W^{1,p}(\Omega)$ such that*

$$w_\varepsilon \rightharpoonup w \quad \text{weakly in } W^{1,p}(\Omega).$$

Then for a subsequence, there exists some $\widehat{w}$ in $L^p(\Omega; W^{1,p}_{per,0}(Y))$ such that

$$\text{(i)} \quad \frac{1}{\varepsilon}\Big(\mathcal{T}_\varepsilon(w_\varepsilon) - \mathcal{M}_\varepsilon(w_\varepsilon)\Big) \rightharpoonup y^c \cdot \nabla w + \widehat{w} \quad \text{weakly in } L^p(\Omega; W^{1,p}(Y)),$$

$$\text{(ii)} \quad \mathcal{T}_\varepsilon(\nabla w_\varepsilon) \rightharpoonup \nabla w + \nabla_y \widehat{w} \quad \text{weakly in } L^p(\Omega \times Y)^N.$$

Moreover

$$\begin{aligned} &\|\widehat{w}\|_{L^p(\Omega; W^{1,p}_{per}(Y))} \leqslant C \limsup_{\varepsilon \to 0} \|w_\varepsilon\|_{W^{1,p}(\Omega)}, \\ &\|\nabla w + \nabla_y \widehat{w}\|_{L^p(\Omega \times Y)} \leqslant |Y|^{1/p} \liminf_{\varepsilon \to 0} \|\nabla w_\varepsilon\|_{L^p(\Omega)}, \end{aligned} \tag{1.46}$$

where the constant C only depends on the Poincaré-Wirtinger constant of Y. For $p = +\infty$, the same convergences hold for the weak-$*$ topologies(7).

Proof. The function Z_ε defined by (1.42) as in the preceding proof, satisfies (1.43), (1.44) and (1.45). In this case,

$$\|\nabla Z_\varepsilon\|_{L^p(\Omega \times Y)} \leqslant C\|\nabla w_\varepsilon\|_{L^p(\Omega)} + \|\nabla w\|_{L^p(\Omega)}\big),$$

so that, by the Poincaré-Wirtinger inequality,

$$\|Z_\varepsilon\|_{L^p(\Omega; W^{1,p}(Y))} \leqslant C\big(\|\nabla w_\varepsilon\|_{L^p(\Omega)} + \|\nabla w\|_{L^p(\Omega)}\big).$$

Consequently, there exists $\widehat{w}$ in $L^p(\Omega; W^{1,p}(Y))$ satisfying (1.46) and such that, up to a subsequence,

$$Z_\varepsilon \rightharpoonup \widehat{w} \quad \text{weakly in } L^p(\Omega; W^{1,p}(Y)).$$

Since by construction, $\mathcal{M}_Y(Z_\varepsilon) = 0$, one has

$$\mathcal{M}_Y(\widehat{w}) = 0.$$

The last convergence above is equivalent to convergences (i) and (ii).

The Y-periodicity of $\widehat{w}$ is proved in a way similar to that in the proof of Theorem 1.36. Recalling that Y' denotes the $N-1$ open parallelotope associated with $(\mathbf{b}_1, \ldots, \mathbf{b}_{N-1})$, evaluate the limit of the function defined for a.e. $(x, y') \in \Omega \times Y'$ as

$$Z_\varepsilon(x, y' + \mathbf{b}_N) - Z_\varepsilon(x, y').$$

On one hand, since

$$(y^c \cdot \nabla w)(x, y' + \mathbf{b}_N) - (y^c \cdot \nabla w)(x, y') = \frac{\partial w}{\partial x_N}(x),$$

(7) See Proposition 2.36 for the case $p = 1$.

one has successively for every ψ in $\mathcal{C}_c^\infty(\Omega \times Y')$,

$$\begin{aligned}
&\int_{\Omega\times Y'} \big[Z_\varepsilon(x, y' + \mathbf{b}_N) - Z_\varepsilon(x, y')\big]\,\psi(x, y')\,dxdy' + \int_{\Omega\times Y'} \frac{\partial w}{\partial x_N}\,\psi(x, y')\,dxdy' \\
&= \int_{\Omega\times Y'} \frac{1}{\varepsilon}\Big\{w_\varepsilon\Big(\varepsilon\Big[\frac{x}{\varepsilon}\Big]_Y + \varepsilon(y' + \mathbf{b}_N)\Big) - w_\varepsilon\Big(\varepsilon\Big[\frac{x}{\varepsilon}\Big]_Y + \varepsilon y'\Big)\Big\}\psi(x, y')\,dxdy' \\
&= \int_{\Omega\times Y'} \Big\{(w_\varepsilon\Big(\varepsilon\Big[\frac{x}{\varepsilon}\Big]_Y + \varepsilon y'\Big)\,\frac{1}{\varepsilon}\big[\psi(x - \varepsilon\,\mathbf{b}_N, y') - \psi(x, y')\big]\Big\}\,dxdy' \\
&= \int_{\Omega\times Y'} \mathcal{T}_\varepsilon(w_\varepsilon)(x, y')\,\frac{1}{\varepsilon}\big[\psi(x - \varepsilon\,\mathbf{b}_N, y') - \psi(x, y')\big]\,dxdy'.
\end{aligned}$$

By Corollary 1.37, $\mathcal{T}_\varepsilon(w_\varepsilon)$ converges weakly to w in $L^p(\Omega; W^{1,p}(Y))$. By the trace theorem in $W^{1,p}(Y)$, the trace of $\mathcal{T}_\varepsilon(w_\varepsilon)$ on $\Omega \times Y'$ converges weakly to w in $L^p(\Omega \times Y')$. Hence the last integral in the equality converges to

$$-\int_{\Omega\times Y'} w(x)\,\frac{\partial \psi}{\partial x_N}(x, y')\,dxdy'.$$

Consequently, passing to the limit in the equality above yields

$$\begin{aligned}
&\int_{\Omega\times Y'} \big[\widehat{w}(x, y') + \mathbf{b}_N) - \widehat{w}(x, y')\big]\,\psi(x, y')\,dxdy \\
&\qquad + \int_{\Omega\times Y'} \frac{\partial w}{\partial x_N}(x)\,\psi(x, y')\,dxdy' = -\int_{\Omega\times Y'} w(x)\,\frac{\partial \psi}{\partial x_N}(x, y')\,dxdy'.
\end{aligned}$$

Since by integration by parts,

$$\int_{\Omega\times Y'} \frac{\partial w}{\partial x_N}(x)\,\psi(x, y')\,dx\,dy' = -\int_{\Omega\times Y'} w(x)\,\frac{\partial \psi}{\partial x_N}(x, y')\,dxdy',$$

what remains is

$$\int_{\Omega\times Y'} \big[\widehat{w}(x, y' + \mathbf{b}_N) - \widehat{w}(x, y')\big]\,\psi(x, y')\,dxdy' = 0,$$

which says precisely that $\widehat{w}$ is periodic in the $\mathbf{b}_N$-direction. By a similar argument, the same holds for the other directions of the period. □

Remark 1.42 (An apparent contradiction).
The notion of periodicity for the limit of a sequence of unfoldings is strictly connected with the space in which the convergence occurs. For the parallelotope Y, the space $L^p_{per}(Y))$ coincides with $L^p(Y)$ whereas $W^{1,p}_{per}(Y)$ is strictly smaller than $W^{1,p}(Y)$. It may happen that the limit of a sequence of unfoldings belong to $L^p(\Omega; W^{1,p}(Y))$ without belonging to $L^p(\Omega; W^{1,p}_{per}(Y))$. This can occur if the original sequence is bounded in $L^p(\Omega)$ but not in $W^{1,p}(\Omega)$.

For example, under the assumptions of Theorem 1.41, the unfolded sequence $\{\frac{1}{\varepsilon}\mathcal{T}_\varepsilon(w_\varepsilon - \mathcal{M}_\varepsilon(w_\varepsilon))\}_\varepsilon$ *of* $\{\frac{1}{\varepsilon}(w_\varepsilon - \mathcal{M}_\varepsilon(w_\varepsilon))\}_\varepsilon$ *converges to a function belonging to the space* $L^p(\Omega; W^{1,p}(Y))$ *but not to* $L^p(\Omega; W^{1,p}_{per}(Y))$.

This is due to the fact that the sequence $\{\frac{1}{\varepsilon}(w_\varepsilon - \mathcal{M}_\varepsilon(w_\varepsilon))\}_\varepsilon$ *is not bounded in* $W^{1,p}(\Omega)$ *(it is not even included in that space !). However, if one considers it as a sequence in* $L^p(\Omega)$, *its unfolding converges in* $L^p(\Omega \times Y) \equiv L^p(\Omega; L^p(Y))$ *(for p finite) which is the same as* $L^p(\Omega; L^p_{per}(Y))$, *so that the limit is indeed* Y*-periodic in that weaker sense.*

The next result is a complement to convergence (1.40).

Corollary 1.43. *Under the assumptions of Theorem 1.41, one has*

$$(i) \quad \frac{1}{\varepsilon}\big(w_\varepsilon \mathbf{1}_{\widehat{\Omega}_\varepsilon} - \mathcal{M}_\varepsilon(w_\varepsilon)\big) \rightharpoonup 0 \quad \text{weakly in } L^p(\Omega),$$

as well as

$$(ii) \quad \frac{1}{\varepsilon}\big(\mathcal{T}_\varepsilon(w_\varepsilon) - w_\varepsilon \mathbf{1}_{\widehat{\Omega}_\varepsilon}\big) \rightharpoonup y^c \cdot \nabla w + \widehat{w} \quad \text{weakly in } L^p(\Omega; W^{1,p}(Y)).$$

Proof. First remark that

$$\mathcal{T}_\varepsilon(w_\varepsilon) - \mathcal{M}_\varepsilon(w_\varepsilon) = \mathcal{T}_\varepsilon\big(w_\varepsilon - \mathcal{M}_\varepsilon(w_\varepsilon)\big) = \mathcal{T}_\varepsilon\big(w_\varepsilon \mathbf{1}_{\widehat{\Omega}_\varepsilon} - \mathcal{M}_\varepsilon(w_\varepsilon)\big).$$

Theorem 1.41(i) implies therefore that

$$\frac{1}{\varepsilon}\mathcal{T}_\varepsilon\big(w_\varepsilon \mathbf{1}_{\widehat{\Omega}_\varepsilon} - \mathcal{M}_\varepsilon(w_\varepsilon)\big) \rightharpoonup y^c \cdot \nabla w + \widehat{w} \quad \text{weakly in } L^p(\Omega; W^{1,p}(Y)).$$

By equality (1.9) and Proposition 1.12(i), convergence (i) follows immediately since the right-hand side above is with zero average on Y. Combining it with the same convergence (i) in Theorem 1.41, gives (ii). □

1.4.3 * Complements

This section completes the statements of Theorems 1.36 and 1.41.

The first result generalizes these theorems. It is used in the proof of Proposition 11.13 concerning the elastic thin plates.

The second result is a version of Theorem 1.41 for sequences converging only in $W^{1,p}_{loc}(\Omega)$. A different setting for unfolding with two parameters is also presented in Section 11.4.

Lemma 1.44. *Suppose* $p \in (1, +\infty)$. *Consider a sequence* $\{w_\varepsilon\}_\varepsilon$ *in* $W^{1,p}(\Omega)$ *converging weakly to some* w *in* $L^p(\Omega)$ *and satisfying for some strictly positive sequence* $\{\delta(\varepsilon)\}_\varepsilon$ *converging to* 0

$$\delta(\varepsilon)\, \|\nabla w_\varepsilon\|_{L^p(\Omega)} \text{ bounded as } \varepsilon \text{ goes to } 0.$$

(i) If $\lim_{\varepsilon\to 0} \varepsilon^{-1}\delta(\varepsilon) = 0$, *then, up to a subsequence, there exist* $\overline{w} \in L^p(\Omega \times Y)$ *with* $\mathcal{M}_Y(\overline{w}) = w$, *and* $\widehat{w}$ *in* $L^p(\Omega; W^{1,p}_{per,0}(Y))$ *such that*

$$\begin{aligned} &\mathcal{T}_\varepsilon(w_\varepsilon) \rightharpoonup \overline{w} \quad \text{weakly in } L^p(\Omega \times Y),\\ &\frac{\delta(\varepsilon)}{\varepsilon}\mathcal{T}_\varepsilon(w_\varepsilon - \mathcal{M}_\varepsilon(w_\varepsilon)) \rightharpoonup \widehat{w} \quad \text{weakly in } L^p(\Omega; W^{1,p}(Y)),\\ &\delta(\varepsilon)\mathcal{T}_\varepsilon(\nabla w_\varepsilon) \rightharpoonup \nabla_y \widehat{w} \quad \text{weakly in } L^p(\Omega \times Y)^N. \end{aligned}$$

(ii) If $\lim_{\varepsilon\to 0} \varepsilon^{-1}\delta(\varepsilon) = +\infty$, *then,*

$$\begin{aligned} &\mathcal{T}_\varepsilon(w_\varepsilon) \rightharpoonup w \text{ weakly in } L^p(\Omega; W^{1,p}(Y)),\\ &\nabla_y \mathcal{T}_\varepsilon(w_\varepsilon) \to 0 \text{ strongly in } L^p(\Omega \times Y)^N. \end{aligned}$$

Furthermore, there exists $\widehat{w}$ *in* $L^p(\Omega; W^{1,p}_{per,0}(Y))$ *such that, up to a subsequence,*

$$\begin{aligned} &\frac{\delta(\varepsilon)}{\varepsilon}\mathcal{T}_\varepsilon(w_\varepsilon - \mathcal{M}_\varepsilon(w_\varepsilon)) \rightharpoonup \widehat{w} \quad \text{weakly in } L^p(\Omega; W^{1,p}(Y)),\\ &\delta(\varepsilon)\mathcal{T}_\varepsilon(\nabla w_\varepsilon) \rightharpoonup \nabla_y \widehat{w} \quad \text{weakly in } L^p(\Omega \times Y)^N. \end{aligned}$$

Remark 1.45. *A "natural" candidate for* $\delta(\varepsilon)$ *is of course* $\|\nabla w_\varepsilon\|^{-1}_{L^p(\Omega)}$. *In many applications, the estimate appears with* $\delta(\varepsilon) = \varepsilon^\gamma$, $\gamma \geqslant 0$. *Theorem 1.36 corresponds to* $\gamma = 1$ *and Theorem 1.41 to* $\gamma = 0$. *The lemma applies to all other cases.*

See also Lemma 11.11 for a more complex situation with two parameters ε *and* δ *going to zero.*

Proof of Lemma 1.44. Applying Theorem 1.41 to the sequence $\{\delta(\varepsilon)\, w_\varepsilon\}_\varepsilon$, which is uniformly bounded in $W^{1,p}(\Omega)$ and converges to 0 in $L^p(\Omega)$, gives a function $\widehat{w}$ in $L^p(\Omega; W^{1,p}_{per,0}(Y))$ such that, up to a subsequence

$$\begin{aligned} &\frac{\delta(\varepsilon)}{\varepsilon}\mathcal{T}_\varepsilon(w_\varepsilon - \mathcal{M}_\varepsilon(w_\varepsilon)) \rightharpoonup y^c \cdot 0 + \widehat{w} \quad \text{weakly in } L^p(\Omega; W^{1,p}(Y)),\\ &\delta(\varepsilon)\mathcal{T}_\varepsilon(\nabla(w_\varepsilon)) \rightharpoonup 0 + \nabla_y \widehat{w} \quad \text{weakly in } L^p(\Omega \times Y)^N. \end{aligned}$$

By Proposition 1.8(iii), $\|\mathcal{T}_\varepsilon(w_\varepsilon)\|_{L^p(\Omega\times Y)}$ is bounded. Therefore, up to a subsequence, there exists $\overline{w}$ in $L^p(\Omega \times Y)$ with $\mathcal{M}_Y(\overline{w}) = w$ and such that

$$\mathcal{T}_\varepsilon(w_\varepsilon) \rightharpoonup \overline{w} \quad \text{weakly in } L^p(\Omega \times Y).$$

This proves (i).

Under the hypothesis $\lim_{\varepsilon\to 0} \varepsilon^{-1}\delta(\varepsilon) = +\infty$, (1.34) implies the convergence

$$\nabla_y \mathcal{T}_\varepsilon(w_\varepsilon) \to 0 \quad \text{strongly in } L^p(\Omega \times Y).$$

Consequently, $\nabla_y \overline{w} \equiv 0$, so that $\mathcal{M}_Y(\overline{w}) = \overline{w}$ which gives $\overline{w} = w$. By the uniqueness of the limit, the whole sequence $\{\mathcal{T}_\varepsilon(w_\varepsilon)\}_\varepsilon$ now converges weakly to w in $L^p(\Omega; W^{1,p}(Y))$. □

Another consequence of Theorem 1.41 concerns sequences converging in $W^{1,p}_{loc}(\Omega)$. To state it, introduce the following notations:

$$\begin{aligned} &d(Y) = \text{the diameter of } Y, \\ &\qquad\qquad (\text{more generally, } d(\mathcal{O}) \text{ is the diameter of } \mathcal{O} \subset \mathbb{R}^N), \\ &\rho(x) = \text{dist}\,(x, \partial\Omega), \quad \forall x \in \mathbb{R}^N. \end{aligned} \tag{1.47}$$

Consider also the Banach spaces,

$$\begin{aligned} &W^{1,p}_\rho(\Omega) \doteq \{\varphi \in L^p(\Omega) \mid \rho\nabla\varphi \in L^p(\Omega)^N\} \;(H^1_\rho(\Omega) \text{ for } p = 2), \\ &L^p_\rho(\Omega; L^p(Y)) \doteq \{\Phi \in L^p_{loc}(\Omega \times Y) \mid \rho\,\Phi \in L^p(\Omega \times Y)\}, \\ &L^p_\rho(\Omega; W^{1,p}(Y)) \doteq \{\Psi \in L^p_\rho(\Omega; L^p(Y)) \mid \rho\,\nabla_y\Psi \in L^p(\Omega \times Y)^N\}, \end{aligned} \tag{1.48}$$

endowed respectively, with the norms

$$\begin{aligned} &\|\varphi\|_{W^{1,p}_\rho(\Omega)} = \Big(\|\varphi\|^p_{L^p(\Omega)} + \|\rho\nabla\varphi\|^p_{L^p(\Omega)}\Big)^{1/p}, \\ &\|\Phi\|_{L^p_\rho(\Omega; L^p(Y))} = \|\rho\,\Phi\|_{L^p(\Omega\times Y)}, \\ &\|\Psi\|_{L^p_\rho(\Omega; W^p(Y))} = \Big(\|\rho\,\Psi\|^p_{L^p(\Omega\times Y)} + \|\rho\,\nabla_y\Psi\|^p_{L^p(\Omega\times Y)}\Big)^{1/p}. \end{aligned}$$

Let $\mathcal{O}$ be an open set in $\mathbb{R}^N$. For $\delta > 0$, we use the following notation:

$$\mathcal{O}^{int}_\delta \doteq \{x \in \mathcal{O} \mid \text{dist}(x, \partial\mathcal{O}) > \delta\}. \tag{1.49}$$

Theorem 1.46. *Suppose $p \in (1, +\infty)$. Let $\{w_\varepsilon\}_\varepsilon$ be a sequence in $W^{1,p}_\rho(\Omega)$, such that*

$$w_\varepsilon \rightharpoonup w \quad \textit{weakly in } W^{1,p}_\rho(\Omega).$$

Then, up to a subsequence, there exists $\widehat{w}$ in $L^p_\rho(\Omega; W^{1,p}_{per,0}(Y))$ such that,

$$\begin{aligned} &(i) \quad \mathcal{T}_\varepsilon(w_\varepsilon) \rightharpoonup w \quad \textit{weakly in } L^p(\Omega \times Y), \\ &(ii) \quad \mathbf{1}_{\Omega^{int}_{2\varepsilon d(Y)}} \mathcal{T}_\varepsilon(\nabla w_\varepsilon) \rightharpoonup \nabla w + \nabla_y\widehat{w} \quad \textit{weakly in } L^p_\rho(\Omega; L^p(Y))^N, \\ &(iii) \quad \mathcal{T}_\varepsilon(\rho\nabla w_\varepsilon) \rightharpoonup \rho\nabla w + \rho\nabla_y\widehat{w} \quad \textit{weakly in } L^p(\Omega \times Y)^N. \end{aligned} \tag{1.50}$$

Moreover, if $\{\varepsilon\nabla w_\varepsilon\}_\varepsilon$ is uniformly bounded in $W^{1,p}(\Omega)$, then

$$\mathcal{T}_\varepsilon(\nabla w_\varepsilon) \rightharpoonup \nabla w + \nabla_y\widehat{w} \quad \textit{weakly in } L^p_\rho(\Omega; L^p(Y))^N. \tag{1.51}$$

Proof. Since ρ is Lipschitz continuous with Lipschitz constant 1,

$$\begin{aligned} &\Big|\rho\Big(\varepsilon\Big[\frac{x}{\varepsilon}\Big]_Y + \varepsilon y\Big) - \rho(x)\Big| \leqslant \varepsilon d(Y) \quad \text{for } (x, y) \in \Omega \times Y, \\ &\frac{\rho(x)}{2} \leqslant \rho\Big(\varepsilon\Big[\frac{x}{\varepsilon}\Big]_Y + \varepsilon y\Big) \leqslant \frac{3\rho(x)}{2} \quad \text{for } (x, y) \in \Omega_{2\varepsilon d(Y)} \times Y. \end{aligned}$$

This implies that

$$\left\{\mathbf{1}_{\Omega^{int}_{2\varepsilon d(Y)}} \mathcal{T}_\varepsilon(\nabla w_\varepsilon)\right\} \text{ is bounded in } L^p_\rho(\Omega; L^p(Y))^N.$$

Then, there exist W in $L^p(\Omega \times Y)$ and F in $L^p_\rho(\Omega; L^p(Y))^N$ such that (for a subsequence of $\{\varepsilon\}$),

$$\begin{aligned} &\mathcal{T}_\varepsilon(w_\varepsilon) \rightharpoonup W \quad \text{weakly in } L^p(\Omega \times Y),\\ &\mathbf{1}_{\Omega^{int}_{2\varepsilon d(Y)}} \mathcal{T}_\varepsilon(\nabla w_\varepsilon) \rightharpoonup F \quad \text{weakly in } L^p_\rho(\Omega; L^p(Y))^N. \end{aligned}$$

We now shall identify W and F. To do so, let ω be a relatively compact open subset in Ω. The sequence $\{w_\varepsilon\}_\varepsilon$ is bounded in $W^{1,p}(\omega)$ and then by hypothesis,

$$w_\varepsilon \rightharpoonup w \quad \text{weakly in } W^{1,p}(\omega).$$

Furthermore, there exists some $\widehat{w}_\omega$ in $L^p(\omega; W^{1,p}_{per,0}(Y))$ such that (for another subsequence of $\{\varepsilon\}$),

$$\begin{aligned} &\mathcal{T}_\varepsilon(w_\varepsilon) \rightharpoonup w \quad \text{weakly in } L^p(\omega; W^{1,p}(Y)),\\ &\mathcal{T}_\varepsilon(\nabla w_\varepsilon) \rightharpoonup \nabla w + \nabla_y \widehat{w}_\omega \quad \text{weakly in } L^p(\omega \times Y)^N. \end{aligned}$$

As a consequence,

$$W = w, \qquad F = \nabla w + \nabla_y \widehat{w}_\omega \quad \text{in} \quad L^p(\omega \times Y),$$

which in turn, imply that $W = w$ in $L^p(\Omega \times Y)$ and that there exists $\widehat{w}$ in $L^p_\rho(\Omega; W^{1,p}_{per,0}(Y))$ such that $F - \nabla w = \nabla_y \widehat{w}$. This completes the proof of assertions (i) and (ii) from (1.50).

It remains to prove convergence (1.50)(iii). At this point observe that from Proposition 1.9(i), for every $q \in [1, +\infty)$,

$$\mathcal{T}_\varepsilon(\rho) \to \rho \quad \text{strongly in } L^q(\Omega \times Y). \tag{1.52}$$

Then (1.50)(iii) is an immediate consequence of (1.50)(ii) taking into account the facts that ρw_ε and ρw belong to $W^{1,p}_0(\Omega)$, and $\rho\widehat{w}$ to $L^p(\Omega; W^{1,p}_{per,0}(Y))$.

From (1.39), we obtain

$$\|\mathcal{T}_\varepsilon(\rho) - \rho\|_{L^\infty(\widehat{\Omega}_\varepsilon)} \leqslant \varepsilon.$$

Therefore,

$$\|\mathcal{T}_\varepsilon(\rho\nabla w_\varepsilon) - \rho\mathcal{T}_\varepsilon(\nabla w_\varepsilon)\|_{L^p(\Omega)} \leqslant C\varepsilon\|\nabla w_\varepsilon\|_{L^p(\Omega)}.$$

Thus, the sequence $\{\rho\mathcal{T}_\varepsilon(\nabla w_\varepsilon)\}_\varepsilon$ is uniformly bounded in $L^p(\Omega \times Y)^N$. Convergences $(1.50)_3$, (1.52) and the above estimates finally give (1.51). □

1.4.4 Case of $\{w_\varepsilon\}_\varepsilon$ bounded in $W^{k,p}(\Omega)$

Theorem 1.41 can be generalized to the case of $W^{k,p}(\Omega)$−spaces with $k \geqslant 2$ and $p \in (1, +\infty)$. In order to do so, for $r = (r_1, \dots, r_N) \in \mathbb{N}^N$ with $|r| = r_1 + \dots + r_N \leqslant k$, introduce the notation D^r and D^r_y,

$$D^r = \frac{\partial^{|r|}}{\partial x_1^{r_1} \dots \partial x_N^{r_N}}, \qquad D^r_y = \frac{\partial^{|r|}}{\partial y_1^{r_1} \dots \partial y_N^{r_N}}.$$

Then the following result holds:

Theorem 1.47. *Suppose $p \in (1, +\infty)$. Let $\{w_\varepsilon\}_\varepsilon$ in be a sequence in $W^{k,p}(\Omega)$ for $k \geqslant 1$, such that*

$$w_\varepsilon \rightharpoonup w \quad \text{weakly in } W^{k,p}(\Omega).$$

Then, up to a subsequence, there exists $\widehat{W}$ in $L^p(\Omega; W^{k,p}_{per}(Y))$ such that

$$\begin{aligned} &\mathcal{T}_\varepsilon(D^l w_\varepsilon) \rightharpoonup D^l w \quad \text{weakly in } L^p(\Omega; W^{k-l,p}(Y)), \quad |l| \leqslant k-1,\\ &\mathcal{T}_\varepsilon(D^l w_\varepsilon) \rightharpoonup D^l w + D^l_y \widehat{W} \quad \text{weakly in } L^p(\Omega \times Y), \quad |l| = k. \end{aligned} \tag{1.53}$$

Furthermore, if $\{w_\varepsilon\}_\varepsilon$ converges strongly to w in $W^{k-1,p}(\Omega)$[(8)]*, the above convergences are strong in $L^p(\Omega; W^{k-l,p}(Y))$ for $|l| \leqslant k-1$.*

Similar statements can be made for the case $p = +\infty$.

Proof. We give a brief proof for $k = 2$. The same argument generalizes for $k > 2$.

If $|l| = 1$, the first convergence in (1.53) and the final statement of the theorem follow directly from Corollary 1.37. We now prove the second convergence in (1.53). Set

$$Z_\varepsilon = \frac{1}{\varepsilon^2}\Big[\mathcal{T}_\varepsilon(w_\varepsilon) - \mathcal{M}_\varepsilon(w_\varepsilon) - y^c \cdot \mathcal{M}_\varepsilon(\nabla w_\varepsilon)\Big].$$

Computing the derivatives of Z_ε up to order 2 gives

$$\nabla_y(Z_\varepsilon) = \frac{1}{\varepsilon}\Big(\mathcal{T}_\varepsilon(\nabla w_\varepsilon) - \mathcal{M}_\varepsilon(\nabla w_\varepsilon)\Big), \tag{1.54}$$

and

$$D^l_y(Z_\varepsilon) = \mathcal{T}_\varepsilon(D^l w_\varepsilon) \quad \text{with} \quad |l| = 2.$$

Note that $\mathcal{M}_Y(Z_\varepsilon) = 0$ and $\mathcal{M}_Y(\nabla_y Z_\varepsilon) = 0$.

The sequence $\{w_\varepsilon\}_\varepsilon$ being bounded in $W^{2,p}(\Omega)$, inequality (1.33) (applied twice) implies that $\{Z_\varepsilon\}_\varepsilon$ is bounded in $L^p(\Omega; W^{2,p}(Y))$. Therefore, up to

(8) Due to the Sobolev embedding theorem (see [2]), this is the case when Ω is bounded with Lipschitz boundary.

a subsequence, there exists Z in $L^p(\Omega; W^{2,p}(Y))$ such that $\mathcal{M}_Y(Z) = 0$, $\mathcal{M}_Y(\nabla_y Z) = 0$, and

$$\begin{aligned} &Z_\varepsilon \rightharpoonup Z \quad \text{weakly in } L^p(\Omega; W^{2,p}(Y)),\\ &\frac{\partial Z_\varepsilon}{\partial y_i} \rightharpoonup \frac{\partial Z}{\partial y_i} \quad \text{weakly in } L^p(\Omega; W^{1,p}(Y)),\\ &D_y^l(Z_\varepsilon) \rightharpoonup D_y^l Z \quad \text{weakly in } L^p(\Omega \times Y) \quad \text{for } |l| = 2. \end{aligned} \tag{1.55}$$

Applying Theorem 1.41 to each derivative $\dfrac{\partial w_\varepsilon}{\partial x_i}$ ($i \in \{1, \dots, N\}$), there exist a function $\widehat{w}_i$ in $L^p(\Omega; W^{1,p}_{per}(Y))$ such that $\mathcal{M}_Y(\widehat{w}_i) \equiv 0$ and

$$\frac{1}{\varepsilon}\Big(\mathcal{T}_\varepsilon\big(\frac{\partial w_\varepsilon}{\partial x_i}\big) - \mathcal{M}_\varepsilon\big(\frac{\partial w_\varepsilon}{\partial x_i}\big)\Big) \rightharpoonup y^c \cdot \nabla\Big(\frac{\partial w}{\partial x_i}\Big) + \widehat{w}_i \quad \text{weakly in } L^p(\Omega \times Y).$$

Then, from (1.54) and (1.55) we derive the equality

$$\frac{\partial Z}{\partial y_i} = y^c \cdot \nabla\Big(\frac{\partial w}{\partial x_i}\Big) + \widehat{w}_i.$$

Set

$$\widehat{W} = Z - \frac{1}{2}\sum_{i,j=1}^N \big(y_i^c\, y_j^c - \mathcal{M}_Y(y_i^c\, y_j^c)\big)\frac{\partial^2 w}{\partial x_i \partial x_j}.$$

By construction, the function $\widehat{W}$ belongs to $L^p(\Omega; W^{2,p}(Y))$ and verifies $\mathcal{M}_Y(\widehat{W}) = 0$. Furthermore,

$$\frac{\partial \widehat{W}}{\partial y_i} = \frac{\partial Z}{\partial y_i} - y^c \cdot \nabla\Big(\frac{\partial w}{\partial x_i}\Big) = \widehat{w}_i,$$

and therefore $\mathcal{M}_Y(\nabla_y \widehat{W}) = 0$ since $\mathcal{M}_Y(\widehat{w}_i) \equiv 0$. This implies that $\widehat{W}$ is Y-periodic, hence belongs to $L^p(\Omega; W^{2,p}_{per}(Y))$. Finally from (1.4.4), one gets

$$D_y^l Z = D^l w + D_y^l \widehat{W} \qquad \text{for} \quad |l| = 2,$$

which together with $(1.55)_3$, prove the last convergence of (1.53). ☐

The next result is easily proved.

Corollary 1.48. *Suppose $p \in (1, +\infty)$. Let $\{w_\varepsilon\}_\varepsilon$ be a sequence in $W^{2,p}(\Omega)$, such that*

$$w_\varepsilon \rightharpoonup w \quad \textit{weakly in } W^{2,p}(\Omega).$$

Then, up to a subsequence, there exists $\widehat{W}$ in $L^p(\Omega; W^{2,p}_{per}(Y))$ satisfying $\mathcal{M}_Y(\widehat{W}) = 0$, and such that, weakly in $L^p(\Omega; W^{2,p}(Y))$,

$$\frac{1}{\varepsilon^2}\Big[\mathcal{T}_\varepsilon(w_\varepsilon) - \mathcal{M}_\varepsilon(w_\varepsilon) - y^c \cdot \mathcal{M}_\varepsilon(\nabla w_\varepsilon)\Big] \rightharpoonup \frac{1}{2}\sum_{i,j=1}^N \Big(y_i^c\, y_j^c - \mathcal{M}_Y(y_i^c\, y_j^c)\Big)\frac{\partial^2 w}{\partial x_i \partial x_j} + \widehat{W}.$$

1.5 Unfolding with parameters and iterated unfolding

In this section, we add a set of parameters $\mathcal{O}$ which is a bounded domain in $\mathbb{R}^d$ (more generally an abstract σ-finite measure space). Consider the space $L^p(\mathcal{O} \times \Omega)$. One can unfold with respect to the second variable $x \in \Omega$ for almost every $t \in \mathcal{O}$ (see Definition 1.2), while the first variable is a parameter.

Definition 1.49 (The partial unfolding operator). *For ϕ a measurable function on $\mathcal{O} \times \Omega$, the unfolding operator $\mathcal{T}_\varepsilon^Y$ is defined as follows:*

$$\mathcal{T}_\varepsilon^Y(\phi)(t,x,y) = \begin{cases} \phi\Big(t, \varepsilon\Big[\dfrac{x}{\varepsilon}\Big]_Y + \varepsilon y\Big) & \text{for a.e.} \quad (t,x,y) \in \mathcal{O} \times \widehat{\Omega}_\varepsilon \times Y, \\ 0 & \text{for a.e.} \quad (t,x,y) \in \mathcal{O} \times \Lambda_\varepsilon \times Y. \end{cases}$$

In the case where $\mathcal{O}$ is a domain in some $\mathbb{R}^M$, $M \in \mathbb{N}^*$, this operator maps $L^p(\mathcal{O} \times \Omega)$ into $L^p(\mathcal{O} \times \Omega \times Y)$ for $p \in [1, +\infty]$.

This is a partial unfolding and it has properties similar to the original unfolding operator without parameters $\mathcal{T}_\varepsilon$. We shall only state a few of them, listed in the next proposition (with a reference to the original statements). Note that the partial unfolding operator commutes with any derivation with respect to the parameter t in the case where $\mathcal{O}$ is a domain in some $\mathbb{R}^M$ (for M in $\mathbb{N}^*$) and the original function ϕ is differentiable. We give some examples in the next proposition.

The approach of partial unfolding is useful in evolution problems (where t is the time variable) as well as in applications where iterated unfolding is used (as exemplified in Subsection 3.1.4).

The approach we present here is also revisited with a slightly different point of view in Chapter 7.

Proposition 1.50.
(i) (cf. Proposition 1.8). For ϕ in $L^1(\mathcal{O} \times \Omega)$,

$$\frac{1}{|Y|}\int_{\mathcal{O}\times\Omega\times Y} \mathcal{T}_\varepsilon^Y(\phi)(t,x,y)\, dxdydt = \int_{\mathcal{O}\times\widehat{\Omega}_\varepsilon} \phi(t,x)\, dxdt.$$

Suppose $p \in [1, +\infty)$.
(ii) (cf. Proposition 1.8). For w in $L^p(\mathcal{O} \times \Omega)$,

$$\|\mathcal{T}_\varepsilon^Y(w)\|_{L^p(\mathcal{O}\times\Omega\times Y)} \leqslant |Y|^{1/p}\, \|w\|_{L^p(\mathcal{O}\times\Omega)}.$$

(iii) (cf. Proposition 1.9). Let $\{w_\varepsilon\}_\varepsilon$ be a sequence in $L^p(\mathcal{O} \times \Omega)$ converging strongly to w. Then

$$\mathcal{T}_\varepsilon^Y(w_\varepsilon) \to w \quad \textit{strongly in} \quad L^p(\mathcal{O} \times \Omega \times Y).$$

If $\{w_\varepsilon\}_\varepsilon$ converges strongly to w in $L^p(\Omega; W^{1,p}(\mathcal{O}))$, then

$$\mathcal{T}_\varepsilon^Y(w_\varepsilon) \to w \quad \text{strongly in } L^p(\Omega \times Y; W^{1,p}(\mathcal{O})).$$

Suppose $p \in (1, +\infty)$.
(iv) (cf. Proposition 1.12). For every relatively weakly compact sequence $\{w_\varepsilon\}_\varepsilon$ in $L^p(\mathcal{O} \times \otimes)$, the corresponding sequence $\{\mathcal{T}_\varepsilon^Y(w_\varepsilon)\}_\varepsilon$ is relatively weakly compact in $L^p(\mathcal{O} \times \otimes \times \mathcal{Y})$.

Furthermore, if

$$\mathcal{T}_\varepsilon^Y(w_\varepsilon) \rightharpoonup \widehat{w} \quad \text{weakly in } L^p(\mathcal{O} \times \Omega \times Y),$$

then

$$w_\varepsilon \rightharpoonup \mathcal{M}_Y(\widehat{w}) \quad \text{weakly in } L^p(\mathcal{O} \times \Omega),$$

and

$$\|\widehat{w}\|_{L^p(\mathcal{O}\times\Omega\times Y)} \leqslant \liminf_{\varepsilon\to 0} |Y|^{1/p}\, \|w_\varepsilon\|_{L^p(\mathcal{O}\times\Omega)}.$$

(v) (cf. Theorem 1.41). Let $\{w_\varepsilon\}_\varepsilon$ be a sequence in $L^p(\mathcal{O}; W^{1,p}(\Omega))$, such that

$$w_\varepsilon \rightharpoonup w \quad \text{weakly in } L^p(\mathcal{O}; W^{1,p}(\Omega)).$$

Then, up to a subsequence, there exists some $\widehat{w}$ in $L^p(\mathcal{O}\times\Omega; W^{1,p}_{per,0}(Y))$ such that

$$\mathcal{T}_\varepsilon^Y(\nabla_x w_\varepsilon) \rightharpoonup \nabla_x w + \nabla_y \widehat{w} \quad \text{weakly in } L^p(\mathcal{O} \times \Omega \times Y)^N.$$

Suppose $p \in (1, +\infty)$ and let $\mathcal{O}$ be a domain in $\mathbb{R}^M$ for $M \in \mathbb{N}^*$.
(vi) (cf. Theorem 1.36). Let $\{w_\varepsilon\}_\varepsilon$ be a sequence in $W^{1,p}(\mathcal{O} \times \Omega)$, such that

$$w_\varepsilon \rightharpoonup w \quad \text{weakly in } L^p(\mathcal{O} \times \Omega)$$

with $\varepsilon\|\nabla_x w_\varepsilon\|_{L^p(\mathcal{O}\times\Omega)} + \|\nabla_t w_\varepsilon\|_{L^p(\mathcal{O}\times\Omega)}$ bounded. Then, up to a subsequence, there exists some $\widehat{w}$ in $L^p(\Omega; W^{1,p}(\mathcal{O} \times Y)) \cap L^p(\mathcal{O} \times \Omega; W^{1,p}_{per}(Y))$ such that

$$\begin{aligned}
&\mathcal{T}_\varepsilon^Y(w_\varepsilon) \rightharpoonup \widehat{w} \quad \text{weakly in } L^p(\Omega; W^{1,p}(\mathcal{O} \times Y)),\\
&\varepsilon\mathcal{T}_\varepsilon^Y(\nabla_x w_\varepsilon) \rightharpoonup \nabla_y \widehat{w} \quad \text{weakly in } L^p(\mathcal{O} \times \Omega \times Y)^N,\\
&\mathcal{T}_\varepsilon^Y(\nabla_t w_\varepsilon) \rightharpoonup \nabla_t \widehat{w} \quad \text{weakly in } L^p(\mathcal{O} \times \Omega \times Y)^M.
\end{aligned}$$

(vii) (cf. Theorem 1.41; compare also with Theorem 7.1). Let $\{w_\varepsilon\}_\varepsilon$ be a sequence in $W^{1,p}(\mathcal{O} \times \Omega)$, such that

$$w_\varepsilon \to w \quad \text{weakly in } W^{1,p}(\mathcal{O} \times \Omega).$$

Then, up to a subsequence, there exists some $\widehat{w}$ in $L^p(\mathcal{O}\times\Omega; W^{1,p}_{per,0}(Y))$ such that

$$\begin{aligned}
&\mathcal{T}_\varepsilon^Y(\nabla_x w_\varepsilon) \rightharpoonup \nabla_x w + \nabla_y \widehat{w} \quad \text{weakly in } L^p(\mathcal{O} \times \Omega \times Y)^N,\\
&\mathcal{T}_\varepsilon^Y(\nabla_t w_\varepsilon) \rightharpoonup \nabla_t w \quad \text{weakly in } L^p(\mathcal{O} \times \Omega \times Y)^M.
\end{aligned}$$

Let $\mathbf{H}$ be a second additive subgroup generated by a basis $\mathbf{C} \doteq (\mathbf{c}_1, .., \mathbf{c}_N)$ of $\mathbb{R}^N$, and Z be the open parallelotope associated with this basis. Let δ be given strictly positive. One can define the partial unfolding operator $\mathcal{T}_\delta^Z$ which unfolds the second variable $y \in Y$ into $(y, z) \in Y \times Z$ (with the definitions for $\widehat{Y}_\delta$ and Λ_δ^Y similar to (1.3)).

Definition 1.51 (The iterated unfolding operator). *For ϕ in $L^p(\Omega \times Y)$, the iterated unfolding operator $\mathcal{T}_\delta^Z$ is defined as follows:*

$$\mathcal{T}_\delta^Z(\phi)(x, y, z) = \begin{cases} \phi\Big(x, \delta\Big[\dfrac{y}{\delta}\Big]_Z + \delta z\Big) & \text{for a.e.} \quad (x, y, z) \in \Omega \times \widehat{Y}_\delta \times Z, \\ 0 & \text{for a.e.} \quad (x, y, z) \in \Omega \times \Lambda_\delta^Y \times Z. \end{cases}$$

Suppose a sequence $\{(\varepsilon, \delta)\}$ in $\mathbb{R}_+^* \times \mathbb{R}_+^*$ going to $(0,0)$ is given. Consider a sequence $\{w_{\varepsilon,\delta}\}_{(\varepsilon,\delta)}$ which is bounded in $W^{1,p}(\Omega)$ $(1 < p < +\infty)$. In this framework, Proposition 1.50(ii) implies the following result:

Proposition 1.52. *Under the above hypotheses, up to a subsequence, there are w_0 in $W^{1,p}(\Omega)$, $\widehat{w}_1$ in $L^p(\Omega; W_{per,0}^{1,p}(Y))$ and $\widehat{w}_2$ in $L^p(\Omega \times Y; W_{per,0}^{1,p}(Z))$, such that*

$$\mathcal{T}_\varepsilon^Y(\nabla w_{\varepsilon,\delta}) \rightharpoonup \nabla w_0 + \nabla_y \widehat{w}_1 \qquad \text{weakly in} \ \ L^p(\Omega \times Y)^N,$$

$$\mathcal{T}_\delta^Z \circ \mathcal{T}_\varepsilon^Y(\nabla w_{\varepsilon,\delta}) \rightharpoonup \nabla w_0 + \nabla_y \widehat{w}_1 + \nabla_z \widehat{w}_2 \qquad \text{weakly in} \ \ L^p(\Omega \times Y \times Z)^N,$$

as well as

$$W_\delta \doteq \frac{1}{\varepsilon}\big(\mathcal{T}_\varepsilon^Y(w_{\varepsilon,\delta} - \mathcal{M}_\varepsilon^Y(w_{\varepsilon,\delta}))\big) \rightharpoonup y^c \cdot \nabla w_0 + \widehat{w}_1 \qquad \text{weakly in} \ \ L^p(\Omega; W^{1,p}(Y)),$$

$$\frac{1}{\delta}\big(\mathcal{T}_\delta^Z(W_\delta) - \mathcal{M}_\delta^Z(W_\delta)\big) \rightharpoonup z^c \cdot (\nabla w_0 + \nabla_y \widehat{w}_1) + \widehat{w}_2 \qquad \text{weakly in} \ \ L^p(\Omega \times Y; W^{1,p}(Z)).$$

Remark 1.53. *Similar results hold for any finite number of microscopic scales ε, $\varepsilon\delta_1$, $\varepsilon\delta_1\delta_2$, $\ldots$, $\varepsilon\delta_1\delta_2\ldots\delta_k$ with $(\varepsilon, \delta_1, \delta_2, \ldots, \delta_k) \to (0, 0, 0, \ldots, 0)$ without any condition of separation of the scales.*

1.6 Macro-micro decomposition and the scale-splitting operators

In this section, we give a different proof of Theorem 1.41 which was the one given originally in [60]. The method will be used also in the case of domains with holes in Chapter 4).

The procedure is based on a splitting of functions ϕ in $W^{1,p}(\Omega)$ (or in $W_0^{1,p}(\Omega)$) for $p \in [1,+\infty]$, in the form

$$\phi = \mathcal{Q}_\varepsilon(\phi) + \mathcal{R}_\varepsilon(\phi),$$

where $\mathcal{Q}_\varepsilon(\phi)$ is an approximation of ϕ having the same behavior as ϕ, while $\mathcal{R}_\varepsilon(\phi)$ is a remainder of order ε.

The results of Section 1.4 can be seen as a zero order scale splitting where the main approximation is piece-wise constant (operator $\mathcal{M}_\varepsilon$). Here $\mathcal{Q}_\varepsilon(\varphi)$ is always in $W^{1,\infty}(\Omega)$ which improves on some estimates.

The construction of $\mathcal{Q}_\varepsilon$ is based on the Q_1–interpolate of some discrete approximation, as is customary in the Finite Element Method ([56,178]). The idea of using these type of interpolate was already present in Griso [116,119] for the study of truss-like structures.

For nodal values of the Q_1–interpolate, one cannot take the values of the considered functions (they do not make sense in general except for $N = 1$ for functions in $W^{1,p}$ since these functions are continuous). Therefore we take the averages on the cells $\varepsilon\xi + \varepsilon Y$(9). The explicit formulas for the Q_1 interpolates (well-known in the Finite Element method) are recalled first.

Without loss of generality, we use the coordinate system on $\mathbb{R}^N$ associated with the basis B, i.e.,

$$z = \sum_{i=1}^N z_i \mathbf{b}_i,$$

and for convenience, we shall write $z = (z_1, \dots, z_N)$. Under this convention,

$$\kappa = (\kappa_1, \dots, \kappa_N) \in \{0,1\}^N \quad \text{is identified with } \sum_{i=1}^N \kappa_i \mathbf{b}_i.$$

Let $\mathbf{Q}_1(Y)$ denote the finite dimensional space of polynomials on $\overline{Y}$ which are the Lagrange interpolate of their values at the vertices of $\overline{Y}$ (it is the set of all polynomials of degree less or equal to 1 separately with respect to each variable). A basis of $\mathbf{Q}_1(Y)$ is obtained by choosing the 2^N elements $\{\mathbf{q}_\kappa \mid \kappa \in \{0,1\}^N\}$ (these are the **shape functions** in the theory of Finite Elements, see for example [56]) such that

$$\begin{aligned}
&\mathbf{q}_\kappa(\kappa) = 1,\\
&\mathbf{q}_\kappa(\kappa') = 0 \ \text{ at all other } 2^N - 1 \text{ vertices } \kappa' \neq \kappa \text{ of } \overline{Y}.
\end{aligned}$$

Consequently, with the above conventions,

$$\mathbf{q}_\kappa(y) \doteq \prod_{\ell=1}^N y_\ell^{\kappa_\ell}(1-y_\ell)^{1-\kappa_\ell} \quad \text{for } \kappa \text{ in } \{0,1\}^N \text{ and } y \text{ in } \overline{Y}.$$

(9) For $N = 1$ and functions in $W^{1,p}$, taking the actual Lagrange interpolate to define the operators $\mathcal{Q}_\varepsilon$ makes for slightly simpler proofs.

At some point, the following notations will also be used, for $1 \leqslant i < j \leqslant N$,

$$\mathbf{q}^i_\kappa(y) \doteq \prod_{\ell=1,\ell\neq i}^{N} y_\ell^{\kappa_\ell}(1-y_\ell)^{1-\kappa_\ell} \qquad \text{for } \kappa \text{ in } \{0,1\}^N \text{ and } y \text{ in } \overline{Y},$$

as well as

$$\mathbf{q}^{ij}_\kappa(y) \doteq \prod_{\ell=1,\ell\neq i,j}^{N} y_\ell^{\kappa_\ell}(1-y_\ell)^{1-\kappa_\ell} \quad \text{for } \kappa \text{ in } \{0,1\}^N \text{ and } y \text{ in } \overline{Y}.$$

Remark 1.54. *It is well known (cf. [56]) that for all y in $\overline{Y}$,*

$$\sum_{\kappa\in\{0,1\}^N} \mathbf{q}_\kappa(y) = 1,$$

and for all y in $\overline{Y}$, and $1 \leqslant i \leqslant N$,

$$\sum_{\kappa\in\mathbf{K}_i} \mathbf{q}^i_\kappa(y) = 1 \quad \textit{where } \mathbf{K}_i = \{\kappa \in \{0,1\}^N \mid \kappa_i = 1\}.$$

It follows that for all y in $\overline{Y}$,

$$\sum_{\kappa\in\{0,1\}^N} \mathbf{q}_\kappa(y)\kappa = y.$$

Similarly, for all y in $\overline{Y}$ and $1 \leqslant i < j \leqslant N$,

$$\sum_{\kappa\in\mathbf{K}_{ij}} \mathbf{q}^{ij}_\kappa(y) = 1 \quad \textit{where } \mathbf{K}_{ij} = \{\kappa \in \{0,1\}^N \mid \kappa_i = \kappa_j = 1\}.$$

Note also that

$$\begin{aligned}
&\int_Y \mathbf{q}_\kappa(y)\,dy = \frac{|Y|}{2^N}, \qquad \int_Y \mathbf{q}^i_\kappa(y)\,dy = \frac{|Y|}{2^{N-1}} \quad \textit{for } 1 \leqslant i \leqslant N,\\
&\int_Y \mathbf{q}^{ij}_\kappa(y)\,dy = \frac{|Y|}{2^{N-2}} \quad \textit{for } 1 \leqslant i < j \leqslant N.
\end{aligned} \tag{1.56}$$

The gradients of elements of $\mathbf{Q}_1(Y)$ are polynomials of degree at most one with respect to each basis variable and furthermore, for every Ψ in $\mathbf{Q}_1(Y)$ and $i = 1,\ldots,N$, $\dfrac{\partial\Psi}{\partial y_i}$ is independent of y_i and separately affine with respect to each $y_j, j \neq i$.

More precisely, the following formulas hold for every y in $\overline{Y}$ and Ψ in $\mathbf{Q}_1(Y)$:

$$\begin{aligned}
\Psi(y) &= \sum_{\kappa\in\{0,1\}^N} \Psi(\kappa)\,\mathbf{q}_\kappa(y),\\
\frac{\partial\Psi}{\partial y_i}(y) &= \Psi(y_1,\ldots,\underset{\mathrm{i}}{1},\ldots,y_N) - \Psi(y_1,\ldots,\underset{\mathrm{i}}{0},\ldots,y_N)\\
&= \sum_{\kappa\in\mathbf{K}_i} \big(\Psi(\kappa+\mathbf{b}_i) - \Psi(\kappa)\big)\mathbf{q}^i_\kappa(y).
\end{aligned} \tag{1.57}$$

Therefore, for every $i \in \{1, \ldots, N\}$,

$$\frac{\partial^2 \Psi}{\partial y_i^2}(y) \equiv 0. \tag{1.58}$$

Similarly, for $1 \leqslant i < j \leqslant N$,

$$\begin{aligned} \frac{\partial^2 \Psi}{\partial y_i \partial y_j}(y) &= \Psi(y_1, \ldots, \underset{\mathrm{i}}{1}, \ldots, \underset{\mathrm{j}}{1}, \ldots, y_N) - \Psi(y_1, \ldots, \underset{\mathrm{i}}{1}, \ldots, \underset{\mathrm{j}}{0}, \ldots, y_N) \\ &\quad - \Psi(y_1, \ldots, \underset{\mathrm{i}}{0}, \ldots, \underset{\mathrm{j}}{1}, \ldots, y_N) + \Psi(y_1, \ldots, \underset{\mathrm{i}}{0}, \ldots, \underset{\mathrm{j}}{0}, \ldots, y_N) \\ &= \sum_{\kappa \in \mathbf{K}_{ij}} \big(\Psi(\kappa + \mathbf{b}_i + \mathbf{b}_j) - \Psi(\kappa + \mathbf{b}_i) - \Psi(\kappa + \mathbf{b}_j) + \Psi(\kappa)\big) \mathbf{q}_\kappa^{ij}(y). \end{aligned} \tag{1.59}$$

Observe that this second derivative is constant with respect to y_i and y_j, and separately affine with respect to the other variables.

Lemma 1.55. *Suppose $p \in [1, +\infty]$. There exists a constant C depending on N and Y (but independent of p), such that for $\phi \in \mathbf{Q}_1(Y)$ and $\psi \in L^p(Y)$,*

$$\begin{aligned} &(i) \quad \|\phi\,\psi\|_{L^p(Y)} \leqslant C\,\|\phi\|_{L^p(Y)} \|\psi\|_{L^p(Y)}, \\ &(ii) \quad \|\phi - \phi(0)\|_{L^p(Y)} \leqslant C\,\|\nabla_y \phi\|_{L^p(Y)}. \end{aligned} \tag{1.60}$$

Proof. Let ϕ be in $\mathbf{Q}_1(Y)$ and $\psi \in L^p(Y)$.
(i) For $p = +\infty$, inequality (1.60)(i) is obvious.

For $p \in [1, +\infty)$, (1.60)(i) follows from the existence of a constant C (depending only on N and Y) such that for every $\phi \in \mathbf{Q}_1(Y)$,

$$\|\phi\|_{L^\infty(Y)} \leqslant C\,\|\phi\|_{L^p(Y)},$$

consequence of the fact that the space $\mathbf{Q}_1(Y)$ is of dimension 2^N so all its norms are equivalent, C can be chosen independent of p by using the constant for $p = 1$ and applying the Hölder inequality

$$\|\phi\|_{L^1(Y)} \leqslant \|\phi\|_{L^p(Y)} |Y|^{1-1/p}.$$

(ii) The subspace of $\mathbf{Q}_1(Y)$ consisting of elements that vanish at the origin (of $\overline{Y}$), is of dimension $2^N - 1$. Hence, all its norms are equivalent. So, there is a constant C (depending only on N and Y) such that for every $\phi \in \mathbf{Q}_1(Y)$ with $\phi(0) = 0$,

$$\|\phi\|_{L^\infty(Y)} \leqslant C\,\|\nabla_y \phi\|_{L^1(Y)}.$$

Consequently, for every ϕ in $\mathbf{Q}_1(Y)$ the following inequality holds:

$$\|\phi - \phi(0)\|_{L^\infty(Y)} \leqslant C\,\|\nabla_y \phi\|_{L^p(Y)}.$$

As before, the constant C can be chosen independently of p. □

1.6.1 The case of the whole space

As a tool (and to make notations easier), we first introduce a kind of averaging map on $L^p(\mathbb{R}^N)$.

Definition 1.56 (Average over a line segment). *Let* $\mathbf{a}$ *be in* $\mathbb{R}^N$. *For* φ *continuous on* $\mathbb{R}^N$, *set*

$$\mathbb{M}_{\mathbf{a}}(\varphi)(x) = \int_0^1 \varphi(x + t\mathbf{a})\,dt \qquad \text{for } x \in \mathbb{R}^N,$$

which itself is continuous.

Lemma 1.57. *For* φ *continuous and bounded on* $\mathbb{R}^N$, *the following inequality holds:*

$$\|\mathbb{M}_{\mathbf{a}}(\varphi)\|_{L^\infty(\mathbb{R}^N)} \leqslant \|\varphi\|_{L^\infty(\mathbb{R}^N)}.$$

If φ *is continuous with compact support, then,*

$$\|\mathbb{M}_{\mathbf{a}}(\varphi)\|_{L^p(\mathbb{R}^N)} \leqslant \|\varphi\|_{L^p(\mathbb{R}^N)} \qquad \text{for all } p \in [1, +\infty).$$

If φ *is in* $C^1(\mathbb{R}^N)$, *then so is* $\mathbb{M}_{\mathbf{a}}(\phi)$ *and*

$$\mathbb{M}_{\mathbf{a}}(\nabla\phi) = \nabla\mathbb{M}_{\mathbf{a}}(\phi) \qquad \text{a.e. in } \mathbb{R}^N. \tag{1.61}$$

Moreover, for ϕ *and* ψ *continuous (with one at least with compact support),*

$$\int_{\mathbb{R}^N} \mathbb{M}_{\mathbf{a}}(\phi)(x)\,\psi(x)\,dx = \int_{\mathbb{R}^N} \phi(x)\,\mathbb{M}_{-\mathbf{a}}(\psi)(x)\,dx. \tag{1.62}$$

Proof. The first inequality is obvious. For $p = 1$, the second inequality follows directly from Fubini's theorem applied to the function $(x,t) \mapsto |\varphi(x + t\mathbf{a})|$. For $1 < p < +\infty$, one uses in the same way Fubini's theorem after applying Hölder's inequality (which implies $|\mathbb{M}_{\mathbf{a}}(\varphi)|^p \leqslant \mathbb{M}_{\mathbf{a}}(|\varphi|^p)$). The third equality is a straightforward derivation under the integral. The last equality (1.62) follows again by Fubini's theorem. □

A density argument (weak-$*$ density for $p = +\infty$) gives

Proposition 1.58. *For* $p \in [1, +\infty]$, *the linear operator* $\mathbb{M}_{\mathbf{a}}$ *has a unique extension, still denoted* $\mathbb{M}_{\mathbf{a}}$, *from* $L^p(\mathbb{R}^N)$ *into itself and with norm 1.*

For every $\phi \in L^p(\mathbb{R}^N)$ *and* $\psi \in L^{p'}(\mathbb{R}^N)$, *one has*

$$\int_{\mathbb{R}^N} \mathbb{M}_{\mathbf{a}}(\phi)(x)\psi(x)\,dx = \int_{\mathbb{R}^N} \phi(x)\mathbb{M}_{-\mathbf{a}}(\psi)(x)\,dx.$$

Furthermore, for every φ *in* $W^{1,p}(\mathbb{R}^N)$, $\mathbb{M}_{\mathbf{a}}(\phi)$ *is also in* $W^{1,p}(\mathbb{R}^N)$ *and*

$$\mathbb{M}_{\mathbf{a}}(\nabla\phi) = \nabla\mathbb{M}_{\mathbf{a}}(\phi) \quad \text{a.e. in } \mathbb{R}^N.$$

The next lemma is proved in a similar fashion as Proposition 1.25.

Lemma 1.59.

Suppose $p \in [1, +\infty)$.

(i) Let $\{w_\varepsilon\}_\varepsilon$ *be a sequence in* $L^p(\mathbb{R}^N)$ *strongly convergent to* w. *Then*

$$\mathrm{M}_{\varepsilon\mathbf{a}}(w_\varepsilon) \to w \quad \textit{strongly in } \ L^p(\mathbb{R}^N).$$

In particular, for every $\phi \in L^p(\mathbb{R}^N)$,

$$\mathrm{M}_{\varepsilon\mathbf{a}}(\phi) \to \phi \quad \textit{strongly in } \ L^p(\mathbb{R}^N).$$

(ii) For any $\phi \in W^{1,p}(\mathbb{R}^N)$, *one has* $\nabla\big(\mathrm{M}_{\mathbf{a}}(\phi)\big) = \mathrm{M}_{\mathbf{a}}(\nabla\phi)$, *and*

$$\|\mathrm{M}_{\mathbf{a}}(\phi)\|_{W^{1,p}(\mathbb{R}^N)} \leqslant \|\phi\|_{W^{1,p}(\mathbb{R}^N)}.$$

Moreover, if $\{w_\varepsilon\}_\varepsilon$ *is a sequence converging strongly to* w *in* $W^{1,p}(\mathbb{R}^N)$, *then*

$$\mathrm{M}_{\varepsilon\mathbf{a}}(w_\varepsilon) \to w \quad \textit{strongly in } \ W^{1,p}(\mathbb{R}^N).$$

Suppose $p \in (1, +\infty)$.

(iii) Let $\{w_\varepsilon\}_\varepsilon$ *be a sequence in* $L^p(\mathbb{R}^N)$ *weakly convergent to* w. *Then*

$$\mathrm{M}_{\varepsilon\mathbf{a}}(w_\varepsilon) \rightharpoonup w \quad \textit{weakly in } \ L^p(\mathbb{R}^N).$$

(iv) Let $\{w_\varepsilon\}_\varepsilon$ *be a sequence in* $W^{1,p}(\mathbb{R}^N)$ *weakly convergent to* w. *Then*

$$\mathrm{M}_{\varepsilon\mathbf{a}}(w_\varepsilon) \rightharpoonup w \quad \textit{weakly in } \ W^{1,p}(\mathbb{R}^N).$$

For $p = +\infty$, *(iii) and (iv) hold with the weak-$*$ topologies.*

Definition 1.60 ($\mathbf{Q_1}$-interpolating operator). *For* $p \in [1, +\infty]$, *the operator*

$$\widetilde{\mathcal{Q}}_\varepsilon : L^p(\mathbb{R}^N) \longrightarrow W^{1,+\infty}(\mathbb{R}^N)$$

(with the convention of Remark 1.22) is defined as follows:

$$\begin{aligned} &\textit{for } \xi \in \mathbf{G}, && \widetilde{\mathcal{Q}}_\varepsilon(\phi)(\varepsilon\xi) = \widetilde{\mathcal{M}}_\varepsilon(\phi)(\varepsilon\xi),\\ &\textit{for } x \in \mathbb{R}^N, && \widetilde{\mathcal{Q}}_\varepsilon(\phi)(x) \textit{ is the } Q_1\textit{-interpolate of the values of } \widetilde{\mathcal{Q}}_\varepsilon(\phi)\\ & && \textit{at the vertices of the cell } \varepsilon\Big[\frac{x}{\varepsilon}\Big]_Y + \varepsilon Y. \end{aligned}$$

One can check the following explicit formulas for $\phi \in L^p(\mathbb{R}^N)$ and for all x in $\mathbb{R}^N$:

$$\widetilde{\mathcal{Q}}_\varepsilon(\phi)(x) = \sum_{\kappa\in\{0,1\}^N} \widetilde{\mathcal{M}}_\varepsilon(\phi)\Big(\varepsilon\Big[\frac{x}{\varepsilon}\Big]_Y + \varepsilon\kappa\Big)\mathbf{q}_\kappa\Big(\Big\{\frac{x}{\varepsilon}\Big\}_Y\Big), \tag{1.63}$$

and from (1.57), for $1 \leqslant i \leqslant N$,

$$\begin{aligned}\frac{\partial \widetilde{\mathcal{Q}}_\varepsilon(\phi)}{\partial x_i}(x) = \sum_{\kappa\in\mathbf{K}_i} \frac{1}{\varepsilon}\Big\{\widetilde{\mathcal{M}}_\varepsilon(\phi)\Big(\varepsilon\Big[\frac{x}{\varepsilon}\Big]_Y + \varepsilon(\kappa+\mathbf{b}_i)\Big)\\ - \widetilde{\mathcal{M}}_\varepsilon(\phi)\Big(\varepsilon\Big[\frac{x}{\varepsilon}\Big]_Y + \varepsilon\kappa\Big)\Big\}\mathbf{q}^i_\kappa\Big(\Big\{\frac{x}{\varepsilon}\Big\}_Y\Big),\end{aligned} \tag{1.64}$$

and from (1.59), for $1 \leqslant i < j \leqslant N$,

$$\begin{aligned}\frac{\partial^2 \widetilde{\mathcal{Q}}_\varepsilon(\phi)}{\partial x_i \partial x_j}(x) = \sum_{\kappa\in\mathbf{K}_{ij}} \frac{1}{\varepsilon^2}\Big\{\widetilde{\mathcal{M}}_\varepsilon(\phi)\Big(\varepsilon\Big[\frac{x}{\varepsilon}\Big]_Y + \varepsilon(\kappa+\mathbf{b}_i+\mathbf{b}_j)\Big)\\ - \widetilde{\mathcal{M}}_\varepsilon(\phi)\Big(\varepsilon\Big[\frac{x}{\varepsilon}\Big]_Y + \varepsilon(\kappa+\mathbf{b}_i)\Big)\\ - \widetilde{\mathcal{M}}_\varepsilon(\phi)\Big(\varepsilon\Big[\frac{x}{\varepsilon}\Big]_Y + \varepsilon(\kappa+\mathbf{b}_j)\Big)\\ + \widetilde{\mathcal{M}}_\varepsilon(\phi)\Big(\varepsilon\Big[\frac{x}{\varepsilon}\Big]_Y + \varepsilon\kappa\Big)\Big\}\mathbf{q}^{ij}_\kappa\Big(\Big\{\frac{x}{\varepsilon}\Big\}_Y\Big).\end{aligned} \tag{1.65}$$

Remark 1.61. *By construction, the function $\widetilde{\mathcal{Q}}_\varepsilon(\phi)$ is separately piece-wise affine (with respect to the variables x_i defined on the basis $\mathbf{B}$) on each cell $\varepsilon(\xi+Y)$. Observe also that for any $i \in \{1,\dots,N\}$, the derivative $\partial\widetilde{\mathcal{Q}_\varepsilon}(\phi)/\partial x_i$ is independent of x_i in each cell $\varepsilon(\xi+Y)$, while it is separately piece-wise affine with respect to the other variables $x_j, j \neq i$. A similar remark can be made for the second order crossed derivatives.*

The next result gives the expressions of $\widetilde{\mathcal{Q}}_\varepsilon(\varphi)$ and of its derivatives in terms of the shape functions.

Lemma 1.62.
(i) For φ in $W^{1,p}(\mathbb{R}^N)$, for $i \in \{1,\dots,N\}$ and for a.e. $x \in \mathbb{R}^N$,

$$\frac{\partial \widetilde{\mathcal{Q}}_\varepsilon(\phi)}{\partial x_i}(x) = \sum_{\kappa\in\mathbf{K}_i} \widetilde{\mathcal{M}}_\varepsilon \circ \mathrm{M}_{\varepsilon\mathbf{b}_i}\Big(\frac{\partial\phi}{\partial x_i}\Big)\Big(\varepsilon\Big[\frac{x}{\varepsilon}\Big]_Y\Big)\mathbf{q}^i_\kappa\Big(\Big\{\frac{x}{\varepsilon}\Big\}_Y\Big) \tag{1.66}$$

and

$$\begin{aligned}&\big(\widetilde{\mathcal{Q}}_\varepsilon - \widetilde{\mathcal{M}}_\varepsilon\big)(\phi)(x)\\ &= \sum_{\kappa\in\{0,1\}^N}\sum_{i=1}^N \widetilde{\mathcal{M}}_\varepsilon \circ \mathrm{M}_{\varepsilon\kappa}\Big(\frac{\partial\phi}{\partial x_i}\Big)\Big(\varepsilon\Big[\frac{x}{\varepsilon}\Big]_Y\Big)\kappa_i\mathbf{q}_\kappa\Big(\Big\{\frac{x}{\varepsilon}\Big\}_Y\Big).\end{aligned} \tag{1.67}$$

(ii) For φ in $W^{2,p}(\mathbb{R}^N)$, for $(i,j) \in \{1,\dots,N\}^2$, $i < j$, and for a.e. $x \in \mathbb{R}^N$,

$$\begin{aligned}&\frac{\partial^2 \widetilde{\mathcal{Q}}_\varepsilon(\phi)}{\partial x_i \partial x_j}(x)\\ &= \sum_{\kappa\in\mathbf{K}_{ij}} \widetilde{\mathcal{M}}_\varepsilon \circ \mathrm{M}_{\varepsilon\mathbf{b}_j} \circ \mathrm{M}_{\varepsilon\mathbf{b}_i}\Big(\frac{\partial^2\phi}{\partial x_i\partial x_j}\Big)\Big(\varepsilon\Big[\frac{x}{\varepsilon}\Big]_Y\Big)\mathbf{q}^{ij}_\kappa\Big(\Big\{\frac{x}{\varepsilon}\Big\}_Y\Big).\end{aligned} \tag{1.68}$$

Proof. *(i)* Let ϕ be in $\mathcal{D}(\mathbb{R}^N)$. For every ξ in $\mathbf{G}$ and every $\kappa \in \{0,1\}^N$,

$$\begin{aligned}
\frac{1}{\varepsilon}\big(\widetilde{\mathcal{M}}_\varepsilon(\phi)(\varepsilon\xi + \varepsilon\kappa) - \widetilde{\mathcal{M}}_\varepsilon(\phi)(\varepsilon\xi)\big)& \\
&= \frac{1}{\varepsilon^N|Y|}\int_{\varepsilon\xi+\varepsilon Y}\int_0^1 \nabla\phi(x+t\varepsilon\kappa)\cdot\kappa\,dx\,dt \\
&= \widetilde{\mathcal{M}}_\varepsilon \circ \mathrm{M}_{\varepsilon\kappa}(\nabla\phi\cdot\kappa)(\varepsilon\xi) \\
&= \sum_{i=1}^N \widetilde{\mathcal{M}}_\varepsilon \circ \mathrm{M}_{\varepsilon\kappa}\Big(\frac{\partial\phi}{\partial\mathbf{b}_i}\kappa_i\Big)(\varepsilon\xi).
\end{aligned} \tag{1.69}$$

Using this formula written for $\kappa = \mathbf{b}_i$ in (1.64), gives (1.66). By density this holds for every φ in $W^{1,p}(\mathbb{R}^N)$ and for a.e. x in $\mathbb{R}^N$.

Similarly, since $\sum\limits_{\kappa\in\{0,1\}^N} \mathbf{q}_\kappa \equiv 1$, formula (1.63) yields,

$$\begin{aligned}
&\frac{1}{\varepsilon}(\widetilde{\mathcal{Q}}_\varepsilon - \widetilde{\mathcal{M}}_\varepsilon)(\phi)(x) \\
&\quad= \sum_{\kappa\in\{0,1\}^N} \frac{1}{\varepsilon}\Big(\widetilde{\mathcal{M}}_\varepsilon(\phi)\Big(\varepsilon\Big[\frac{x}{\varepsilon}\Big]_Y + \varepsilon\kappa\Big) - \widetilde{\mathcal{M}}_\varepsilon(\phi)\Big(\varepsilon\Big[\frac{x}{\varepsilon}\Big]_Y\Big)\Big)\mathbf{q}_\kappa\Big(\Big\{\frac{x}{\varepsilon}\Big\}_Y\Big).
\end{aligned}$$

This, together with (1.69), imply (1.67). By density, it holds for every φ in $W^{1,p}(\mathbb{R}^N)$ and for a.e. x in $\mathbb{R}^N$.

(ii) Now, consider again ϕ in $\mathcal{D}(\mathbb{R}^N)$. For $\xi \in \mathbf{G}$ and for $1 \leqslant i < j \leqslant N$, from (1.69), after multiple applications of (1.61), we easily obtain

$$\begin{aligned}
&\frac{1}{\varepsilon^2}\Big[\big(\widetilde{\mathcal{M}}_\varepsilon(\phi)(\varepsilon\xi + \varepsilon\mathbf{b}_i + \varepsilon\mathbf{b}_j) - \widetilde{\mathcal{M}}_\varepsilon(\phi)(\varepsilon\xi + \varepsilon\mathbf{b}_j)\big) \\
&\qquad\qquad - \big(\widetilde{\mathcal{M}}_\varepsilon(\phi)(\varepsilon\xi + \varepsilon\mathbf{b}_i) - \widetilde{\mathcal{M}}_\varepsilon(\phi)(\varepsilon\xi)\big)\Big] \\
&= \frac{1}{\varepsilon}\Big(\widetilde{\mathcal{M}}_\varepsilon \circ \mathrm{M}_{\varepsilon\mathbf{b}_i}\Big(\frac{\partial\phi}{\partial x_i}\Big)(\varepsilon\xi + \varepsilon\mathbf{b}_j) - \widetilde{\mathcal{M}}_\varepsilon \circ \mathrm{M}_{\varepsilon\mathbf{b}_i}\Big(\frac{\partial\phi}{\partial x_i}\Big)(\varepsilon\xi)\Big) \\
&= \frac{1}{\varepsilon}\Big(\widetilde{\mathcal{M}}_\varepsilon\Big(\frac{\partial\mathrm{M}_{\varepsilon\mathbf{b}_i}(\phi)}{\partial x_i}\Big)(\varepsilon\xi + \varepsilon\mathbf{b}_j) - \widetilde{\mathcal{M}}_\varepsilon\Big(\frac{\partial\mathrm{M}_{\varepsilon\mathbf{b}_i}(\phi)}{\partial x_i}\Big)(\varepsilon\xi)\Big) \\
&\qquad\qquad = \widetilde{\mathcal{M}}_\varepsilon \circ \mathrm{M}_{\varepsilon\mathbf{b}_j}\Big(\frac{\partial^2\mathrm{M}_{\varepsilon\mathbf{b}_i}(\phi)}{\partial x_i\partial x_j}\Big)(\varepsilon\xi) \\
&\qquad\qquad = \widetilde{\mathcal{M}}_\varepsilon \circ \mathrm{M}_{\varepsilon\mathbf{b}_j} \circ \mathrm{M}_{\varepsilon\mathbf{b}_i}\Big(\frac{\partial^2\phi}{\partial x_i\partial x_j}\Big)(\varepsilon\xi).
\end{aligned} \tag{1.70}$$

Consequently, equality (1.68) holds for $\phi \in \mathcal{D}(\mathbb{R}^N)$, and by density, for φ in $W^{2,p}(\mathbb{R}^N)$ and a.e. for x in $\mathbb{R}^N$. □

We use in the sequel the following notation:

Notation 1.63. *For $\delta > 0$, and $\mathcal{O}$ a subset of $\mathbb{R}^N$, $\mathcal{V}_\delta(\mathcal{O})$ denotes the δ-neighborhood of $\mathcal{O}$, that is,*

$$\mathcal{V}_\delta(\mathcal{O}) = \{x \in \mathbb{R}^N \backslash \mathcal{O}) \mid dist\,(x, \partial\Omega) \leqslant \delta.\}.$$

Proposition 1.64. *Suppose p is in $[1, +\infty]$ and recall that $d(Y)$ is the diameter of Y.*

If ϕ belongs to $L^p(\mathbb{R}^N)$, then

$$(i)\quad \|\widetilde{\mathcal{Q}}_\varepsilon(\phi)\|_{L^p(\mathbb{R}^N)} \leqslant \|\phi\|_{L^p(\mathbb{R}^N)}, \quad \|\widetilde{\mathcal{Q}}_\varepsilon(\phi)\|_{L^p(\mathcal{O})} \leqslant \|\phi\|_{L^p(\mathcal{V}_{2\varepsilon d(Y)}(\mathcal{O}))}.$$

$$(ii)\quad \begin{aligned} &\|\nabla\widetilde{\mathcal{Q}}_\varepsilon(\phi)\|_{L^p(\mathbb{R}^N)} \leqslant \frac{2}{\varepsilon}\|\phi\|_{L^p(\mathbb{R}^N)},\\ &\|\nabla\widetilde{\mathcal{Q}}_\varepsilon(\phi)\|_{L^p(\mathcal{O})} \leqslant \frac{2}{\varepsilon}\|\phi\|_{L^p(\mathcal{V}_{2\varepsilon d(Y)}(\mathcal{O}))}.\end{aligned}$$

If ϕ belongs to $W^{1,p}(\mathbb{R}^N)$ and $1 \leqslant i < j \leqslant N$, then

$$(iii)\quad \begin{aligned} &\|\nabla\widetilde{\mathcal{Q}}_\varepsilon(\phi)\|_{L^p(\mathbb{R}^N)} \leqslant \|\nabla\phi\|_{L^p(\mathbb{R}^N)},\\ &\|\nabla\widetilde{\mathcal{Q}}_\varepsilon(\phi)\|_{L^p(\mathcal{O})} \leqslant \|\nabla\phi\|_{L^p(\mathcal{V}_{2\varepsilon d(Y)}(\mathcal{O}))},\end{aligned}$$

$$(iv)\quad \begin{aligned} &\|\widetilde{\mathcal{Q}}_\varepsilon(\phi) - \widetilde{\mathcal{M}}_\varepsilon(\phi)\|_{L^p(\mathbb{R}^N)} \leqslant \varepsilon\|\nabla\phi\|_{L^p(\mathbb{R}^N)},\\ &\|\widetilde{\mathcal{Q}}_\varepsilon(\phi) - \widetilde{\mathcal{M}}_\varepsilon(\phi)\|_{L^p(\mathcal{O})} \leqslant \varepsilon\|\nabla\phi\|_{L^p(\mathcal{V}_{3\varepsilon d(Y)}(\mathcal{O}))},\end{aligned}$$

and

$$(v)\quad \begin{aligned} &\Big\|\frac{\partial^2\widetilde{\mathcal{Q}}_\varepsilon(\phi)}{\partial x_i\partial x_j}\Big\|_{L^p(\mathbb{R}^N)} \leqslant \frac{2}{\varepsilon}\|\nabla\phi\|_{L^p(\mathbb{R}^N)},\\ &\Big\|\frac{\partial^2\widetilde{\mathcal{Q}}_\varepsilon(\phi)}{\partial x_{ji}\partial x_j}\Big\|_{L^p(\mathcal{O})} \leqslant \frac{2}{\varepsilon}\|\nabla\phi\|_{L^p(\mathcal{V}_{3\varepsilon d(Y)}(\mathcal{O}))}.\end{aligned}$$

If ϕ belongs to $W^{2,p}(\mathbb{R}^N)$, then for $1 \leqslant i < j \leqslant N$,

$$(vi)\quad \begin{aligned} &\Big\|\frac{\partial^2\widetilde{\mathcal{Q}}_\varepsilon(\phi)}{\partial x_i\partial x_j}\Big\|_{L^p(\mathbb{R}^N)} \leqslant \|\nabla^2\phi\|_{L^p(\mathbb{R}^N)},\\ &\Big\|\frac{\partial^2\widetilde{\mathcal{Q}}_\varepsilon(\phi)}{\partial x_i\partial x_j}\Big\|_{L^p(\mathcal{O})} \leqslant \|\nabla^2\phi\|_{L^p(\mathcal{V}_{3\varepsilon d(Y)}(\mathcal{O}))}.\end{aligned}$$

$$(vii)\quad \begin{aligned} &\big\|(Id - \widetilde{\mathcal{M}}_\varepsilon)(\nabla\widetilde{\mathcal{Q}}_\varepsilon(\phi))\big\|_{L^p(\mathbb{R}^N)} \leqslant C\varepsilon\|\nabla^2\phi\|_{L^p(\mathbb{R}^N)},\\ &\big\|(Id - \widetilde{\mathcal{M}}_\varepsilon)(\nabla\widetilde{\mathcal{Q}}_\varepsilon(\phi))\big\|_{L^p(\mathcal{O})} \leqslant C\varepsilon\|\nabla^2\phi\|_{L^p(\mathcal{V}_{3\varepsilon d(Y)}(\mathcal{O}))}.\end{aligned}$$

The constant C depends only on N and Y.

Remark 1.65. *Using $(1.60)_2$, applied to $\widetilde{\mathcal{Q}}_\varepsilon(\phi)$ after scaling and summation, it follows that*

$$\begin{aligned} &\|\widetilde{\mathcal{Q}}_\varepsilon(\phi) - \widetilde{\mathcal{M}}_\varepsilon(\phi)\|_{L^p(\mathbb{R}^N)} \leqslant C\,\varepsilon\|\nabla\widetilde{\mathcal{Q}}_\varepsilon(\phi)\|_{L^p(\mathbb{R}^N)},\\ &\|\widetilde{\mathcal{Q}}_\varepsilon(\phi) - \widetilde{\mathcal{M}}_\varepsilon(\phi)\|_{L^p(\mathcal{O})} \leqslant C\,\varepsilon\|\nabla\widetilde{\mathcal{Q}}_\varepsilon(\phi)\|_{L^p(\mathcal{V}_{3\varepsilon d(Y)}(\mathcal{O}))},\end{aligned}$$

with a constant C depending only on N and Y. In view of (iii), this gives another proof of (iv), with the extra constant C, but is of interest in itself.

Proof of Proposition 1.64. Suppose first p finite.
(i) and (ii) Let ϕ be in $L^p(\mathbb{R}^N)$. Using formula (1.63), the fact that

$$\mathbf{q}_\kappa(y) \geqslant 0 \quad \text{and} \quad \sum_{\kappa\in\{0,1\}^N} \mathbf{q}_\kappa(y) = 1 \quad \text{for every } y \in \overline{Y},$$

implies that for all ξ in $\mathbf{G}$, and for every x in $\varepsilon(\xi + \overline{Y})$,

$$|\widetilde{\mathcal{Q}}_\varepsilon(\phi)(x)|^p \leqslant \sum_{\kappa\in\{0,1\}^N} |\widetilde{\mathcal{M}}_\varepsilon(\phi)(\varepsilon\xi + \varepsilon\kappa)|^p \, \mathbf{q}_\kappa\Big(\Big\{\frac{x}{\varepsilon}\Big\}_Y\Big),$$

where we used the convexity of the function $t \mapsto |t|^p$. Integrating with respect to x and recalling the equality (see Remark 1.54),

$$\int_Y \mathbf{q}_\kappa(y) = \frac{|Y|}{2^N},$$

yields

$$\|\widetilde{\mathcal{Q}}_\varepsilon(\phi)\|^p_{L^p(\varepsilon(\xi+Y))} \leqslant \sum_{\kappa\in\{0,1\}^N} \frac{\varepsilon^N|Y|}{2^N} |\widetilde{\mathcal{M}}_\varepsilon(\phi)(\varepsilon\xi + \varepsilon\kappa)|^p.$$

Summation over $\mathbf{G}$ (or over $\mathbf{G} \cap \mathcal{O}$) gives

$$\begin{aligned}
\|\widetilde{\mathcal{Q}}_\varepsilon(\phi)\|^p_{L^p(\mathbb{R}^N)} &\leqslant \|\widetilde{\mathcal{M}}_\varepsilon(\phi)\|^p_{L^p(\mathbb{R}^N)},\\
\|\widetilde{\mathcal{Q}}_\varepsilon(\phi)\|^p_{L^p(\mathcal{O})} &\leqslant \|\widetilde{\mathcal{Q}}_\varepsilon(\phi)\|^p_{L^p(\mathcal{V}_{\varepsilon d(Y)}(\mathcal{O}))} \leqslant \|\widetilde{\mathcal{M}}_\varepsilon(\phi)\|^p_{L^p(\mathcal{V}_{2\varepsilon d(Y)}(\mathcal{O}))}.
\end{aligned}$$

Then inequalities in (i) are immediate consequences of Proposition 1.24(i) applied to $\Omega = \mathbb{R}^N$ or to $\Omega = \mathcal{V}_{2\varepsilon d(Y)}(\mathcal{O})$.

Due to the expression of the partial derivatives (1.64), and proceeding as above, the inequalities in (ii) follow easily.

(iii) Let ϕ be in $\mathcal{D}(\mathbb{R}^N)$. Reasoning as in the proof of (i) and using formulas (1.66) and (1.56) from Remark 1.54, one has for every ξ in $\mathbf{G}$ and x in $\varepsilon(\xi + \overline{Y})$,

$$\Big|\frac{\partial \widetilde{\mathcal{Q}}_\varepsilon(\phi)}{\partial x_i}(x)\Big|^p \leqslant \sum_{\kappa\in\mathbf{K}_i} \Big|\widetilde{\mathcal{M}}_\varepsilon \circ \mathrm{M}_{\varepsilon\mathbf{b}_i}\Big(\frac{\partial\phi}{\partial x_i}\Big)(\varepsilon\xi + \varepsilon\kappa)\Big|^p \mathbf{q}^i_\kappa\Big(\Big\{\frac{x}{\varepsilon}\Big\}_Y\Big).$$

Then, by integrating with respect to x,

$$\Big\|\frac{\partial \widetilde{\mathcal{Q}}_\varepsilon(\phi)}{\partial x_i}\Big\|^p_{L^p(\varepsilon\xi+\varepsilon Y)} \leqslant \frac{\varepsilon^N|Y|}{2^{N-1}} \sum_{\kappa\in\mathbf{K}_i} \Big|\widetilde{\mathcal{M}}_\varepsilon \circ \mathrm{M}_{\varepsilon\mathbf{b}_i}\Big(\frac{\partial\phi}{\partial x_i}\Big)(\varepsilon\xi + \varepsilon\kappa)\Big|^p.$$

Due to Proposition 1.24(i) (with $\Omega = \mathbb{R}^N$ or $\Omega = \mathcal{V}_{\varepsilon d(Y)}(\mathcal{O})$), and to Lemma 1.59(i), summation over $\mathbf{G}$ (or over $\mathbf{G} \cap \mathcal{O}$) gives the inequalities in (iii) for p finite. By density, they hold true for $\phi \in L^p(\mathbb{R}^N)$ with $\dfrac{\partial \phi}{\partial x_i} \in L^p(\mathbb{R}^N)$.

(iv) Proceeding as in the above proofs and recalling formula (1.67), gives

$$\Big\|\frac{1}{\varepsilon}(\widetilde{\mathcal{Q}}_\varepsilon - \widetilde{\mathcal{M}}_\varepsilon)(\phi)\Big\|^p_{L^p(\varepsilon\xi+\varepsilon Y)} \leqslant \sum_{\kappa \in \{0,1\}^N} \frac{\varepsilon^N |Y|}{2^N} \sum_{i=1}^N \Big|\widetilde{\mathcal{M}}_\varepsilon \circ \mathrm{M}_{\varepsilon\kappa}\Big(\frac{\partial \phi}{\partial x_i}\Big)(\varepsilon\xi)\Big|^p.$$

Summing over $\mathbf{G}$ (and $\mathbf{G} \cap \mathcal{O}$, resp.) leads to

$$\|\widetilde{\mathcal{Q}}_\varepsilon(\phi) - \widetilde{\mathcal{M}}_\varepsilon(\phi)\|^p_{L^p(\mathbb{R}^N)} \leqslant \varepsilon^p \sum_{i=1}^N \Big\|\widetilde{\mathcal{M}}_\varepsilon \circ \mathrm{M}_{\varepsilon\kappa}\Big(\frac{\partial \phi}{\partial x_i}\Big)\Big\|^p_{L^p(\mathbb{R}^N)},$$

$$\|\widetilde{\mathcal{Q}}_\varepsilon(\phi) - \widetilde{\mathcal{M}}_\varepsilon(\phi)\|^p_{L^p(\mathcal{O})} \leqslant \varepsilon^p \sum_{i=1}^N \Big\|\widetilde{\mathcal{M}}_\varepsilon \circ \mathrm{M}_{\varepsilon\kappa}\Big(\frac{\partial \phi}{\partial x_i}\Big)\Big\|^p_{L^p(\mathcal{V}_{\varepsilon d(Y)}(\mathcal{O}))}.$$

Proposition 1.24(i) (with $\Omega = \mathbb{R}^N$ or $\Omega = \mathcal{V}_{\varepsilon d(Y)}(\mathcal{O})$) and Lemma 1.59(ii) end the proof of (iv)[10].

(v) and (vi) Proceed as in the proof of (iii) with ϕ in $\mathcal{D}(\mathbb{R}^N)$ and make use of (1.68) and again of (1.56), to obtain for $\xi \in \mathbf{G}$ and $1 \leqslant i < j \leqslant N$, the following estimates:

$$\Big|\frac{\partial^2 \widetilde{\mathcal{Q}}_\varepsilon(\phi)}{\partial x_i \partial x_j}(x)\Big|^p \leqslant \frac{2^{p-1}}{\varepsilon^p} \sum_{\kappa \in \mathbf{K}_{ij}} \Big(\Big|\widetilde{\mathcal{M}}_\varepsilon \circ \mathrm{M}_{\varepsilon \mathbf{b}_i}\Big(\frac{\partial \phi}{\partial x_i}\Big)(\varepsilon\xi + \varepsilon\kappa + \varepsilon \mathbf{b}_j)\Big|^p$$
$$+ \Big|\widetilde{\mathcal{M}}_\varepsilon \circ \mathrm{M}_{\varepsilon \mathbf{b}_i}\Big(\frac{\partial \phi}{\partial x_i}\Big)(\varepsilon\xi + \varepsilon\kappa)\Big|^p\Big) \mathbf{q}^{ij}_\kappa\Big(\Big\{\frac{x}{\varepsilon}\Big\}_Y\Big),$$

$$\Big|\frac{\partial^2 \widetilde{\mathcal{Q}}_\varepsilon(\phi)}{\partial x_i \partial x_j}(x)\Big|^p$$
$$\leqslant \sum_{\kappa \in \mathbf{K}_{ij}} \Big|\widetilde{\mathcal{M}}_\varepsilon \circ \mathrm{M}_{\mathbf{b}_j,\varepsilon} \circ \mathrm{M}_{\varepsilon \mathbf{b}_i}\Big(\frac{\partial^2 \phi}{\partial x_i \partial x_j}\Big)(\varepsilon\xi + \varepsilon\kappa)\Big|^p \mathbf{q}^{ij}_\kappa\Big(\Big\{\frac{x}{\varepsilon}\Big\}_Y\Big).$$

Now, with the same arguments as before, it is easily seen that the last two inequalities imply (v) and (vi), respectively.

(vii) For $\phi \in W^{2,p}(\mathbb{R}^N)$ consider the function $\widetilde{\mathcal{T}}_\varepsilon(\nabla \widetilde{\mathcal{Q}}_\varepsilon(\phi))$. In view of (1.58),

$$\frac{\partial}{\partial y_i}\Big(\widetilde{\mathcal{T}}_\varepsilon\Big(\frac{\partial \widetilde{\mathcal{Q}}_\varepsilon(\phi)}{\partial x_i}\Big)\Big) = 0. \tag{1.71}$$

Recall that for $1 \leqslant i < j \leqslant N$,

$$\frac{\partial}{\partial y_i}\Big(\widetilde{\mathcal{T}}_\varepsilon\Big(\frac{\partial \widetilde{\mathcal{Q}}_\varepsilon(\phi)}{\partial x_j}\Big)\Big) = \varepsilon \widetilde{\mathcal{T}}_\varepsilon\Big(\frac{\partial^2 \widetilde{\mathcal{Q}}_\varepsilon(\phi)}{\partial x_i \partial x_j}\Big).$$

(10) An alternative proof could be given by using (1.60)(ii).

Therefore by estimates (vi), $\widetilde{\mathcal{T}}_\varepsilon(\nabla\widetilde{\mathcal{Q}}_\varepsilon(\phi))$ belongs to $L^p(\mathbb{R}^N; W^{1,p}(Y))^N$ and moreover,

$$\begin{aligned}
\|\nabla_y(\widetilde{\mathcal{T}}_\varepsilon(\nabla\widetilde{\mathcal{Q}}_\varepsilon(\phi)))\|_{L^p(\mathbb{R}^N\times Y)} &\leqslant C\varepsilon|Y|^{1/p}\|\nabla^2\phi\|_{L^p(\mathbb{R}^N)},\\
\|\nabla_y(\widetilde{\mathcal{T}}_\varepsilon(\nabla\widetilde{\mathcal{Q}}_\varepsilon(\phi)))\|_{L^p(\mathcal{O}\times Y)} &\leqslant C\varepsilon|Y|^{1/p}\|\nabla^2\phi\|_{L^p(\mathcal{V}_{2\varepsilon d(Y)}(\mathcal{O}))}.
\end{aligned}$$

Now, applying inequality (1.60) leads to

$$\begin{aligned}
&\|\widetilde{\mathcal{T}}_\varepsilon(\nabla\widetilde{\mathcal{Q}}_\varepsilon(\phi)) - \mathcal{M}_\varepsilon(\nabla\widetilde{\mathcal{Q}}_\varepsilon(\phi))\|_{L^p(\mathbb{R}^N\times Y)}\\
&\qquad\qquad \leqslant C\|\nabla_y(\widetilde{\mathcal{T}}_\varepsilon(\nabla\widetilde{\mathcal{Q}}_\varepsilon(\phi)))\|_{L^p(\mathbb{R}^N\times Y)}\\
&\qquad\qquad \leqslant C\varepsilon|Y|^{1/p}\|\nabla^2\phi\|_{L^p(\mathbb{R}^N)},\\
&\|\widetilde{\mathcal{T}}_\varepsilon(\nabla\widetilde{\mathcal{Q}}_\varepsilon(\phi)) - \mathcal{M}_\varepsilon(\nabla\widetilde{\mathcal{Q}}_\varepsilon(\phi))\|_{L^p(\mathcal{O}\times Y)}\\
&\qquad\qquad \leqslant C\|\nabla_y(\widetilde{\mathcal{T}}_\varepsilon(\nabla\widetilde{\mathcal{Q}}_\varepsilon(\phi)))\|_{L^p(\mathcal{O}\times Y)}\\
&\qquad\qquad \leqslant C\varepsilon|Y|^{1/p}\|\nabla^2\phi\|_{L^p(\mathcal{V}_{2\varepsilon d(Y)}(\mathcal{O}))}.
\end{aligned}\tag{1.72}$$

Then the estimates from (vii) in the case of the whole space, are immediate by virtue of Proposition 1.8(i) applied in $\mathbb{R}^N$.
In the case of a general open set $\mathcal{O}$, the same proposition shows that

$$\begin{aligned}
&\|\nabla\widetilde{\mathcal{Q}}_\varepsilon(\phi) - \mathcal{M}_\varepsilon(\nabla\widetilde{\mathcal{Q}}_\varepsilon(\phi))\|_{L^p(\mathcal{O})}\\
&\qquad \leqslant |Y|^{-1/p}\|\widetilde{\mathcal{T}}_\varepsilon(\nabla\widetilde{\mathcal{Q}}_\varepsilon(\phi)) - \mathcal{M}_\varepsilon(\nabla\widetilde{\mathcal{Q}}_\varepsilon(\phi))\|_{L^p(\mathcal{V}_{2\varepsilon d(Y)}(\mathcal{O})\times Y)}.
\end{aligned}$$

This, together with (1.72) applied in the open set $\mathcal{V}_{\varepsilon d(Y)}(\mathcal{O})$ instead of $\mathcal{O}$, concludes the proof of estimate (vii).

All the inequalities from the statement of theorem are also satisfied in the case $p = +\infty$, by passing to the limit. □

Corollary 1.66. *Suppose $p \in [1, +\infty)$. For ϕ in $L^p(\mathbb{R}^N)$, the following convergences hold:*

$$\begin{aligned}
&(i) \quad \widetilde{\mathcal{Q}}_\varepsilon(\phi) \to \phi \quad \text{strongly in } L^p(\mathbb{R}^N),\\
&(ii) \quad \varepsilon\nabla\widetilde{\mathcal{Q}}_\varepsilon(\phi) \to 0 \quad \text{strongly in } L^p(\mathbb{R}^N)^N.
\end{aligned}\tag{1.73}$$

For ϕ in $W^{1,p}(\mathbb{R}^N)$, one has the convergence

$$\widetilde{\mathcal{Q}}_\varepsilon(\phi) \to \phi \quad \text{strongly in } W^{1,p}(\mathbb{R}^N).\tag{1.74}$$

For ϕ in $W^{2,p}(\mathbb{R}^N)$, and $1 \leqslant i \neq j \leqslant N$,

$$\frac{\partial^2\widetilde{\mathcal{Q}}_\varepsilon(\phi)}{\partial x_i\partial x_j} \to \frac{\partial^2\phi}{\partial x_i\partial x_j} \quad \text{strongly in } L^p(\mathbb{R}^N).\tag{1.75}$$

Proof. The first convergence in (1.73) comes from the inequalities (iv) of Proposition 1.64 and $(1.39)_1$ when ϕ belongs to $W^{1,p}(\mathbb{R}^N)$, and by a density argument (of $W^{1,p}(\mathbb{R}^N)$ in $L^p(\mathbb{R}^N)$) together with estimate (i) of the same Proposition 1.64.

The second convergence in (1.73) is obtained in a similar way, by using first estimates (iii) of Proposition 1.64 for ϕ in $W^{1,p}(\mathbb{R}^N)$, and afterwards estimates (ii) from the same proposition.

For $\psi \in \mathcal{C}_c^\infty(\mathbb{R}^N)$, formula (1.64) allows to obtain the strong convergence

$$\frac{\partial \widetilde{\mathcal{Q}}_\varepsilon(\psi)}{\partial x_i} \to \frac{\partial \psi}{\partial x_i} \quad \text{strongly in } L^\infty(\mathbb{R}^N),$$

and then in $L^p(\mathbb{R}^N)$ for p finite, since these two partial derivatives vanish outside a fixed bounded closed set. Again by a density argument, estimates (iii) in Proposition 1.64 imply for $\phi \in W^{1,p}(\mathbb{R}^N)$, the convergence

$$\nabla \widetilde{\mathcal{Q}}_\varepsilon(\phi) \to \nabla \phi \quad \text{strongly in } L^p(\mathbb{R}^N)^N,$$

for p finite. Using $(1.73)_1$, convergence (1.74) follows immediately.

As regarding convergence (1.75), it is obtained by using formula (1.68), Lemma 1.59 and proceeding as above. □

Definition 1.67 (Remainder in the scale-splitting). *The remainder* $\widetilde{\mathcal{R}}_\varepsilon(\phi)$ *is given by*

$$\widetilde{\mathcal{R}}_\varepsilon(\phi) = \phi - \widetilde{\mathcal{Q}}_\varepsilon(\phi) \quad \textit{for every } \phi \in W^{1,p}(\mathbb{R}^N).$$

Corollary 1.68. *Suppose* $p \in [1, +\infty]$. *For* $\phi \in W^{1,p}(\mathbb{R}^N)$, *one has*

$$\begin{aligned} &\|\widetilde{\mathcal{R}}_\varepsilon(\phi)\|_{L^p(\mathbb{R}^N)} \leqslant \varepsilon C \|\nabla \phi\|_{L^p(\mathbb{R}^N)}, \\ &\|\nabla \widetilde{\mathcal{R}}_\varepsilon(\phi)\|_{L^p(\mathbb{R}^N)} \leqslant C \|\nabla \phi\|_{L^p(\mathbb{R}^N)}. \end{aligned}$$

Moreover, for $p \in [1, +\infty)$,

$$\begin{aligned} &\frac{1}{\varepsilon}\widetilde{\mathcal{T}}_\varepsilon \circ \widetilde{\mathcal{R}}_\varepsilon(\phi) \to -\sum_{i=1}^N \mathcal{M}_Y(y_i)\frac{\partial \phi}{\partial x_i} \quad \textit{strongly in } L^p(\mathbb{R}^N \times Y), \\ &\frac{1}{\varepsilon}\widetilde{\mathcal{R}}_\varepsilon(\phi) \to -\sum_{i=1}^N \mathcal{M}_Y(y_i)\frac{\partial \phi}{\partial x_i} \quad \textit{strongly in } L^p(\mathbb{R}^N). \end{aligned} \tag{1.76}$$

The constant C *does not depend on* ε *and* p.

Proof. The two estimates are obtained from Proposition 1.64(iv) and (iii) respectively, together with $(1.39)_1$ (with $\mathbb{R}^N$ in place of $\widehat{\Omega}_\varepsilon$).

For $\phi \in W^{1,p}(\mathbb{R}^N)$ formula (1.67) reads for a.e. $(x,y) \in \mathbb{R}^N \times Y$,

$$\frac{1}{\varepsilon}\widetilde{\mathcal{T}}_\varepsilon \circ (\widetilde{\mathcal{Q}}_\varepsilon - \widetilde{\mathcal{M}}_\varepsilon)(\phi)(x,y) = \sum_{\kappa \in \{0,1\}^N} \mathbf{q}_\kappa(y) \sum_{i=1}^N \widetilde{\mathcal{M}}_\varepsilon \circ \mathrm{M}_{\varepsilon\kappa}\Big(\frac{\partial \phi}{\partial x_i}\Big)(x)\kappa_i. \tag{1.77}$$

For p finite, Proposition 1.25(i) and Lemma 1.59(ii) allow to pass to the limit and obtain the following strong convergence in $L^p(\mathbb{R}^N \times Y)$:

$$\frac{1}{\varepsilon}\widetilde{\mathcal{T}}_\varepsilon \circ (\widetilde{\mathcal{Q}}_\varepsilon - \widetilde{\mathcal{M}}_\varepsilon)(\psi)(\cdot\,, y) \to \sum_{\kappa\in\{0,1\}^N} \mathbf{q}_\kappa(y)\Big(\sum_{i=1}^N \frac{\partial\psi}{\partial x_i}\kappa_i\Big) = \sum_{i=1}^N \frac{\partial\psi}{\partial x_i} y_i.$$

Then, according to Proposition 1.39(ii) with $\mathbb{R}^N$ in place of Ω, $(1.76)_1$ is straightforward. Proposition 1.31(vi) then easily gives $(1.76)_2$. □

We now state convergence results concerning the sequences $\{\widetilde{\mathcal{Q}}_\varepsilon(w_\varepsilon)\}_\varepsilon$, $\{\widetilde{\mathcal{R}}_\varepsilon(w_\varepsilon)\}_\varepsilon$ and $\{\widetilde{\mathcal{T}}_\varepsilon \circ \widetilde{\mathcal{Q}}_\varepsilon(w_\varepsilon)\}_\varepsilon$ for a sequence $\{w_\varepsilon\}_\varepsilon$ converging weakly in $W^{1,p}(\mathbb{R}^N)$. Note that by construction $\widetilde{\mathcal{T}}_\varepsilon \circ \widetilde{\mathcal{Q}}_\varepsilon(w_\varepsilon)$ belongs to the closed subspace $L^p(\mathbb{R}^N; \mathbf{Q}_1(Y))$ of the space $L^p(\mathbb{R}^N; W^{1,p}(Y))$.

Proposition 1.69. *Suppose $p \in (1, +\infty)$. Let $\{w_\varepsilon\}_\varepsilon$ be a sequence such that*

$$w_\varepsilon \rightharpoonup w \quad \text{weakly in } W^{1,p}(\mathbb{R}^N).$$

Then, the following convergences hold true:

$$\begin{aligned}
&(i)\ \widetilde{\mathcal{R}}_\varepsilon(w_\varepsilon) \to 0 \quad \text{strongly in } L^p(\mathbb{R}^N) \text{ and weakly in } W^{1,p}(\mathbb{R}^N),\\
&(ii)\ \widetilde{\mathcal{Q}}_\varepsilon(w_\varepsilon) \to w \quad \text{strongly in } L^p_{loc}(\mathbb{R}^N) \text{ and weakly in } W^{1,p}(\mathbb{R}^N),\\
&(iii)\ \widetilde{\mathcal{T}}_\varepsilon \circ \widetilde{\mathcal{Q}}_\varepsilon(w_\varepsilon) \to w \quad \text{strongly in } L^p_{loc}(\mathbb{R}^N; \mathbf{Q}_1(Y)),\\
&(iv)\ \widetilde{\mathcal{T}}_\varepsilon(\nabla\widetilde{\mathcal{Q}}_\varepsilon(w_\varepsilon)) \rightharpoonup \nabla w \quad \text{weakly in } L^p(\mathbb{R}^N \times Y)^N.
\end{aligned}$$

For $p = +\infty$, these strong convergences are the same with weak-$*$ in place of weak in (i), (ii) and (iv)[11].

Proof. (i) Convergences (i) are direct consequences of the two inequalities given in Corollary 1.68.
(ii) Convergences (ii) follow from (i) and the weak convergence of $\{w_\varepsilon\}_\varepsilon$ in $W^{1,p}(\mathbb{R}^N)$, which implies the strong convergence $w_\varepsilon \to w$ in $L^p_{loc}(\mathbb{R}^N)$.
(iii) The third convergence is obtained from (ii) and Corollary 1.37.
(iv) By Propositions 1.8(iii) and 1.64(iii), and with C independent of ε,

$$\Big\|\widetilde{\mathcal{T}}_\varepsilon\Big(\frac{\partial\widetilde{\mathcal{Q}}_\varepsilon(w_\varepsilon)}{\partial x_i}\Big)\Big\|_{L^p(\mathbb{R}^N\times Y)} \leqslant C,$$

and by Propositions 1.35 and 1.64(v), and with equality (1.71),

$$\Big\|\nabla_y\Big\{\widetilde{\mathcal{T}}_\varepsilon\Big(\frac{\partial\widetilde{\mathcal{Q}}_\varepsilon(w_\varepsilon)}{\partial x_i}\Big)\Big\}\Big\|_{L^p(\mathbb{R}^N\times Y)} \leqslant C.$$

[11] See Proposition 2.38 for the case $p = 1$.

As a consequence, there exist a subsequence, still denoted ε, and W_i in $L^p(\mathbb{R}^N; W^{1,p}(Y))$, such that

$$\widetilde{\mathcal{T}}_\varepsilon\Big(\frac{\partial \widetilde{\mathcal{Q}}_\varepsilon(w_\varepsilon)}{\partial x_i}\Big) \rightharpoonup W_i \quad \text{weakly in } \; L^p(\mathbb{R}^N \times Y).$$

By Remark 1.61, $\dfrac{\partial \widetilde{\mathcal{Q}}_\varepsilon(w_\varepsilon)}{\partial x_i}$ is piecewise constant with respect to x_i and piecewise affine with respect to the other x_j. So, $W_i(\cdot, y)$ is independent of y_i and separately affine with respect to the other y_j $(j \neq i)$.

On the other hand, by Proposition 1.64, one has

$$\Big\|\varepsilon\frac{\partial}{\partial x_j}\Big(\frac{\partial \widetilde{\mathcal{Q}}_\varepsilon(w_\varepsilon)}{\partial x_i}\Big)\Big\|_{L^p(\mathbb{R}^N)} \leqslant C,$$

for $j \neq i$, and with C independent of ε. Then one can show, as in the proof of Theorem 1.36, that $W_i(\cdot, y)$ is periodic with respect to the variables y_j for $j \neq i$, as well as affine. This obviously implies that it is independent of y_j, for all $j \in \{1, \dots, N\}$. Consequently, W_i does not depend on y so that $W_i = \mathcal{M}_Y(W_i)$.
But Proposition 1.12(iv) together with convergence (ii) above imply that

$$\mathcal{M}_Y(W_i) = \frac{\partial w}{\partial x_i}.$$

Now, the whole sequence $\Big\{\widetilde{\mathcal{T}}_\varepsilon\Big(\dfrac{\partial \widetilde{\mathcal{Q}}_\varepsilon(w_\varepsilon)}{\partial x_i}\Big)\Big\}_\varepsilon$ having a unique weak limit point, converges weakly in the space $L^p(\mathbb{R}^N \times Y)$. This proves the claim.

The case $p = +\infty$ is treated in exactly the same fashion. □

Corollary 1.70. *Suppose $p \in (1, +\infty)$. Let $\{w_\varepsilon\}_\varepsilon$ be a sequence converging weakly in $W^{1,p}(\mathbb{R}^N)$ to w. Then, up to a subsequence, there exists some $\widehat{w}'$ in the space $L^p(\mathbb{R}^N; W^{1,p}_{per}(Y))$ such that*

$$\begin{aligned}
&(i) \quad \frac{1}{\varepsilon}\widetilde{\mathcal{T}}_\varepsilon \circ \widetilde{\mathcal{R}}_\varepsilon(w_\varepsilon) \rightharpoonup \widehat{w}' \quad \textit{weakly in } \; L^p(\mathbb{R}^N; W^{1,p}(Y)),\\
&(ii) \quad \frac{1}{\varepsilon}\widetilde{\mathcal{T}}_\varepsilon(\nabla\widetilde{\mathcal{R}}_\varepsilon(w_\varepsilon)) \rightharpoonup \nabla_y\widehat{w}' \quad \textit{weakly in } \; L^p(\mathbb{R}^N \times Y)^N,\\
&(iii) \quad \frac{1}{\varepsilon}\widetilde{\mathcal{T}}_\varepsilon(\nabla w_\varepsilon) \rightharpoonup \nabla w + \nabla_y\widehat{w}' \quad \textit{weakly in } \; L^p(\mathbb{R}^N \times Y)^N.
\end{aligned}$$

The connection with $\widehat{w}$ of Theorem 1.41 (in the case $\Omega = \mathbb{R}^N$), is given by

$$\widehat{w} = \widehat{w}' - \mathcal{M}_Y(\widehat{w}') \quad \textit{and} \quad \mathcal{M}_Y(\widehat{w}') = -\nabla w \cdot \mathcal{M}_Y(y)^{(12)}, \tag{1.78}$$

provided the same subsequence is used.
For $p = +\infty$, the same convergences hold for the weak- topologies*(13).

(12) with the notation convention introduced before Theorem 1.41.
(13) See Corollary 2.39 for the case $p = 1$.

Proof. Convergences (i), (ii) and (iii) follow from the estimates of Corollary 1.68, from the convergence in Proposition 1.69(iv), and from Theorem 1.36 (for $\Omega = \mathbb{R}^N$).

The relation between $\widehat{w}'$ and $\widehat{w}$ follows from (1.78). The first equality is straightforward since $\widehat{w}$ and $\widehat{w}'$ having the same gradient with respect to y, they differ by a function of x only which is determined as $\mathcal{M}_Y(\widehat{w}')$, because $\mathcal{M}_Y(\widehat{w}) \equiv 0$.

To obtain the value of $\mathcal{M}_Y(\widehat{w}')$, write

$$\frac{1}{\varepsilon}\widetilde{\mathcal{T}}_\varepsilon \circ \widetilde{\mathcal{R}}_\varepsilon(w_\varepsilon) = \frac{1}{\varepsilon}\widetilde{\mathcal{T}}_\varepsilon\big(w_\varepsilon - \widetilde{\mathcal{M}}_\varepsilon(w_\varepsilon)\big) - \frac{1}{\varepsilon}\widetilde{\mathcal{T}}_\varepsilon\big(\widetilde{\mathcal{Q}}_\varepsilon(w_\varepsilon) - \widetilde{\mathcal{M}}_\varepsilon(w_\varepsilon)\big).$$

The average over Y of the first term in the right-hand side is always 0. By Proposition 1.25(ii), the average over Y of the second term converges to $\mathcal{M}_Y(\widehat{w}')$. Use Formula (1.77) to get for a.e. $(x,y) \in \mathbb{R}^N \times Y$,

$$\frac{1}{\varepsilon}\widetilde{\mathcal{T}}_\varepsilon \circ (\widetilde{\mathcal{Q}}_\varepsilon - \widetilde{\mathcal{M}}_\varepsilon)(w_\varepsilon)(x,y) = \sum_{\kappa\in\{0,1\}^N} \mathbf{q}_\kappa(y) \sum_{i=1}^N \widetilde{\mathcal{M}}_\varepsilon \circ \mathbb{M}_{\varepsilon\kappa}\Big(\frac{\partial w_\varepsilon}{\partial x_i}\Big)(x)\kappa_i.$$

Passing to the limit with Lemma 1.59(iii) gives the following weak convergence in $L^p(\mathbb{R}^N \times Y)$:

$$\begin{aligned}\frac{1}{\varepsilon}\widetilde{\mathcal{T}}_\varepsilon \circ (\widetilde{\mathcal{Q}}_\varepsilon - \widetilde{\mathcal{M}}_\varepsilon)(w_\varepsilon)(\cdot,y) \rightharpoonup & \sum_{\kappa\in\{0,1\}^N} \mathbf{q}_\kappa(y)\Big(\sum_{i=1}^N \frac{\partial w}{\partial x_i}\kappa_i\Big)\\ &= \sum_{i=1}^N \frac{\partial w}{\partial x_i} y_i \equiv \nabla w \cdot y.\end{aligned}$$

Taking the average over Y, finally implies (1.78). □

We close this subsection with some remarkable estimates which generalize Proposition 1.64.

Proposition 1.71. *Suppose $p \in [1,+\infty]$. For ψ in $L^p(Y)$ and Φ in $L^p(\mathbb{R}^N)$, with $\Phi_{|\varepsilon(\xi+Y)}$ in $\mathbf{Q}_1(\varepsilon(\xi+Y))$ for every $\xi \in \mathbf{G}$,*

$$\Big\|\psi\Big(\Big\{\frac{\cdot}{\varepsilon}\Big\}_Y\Big)\Phi\Big\|_{L^p(\mathbb{R}^N)} \leqslant C\|\psi\|_{L^p(Y)}\|\Phi\|_{L^p(\mathbb{R}^N)}.$$

Similarly, for every open subset $\mathcal{O}$,

$$\Big\|\psi\Big(\Big\{\frac{\cdot}{\varepsilon}\Big\}_Y\Big)\Phi\Big\|_{L^p(\mathcal{O})} \leqslant C\|\psi\|_{L^p(Y)}\|\Phi\|_{L^p(\mathcal{V}_{2\varepsilon d(Y)}(\mathcal{O}))}.$$

The constants depend only on N and Y.

Proof. As a consequence of (1.60)(i), for $p \in [1,+\infty)$,

$$\Big\|\psi\Big(\Big\{\frac{\cdot}{\varepsilon}\Big\}_Y\Big)\Phi\Big\|^p_{L^p(\varepsilon(\xi+Y))} \leqslant C^p\|\Phi\|^p_{L^p(\varepsilon(\xi+Y))}\|\psi\|^p_{L^p(Y)}.$$

Summation over $\mathbf{G}$ (and over $\mathbf{G}\cap\mathcal{O}$, resp.) gives the result. For $p = +\infty$ the result is straightforward. □

Corollary 1.72. *Suppose $p \in [1, +\infty]$. For ψ in $L^p(Y)$ and ϕ in $L^p(\mathbb{R}^N)$,*

$$\left\| \psi\left(\left\{\frac{\cdot}{\varepsilon}\right\}_Y\right) \widetilde{\mathcal{Q}}_\varepsilon(\phi) \right\|_{L^p(\mathbb{R}^N)} \leqslant C \, \|\psi\|_{L^p(Y)} \|\phi\|_{L^p(\mathbb{R}^N)},$$

$$\left\| \psi\left(\left\{\frac{\cdot}{\varepsilon}\right\}_Y\right) \widetilde{\mathcal{Q}}_\varepsilon(\phi) \right\|_{L^p(\mathcal{O})} \leqslant C \, \|\psi\|_{L^p(Y)} \|\phi\|_{L^p(\mathcal{V}_{2\varepsilon d(Y)}(\mathcal{O}))},$$

and

$$\left\| \psi\left(\left\{\frac{\cdot}{\varepsilon}\right\}_Y\right) \nabla \widetilde{\mathcal{Q}}_\varepsilon(\phi) \right\|_{L^p(\mathbb{R}^N)} \leqslant \frac{C}{\varepsilon} \, \|\psi\|_{L^p(Y)} \|\phi\|_{L^p(\mathbb{R}^N)},$$

$$\left\| \psi\left(\left\{\frac{\cdot}{\varepsilon}\right\}_Y\right) \nabla \widetilde{\mathcal{Q}}_\varepsilon(\phi) \right\|_{L^p(\mathcal{O})} \leqslant \frac{C}{\varepsilon} \, \|\psi\|_{L^p(Y)} \|\phi\|_{L^p(\mathcal{V}_{2\varepsilon d(Y)}(\mathcal{O}))}.$$

Furthermore, if ϕ belongs to $W^{1,p}(\mathbb{R}^N)$, then

$$\left\| \psi\left(\left\{\frac{\cdot}{\varepsilon}\right\}_Y\right) \nabla \widetilde{\mathcal{Q}}_\varepsilon(\phi) \right\|_{L^p(\mathbb{R}^N)} \leqslant C \, \|\psi\|_{L^p(Y)} \|\nabla\phi\|_{L^p(\mathbb{R}^N)},$$

$$\left\| \psi\left(\left\{\frac{\cdot}{\varepsilon}\right\}_Y\right) \nabla \widetilde{\mathcal{Q}}_\varepsilon(\phi) \right\|_{L^p(\mathcal{O})} \leqslant C \, \|\psi\|_{L^p(Y)} \|\nabla\phi\|_{L^p(\mathcal{V}_{2\varepsilon d(Y)}(\mathcal{O}))}.$$

The constants C depend only on N and Y.

Proof. These estimates follow from Lemma 1.55 and Proposition 1.64. □

1.6.2 The case of a subdomain

We now turn to the definitions of the maps $\mathcal{Q}_\varepsilon$ and $\mathcal{R}_\varepsilon$ on the spaces $W_0^{1,p}(\Omega)$ and $W^{1,p}(\Omega)$. The case of $W^{1,p}(\Omega)$ is considered with the extra assumption that Ω is bounded with Lipschitz boundary.

Definition 1.73 ($\mathbf{Q_1}$-interpolating operator in $L^p(\Omega)$ and $W_0^{1,p}(\Omega)$). *On $L^p(\Omega)$, the operator*

$$\mathcal{Q}^o_\varepsilon : L^p(\Omega) \longrightarrow W^{1,\infty}(\Omega),$$

is defined as

$$\mathcal{Q}^o_\varepsilon(\phi) = \widetilde{\mathcal{Q}}_\varepsilon(\widetilde{\phi})_{|\Omega},$$

where $\widetilde{\phi}$ is the extension of ϕ by zero to $\mathbb{R}^N \setminus \overline{\Omega}$. The same operator is used for ϕ in $W_0^{1,p}(\Omega)$. The map $\mathcal{R}^o_\varepsilon$ is defined as $Id - \mathcal{Q}^o_\varepsilon$.

Note that $\widetilde{\mathcal{Q}}_\varepsilon(\widetilde{\phi})$ vanishes outside of the $2\varepsilon d(Y)$-neighborhood of Ω. In particular $\mathcal{Q}^o_\varepsilon(\phi)$ does not necessarily belong to $W_0^{1,\infty}(\Omega)$ for ϕ in $W_0^{1,p}(\Omega)$.

Concerning the case of $W^{1,p}(\Omega)$, as Ω is bounded with Lipschitz boundary, it is then classical (see for instance [2, 37]) that there exists a continuous extension operator

$$\widetilde{\mathcal{P}} : W^{1,p}(\Omega) \to W^{1,p}(\mathbb{R}^N),$$

and a constant C satisfying, for all $\phi \in W^{1,p}(\Omega)$,

$$\begin{aligned} \|\widetilde{\mathcal{P}}\phi\|_{L^p(\mathbb{R}^N)} &\leqslant C\|\phi\|_{L^p(\Omega)}, \\ \|\widetilde{\mathcal{P}}\phi\|_{W^{1,p}(\mathbb{R}^N)} &\leqslant C\|\phi\|_{W^{1,p}(\Omega)}. \end{aligned} \tag{1.79}$$

Proposition 1.74. *There exists an extension operator $\mathcal{P}$ from $W^{1,p}(\Omega)$ to $W^{1,p}(\mathbb{R}^N)$ and a constant C satisfying for all $\phi \in W^{1,p}(\Omega)$,*

$$\begin{aligned} &\|\mathcal{P}\phi\|_{L^p(\mathbb{R}^N)} \leqslant C\|\phi\|_{L^p(\Omega)}, \\ &\|\mathcal{P}\phi\|_{W^{1,p}(\mathbb{R}^N)} \leqslant C\|\phi\|_{W^{1,p}(\Omega)}, \\ &\|\nabla(\mathcal{P}\phi)\|_{L^p(\mathcal{V}_1(\Omega))} \leqslant C\|\nabla\phi\|_{L^p(\Omega)}, \end{aligned} \tag{1.80}$$

where $\mathcal{V}_1(\Omega)$ is the 1-neighborhood of Ω (cf. notation 1.63).

Proof. Let θ be positive in $\mathcal{D}(\mathbb{R}^N)$ with value 1 on $\mathcal{V}_1(\Omega)$. Set

$$\mathcal{P}\phi \doteq \big(\widetilde{\mathcal{P}}(\phi - \mathcal{M}_\Omega(\phi)) + \mathcal{M}_\Omega(\phi)\big)\theta.$$

Inequality $(1.80)_1$ follows from the fact that $|\mathcal{M}_\Omega(\phi)| \leqslant |\Omega|^{1/p'}\|\phi\|_{L^p(\Omega)}$ and from $(1.79)_1$.

The Poincaré-Wirtinger inequality (2.23) (see Subsection 2.3.1), applied to ϕ in Ω gives

$$\|\phi - \mathcal{M}_\Omega(\phi)\|_{W^{1,p}(\Omega)} \leqslant C\|\nabla\phi\|_{L^p(\Omega)}.$$

By $(1.79)_2$,

$$\|\widetilde{\mathcal{P}}(\phi - \mathcal{M}_\Omega(\phi))\|_{W^{1,p}(\mathbb{R}^N)} \leqslant C\|\nabla\phi\|_{L^p(\Omega)},$$

so that $(1.80)_2$ holds.

Finally, in view of the identity

$$\nabla(\mathcal{P}\phi)_{|\mathcal{V}_1(\Omega)} = \nabla\widetilde{\mathcal{P}}(\phi - \mathcal{M}_\Omega(\phi))_{|\mathcal{V}_1(\Omega)},$$

the last inequality $(1.80)_3$ is straightforward. □

Definition 1.75 ($\mathbf{Q}_1$-interpolating operator in $W^{1,p}(\Omega)$). *On $W^{1,p}(\Omega)$, the operator*

$$\mathcal{Q}_\varepsilon : W^{1,p}(\Omega) \to W^{1,+\infty}(\Omega),$$

is defined as follows:

$$\mathcal{Q}_\varepsilon(\phi) = \widetilde{\mathcal{Q}}_\varepsilon(\mathcal{P}\phi)_{|\Omega}.$$

The map $\mathcal{R}_\varepsilon$ is defined as $Id - \mathcal{Q}_\varepsilon$.

In the case of a function in $W_0^{1,p}(\Omega)$, the operators $\mathcal{Q}_\varepsilon^o$ and $\mathcal{Q}_\varepsilon$ can both be used. Their values differ only in the boundary layer $\Omega^{bl}_{2\varepsilon d(Y)}$ (of thickness $2\varepsilon d((Y))$ and for practical purposes, either can be used when the boundary of Ω is Lipschitz. If $\partial\Omega$ is not Lipschitz, only $\mathcal{Q}_\varepsilon^o$ can be used.

They essentially enjoy the same properties which are summarized below (we only use the notation $\mathcal{Q}_\varepsilon$ but the same results hold for $\mathcal{Q}_\varepsilon^o$ when applied on $W_0^{1,p}(\Omega)$).

Lemma 1.76. *Suppose $p \in [1,+\infty]$. There exists a constant C depending only on N, Y and $\mathcal{P}$, such that for every ϕ in $W^{1,p}(\Omega)$, the following estimates hold (for $2\varepsilon d(Y) \leqslant 1$):*

$$\begin{aligned} &\|\mathcal{Q}_\varepsilon(\phi)\|_{L^p(\Omega)} \leqslant C\|\phi\|_{L^p(\Omega)},\\ &\|\mathcal{R}_\varepsilon(\phi)\|_{L^p(\Omega)} \leqslant C\varepsilon\|\nabla\phi\|_{L^p(\Omega)},\\ &\|\nabla\mathcal{Q}_\varepsilon(\phi)\|_{L^p(\Omega)} + \|\nabla\mathcal{R}_\varepsilon(\phi)\|_{L^p(\Omega)} \leqslant C\|\nabla\phi\|_{L^p(\Omega)}. \end{aligned} \tag{1.81}$$

Proof. These inequalities follow from the estimates of Proposition 1.64 and from estimate $(1.39)_1$. □

Corollary 1.77. *Suppose $p \in [1,+\infty)$. For ϕ in $L^p(\Omega)$,*

$$\begin{aligned} &(i) \quad \mathcal{Q}^o_\varepsilon(\phi) \to \phi \quad \textit{strongly in } L^p(\Omega),\\ &(ii) \quad \varepsilon\nabla\mathcal{Q}^o_\varepsilon(\phi) \to 0 \quad \textit{strongly in } L^p(\Omega)^N. \end{aligned} \tag{1.82}$$

For ϕ in $W^{1,p}(\Omega)$, one has the convergence

$$\mathcal{Q}_\varepsilon(\phi) \to \phi \quad \textit{strongly in } W^{1,p}(\Omega). \tag{1.83}$$

For ϕ in $W^{2,p}(\Omega)$ and $1 \leqslant i \neq j \leqslant N$,

$$\frac{\partial^2\mathcal{Q}_\varepsilon(\phi)}{\partial x_i\partial x_j} \to \frac{\partial^2\phi}{\partial x_i\partial x_j} \quad \textit{strongly in } L^p_{loc}(\Omega)^{(14)}. \tag{1.84}$$

Proof. Convergences (1.82)-(1.83) are the immediate consequences of convergences (1.73)-(1.74) of Corollary 1.66.

To prove (1.84) let ω be an open set strictly included in Ω. From Lemma 1.59(i) and Proposition 1.25(ii) one has, for every $\kappa \in \mathbf{K}_{ij}$,

$$\mathcal{M}_\varepsilon \circ \mathrm{M}_{\varepsilon\mathbf{b_j}} \circ \mathrm{M}_{\varepsilon\mathbf{b_i}}\Big(\frac{\partial^2\phi}{\partial x_i\partial x_j}\Big)(\cdot + \varepsilon\kappa)\mathbf{1}_\omega \to \frac{\partial^2\phi}{\partial x_i\partial x_j} \quad \text{strongly in} \quad L^p(\omega).$$

Then using the explicit expression (1.65) of $\dfrac{\partial^2\widetilde{\mathcal{Q}}_\varepsilon(\phi)}{\partial x_i\partial x_j}$ and equality (1.68), the above strong convergence implies that

$$\frac{\partial^2\widetilde{\mathcal{Q}}_\varepsilon(\phi)}{\partial x_i\partial x_j}\mathbf{1}_\omega \to \frac{\partial^2\phi}{\partial x_i\partial x_j} \quad \text{strongly in} \quad L^p(\omega),$$

and this yields convergence (1.84). □

(14) If the extension operator $\mathcal{P}$ also maps $W^{2,p}(\Omega)$ continuously into $W^{2,p}(\mathbb{R}^N)$ (see [2, Chapter 5], [184, Chapter 6]) and using convergence (1.75), this convergence in $L^p_{loc}(\Omega)$ can be replaced by a strong convergence in $L^p(\Omega)$.

Proposition 1.78. *(Theorem 1.41 revisited) Suppose* $p \in (1, +\infty)$. *Let* $\{w_\varepsilon\}_\varepsilon$ *be a sequence such that*

$$w_\varepsilon \rightharpoonup w \quad \text{weakly in } W^{1,p}(\Omega).$$

Then there exists $\widehat{w}'$ *in* $L^p(\Omega; W^{1,p}_{per}(Y))$ *such that*

$$\begin{aligned}
&(i) && \mathcal{R}_\varepsilon(w_\varepsilon) \to 0 \quad \text{strongly in } L^p(\Omega),\\
&(ii) && \mathcal{Q}_\varepsilon(w_\varepsilon) \rightharpoonup w \quad \text{weakly in } W^{1,p}(\Omega),\\
&(iii) && \mathcal{T}_\varepsilon(\nabla \mathcal{Q}_\varepsilon(w_\varepsilon)) \rightharpoonup \nabla w \quad \text{weakly in } L^p(\Omega \times Y)^N,\\
&(iv) && \frac{1}{\varepsilon}\mathcal{T}_\varepsilon \circ \mathcal{R}_\varepsilon(w_\varepsilon) \rightharpoonup \widehat{w}' \quad \text{weakly in } L^p(\Omega; W^{1,p}(Y)),\\
&(v) && \mathcal{T}_\varepsilon(\nabla \mathcal{R}_\varepsilon(w_\varepsilon)) \rightharpoonup \nabla_y \widehat{w}' \quad \text{weakly in } L^p(\Omega \times Y)^N,\\
&(vi) && \mathcal{T}_\varepsilon(\nabla w_\varepsilon) \rightharpoonup \nabla w + \nabla_y \widehat{w}' \quad \text{weakly in } L^p(\Omega \times Y)^N.
\end{aligned}$$

The same results are true for w_ε *in* $W^{1,p}_0(\Omega)$ *with the operators* $\mathcal{Q}^o_\varepsilon$ *and* $\mathbb{R}^o$.

Similar results hold for $p = +\infty$ *with the weak-∗ topology on the corresponding spaces*(15).

The connection with the $\widehat{w}$ *of Theorem 1.41 is given by*

$$\widehat{w} = \widehat{w}' - \mathcal{M}_Y(\widehat{w}') \quad \text{and} \quad \mathcal{M}_Y(\widehat{w}') = -\nabla w \cdot \mathcal{M}_Y(y), \tag{1.85}$$

provided the same subsequence is used.

Proof. Convergences (i) and (ii) follow from the preceding lemma.

The function $\mathcal{T}_\varepsilon(\nabla \mathcal{Q}_\varepsilon(w_\varepsilon)$ is just the restriction to $\widehat{\Omega}_\varepsilon$ of $\widetilde{\mathcal{T}}_\varepsilon(\nabla \widetilde{\mathcal{Q}}_\varepsilon(\widetilde{w}_\varepsilon))$ (and $\widetilde{\mathcal{T}}_\varepsilon(\nabla(\widetilde{\mathcal{Q}}_\varepsilon \circ \mathcal{P}(w_\varepsilon))$, respectively). Since $\widetilde{w}_\varepsilon$ (resp. $\mathcal{P}(w_\varepsilon)$) converges weakly to $\widetilde{w}$ (resp. $\mathcal{P}(w)$) in $W^{1,p}(\mathbb{R}^N)$, the result follows from (iv) of Proposition 1.69.

Due to Lemma 1.76 and Theorem 1.36, there exists $\widehat{w}'$ in $L^p(\Omega; W^{1,p}_{per}(Y))$ such that, up to a subsequence, convergences (iv) and (v) hold. Combining with convergence (iii), shows (vi). Relations (1.85) come from (1.78). □

(15) See Proposition 2.40 for the case $p = 1$.

Chapter 2

Advanced topics for unfolding

Extending the results of Chapter 1, this chapter presents various advanced topics related to the unfolding method.

The unfolding method appears to depend upon the choice of the basis **B** of macroscopic periods. But in physical problems, it is the group of periods **G** which is given. So there are many way to define a basis and the corresponding unfolding. Section 2.1 shows that the final result is actually independent of the choice of the set Y as long as it has the paving property (equivalently is a fundamental domain) with respect to the action of **G** and is connected with Lipschitz boundary. Thus, up to a null set, Y is simply a realization of the flat torus, i.e., the quotient of $\mathbb{R}^N$ under the action of **G** (see Proposition 2.5). This section concludes with a general proof for the periodicity of some limits of sequences of unfoldings which also applies to the case of problems with holes in Part II.

In Section 2.2, unfolding with partial information on derivatives is studied. The case of fields with bounded divergence or curl or symmetric gradients are obviously of interest in Mechanics. The case of a single derivative with respect to a direction which is not that of a period is detailed.

Section 2.3 gives the proofs of the main results for the unfolding operators for the cases of functions and sequences of functions in L^1 or $W^{1,1}$ as well as for bounded measures and sequences of bounded measures and functions in BV spaces (the latter generalizing some attempts to define two-scale convergence for measures and BV functions).

In the Lemmas, Propositions and Theorems of Sections 2.1 and 2.2, the Lebesgue exponent p is restricted to $(1, +\infty)$.

D. Cioranescu et al., *The Periodic Unfolding Method*, Series in Contemporary Mathematics 3, https://doi.org/10.1007/978-981-13-3032-2_2

2.1 Effect of a change of the periodicity cell

This section investigates the effect of changing the reference cell on the unfolding process. Indeed, one starts with the group $\mathbf{G}$ which represents the macroscopic periods of the system. From this a basis $\mathbf{B}$ and the set Y are defined (the latter is the parallelotope generated by the vectors of the basis).

The same construction can be carried out when Y is chosen in a larger class, the set having the paving property with respect to $\mathbf{G}$[1].

Definition 2.1 (Set with the paving property, also known as *fundamental domains*). *The bounded open set Y has the paving property with respect to the group $\mathbf{G}$ when it is connected, its boundary ∂Y is a null set and*

$$\mathbb{R}^N = \bigcup_{\xi\in\mathbf{G}} (\xi + \overline{Y}), \ \forall(\xi_1,\xi_2) \in \mathbf{G}^2, \ \xi_1 \neq \xi_2 \Longrightarrow (\xi_1 + Y) \cap (\xi_2 + Y) = \varnothing.$$

Observe that any translate of Y satisfies the same property and that εY has the same paving property as Y but with $\mathbf{G}$ replaced by $\varepsilon\mathbf{G}$.

From now on, we reserve the notation Y (the reference cell) for a bounded open set having the paving property with respect to the group $\mathbf{G}$. As for the open parallelotope generated by a basis $\mathbf{B}$ generating $\mathbf{G}$, which was denoted Y in Chapter 1, we use the notation

$$\mathcal{P} \doteq \Big\{ y \in \mathbb{R}^N \mid y = \sum_{i=1}^{N} y_i \mathbf{b}_i, \ (y_1,\dots,y_n) \in (0,1)^N \Big\}. \tag{2.1}$$

The parallelotope $\mathcal{P}$ plays an important role, in particular in Subsection 4.3.2.2 for the definition of the macro-micro operators $\mathcal{Q}^*_\varepsilon$ and $\mathcal{R}^*_\varepsilon$ (the analogues of $\mathcal{Q}_\varepsilon$ and $\mathcal{R}_\varepsilon$ of Section 1.6) for perforated domains.

At this point we introduce the following notations:

Notation 2.2. *The space $L^p_{\mathbf{G}\text{-}per}(\mathbb{R}^N)$, $p \in [1,+\infty]$, is defined as*

$$L^p_{\mathbf{G}\text{-}per}(\mathbb{R}^N) \doteq \{\varphi \in L^p_{loc}(\mathbb{R}^N) \,|\, \varphi \text{ is } \mathbf{G}\text{-periodic}\}.$$

It can be identified with $L^p(\mathbb{R}^N/\mathbf{G})$ (where $\mathbb{R}^N/\mathbf{G}$ denotes the quotient flat torus) and as such to $L^p(Y)$ for every Y having the paving property.

Similarly, the space $W^{1,p}_{\mathbf{G}\text{-}per}(\mathbb{R}^N)$ is

$$W^{1,p}_{\mathbf{G}\text{-}per}(\mathbb{R}^N) \doteq \{\varphi \in W^{1,p}_{loc}(\mathbb{R}^N) \,|\, \varphi \text{ is } \mathbf{G}\text{-periodic}\}.$$

It can be identified with $W^{1,p}(\mathbb{R}^N/\mathbf{G})$ and as such to $W^{1,p}_{per}(Y)$) for every Y having the paving property (the latter space is a strict closed subspace of $W^{1,p}(Y)$).

(1) These sets will play an important role in Part II (unfolding for perforated domains).

For a set Y with the paving property, the notations are the same: $[x]_Y$ is the ("unique" for a.e. $x \in \mathbb{R}^N$) element ξ in $\mathbf{G}$ such that x belongs to $\xi + Y$, while $\{x\}_Y$ is $x - [x]_Y$ as in Chapter 1 (see (1.1)).

The construction of the unfolding operator is the same as in the previous chapter, except for the proof of periodicity in Theorem 1.36 which only works for ∂Y-Lipschitz.

2.1.1 General proof of periodicity for unfoldings limits

Here is another proof of periodicity which is valid in the general case. It requires a slight generalization of the unfolding operator. For simplicity, we consider only the case $\Omega = \mathbb{R}^N$. All the results extend easily to the case of Ω an open subset of $\mathbb{R}^N$ (see for example, how this is done in Section 4.3).

Definition 2.3. *For* $\mathbf{G}$ *and* Y *as above, consider the following extension* $\overline{\mathcal{T}_\varepsilon}$ *of the unfolding operator* $\widetilde{\mathcal{T}}_\varepsilon$ *to* $\mathbb{R}^N \times \mathbb{R}^N$ *defined for every measurable function* u *in* $\mathbb{R}^N$ *, by*

$$\overline{\mathcal{T}_\varepsilon} u(x,y) = u\Big(\varepsilon\Big[\frac{x}{\varepsilon}\Big]_Y + \varepsilon y\Big) \quad \text{for a.e.} \quad (x,y) \in \mathbb{R}^N \times \mathbb{R}^N. \tag{2.2}$$

Clearly, $\widetilde{\mathcal{T}}_\varepsilon u \equiv \overline{\mathcal{T}_\varepsilon} u_{\,|\mathbb{R}^N \times Y}$ and if $u \in L^p(\mathbb{R}^N)$ ($p \in [1,+\infty)$) then $\overline{\mathcal{T}_\varepsilon} u$ is bounded in $L^p(\mathbb{R}^N; L^p_{loc}(\mathbb{R}^N))$[(2)] uniformly with respect to $\varepsilon \in (0,1]$. More precisely, it is bounded in the subspace $L^p(\mathbb{R}^N; L^p_{loc\,unif}(\mathbb{R}^N))$, where $L^p_{loc\,unif}(\mathbb{R}^N)$ is defined as

$$L^p_{loc\,unif}(\mathbb{R}^N) \doteq \big\{w \in L^p_{loc}(\mathbb{R}^N) \mid \sup_{t \in \mathbb{R}^N} \|w\, \mathbf{1}_{(t+\mathcal{B})}\|_{L^p(\mathbb{R}^N)} < +\infty\big\},$$

where $\mathcal{B}$ is the unit ball in $\mathbb{R}^N$ (but can equivalently be any non empty bounded open subset of $\mathbb{R}^N$). One can reformulate Theorem 1.36 in this more general framework.

Theorem 2.4 (Theorem 1.36 revisited)**.** *Let* $\{w_\varepsilon\}_\varepsilon$ *be a sequence satisfying*

$$w_\varepsilon \rightharpoonup w \quad \textit{weakly in} \quad L^p(\mathbb{R}^N).$$

Then there exists $\widehat{w}$ *in* $L^p(\mathbb{R}^N; L^p_{\mathbf{G}\text{-}per}(\mathbb{R}^N))$ *such that, up to a subsequence,*

$$\overline{\mathcal{T}_\varepsilon}(w_\varepsilon) \rightharpoonup \widehat{w} \quad \textit{weakly in } L^p(\mathbb{R}^N; L^p_{loc\,unif}(\mathbb{R}^N)).$$

Moreover, if

$$\varepsilon \|\nabla w_\varepsilon\|_{L^p(\mathbb{R}^N)} \leqslant C,$$

with C *independent of* ε*, then* $\widehat{w}$ *belongs to* $L^p(\mathbb{R}^N; W^{1,p}_{\mathbf{G}\text{-}per}(\mathbb{R}^N))$.

Similar results hold for $p = +\infty$*, replacing the weak topology by the weak-* topology on the corresponding spaces.*

(2) It also belongs to $L^p_{loc}(\mathbb{R}^N; L^p(\mathbb{R}^N))$, but is not bounded uniformly with respect to $\varepsilon \in (0,1]$.

Proof. The proof follows along the lines of that of Theorem 1.36, except for the **G**-periodicity of $\widehat{w}$.

To prove for example, the $\mathbf{b}_1$-periodicity for a function φ in $C_c^\infty(\mathbb{R}^N \times \mathbb{R}^N)$, recalling definition (2.2) write successively,

$$\begin{aligned}
&\int_{\mathbb{R}^N\times\mathbb{R}^N} \big(\mathcal{T}_\varepsilon(w_\varepsilon)(x, y+\mathbf{b}_1) - \mathcal{T}_\varepsilon(w_\varepsilon)(x,y)\big)\,\varphi(x,y)\,dxdy\\
&= \int_{\mathbb{R}^N\times\mathbb{R}^N} \Big(w_\varepsilon\Big(\varepsilon\Big[\frac{x}{\varepsilon}\Big]_Y + \varepsilon y + \varepsilon\mathbf{b}_1\Big) - w_\varepsilon\Big(\varepsilon\Big[\frac{x}{\varepsilon}\Big]_Y + \varepsilon y\Big)\Big)\,\varphi(x,y)\,dxdy\\
&= \int_{\mathbb{R}^N\times\mathbb{R}^N} \Big(w_\varepsilon\Big(\varepsilon\Big[\frac{x+\varepsilon\mathbf{b}_1}{\varepsilon}\Big]_Y + \varepsilon y\Big) - w_\varepsilon\Big(\varepsilon\Big[\frac{x}{\varepsilon}\Big]_Y + \varepsilon y\Big)\Big)\,\varphi(x,y)\,dxdy\\
&= \int_{\mathbb{R}^N\times\mathbb{R}^N} w_\varepsilon\Big(\varepsilon\Big[\frac{x}{\varepsilon}\Big]_Y + \varepsilon y\Big)\,\big(\varphi(x-\varepsilon\mathbf{b}_1, y) - \varphi(x,y)\big)\,dxdy\\
&= \int_{\mathbb{R}^N\times\mathbb{R}^N} \mathcal{T}_\varepsilon(w_\varepsilon)(x,y)\big(\varphi(x-\varepsilon\mathbf{b}_1,y) - \varphi(x,y)\big)\,dxdy,
\end{aligned}$$

which clearly converges to 0 by the regularity of φ. □

We can now compare the unfolding results for two choices of cells Y and Z having the paving property with respect to the same group **G**. To distinguish them, we use the notations $\mathcal{T}_\varepsilon^{\,Y}$ and $\mathcal{T}_\varepsilon^{\,Z}$ respectively. The next proposition shows that the results are independent of Y and Z but depend only upon the group **G**.

Proposition 2.5. *Let $\{w_\varepsilon\}_\varepsilon$ be a sequence in $L^p_{loc}(\mathbb{R}^N)$ whose unfolding is such that,*

$$\mathcal{T}_\varepsilon^{\,Z}(w_\varepsilon) \rightharpoonup \widehat{w} \quad \textit{weakly in} \quad L^p(\mathbb{R}^N; L^p_{loc}(\mathbb{R}^N)).$$

Then

$$\mathcal{T}_\varepsilon^{\,Y}(w_\varepsilon) \rightharpoonup \widehat{w} \quad \textit{weakly in} \quad L^p(\mathbb{R}^N; L^p_{loc}(\mathbb{R}^N)).$$

Similar results hold for $p = +\infty$, replacing the weak topology by the weak-∗ topology on the corresponding spaces.

In other words, even though the restrictions $\widetilde{\mathcal{T}}_\varepsilon^{Y}$ of $\mathcal{T}_\varepsilon^{\,Y}$ to Y, and $\widetilde{\mathcal{T}}_\varepsilon^{Z}$ of $\mathcal{T}_\varepsilon^{\,Z}$ to Z, are defined on the two different sets $\mathbb{R}^N \times Y$ and $\mathbb{R}^N \times Z$, the convergence of the sequence $\{\widetilde{\mathcal{T}}_\varepsilon^{Y}(w_\varepsilon)\}_\varepsilon$ weakly in $L^p(\mathbb{R}^N \times Y)$ is equivalent to that of the sequence $\{\widetilde{\mathcal{T}}_\varepsilon^{Z}(w_\varepsilon)\}_\varepsilon$ weakly in $L^p(\mathbb{R}^N \times Z)$, the limits being identified when extended by **G**-periodicity to $L^p(\mathbb{R}^N; L^p_{\mathbf{G}\text{-}per}(\mathbb{R}^N))$.

Proof of Proposition 2.5. By definition, there is a finite subset J of **G** such that $\bigcup_{j\in J}(j+Y)$ contains Z, as usual, up to a null set (the sets $j+Y$ are also pair-wise disjoint up to a null set). Set

$$Z_j = Z \cap (j+Y), \quad \forall j \in J.$$

Then for a.e. z in Z,

$$[z]_Y = j \iff z \in Z_j.$$

Consequently for a.e. $x \in \mathbb{R}^N$,

$$\begin{aligned}\{x\}_Z \in Z_j \Longleftrightarrow \{x\}_Z = j + \{x\}_Y \Longleftrightarrow x = [x]_Z + j + \{x\}_Y \\ \Longleftrightarrow [x]_Y = j + [x]_Z.\end{aligned} \tag{2.3}$$

Let $\mathbb{R}^N_j$ be the set defined as

$$\mathbb{R}^N_j \doteq \{z \in \mathbb{R}^N \mid \{z\}_Z \in Z_j\},$$

so that up to a null set, $\mathbb{R}^N$ is the disjoint union of the $\mathbb{R}^N_j$'s for $j \in J$. With this notation, and taking into account (2.3), compute $\mathcal{T}_\varepsilon^Y(u)$ as follows:

$$\begin{aligned}\mathcal{T}_\varepsilon^Y(u)(x,y) &= \sum_{j\in J} u\Big(\varepsilon\Big[\frac{x}{\varepsilon}\Big]_Y + \varepsilon y\Big)\, \mathbf{1}_{\mathbb{R}^N_j}\Big(\frac{x}{\varepsilon}\Big) \\ &= \sum_{j\in J} u\Big(\varepsilon\Big[\frac{x}{\varepsilon}\Big]_Z + \varepsilon j + \varepsilon y\Big)\, \mathbf{1}_{\mathbb{R}^N_j}\Big(\frac{x}{\varepsilon}\Big) \\ &= \sum_{j\in J} \mathcal{T}_\varepsilon^Z(u)(x+\varepsilon j, y)\, \mathbf{1}_{\mathbb{R}^N_j}\Big(\frac{x}{\varepsilon}\Big) \qquad \text{for } x,\, y \in \mathbb{R}^N.\end{aligned}$$

Testing with a function φ in $L^{p'}(\mathbb{R}^N \times \mathbb{R}^N)$ with compact support gives

$$\begin{aligned}&\int_{\mathbb{R}^N\times\mathbb{R}^N} \mathcal{T}_\varepsilon^Y(u)(x,y)\, \varphi(x,y)\, dxdy \\ &\qquad = \sum_{j\in J}\int_{\mathbb{R}^N\times\mathbb{R}^N} \mathcal{T}_\varepsilon^Z(u)(x,y)\, \mathbf{1}_{\mathbb{R}^N_j}\Big(\frac{x}{\varepsilon} - j\Big)\, \varphi(x-\varepsilon j, y)\, dxdy.\end{aligned}$$

But

$$\mathbf{1}_{\mathbb{R}^N_j}\Big(\frac{x}{\varepsilon} - j\Big) = \mathbf{1}_{\mathbb{R}^N_j}\Big(\frac{x}{\varepsilon}\Big) \quad \text{and} \quad \sum_{j\in J} \mathbf{1}_{\mathbb{R}^N_j} \equiv 1,$$

so that

$$\begin{aligned}&\int_{\mathbb{R}^N\times\mathbb{R}^N} \Big(\mathcal{T}_\varepsilon^Y(u)(x,y) - \mathcal{T}_\varepsilon^Z(u)(x,y)\Big)\varphi(x,y)\, dxdy \\ &\qquad = \sum_{j\in J}\int_{\mathbb{R}^N\times\mathbb{R}^N} \mathcal{T}_\varepsilon^Z(u)(x,y)\, \mathbf{1}_{\mathbb{R}^N_j}\Big(\frac{x}{\varepsilon}\Big)\, \big(\varphi(x-\varepsilon j, y) - \varphi(x,y)\big)\, dxdy.\end{aligned}$$

Thus,

$$\begin{aligned}&\Big|\int_{\mathbb{R}^N\times\mathbb{R}^N} \Big(\mathcal{T}_\varepsilon^Y(u)(x,y) - \mathcal{T}_\varepsilon^Z(u)(x,y)\Big)\varphi(x,y)\, dxdy\Big| \\ &\qquad \leqslant \|\mathcal{T}_\varepsilon^Z(u)\|_{L^p(K)} \sum_{j\in J} \|\varphi(\cdot - \varepsilon j, \cdot) - \varphi(\cdot, \cdot)\|_{L^{p'}(K)},\end{aligned}$$

for K a compact containing a fixed neighborhood of the support of φ. The right-hand side converge to 0 with ε, and this concludes the proof. ☐

Remark 2.6. *Note that by restriction to Y, Proposition 2.5 gives another proof of the periodicity in Theorem 1.36.*

2.2 Unfolding with partial information

The results of this section are useful when studying classical problems arising in Fluid Mechanics, Electromagnetism and Solid Mechanics, where some first order operators play a crucial role (divergence, curl and symmetric gradient, see e.g. [30, 31]). The last paragraph is dedicated to the more complex case of a single directional derivative.

Here, Y is the parallelotope associated with a basis of $\mathbf{G}$ and in order to simplify the presentation, we use the notation

$$\langle \cdot, \cdot \rangle_{\Omega \times Y'} \doteq \langle \cdot, \cdot \rangle_{L^p(\Omega,(W^{1-1/p,p}(Y'))'),L^p(\Omega,W^{1-1/p,p}(Y'))}. \tag{2.4}$$

2.2.1 Unfolding with information on the divergence

We use the following notations of classical Banach spaces for a given domain $\mathcal{O} \in \mathbb{R}^N$($p \in [1, +\infty]$):

$$L^p_{div}(\mathcal{O}) \doteq \{\varphi \in L^p(\mathcal{O})^N \mid \text{div } \varphi \in L^p(\mathcal{O})\},$$

endowed with the norm

$$\|\varphi\|_{L^p_{div}(\mathcal{O})} = \big(\|\varphi\|^p_{L^p(\mathcal{O})} + \|\text{div } \varphi\|^p_{L^p(\mathcal{O})}\big)^{1/p},$$

for p finite (and the similar norm for $p = +\infty$).

Recall that for a bounded set $\mathcal{O}$ with Lipschitz boundary, all elements of $L^p_{div}(\mathcal{O})$ have normal traces on $\partial\mathcal{O}$. This trace operator is continuous with values in $W^{-1/p,p}(\partial\mathcal{O})$ (see [151]).

In the case of Y, let $L^p_{\text{div},per}(Y)$ be the subspace of $L^p_{\text{div}}(Y)$, defined as

$$L^p_{\text{div},per}(Y) \doteq \Big\{\varphi_{|Y} \mid \varphi \in L^p_{loc}(\mathbb{R}^N)^N, \ \varphi \ \mathbf{G}\text{-periodic}, \ \text{div}_y \varphi \in L^p_{loc}(\mathbb{R}^N)\Big\},$$

$$L^p_{\text{div}\,0,per}(Y) \doteq \Big\{\varphi \in L^p_{\text{div},per}(Y) \mid \ \text{div}_y \varphi = 0\Big\}.$$

The space $L^p_{\text{div},per}(Y)$ is actually the subspace of $L^p_{\text{div}}(Y)$ whose elements have matching normal traces on opposite faces of Y (up to sign, since the outward unit normals of opposite faces are opposite).

Proposition 2.7. *Let w_ε be in $L^p_{\text{div}}(\Omega)$ satisfying*

$$\|w_\varepsilon\|_{L^p(\Omega)} + \varepsilon\|\text{div } w_\varepsilon\|_{L^p(\Omega)} \leqslant C.$$

Then there exist a subsequence (still denoted ε) and $\widehat{w} \in L^p(\Omega; L^p_{\text{div},per}(Y))$ such that,

$$\begin{aligned}
&\mathcal{T}_\varepsilon(w_\varepsilon) \rightharpoonup \widehat{w} \quad \text{weakly in } L^p(\Omega \times Y)^N,\\
&\text{div}_y(\mathcal{T}_\varepsilon(w_\varepsilon)) = \varepsilon\mathcal{T}_\varepsilon(\text{div}_y\, w_\varepsilon) \rightharpoonup \text{div}_y\, \widehat{w} \quad \text{weakly in } L^p(\Omega \times Y).
\end{aligned}$$

Similar results hold for $p = +\infty$ replacing the weak topology by the weak-$*$ topology on the corresponding spaces.

Proof. The proof of the existence of $\widehat{w}$ in $L^p(\Omega \times Y)^N$ with $\mathrm{div}_y\, \widehat{w}$ in $L^p(\Omega \times Y)$, follows from the fact that, by hypotheses, $\{\mathcal{T}_\varepsilon(w_\varepsilon)\}_\varepsilon$ is bounded in $L^p(\Omega \times Y)^N$ and such that $\{\mathrm{div}_y \mathcal{T}_\varepsilon(w_\varepsilon)\}_\varepsilon$ is bounded in $L^p(\Omega \times Y)$.

In order to show that the normal traces on opposite faces coincide (up to their sign), consider the case of the face pair $(Y', Y' + \mathbf{b}_N)$ using the *same normal unit vector* $\mathbf{b}_N^{*}$(3). Introduce a smooth test function ψ in $\mathcal{D}(\Omega \times Y')$ and make the following computation using the notation (2.4):

$$\begin{aligned}
&\big\langle \widehat{w} \cdot \mathbf{b}^*_{N|Y'+\mathbf{b}_N} - \widehat{w} \cdot \mathbf{b}^*_{N|Y'}, \psi \big\rangle_{\Omega \times Y'} \\
&\qquad = \lim_{\varepsilon \to 0} \big\langle \mathcal{T}_\varepsilon(w_\varepsilon) \cdot \mathbf{b}^*_{N|Y'+\mathbf{b}_N} - \mathcal{T}_\varepsilon(w_\varepsilon) \cdot \mathbf{b}^*_{N|Y'}, \psi \big\rangle_{\Omega \times Y'}.
\end{aligned} \tag{2.5}$$

Expression (2.5) can be computed as it was in the proof of Theorem 1.36. We have successively,

$$\begin{aligned}
&\big\langle \mathcal{T}_\varepsilon(w_\varepsilon) \cdot \mathbf{b}^*_{N|Y'+\mathbf{b}_N} - \mathcal{T}_\varepsilon(w_\varepsilon) \cdot \mathbf{b}^*_{N|Y'}, \psi \big\rangle_{\Omega \times Y'} \\
&\quad = \Big\langle w_\varepsilon\Big(\varepsilon\Big[\frac{x}{\varepsilon}\Big]_Y + \varepsilon \mathbf{b}_N + \varepsilon y'\Big) \cdot \mathbf{b}^*_{N|Y'+\mathbf{b}_N} \\
&\qquad\qquad - w_\varepsilon\Big(\Big[\frac{x}{\varepsilon}\Big]_Y + \varepsilon y'\Big) \cdot \mathbf{b}^*_{N|Y'}, \psi(x, y')\Big\rangle_{\Omega \times Y'} \\
&\quad = \Big\langle w_\varepsilon\Big(\varepsilon\Big[\frac{x}{\varepsilon}\Big]_Y + \varepsilon y'\Big) \cdot \mathbf{b}^*_{N|Y'}, \psi(x - \varepsilon \mathbf{b}_N, y') - \psi(x, y')\Big\rangle_{\Omega \times Y'} \\
&\quad = \big\langle \mathcal{T}_\varepsilon(w_\varepsilon) \cdot \mathbf{b}^*_{N|Y'}, \psi(x - \varepsilon \mathbf{b}_N, y') - \psi(x, y')\big\rangle_{\Omega \times Y'}.
\end{aligned}$$

The last term goes to 0 with ε so that the normal traces of $\widehat{w}$ on Y' and $Y' + b_N$ coincide, i.e., $\widehat{w}$ belongs to $L^p(\Omega; L^p_{\mathrm{div},per}(Y))$. □

Proposition 2.8. *Let $\{w_\varepsilon\}_\varepsilon$ be a sequence converging weakly to some w in $L^p_{\mathrm{div}}(\Omega)$. Then, up to a subsequence, there exist $\overline{w}$ in $L^p(\Omega; L^p_{\mathrm{div}\,0,per}(Y))$ with $\mathcal{M}_Y(\overline{w}) = w$ and $\widehat{w}$ in $L^p(\Omega; L^p_{\mathrm{div},per}(Y))$ with $\mathcal{M}_Y(\widehat{w}) = 0$, such that*

$$\begin{aligned}
&\mathcal{T}_\varepsilon(w_\varepsilon) \rightharpoonup \overline{w} \quad \textit{weakly in } L^p(\Omega \times Y)^N, \\
&\mathcal{T}_\varepsilon(\mathrm{div}\, w_\varepsilon) \rightharpoonup \mathrm{div}\, w + \mathrm{div}_y\, \widehat{w} \quad \textit{weakly in } L^p(\Omega \times Y).
\end{aligned}$$

Proof. The existence of $\overline{w}$ follows from Proposition 2.7 and the fact that $\mathrm{div}_y \mathcal{T}_\varepsilon(w_\varepsilon)$ is now bounded in $L^p(\Omega \times Y)$ by $C\varepsilon$.

To construct $\widehat{w}$, start by lifting the sequence $\{\mathrm{div}\, w_\varepsilon\}_\varepsilon$ to a sequence of functions $\{\varphi_\varepsilon\}_\varepsilon$ bounded in $W^{2,p}(\Omega)$ such that $-\Delta \varphi_\varepsilon + \varphi_\varepsilon = \mathrm{div}\, w_\varepsilon$. This can be achieved for example, by taking the restriction to Ω of the solution in $W^{2,p}(\mathbb{R}^N)$ of

$$-\Delta \varphi_\varepsilon + \varphi_\varepsilon = \begin{cases} \mathrm{div}\, w_\varepsilon & \text{in } \Omega, \\ 0 & \text{in } \mathbb{R}^N \setminus \overline{\Omega}. \end{cases}$$

It satisfies the estimate (see [114, Theorem 9.11] with proper modifications)

$$\|\varphi_\varepsilon\|_{W^{2,p}(\mathbb{R}^N)} \leqslant C \|\mathrm{div}\, w_\varepsilon\|_{L^p(\Omega)} \leqslant C.$$

(3) This normal vector is not necessarily $\mathbf{b}_N$ nor colinear to it.

By Theorem 1.47, there exist $\varphi \in W^{2,p}(\Omega)$, $\Phi \in L^p(\Omega; W^{2,p}_{per}(Y))$ such that

$$\begin{aligned}
&\mathcal{T}_\varepsilon(\varphi_\varepsilon) \rightharpoonup \varphi \quad \text{weakly in } L^p(\Omega; W^{2,p}(Y)),\\
&\mathcal{T}_\varepsilon(\nabla\varphi_\varepsilon) \rightharpoonup \nabla\varphi \quad \text{weakly in } L^p(\Omega; W^{1,p}(Y))^N,\\
&\mathcal{T}_\varepsilon(\Delta\varphi_\varepsilon) \rightharpoonup \Delta\varphi + \Delta_y\Phi \quad \text{weakly in } L^p(\Omega \times Y).
\end{aligned}$$

Since $-\Delta\varphi_\varepsilon + \varphi_\varepsilon = \operatorname{div} w_\varepsilon$, $-\Delta\varphi + \varphi = \operatorname{div} w$ a.e. in Ω and

$$\mathcal{T}_\varepsilon(\operatorname{div} w_\varepsilon) = \mathcal{T}_\varepsilon(\varphi_\varepsilon) - \mathcal{T}_\varepsilon(\Delta\varphi_\varepsilon),$$

passing to the limit and setting $\widehat{w} \doteq -\nabla_y\Phi$ ends the proof. Note that, as $\widehat{w}$ is the gradient with respect to y of the Y-periodic function Φ, it satisfies $\mathcal{M}_Y(\widehat{w}) = 0$. □

2.2.2 Unfolding with information on the curl

We use the following notation of classical Banach spaces for a given domain $\mathcal{O}$ in R^N:

$$L^p_{\text{curl}}(\mathcal{O}) \doteq \{\varphi \in L^p(\mathcal{O})^N \mid \text{curl } \varphi \in L^p(\mathcal{O})^{N(N-1)/2}\},$$

endowed with the norm

$$\|\varphi\|_{L^p_{\text{curl}}(\mathcal{O})} = \big(\|\varphi\|^p_{L^p(\mathcal{O})} + \|\text{curl } \varphi\|^p_{L^p(\mathcal{O})}\big)^{1/p},$$

for p finite, and the similar norm for $p = +\infty$.

When $\mathcal{O}$ is bounded with Lipschitz boundary, all the elements of $L^p_{\text{curl}}(\mathcal{O})$ have tangential traces on $\partial\mathcal{O}$. This trace operator is continuous with values in $W^{-1/p,p}(\partial\mathcal{O})^N$ (see [151]).

In the case of Y, let the subspace $L^p_{\text{curl},per}(Y)$ of $L^p_{\text{curl}}(Y)$ be defined as

$$\begin{aligned}
L^p_{\text{curl},per}(Y) \doteq \Big\{\varphi_{|Y} \mid \varphi \in L^p_{loc}(\mathbb{R}^N)^N,\ \varphi\ \mathbf{G}\text{-periodic},\\
\text{curl}_y\, \varphi \in L^p_{loc}(\mathbb{R}^N)^{N(N-1)/2}\Big\}.
\end{aligned}$$

It is the same as the subspace of $L^p_{\text{curl}}(Y)$ whose elements have matching tangential traces on opposite faces of Y.

Proposition 2.9. *Let $\{w_\varepsilon\}_\varepsilon$ be a sequence in $L^p_{\text{curl}}(\Omega)$ satisfying*

$$\|w_\varepsilon\|_{L^p(\Omega)} + \varepsilon\|\text{curl } w_\varepsilon\|_{L^p(\Omega)} \leqslant C.$$

Then, up to a subsequence, there exists $\widehat{w} \in L^p(\Omega; L^p_{\text{curl},per}(Y))$ such that the following convergences hold:

$$\begin{aligned}
&\mathcal{T}_\varepsilon(w_\varepsilon) \rightharpoonup \widehat{w} \quad \textit{weakly in } L^p(\Omega \times Y)^N,\\
&\text{curl}_y(\mathcal{T}_\varepsilon(w_\varepsilon)) = \varepsilon\mathcal{T}_\varepsilon(\text{curl } w_\varepsilon) \rightharpoonup \text{curl}_y\widehat{w} \quad \textit{weakly in } L^p(\Omega \times Y)^{N(N-1)/2}.
\end{aligned}$$

Similar results hold for $p = +\infty$, replacing the weak topology by the weak- topology on the corresponding spaces.*

Proof. The proof of the existence of $\widehat{w}$ in $L^p(\Omega \times Y)^N$ with $\text{curl}_y \widehat{w}$ belonging to $L^p(\Omega \times Y)^{N(N-1)/2}$, follows from the facts that under the hypothesis, the sequence $\{\mathcal{T}_\varepsilon(w_\varepsilon)\}_\varepsilon$ is bounded in $L^p(\Omega \times Y)^N$, and the sequence $\{\text{curl}_y \mathcal{T}_\varepsilon(w_\varepsilon))\}_\varepsilon$ is bounded in $L^p(\Omega \times Y)^{N(N-1)/2}$.

In order to show that the tangential traces of $\widehat{w}$ on opposite faces coincide, consider the case of the pair $(Y', Y' + \mathbf{b}_N)$ of two opposite faces, using a fixed tangential unit vector $\mathbf{t}$. Introduce a smooth test function ψ in $\mathcal{D}(\Omega \times Y')$ and perform the same computations on the duality pairing as in Proposition 2.7 (by using again notation (2.4)),

$$\begin{aligned}\Big\langle \widehat{w} \cdot \mathbf{t}_{|Y'+\mathbf{b}_N} &- \widehat{w} \cdot \mathbf{t}_{|Y'}, \psi \Big\rangle_{\Omega \times Y'} \\ &= \lim_{\varepsilon \to 0} \Big\langle \mathcal{T}_\varepsilon(w_\varepsilon) \cdot \mathbf{t}_{|Y'+\mathbf{b}_N} - \mathcal{T}_\varepsilon(w_\varepsilon) \cdot \mathbf{t}_{|Y'}, \psi \Big\rangle_{\Omega \times Y'}.\end{aligned}$$

The latter limit can be computed as it was done in the proofs of Theorem 1.36 and of Proposition 2.7, and the result is

$$\begin{aligned}&\Big\langle \mathcal{T}_\varepsilon(w_\varepsilon) \cdot \mathbf{t}_{|Y'+\mathbf{b}_N} - \mathcal{T}_\varepsilon(w_\varepsilon) \cdot \mathbf{t}_{|Y'}, \psi \Big\rangle_{\Omega \times Y'} \\ &\quad = \Big\langle w_\varepsilon\Big(\varepsilon\Big[\frac{x}{\varepsilon}\Big]_Y + \varepsilon \mathbf{b}_N + \varepsilon y'\Big) \cdot \mathbf{t}_{|Y'+\mathbf{b}_N} \\ &\qquad\qquad - w_\varepsilon\Big(\varepsilon\Big[\frac{x}{\varepsilon}\Big]_Y + \varepsilon y'\Big) \cdot \mathbf{t}_{|Y'}, \psi(x, y')\Big\rangle_{\Omega \times Y'} \\ &\quad = \Big\langle w_\varepsilon\Big(\varepsilon\Big[\frac{x}{\varepsilon}\Big]_Y + \varepsilon y'\Big) \cdot \mathbf{t}_{|Y'}, \psi(x - \varepsilon \mathbf{b}_N, y') - \psi(x, y')\Big\rangle_{\Omega \times Y'} \\ &\quad = \big\langle \mathcal{T}_\varepsilon(w_\varepsilon) \cdot \mathbf{t}_{|Y'}, \psi(x - \varepsilon \mathbf{b}_N, y') - \psi(x, y')\big\rangle_{\Omega \times Y'}.\end{aligned}$$

The last term goes to zero, which completes the proof. □

Proposition 2.10. *Let $\{w_\varepsilon\}_\varepsilon$ be a sequence converging weakly to some w in $L^p_{\text{curl}}(\Omega)$. Then, up to a subsequence, there exist $\overline{w}$ in $L^p(\Omega; W^{1,p}_{per}(Y))$ with $\mathcal{M}_Y(\overline{w}) = 0$, and $\widehat{w}$ in $L^p(\Omega; W^{1,p}_{per}(Y))^N$ with $\mathcal{M}_Y(\widehat{w}) = 0$ and $\text{div}_y\, \widehat{w} = 0$, such that*

$$\begin{aligned}&\mathcal{T}_\varepsilon(w_\varepsilon) \rightharpoonup w + \nabla_y \overline{w} \quad \textit{weakly in } L^p(\Omega \times Y)^N, \\ &\mathcal{T}_\varepsilon(\text{curl}\, w_\varepsilon) \rightharpoonup \text{curl}\, w + \text{curl}_y \widehat{w} \quad \textit{weakly in } L^p(\Omega \times Y)^{N(N-1)/2}.\end{aligned} \tag{2.6}$$

Proof. There exist two functions W in $L^p(\Omega; L^p_{\text{curl},per}(Y))$ with $\text{curl}_y(W) = 0$ (by Proposition 2.9) and Z in $L^p(\Omega \times Y)^{N(N-1)/2}$ (by Proposition 1.12), such that, up to a subsequence,

$$\begin{aligned}&\mathcal{T}_\varepsilon(w_\varepsilon) \rightharpoonup W \quad \text{weakly in } L^p(\Omega \times Y)^N, \\ &\mathcal{T}_\varepsilon(\text{curl}\, w_\varepsilon) \rightharpoonup Z \quad \text{weakly in } L^p(\Omega \times Y)^{N(N-1)/2}.\end{aligned} \tag{2.7}$$

In order to prove convergences (2.6), it remains to identify the limits W and Z. That will be done in two steps, first for Ω bounded and then, for Ω unbounded.

Step 1. Assume Ω is a bounded domain.

As in the proof of Proposition 2.8, start by lifting the sequence $\{\operatorname{div} w_\varepsilon\}_\varepsilon$ which is bounded in $W^{-1,p}(\Omega)$, to a sequence $\{\psi_\varepsilon\}_\varepsilon$ bounded in $W^{1,p}(\Omega)$ by taking the Newtonian potential of $\operatorname{div} w_\varepsilon$, that is

$$\Delta\psi_\varepsilon = \operatorname{div} w_\varepsilon \quad \text{in} \quad \Omega.$$

It is well-known that (see [114, Chapter 9])

$$\|\psi_\varepsilon\|_{W^{1,p}(\Omega)} \leqslant C\|\operatorname{div} w_\varepsilon\|_{W^{-1,p}(\Omega)} \leqslant C\|w_\varepsilon\|_{L^p(\Omega)}.$$

Then, due to the above estimate and by Theorem 1.41, up to a subsequence, there exist ψ in $W^{1,p}(\Omega)$ and $\overline{w}$ in $L^p(\Omega; W^{1,p}_{per,0}(Y))$ such that

$$\begin{aligned} &\psi_\varepsilon \rightharpoonup \psi \quad \text{weakly in} \quad W^{1,p}(\Omega), \\ &\mathcal{T}_\varepsilon(\nabla\psi_\varepsilon) \rightharpoonup \nabla\psi + \nabla_y\overline{w} \quad \text{weakly in} \quad L^p(\Omega \times Y)^N. \end{aligned} \tag{2.8}$$

Set now,

$$u_\varepsilon = -\nabla\psi_\varepsilon + w_\varepsilon \quad \text{in} \quad \Omega. \tag{2.9}$$

The field u_ε is bounded in $L^p(\Omega)$ and satisfies

$$\operatorname{div} u_\varepsilon = 0, \quad \operatorname{curl} u_\varepsilon = \operatorname{curl} w_\varepsilon \quad \text{and} \quad \|u_\varepsilon\|_{L^p(\Omega)} \leqslant C\|w_\varepsilon\|_{L^p(\Omega)}.$$

Convergences (2.8) and equalities above give

$$u_\varepsilon \rightharpoonup u \quad \text{weakly in} \quad L^p_{\text{curl}}(\Omega).$$

Due to the following identities, true for $k \in \{1, \dots, N\}$,

$$\begin{aligned} \Delta u_{\varepsilon,k} = \Delta u_{\varepsilon,k} - \frac{\partial(\operatorname{div} u_\varepsilon)}{\partial x_k} &= \sum_{i=1}^N \frac{\partial}{\partial x_i}\Big(\frac{\partial u_{\varepsilon,k}}{\partial x_i} - \frac{\partial u_{\varepsilon,i}}{\partial x_k}\Big) \\ &= \sum_{i=1}^N \frac{\partial}{\partial x_i}\Big(\frac{\partial w_{\varepsilon,k}}{\partial x_i} - \frac{\partial w_{\varepsilon,i}}{\partial x_k}\Big), \end{aligned} \tag{2.10}$$

one has

$$\|\Delta u_\varepsilon\|_{W^{-1,p}(\Omega)} \leqslant N^2\|\nabla(\operatorname{curl} u_\varepsilon)\|_{W^{-1,p}(\Omega)} = N^2\|\operatorname{curl} w_\varepsilon\|_{L^p(\Omega)}.$$

As a consequence, $\{u_\varepsilon\}_\varepsilon$ belongs to $W^{1,p}_{loc}(\Omega)^N$ (4) and for every $\omega \Subset \Omega$, there exists a constant C_ω such that

$$\begin{aligned} \|u_\varepsilon\|_{W^{1,p}(\omega)} &\leqslant C_\omega\big(\|\Delta u_\varepsilon\|_{W^{-1,p}(\Omega)} + \|u_\varepsilon\|_{L^p(\Omega)}\big) \\ &\leqslant C_\omega\big(\|u_\varepsilon\|_{L^p(\Omega)} + \|\operatorname{curl} w_\varepsilon\|_{L^p(\Omega)}\big). \end{aligned}$$

(4) If Ω is a bounded domain with Lipschitz boundary, an alternative way for ending this step, would be to use the following estimate: $\|\rho\nabla u_\varepsilon\|_{L^p(\Omega)} \leqslant C\big(\|u_\varepsilon\|_{L^p(\Omega)} + \|\operatorname{curl} w_\varepsilon\|_{L^p(\Omega)}\big)$, where the constant does not depend on ε, and then using Theorem (1.46).

Then, due to the above estimate of u_ε and by Theorem 1.41, up to a subsequence, one has u in $W^{1,p}(\omega)^N$ and there exists $\widehat{w}_\omega$ in $L^p(\omega; W^{1,p}_{per,0}(Y))^N$ such that

$$\begin{aligned} &u_\varepsilon \rightharpoonup u \quad \text{weakly in } W^{1,p}(\omega),\\ &\mathcal{T}_\varepsilon(u_\varepsilon) \rightharpoonup u \quad \text{weakly in } L^p(\omega \times Y)^N,\\ &\mathcal{T}_\varepsilon(\nabla u_\varepsilon) \rightharpoonup \nabla u + \nabla_y \widehat{w}_\omega \quad \text{weakly in } L^p(\omega \times Y)^{N\times N}. \end{aligned} \tag{2.11}$$

Since $\operatorname{div} u_\varepsilon = 0$, it yields $\operatorname{div}_y \widehat{w}_\omega = 0$.

Now, recalling the first convergence from (2.7), thanks to (2.9) one gets,

$$\begin{aligned} &w = u + \nabla\psi \quad \text{in } L^p(\omega)^N,\\ &W = u + \nabla\psi + \nabla_y \overline{w}_\omega = w + \nabla_y \overline{w}_\omega \quad \text{in } L^p(\omega \times Y)^N. \end{aligned}$$

We now claim that there exits $\widehat{w}$ in $L^p(\Omega; W^{1,p}_{per}(Y))^N$ such that $\widehat{w}_{|\omega} = \widehat{w}_\omega$ for every subdomain relatively compact in Ω. In order to do so, let ω_1 and ω_2 be two balls included in Ω. Assume that they intersect, and that the same subsequence is used for both to obtain $\widehat{w}_{\omega_1}$ and $\widehat{w}_{\omega_2}$. The difference $\widehat{w}_{\omega_1} - \widehat{w}_{\omega_2}$, whose restriction to $\omega_1 \cap \omega_2$ belongs to $L^p(\omega_1 \cap \omega_2; W^{1,p}_{per,0}(Y))^N$, satisfies

$$\operatorname{curl}_y(\widehat{w}_{\omega_1} - \widehat{w}_{\omega_2}) = 0 \text{ and } \operatorname{div}_y(\widehat{w}_{\omega_1} - \widehat{w}_{\omega_2}) = 0$$

a.e. in $(\omega_1 \cap \omega_2) \times Y$. Then using the same identities as (2.10), one obtains $\Delta_y \widehat{w}(x,\cdot) = 0$ in $(W^{1,p}_{per}(Y))'$ for a.e. $x \in \omega_1 \cap \omega_2$. Thus

$$\widehat{w}_{\omega_1} = \widehat{w}_{\omega_2} \quad \text{in} \quad (\omega_1 \cap \omega_2) \times Y.$$

Using a countable set of balls included in Ω whose union is Ω and using a diagonal subsequence argument, proves that there exits $\widehat{w}$ in $L^p_{loc}(\Omega; W^{1,p}_{per,0}(Y))^N$ with $\operatorname{div}_y \widehat{w} = 0$ such that $\widehat{w}_{|\omega} = \widehat{w}_\omega$ for every subdomain relatively compact in Ω.

At this point, the last line of (2.11) together with $(2.7)_2$ imply

$$Z = \operatorname{curl} w + \operatorname{curl}_y \widehat{w} \quad \text{in } L^p(\omega \times Y)^{N(N-1)/2}.$$

Both fields Z and $\operatorname{curl} w$ are in $L^p(\Omega \times Y)^{N(N-1)/2}$, it yields that $\operatorname{curl}_y \widehat{w}$ is in the same space. Then, using the same identities as (2.10), one obtains

$$\|\Delta_y \widehat{w}(x,\cdot)\|_{(W^{1,p}_{per}(Y))'} \leqslant N^2 \|\operatorname{curl}_y \widehat{w}(x,\cdot)\|_{L^p(Y)} \quad \text{for a.e. } x \text{ in } \Omega.$$

Hence (recall that $\mathcal{M}_Y(\widehat{w}) = 0$ a.e. in Ω), there exists a constant C which only depends on N, p and Y (but not on Ω) such that

$$\|\widehat{w}(x,\cdot)\|_{W^{1,p}(Y)} \leqslant C \|\operatorname{curl}_y \widehat{w}(x,\cdot)\|_{L^p(Y)} \quad \text{for a.e. } x \text{ in } \Omega.$$

Therefore,

$$\|\widehat{w}\|_{L^p(\Omega; W^{1,p}_{per}(Y))} \leqslant C \limsup_{\varepsilon\to 0} \|\operatorname{curl} w_\varepsilon\|_{L^p(\Omega)}.$$

Step 2. Assume Ω unbounded.

Again using a countable set of balls included in Ω whose union is Ω, a diagonal subsequence argument and the results of Step 1, completes the proof. □

2.2.3 Unfolding with information on the symmetric gradient

For a given domain Ω in $\mathbb{R}^N$, we consider sequences of vector fields w which are bounded in $L^p(\Omega)^N$, $p \in [1, +\infty]$, with their symmetric gradient

$$e(w) \doteq \frac{1}{2}\big(\nabla w + (\nabla w)^T\big) \text{ in } L^p(\Omega)^{N\times N}.$$

For fields depending upon the variable y, the corresponding symmetric gradient operator is denoted e_y.

The following Korn-type inequalities, valid for bounded domains ω with Lipschitz boundary, will be used below and also in Chapter 11 (see for example [57, 115, 168]):

Proposition 2.11. *There exists a constant $C(\omega, p)$ such that, for every vector field φ in $L^p(\omega)^N$ with $e(\varphi)$ in $L^p(\omega)^{N\times N}$, there exists a rigid displacement R ($R(x) = \mathbf{a} + \mathbf{A}\,x$, $x \in \mathbb{R}^N$, where $\mathbf{a}$ is in $\mathbb{R}^N$ and $\mathbf{A}$ is an antisymmetric $N \times N$ matrix) the following inequality holds:*

$$\|\varphi - R\|_{W^{1,p}(\omega)} \leqslant C(\omega, p)\|e(\varphi)\|_{L^p(\omega)}. \tag{2.12}$$

Consequently, one also has

$$\|\varphi\|_{W^{1,p}(\omega)} \leqslant C(\omega, p)\big(\|\varphi\|_{L^p(\omega)} + \|e(\varphi)\|_{L^p(\omega)}\big), \tag{2.13}$$

and, if $\varphi = 0$ on a part of $\partial\omega$ with non zero measure, then

$$\|\varphi\|_{W^{1,p}(\omega)} \leqslant C(\omega, p)\|e(\varphi)\|_{L^p(\omega)}.$$

If ω is a parallelotope $\mathcal{P}$, then, for every vector field ψ in $W^{1,p}_{per}(\mathcal{P})^N$,

$$\|\psi - \mathcal{M}_Y(\psi)\|_{W^{1,p}(\mathcal{P})} \leqslant C(\mathcal{P}, p)\|e_y(\psi)\|_{L^p(\mathcal{P})}. \tag{2.14}$$

Proposition 2.12. *Let $\{w_\varepsilon\}_\varepsilon$ be a sequence in $L^p(\Omega)^N$ satisfying*

$$\|w_\varepsilon\|_{L^p(\Omega)} + \varepsilon\|e(w_\varepsilon)\|_{L^p(\Omega)} \leqslant C. \tag{2.15}$$

Then there exist a subsequence (still denoted ε) and $\widehat{w} \in L^p(\Omega; W^{1,p}_{per}(Y))^N$, such that

$$\begin{aligned}
&\mathcal{T}_\varepsilon(w_\varepsilon) \rightharpoonup \widehat{w} \quad \text{weakly in } L^p(\Omega \times Y)^N,\\
&e_y(\mathcal{T}_\varepsilon(w_\varepsilon)) = \varepsilon\mathcal{T}_\varepsilon(e(w_\varepsilon)) \rightharpoonup e_y(\widehat{w}) \quad \text{weakly in } L^p(\Omega \times Y)^{N\times N}.
\end{aligned} \tag{2.16}$$

Similar results hold for $p = +\infty$ in the weak-$*$ topology.

Proof. Proceeding as in Proposition 1.35, one obtains the equality

$$e_y(\mathcal{T}_\varepsilon(w_\varepsilon)) = \varepsilon\mathcal{T}_\varepsilon(e(w_\varepsilon)).$$

Then, from (2.15) it follows that

$$\|\mathcal{T}_\varepsilon(w_\varepsilon)\|_{L^p(\Omega\times Y)} + \|e_y(\mathcal{T}_\varepsilon(w_\varepsilon))\|_{L^p(\Omega\times Y)} \leqslant C,$$

with C independent of ε.

By (2.13), the sequence $\{\mathcal{T}_\varepsilon(w_\varepsilon)\}_\varepsilon$ is bounded in $L^p(\Omega; W^{1,p}(Y))^N$. So, there exists a function $\widehat{w}$ in $L^p(\Omega; W^{1,p}(Y))^N$ such that, up to a subsequence, convergences (2.16) hold.

It remains to show that $\widehat{w}$ is Y-periodic. To do so, let ω be an arbitrary bounded and Lipschitz subdomain of Ω. It is enough to show that the restriction of $\widehat{w}$ to $\omega \times Y$ is Y-periodic. Again by (2.13), there exists a constant C_ω such that the following estimate holds in ω:

$$\|\nabla w_\varepsilon\|_{L^p(\omega)} \leqslant C_\omega\big(\|w_\varepsilon\|_{L^p(\omega)} + \|e(w_\varepsilon)\|_{L^p(\omega)}\big).$$

Together with (2.15), this shows that $\{\varepsilon\nabla w_\varepsilon\}_\varepsilon$ is a bounded sequence in $L^p(\omega)^{N\times N}$. Then the Y-periodicity is obtained by applying Theorem 1.36 and this completes the proof. □

Proposition 2.13. *Let $\{w_\varepsilon\}_\varepsilon$ be a bounded sequence in $L^p(\Omega)^N$. Assume that $\{e(w_\varepsilon)\}_\varepsilon$ is also bounded in $L^p(\Omega)^{N\times N}$. Then, up to a subsequence, there exist w in $(L^p(\Omega) \cap W^{1,p}_{loc}(\Omega))^N$ and $\widehat{w}$ in $L^p(\Omega; W^{1,p}_{per,0}(Y))^N$, such that*

$$\begin{aligned}
&w_\varepsilon \rightharpoonup w \quad \textit{weakly in } (L^p(\Omega) \cap W^{1,p}_{loc}(\Omega))^N,\\
&\mathcal{T}_\varepsilon(w_\varepsilon) \rightharpoonup w \quad \textit{weakly in } L^p(\Omega \times Y)^N,\\
&\mathcal{T}_\varepsilon(e(w_\varepsilon)) \rightharpoonup e(w) + e_y(\widehat{w}) \quad \textit{weakly in } L^p(\Omega \times Y)^{N\times N}.
\end{aligned}$$

Furthermore, if the domain Ω is bounded with Lipschitz boundary (so that the Korn inequality (2.13) *holds in Ω), the function w belongs to $W^{1,p}(\Omega)^N$ and the first convergence above holds in that space.*

Proof. The hypotheses on boundedness imply that there exist three vector fields namely, w in $L^p(\Omega)^N$, W in $L^p(\Omega\times Y)^N$ and E in $L^p(\Omega\times Y)^{N\times N}$ such that, up to a subsequence,

$$\begin{aligned}
&w_\varepsilon \rightharpoonup w \quad \text{weakly in } L^p(\Omega)^N,\\
&\mathcal{T}_\varepsilon(w_\varepsilon) \rightharpoonup W \quad \text{weakly in } L^p(\Omega \times Y)^N,\\
&\mathcal{T}_\varepsilon(e(w_\varepsilon)) \rightharpoonup E \quad \text{weakly in } L^p(\Omega \times Y)^{N\times N}.
\end{aligned}$$

Due to (2.13), for every bounded subdomain ω of Ω with smooth boundary the sequence $\{w_\varepsilon\}_\varepsilon$ is bounded in $W^{1,p}(\omega)^N$. Then Theorem 1.41 implies that the vector field W satisfies

$$W = w \quad \text{in } \omega \times Y \; quad \text{with } w \in W^{1,p}(\omega).$$

By the same theorem, there exists a vector field $\widehat{w}_\omega$ in $L^p(\omega, W^{1,p}_{per,0}(Y))^N$ such that

$$E = e(w) + e_y(\widehat{w}_\omega) \quad \text{in } \omega \times Y.$$

From this identity and due to the Korn inequality for periodic fields (2.14), one obtains a global field $\widehat{w}$ in the same way as in the first part of the proof of Proposition 2.10. □

2.2.4 * Unfolding with information on one derivative

In this paragraph, we present the case where $\Omega = \mathbb{R}^N$ (the case $\Omega \subset \mathbb{R}^N$ follows by restriction). In Section 1.4 of Chapter 1, the case of N directional derivatives was considered. Here, we present the more intricate situation associated to a single directional derivative. The cases of a domain Ω included in $\mathbb{R}^N$ and with k directional derivatives can be treated in the same spirit.

Definition 2.14 (Directional derivatives). *For an open subset $\mathcal{O}$ of $\mathbb{R}^N$, a non zero vector* **a** *in $\mathbb{R}^N$ and p in $[1, +\infty]$, the* **a**-*directional derivative for a function f in $L^p(\mathcal{O})$ is defined in the distributional sense as*

$$\langle \partial_{\mathbf{a}} f, \varphi \rangle_{\mathcal{D}'(\Omega),\mathcal{D}(\Omega)} = - \int_{\mathcal{O}} f\, \mathbf{a} \cdot \nabla\varphi\, dx, \quad \textit{for every } \varphi \in \mathcal{D}(\mathcal{O}).$$

For functions defined on $\mathcal{O} \times Y$ or $\mathcal{O} \times \mathbb{R}^N$, $\partial_{\mathbf{a}}$ is the directional derivative with respect to x while $\partial_{\mathbf{a},y}$ is the directional derivative with respect to the variable y.

Definition 2.15. *The Banach spaces $W^{1,p}_{\mathbf{a}}(\mathcal{O})$ and $W^{1,p}_{\mathbf{a},loc}(\mathbb{R}^N)$ are respectively, defined as*

$$W^{1,p}_{\mathbf{a}}(\mathcal{O}) = \{\varphi \in L^p(\mathcal{O}) \,|\, \partial_{\mathbf{a}}\varphi \in L^p(\mathcal{O})\},$$
$$W^{1,p}_{\mathbf{a},\,loc}(\mathbb{R}^N) = \{\varphi \in L^p_{loc}(\mathbb{R}^N) \,|\, \partial_{\mathbf{a}}\varphi \in L^p_{loc}(\mathbb{R}^N)\}.$$

Define also the space

$$W^{1,p}_{\mathbf{a},\, \mathbf{G}\text{-}per}(\mathbb{R}^N) = W^{1,p}_{\mathbf{a},\,loc}(\mathbb{R}^N) \cap L^p_{\mathbf{G}\text{-}per}(\mathbb{R}^N).$$

In the above definition, as in the remainder of this section, it is obvious that the vector **a** can be replaced by any of its non zero real multiples. The case where **a** is part of a basis of **G** is easily obtained by going back to the proofs of Theorems 1.36 and 1.41, choosing $\mathbf{b_n} \equiv \mathbf{a}$. When the hypotheses hold for several vectors $\mathbf{a_1}, \mathbf{a_2}, \ldots$, the results below can be combined.

We now consider the case where **a** is not colinear to a basic period.

Remark 2.16. *The functions in $W^{1,p}_{\mathbf{a},\, \mathbf{G}\text{-}per}(\mathbb{R}^N)$ have traces on every hyperplane $\mathcal{H}$ which is transverse to the vector. These traces are in $L^p_{loc\, unif}(\mathcal{H})$. If Y is a polyhedral cell with Lipschitz boundary and having the paving property with respect to* **G** *(such as a parallelotope), the restrictions to Y of the functions of $W^{1,p}_{\mathbf{a},\, \mathbf{G}\text{-}per}(\mathbb{R}^N)$ are in $W^{1,p}_{\mathbf{a}}(Y)$, and have the same traces on each set of opposite faces of Y provided these faces are transverse to* **a**.

Suppose Y is a parallelotope. A function in $W^{1,p}_{\mathbf{a}}(Y)$ is the restriction to Y of an element of $W^{1,p}_{\mathbf{a},\, \mathbf{G}\text{-}per}(\mathbb{R}^N)$ if and only if it has the same traces on each set of opposite faces of Y which are transverse to **a**.

Proposition 2.17. *Let $\{w_\varepsilon\}_\varepsilon$ be a bounded sequence in $L^p(\mathbb{R}^N)$, with $\partial_{\mathbf{a}} w_\varepsilon$ in $L^p(\mathbb{R}^N)$ and satisfying*

$$\varepsilon \, \|\partial_{\mathbf{a}} w_\varepsilon\|_{L^p(\mathbb{R}^N)} \leqslant C.$$

Then, up to a subsequence, there is a function $\widehat{w}$ in $L^p(\mathbb{R}^N; W^{1,p}_{\mathbf{a},\,\mathbf{G}\text{-}per}(\mathbb{R}^N))$ such that, for $1 < p < +\infty$ and for every Y having the paving property with respect to $\mathbf{G}$,

$$\overline{\mathcal{T}_\varepsilon}(w_\varepsilon) \rightharpoonup \widehat{w} \quad \text{weakly in } \ L^p(\mathbb{R}^N; W^{1,p}_{\mathbf{a},loc}(\mathbb{R}^N)), \tag{2.17}$$

and by restriction to $\mathbb{R}^N \times Y$,

$$\widetilde{\mathcal{T}}^Y_\varepsilon(w_\varepsilon) \rightharpoonup \widehat{w} \quad \text{weakly in } \ L^p(\mathbb{R}^N; W^{1,p}_{\mathbf{a}}(Y)).$$

For $p = +\infty$, the convergence takes place in the weak-$*$ topology of the space $L^\infty(\mathbb{R}^N; W^{1,\infty}_{\mathbf{a},loc}(\mathbb{R}^N))$.

Proof. From the hypotheses, as in Proposition 1.35, it follows that

$$\|\overline{\mathcal{T}_\varepsilon}^Y(w_\varepsilon)\|_{L^p(\mathbb{R}^N; W^{1,p}_{\mathbf{a},loc}(\mathbb{R}^N))} \leqslant C,$$

with C independent of ε. Hence, up to a subsequence, there exists some $\widehat{w}$ in $L^p(\mathbb{R}^N; W^{1,p}_{\mathbf{a},loc}(\mathbb{R}^N))$, such that convergences (2.17) hold. By the first statement of Theorem 2.4, this convergence takes place in $L^p(\mathbb{R}^N; L^p_{\mathbf{G}\text{-}per}(\mathbb{R}^N)))$.

The proof can be easily adapted to the case $p = +\infty$. ☐

We now turn to the case in which the sequence of directional derivatives $\partial_{\mathbf{a}} w_\varepsilon$ is bounded in $L^p(\mathbb{R}^N)$. The result is stated in Theorem 2.23 below and requires some preliminary notations and results. To do so, we shall make a small detour via *Algebra*.

The minimal representation of a vector as linear combinations of elements of G.

Proposition 2.18. *Let $\mathbf{a}$ be a given non zero vector in $\mathbb{R}^N$ and let $\mathbf{G}$ be as before. Then there exits a unique minimal subspace $\mathbf{V}$ of $\mathbb{R}^N$ generated by a subgroup of $\mathbf{G}$ and which contains $\mathbf{a}$. The set $\mathbf{H} \doteq \mathbf{V} \cap \mathbf{G}$ is a subgroup of $\mathbf{G}$ and it generates $\mathbf{V}$.*

Proof. Consider all the vector subspaces of $\mathbb{R}^N$ containing $\mathbf{a}$ and generated by subgroups of $\mathbf{G}$. Let r be the minimum of the dimensions of such subspaces and $\mathbf{V}$ be one of these subspaces with dimension r.

By construction, there is a subgroup $\mathbf{H}_1$ of $\mathbf{G}$ with rank r (its dimension as a $\mathbb{Z}$-module), whose span in $\mathbb{R}^N$ is $\mathbf{V}$ so that $\mathbf{H}_1$ is included in $\mathbf{H} \doteq \mathbf{V} \cap \mathbf{G}$, itself a subgroup of $\mathbf{G}$. By construction, the span of $\mathbf{H}$ also $\mathbf{V}$ and $\mathbf{H}$ is also of rank r. Consequently, $\mathbf{H}$ is the largest subgroup of $\mathbf{G}$ which spans $\mathbf{V}$[(5)].

(5) Every subgroup of $\mathbf{H}$ with rank r, such as $\mathbf{H}_1$, also generates $\mathbf{V}$.

The space $\mathbf{V}$ is unique. Indeed, let $\mathbf{V}'$ be another subspace having the same dimension r, containing $\mathbf{a}$ and generated by the subgroup $\mathbf{H}' \doteq \mathbf{V}' \cap \mathbf{G}$. The subspace $\mathbf{V} + \mathbf{V}'$ is generated by $\mathbf{H} + \mathbf{H}'$, while the span of $\mathbf{H} \cap \mathbf{H}'$ is included in $\mathbf{V} \cap \mathbf{V}'$. We claim that latter inclusion is an equality. Indeed,

$$\begin{aligned}\dim_{\mathbb{R}}(\mathbf{V} \cap \mathbf{V}') &= \dim_{\mathbb{R}}(\mathbf{V}) + \dim_{\mathbb{R}}(\mathbf{V}') - \dim_{\mathbb{R}}(\mathbf{V} + \mathbf{V}') \\ &= \operatorname{rank}(\mathbf{H}) + \operatorname{rank}(\mathbf{H}') - \operatorname{rank}(\mathbf{H} + \mathbf{H}') \\ &= \operatorname{rank}(\mathbf{H} \cap \mathbf{H}'),\end{aligned}$$

which implies the claim that $\mathbf{H} \cap \mathbf{H}'$ spans $\mathbf{V} \cap \mathbf{V}'$. Since $\mathbf{V} \cap \mathbf{V}'$ contains $\mathbf{a}$, the minimality of r gives $\dim_{\mathbb{R}}(\mathbf{V} \cap \mathbf{V}') = r$ so that $\mathbf{V} \cap \mathbf{V}' = \mathbf{V}$, i.e., $\mathbf{V} = \mathbf{V}'$. □

This result implies the existence of a subset $(\mathbf{p}_1, \ldots, \mathbf{p}_r) \in \mathbf{H}^r$ which is a basis for the group $\mathbf{H}$ as well as a basis for its span $\mathbf{V}$. Thus

$$\begin{aligned}&(i) \quad \mathbf{H} = \sum_{i=1}^{r} \mathbb{Z}\mathbf{p}_i, \qquad \mathbf{V} \doteq \sum_{i=1}^{r} \mathbb{R}\mathbf{p}_i, \\ &(ii) \quad \mathbf{a} \doteq \sum_{i=1}^{r} \lambda_i \, \mathbf{p}_i \quad \text{with} \quad (\lambda_1, \ldots, \lambda_r) \in \mathbb{R}^r.\end{aligned} \tag{2.18}$$

Lemma 2.19.
(i) *The real numbers λ_i's of the decomposition* (2.18)*(ii) of* $\mathbf{a}$ *are rationally independent (i.e. linearly independent over $\mathbb{Q}$).*
(ii) *The subspace $\mathbf{V}$ is the closure $\overline{\mathbf{H} + \mathbb{R}\mathbf{a}}$ of the subgroup $\mathbf{H} + \mathbb{R}\mathbf{a}$.*
(iii) *There exists $\mathbf{p}_{r+1}, \ldots, \mathbf{p}_N$ in $\mathbf{G}$ such that $(\mathbf{p}_1, \ldots, \mathbf{p}_N)$ is a basis of $\mathbf{G}$.*

Proof.
(i) The case $r = 1$ is obvious since $\mathbf{a}$ is not zero. For $r \geqslant 2$, if the λ_i's were linearly dependent, one would write one of them, say λ_r (by a reordering if needed), as a rational combination of the others,

$$\lambda_r = \sum_{i=1}^{r-1} \mu_i \, \lambda_i,$$

which implies that

$$\mathbf{a} = \sum_{i=1}^{r-1} \lambda_i \, (\mathbf{p}_i + \mu_i \, \mathbf{p}_r).$$

Denoting by m the product of the denominators of the μ_i's, this is just a representation of the form

$$\mathbf{a} = \sum_{i=1}^{r-1} \frac{\lambda_i}{m} (m \, \mathbf{p}_i + m \, \mu_i \, \mathbf{p_r}).$$

As $m\mathbf{p}_i + m\mu_i\mathbf{p_r}$ belongs to $\mathbf{G}$, this contradicts the minimality of r.

(ii) The subspace $\mathbf{V}$ contains both $\mathbf{H}$ and $\mathbb{R}\mathbf{a}$ and is a closed in $\mathbb{R}^N$, hence,

$$\mathbf{H} + \mathbb{R}\mathbf{a} \subset \overline{\mathbf{H} + \mathbb{R}\mathbf{a}} \subset \mathbf{V}. \tag{2.19}$$

Then (2.18)(ii) implies that for every $(k, \theta) \in \mathbb{Z} \times \mathbb{R}$, the sum $\sum_{i=1}^{r} k\,\theta\,\lambda_i \mathbf{p}_i$ belongs to $\mathbb{R}\mathbf{a}$. Using the notation $\{\tau\}$ for the fractional part of the real number τ, it follows that x

$$\sum_{i=1}^{r} \{k\theta\lambda_i\}\mathbf{p}_i \text{ belongs to } \mathbf{H} + \mathbb{R}\mathbf{a}, \qquad \forall (k, \theta) \in \mathbb{Z} \times \mathbb{R}.$$

Now, choose θ such that

$$\frac{1}{\theta} \in \mathbb{R} \setminus (\mathbb{Q}\lambda_1 \oplus \ldots \oplus \mathbb{Q}\lambda_r).$$

With this choice, the $r + 1$ real numbers $1, \theta\lambda_1, \ldots, \theta\lambda_r$ are rationally independent because the r real numbers $\lambda_1, \ldots, \lambda_r$ are rationally independent. Therefore, the sequence $(\{k\theta\lambda_1\}, \ldots, \{k\theta\lambda_r\})(k \in \mathbb{Z})$ is uniformly distributed in $(0, 1)^r$ (see [143]), hence dense in $[0, 1]^r$. Thus

$$\sum_{i=1}^{r} [0, 1]\mathbf{p}_i \in \overline{\mathbf{H} + \mathbb{R}\mathbf{a}} \quad \text{and by scaling} \quad \mathbf{V} \in \overline{\mathbf{H} + \mathbb{R}\mathbf{a}}.$$

This together with (2.19) gives the claim (ii).

(iii) In the case $r = N$, one can choose $\mathbf{H} \equiv \mathbf{G}$, so the result is obvious. In the case where $r < N$, the abelian group $\mathbf{G}/\mathbf{H}$ is a free $\mathbb{Z}$-module since, if $g \in \mathbf{G}$ is such that $ng \in \mathbf{H}$ for some $n \in \mathbb{Z}^*$, then g belongs to $\mathbf{G} \cap \mathbf{V}$, hence to $\mathbf{H}$ (by the definition of the latter).

To end the proof, observe that if $\mathbf{p}_{r+1} + \mathbf{H}, \ldots, \mathbf{p}_N + \mathbf{H}$ is a basis of $\mathbf{G}/\mathbf{H}$, the set $(\mathbf{p}_1, \ldots, \mathbf{p}_N)$ is a basis of $\mathbf{G}$. □

Remark 2.20. *In reference to Remark 2.16, with the notations of* (2.18), *suppose that Y is the parallelotope constructed on the basis $\mathbf{p}_1, \ldots, \mathbf{p}_N$, the faces of Y which are transverse are exactly the r faces generated by the sets* $\{\mathbf{p}_2, \ldots, \mathbf{p}_N\}$, $\{\mathbf{p}_1, \mathbf{p}_3, \ldots, \mathbf{p}_N\}$, $\ldots$, $\{\mathbf{p}_1, \ldots, \mathbf{p}_{r-1}, \mathbf{p}_{r+1}, \ldots, \mathbf{p}_N\}$.

Properties of periodic functions in the kernel of $\partial_{\mathbf{a},y}$.

In this context, being in the kernel of $\partial_{\mathbf{a},y}$ implies a lot more for functions of $W^{1,p}_{\mathbf{a},\,\mathbf{G}\text{-}per}(\mathbb{R}^N)$.

Proposition 2.21. *Let* $\mathbf{a}$ *and* $\mathbf{V}$ *be defined as above. Suppose Ψ belongs to $W^{1,p}_{\mathbf{a},\,\mathbf{G}\text{-}per}(\mathbb{R}^N)$ and is such that $\partial_{\mathbf{a},y}\Psi \equiv 0$. Then, Ψ satisfies $\partial_{\mathbf{v},y}\Psi \equiv 0$ for all $\mathbf{v} \in \mathbf{V}$.*

The same result holds for functions in $L^p(\mathbb{R}^N, W^{1,p}_{\mathbf{a},\,\mathbf{G}\text{-}per}(\mathbb{R}^N))$. In particular, every weak limit point of a sequence of unfolding of a bounded sequence in the space $W^{1,p}_{\mathbf{a}}(\Omega)$ is in the kernel of all $\partial_{\mathbf{v},y}\Psi = 0$ for $\mathbf{v} \in \mathbf{V}$.

Lemma 2.22. *The group of periods with respect to the variable y of an element of $L^p_{\mathbf{G}-per}(\mathbb{R}^N)$ is closed in $\mathbb{R}^N$. Similarly, the group of periods with respect to y of an element of $L^p(\mathbb{R}^N; L^p_{\mathbf{G}-per}(\mathbb{R}^N))$ is closed in $\mathbb{R}^N$.*

Proof. We give the proof for the second case. Let Y be a periodicity cell for $\mathbf{G}$. The vector $\mathbf{p}$ is a period of $\widehat{w}$ if and only if

$$h(\mathbf{p}) \doteq \|\widehat{w} - \widehat{w}(\cdot, \cdot + \mathbf{p})\|_{L^p(\mathbb{R}^N \times Y)} = 0.$$

The result follows because h is continuous. □

Proof of Proposition 2.21. Under the hypotheses, the group of periods of Ψ contains $\mathbb{R}\mathbf{a}$ and $\mathbf{G}$. It is closed by Lemma 2.22. Therefore, Lemma 2.19(ii) implies that its group of periods contains $\mathbf{V}$. One concludes by the fact that it is equivalent for a function to have $\mathbb{R}\mathbf{v}$ in its group of periods, and to be in the kernel of $\partial_{\mathbf{v}}$. □

We can now state the main result, with the same notations as above.

Theorem 2.23. *Suppose $p \in (1, +\infty]$ and let $\mathbf{a} \neq 0$ be a vector in $\mathbb{R}^N$ with an associated subspace $\mathbf{V}$. Let $\{w_\varepsilon\}_\varepsilon$ be a sequence in $W^{1,p}_{\mathbf{a}}(\mathbb{R}^N)$ such that*

$$w_\varepsilon \rightharpoonup w \quad \text{weakly in } W^{1,p}_{\mathbf{a}}(\mathbb{R}^N).$$

Then, up to a subsequence and for every cell Y having the paving property with respect to $\mathbf{G}$, there exist two functions, $\widehat{w}$ in $W^{1,p}_{\mathbf{a}}(\mathbb{R}^N; W^{1,p}_{\mathbf{a},\mathbf{G}\text{-}per}(\mathbb{R}^N))$ and $\widehat{\widehat{w}}$ in $L^p(\mathbb{R}^N; W^{1,p}_{\mathbf{a}}(Y))$, (6) *such that for $1 < p < +\infty$,*

$$\begin{aligned}
&(i) \quad \mathcal{T}_\varepsilon(w_\varepsilon) \rightharpoonup \widehat{w} \quad \text{weakly in } L^p(\mathbb{R}^N; W^{1,p}_{\mathbf{a},loc}(\mathbb{R}^N)) \\
&\qquad (\text{by restriction, } \widetilde{\mathcal{T}}^Y_\varepsilon(w_\varepsilon) \rightharpoonup \widehat{w} \text{ weakly in } L^p(\mathbb{R}^N; W^{1,p}_{\mathbf{a}}(Y))), \\
&(ii) \quad \mathcal{M}_Y(\widehat{w}) \equiv w \quad \text{and} \quad \partial_{\mathbf{a},y}\widehat{w} \equiv 0, \\
&(iii) \quad \partial_{\mathbf{v},y}\widehat{w} \equiv 0 \text{ for every } \mathbf{v} \in \mathbf{V}, \\
&(iv) \quad \partial_{\mathbf{a}}\widehat{w} \in L^p(\mathbb{R}^N \times Y), \\
&(v) \quad \widetilde{\mathcal{T}}^Y_\varepsilon(\partial_{\mathbf{a}} w_\varepsilon) \rightharpoonup \partial_{\mathbf{a}}\widehat{w} + \partial_{\mathbf{a},y}\widehat{\widehat{w}} \quad \text{weakly in } L^p(\mathbb{R}^N \times Y).
\end{aligned} \tag{2.20}$$

For $p = +\infty$, the convergences are in the weak-$*$ topology of $L^\infty(\mathbb{R}^N \times Y)$ (as well as weakly in every $L^q_{loc}(\mathbb{R}^N; L^q(Y))$ for q finite).

Proof. The proof for p finite is given below, the case $p = +\infty$ can be treated in the same way. By Proposition 2.5, one can choose for Y the parallelotope associated with the basis of $\mathbf{G}$ obtained in Lemma 2.19 (iii).

Convergence (2.20)(i) for some $\widehat{w} \in L^p(\mathbb{R}^N; W^{1,p}_{\mathbf{a},\mathbf{G}\text{-}per}(\mathbb{R}^N))$, is a direct consequence of convergence (2.17) from Proposition 2.17. The fact that

(6) One would expect that the function $\widehat{\widehat{w}}$ belong to the space $L^p(\mathbb{R}^N; W^{1,p}_{\mathbf{a},\,\mathbf{G}\text{-}per}(\mathbb{R}^N))$ but this is an open question.

$\mathcal{M}_Y(\widehat{w}) \equiv w$ follows from assertion (iii) of Proposition 1.9. Furthermore, due to the identity

$$\partial_{\mathbf{a},y}\widetilde{\mathcal{T}}_\varepsilon^Y(w_\varepsilon) = \varepsilon\widetilde{\mathcal{T}}_\varepsilon^Y(\partial_{\mathbf{a}}w_\varepsilon),$$

it is immediate that $\partial_{\mathbf{a},y}\widehat{w} \equiv 0$, whence (2.20)(ii).

The function $\widehat{w}$ belongs to $L^p(\mathbb{R}^N; W^{1,p}_{\mathbf{a},\,\mathbf{G}\text{-}per}(\mathbb{R}^N))$ with $\partial_{\mathbf{a},y}\widehat{w} \equiv 0$. By Proposition 2.21 it satisfies (2.20)(iii).

The proof of (2.20)(iv) for $r = N$ is straightforward. Indeed, by (2.20)(iii), $\widehat{w}$ is independent of y, so that $\widehat{w} = w$ and

$$\partial_{\mathbf{a}}\widehat{w} = \partial_{\mathbf{a}}w \in L^p(\mathbb{R}^N).$$

The proof of (2.20)(iv) for $1 \leqslant r < N$ is more complicated and requires extra notations.

Let Y_1 and Y_2 be the parallelotopes associated with the vectors $\mathbf{p}_1, \ldots, \mathbf{p}_r$ and $\mathbf{p}_{r+1}, \ldots, \mathbf{p}_N$, thus $Y = Y_1 \times Y_2$. Let also $\mathbf{V}'$ denote the subspace $\mathbb{R}\mathbf{p}_{r+1} \oplus \ldots \oplus \mathbb{R}\mathbf{p}_N$.

In the remainder of the proof, set

$$\begin{aligned} &x \in \mathbb{R}^N, \qquad x = x_{\mathbf{V}} + x_{\mathbf{V}'}, \quad (x_{\mathbf{V}}, x_{\mathbf{V}'}) \in \mathbf{V} \times \mathbf{V}',\\ &y \in Y, \qquad y = y_{\mathbf{V}} + y_{\mathbf{V}'}, \quad (y_{\mathbf{V}}, y_{\mathbf{V}'}) \in Y_1 \times Y_2. \end{aligned}$$

By (2.20)(ii), $\widehat{w}$ does not depend upon the variables $y_{\mathbf{V}}$.

We now split the unfolding with respect to Y_1 and Y_2 (these are partial unfoldings)[(7)]. Let $\widetilde{\mathcal{T}}_\varepsilon^1 : L^p(\mathbb{R}^N \times Y_2) \mapsto L^p(\mathbb{R}^N \times Y)$ be the unfolding operator which unfolds functions of $x_{\mathbf{V}}$ to functions of $(x_{\mathbf{V}}, y_{\mathbf{V}})$ ($x_{\mathbf{V}'}$ *and* $y_{\mathbf{V}'}$ being parameters):

$$\forall\Phi \in L^p(\mathbb{R}^N \times Y_2), \qquad \widetilde{\mathcal{T}}_\varepsilon^1(\Phi)(x, y_{\mathbf{V}}, y_{\mathbf{V}'}) = \Phi\Big(\varepsilon\Big[\frac{x_{\mathbf{V}}}{\varepsilon}\Big]_Y + \varepsilon y_{\mathbf{V}} + x_{\mathbf{V}'}, y_{\mathbf{V}'}\Big).$$

Similarly, $\widetilde{\mathcal{T}}_\varepsilon^2 : L^p(\mathbb{R}^N \times Y_1) \mapsto L^p(\mathbb{R}^N \times Y)$ unfolds functions of $x_{\mathbf{V}'}$ to functions of $(x_{\mathbf{V}'}, y_{\mathbf{V}'})$:

$$\forall\Psi \in L^p(\mathbb{R}^N \times Y_1), \qquad \widetilde{\mathcal{T}}_\varepsilon^2(\Psi)(x, y_{\mathbf{V}}, y_{\mathbf{V}'}) = \Psi\Big(x_{\mathbf{V}} + \varepsilon\Big[\frac{x_{\mathbf{V}'}}{\varepsilon}\Big]_Y + \varepsilon y_{\mathbf{V}'}, y_{\mathbf{V}}\Big).$$

For $\phi \in L^p(\mathbb{R}^N)$, clearly,

$$\begin{aligned} \widetilde{\mathcal{T}}_\varepsilon^Y(\phi)(x, y) = \phi\Big(\varepsilon\Big[\frac{x}{\varepsilon}\Big]_Y + \varepsilon y\Big) &= \phi\Big(\varepsilon\Big[\frac{x_{\mathbf{V}}}{\varepsilon}\Big]_Y + \varepsilon\Big[\frac{x_{\mathbf{V}'}}{\varepsilon}\Big]_Y + \varepsilon y_{\mathbf{V}} + \varepsilon y_{\mathbf{V}'}\Big)\\ &= \widetilde{\mathcal{T}}_\varepsilon^1 \circ \widetilde{\mathcal{T}}_\varepsilon^2(\phi)(x, y) \quad \text{for a.e. } (x, y) \in \mathbb{R}^N \times Y. \end{aligned}$$

Set $\Theta_\varepsilon \doteq \widetilde{\mathcal{T}}_\varepsilon^2(w_\varepsilon)$. By hypothesis,

$$\widetilde{\mathcal{T}}_\varepsilon^1(\Theta_\varepsilon) \doteq \widetilde{\mathcal{T}}_\varepsilon(w_\varepsilon) \rightharpoonup \widehat{w} \quad \text{weakly in } L^p(\mathbb{R}^N \times Y).$$

---

(7) See Part III for a general discussion of partial unfoldings.

So, using the equivalent of Proposition 1.9(iii) for the partial unfolding $\widetilde{\mathcal{T}}_\varepsilon^1$,

$$\Theta_\varepsilon \rightharpoonup \mathcal{M}_{Y_1}(\widehat{w}) \text{ weakly in } L^p(\mathbb{R}^N \times Y_2).$$

But $\widehat{w}$ does not depend on $y_{\mathbf{V}}$, hence $\mathcal{M}_{Y_1}(\widehat{w}) \equiv (\widehat{w})$ and

$$\widetilde{\mathcal{T}}_\varepsilon^2(w_\varepsilon) \rightharpoonup \widehat{w} \text{ weakly in } L^p(\mathbb{R}^N \times Y_2).$$

Since $\mathbf{a}$ belongs to $\mathbf{V}$, $\partial_{\mathbf{a}}\widetilde{\mathcal{T}}_\varepsilon^2(w_\varepsilon) \equiv \widetilde{\mathcal{T}}_\varepsilon^2(\partial_{\mathbf{a}} w_\varepsilon)$ and by the hypotheses, the latter is bounded in $L^p(\mathbb{R}^N \times Y_2)$. Therefore,

$$\partial_{\mathbf{a}}\widetilde{\mathcal{T}}_\varepsilon^2(w_\varepsilon) \rightharpoonup \partial_{\mathbf{a}}\widehat{w} \quad \text{weakly in } L^p(\mathbb{R}^N \times Y_2).$$

Taking into account that $\partial_{\mathbf{a}}\widehat{w}$ does not depend upon $y_{\mathbf{V}}$, it follows that $\partial_{\mathbf{a}}\widehat{w}$ belongs to $L^p(\mathbb{R}^N \times Y)$, which completes the proof of (2.20)(iv).

To prove (2.20)(v), first let $\mathbf{W}$ be the $(N-1)$-dimensional subspace of $\mathbb{R}^N$ generated by $\mathbf{p}_2, \ldots, \mathbf{p}_N$. Without loss of generality, one can assume that $\mathbf{a}-\mathbf{p}_1$ belongs to $\mathbf{W}$. Indeed, recalling the representation of $\mathbf{a}$ in (2.18), where all the λ_j's are different from zero, one can replace $\mathbf{a}$ by $(1/\lambda_1)\mathbf{a}$.

For $\varphi \in L^p(\mathbb{R}^N)$, define the function $\widetilde{\mathcal{M}}_{\varepsilon,\mathbf{a}}(\varphi) \in L^p(\mathbb{R}^N)$ by

$$\widetilde{\mathcal{M}}_{\varepsilon,\mathbf{a}}(\varphi)(x' + \varepsilon k\mathbf{b}_1) \doteq \int_0^1 \varphi\big(x' + \varepsilon k\mathbf{b}_1 + \varepsilon t\mathbf{a}\big)\, dt, \quad k \in \mathbb{Z}, \text{ for a.e. } x' \in \mathbf{W},$$

and then extend it to be constant on the line intervals

$$[x' + \varepsilon k\mathbf{b}_1, x' + \varepsilon k\mathbf{b}_1 + \varepsilon\mathbf{a}), \quad k \in \mathbb{Z} \text{ and for a.e. } x' \in \mathbf{W}.$$

As in the proof of Proposition 1.24(i), $\widetilde{\mathcal{M}}_{\varepsilon,\mathbf{a}}(\varphi)$ belongs to $L^p(\mathbb{R}^N)$ and

$$\|\widetilde{\mathcal{M}}_{\varepsilon,\mathbf{a}}(\varphi)\|_{L^p(\mathbb{R}^N)} \leqslant \|\varphi\|_{L^p(\mathbb{R}^N)}. \tag{2.21}$$

If φ belongs to $W^{1,p}_{\mathbf{a}}(\mathbb{R}^N)$, the one dimensional Poincaré-Wirtinger inequality implies

$$\|\varphi - \widetilde{\mathcal{M}}_{\varepsilon,\mathbf{a}}(\varphi)\|_{L^p(\mathbb{R}^N)} \leqslant C\varepsilon\|\partial_{\mathbf{a}}\varphi\|_{L^p(\mathbb{R}^N)}{}^{(8)}, \tag{2.22}$$

with a constant independent of ε. Consider the function

$$Z_\varepsilon = \widetilde{\mathcal{T}}_\varepsilon^Y\Big(\frac{w_\varepsilon - \widetilde{\mathcal{M}}_{\varepsilon,\mathbf{a}}(w_\varepsilon)}{\varepsilon}\Big) - \frac{y^c \cdot \mathbf{a}}{|\mathbf{a}|^2}\partial_{\mathbf{a}}\widehat{w}.$$

From (2.22), it follows that $\{Z_\varepsilon\}_\varepsilon$ is uniformly bounded in $L^p(\mathbb{R}^N \times Y)$. Recall that $\partial_{\mathbf{a}}\widehat{w}$ is independent of $y_{\mathbf{V}}$ therefore $\partial_{y,\mathbf{a}}\partial_{\mathbf{a}}\widehat{w} = 0$, so a direct computation shows that

$$\partial_{\mathbf{a},y} Z_\varepsilon = \widetilde{\mathcal{T}}_\varepsilon^Y(\partial_{\mathbf{a}} w_\varepsilon) - \partial_{\mathbf{a}}\widehat{w} \quad \text{a.e. in } \mathbb{R}^N \times Y,$$

which is bounded in $L^p(\mathbb{R}^N \times Y)$.

(8) To prove (2.21) and (2.22), one can proceed as in Lemma 1.57 and Proposition 1.58 starting with a function $\varphi \in \mathcal{D}(\mathbb{R}^N)$.

Consequently, $\|Z_\varepsilon\|_{L^p(\mathbb{R}^N;W^{1,p}_{\mathbf{a}}(Y))}$ is bounded independently of ε. Therefore, up to a subsequence, there exists $\widehat{\overline{w}}$ in $L^p(\mathbb{R}^N; W^{1,p}_{\mathbf{a}}(Y))$ such that

$$\begin{aligned} Z_\varepsilon \rightharpoonup \widehat{\overline{w}} \quad & \text{weakly in } L^p(\mathbb{R}^N; W^{1,p}_{\mathbf{a}}(Y)), \\ \partial_{\mathbf{a},y} Z_\varepsilon \rightharpoonup \partial_{\mathbf{a},y}\widehat{\overline{w}} \quad & \text{weakly in } L^p(\mathbb{R}^N \times Y), \end{aligned}$$

and this proves (2.20)(v). □

2.3 Unfolding in L^1 and for measures

This section gives complements concerning the various unfolding operators for the case of the space $L^1(\Omega)$, as well as some results concerning the space of bounded measures $\mathcal{M}^1(\Omega)$ and the corresponding $L^1(\Omega \times Y)$ and $\mathcal{M}^1(\Omega \times Y)$. For the case of weak compactness in L^1 in particular, the Poincaré-Wirtinger inequality is used to its fullest.

2.3.1 About the Poincaré-Wirtinger inequality

Recall the definition of the mean value over a bounded domain:

Definition 2.24. *Let $\mathcal{O}$ be a bounded domain in $\mathbb{R}^N$ and φ integrable over $\mathcal{O}$. The average (also called the mean value) of φ over $\mathcal{O}$ is defined as*

$$\mathcal{M}_{\mathcal{O}}(\varphi) = \frac{1}{|\mathcal{O}|}\int_{\mathcal{O}} \varphi(x)\,dx,$$

where $|\mathcal{O}|$ = measure($\mathcal{O}$) is the Lebesgue measure of the set $\mathcal{O}$.

The following inequality was recalled in (1.33):

Definition 2.25 (The Poincaré-Wirtinger inequality). *A bounded open set $\mathcal{O}$ in $\mathbb{R}^N$ satisfies the Poincaré-Wirtinger inequality for the exponent $p \in [1, +\infty]$ if there exists a constant C_p such that*

$$\forall u \in W^{1,p}(\mathcal{O}), \quad \|u - \mathcal{M}_{\mathcal{O}}(u)\|_{L^p(\mathcal{O})} \leqslant C_p \|\nabla u\|_{L^p(\mathcal{O})}. \tag{2.23}$$

If this holds, we say that $\mathcal{O}$ is a PW-domain for the exponent p.

Obviously, for $\mathcal{O}$ to satisfy the above condition for some p, it has to be connected. But this condition is far from being sufficient. A simple sufficient condition is that the Sobolev embedding from $W^{1,p}(\mathcal{O})$ to $L^p(\mathcal{O})$ be compact, the proof of the inequality in this situation is obtained by contradiction.

A large class of domains for which this compactness holds is that of bounded domains $\mathcal{O} \subset \mathbb{R}^N$ with the property that there exists a continuous linear extension operator from $W^{1,p}(\mathcal{O})$ to $W^{1,p}(\mathbb{R}^N)$ (by Rellich's theorem). This class includes the so-called Jones-domains and in particular, the domains with Lipschitz boundary (see [2, 37]).

A larger class of domains which satisfy the Poincaré-Wirtinger inequality is that of John-domains. These domains were introduced by Fritz John in one of his important papers on elasticity in particular, in connection with the Korn inequality, see [157]. The class of John-domains is strictly larger than the class of Jones-domains. Indeed, there are examples of John-domains without Lipschitz boundary. They can be constructed explicitly using sequences of "rooms and corridors" (see a discussion in [88] and references therein).

It is remarkable that the Poincaré-Wirtinger inequality is very stable. Indeed, it is equivalent to many variants as explained in the next proposition.

Proposition 2.26. *Let $\mathcal{O}$ be a bounded domain. If it is a PW-domain-domain for the exponent $p \in [1, +\infty]$, then for every measure μ such that $\mu(\mathcal{O}) = 1$ and with finite energy (i.e. μ belonging to $\mathcal{M}^1(\mathcal{O}) \cap W^{-1,p'}(\mathcal{O})$), the following variant of the Poincaré-Wirtinger inequality holds for some other constant C':*

$$\forall u \in W^{1,p}(\mathcal{O}), \quad \Big\| u - \int_{\mathcal{O}} u \, d\mu \Big\|_{L^p(\mathcal{O})} \leqslant C' \|\nabla u\|_{L^p(\mathcal{O})}. \tag{2.24}$$

Conversely, if inequality (2.24) holds, the domain $\mathcal{O}$ is a PW-domain for the exponent p.

Proof. Assuming inequality (2.23) holds, by integration with respect to μ,

$$\Big| \int_{\mathcal{O}} u \, d\mu - \mathcal{M}_{\mathcal{O}}(u) \Big| \leqslant C_p \|\nabla u\|_{L^p(\mathcal{O})}.$$

Combining this inequality with (2.23) itself implies (2.24). The converse is proved by a similar argument. □

Remark 2.27. *The previous result can be used to show that if two bounded open sets with non empty intersection $\mathcal{O}_1$ and $\mathcal{O}_2$ are PW-domains for the same exponent p, then their union $\mathcal{O}_1 \cup \mathcal{O}_2$ is also a PW-domain for p.*

More generally, let two bounded connected open sets $\mathcal{O}_1$ and $\mathcal{O}_2$ be PW-domains for the same exponent p. Assume that $(\overline{\mathcal{O}_1} \cap \overline{\mathcal{O}_2})$ contains a portion of a Lipschitz graph. Then the domain $\mathcal{O} \doteq \operatorname{interior}(\overline{\mathcal{O}_1 \cup \mathcal{O}_2})$ is a PW-domain for the same exponent p.

The Poincaré-Wirtinger inequality has several deep consequences. To list some of them, for $p \in [1, +\infty]$, consider the spaces,

$$\begin{aligned} W^{1,p}_{\mathcal{M}}(\mathcal{O}) &= \{\psi \in W^{1,p}(\mathcal{O}) \mid \mathcal{M}_{\mathcal{O}}(\psi) = 0\}, \\ \mathcal{Z}_p(\mathcal{O}) &= \{\nabla \psi \mid \psi \in W^{1,p}_{\mathcal{M}}(\mathcal{O})\} \subset L^p(\mathcal{O})^N\}. \end{aligned}$$

Proposition 2.28. *Assume that $\mathcal{O}$ is a bounded connected domain in $\mathbb{R}^N$ and that p is given in $[1, +\infty]$.*

(i) The following assertions are equivalent:

(a) $\mathcal{O}$ is a PW-domain for the exponent p,

(b) $\mathcal{Z}_p(\mathcal{O})$ is closed in $L^p(\mathcal{O})^N$.

(ii) On the space $W^{1,p}_{\mathcal{M}}(\mathcal{O})$, the function $\psi \mapsto \|\nabla\psi\|_{L^p(\mathcal{O})}$ is an equivalent norm.
(iii) The map ∇ is an isomorphism from $W^{1,p}_{\mathcal{M}}(Y)$ to $\mathcal{Z}_p(\mathcal{O})$ and it has a continuous linear inverse $\mathcal{L}_p$. Actually, if $\mathcal{L}_p$ and $\mathcal{L}_q$ exist for two values $q \leqslant p$, the map $\mathcal{L}_p$ is the restriction to $\mathcal{Z}_p$ of the map $\mathcal{L}_q$.

Proof. (i) Clearly,

$$\|\nabla\psi\|_{L^p(Y)} \leqslant C\,\|\psi\|_{W^{1,p}(\mathcal{O})}, \tag{2.25}$$

so that the gradient operator is linear continuous from $W^{1,p}(\mathcal{O})$ to $\mathcal{Z}_p(\mathcal{O})$. Since $\mathcal{O}$ is connected, the gradient map is continuous bijective from $W^{1,p}_{\mathcal{M}}(\mathcal{O})$ to $\mathcal{Z}_p(\mathcal{O})$. If the latter is closed in $L^p(\mathcal{O})^N$, it is a Banach space, and by the Banach isomorphism theorem, ∇ has a continuous inverse $\mathcal{L}_p$, which implies that there is a constant C such that

$$\forall u \in W^{1,p}_{\mathcal{M}}(\mathcal{O}), \quad \|u\|_{W^{1,p}(\mathcal{O})} \leqslant C\,\|\nabla u\|_{L^p(\mathcal{O})}.$$

For v in $W^{1,p}(\mathcal{O})$, applying the previous inequality to $u \doteq v - \mathcal{M}_{\mathcal{O}}(v)$, entails that $\mathcal{O}$ is a PW-domain for the exponent p.

The converse follows from the considerations below where it is assumed that $\mathcal{O}$ is a PW-domain for the exponent p.
(ii) For ψ in $W^{1,p}_{\mathcal{M}}(\mathcal{O})$, by the Poincaré-Wirtinger inequality

$$\|\psi\|_{W^{1,p}(\mathcal{O})} \leqslant \|\psi\|_{L^p(\mathcal{O})} + \|\nabla\psi\|_{L^p(\mathcal{O})} \leqslant (1 + C_p)\|\nabla\psi\|_{L^p(\mathcal{O})}.$$

This, together with (2.25) gives the equivalence of the norms.
(iii) Assertion (ii) implies that the space $\mathcal{Z}_p(\mathcal{O})$ is complete since a Cauchy sequence in $\mathcal{Z}_p(\mathcal{O})$ is the image by the gradient map of a Cauchy sequence in $W^{1,p}_{\mathcal{M}}(\mathcal{O})$. Its limit is therefore the gradient of the corresponding limit in $W^{1,p}(\mathcal{O})$.

The fact that, for $1 \leqslant q \leqslant p \leqslant +\infty$, the restriction of $\mathcal{L}_q$ to $\mathcal{Z}_p(\mathcal{O})$ is $\mathcal{L}_p$, follows from the same property of the gradient map. □

Remark 2.29. *In [191] a more general situation is proposed with the notion of p-connectedness.*

We conclude this paragraph with a remark concerning another kind of Poincaré-Wirtinger inequality with a shift in the order of derivation (see for example [183] Theorem 14).

Proposition 2.30. *Let $\mathcal{O}$ be a connected bounded domain in $\mathbb{R}^N$ with Lipschitz boundary. Then for every $p \in (1, +\infty)$, there is a constant $C(\mathcal{O}, p)$ such that*

$$\|u - \mathcal{M}_{\mathcal{O}}(u)\|_{L^p(\mathcal{O})} \leqslant C(\mathcal{O}, p)\|\nabla u\|_{W^{-1,p}(\mathcal{O})}, \quad \forall u \in L^p(\mathcal{O}).$$

As a consequence, for g in $L^{p'}(\mathcal{O})$ such that $\displaystyle\int_{\mathcal{O}} g(x)\,dx = 1$, and for u in $L^p(\mathcal{O})$

$$\Big\| u - \int_{\mathcal{O}} u(x)g(x)\,dx \Big\|_{L^p(\mathcal{O})} \leqslant C(\mathcal{O},p)\Big(1 + \|g\|_{L^{p'}(\mathcal{O})}|\mathcal{O}|^{1/p}\Big)\|\nabla u\|_{W^{-1,p}(\mathcal{O})}.$$

This last inequality is obtained in the same way as the variant inequality (2.24) was derived from the Poincaré-Wirtinger inequality (2.23).

2.3.2 Unfolding in L^1

Concerning weak-compactness and weakly convergent sequences, one needs a characterization of weakly compact sets in the space $L^1(O)$ for a domain O in $\mathbb{R}^N$. Here we shall use the so-called "de la Vallée-Poussin criterion", defined as follows:

Definition 2.31 (De la Vallée-Poussin criterion). *A sequence $\{w_\varepsilon\}_\varepsilon$(or a set) of measurable functions on O satisfies the de la Vallée-Poussin criterion whenever there exists a non-decreasing convex function $\mathcal{F} : \mathbb{R}^+ \mapsto \mathbb{R}^+$, such that*

$$\mathcal{F}(0) = 0, \quad \lim_{t\to+\infty} \frac{\mathcal{F}(t)}{t} = +\infty \quad \textit{and the set} \left\{ \int_O \mathcal{F} \circ |w_\varepsilon|(x)\,dx \right\} \textit{ is bounded.}$$

The function $\mathcal{F}$ is a de la Vallée-Poussin function for the sequence $\{w_\varepsilon\}_\varepsilon$.

One of the main classical results concerning this criterion is the next proposition.

Proposition 2.32 (cf. [97, 105]). *If $|O|$ is finite, the de la Vallée-Poussin criterion is equivalent to the relative weak compactness in $L^1(O)$.*

If $|O| = +\infty$, the de la Vallée-Poussin criterion is equivalent to the relative weak compactness in $L^1_{loc}(O)$. Furthermore, together with the equi-integrability at $+\infty$, which is defined as

$$\lim_{R\to+\infty} \Big(\sup_\varepsilon \int_{O\setminus O\cap B(0,R)} |w_\varepsilon(x)|dx \Big) = 0,$$

it implies the relative weak compactness in $L^1(O)$. In this definition, $B(0,R)$ stands for the ball in $\mathbb{R}^N$ with center 0 and radius R.

Proposition 2.33 (Proposition 1.9 for $p = 1$). *For every relatively weakly compact sequence $\{w_\varepsilon\}_\varepsilon$ in $L^1(\Omega)$, the corresponding $\{\mathcal{T}_\varepsilon(w_\varepsilon)\}_\varepsilon$ is relatively weakly compact in $L^1(\Omega \times Y)$. Furthermore, if*

$$\mathcal{T}_\varepsilon(w_\varepsilon) \rightharpoonup \widehat{w} \quad \textit{weakly in } L^1(\Omega \times Y),$$

then

$$w_\varepsilon \rightharpoonup \mathcal{M}_Y(\widehat{w}) \quad \textit{weakly in } L^1(\Omega)$$

and

$$\|\widehat{w}\|_{L^1(\Omega\times Y)} \leqslant |Y| \liminf \|w_\varepsilon\|_{L^1(\Omega)}.$$

Proof. Let $\mathcal{F}$ be a de la Vallée-Poussin function for the sequence $\{w_\varepsilon\}_\varepsilon$. Unfolding the integral $\displaystyle\int_\Omega \mathcal{F}\circ|w_\varepsilon|(x)\,dx$ and using Proposition 1.8(ii), give

$$\begin{aligned}\frac{1}{|Y|}\int_{\Omega\times Y}\mathcal{F}\big(|\mathcal{T}_\varepsilon(w_\varepsilon)(x,y)|\big)\,dxdy &= \frac{1}{|Y|}\int_{\Omega\times Y}\mathcal{T}_\varepsilon\big(\mathcal{F}\circ|w_\varepsilon|\big)(x,y)\,dxdy\\ &\leqslant \int_\Omega \mathcal{F}\circ|w_\varepsilon|(x)\,dx.\end{aligned}$$

Hence the sequence

$$\Big\{\int_{\Omega\times Y}\mathcal{F}\big(|\mathcal{T}_\varepsilon(w_\varepsilon)(x,y)|\big)\,dxdy\Big\}_\varepsilon \quad\text{is bounded,}$$

and this completes the proof of weak compactness of $\{\mathcal{T}_\varepsilon(w_\varepsilon)\}_\varepsilon$ in $L^1(\Omega\times Y)$ in the case of Ω with finite measure.

For the case where the measure of Ω is not finite and recalling that $d(Y)$ is the diameter of Y (see (1.47)), the inequality given in Proposition 1.8(iii) applied to

$$|w_\varepsilon|\mathbf{1}_{\Omega\setminus\Omega\cap B(0,R-\varepsilon d(Y))},$$

shows that the equi-integrability at infinity of the sequence $\{w_\varepsilon\}_\varepsilon$ carries over to the sequence $\{\mathcal{T}_\varepsilon(w_\varepsilon)\}_\varepsilon$.

Now, suppose the following convergence holds:

$$\mathcal{T}_\varepsilon(w_\varepsilon)\rightharpoonup\widehat{w}\quad\text{weakly in } L^1(\Omega\times Y),$$

and let ψ be in $C_c^\infty(\Omega)$. For ε sufficiently small, one has

$$\int_\Omega w_\varepsilon(x)\,\psi(x)\,dx = \frac{1}{|Y|}\int_{\Omega\times Y}\mathcal{T}_\varepsilon(w_\varepsilon)(x,y)\,\mathcal{T}_\varepsilon(\psi)(x,y)\,dxdy.$$

In view of Proposition 1.9(i), one can pass to the limit in the right-hand side to obtain

$$\lim_{\varepsilon\to 0}\int_\Omega w_\varepsilon(x)\,\psi(x)\,dx = \int_\Omega\Big\{\frac{1}{|Y|}\int_Y\widehat{w}(x,y)\,dy\Big\}\psi(x)\,dx,$$

which is the claim. The last statement follows from the weak lower semicontinuity of the L^1-norm and from the inequality (iii) of Proposition 1.8 applied for $p=1$. □

Proposition 2.34 (Proposition 1.25(ii) for $p=1$). *If $\{v_\varepsilon\}_\varepsilon$ is a sequence included in a weakly compact subset of $L^1(\Omega)$, the following convergences are equivalent:*

$$(a)\quad v_\varepsilon\rightharpoonup v\quad\textit{weakly in } L^1(\Omega),$$
$$(b)\quad \mathcal{M}_\varepsilon(v_\varepsilon)\rightharpoonup v\quad\textit{weakly in } L^1(\Omega)^{(9)}.$$

(9) If the sequence $\{v_\varepsilon\}_\varepsilon$ is only bounded in $L^1(\Omega)$, Proposition 2.51 applies.

Proof. We first prove that if $\{v_\varepsilon\}_\varepsilon$ is included in a weakly compact subset of $L^1(\Omega)$, so is $\{\mathcal{M}_\varepsilon(v_\varepsilon)\}_\varepsilon$. More precisely, if the de la Vallée-Poussin criterion is satisfied by the sequence $\{v_\varepsilon\}_\varepsilon$ with a function $\mathcal{F}$, it is also satisfied by the sequence $\{\mathcal{M}_\varepsilon(v_\varepsilon)\}_\varepsilon$, with the same $\mathcal{F}$, since for $\mathcal{F}$ non-decreasing and convex, the monotonicity and Jensen's inequality imply for a.e. $x \in \Omega$,

$$\mathcal{F}(|\mathcal{M}_\varepsilon(v_\varepsilon)(x)|) \leqslant \mathcal{F}(\mathcal{M}_\varepsilon(|v_\varepsilon|)(x)) \leqslant \mathcal{M}_\varepsilon(\mathcal{F}(|v_\varepsilon(x)|)).$$

Hence, from Proposition 1.24(i) one gets

$$\int_\Omega \mathcal{F}(|\mathcal{M}_\varepsilon(v_\varepsilon)(x)|)dx \leqslant \int_\Omega \mathcal{M}_\varepsilon(\mathcal{F}(|v_\varepsilon(x)|))dx \leqslant \int_\Omega \mathcal{F}(|v_\varepsilon(x)|)dx.$$

The condition of equi-integrability near $+\infty$ is also easily checked in the case where $|\Omega| = \infty$.

At this point, the equivalence of (a) and (b) is obtained by testing with a function in $\mathcal{D}(\Omega)$ and using the same arguments as in the proof of Proposition 1.25(ii). □

Proposition 2.35 (Proposition 1.31(i)-(ii) for $p = 1$). *Let $\{\Phi_\varepsilon\}_\varepsilon$ be a sequence in $L^1(\Omega \times Y)$ such that*

$$\Phi_\varepsilon \rightharpoonup \Phi \quad \textit{weakly in } L^1(\Omega \times Y).$$

Then

$$\begin{aligned} &(i) \quad \mathcal{U}_\varepsilon(\Phi_\varepsilon) \rightharpoonup \mathcal{M}_Y(\Phi) = \frac{1}{|Y|}\int_Y \Phi(\,\cdot\,, y)\,dy \quad \textit{weakly in } L^1(\Omega),\\ &(ii) \quad \mathcal{T}_\varepsilon \circ \mathcal{U}_\varepsilon(\Phi_\varepsilon) \rightharpoonup \Phi \quad \textit{weakly in } L^1(\Omega \times Y). \end{aligned}$$

Proof. Let $\mathcal{F}$ be a de la Vallée-Poussin function for the sequence $\{\Phi_\varepsilon\}_\varepsilon$. The monotonicity and the Jensen inequality give for a.e. $x \in \widehat{\Omega}_\varepsilon$

$$\begin{aligned} \mathcal{F}\Big(\Big|\frac{1}{|Y|}\int_Y \Phi_\varepsilon\Big(\varepsilon\Big[\frac{x}{\varepsilon}\Big]_Y + \varepsilon z, \Big\{\frac{x}{\varepsilon}\Big\}_Y\Big)\,dz\Big|\Big)&\\ \leqslant \frac{1}{|Y|}\int_Y \mathcal{F}\Big(\Big|\Phi_\varepsilon\Big(\varepsilon\Big[\frac{x}{\varepsilon}\Big]_Y + \varepsilon z, \Big\{\frac{x}{\varepsilon}\Big\}_Y\Big)\Big|\Big)\,dz.& \end{aligned}$$

Hence,

$$\mathcal{F}\big(|\mathcal{U}_\varepsilon(\Phi_\varepsilon)(x)|\big) \leqslant \mathcal{U}_\varepsilon(\mathcal{F} \circ |\Phi_\varepsilon|)(x) \quad \text{for a.e. } x \text{ in } \Omega.$$

By integration and using inequality (1.28), it follows that

$$\int_\Omega \mathcal{F}(|\mathcal{U}_\varepsilon(\Phi_\varepsilon)(x)|)dx \leqslant \int_\Omega \mathcal{U}_\varepsilon(\mathcal{F} \circ |\Phi_\varepsilon|)(x)dx \leqslant \frac{1}{|Y|}\|\mathcal{F} \circ |\Phi_\varepsilon|\|_{L^1(\Omega\times Y)}.$$

The condition of equi-integrability near $+\infty$ is also easily checked in the case $|\Omega| = \infty$. This proves (i). The proof of (ii) is similar making use of (1.27). □

Proposition 2.36 (Theorem 1.41 for $p = 1$). *Let* $\{w_\varepsilon\}_\varepsilon$ *be a sequence in* $W^{1,1}(\Omega)$, *such that*

$$w_\varepsilon \rightharpoonup w \quad \text{weakly in } W^{1,1}(\Omega).$$

Then, up to a subsequence, there exists some $\widehat{w}$ *in* $L^1(\Omega; W^{1,1}_{per,0}(Y))$ *such that*

$$(i) \quad \frac{1}{\varepsilon}\Big(\mathcal{T}_\varepsilon(w_\varepsilon) - \mathcal{M}_\varepsilon(w_\varepsilon)\Big) \rightharpoonup \widehat{w} + y^c \cdot \nabla w \quad \text{weakly in } L^1(\Omega; W^{1,1}(Y)),$$

$$(ii) \quad \mathcal{T}_\varepsilon(\nabla w_\varepsilon) \rightharpoonup \nabla w + \nabla_y \widehat{w} \quad \text{weakly in } L^1(\Omega \times Y)^N.$$

Proof. The proof is essentially the same as that of Theorem 1.41. The only difference is in how to establish the weak compactness, which, for L^1-based spaces, does not simply follow from bounds.

By hypothesis, $\{\nabla w_\varepsilon\}_\varepsilon$ is weakly convergent in $L^1(\mathbb{R}^N)^N$. Therefore by Proposition 2.33, the corresponding sequence $\{\widetilde{\mathcal{T}}_\varepsilon(\nabla w_\varepsilon)\}_\varepsilon$ is in a weakly compact subset of $L^1(\mathbb{R}^N \times Y)^N$. Up to a subsequence, we can assume that it is weakly convergent.

Now, $\widetilde{\mathcal{T}}_\varepsilon(\nabla w_\varepsilon) = \nabla_y Z_\varepsilon$, with

$$Z_\varepsilon = \frac{1}{\varepsilon}\big(\mathcal{T}_\varepsilon(w_\varepsilon) - \mathcal{M}_\varepsilon(w_\varepsilon)\big),$$

(see proof of Theorem 1.41). By Proposition 2.28, the map ∇_y is an isomorphism between the space $L^1(\mathbb{R}^N; W^{1,1}_{\mathcal{M}}(Y))$ and $L^1(\mathbb{R}^N; \mathcal{X}_1(Y))$. The latter space is itself a closed subspace of $L^1(\mathbb{R}^N; L^1(Y))$, which, by Fubini's theorem, is the same as $L^1(\mathbb{R}^N \times Y)$.

Consequently, the weak convergence of $\widetilde{\mathcal{T}}_\varepsilon(\nabla w_\varepsilon)$ in $L^1(\mathbb{R}^N \times Y)^N$ is equivalent to that of the sequence $\big\{\widetilde{\mathcal{T}}_\varepsilon(Z_\varepsilon)\big\}_\varepsilon$ in $L^1(\mathbb{R}^N; W^{1,1}_{\mathcal{M}}(Y))$. We can define $\widehat{w}$ by (i) which is equivalent to (ii).

From these weak convergences, the Y-periodicity of $\widehat{w}'$ follows is the same way as in the proof of Theorem 1.41. □

Corollary 2.37 (Corollary 1.37 for $p = 1$). *Let* $\{w_\varepsilon\}_\varepsilon$ *be a sequence in* $W^{1,1}(\Omega)$, *such that*

$$w_\varepsilon \rightharpoonup w \quad \text{weakly in } W^{1,1}(\Omega).$$

Then

$$\mathcal{T}_\varepsilon(w_\varepsilon) \rightharpoonup w \quad \text{weakly in } L^1(\Omega; W^{1,1}(Y)).$$

Furthermore, if

$$w_\varepsilon \to w \quad \text{strongly in } L^1(\Omega)^{(10)},$$

then

$$\mathcal{T}_\varepsilon(w_\varepsilon) \to w \quad \text{strongly in } L^1(\Omega; W^{1,p}(Y)).$$

(10) This is automatically the case when Ω is bounded with Lipschitz boundary.

Proof. Indeed, by (i) of Proposition 2.36,

$$\mathcal{T}_\varepsilon(w_\varepsilon) - \mathcal{M}_\varepsilon(w_\varepsilon) \to 0 \quad \text{strongly in } L^1(\Omega; W^{1,1}(Y)).$$

If $\{w_\varepsilon\}_\varepsilon$ converges weakly, Proposition 2.34 implies that

$$\mathcal{M}_\varepsilon(w_\varepsilon) \rightharpoonup w \quad \text{weakly in } L^1(\Omega),$$

hence in $L^1(\Omega; W^{1,1}(Y))$, and this gives the first result.

If $\{w_\varepsilon\}_\varepsilon$ converges strongly, the second result follows in the same way from Proposition 1.25(i). □

Proposition 2.38 (Proposition 1.69 for $p = 1$). *Let $\{w_\varepsilon\}_\varepsilon$ be a sequence in $W^{1,1}(\mathbb{R}^N)$, such that*

$$w_\varepsilon \rightharpoonup w \quad \textit{weakly in } W^{1,1}(\mathbb{R}^N).$$

Then the following convergences hold:

(i) $\widetilde{\mathcal{R}}_\varepsilon(w_\varepsilon) \to 0$ *strongly in* $L^1(\mathbb{R}^N)$,

(ii) $\widetilde{\mathcal{Q}}_\varepsilon(w_\varepsilon) \rightharpoonup w$ *strongly in* $L^1_{loc}(\mathbb{R}^N)$ *and weakly in* $W^{1,1}(\mathbb{R}^N)$,

(iii) $\widetilde{\mathcal{T}}_\varepsilon \circ \widetilde{\mathcal{Q}}_\varepsilon(w_\varepsilon) \to w$ *strongly in* $L^1_{loc}(\mathbb{R}^N; \mathbf{Q}_1(Y))$,

(iv) $\widetilde{\mathcal{T}}_\varepsilon(\nabla \widetilde{\mathcal{Q}}_\varepsilon(w_\varepsilon)) \rightharpoonup \nabla w$ *weakly in* $L^1(\mathbb{R}^N \times Y)^N$.

Proof. The proofs of (i), (iii) and of the first statement of (ii) are similar to those of Proposition 1.69 in view of the compact embedding of $W^{1,1}(\mathbb{R}^N)$ in $L^1_{loc}(\mathbb{R}^N)$ (the sequence $\{w_\varepsilon\}_\varepsilon$ being bounded in $W^{1,1}(\mathbb{R}^N)$).

In view of the first statement in (ii), to prove the second one it remains to show that $\nabla \widetilde{\mathcal{Q}}_\varepsilon(w_\varepsilon)$ is relatively weakly compact in $L^1(\mathbb{R}^N)^N$. Let $\mathcal{F}$ be a de la Vallée-Poussin function for which $\|\mathcal{F} \circ |\nabla w_\varepsilon|\|_{L^1(\Omega\times Y)}$ is bounded. Due to the monotonicity and the convexity of $\mathcal{F}$, by Jensen's inequality, one has for every $j \in \{1, \dots, N\}$,

$$\begin{aligned}
\int_{\mathbb{R}^N} \mathcal{F}(|\nabla \mathcal{Q}_\varepsilon(w_\varepsilon)\cdot \mathbf{b}_j|(x))dx &= \int_{\mathbb{R}^N} \mathcal{F} \circ \Big|\frac{\mathcal{M}_\varepsilon(w_\varepsilon)(\cdot + \varepsilon \mathbf{b}_j) - \mathcal{M}_\varepsilon(w_\varepsilon)}{\varepsilon}\Big|(x)dx \\
&\leqslant \int_{\mathbb{R}^N} \mathcal{M}_\varepsilon\Big(\mathcal{F} \circ \Big|\frac{w_\varepsilon(\cdot + \varepsilon \mathbf{b}_j) - w_\varepsilon}{\varepsilon}\Big|\Big)(x)dx \\
&\leqslant \int_{\mathbb{R}^N} \mathcal{F} \circ \Big|\frac{w_\varepsilon(\cdot + \varepsilon \mathbf{b}_j) - w_\varepsilon}{\varepsilon}\Big|(x)dx \\
&\leqslant \int_{\mathbb{R}^N} \mathcal{F} \circ |\nabla w_\varepsilon \cdot \mathbf{b}_j|(x)dx.
\end{aligned}$$

The condition of equi-integrability near $+\infty$ is also easily checked. This completes the proof of (ii).

For the proof of (iv), as was just shown, $\nabla \widetilde{\mathcal{Q}}_\varepsilon(w_\varepsilon)$ converges weakly to ∇w in $L^1(\mathbb{R}^N)^N$. Hence, by Proposition 2.33, $\widetilde{\mathcal{T}}_\varepsilon(\nabla \widetilde{\mathcal{Q}}_\varepsilon(w_\varepsilon))$ is weakly compact in $L^1(\mathbb{R}^N \times Y)^N$. The remainder of the proof is the same as that of (iv) of Proposition 1.69. □

Corollary 2.39 (Corollary 1.70 for $p = 1$). *Let $\{w_\varepsilon\}_\varepsilon$ be a sequence in $W^{1,1}(\mathbb{R}^N)$, such that*

$$w_\varepsilon \rightharpoonup w \quad \text{weakly in } W^{1,1}(\mathbb{R}^N).$$

Then, up to a subsequence, there exists some $\widehat{w}'$ in the space $L^1(\mathbb{R}^N; W^{1,1}_{per}(Y))$ such that the following convergences hold:

$$(i) \quad \widetilde{\mathcal{T}}_\varepsilon(\nabla \widetilde{\mathcal{R}}_\varepsilon(w_\varepsilon)) \rightharpoonup \nabla_y \widehat{w}' \quad \text{weakly in } L^1(\mathbb{R}^N \times Y)^N,$$

$$(ii) \quad \widetilde{\mathcal{T}}_\varepsilon(\nabla w_\varepsilon) \rightharpoonup \nabla w + \nabla_y \widehat{w}' \quad \text{weakly in } L^1(\mathbb{R}^N \times Y)^N.$$

Proof. Since

$$\widetilde{\mathcal{T}}_\varepsilon(\nabla \widetilde{\mathcal{R}}_\varepsilon(w_\varepsilon)) = \widetilde{\mathcal{T}}_\varepsilon(\nabla w_\varepsilon) - \widetilde{\mathcal{T}}_\varepsilon(\nabla \widetilde{\mathcal{Q}}_\varepsilon(w_\varepsilon)),$$

the sequence $\{\widetilde{\mathcal{T}}_\varepsilon(\nabla \widetilde{\mathcal{R}}_\varepsilon(w_\varepsilon))\}_\varepsilon$ is weakly compact in $L^1(\mathbb{R}^N \times Y)$ as difference of two weakly compact sequences (the former by Proposition 2.33 and the latter by Proposition 2.38 (iv)). Therefore, there exists a $\widehat{W}'$ in $L^1(\mathbb{R}^N \times Y))$ such that, up to a subsequence,

$$\widetilde{\mathcal{T}}_\varepsilon(\nabla \widetilde{\mathcal{R}}_\varepsilon(w_\varepsilon)) \equiv \nabla_y\Big(\widetilde{\mathcal{T}}_\varepsilon\big(\frac{1}{\varepsilon}\widetilde{\mathcal{R}}_\varepsilon(w_\varepsilon)\big)\Big) \rightharpoonup \widehat{W}' \quad \text{weakly in } L^1(\mathbb{R}^N \times Y)^N. \tag{2.26}$$

The same argument as in the proof of Proposition 2.36 applied here, shows that convergence (2.26) is equivalent to the following weak convergence:

$$\widetilde{\mathcal{T}}_\varepsilon\Big(\frac{1}{\varepsilon}\widetilde{\mathcal{R}}_\varepsilon(w_\varepsilon)\Big) - \mathcal{M}_Y\Big(\widetilde{\mathcal{T}}_\varepsilon\Big(\frac{1}{\varepsilon}\widetilde{\mathcal{R}}_\varepsilon(w_\varepsilon)\Big)\Big) \rightharpoonup \widehat{w}' \quad \text{weakly in } L^1(\mathbb{R}^N; W^{1,1}_{\mathcal{M}}(Y)).$$

From this fact, the Y-periodicity of $\widehat{w}'$ is obtained proceeding as in the proof of Theorem 1.41. □

As a consequence, the following result for the case of a domain Ω holds:

Proposition 2.40 (Proposition 1.78 for $p = 1$). *Let $\{w_\varepsilon\}$ be a sequence converging weakly to w in $W^{1,1}(\Omega)$ or in $W^{1,1}_0(\Omega)$. Then*

$$\begin{aligned} &(i) && \mathcal{R}_\varepsilon(w_\varepsilon) \to 0 \quad \text{strongly in } L^1(\Omega),\\ &(ii) && \mathcal{Q}_\varepsilon(w_\varepsilon) \rightharpoonup w \quad \text{weakly in } W^{1,1}(\Omega),\\ &(iii) && \mathcal{T}_\varepsilon(\nabla \mathcal{Q}_\varepsilon(w_\varepsilon)) \rightharpoonup \nabla w \quad \text{weakly in } L^1(\Omega \times Y)^N. \end{aligned}$$

2.3.3 * Unfolding for measures

In this section, we denote by $\widetilde{Y}$ the semi-closed parallelotope defined as

$$\widetilde{Y} \doteq \Big\{ \sum_{j=1}^N y_j \mathbf{b}_j \,\big|\, 0 \leqslant y_j < 1 \Big\},$$

so that $\mathbb{R}^N$ is exactly the disjoint union $\bigcup_{\xi \in \mathbf{G}} (\xi + \widetilde{Y})$ (not only up to a set of Lebesgue measure 0 as with Y).

The notations $[x]_{\widetilde{Y}}$ and $\{x\}_{\widetilde{Y}}$ are now meant for every x in $\mathbb{R}^N$ without ambiguity. In particular, (1.2) becomes

$$x = \varepsilon\Big(\Big[\frac{x}{\varepsilon}\Big]_{\widetilde{Y}} + \Big\{\frac{x}{\varepsilon}\Big\}_{\widetilde{Y}}\Big)$$

for every $x \in \mathbb{R}^N$. Consequently, the unfolding operator (still denoted $\widetilde{\mathcal{T}}_\varepsilon$) is defined everywhere for a function defined everywhere. For example, it can be applied to a Borel function on $\mathbb{R}^N$ to obtain a Borel function on $\mathbb{R}^N \times \widetilde{Y}$. The definition for the modified local average operator $\widetilde{\mathcal{M}}_\varepsilon$ follows easily.

The modified averaging operator $\widetilde{\mathcal{U}}_\varepsilon$ is defined in a similar way for Borel functions on $\mathbb{R}^N \times Y$, and yields a Borel function on $\mathbb{R}^N$. The algebraic properties of these operators (proved above) still hold with these modified version.

Here, we mostly present the results for bounded Borel measures on $\mathbb{R}^N$ (the corresponding space is denoted $\mathcal{M}^1$). Similar results hold for positive σ-finite Borel measures and for signed Radon measures on $\mathbb{R}^N$. The case of measures on a Borel subset Ω of $\mathbb{R}^N$ is treated simply by considering the corresponding measures on $\mathbb{R}^N$ which are supported in Ω.

The notion of unfolding for measures extends the two-scale convergence for measures for which we refer to [32], where only Example 2.53 is used (the abstract results of Section 2 of [32] follow easily from Proposition 2.47 below). It also extends the "generalized " two-scale convergence of [192].

Definition 2.41. *The unfolding $\widetilde{\mathcal{T}}_\varepsilon(\mu)$ of a Borel measure $\mu \in \mathcal{M}^1(\mathbb{R}^N)$ is defined by duality for every bounded Borel function Φ on $\Omega \times Y$, as*

$$\langle \widetilde{\mathcal{T}}_\varepsilon(\mu), \Phi \rangle \doteq |Y| \langle \mu, \widetilde{\mathcal{U}}_\varepsilon(\Phi) \rangle.$$

Proposition 2.42. *Let μ be in $\mathcal{M}^1(\mathbb{R}^N)$. Then*

$$\widetilde{\mathcal{T}}_\varepsilon(\mu) \in L^1 \cap L^\infty(\mathbb{R}^N; \mathcal{M}^1(\widetilde{Y})), \tag{2.27}$$

is piecewise constant with respect to $x \in \mathbb{R}^N$, and satisfies

$$\begin{aligned} \|\widetilde{\mathcal{T}}_\varepsilon(\mu)\|_{L^1(\mathbb{R}^N;\mathcal{M}^1(\widetilde{Y}))} &\leqslant |Y|\; \|\mu\|_{\mathcal{M}^1(\Omega)} \\ \|\widetilde{\mathcal{T}}_\varepsilon(\mu)\|_{L^\infty(\mathbb{R}^N;\mathcal{M}^1(\widetilde{Y}))} &\leqslant \frac{1}{\varepsilon^N}\; \|\mu\|_{\mathcal{M}^1(\Omega)}. \end{aligned} \tag{2.28}$$

More precisely, for every $x \in \mathbb{R}^N$,

$$\widetilde{\mathcal{T}}_\varepsilon(\mu)(x) = \mu_{\varepsilon,[\frac{x}{\varepsilon}]_{\widetilde{Y}}}, \tag{2.29}$$

where $\mu_{\varepsilon,\xi} \in \mathcal{M}^1(\widetilde{Y})$ denotes the translate of the scaled restriction to $\varepsilon(\xi + \widetilde{Y})$ of μ, defined, for φ a bounded Borel function on $\widetilde{Y}$, as

$$\mu_{\varepsilon,\xi}(\varphi) = \frac{1}{\varepsilon^N} \int \varphi\Big(\Big\{\frac{x}{\varepsilon}\Big\}_{\widetilde{Y}}\Big) \mathbf{1}_{\varepsilon\xi + \varepsilon\widetilde{Y}}(x)\, \mu(dx). \tag{2.30}$$

A similar result holds for μ a Radon measure on $\mathbb{R}^N$, in which case $\widetilde{\mathcal{T}}_\varepsilon(\mu)$ belongs to $L^\infty_{loc}(\mathbb{R}^N;\mathcal{M}^1(\widetilde{Y}))$.

For μ a σ-finite positive Borel measure on $\mathbb{R}^N$, $\widetilde{\mathcal{T}}_\varepsilon(\mu)$ is a piece-wise constant map from $\mathbb{R}^N$ to the set of σ-finite positive Borel measures on $\widetilde{Y}$, taking for each x the value $\mu_{\varepsilon,x}$ which is the σ-finite Borel positive measure defined on $\widetilde{Y}$ by the following formula for each Borel subset A of $\widetilde{Y}$:

$$\mu_{\varepsilon,x}(A) = \frac{1}{\varepsilon^N}\mu\Big(\varepsilon\Big[\frac{x}{\varepsilon}\Big]_{\widetilde{Y}} + \varepsilon A\Big).$$

Example 2.43. *The formulas above show that $\widetilde{\mathcal{T}}_\varepsilon(dx)$ is simply the product Lebesgue measure $\dfrac{1}{|Y|}dx \otimes dy$ on $\mathbb{R}^N \times \widetilde{Y}$. More generally, if μ is with density ρ with respect to dx, its unfolding is with density $\widetilde{\mathcal{T}}_\varepsilon(\rho)$ with respect to $\dfrac{1}{|Y|}dx \otimes dy$. This is in agreement with the unfolding of functions.*

Proof of Proposition 2.42. The following estimate is straightforward for μ positive and extends to signed μ:

$$\|\mu_{\varepsilon,\xi}\|_{\mathcal{M}^1(\widetilde{Y})} \leqslant \frac{1}{\varepsilon^N}|\mu|(\varepsilon\xi + \varepsilon\widetilde{Y}). \tag{2.31}$$

This implies that the map $x \mapsto \mu_{\varepsilon,[\frac{x}{\varepsilon}]_{\widetilde{Y}}}$ is in $L^1 \cap L^\infty(\mathbb{R}^N;\mathcal{M}^1(\widetilde{Y}))$ with the bound (2.28) above (since the Lebesgue measure of $\varepsilon\xi + \varepsilon\widetilde{Y}$ is $\varepsilon^N|Y|$).

It remains to show that it is actually $\widetilde{\mathcal{T}}_\varepsilon(\mu)$. To do so, it is enough to consider μ positive. Testing against a bounded Borel function $\Phi(x,y)$, making multiple use of Fubini's theorem, the result is obtained from the following computation:

$$\begin{aligned}
\langle \widetilde{\mathcal{T}}_\varepsilon(\mu), \Phi\rangle &= |Y|\int_{\mathbb{R}^N} \widetilde{\mathcal{U}}_\varepsilon(\Phi)(x)\mu(dx)\\
&= \int_{\mathbb{R}^N\times\widetilde{Y}} \Phi\Big(\varepsilon\Big[\frac{x}{\varepsilon}\Big]_{\widetilde{Y}} + \varepsilon z, \Big\{\frac{x}{\varepsilon}\Big\}_{\widetilde{Y}}\Big)\,\mu(dx)dz\\
&= \sum_{\xi\in\mathbf{G}}\int_{\varepsilon(\xi+\widetilde{Y})\times\widetilde{Y}} \Phi\Big(\varepsilon\xi + \varepsilon z, \Big\{\frac{x}{\varepsilon}\Big\}_{\widetilde{Y}}\Big)\,\mu(dx)dz\\
&= \varepsilon^N\sum_{\xi\in\mathbf{G}}\int_{\widetilde{Y}\times\widetilde{Y}} \Phi(\varepsilon\xi + \varepsilon z, y)\,\mu_{\varepsilon,\xi}(dy)dz\\
&= \sum_{\xi\in\mathbf{G}}\int_{\widetilde{Y}\times\varepsilon(\xi+\widetilde{Y})} \Phi(x,y)\,\mu_{\varepsilon,\xi}(dy)dx\\
&= \int_{\mathbb{R}^N\times\widetilde{Y}} \Phi(x,y)\,d\mu_{\varepsilon,[\frac{x}{\varepsilon}]_{\widetilde{Y}}}(y)dx\\
&= \int_{\mathbb{R}^N} \mu_{\varepsilon,[\frac{x}{\varepsilon}]_{\widetilde{Y}}}(\Phi(x,\cdot))dx.
\end{aligned} \tag{2.32}$$

For μ a Radon measure, the same formulas hold for Φ compactly supported in $\mathbb{R}^N$. The fact that $\widetilde{\mathcal{T}}_\varepsilon(\mu)$ belongs to $L^\infty_{loc}(\mathbb{R}^N; \mathcal{M}^1(\widetilde{Y}))$ follows by formula (2.31). A similar proof holds for μ *sigma*-finite. □

For each $\varepsilon > 0$, introduce a special involution $\mathcal{X}_\varepsilon$ of $\mathbb{R}^N \times \widetilde{Y}$, which is piecewise affine (hence Borel),

$$\mathcal{X}_\varepsilon(x,y) = (x',y') \quad \text{where } x' = \varepsilon\Big[\frac{x}{\varepsilon}\Big]_{\widetilde{Y}} + \varepsilon y \quad \text{and } y' = \Big\{\frac{x}{\varepsilon}\Big\}_{\widetilde{Y}}.$$

It is straightforward that $\mathcal{X}_\varepsilon^{-1} = \mathcal{X}_\varepsilon$.

One can describe the unfolding of the measure μ with the use of the involution $\mathcal{X}_\varepsilon$.

Proposition 2.44. *The following identity holds:*

$$\widetilde{\mathcal{T}}_\varepsilon(\mu) \equiv \mathcal{X}_\varepsilon^*(\mu \otimes dy).$$

Proof. The second line of formula (2.32) reads

$$\int_{\mathbb{R}^N\times\widetilde{Y}} \Phi\Big(\varepsilon\Big[\frac{x}{\varepsilon}\Big]_{\widetilde{Y}} + \varepsilon y, \Big\{\frac{x}{\varepsilon}\Big\}_{\widetilde{Y}}\Big)\, d\mu(x)dy = \int_{\mathbb{R}^N\times\widetilde{Y}} \Phi\circ\mathcal{X}_\varepsilon(x,y)\, d\mu(x)dy,$$

whence the claim. □

Definition 2.45. *The averaging operator $\widetilde{\mathcal{U}}_\varepsilon$, applied to a Borel measure Σ in $\mathcal{M}^1(\mathbb{R}^N \times \widetilde{Y})$, is the Borel measure on $\mathbb{R}^N$ defined by duality for every bounded Borel function ϕ, as*

$$\big\langle \widetilde{\mathcal{U}}_\varepsilon(\Sigma), \phi \big\rangle \doteq \frac{1}{|Y|}\big\langle \Sigma, \widetilde{\mathcal{T}}_\varepsilon(\phi)\big\rangle = \frac{1}{|Y|}\int_{\mathbb{R}^N\times\widetilde{Y}} \phi\Big(\varepsilon\Big[\frac{x}{\varepsilon}\Big]_{\widetilde{Y}} + \varepsilon y\Big)\Sigma(dx,dy).$$

Consequently, from (iii) of Proposition 1.8 applied to bounded Borel functions, it follows that

$$\|\widetilde{\mathcal{U}}_\varepsilon(\Sigma)\|_{\mathcal{M}^1(\mathbb{R}^N)} \leqslant \frac{1}{|Y|}\|\Sigma\|_{\mathcal{M}^1(\mathbb{R}^N\times\widetilde{Y})}.$$

As for the unfolding operator, one can describe the averaging operator by using the map $\mathcal{X}_\varepsilon$.

Proposition 2.46. $\displaystyle \widetilde{\mathcal{U}}_\varepsilon(\Sigma) = \frac{1}{|Y|}\int_{\widetilde{Y}} \mathcal{X}_\varepsilon^*(\Sigma)(\,\cdot\,,dy).$

Proposition 2.47. *The following properties hold for μ in $\mathcal{M}^1(\mathbb{R}^N)$ and Σ in $\mathcal{M}^1(\mathbb{R}^N \times \widetilde{Y})$:*

(i) $\widetilde{\mathcal{T}}_\varepsilon(\phi\mu)(x) = \widetilde{\mathcal{T}}_\varepsilon(\phi)\widetilde{\mathcal{T}}_\varepsilon(\mu)$ for every ϕ bounded Borel function on $\mathbb{R}^N$ or more generally locally integrable with respect to μ,

(ii) $\frac{1}{|Y|}\langle \widetilde{\mathcal{T}}_\varepsilon(\mu), \widetilde{\mathcal{T}}_\varepsilon(\phi)\rangle = \langle \mu, \phi\rangle$ *for every* ϕ *bounded Borel function on* $\mathbb{R}^N$ *or more generally integrable with respect to* μ,

(iii) $\widetilde{\mathcal{U}}_\varepsilon \circ \widetilde{\mathcal{T}}_\varepsilon(\mu) = \mu$,

(iv) $\langle \widetilde{\mathcal{T}}_\varepsilon \circ \widetilde{\mathcal{U}}_\varepsilon(\Sigma), \Phi\rangle = \frac{1}{|Y|}\int_Y dz \int_{\mathbb{R}^N\times\widetilde{Y}} \Phi(\varepsilon\big[\frac{x}{\varepsilon}\big]_{\widetilde{Y}} + z, y)\,\Sigma(dx, dy)$ *for every* Φ *bounded Borel function on* $\mathbb{R}^N \times \widetilde{Y}$.

Similar results hold for positive Borel measures and for Radon measures.

Proof.
(i) Computing formulas (2.29) and (2.30) for the Radon measure $\phi\mu$ on $\mathbb{R}^N$, leads to

$$
\begin{aligned}
\langle \widetilde{\mathcal{T}}_\varepsilon(\phi\mu)(x), \varphi\rangle &= \frac{1}{\varepsilon^N}\int \varphi\Big(\Big\{\frac{x'}{\varepsilon}\Big\}_{\widetilde{Y}}\Big)\mathbf{1}_{\varepsilon[\frac{x}{\varepsilon}]_{\widetilde{Y}}+\varepsilon\widetilde{Y}}(x')\phi(x')\,\mu(dx')\\
&= \frac{1}{\varepsilon^N}\int \varphi\Big(\Big\{\frac{x'}{\varepsilon}\Big\}_{\widetilde{Y}}\Big)\mathbf{1}_{\varepsilon[\frac{x}{\varepsilon}]_{\widetilde{Y}}+\varepsilon\widetilde{Y}}(x')\widetilde{\mathcal{T}}_\varepsilon(\phi)\Big(\varepsilon\Big[\frac{x}{\varepsilon}\Big]_{\widetilde{Y}} + \varepsilon\Big(\Big\{\frac{x'}{\varepsilon}\Big\}_{\widetilde{Y}}\Big)\,\mu(dx')\\
&= \Big\langle \widetilde{\mathcal{T}}_\varepsilon(\mu)(x), \phi\Big(\varepsilon\Big[\frac{x}{\varepsilon}\Big]_{\widetilde{Y}} + \varepsilon\,\cdot\Big)\varphi(\cdot)\Big\rangle,
\end{aligned}
$$

for every φ bounded Borel function on $\widetilde{Y}$, and for every $x \in \mathbb{R}^N$. This gives the result.
(ii) By definition,

$$
\frac{1}{|Y|}\langle \widetilde{\mathcal{T}}_\varepsilon(\mu), \widetilde{\mathcal{T}}_\varepsilon(\phi)\rangle = \langle \mu, \widetilde{\mathcal{U}}_\varepsilon \circ \widetilde{\mathcal{T}}_\varepsilon(\phi)\rangle = \langle \mu, \phi\rangle,
$$

where (1.26) is used for the Borel function ϕ.
(iii) From Propositions 2.44 and 2.46, one gets

$$
\widetilde{\mathcal{U}}_\varepsilon \circ \widetilde{\mathcal{T}}_\varepsilon(\mu) = \frac{1}{|Y|}\int_{\widetilde{Y}} \mathcal{X}_\varepsilon^*(\widetilde{\mathcal{T}}_\varepsilon(\mu))(\cdot, dy) = \frac{1}{|Y|}\int_{\widetilde{Y}} \mathcal{X}_\varepsilon^* \circ \mathcal{X}_\varepsilon^*(\mu(dx)\otimes dy)(\cdot, dy).
$$

Since $\mathcal{X}_\varepsilon$ is an involution, so is $\mathcal{X}_\varepsilon^*$, and the last integral becomes

$$
\frac{1}{|Y|}\int_{\widetilde{Y}} \mu(\cdot)\otimes dy = \mu.
$$

(iv) By definition,

$$
\langle \widetilde{\mathcal{T}}_\varepsilon \circ \widetilde{\mathcal{U}}_\varepsilon(\Sigma), \Phi\rangle = \langle \Sigma, \widetilde{\mathcal{T}}_\varepsilon \circ \widetilde{\mathcal{U}}_\varepsilon(\Phi)\rangle.
$$

The result follows by Fubini's theorem and formula (1.27) which holds for bounded Borel maps on $\mathbb{R}^N \times \widetilde{Y}$. ☐

Remark 2.48. *Proposition 2.47(ii) implies that $\widetilde{\mathcal{T}}_\varepsilon$ defines a continuous linear map from $L^p(\mathbb{R}^N,\mu)$ to $L^p(\mathbb{R}^N \times \widetilde{Y}, \widetilde{\mathcal{T}}_\varepsilon(\mu))$ with a norm bounded above by $|Y|^{1/p}$. This is also true for μ a non negative Borel measure on $\mathbb{R}^N$, and generalizes the result (iii) of Proposition 1.8.*

Definition 2.49 (Average of a measure over Y). *The average $\mathcal{M}_Y(\nu)$ of a measure $\nu \in \mathcal{M}^1(\mathbb{R}^N \times \widetilde{Y})$ on $\mathbb{R}^N$, is the element of $\mathcal{M}^1(\mathbb{R}^N)$ defined as the linear map*

$$\varphi \in C_c^\infty(\mathbb{R}^N) \mapsto \frac{1}{|Y|}\int_{R^N \times Y} \varphi(x)\, d\nu(x,y).$$

It is the projection of ν over $\mathbb{R}^N$ up to the factor $1/|Y|$.

Definition 2.50 (Local average of a measure). *For $\mu \in \mathcal{M}^1(\mathbb{R}^N)$, the local average $\widetilde{\mathcal{M}}_\varepsilon(\mu)$ is the average over Y of its unfolding:*

$$\widetilde{\mathcal{M}}_\varepsilon(\mu) \doteq \mathcal{M}_Y(\widetilde{\mathcal{T}}_\varepsilon(\mu)) = \frac{1}{\varepsilon^N|Y|}\,\mu\Big(\varepsilon\Big[\frac{x}{\varepsilon}\Big]_{\widetilde{Y}} + \varepsilon\widetilde{Y}\Big),$$

which belongs to $L^1 \cap L^\infty(\mathbb{R}^N)$ by (2.27).

The next result corresponds to Proposition 1.9 for bounded measures, weak-$*$ convergence here is also called vague convergence.

Proposition 2.51.
(i) For μ in $\mathcal{M}^1(\mathbb{R}^N)$, the following convergence holds:

$$\widetilde{\mathcal{M}}_\varepsilon(\mu) \rightharpoonup \mu \quad \text{weakly-}* \text{ in } \mathcal{M}^1(\mathbb{R}^N).$$

(ii) If $\{\mu_\varepsilon\}_\varepsilon$ is a bounded sequence in $\mathcal{M}^1(\mathbb{R}^N)$, the following convergences are equivalent:

$$(a)\quad \mu_\varepsilon \rightharpoonup \mu \quad \text{weakly-}* \text{ in } \mathcal{M}^1(\mathbb{R}^N),$$
$$(b)\quad \widetilde{\mathcal{M}}_\varepsilon(\mu_\varepsilon) \rightharpoonup \mu \quad \text{weakly-}* \text{ in } \mathcal{M}^1(\mathbb{R}^N).$$

(iii) Let $\{\mu_\varepsilon\}_\varepsilon$ be a bounded sequence of Borel measures such that $\{\widetilde{\mathcal{T}}_\varepsilon(\mu_\varepsilon)\}_\varepsilon$ converges vaguely to some measure ν in $\mathcal{M}^1(\mathbb{R}^N \times \widetilde{Y})$. Then, the sequence $\{\mu_\varepsilon\}_\varepsilon$ itself converges vaguely to $\mathcal{M}_Y(\nu)$ on Ω.

Similar results hold for sequences of Radon measures.

Proof.
(i) Clearly, (ii) implies (i).
(ii) *(a)* $\Rightarrow$ *(b)* Let φ be in $C_c^\infty(\mathbb{R}^N)$. Then, by the definitions,

$$\int_{\mathbb{R}^N} \widetilde{\mathcal{M}}_\varepsilon(\mu_\varepsilon)(x)\varphi(x)\, dx \equiv \frac{1}{\varepsilon^N|Y|}\sum_{\xi\in\mathbf{G}}\int_{\varepsilon\xi+\varepsilon Y} \mu_\varepsilon(\varepsilon\xi + \varepsilon\widetilde{Y})\varphi(x)\, dx$$
$$= \sum_{\xi\in\mathbf{G}} \mu_\varepsilon(\varphi \mathbf{1}_{\varepsilon\xi+\varepsilon\widetilde{Y}}) = \big\langle \mu_\varepsilon, \widetilde{\mathcal{M}}_\varepsilon(\varphi)\big\rangle.$$

Since $\widetilde{\mathcal{M}}_\varepsilon(\varphi)$ converges uniformly to φ in $\mathbb{R}^N$, one concludes using the weak-$*$ convergence of μ_ε to μ. The converse is obtained as in the cases $p \in [1, +\infty)$ from the uniqueness of the weak-$*$ limit point for the sequence $\{\mu_\varepsilon\}_\varepsilon$.
(iii) Similarly, the following equalities hold:

$$\begin{aligned}\frac{1}{|Y|}\big\langle \nu, \varphi \big\rangle &= \frac{1}{|Y|}\lim \big\langle \widetilde{\mathcal{T}}_\varepsilon(\mu_\varepsilon), \varphi \big\rangle \\ &= \lim_{\varepsilon\to 0} \big\langle \mu_\varepsilon, \widetilde{\mathcal{U}}_\varepsilon(\varphi) \big\rangle = \lim\big\langle \mu_\varepsilon, \varphi \big\rangle,\end{aligned}$$

and this concludes the proof. □

Proposition 2.52 (Proposition 1.31 for measures).
(i) Let $\{\Sigma_\varepsilon\}$ be a bounded sequence in $\mathcal{M}^1(\mathbb{R}^N \times \widetilde{Y})$ converging vaguely to some measure Σ in $\mathcal{M}^1(\mathbb{R}^N \times \widetilde{Y})$. Then, for every bounded uniformly continuous function ϕ on $\mathbb{R}^N$,

$$\lim_{\varepsilon\to 0}\big\langle \widetilde{\mathcal{U}}_\varepsilon(\Sigma_\varepsilon), \phi \big\rangle = \frac{1}{|Y|}\big\langle \Sigma, \phi \big\rangle,$$

which express the fact that

$$\widetilde{\mathcal{U}}_\varepsilon(\Sigma_\varepsilon) \quad \textit{converges vaguely to } \frac{1}{|Y|}\int_{\widetilde{Y}} \Sigma(\,\cdot\,, dy).$$

(ii) Under the same hypotheses, and for every bounded continuous function Φ on $\mathbb{R}^N \times \widetilde{Y}$,

$$\lim_{\varepsilon\to 0}\big\langle \widetilde{\mathcal{T}}_\varepsilon \circ \widetilde{\mathcal{U}}_\varepsilon(\Sigma_\varepsilon), \Phi \big\rangle = \big\langle \Sigma, \Phi \big\rangle,$$

i.e.,

$$\widetilde{\mathcal{T}}_\varepsilon \circ \widetilde{\mathcal{U}}_\varepsilon(\Sigma_\varepsilon) \quad \textit{converges vaguely in the sense of measures to } \Sigma.$$

Proof. The statements follow directly from formulas (ii) and (iii) of Proposition 2.47 and from the fact that for every bounded uniformly continuous function ϕ on $\mathbb{R}^N$,

$$|\phi(x) - \widetilde{\mathcal{T}}_\varepsilon(\phi)(x, y)| \leqslant \varpi(\varepsilon d(Y)) \to 0,$$

where ϖ is the modulus of continuity of ϕ. □

Example 2.53. *Let ν be a given element of $\mathcal{M}^1(\widetilde{Y})$. It can be extended by periodicity to $\mathbb{R}^N$ as a Radon measure $\widetilde{\nu}$. Consider its image ν^ε via the scaling ε, which is the $\varepsilon\widetilde{Y}$-periodic Radon measure defined for ψ continuous and with compact support in $\mathbb{R}^N$ as*

$$\big\langle \nu^\varepsilon, \psi \big\rangle = \varepsilon^N \big\langle \widetilde{\nu}, \psi(\varepsilon\,\cdot\,) \big\rangle = \varepsilon^N \sum_{\xi\in\mathbf{G}} \big\langle \nu, \psi(\varepsilon\xi + \varepsilon\,\cdot\,) \big\rangle.$$

We now unfold the measure ν^ε. *By formula* (2.30), *for every* $x \in \mathbb{R}^N$,

$$\langle \widetilde{\mathcal{T}}_\varepsilon(\nu^\varepsilon)(x), \varphi \rangle = \frac{1}{\varepsilon^N} \int \varphi\Big(\Big\{\frac{x'}{\varepsilon}\Big\}_{\widetilde{Y}}\Big) \mathbf{1}_{\varepsilon[\frac{x}{\varepsilon}]_{\widetilde{Y}}+\varepsilon\widetilde{Y}}(x')\, \nu^\varepsilon(dx') = \langle \nu, \varphi \rangle.$$

Hence $\widetilde{\mathcal{T}}_\varepsilon(\nu^\varepsilon)$ *is the constant map* ν *from* $\mathbb{R}^N$ *to* $\mathcal{M}^1(\widetilde{Y})$ *(it can be seen also as the measure* $dx \otimes \nu$ *on* $\mathbb{R}^N \times \widetilde{Y}$*).*

Note that these are the sequences of measures used in [32,192] with respect to which two-scale convergence is defined for sequences of functions. With the viewpoint of the unfolding method, this two-scale convergence is just the weak convergence of the sequence of unfolded functions with respect to the fixed measure ν. *In Section 4.4, there is an example of such measures, where* ν *is a restriction of the Hausdorff measure of codimension 1.*

Remark 2.54. *With the unfolding point of view, it is very easy to consider simultaneously sequences of measures and functions depending upon* ε, *making use of (i) of Proposition 2.47.*

Remark 2.55 (Generalization to Lipschitz domains Ω in $\mathbb{R}^N$)**.** *To consider the unfolding of a measure* $\mu \in \mathcal{M}^1(\Omega)$ *for* Ω *a strict subdomain of* $\mathbb{R}^N$, *according to (i) of Proposition 2.47, one can set*

$$\mathcal{T}_\varepsilon(\mu) \doteq \widetilde{\mathcal{T}}_\varepsilon(\mathbf{1}_{\widehat{\Omega}_\varepsilon}\mu) = \mathbf{1}_{\widehat{\Omega}_\varepsilon}\widetilde{\mathcal{T}}_\varepsilon(\mu),$$

which belongs to $L^1 \cap L^\infty(\widehat{\Omega}_\varepsilon; \mathcal{M}^1(\widetilde{Y}))$. *Here* $\widehat{\Omega}_\varepsilon$ *denotes precisely the set*

$$\bigcup_{\xi\in\mathbf{G}} \big\{\varepsilon\xi + \varepsilon\widetilde{Y} \;\big|\; \varepsilon\xi + \varepsilon\widetilde{Y} \subset \Omega\big\}.$$

The results concerning the unfolding of measures on $\mathbb{R}^N$ *can then be applied, using test functions* φ *having compact supports in* Ω *so that, for* ε *small enough,* $\mathbf{1}_{\widehat{\Omega}_\varepsilon}\varphi \equiv \varphi$.

We close this section with the equivalent of Theorem 1.41 for the case of the space $BV(\Omega)$ for a bounded open subset Ω of $\mathbb{R}^N$ with Lipschitz boundary. Recall that in such a situation, the embedding from $BV(\Omega)$ to $L^1(\Omega)$ is compact. See, for example, [18] for general properties of BV spaces.

Recall that the weak-$*$ convergence of a sequence in $\mathcal{M}^1(\Omega; BV(Y))$ is actually defined by the weak-$*$ convergence in $\mathcal{M}^1(\Omega \times Y)^{N+1}$ of the image of the sequence by the map $\phi \mapsto (\phi, \nabla\phi)$.

Theorem 2.56. *Let* $\{w_\varepsilon\}_\varepsilon$ *be a bounded sequence in* $BV(\Omega)$. *For a subsequence, there exist* w *in* $BV(\Omega)$, $\widehat{w}$ *in* $\mathcal{M}^1(\Omega; BV_{per}(Y))$ *with* $\mathcal{M}_Y(\widehat{w}) = 0$, *such that*

$$\frac{1}{\varepsilon}\big(\mathcal{T}_\varepsilon(w_\varepsilon) - \mathcal{M}_Y \circ \mathcal{T}_\varepsilon(w_\varepsilon)\big) \rightharpoonup y^c \cdot \nabla w + \widehat{w} \text{ weakly-}* \text{ in } \mathcal{M}^1(\Omega; BV(Y)),$$

and

$$\mathcal{T}_\varepsilon(\nabla w_\varepsilon) \rightharpoonup \nabla w + \nabla_y \widehat{w} \quad \text{weakly-}* \text{ in } \mathcal{M}^1(\Omega \times Y)^N.$$

Proof. By the compact embedding of $BV(\Omega)$ in $L^1(\Omega)$, one can assume that w_ε converges strongly to some w in $L^1(\Omega)$. Therefore, up to a subsequence, its unfolding converges strongly to the same limit w in $L^1(\Omega \times Y)$.

Since Y is with Lipschitz boundary, it satisfies the Poincaré-Wirtinger inequality for the exponent 1. By the density of the regular functions in $BV(Y)$ for the intermediate convergence (i.e. weak-$*$ density plus convergence of the norms), it follows that the Poincaré-Wirtinger inequality holds for the space $BV(Y)$. Therefore, it is straightforward that

$$\Big\|\frac{1}{\varepsilon}\big(\mathcal{T}_\varepsilon(w_\varepsilon) - \mathcal{M}_\varepsilon(w_\varepsilon)\big)\Big\|_{L^1(\Omega;BV(Y))} \leqslant C\|\nabla w_\varepsilon\|_{\mathcal{M}^1(\Omega)},$$

which is bounded. The first convergence then follows for a subsequence.

Applying the gradient operator gives the second convergence. As

$$\mathcal{M}_Y\big(\mathcal{T}_\varepsilon(w_\varepsilon) - \mathcal{M}_Y \circ \mathcal{T}_\varepsilon(w_\varepsilon)\big) = 0,$$

by the weak-$*$ continuity of $\mathcal{M}_Y$, one gets $\mathcal{M}_Y(\widehat{w}) = 0$.

It remains to show that the measure $\widehat{w}$ is Y periodic. This cannot be proved by comparing traces because a periodic $BV(\Omega)$ function does not necessarily have the same traces on opposite faces of Y. The proof can be given in the same way as for the case of perforated domains in Part II Theorem 4.28. □

Remark 2.57 (Generalization to $\mathbb{R}^N$ or unbounded domains therein). *The previous result holds provided the L^1-convergence is replaced by the L^1_{loc}-convergence.*

Chapter 3

Homogenization in fixed domains

This chapter is devoted to examples of the application of the unfolding method to the homogenization of some partial differential equations with highly oscillating coefficients.

All examples are elliptic problems, but the method applies readily to evolution problems with coefficients oscillating in space (using partial unfolding in space). The case of oscillating coefficients in space-time can be also treated by a global unfolding.

The homogenization of linear second order elliptic diffusion equations is presented in details in Section 3.1, for the case of homogeneous Dirichlet conditions on the boundary (subsection 3.1.1). Nonhomogeneous Dirichlet conditions as well as Neumann boundary conditions are easily adapted.The case of the Fredholm alternative with homogeneous Neumann boundary condition is also discussed (subsection 3.1.2). The notion of frame-periodic oscillating coefficients is introduced as a generalization of oscillating periodic coefficients and their homogenization is obtained again via unfolding (subsection 3.1.3).

Subsection 3.1.4 shows how to use the unfolding method for an example of multiscale linear diffusion (this is an example of iterated unfolding) and subsection 3.1.5 gives an example of treatment when the right-hand side only converges weakly. This section closes with a first result for correctors using the unfolding operators (these are much improved upon in Chapter 14).

In Section 3.2, homogenization of nonlinear diffusion problems involving Višik-Leray-Lions type operators is treated using the unfolding method (with a brief presentation of the necessary convergence theory for maximal monotone operators).

Finally, Section 3.3 shows how the unfolding method can be used to study the Γ-convergence of oscillating functionals of gradients.

Other types of transmission conditions on the boundary of ε-periodically distributed subdomains, themselves of size ε, have been treated with the

D. Cioranescu et al., *The Periodic Unfolding Method*, Series in Contemporary Mathematics 3, https://doi.org/10.1007/978-981-13-3032-2_3

unfolding method. An example of such conditions concerns the homogenization of problems with equi-valued surfaces when these surfaces are small and ε-periodically distributed – e.g., the torsion problem (see [67] for details). Another example concerns imperfect heat transmission conditions (see [101, 102]).

3.1 Homogenization of linear diffusion problems

Definition 3.1. *Let $\alpha, \beta \in \mathbb{R}$, such that $0 < \alpha < \beta$ and $\mathcal{O}$ be an open subset of $\mathbb{R}^N$. $M(\alpha, \beta, \mathcal{O})$ is the set of the $N \times N$ matrices $A = (a_{ij})_{1 \leqslant i,j \leqslant N}$ with coefficients in $L^\infty(\mathcal{O})$ such that for any $\lambda \in \mathbb{R}^N$ and for a.e. x in $\mathcal{O}$, the following two inequalities hold:*

$$\begin{aligned} &(i) \quad (A(x)\lambda, \lambda) \geqslant \alpha|\lambda|^2, \\ &(ii) \quad |A(x)\lambda|^2 \leqslant \beta\,(A(x)\lambda, \lambda). \end{aligned} \tag{3.1}$$

Remark 3.2. *Inequality* (3.1)*(ii) implies that*

$$|A(x)\lambda| \leqslant \beta|\lambda|. \tag{3.2}$$

The converse is true in case of positive symmetric matrices (and is often taken as an alternate second condition, even in the general case). It will be shown that the sets $M(\alpha, \beta, \mathcal{O})$ (with the definition used above) are stable under homogenization. This is not be the case with the alternate definition (3.2).

Let Ω be a bounded domain in $\mathbb{R}^N$ and $A^\varepsilon = (a^\varepsilon_{ij})_{1 \leqslant i,j \leqslant N}$ a sequence of matrices in $M(\alpha, \beta, \Omega)$. Many different diffusion problems can be set in Ω with diffusion matrix field A^ε. We will consider the classical Dirichlet and Neumann problems.

For f_ε given in $H^{-1}(\Omega)$, consider the homogeneous Dirichlet problem

$$\begin{cases} -\text{div}\,(A^\varepsilon \nabla u_\varepsilon) = f_\varepsilon & \text{in } \Omega, \\ u_\varepsilon = 0 & \text{on } \partial\Omega. \end{cases} \tag{3.3}$$

Its variational formulation is: *Find u_ε in $H^1_0(\Omega)$ such that*

$$\int_\Omega A^\varepsilon \nabla u_\varepsilon \, \nabla v \, dx = \langle f_\varepsilon, v \rangle_{H^{-1}(\Omega), H^1_0(\Omega)}, \quad \forall v \in H^1_0(\Omega). \tag{3.4}$$

It admits a unique solution by the Lax-Milgram theorem. Using the standard Poincaré inequality (which holds in Ω), there is a uniform estimate (i.e. independent of ε), which states that

$$\|u_\varepsilon\|_{H^1_0(\Omega)} \leqslant \frac{1}{\alpha} \|f_\varepsilon\|_{H^{-1}(\Omega)}. \tag{3.5}$$

Consequently, assuming the sequence $\{f_\varepsilon\}_\varepsilon$ is bounded in $H^{-1}(\Omega)$, there exist u_0 in $H^1_0(\Omega)$ and a subsequence, still denoted ε, such that

$$u_\varepsilon \rightharpoonup u_0 \quad \text{weakly in } H^1_0(\Omega).$$

The homogeneous Neumann problem for f_ε in $L^2(\Omega)$ with $\mathcal{M}_\Omega(f_\varepsilon) = 0$ (known as the Fredholm condition), is

$$\begin{cases} -\text{div}\,(A^\varepsilon \nabla u_\varepsilon) = f_\varepsilon & \text{in } \Omega, \\ A^\varepsilon \nabla u_\varepsilon \cdot n = 0 & \text{on } \partial\Omega, \\ \mathcal{M}_\Omega(u_\varepsilon) = 0. \end{cases} \tag{3.6}$$

Its variational formulation

$$\int_\Omega A^\varepsilon \nabla u_\varepsilon \, \nabla v \, dx = \int_\Omega f_\varepsilon \, v \, dx, \quad \forall v \in H^1(\Omega),$$

admits a unique solution by the standard Poincaré-Wirtinger inequality in Ω and the Lax-Milgram theorem. Moreover, there is a uniform estimate,

$$\|u_\varepsilon\|_{H^1(\Omega)} \leqslant \frac{1}{\alpha} \|f_\varepsilon\|_{L^2(\Omega)}.$$

So, assuming the sequence $\{f_\varepsilon\}_\varepsilon$ is bounded in $L^2(\Omega)$, there exist u_0 in $H^1(\Omega)$ such that, up to a subsequence,

$$u_\varepsilon \rightharpoonup u_0 \quad \text{weakly in } H^1(\Omega).$$

In each case, assuming that the matrix A^ε is ε-periodic, the classical problem of homogenization is to find an equation satisfied by every weak limit-point of the corresponding sequence of solutions. If the homogenized problem has a unique solution, this also implies that the whole sequence converges weakly.

Remark 3.3 (A general remark). *Second order variational problems exhibit invariance under* **bi-Lipschitz diffeomorphisms**[(1)] *(the coefficients are obviously modified but stay in the same general class; this also holds for higher order problems with higher regularity for the diffeomorphism used as change of variable). Therefore, every result stated in the periodic case can be extended to the case of* **bi-Lipschitz equivalent to periodic**.

We do not give any explicit examples of such problems (as for instance, problems which are periodic with respect to polar or cylindric coordinates). Note that these global bi-Lipschitz diffeomorphisms can be combined with the local bi-Lipschitz change of variables which we introduce in Subsection 3.1.3 under the terminology of "frame-periodic".

3.1.1 Homogenization of the Dirichlet problem

We are now interested to find a limit problem, the "homogenized" problem, satisfied by u_0 in the case of (3.3). This is called standard homogenization

(1) A bi-Lipschitz diffeomorphism is a bijection which is Lipschitz continuous as well as its inverse.

and the answer, for some classes of A^ε, can be found in many works, starting with the classical book of Bensoussan, Lions and Papanicolaou [21] (see, for instance Cioranescu and Donato [70] and the references therein). We recall this result below (Theorem 3.4) for the case of a single scale (i.e. there exists a single small scale of order ε in the problem), then give the proof in a more general setting using the unfolding method.

Theorem 3.4. *(Standard single scale periodic homogenization).*
Let $A = (a_{ij})_{1\leqslant i,j\leqslant N}$ belong to $M(\alpha, \beta, Y)$, and be extended to the whole $\mathbb{R}^N$ by Y-periodicity. Set

$$A^\varepsilon(x) = \Big(a_{ij}\Big(\frac{x}{\varepsilon}\Big)\Big)_{1\leqslant i,j\leqslant N} \qquad \text{for a.e. } x \text{ in } \Omega. \tag{3.7}$$

Let f in $H^{-1}(\Omega)$ and consider the problem

$$\begin{cases} -\operatorname{div}(A^\varepsilon \nabla u_\varepsilon) = f & \text{in } \Omega, \\ u_\varepsilon = 0 & \text{on } \partial\Omega, \end{cases}$$

Then the whole sequence $\{u_\varepsilon\}_\varepsilon$ converges weakly in $H^1_0(\Omega)$ to a limit u_0 which is the unique solution of the homogenized problem

$$\begin{cases} -\displaystyle\sum_{i,j=1}^{N} a^0_{ij}\frac{\partial^2 u_0}{\partial x_i \partial x_j} = f & \text{in } \Omega, \\ u_0 = 0 & \text{on } \partial\Omega, \end{cases} \tag{3.8}$$

where the homogenized constant matrix $A^0 = (a^0_{ij})_{1\leqslant i,j\leqslant N}$ is elliptic and given by

$$a^0_{ij} = \mathcal{M}_Y\Big(a_{ij} + \sum_{k=1}^{N} a_{ik}\frac{\partial \widehat{\chi}_j}{\partial y_k}\Big). \tag{3.9}$$

In (3.9), the functions $\widehat{\chi}_j$ $(j = 1, \dots, N)$, often referred to as **correctors,** *are the solutions of the cell problems*

$$\begin{cases} -\displaystyle\sum_{i,k=1}^{N} \frac{\partial}{\partial y_i}\Big(a_{ik}(y)\frac{\partial(\widehat{\chi}_j(y) + y_j)}{\partial y_k}\Big) = 0 & \text{in } \mathcal{D}'(\mathbb{R}^N), \\ \mathcal{M}_Y(\widehat{\chi}_j) = 0, \quad \widehat{\chi}_j \ \mathbf{G}\text{-periodic.} \end{cases} \tag{3.10}$$

The correctors belong to $H^1_{per,0}(Y) \cap L^\infty(\mathbb{R}^N)$ (see Proposition 3.7 below).

As will be seen below, using the periodic unfolding method, the proof of this theorem is elementary! Actually, with the same proof, a more general result can be obtained, with a general sequence of matrices $\{A^\varepsilon\}_\varepsilon$ in $M(\alpha, \beta, Y)$ (that is, such that $\mathcal{T}_\varepsilon(A^\varepsilon)$ converges a.e. in $\Omega \times Y$). The proof also works for problem (3.4) with strongly convergent right-hand sides f_ε in $H^{-1}(\Omega)$.

The case of **weakly convergent** right-hand sides in $H^{-1}(\Omega)$ is more elaborate. It is treated in Subsection 3.1.5.

Theorem 3.5 (Single scale periodic homogenization via unfolding). *Let Ω be a bounded domain in $\mathbb{R}^N$ and u_ε be the solution of problem (3.3) with f_ε in $H^{-1}(\Omega)$ and $A^\varepsilon = (a^\varepsilon_{ij})_{1\leqslant i,j\leqslant N}$ in $M(\alpha,\beta,\Omega)$. Suppose that*

$$f_\varepsilon \to f \quad \text{strongly in } H^{-1}(\Omega),$$

and that there exists a matrix B such that

$$B^\varepsilon \doteq \mathcal{T}_\varepsilon(A^\varepsilon) \to B \quad \text{a.e. in } \Omega \times Y^{(2)}. \tag{3.11}$$

Then there exists $u_0 \in H^1_0(\Omega)$ and $\widehat{u} \in L^2(\Omega; H^1_{per,0}(Y))$ such that

$$\begin{aligned} &u_\varepsilon \rightharpoonup u_0 \quad \text{weakly in } H^1_0(\Omega),\\ &\mathcal{T}_\varepsilon(u_\varepsilon) \to u_0 \quad \text{strongly in } L^2(\Omega; H^1(Y)),\\ &\mathcal{T}_\varepsilon(\nabla u_\varepsilon) \rightharpoonup \nabla u_0 + \nabla_y \widehat{u} \quad \text{weakly in } L^2(\Omega \times Y)^N, \end{aligned} \tag{3.12}$$

and the pair $(u_0, \widehat{u})$ is the unique solution of the problem

$$\begin{cases} \dfrac{1}{|Y|}\displaystyle\int_{\Omega\times Y} B(x,y)\big[\nabla u_0(x) + \nabla_y \widehat{u}(x,y)\big]\big[\nabla\Psi(x) + \nabla_y\Phi(x,y)\big]dxdy \\ \qquad\qquad = \langle f, \Psi\rangle_{H^{-1}(\Omega),H^1_0(\Omega)}, \\ \forall \Psi \in H^1_0(\Omega),\ \forall\Phi \in L^2(\Omega;\ H^1_{per,0}(Y)). \end{cases} \tag{3.13}$$

Problem (3.13) is the unfolded formulation of the homogenized problem. It is of standard variational form in the space

$$\mathcal{H} = H^1_0(\Omega) \times L^2(\Omega;\ H^1_{per,0}(Y))$$

endowed with the norm $\|(v,\widehat{v})\|_{\mathcal{H}} = (\|\nabla v\|^2_{L^2(\Omega)} + \|\nabla_y \widehat{v}\|^2_{L^2(\Omega\times Y)})^{1/2}$. Note that for every $(v,\widehat{v}) \in \mathcal{H}$, due to the Y-periodicity of $\widehat{v}$, $\mathcal{M}_Y(\nabla_y\widehat{v}) = 0$ and the following equality holds:

$$\|\nabla v + \nabla_y \widehat{v}\|^2_{L^2(\Omega\times Y)} = |Y|\,\|\nabla v\|^2_{L^2(\Omega)} + \|\nabla_y\widehat{v}\|^2_{L^2(\Omega\times Y)}. \tag{3.14}$$

Remark 3.6.
1. Hypothesis (3.11) implies that B belongs to $M(\alpha,\beta,\Omega\times Y)$.
2. If A^ε is of the form (3.7), then $B(x,y) = A(y)$. In the case where

$$A^\varepsilon(x) = A_1(x)A_2\Big(\frac{x}{\varepsilon}\Big) \qquad \text{one has} \qquad B(x,y) = A_1(x)A_2(y).$$

3. Note that every matrix $B \in M(\alpha,\beta,\Omega\times Y)$ can be approached by a sequence of matrices $\{\mathcal{T}_\varepsilon(A^\varepsilon)\}_\varepsilon$ with A^ε in $M(\alpha,\beta,\Omega)$. One of many such matrices can be defined as follows:

$$A^\varepsilon = \begin{cases} \mathcal{U}_\varepsilon(B) & \text{in } \widehat{\Omega}_\varepsilon \\ \alpha I_n & \text{in } \Lambda_\varepsilon, \end{cases}$$

where $\mathcal{U}_\varepsilon$ is the averaging operator from Definition 1.26.

(2) This condition can be generalized to the convergence in measure in $\Omega\times Y$ of $\mathcal{T}_\varepsilon(A^\varepsilon)$ to B, see footnote of Lemma 1.17.

Proof of Theorem 3.5. By estimate (3.5), the sequence $\{u_\varepsilon\}_\varepsilon$ is uniformly bounded in $H_0^1(\Omega)$. By weak compactness, there exists u_0 in $H_0^1(\Omega)$ such that, up to a subsequence, convergence $(3.12)_1$ holds. Since Ω is bounded, this convergence is strong in $L^2(\Omega)$. Corollary 1.37 implies convergence $(3.12)_2$. By Theorem 1.41, there exist a sub-subsequence and $\widehat{u}$ in $L^2(\Omega; H^1_{per,0}(Y))$, such that convergence $(3.12)_3$ holds.

Choose Ψ in $H_0^1(\Omega)$ as test function in (3.4). By Corollary 1.18, using hypothesis (3.11), the convergences of (3.12) and passing to the limit gives

$$\frac{1}{|Y|}\int_{\Omega\times Y} B(x,y)\big[\nabla u_0(x)+\nabla_y\widehat{u}(x,y)\big]\nabla\Psi(x)\,dxdy \\ = \langle f,\Psi\rangle_{H^{-1}(\Omega),H_0^1(\Omega)}. \tag{3.15}$$

For given Ψ in $C_c^\infty(\Omega)$ and ψ in $C^1_{per}(\overline{Y})$, consider now the test function

$$v_\varepsilon(x)=\varepsilon\Psi(x)\psi\Big(\frac{x}{\varepsilon}\Big).$$

The sequence $\{v_\varepsilon\}_\varepsilon$ converges to 0 uniformly in Ω. Evidently, its gradient

$$\nabla v_\varepsilon(x)=\varepsilon\nabla\Psi(x)\psi\Big(\frac{x}{\varepsilon}\Big)+\Psi(x)\nabla_y\psi\Big(\frac{x}{\varepsilon}\Big), \tag{3.16}$$

is bounded in $L^2(\Omega)^N$. Therefore,

$$v_\varepsilon \rightharpoonup 0 \quad \text{weakly in } H_0^1(\Omega).$$

Unfolding the expression (3.16) leads to the convergence

$$\mathcal{T}_\varepsilon(\nabla v_\varepsilon)\to\Psi\,\nabla_y\psi \quad \text{strongly in } L^2(\Omega\times Y)^N.$$

Taking v_ε as test function in (3.4) and passing to the limit, by Corollary 1.37 one gets

$$\frac{1}{|Y|}\int_{\Omega\times Y} B(x,y)\big[\nabla u_0(x)+\nabla_y\widehat{u}(x,y)\big]\Psi(x)\nabla_y\psi(y)\,dxdy=0. \tag{3.17}$$

The density of $C_c^\infty(\Omega)\otimes C^1_{per}(\overline{Y})$ in $L^2(\Omega;H^1_{per}(Y))$ implies

$$\frac{1}{|Y|}\int_{\Omega\times Y} B(x,y)\big[\nabla u_0(x)+\nabla_y\widehat{u}(x,y)\big]\nabla_y\Phi(x,y)\,dxdy=0,$$

for every Φ in $L^2(\Omega;H^1_{per}(Y))$. This, together with (3.15), yields (3.13).

Observe that due to the properties of the matrix B (see Remark 3.6.1) and equality (3.14), the Lax-Milgram theorem applies to problem (3.13) which, therefore, has a unique solution. Consequently, the whole sequence $\{(u_\varepsilon,\mathcal{T}_\varepsilon(\nabla u_\varepsilon)\}_\varepsilon$ converges weakly in the space $H_0^1(\Omega)\times L^2(\Omega\times Y)^N$ to the limit $(u_0,\nabla u_0+\nabla_y\widehat{u})$, and this concludes the proof. □

One can now give the standard form of the homogenized equation.

Proposition 3.7. *The function u_0 is the unique solution of the following homogenized problem:*

$$\int_\Omega A^{hom}\nabla u_0\,\nabla v\,dx = \langle f,v\rangle_{H^{-1}(\Omega),H_0^1(\Omega)},\qquad \forall v\in H_0^1(\Omega), \tag{3.18}$$

The matrix field A^{hom}, given by its coefficients

$$a_{ij}^{hom}(x) = \mathcal{M}_Y\Big(b_{ij}(x,\cdot) + \sum_{k=1}^N b_{ik}(x,\cdot)\frac{\partial\widehat{\chi}_j}{\partial y_k}(x,\cdot)\Big), \tag{3.19}$$

for a.e. $x\in\Omega$, belongs to $\mathcal{M}(\alpha,\beta,\Omega)$.

Here the correctors $\widehat{\chi}_j$ for $j = 1,\dots,N$, are the unique solutions in $L^\infty(\Omega;H^1_{per,0}(Y))$ of the cell-problems

$$\begin{cases}\displaystyle\int_Y \sum_{i,k=1}^N b_{ik}(\cdot,y)\Big(\frac{\partial\widehat{\chi}_j}{\partial y_k}(\cdot,y)+\delta_{jk}\Big)\frac{\partial\varphi}{\partial y_i}(y)\,dy = 0,\\ \forall\varphi\in H^1_{per,0}(Y).\end{cases} \tag{3.20}$$

Proof. Step 1. Proof of (3.18)-(3.19).

By the Lax-Milgram Theorem applied in $L^2(\Omega;H^1_{per,0}(Y))$, problem (3.20) has a unique solution in that space for each $j = 1,\dots,N$. Since B belongs to $\mathcal{M}(\alpha,\beta,\Omega\times Y)$, this implies that $\|\nabla_y\widehat{\chi}_j(x,\cdot)\|_{L^2(Y)}$ is bounded for a.e. x in Ω by a constant independent of x and j. Consequently, all the $\widehat{\chi}_j$'s are in $L^\infty(\Omega;H^1_{per,0}(Y))$.(3)

Equation (3.13) with $\Psi\equiv 0$ now gives $\widehat{u}$ in terms of ∇u_0 and these correctors:

$$\widehat{u}(x,y) = \sum_{i=1}^N \frac{\partial u_0}{\partial x_i}(x)\widehat{\chi}_i(x,y)\quad\text{for a.e. } (x,y)\in\Omega\times Y. \tag{3.21}$$

Using formula (3.21) in (3.13) with $\Phi\equiv 0$ gives (3.18) with

$$A^{hom}\nabla u_0 \doteq \frac{1}{|Y|}\int_Y B(\cdot,y)\big[\nabla u_0 + \nabla_y\widehat{u}(\cdot,y)\big]\,dy\quad\text{for a.e. in } \Omega. \tag{3.22}$$

Formula (3.19) follows and shows that A^{hom} is in $L^\infty(\Omega)^{N\times N}$.

Step 2. Consider the map G which to f in $H^{-1}(\Omega)$ assigns u_0, the first component of the solution of problem (3.13). We claim that its image is dense in $H_0^1(\Omega)$.(4)

(3) Lemma 3.29 further implies that $\|\widehat{\chi}_j(x,\cdot)\|_{L^\infty(Y)}$ is also uniformly bounded for a.e. x in Ω and every $j = 1,\dots,N$, which implies that all the norms $\|\widehat{\chi}_j\|_{L^\infty(\Omega\times Y)}$ are bounded.

(4) Actually, at the end of this proof, one can see that this image is the whole space $H_0^1(\Omega)$, since G is also the solution map of the homogenized problem (3.18).

To do so, let $g \in H^{-1}(\Omega)$ be such that

$$\langle g, G(f)\rangle_{H^{-1}(\Omega),H_0^1(\Omega)} \equiv 0, \quad \forall f \in H^{-1}(\Omega).$$

Let $(v_0, \widehat{v})$ be in $H_0^1(\Omega) \times L^2(\Omega; H^1_{per,0}(Y))$ the unique solution of the adjoint problem

$$\begin{cases} \dfrac{1}{|Y|}\displaystyle\int_{\Omega\times Y} B(x,y)\big[\nabla\Psi(x) + \nabla_y\Phi(x,y)\big]\big[\nabla v_0(x) + \nabla_y\widehat{v}(x,y)\big]dxdy \\ \qquad\qquad\qquad\qquad = \langle g, \Psi\rangle_{H^{-1}(\Omega),H_0^1(\Omega)}, \\ \forall \Psi \in H_0^1(\Omega), \quad \forall\Phi \in L^2(\Omega; H^1_{per,0}(Y)). \end{cases} \tag{3.23}$$

The Lax-Milgram theorem implies that this problem has a unique solution.

Let f be in $H^{-1}(\Omega)$ and $(u_0, \widehat{u})$ be the solution of problem (3.13). Choosing in (3.23) $\Psi = u_0$ and $\Phi = \widehat{u}$ implies

$$\forall f \in H^{-1}(\Omega), \quad \langle f, v_0\rangle_{H^{-1}(\Omega),H_0^1(\Omega)} = \langle g, G(f)\rangle_{H^{-1}(\Omega),H_0^1(\Omega)} \equiv 0,$$

hence $v_0 \equiv 0$. Consequently, $\widehat{v} = 0$ (choose $\Psi = 0$ and $\Phi = \widehat{v}$ in (3.23)), so that $g = 0$. This implies the density of the range of G.

Step 3. A^{hom} belongs to $\mathcal{M}(\alpha, \beta, \Omega)$.

In (3.13), let $\Psi = u_0$ and $\Phi = \widehat{u}$. This gives

$$\begin{aligned} \frac{1}{|Y|}\int_{\Omega\times Y} B(x,y)\big[\nabla u_0(x) + \nabla_y\widehat{u}(x,y)\big]\big[\nabla u_0(x) + \nabla_y\widehat{u}(x,y)\big]\, dxdy \\ = \langle f, u_0\rangle_{H^{-1}(\Omega),H_0^1(\Omega)} = \int_\Omega A^{hom}\nabla u_0\, \nabla u_0\, dx. \end{aligned}$$

As noted in Remark 3.2, $B \in \mathcal{M}(\alpha, \beta, \Omega \times Y)$, so it is α-elliptic. Thus,

$$\frac{\alpha}{|Y|}\|\nabla u_0 + \nabla_y\widehat{u}\|^2_{L^2(\Omega\times Y)} \leqslant \int_\Omega A^{hom}\nabla u_0\, \nabla u_0\, dx.$$

Since by periodicity $\displaystyle\int_Y \nabla_y\widehat{u}(\,\cdot\,, y)\, dy \equiv 0$,

$$\frac{1}{|Y|}\|\nabla u_0 + \nabla_y\widehat{u}\|^2_{L^2(\Omega\times Y)} = \|\nabla u_0\|^2_{L^2(\Omega)} + \frac{1}{|Y|}\|\nabla_y\widehat{u}\|^2_{L^2(\Omega\times Y)}.$$

Consequently, for u_0 in $H_0^1(\Omega)$,

$$\alpha\|\nabla u_0\|^2_{L^2(\Omega)} \leqslant \int_\Omega A^{hom}\nabla u_0\, \nabla u_0\, dx. \tag{3.24}$$

By the Cauchy-Schwarz inequality in (3.22)

$$\begin{aligned} \|\, A^{hom}\nabla u_0\|^2_{L^2(\Omega)} &= \Big\|\frac{1}{|Y|}\int_Y B(\,\cdot\,,y)\big[\nabla u_0 + \nabla_y\widehat{u}(\,\cdot\,,y)\big]dy\Big\|^2_{L^2(\Omega)} \\ &\leqslant \frac{1}{|Y|}\|B\big[\nabla u_0 + \nabla_y\widehat{u}\big]\|^2_{L^2(\Omega\times Y)}. \end{aligned} \tag{3.25}$$

Inequality (3.1)(ii) from Definition 3.1 of $M(\alpha, \beta, \Omega \times Y)$, written for B gives

$$\big|B(x,y)\big[\nabla u_0(x) + \nabla_y \widehat{u}(x,y)\big]\big|^2 \leqslant \beta B(x,y)\big(\nabla u_0(x) + \nabla_y \widehat{u}(x,y)\big)\,\big(\nabla u_0(x) + \nabla_y \widehat{u}(x,y)\big),$$

for a.e. $(x,y) \in \Omega \times Y$. This, used in (3.25) yields

$$\| \, A^{hom} \nabla u_0 \|^2_{L^2(\Omega)} \leqslant \beta \int_\Omega A^{hom} \nabla u_0 \, \nabla u_0 \, dx. \tag{3.26}$$

The conclusion, namely $A^{hom} \in M(\alpha, \beta, \Omega \times Y)$, now follows by applying Lemma 3.8 below, successively for $F \doteq A^{hom} - \alpha I$ (using (3.24)), and for $F \doteq (\beta I - A^{hom^T}) A^{hom}$ (using (3.26)). □

Lemma 3.8. *Let F be a matrix field in $L^\infty(\Omega)^{N \times N}$ and suppose that for every θ in a dense subset of $H^1_0(\Omega)$ the following inequality holds:*

$$\int_\Omega \big(F(x)\nabla\theta(x), \nabla\theta(x)\big)\, dx \geqslant 0. \tag{3.27}$$

Then, for almost every x in Ω and every λ in $\mathbb{R}^N$, $(F(x)\lambda, \lambda) \geqslant 0$.

Proof. First, note that by density, (3.27) holds for every θ in $H^1_0(\Omega)$.

Now, let $\lambda \neq 0$ be given in $\mathbb{R}^N$, φ in $\mathcal{C}^\infty_c(\Omega)$, and $\psi(t) = |t - 1/2|$ defined on $[0,1]$ and extended by periodicity to $\mathbb{R}$. We choose as reference cell

$$Y = (0, \mu_1) \times \ldots \times (0, \mu_N),$$

where for $i = 1, \ldots, N$,

$$\mu_i = \begin{cases} |\lambda_i|^{-1} & \text{if } \lambda_i \neq 0, \\ 1 & \text{otherwise.} \end{cases}$$

In (3.27), taking

$$\theta_\varepsilon(x) = \varepsilon \psi\Big(\lambda \cdot \frac{x}{\varepsilon}\Big)\varphi(x),$$

unfolding and letting $\varepsilon \to 0$, give

$$\frac{1}{|Y|} \int_{\Omega \times Y} \big(F(x)\lambda, \lambda\big) \psi'(\lambda \cdot y)^2 \varphi(x)^2 \, dx dy \geqslant 0.$$

Since $\psi'^2 \equiv 1$, this is simply

$$\int_\Omega \big(F(x)\lambda, \lambda\big)\, \varphi(x)^2 \, dx \geqslant 0 \quad \text{for every } \varphi \in \mathcal{C}^\infty_c(\Omega).$$

By density, we obtain

$$\int_\Omega \big(F(x)\lambda, \lambda\big)\, \Phi(x) \, dx \geqslant 0 \quad \text{for every } \Phi \in L^1(\Omega)^+.$$

Choosing $\Phi(x) = (F(x)\lambda, \lambda)^-$ concludes the proof. □

Proposition 3.9 (Convergence of the energy). *Under the hypotheses of Theorem 3.5,*

$$\lim_{\varepsilon\to 0}\int_\Omega A^\varepsilon \nabla u_\varepsilon \nabla u_\varepsilon \, dx = \int_\Omega A^{hom} \nabla u_0 \nabla u_0 \, dx. \tag{3.28}$$

Proof. Since

$$\int_\Omega A^{hom} \nabla u_0 \nabla u_0 \, dx = \langle f, u_0 \rangle_{H^{-1}(\Omega), H_0^1(\Omega)},$$

the result is straightforward from the equality,

$$\int_\Omega A^\varepsilon \nabla u_\varepsilon \nabla u_\varepsilon \, dx = \langle f_\varepsilon, u_\varepsilon \rangle_{H^{-1}(\Omega), H_0^1(\Omega)},$$

and the convergence of its right-hand side to $\langle f, u_0 \rangle_{H^{-1}(\Omega), H_0^1(\Omega)}$. □

Corollary 3.10. *The following convergences hold:*

$$\begin{aligned} &(i) \quad \mathcal{T}_\varepsilon(\nabla u_\varepsilon) \to \nabla u_0 + \nabla_y \widehat{u} \quad \textit{strongly in } L^2(\Omega \times Y)^N, \\ &(ii) \quad \int_{\Lambda_\varepsilon} |\nabla u_\varepsilon|^2 dx \to 0. \end{aligned} \tag{3.29}$$

For the proof, we use the classical result stated (and proved) below.

Lemma 3.11. *Let $\{D_\varepsilon\}_\varepsilon$ be a sequence of $N \times N$ matrix fields in $M(\alpha, \beta, \mathcal{O})$ for some open set $\mathcal{O}$, such that $D_\varepsilon \to D$ a.e. on $\mathcal{O}$ (or more generally, in measure in $\mathcal{O}$).*

If the sequence of vector fields $\{\zeta_\varepsilon\}_\varepsilon$ converges weakly to ζ in $L^2(\mathcal{O})^N$, then

$$\liminf_{\varepsilon\to 0}\int_{\mathcal{O}} D_\varepsilon \zeta_\varepsilon \, \zeta_\varepsilon \, dx \geqslant \int_{\mathcal{O}} D\zeta \, \zeta \, dx.$$

Furthermore, if

$$\limsup_{\varepsilon\to 0}\int_{\mathcal{O}} D_\varepsilon \zeta_\varepsilon \, \zeta_\varepsilon \, dx \leqslant \int_{\mathcal{O}} D\zeta \, \zeta \, dx,$$

then

$$\int_{\mathcal{O}} D\zeta \, \zeta \, dx = \lim_{\varepsilon\to 0}\int_{\mathcal{O}} D_\varepsilon \zeta_\varepsilon \, \zeta_\varepsilon \, dx \quad \textit{and} \quad \zeta_\varepsilon \to \zeta \quad \textit{strongly in } L^2(\mathcal{O})^N.$$

Proof. In the statements, it is clear that only the symmetric part of the matrix fields D_ε and D play a role. It is therefore enough to assume that D_ε and D are symmetric. Set

$$\Phi_\varepsilon(\zeta) \doteq \frac{1}{2}\int_{\mathcal{O}} D_\varepsilon \zeta \, \zeta \, dx \quad \text{and} \quad \Phi(\zeta) \doteq \frac{1}{2}\int_{\mathcal{O}} D\zeta \, \zeta \, dx.$$

They are equivalent Hilbert norms and convex continuous functions on the space $L^2(\mathcal{O})^N$. By a classical computation, their conjugate convex functions are respectively,

$$\Phi_\varepsilon^*(\eta) \doteq \frac{1}{2}\int_{\mathcal{O}} D_\varepsilon^{-1}\eta\,\eta\,dx \quad \text{and} \quad \Phi^*(\eta) \doteq \frac{1}{2}\int_{\mathcal{O}} D^{-1}\eta\,\eta\,dx.$$

In particular (biconjugation is the identity for l.s.c. convex functions, so it is in this case),

$$\Phi_\varepsilon(\zeta_\varepsilon) = \sup_{\eta\in L^2(\mathcal{O})}\Big(\int_{\mathcal{O}} \zeta_\varepsilon\,\eta\,dx - \Phi_\varepsilon^*(\eta)\Big).$$

Going to the limit yields

$$\liminf_{\varepsilon\to 0} \Phi_\varepsilon(\zeta_\varepsilon) \geqslant \sup_{\eta\in L^2(\mathcal{O})}\Big(\int_{\mathcal{O}} \zeta\,\eta\,dx - \Phi_\varepsilon^*(\eta)\Big) = \Phi(\zeta),$$

whence the first inequality in the statement of the lemma.

Assuming the second inequality holds, it obviously implies the following equality at the limit:

$$\Phi(\zeta) = \lim_{\varepsilon\to 0} \Phi_\varepsilon(\zeta_\varepsilon).$$

Due to the fact that Φ_ε is a quadratic form and D_ε is in $\mathcal{M}(\alpha,\beta,\mathcal{O})$,

$$\alpha\|\zeta_\varepsilon - \zeta\|_{L^2(\mathcal{O})} \leqslant 2\Phi_\varepsilon(\zeta_\varepsilon - \zeta) = 2\Phi_\varepsilon(\zeta_\varepsilon) + 2\Phi_\varepsilon(\zeta) - 4\int_{\mathcal{O}} D_\varepsilon\zeta\,\zeta_\varepsilon\,dx. \quad (3.30)$$

The convergence a.e. (or in measure) in $\mathcal{O}$ of D_ε to D, and the fact that they are all uniformly (essentially) bounded, imply that

$$\Phi_\varepsilon(\zeta) \to \Phi(\zeta) \quad \text{and} \quad \int_{\mathcal{O}} D_\varepsilon\zeta\,\zeta_\varepsilon\,dx \to \int_{\mathcal{O}} D\zeta\,\zeta\,dx.$$

Using these convergences in the right-hand side of (3.30) shows that ζ_ε converges strongly to ζ in $L^2(\mathcal{O})^N$. □

Proof of Corollary 3.10. Making use of Lemma 3.11, it is enough to show that

$$\lim_{\varepsilon\to 0}\frac{1}{|Y|}\int_{\Omega\times Y} B^\varepsilon\,\mathcal{T}_\varepsilon(\nabla u_\varepsilon)\mathcal{T}_\varepsilon(\nabla u_\varepsilon)\,dx\,dy$$
$$= \frac{1}{|Y|}\int_{\Omega\times Y} B\big[\nabla u_0 + \nabla_y\widehat{u}\big]\big[\nabla u_0 + \nabla_y\widehat{u}\big]\,dx\,dy.$$

Indeed, one successively obtains (note that the fourth inequality comes from the α-coercivity of the matrix fields A^ε),

$$
\begin{aligned}
\frac{1}{|Y|}\int_{\Omega\times Y} & B\big[\nabla u_0+\nabla_y\widehat{u}\big]\big[\nabla u_0+\nabla_y\widehat{u}\big]\,dx\,dy\\
&\leqslant \liminf_{\varepsilon\to 0}\frac{1}{|Y|}\int_{\Omega\times Y} B^\varepsilon\,\mathcal{T}_\varepsilon(\nabla u_\varepsilon)\mathcal{T}_\varepsilon(\nabla u_\varepsilon)\,dx\,dy\\
&=\liminf_{\varepsilon\to 0}\Big(\int_\Omega A^\varepsilon\,\nabla u_\varepsilon\nabla u_\varepsilon\,dx-\int_{\Lambda_\varepsilon} A^\varepsilon\,\nabla u_\varepsilon\nabla u_\varepsilon\,dx\Big)\\
&\leqslant \limsup_{\varepsilon\to 0}\frac{1}{|Y|}\int_{\Omega\times Y} B^\varepsilon\,\mathcal{T}_\varepsilon(\nabla u_\varepsilon)\mathcal{T}_\varepsilon(\nabla u_\varepsilon)\,dx\,dy\\
&=\limsup_{\varepsilon\to 0}\Big(\int_\Omega A^\varepsilon\,\nabla u_\varepsilon\nabla u_\varepsilon\,dx-\int_{\Lambda_\varepsilon} A^\varepsilon\,\nabla u_\varepsilon\nabla u_\varepsilon\,dx\Big)\\
&\leqslant \lim_{\varepsilon\to 0}\int_\Omega A^\varepsilon\,\nabla u_\varepsilon\nabla u_\varepsilon\,dx=\int_\Omega A^{hom}\nabla u_0\nabla u_0\,dx\ \text{(by (3.28))}\\
&=\frac{1}{|Y|}\int_{\Omega\times Y} B\big[\nabla u_0+\nabla_y\widehat{u}\big]\big[\nabla u_0+\nabla_y\widehat{u}\big]\,dx\,dy,
\end{aligned}
$$

the last equality coming from the definition of A^{hom} (see (3.22)). Hence all inequalities above are equalities. This, by Lemma 3.11, gives convergence (3.29)(i) and implies that

$$
\lim_{\varepsilon\to 0}\int_{\Lambda_\varepsilon} A^\varepsilon\,\nabla u_\varepsilon\nabla u_\varepsilon\,dx=0.
$$

Then, convergence (3.29)(ii) follows from the α-coerciveness of A^ε. □

The convergence of the energy is the main tool used to obtain corrector results which are described in subsection 3.1.6 below.

3.1.2 Homogenization of a Neumann problem

Let Ω be a bounded domain in $\mathbb{R}^N$ on which we consider a pure Neumann problem (i.e., without zero order term) which is traditionally associated with the Fredholm alternative.

For f_ε given in $L^2(\Omega)$, consider the homogeneous Neumann problem (3.6)

$$
\begin{cases}
-\mathrm{div}\,(A^\varepsilon\nabla u_\varepsilon)=f_\varepsilon & \text{in }\Omega,\\
A^\varepsilon\nabla u_\varepsilon\cdot n=0 & \text{on }\partial\Omega,\\
\displaystyle\int_\Omega u_\varepsilon\,dx=0,
\end{cases}
$$

where the sequence of matrix fields $\{A^\varepsilon\}_\varepsilon$ in $\mathcal{M}(\alpha,\beta,\Omega)$.

The Fredholm alternative implies that this problem has a unique solution (up to a constant), provided Ω is bounded, connected with Lipschitz boundary and *if and only if* $\mathcal{M}_\Omega(f_\varepsilon)=0$ (the Fredholm condition).[5]

[5] More generally, Ω can be multiply connected provided f is with zero average on each of its connected components.

Without this condition, an alternative is to consider the Hilbert space

$$V(\Omega) \doteq \{w \in H^1(\Omega) \mid \mathcal{M}_\Omega(w) = 0\}, \tag{3.31}$$

endowed with the norm $\|w\|_{V(\Omega)} \doteq \|\nabla w\|_{L^2(\Omega)}$ (which is equivalent to the $H^1(\Omega)$ norm by the Poincaré-Wirtinger inequality on Ω for $p = 2$; cf. Section 2.3.1). Then, by the Lax-Milgram theorem, the corresponding variational formulation

$$\int_\Omega A^\varepsilon \nabla u_\varepsilon \nabla\varphi \, dx = \int_\Omega f_\varepsilon \varphi \, dx, \quad \forall \varphi \in V(\Omega), \tag{3.32}$$

has a unique solution irrespective of the Fredholm condition ! But the strong formulation is actually

$$\begin{cases} -\mathrm{div}\,(A^\varepsilon \nabla u) = f_\varepsilon - \mathcal{M}_\Omega(f_\varepsilon) & \text{in } \Omega, \\ A^\varepsilon \nabla u \cdot n = 0 & \text{on } \partial\Omega. \end{cases}$$

The homogenization result for problem (3.32) is stated below. We skip its proof since it is entirely similar to that of the Dirichlet case.(6) Note that the correctors are the same as in the Dirichlet case.

Theorem 3.12. *Let Ω be a bounded domain in $\mathbb{R}^N$ with a Lipschitz boundary and u_ε be the solution of problem (3.32). Assume that f_ε converges weakly to f in $L^2(\Omega)$. Then there exists $u_0' \in V(\Omega)$ and $\widehat{u}' \in L^2(\Omega; H^1_{per,0}(Y))$ such that*

$$\begin{cases} u_\varepsilon \rightharpoonup u_0' & \textit{weakly in } V(\Omega), \\ \mathcal{T}_\varepsilon(u_\varepsilon) \to u_0' & \textit{strongly in } L^2(\Omega; H^1(Y)), \\ \mathcal{T}_\varepsilon(\nabla u_\varepsilon) \to \nabla u_0' + \nabla_y \widehat{u}' & \textit{strongly in } L^2(\Omega \times Y)^N, \end{cases}$$

and the pair $(u_0', \widehat{u}')$ is the unique solution of the following problem: Find $(u_0', \widehat{u}')$ in $V(\Omega) \times L^2(\Omega; H^1_{per,0}(Y))$ satisfying

$$\begin{cases} \dfrac{1}{|Y|} \displaystyle\int_{\Omega \times Y} A(y)\big[\nabla u_0'(x) + \nabla_y \widehat{u}'(x,y)\big]\big[\nabla\Psi(x) + \nabla_y \Phi(x,y)\big] dx dy \\ \qquad\qquad\qquad\qquad = \displaystyle\int_\Omega f\,\Psi \, dx, \\ \forall \Psi \in H^1(\Omega), \ \ \forall \Phi \in L^2(\Omega;\, H^1_{per,0}(Y)). \end{cases}$$

The function $u_0' \in V(\Omega)$ is the unique solution of the homogenized problem

$$\int_\Omega A^{hom} \nabla u_0' \nabla\varphi \, dx = \int_\Omega f\,\varphi \, dx \quad \forall \varphi \in V(\Omega), \tag{3.33}$$

where the homogenized matrix A^{hom} is given by the same formulas (3.19).

(6) Equality (3.14) holds also for every $(v, \widehat{v})$ in $V(\Omega) \times L^2(\Omega; H^1_{per,0}(Y))$.

3.1.3 The case of frame-periodic coefficients

In this section we define frame-periodic (or approximate periodic) coefficients. They appear as "local perturbations" of periodic coefficients. We introduce a condition so they satisfy hypothesis (3.11). Recall that as in the case of one dimension, the functions $[\cdot]_Y$ and $\{\cdot\}_Y$ can be defined everywhere in $\mathbb{R}^N$ (instead of only almost everywhere).

In this section, $Diff_{Lip}(Y)$ denotes the group of Lipschitz continuous homeomorphisms of Y with Lipschitz inverse[(7)], and $Diff_{Lip}(\overline{Y}; \partial Y)$ the set of Lipschitz diffeomorphisms of $\overline{Y}$ leaving each point of the boundary ∂Y invariant.

Definition 3.13 (Convergence in the group $Diff_{Lip}(Y)$). *We say that a sequence in $\{\theta_\varepsilon\}_\varepsilon$ converge to an element θ_0 in $Diff_{Lip}(Y)$ if it converges to θ_0 in $C^{0,1}(Y)^N$ and its inverse sequence converges to θ_0^{-1} in $C^{0,1}(Y)^N$.*

This implies that the sequence of jacobians $\{Jac(\theta_\varepsilon)\}_\varepsilon$ converges for almost every $y \in Y$ to $Jac(\theta_0)$ with the same property for the sequence of the Jacobian of their inverses.

Definition 3.14 (Sequences of frame-periodic deformations). *Let $\mathfrak{M}_\varepsilon$ be the set of maps Θ in $L^\infty(\Omega; Diff_{Lip}(Y))$ such that,*[(8)]

(i) $\Theta(x, \cdot) = Id_Y$ *(the identity of Y) for a.e. x in Λ_ε,*

(ii) for every ξ in Ξ_ε, $\Theta(x, \cdot)$ is independent of x for a.e. x in $\varepsilon(\xi + Y)$.

To each element Θ in $\mathfrak{M}_\varepsilon$ is associated a map $\mathbf{Fpd}(\Theta)$ *defined by*

$$\mathbf{Fpd}(\Theta) \,:\, \Big\{x \in \Omega \;|\; \Big\{\frac{x}{\varepsilon}\Big\}_Y \in Y\Big\} \longmapsto \Big\{x \in \Omega \;|\; \Big\{\frac{x}{\varepsilon}\Big\}_Y \in Y\Big\},$$
$$\mathbf{Fpd}(\Theta)(x) \doteq \varepsilon\Big[\frac{x}{\varepsilon}\Big]_Y + \varepsilon\,\Theta\Big(x, \Big\{\frac{x}{\varepsilon}\Big\}_Y\Big),$$

which we call an **ε-frame-periodic deformation** *of Ω (the set where it is defined differs from Ω by a set of zero Lebesgue measure). If $\Theta(x, \cdot) \equiv Id_Y$ for a. e. $x \in \Omega$,* $\mathbf{Fpd}(\Theta)(x) = x$ *a.e. in Ω.*

Let $\mathfrak{S}$ denote the set of all sequences $\{\Theta_\varepsilon\}_\varepsilon$ of maps from Ω to $Diff_{Lip}(Y)$ such that for each ε, Θ_ε belongs to $\mathfrak{M}_\varepsilon$. A sequence of ε-frame-periodic deformations of Ω is therefore naturally associated with every element of $\mathfrak{S}$.

Note that an ε-frame-periodic deformation is not necessarily continuous across the boundaries of the ε-cells unless the corresponding Θ's all have their image in $Diff_{Lip}(\overline{Y}; \partial Y)$, in which case the deformation is a bi-Lipschitz diffeomorphism of Ω. Such global deformations will be used in Section 5.3.

---

(8) Note that every bi-Lipschitz diffeomorphism of Y is actually the restriction to Y of a bi-Lipschitz diffeomorphism of $\overline{Y}$. The definition could therefore be expressed in terms of diffeomorphisms of $\overline{Y}$.

Example. To a single map θ belonging to $\mathcal{C}(\overline{\Omega}; Diff_{Lip}(Y))$, one can associate a element of $\mathfrak{S}$ by setting

$$\Theta_\varepsilon(x,\,\cdot\,) = \begin{cases} \theta\Big(\varepsilon\Big[\dfrac{x}{\varepsilon}\Big]_Y,\,\cdot\Big) & \text{for } x \text{ in } \widehat{\Omega}_\varepsilon, \\ Id_Y & \text{for } x \text{ in } \Lambda_\varepsilon. \end{cases} \tag{3.34}$$

We now show that using the sequences $\{\Theta_\varepsilon\}_\varepsilon$, one can construct more examples of sequences of coefficients satisfying hypothesis (3.11).

Proposition 3.15. *Suppose the following properties are satisfied by the sequence $\{\Theta_\varepsilon\}_\varepsilon$ in $\mathfrak{S}$:*

(i) there exists a map Θ_0 in $L^\infty(\Omega; Diff_{Lip}(Y))$ such that for a.e. $x \in \Omega$, $\Theta_\varepsilon(x,\cdot)$ converges to $\Theta_0(x,\,\cdot\,)$ in $Diff_{Lip}(Y)$,

(ii) the sequences of functions $\{|Jac(\Theta_\varepsilon|\}_\varepsilon$ and $\{|Jac^{-1}(\Theta_\varepsilon)|\}_\varepsilon$ are bounded above by a fixed function in $L^1(\Omega \times Y)$.

Let A^ε be in $M(\alpha,\beta,\Omega)$. Assume that $\mathcal{T}_\varepsilon(A^\varepsilon)$ converges in measure to A^0 in $\Omega \times Y$. Consider the modified sequence $x \mapsto B^\varepsilon \doteq A^\varepsilon \circ \mathbf{Fpd}(\Theta_\varepsilon)$. Then $\mathcal{T}_\varepsilon(B^\varepsilon)$ converges in measure to $(x,y) \mapsto B(x,y) \doteq A^0\big(x, \Theta_0(x,y)\big)$.

Proof. Reasoning component-wise reduces the proof to the scalar case, so that without changing notations, we assume A^ε to be a scalar valued function. Since Θ_ε is constant with respect to x on each ε-cell, the unfolding of B^ε is given by

$$\mathcal{T}_\varepsilon(B^\varepsilon)(x,y) = A^\varepsilon\Big(\varepsilon\Big[\frac{x}{\varepsilon}\Big]_Y + \varepsilon\,\Theta_\varepsilon(x,y)\Big)\mathbf{1}_{\widehat{\Omega}_\varepsilon}.$$

We will show first that it converges weakly to B in $L^2(\Omega \times Y)$, then that the convergence is actually strong. This will imply the claimed convergence in measure.

Let φ be a smooth function on $\Omega \times Y$ with compact support. Performing the bi-Lipschitz change of variable $\Theta_\varepsilon(x,\cdot)$ on each ε-cell, and denoting $\Theta_\varepsilon^{-1}(x,\cdot)$ the inverse map of $\Theta_\varepsilon(x,\cdot)$ in the space $Diff_{Lip}(Y)$, we get

$$\begin{aligned}
&\int_{\Omega\times Y} A^\varepsilon\Big(\varepsilon\Big[\frac{x}{\varepsilon}\Big]_Y + \varepsilon\,\Theta_\varepsilon(x,y)\Big)\mathbf{1}_{\widehat{\Omega}_\varepsilon}(x)\,\varphi(x,y)dxdy \\
&= \sum_{\xi\in\Xi_\varepsilon}\int_{\varepsilon(\xi+Y)\times Y} A^\varepsilon\Big(\varepsilon\xi + \varepsilon\,\Theta_\varepsilon(\varepsilon\xi,y)\Big)\,\varphi(x,y)dxdy \\
&= \sum_{\xi\in\Xi_\varepsilon}\int_{\varepsilon(\xi+Y)\times Y} A^\varepsilon\Big(\varepsilon\xi + \varepsilon y'\Big)\,\varphi(x,\Theta_\varepsilon^{-1}(\varepsilon\xi,y'))\big|Jac(\Theta_\varepsilon^{-1})(\varepsilon\xi,y')\big|dxdy' \\
&= \sum_{\xi\in\Xi_\varepsilon}\int_{\varepsilon(\xi+Y)\times Y} A^\varepsilon\Big(\varepsilon\Big[\frac{x}{\varepsilon}\Big]_Y + \varepsilon y'\Big)\,\varphi(x,\Theta_\varepsilon^{-1}(x,y'))\big|Jac(\Theta_\varepsilon^{-1})(x,y')\big|dxdy' \\
&= \int_{\Omega\times Y} \mathcal{T}_\varepsilon(A^\varepsilon)(x,y')\,\varphi(x,\Theta_\varepsilon^{-1}(x,y'))\big|Jac(\Theta_\varepsilon^{-1})(x,y')\big|\mathbf{1}_{\widehat{\Omega}_\varepsilon}(x)dxdy'.
\end{aligned}$$

Under the hypotheses, $\{\varphi(x, \Theta_\varepsilon^{-1}(x, y'))\}_\varepsilon$ converges to $\varphi(x, \Theta_0^{-1}(x, y'))$, $\{Jac(\Theta_\varepsilon^{-1})(x, y')\}_\varepsilon$ converges to $Jac(\Theta_0^{-1})(x, y')$ whereas $\{\mathcal{T}_\varepsilon(A^\varepsilon)(x, y')\}_\varepsilon$ converges to $A^0(x, y')$ a.e. in $\Omega \times Y$. The integrand is bounded by a fixed $L^1(\Omega \times Y)$ function, so by the Lebesgue theorem,

$$\lim_{\varepsilon\to 0}\int_{\Omega\times Y} \mathcal{T}_\varepsilon(B^\varepsilon)(x, y)\,\varphi(x, y)dxdy = \int_{\Omega\times Y} A^0(x, y')\varphi(x, \Theta_0^{-1}(x, y'))\big|Jac(\Theta_0^{-1})(x, y')\big|dxdy'.$$

Performing the change of variable Θ_0^{-1} with respect to the variable y' yields

$$\lim_{\varepsilon\to 0}\int_{\Omega\times Y} \mathcal{T}_\varepsilon(B^\varepsilon)(x, y)\,\varphi(x, y)dxdy = \int_{\Omega\times Y} A^0(x, \Theta_0(x, y))\,\varphi(x, y)dxdy,$$

which gives the claimed weak convergence in every $L^p(\Omega\times Y)$ for p in $(1, +\infty)$, and in the weak-$*$ topology of $L^\infty(\Omega \times Y)$. A similar computation shows that the $L^2(\Omega \times Y)$-norm of $\mathcal{T}_\varepsilon(B^\varepsilon)$ converges to that of its weak limit $A^0(x, \Theta_0(x, y))$, whence the strong convergence in L^2 and the convergence in measure. □

Remark 3.16. *In the case of the sequence generated by a single map as in* (3.34)*, the hypothesis of Proposition 3.15 is satisfied with limit* θ.

Example. Let A be an element in $M(\alpha, \beta, Y)$ extended by Y-periodicity and set $A^\varepsilon(x) \doteq A(\dfrac{x}{\varepsilon})$. In this case, given an element $\{\Theta_\varepsilon\}_\varepsilon$ of $\mathfrak{S}$, the previous construction reduces to $B^\varepsilon(x) \doteq A \circ \Theta_\varepsilon(x, \{\frac{x}{\varepsilon}\}_Y)$.

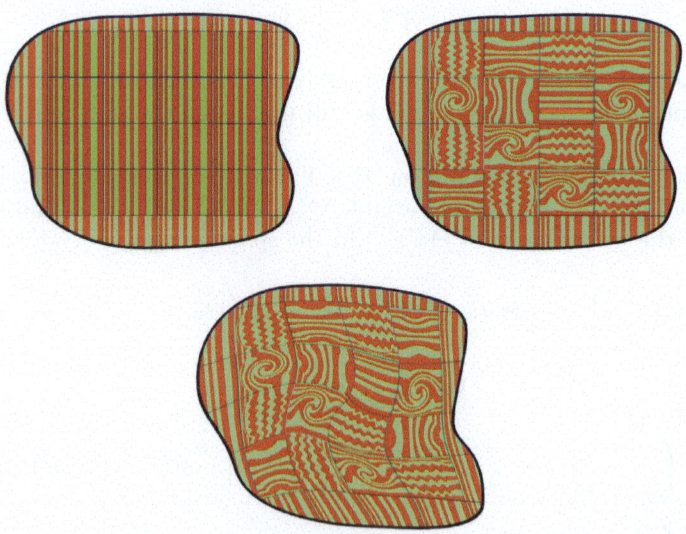

Figure 3.1: A two-component layered material under a frame-periodic deformation and a global bi-Lipschitz transformation

Note that this sequence has an εY- periodic frame, but varies "slowly" from ε-cell to neighboring ε-cells (see Figure 3.1 for an example). Under the hypothesis of Proposition 3.15 for Θ_ε, it follows that $\mathcal{T}_\varepsilon(B^\varepsilon)$ converges in measure to $A \circ \Theta_0$ in $\Omega \times Y$.

Remark 3.17. *For variational problems of second order, one can furthermore apply a fixed global bi-Lipschitz diffeomorphism to obtain a situation without apparent periodicity. For higher order problems, the diffeomorphism has to be more regular.*

3.1.4 Homogenization of multiscale diffusion

The periodic unfolding method turns out to be particularly well-adapted to multi-scales problems as it can be iterated. A simple example of a problem with two different small scales is presented here. This a particular case of the use of unfolding with parameters discussed in Section 1.5. Consider two periodicity cells Y and Z (parallelotopes each associated to its set of periods). Suppose that Y is "partitioned" in two non-empty disjoint open subsets Y_1 and Y_2, i.e. such that $Y_1 \cap Y_2 = \varnothing$ and $\overline{Y} = \overline{Y}_1 \cup \overline{Y}_2$. Let $A^{\varepsilon,\delta}$ be a matrix field defined by

$$A^{\varepsilon,\delta}(x) = \begin{cases} A_1\Big(\Big\{\dfrac{x}{\varepsilon}\Big\}_Y\Big) & \text{for } \Big\{\dfrac{x}{\varepsilon}\Big\}_Y \in Y_1 \\ A_2\Bigg(\Bigg\{\dfrac{\{\frac{x}{\varepsilon}\}_Y}{\delta}\Bigg\}_Z\Bigg) & \text{for } \Big\{\dfrac{x}{\varepsilon}\Big\}_Y \in Y_2, \end{cases}$$

where A_1 is in $M(\alpha, \beta, Y_1)$ and A_2 in $M(\alpha, \beta, Z)$ (cf. Definition 3.1).

Here, there are two small scales, namely ε and $\varepsilon\delta$, associated respectively to the cells Y and Z (see Figure 3.2).

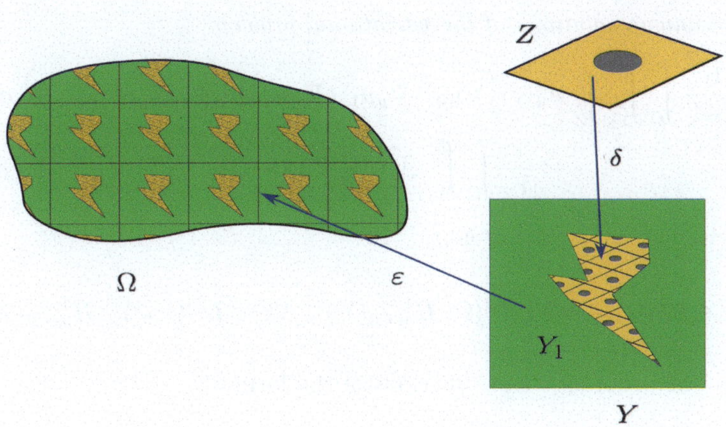

Figure 3.2: A domain with two small periodic scales ε and $\varepsilon\delta$.

Consider the problem

$$\begin{cases} \displaystyle\int_\Omega A^{\varepsilon,\delta}\nabla u_{\varepsilon,\delta}\nabla w\, dx = \int_\Omega f\, w\, dx, \\ \forall w \in H_0^1(\Omega), \end{cases}$$

with f in $L^2(\Omega)$. As before, the Lax-Milgram theorem gives the existence and uniqueness of $u_{\varepsilon,\delta}$ in $H_0^1(\Omega)$ satisfying the estimate

$$\|u_{\varepsilon,\delta}\|_{H_0^1(\Omega)} \leqslant \frac{1}{\alpha}\|f\|_{L^2(\Omega)}.$$

So, by Proposition 1.52, for any sequence $\varepsilon(\delta)$ going to 0 with δ (for simplicity we will suppress the dependence of ε upon δ in the formulas), there is a subsequence extracted from δ and some $u_0 \in H_0^1(\Omega)$, $u_1 \in L^2(\Omega; H^1_{per,0}(Y))$ and $u_2 \in L^2(\Omega \times Y; H^1_{per,0}(Z))$ such that

$$\begin{aligned} & u_{\varepsilon,\delta} \rightharpoonup u_0 \quad \text{weakly in } H_0^1(\Omega) \\ & \mathcal{T}_\varepsilon^Y(u_{\varepsilon,\delta}) \rightharpoonup u_0 \quad \text{weakly in } L^2(\Omega;\, H^1(Y)), \\ & \mathcal{T}_\delta^Z \circ \mathcal{T}_\varepsilon^Y(u_{\varepsilon,\delta}) \rightharpoonup u_0 \quad \text{weakly in } L^2(\Omega \times Y;\, H^1(Z)), \\ & \mathcal{T}_\varepsilon^Y\big(\nabla u_{\varepsilon,\delta}\big) \rightharpoonup \nabla u_0 + \nabla_y u_1 \quad \text{in } L^2(\Omega \times Y)^N, \\ & \mathcal{T}_\delta^Z \circ \mathcal{T}_\varepsilon^Y\big(\nabla u_{\varepsilon,\delta}\big) \rightharpoonup \nabla u_0 + \nabla_y u_1 + \nabla_z u_2 \quad \text{in } L^2(\Omega \times Y \times Z)^N. \end{aligned} \tag{3.35}$$

The limit functions u_0, u_1 and u_2 are characterized in the next result.

Theorem 3.18. *Under the assumptions above,*

$$u_0 \in H_0^1(\Omega), \quad u_1 \in L^2(\Omega; H^1_{per,0}(Y)), \quad u_2 \in L^2(\Omega \times Y_2; H^1_{per,0}(Z))$$

are the unique solutions of the variational problem

$$\begin{cases} \displaystyle\frac{1}{|Y||Z|}\int_\Omega\int_{Y_2}\int_Z A_2(z)\Big\{\nabla u_0 + \nabla_y u_1 + \nabla_z u_2\Big\}\Big\{\nabla\Psi + \nabla_y\Phi + \nabla_z\Theta\Big\}\, dxdydz \\ \displaystyle\qquad + \frac{1}{|Y|}\int_\Omega\int_{Y_1} A_1(y)\Big\{\nabla u_0 + \nabla_y u_1\Big\}\Big\{\nabla\Psi + \nabla_y\Phi\Big\}\, dxdy \\ \displaystyle\qquad = \int_\Omega f\,\Psi\, dx, \\ \forall\Psi \in H_0^1(\Omega), \quad \forall\Phi \in L^2(\Omega;\, H^1_{per,0}(Y)), \ \forall\Theta \in L^2(\Omega \times Y_2; H^1_{per,0}(Z)). \end{cases}$$

Proof. The proof uses test functions of the form

$$\Psi(x) + \varepsilon\Phi\Big(x, \frac{x}{\varepsilon}\Big) + \varepsilon\delta\ \Theta\Big(x, \frac{x}{\varepsilon}, \frac{1}{\varepsilon\delta}\, x\Big),$$

where Ψ is in $\mathcal{D}(\Omega)$, Φ in $\mathcal{D}(\Omega; H^1_{per,0}(Y))$ and Θ in $\mathcal{D}(\Omega \times Y_2; H^1_{per,0}(Z))$.

Unfolding with the operators from (3.35) and passing to the limit, gives

$$\frac{1}{|Y||Z|}\int_\Omega\int_{Y_1}\int_Z A_1(y)\Big\{\nabla u_0+\nabla_y u_1+\nabla_z u_2\Big\}\Big\{\nabla\Psi+\nabla_y\Phi+\nabla_z\Theta\Big\}\,dx\,dy\,dz$$
$$+\frac{1}{|Y||Z|}\int_\Omega\int_{Y_2}\int_Z A_2(z)\Big\{\nabla u_0+\nabla_y u_1+\nabla_z u_2\Big\}\Big\{\nabla\Psi+\nabla_y\Phi+\nabla_z\Theta\Big\}\,dx\,dy\,dz$$
$$=\int_\Omega f\,\Psi\,dx.$$

By density, this holds for all Ψ in $H_0^1(\Omega)$, Φ in $L^2(\Omega;\,H^1_{per,0}(Y))$ and Θ in $L^2(\Omega\times Y_2;H^1_{per,0}(Z))$.
Choosing in particular $\Psi=0,\Phi=0$, and $\Theta=u_2\mathbf{1}_{Y_1}$ implies that $\nabla_z u_2\equiv 0$ in $\Omega\times Y_1\times Z$.

Since the limit problem does not depend on the δ-sequence nor on the sequence $\varepsilon(\delta)$, this implies the convergence for the full $(\varepsilon,\delta)\to(0,0)$. □

Proposition 3.9 (convergence of the energy) and Corollary 3.10 extend without any difficulty to the multiscale case.

Proposition 3.19. *The convergence for the energy is given by*

$$\lim_{\varepsilon,\delta\to 0}\int_\Omega A^{\varepsilon,\delta}\nabla u_{\varepsilon,\delta}\nabla u_{\varepsilon,\delta}\,dx$$
$$=\frac{1}{|Y||Z|}\int_\Omega\int_{Y_2}\int_Z A_2(z)\Big[\nabla u_0+\nabla_y u_1+\nabla_z u_2\Big]\Big[\nabla u_0+\nabla_y u_1+\nabla_z u_2\Big]dx\,dy\,dz$$
$$+\frac{1}{|Y|}\int_\Omega\int_{Y_1} A_1(y)\Big[\nabla u_0+\nabla_y u_1\Big]\Big[\nabla u_0+\nabla_y u_1\Big]\,dx\,dy.$$

Corollary 3.20. *The following strong convergences hold true:*

$$\mathcal{T}^Y_\varepsilon(\nabla u_{\varepsilon,\delta})\rightharpoonup\nabla u_0+\nabla_y u_1\quad\textit{strongly in } L^2(\Omega\times Y_1)^N,$$
$$\mathcal{T}^Z_\delta\circ\mathcal{T}^Y_\varepsilon(\nabla u_{\varepsilon,\delta})\rightharpoonup\nabla u_0+\nabla_y u_1+\nabla_z u_2\quad\textit{strongly in } L^2(\Omega\times Y_2\times Z)^N.$$

Remark 3.21. *Theorem 3.18 and Proposition 3.9 can be extended to the case of any finite number of distinct scales, by a simple reiteration.*

Remark 3.22. *A corrector result, similar to that of Theorem 3.27 below, can be obtained in the case of finitely many scales.*

3.1.5 The case of weakly convergent right-hand side

In this subsection, we again consider Problem (3.3) with a right-hand side f_ε which converges weakly to f in $H^{-1}(\Omega)$. All the other hypotheses are the same as before. Of course, f_ε is bounded in $H^{-1}(\Omega)$, so that the solutions u_ε are bounded in $H_0^1(\Omega)$. We shall see however, that the weak convergence for the full sequence of solutions does not necessarily hold.

To understand this phenomenon, we have to digress into how to treat the unfolding of elements of $H^{-1}(\Omega)$.

Recall that due to the Poincaré inequality in $H_0^1(\Omega)$, every element f of $H^{-1}(\Omega)$ can be written uniquely in the form $f = -\Delta w$ with $w$in $H_0^1(\Omega)$ and $\|w\|_{H_0^1(\Omega)} = \|f\|_{H^{-1}(\Omega)}$. The weak convergence to w of a sequence $\{w_\varepsilon\}_\varepsilon$ in $H_0^1(\Omega)$ is equivalent to the weak convergence to Δw of the corresponding $f_\varepsilon \doteq \Delta w_\varepsilon$ in $H^{-1}(\Omega)$. Instead of detailing the unfolding for f_ε in $H^{-1}(\Omega)$, we will unfold the corresponding w_ε in $H_0^1(\Omega)$.

The question of convergence of the whole sequence u_ε is answered in the next theorem.

Theorem 3.23. *Consider the following problem:*

$$\begin{cases} -\mathrm{div}\,(A^\varepsilon \nabla u_\varepsilon) = f_\varepsilon & \text{in } \Omega, \\ u_\varepsilon = 0 & \text{on } \partial\Omega. \end{cases}$$

Assume that

$$f_\varepsilon \rightharpoonup f \quad \text{weakly in } H^{-1}(\Omega).$$

Let w_ε be defined as above as the solution in $H_0^1(\Omega)$ of

$$-\Delta w_\varepsilon = f_\varepsilon.$$

Thus, $w_\varepsilon \rightharpoonup w$ in $H_0^1(\Omega)$ where w is the solution in $H_0^1(\Omega)$ of $-\Delta w = f$. Assume furthermore that there exists a $\widehat{w}$ in $L^2(\Omega; H^1_{per,0}(Y))$ such that

$$\mathcal{T}_\varepsilon(\nabla w_\varepsilon) \rightharpoonup \nabla w + \nabla_y \widehat{w} \quad \text{weakly in} \quad L^2(\Omega \times Y)^{N\,(9)}. \tag{3.36}$$

Then the sequence $\{u_\varepsilon\}_\varepsilon$ converges to the solution u of the problem

$$\begin{cases} \displaystyle\int_\Omega \big(A^{hom}(x)\nabla u_0(x) + \mathcal{M}_Y\big(B(x,\cdot)\nabla_y \widehat{\chi}_0(x,\cdot)\big)\big)\nabla v(x)\,dx \\ \qquad\qquad = \langle f, v\rangle_{H^{-1}(\Omega),H_0^1(\Omega)}, \\ \forall v \in H_0^1(\Omega). \end{cases} \tag{3.37}$$

In this formulation, A^{hom} is the same as in Proposition 3.7, while $\widehat{\chi}_0$ is the solution in $L^2(\Omega; H^1_{per,0}(Y))$ of the following problem for a.e x in Ω :

$$\begin{cases} \displaystyle\int_Y B(x,y)\nabla_y \widehat{\chi}_0(x,y)\,\nabla_y \psi(y)\,dy = \int_Y \nabla_y \widehat{w}(x,y)\nabla_y \psi(y)\,dy \\ \forall \psi \in H^1_{per,0}(Y). \end{cases} \tag{3.38}$$

(9) This hypothesis is not generally satisfied for the full sequence but holds up to a subsequence by Theorem 1.41.

Proof. The proof follows along the lines of that of Theorem 3.4, up to formula (3.18). The change arises when using the test function

$$v^{\varepsilon}(x) = \varepsilon\Psi(x)\psi\Big(\frac{x}{\varepsilon}\Big), \quad \Psi \in \mathcal{C}_c^{\infty}(\Omega) \text{ and } \psi \in H^1_{per,0}(Y).$$

Indeed, going to the limit in the left-hand side of the variational formulation of (3.37) remains unchanged. But in order to go to the limit in its right-hand side, more computations have to be performed. First, one uses unfolding (for ε small enough so that Λ_ε does not intersect the support of Ψ) to compute the right-hand side as follows:

$$\begin{aligned}\langle f_\varepsilon, v^\varepsilon\rangle_{H^{-1}(\Omega),H_0^1(\Omega)} &= \int_\Omega \nabla w_\varepsilon \nabla v^\varepsilon \, dx\\ &= \frac{1}{|Y|}\int_{\Omega\times Y} \mathcal{T}_\varepsilon(\nabla w_\varepsilon)\mathcal{T}_\varepsilon(\nabla v^\varepsilon)\, dxdy.\end{aligned}$$

The last formula passes to the limit by (3.36) and the fact (already used to pass to the limit in the left-hand side) that $\mathcal{T}_\varepsilon(\nabla v^\varepsilon)$ converges strongly to $\Psi(x)\nabla_y\psi(y)$, gives

$$\begin{aligned}&\frac{1}{|Y|}\int_{\Omega\times Y} B(x,y)\big[\nabla u_0(x) + \nabla_y\widehat{u}(x,y)\big]\Psi(x)\nabla_y\psi(y)\, dxdy\\ &\qquad = \frac{1}{|Y|}\int_{\Omega\times Y}\Big(\nabla w(x) + \nabla_y\widehat{w}(x,y)\Big)\nabla_y\psi(y)\Psi(x)\, dxdy,\end{aligned}$$

which replaces (3.17) in this setting.

Now, by the same density argument, the limit unfolded formulation becomes: the pair $(u_0, \widehat{u}) \in H_0^1(\Omega) \times L^2(\Omega; H^1_{per,0}(Y))$ is the unique solution of the problem

$$\begin{cases}\dfrac{1}{|Y|}\displaystyle\int_{\Omega\times Y} B(x,y)\big[\nabla u_0(x) + \nabla_y\widehat{u}(x,y)\big]\big[\nabla\Psi(x) + \nabla_y\Phi(x,y)\big]dxdy\\ \quad = \langle f, \Psi\rangle_{H^{-1}(\Omega),H_0^1(\Omega)} + \dfrac{1}{|Y|}\displaystyle\int_{\Omega\times Y}(\nabla w(x) + \nabla_y\widehat{w}(x,y))\nabla_y\Phi(x,y)\, dxdy,\\ \forall\Psi \in H_0^1(\Omega),\ \forall\Phi \in L^2(\Omega; H^1_{per,0}(Y)).\end{cases}$$

To obtain the corresponding homogenized formulation, one expresses $\widehat{u}$ as in (3.21), but with an extra term as follows:

$$\widehat{u} = \widehat{\chi}_0 + \sum_{i=1}^N \frac{\partial u_0}{\partial x_i}\widehat{\chi}_i,$$

where the $\widehat{\chi}_i$'s are, as before, given by (3.20), while $\widehat{\chi}_0$ is the solution for a.e. $x \in \Omega$, of problem (3.38). Lax-Milgram's theorem applies again. Finally, the variational formulation for the homogenized problem becomes (3.37). □

Remark 3.24. *A new term has appeared in the right-hand side. In the strong formulation, it is written as*

$$-\mathrm{div}\,\big(\mathcal{M}_Y\big(B(\cdot,y)\nabla_y\widehat{\chi}_0(\cdot,y)\big)\big).$$

It belongs to $H^{-1}(\Omega)$ and depends explicitly upon both B and $\widehat{w}$ but not on w, that is not on f. If $\widehat{\chi}_0 = 0$, this term disappears. By (3.38) this is equivalent to $\widehat{w} \equiv 0$. There are no other obvious sufficient conditions for it to vanish.

As a consequence, the weak convergence of $\{f_\varepsilon\}_\varepsilon$ in $H^{-1}(\Omega)$ alone does not imply in general, the weak convergence of the whole sequence $\{u_\varepsilon\}_\varepsilon$ in $H^1_0(\Omega)$. This fact is made very clear by the use of the unfolding method.

Remark 3.25. *If in Theorem 3.23 one makes the extra assumption*[10]*:*

$$\mathcal{T}_\varepsilon(\nabla w_\varepsilon) \to \nabla w + \nabla_y\widehat{w} \quad \textit{strongly in } \; L^2(\Omega\times Y)^N \textit{and} \; \int_{\Lambda_\varepsilon} |\nabla w_\varepsilon|^2 dx \to 0,$$

one can show that

$$\lim_{\varepsilon\to 0}\langle f_\varepsilon, u_\varepsilon\rangle_{H^{-1},H^1_0} = \langle f, u_0\rangle_{H^{-1},H^1_0} + \frac{1}{|Y|}\int_{\Omega\times Y} \nabla_y\widehat{w}(x,y)\nabla_y\widehat{u}(x,y)\,dxdy,$$

which implies that the convergence of the energy (Proposition 3.9) still holds true. Therefore Corollary 3.10, as well as the corresponding parts of Theorem 3.27, hold true also. This is a case where a corrector result can be obtained despite the weak convergence of the right-hand side.

Example 3.26. *Suppose that $\{f_\varepsilon\}_\varepsilon$ is a sequence in $L^2(\Omega)$ such that $\{\varepsilon f_\varepsilon\}_\varepsilon$ and $\mathcal{M}_\varepsilon(f_\varepsilon)$ are bounded in $L^2(\Omega)$. We claim that $\{f_\varepsilon\}_\varepsilon$ is bounded in $H^{-1}(\Omega)$. Indeed,*

$$\begin{aligned}\langle f_\varepsilon,\varphi\rangle_{H^{-1}(\Omega),H^1_0(\Omega)} &= \int_\Omega f_\varepsilon(x)\varphi(x)\,dx\\ &= \int_{\widehat{\Omega}_\varepsilon} f_\varepsilon(x)\varphi(x)\,dx + \int_{\Lambda_\varepsilon}(\varepsilon f_\varepsilon)(x)(\varepsilon^{-1}\varphi)(x)\,dx.\end{aligned}$$

The last integral is bounded by $C\|\varepsilon f_\varepsilon\|_{L^2(\Omega)}\|\nabla\varphi\|_{L^2(\Omega^{bl}_{\varepsilon d(Y)})}$ thanks to the Poincaré inequality applied in $\Omega^{bl}_{\varepsilon d(Y)}$ (which contains Λ_ε). Consequently it converges to 0.

The first integral in the right-and side is computed as follows:

$$\begin{aligned}\int_{\widehat{\Omega}_\varepsilon} f_\varepsilon(x)\varphi(x)\,dx &= \int_{\widehat{\Omega}_\varepsilon} f_\varepsilon(x)\big(\varphi(x) - \mathcal{M}_\varepsilon(\varphi)(x)\big)\,dx + \int_\Omega \mathcal{M}_\varepsilon(f_\varepsilon)(x)\;\varphi(x)\,dx\\ &= \int_{\widehat{\Omega}_\varepsilon} [\varepsilon f_\varepsilon(x)\varepsilon^{-1}\big(\varphi(x) - \mathcal{M}_\varepsilon(\varphi)(x)\big)\,dx + \int_\Omega \mathcal{M}_\varepsilon(f_\varepsilon)(x)\;\varphi(x)\,dx.\end{aligned}$$

[10] In view of Proposition 1.29(ii), this is equivalent to $\nabla w_\varepsilon - \nabla w - \mathcal{U}_\varepsilon(\nabla_y\widehat{w}) \to 0$ strongly in $L^2(\Omega)$.

By $(1.39)_1$, *it is bounded by* $C\|\varepsilon f_\varepsilon\|_{L^2(\Omega)}\|\nabla\varphi\|_{L^2(\Omega)} + \|\mathcal{M}_\varepsilon(f_\varepsilon)\|_{L^2(\Omega)}\|\varphi\|_{L^2(\Omega)}$. *So,* f_ε *is bounded in* $H^{-1}(\Omega)$ *by* $C\|\varepsilon f_\varepsilon\|_{L^2(\Omega)} + \|\mathcal{M}_\varepsilon(f_\varepsilon)\|_{L^2(\Omega)}$.

Now, assuming that $\mathcal{M}_\varepsilon(f_\varepsilon) \rightharpoonup F$ *in* $L^2(\Omega)$ *and* $\mathcal{T}_\varepsilon(\varepsilon f_\varepsilon) \rightharpoonup \widehat{f}$ *in* $L^2(\Omega\times Y)$, *unfolding the previous equality implies that* $f_\varepsilon \rightharpoonup f$ *in* $H^{-1}(\Omega)$ *where*

$$\langle f, \varphi\rangle_{H^{-1}(\Omega),H_0^1(\Omega)} = \int_\Omega \Big(\mathcal{M}_Y(\widehat{f}(x,y)\, y^c)\cdot\nabla\varphi(x) + F(x)\,\varphi(x)\Big)\,dx.$$

Up to a subsequence,

$$\mathcal{T}_\varepsilon(\nabla w_\varepsilon) \rightharpoonup \nabla w + \nabla_y\widehat{w} \quad \textit{in } L^2(\Omega\times Y)^N.$$

Theorem 3.23 now applies.

The equality $f_\varepsilon = -\Delta w_\varepsilon$ *implies that*

$$\operatorname{div}_y \mathcal{T}_\varepsilon(\nabla w_\varepsilon) = \varepsilon\mathcal{T}_\varepsilon(-\Delta w_\varepsilon) = \mathcal{T}_\varepsilon(\varepsilon f_\varepsilon),$$

which at the limit, gives

$$\mathcal{T}_\varepsilon(\varepsilon f_\varepsilon) \rightharpoonup \widehat{f} \doteq -\operatorname{div}_y(\nabla w + \nabla_y\widehat{w}) = -\Delta_y\widehat{w} \textit{ weakly in } L^2(\Omega\times Y).$$

Hence $\widehat{w}$ *belongs to* $L^2(\Omega; H^1_{per,0}(Y)\cap H^2(Y))$ *and*

$$-\Delta w = f = F - \operatorname{div}\left(\mathcal{M}_Y(\widehat{f}(\cdot,y)\, y^c\right) = F + \operatorname{div}\left(\mathcal{M}_Y(\Delta_y\widehat{w}(\cdot,y)\, y^c\right),$$

in $H^{-1}(\Omega)$. *This is an additional information relating* Δw *and* $\Delta_y\widehat{w}$.

3.1.6 A general corrector result

We return to the setting of the beginning of this section. Under additional regularity assumptions on the homogenized solution u_0 and the cell-functions $\widehat{\chi}_j$, the strong convergence for the gradient of u_0 with a corrector is known (cf. [21, Chapter 1, Section 5], [70, Chapter 8, Section 3], and references therein). More precisely, suppose that $\nabla_y\widehat{\chi}_j \in L^r(Y)^N$, $j = 1,\dots,N$ and $\nabla u^0 \in L^s(\Omega)$ with $1 \leqslant r, s < +\infty$ and such that $1/r + 1/s = 1/2$. Then

$$\nabla u_\varepsilon - \nabla u_0 - \sum_{j=1}^N \frac{\partial u_0}{\partial x_j}\,\nabla_y\widehat{\chi}_j\Big(\frac{\cdot}{\varepsilon}\Big) \to 0 \quad \text{strongly in } L^2(\Omega)^N.$$

However, making use of the unfolding operators, one can obtain a general corrector result **without any regularity assumption** on $\widehat{\chi}_j$ and with a very short proof.

Theorem 3.27. *Under the hypotheses of Theorem 3.5, one has*

$$\nabla u_\varepsilon - \nabla u_0 - \sum_{i=1}^N \mathcal{M}_\varepsilon\Big(\frac{\partial u_0}{\partial x_i}\Big)\,\mathcal{U}_\varepsilon(\nabla_y\widehat{\chi}_i) \to 0 \quad \textit{strongly in } L^2(\Omega)^N. \tag{3.39}$$

If $A^\varepsilon(x) = A\Big(\dfrac{x}{\varepsilon}\Big)$, *then* $u_0 + \varepsilon \sum_{i=1}^N \mathcal{Q}_\varepsilon\Big(\dfrac{\partial u_0}{\partial x_i}\Big)\widehat{\chi}_i\Big(\dfrac{\cdot}{\varepsilon}\Big)$ *belongs to* $H^1(\Omega)$, *and*

$$u_\varepsilon - u_0 - \varepsilon \sum_{i=1}^N \mathcal{Q}_\varepsilon\Big(\frac{\partial u_0}{\partial x_i}\Big)\widehat{\chi}_i\Big(\frac{\cdot}{\varepsilon}\Big) \to 0 \quad \textit{strongly in } H^1(\Omega). \tag{3.40}$$

Proof. From (3.29) and Proposition 1.29(ii), one immediately has

$$\nabla u_\varepsilon - \mathcal{U}_\varepsilon(\nabla u_0) - \mathcal{U}_\varepsilon(\nabla_y \widehat{u}) \to 0 \quad \text{strongly in } L^2(\Omega)^N.$$

Since ∇u_0 belongs to $L^2(\Omega)^N$, by (1.24)

$$\mathcal{U}_\varepsilon(\nabla u_0) \to \nabla u_0 \quad \text{strongly in } L^2(\Omega)^N.$$

On the other hand, thanks to (3.21),

$$\mathcal{U}_\varepsilon(\nabla_y \widehat{u}) = \sum_{i=1}^N \mathcal{U}_\varepsilon\Big(\frac{\partial u_0}{\partial x_i}\nabla_y \widehat{\chi}_i\Big).$$

We are now in position to apply Proposition 1.34 with

$$\alpha_i = \frac{\partial u_0}{\partial x_i} \quad \text{and} \quad \beta_i = \nabla_y \widehat{\chi}_i, \quad \forall i \in \{1, \dots, N\},$$

to get

$$\mathcal{U}_\varepsilon(\nabla_y \widehat{u}) - \sum_{i=1}^N \mathcal{U}_\varepsilon\Big(\frac{\partial u_0}{\partial x_i}\Big)\mathcal{U}_\varepsilon(\nabla_y \widehat{\chi}_i) \to 0 \quad \text{strongly in } L^2(\Omega)^N.$$

This together with (1.23) (applied to $\dfrac{\partial u_0}{\partial x_i}$) proves (3.39).

From Corollary 1.72 (applied three times!), it follows that

$$u_0 + \varepsilon \sum_{i=1}^N \mathcal{Q}_\varepsilon\Big(\frac{\partial u_0}{\partial x_i}\Big)\widehat{\chi}_i\Big(\frac{\cdot}{\varepsilon}\Big) \in H^1(\Omega).$$

Then, (3.21) implies

$$\begin{aligned}
\nabla u_0 + \mathcal{U}_\varepsilon(\nabla_y \widehat{u}) - \nabla\Big[u_0 + \varepsilon \sum_{i=1}^N \mathcal{Q}_\varepsilon\Big(\frac{\partial u_0}{\partial x_i}\Big)\widehat{\chi}_i\Big(\frac{\cdot}{\varepsilon}\Big)\Big]& \\
= - \sum_{i=1}^N \Big[\mathcal{Q}_\varepsilon\Big(\frac{\partial u_0}{\partial x_i}\Big) - \mathcal{M}_\varepsilon\Big(\frac{\partial u_0}{\partial x_i}\Big)\Big]\nabla_y \widehat{\chi}_i\Big(\frac{\cdot}{\varepsilon}\Big)& \\
- \sum_{i=1}^N \varepsilon \nabla\Big[\mathcal{Q}_\varepsilon\Big(\frac{\partial u_0}{\partial x_i}\Big)\Big]\widehat{\chi}_i\Big(\frac{\cdot}{\varepsilon}\Big).&
\end{aligned}$$

Using Corollary 1.72, Proposition 1.25(i) and Corollary 1.77, one immediately gets the strong convergence to 0 in $L^2(\Omega)$ of the right-hand side in the above equality. Together with (3.39), this implies (3.40). □

The next result concerns some estimates for the solutions of cell-problems of the type (3.20) defining the correctors $\widehat{\chi}_i$ $(i = 1, \dots, N)$. It will be used below in order to improve convergence (3.40) (see Corollary 3.31). It is also a key ingredient in the proof of Theorem 14.3 giving error estimates which show the speed of convergences (3.40) with respect to ε, as well as in the proof of Corollary 3.31.

For $\delta > 0$, set (see also Subsection 12.1)

$$\Omega_\delta^{bl} \doteq \{x \in \Omega \mid \rho(x) < \delta\}.$$

Recall that for every $\phi \in H^1(\Omega)$ and $\delta > 0$,

$$\|\phi\|_{L^2(\Omega_\delta^{bl})} \leqslant C\big(\delta^{1/2}\|\phi\|_{L^2(\partial\Omega)} + \delta\|\nabla\phi\|_{L^2(\Omega)}\big). \tag{3.41}$$

The constant does not depend on δ (in Proposition 12.2 we detail several estimates of this type).

Lemma 3.28. *Let d and f be in elements of $L^2(Y)^N$ and $L^2(Y)$, respectively, extended by Y-periodicity to the whole of $\mathbb{R}^N$. Suppose they satisfy for every φ in $H^1_{per}(Y)$*

$$\int_Y d(y) \cdot \nabla_y \varphi(y)\, dy = \int_Y f(y)\, \varphi(y)\, dy.^{(11)}$$

For $\varepsilon > 0$, let

$$d_\varepsilon \doteq d\Big(\frac{\cdot}{\varepsilon}\Big) \quad \textit{and} \quad f_\varepsilon \doteq f\Big(\frac{\cdot}{\varepsilon}\Big).$$

Then,

$$-\operatorname{div} d_\varepsilon = \frac{1}{\varepsilon} f_\varepsilon \quad \textit{in} \ \ \mathcal{D}'(\mathbb{R}^N).$$

Proof. Let ϕ be in $\mathcal{D}(\mathbb{R}^N)$. Using the unfolding identity, one obtains

$$\begin{aligned}
&\int_{\mathbb{R}^N} \big(\varepsilon\, d_\varepsilon(x) \cdot \nabla\phi(x) + f_\varepsilon(x)\phi(x)\big)\, dx \\
&= \frac{1}{|Y|} \int_{\mathbb{R}^N \times Y} \Big(d(y) \cdot \nabla_y \widetilde{\mathcal{T}}_\varepsilon(\phi)(x,y) + f(y)\widetilde{\mathcal{T}}_\varepsilon(\phi)(x,y)\Big)\, dxdy \\
&= \frac{1}{|Y|} \int_Y \Big[d(y) \cdot \nabla_y\Big(\int_{\mathbb{R}^N} \widetilde{\mathcal{T}}_\varepsilon(\phi)(x,y) dx\Big) + f(y) \int_{\mathbb{R}^N} \widetilde{\mathcal{T}}_\varepsilon(\phi)(x,y) dx\Big] dy.
\end{aligned}$$

The last integral vanishes since

$$\int_{\mathbb{R}^N} \mathcal{T}_\varepsilon(\phi)(x,y) dx = \varepsilon^N \sum_{\xi \in \mathbf{G}} \phi(\varepsilon\xi + \varepsilon y) \in H^1_{per}(Y),$$

the summation being actually finite. □

(11) By Green's formula, this equivalent to the following strong formulation:

$$\begin{cases} -\text{div } d = f \ \ \textit{in} \ \ \mathcal{D}'(Y), \\ \text{the normal trace } d \cdot n \text{ belongs to } H^{-1/2}(\partial Y) \text{ and} \\ \text{it takes opposite values on the } N \text{ pairs of opposite faces of } Y. \end{cases}$$

Lemma 3.29. *Let A be in $M(\alpha,\beta,Y)$, F in $L^\infty(Y)^N$ and χ in $H^1_{per,0}(Y)$ the solution of*

$$\begin{cases} \displaystyle\int_Y (A(y)\nabla_y\chi(y) + F(y))\nabla_y\varphi(y)\,dy = 0, \\ \forall\varphi \in H^1_{per,0}(Y). \end{cases} \tag{3.42}$$

Then the following estimates hold:

$$\begin{aligned} &\|\chi\|_{L^2(Y)} \leqslant C\|F\|_{L^2(Y)}, \\ &\|\nabla\chi\|_{L^2(Y)} \leqslant C\|F\|_{L^2(Y)}, \\ &\|\chi\|_{L^\infty(Y)} \leqslant C\|F\|_{L^\infty(Y)}. \end{aligned} \tag{3.43}$$

Furthermore, for every w in $H^1(\mathbb{R}^N)$,

$$\begin{aligned} &\int_{\mathbb{R}^N} \Big|\nabla_y\chi\Big(\frac{x}{\varepsilon}\Big)\,w(x)\Big|^2 dx \\ &\qquad \leqslant C\|F\|^2_{L^\infty(Y)}\big(\|w\|^2_{L^2(\mathbb{R}^N)} + \varepsilon^2\|\nabla w\|^2_{L^2(\mathbb{R}^N)}\big). \end{aligned} \tag{3.44}$$

All the constants depend only upon Y, α and β and are independent of ε.

If Ω is a bounded domain with a Lipschitz boundary, then

$$\int_\Omega \Big|\nabla_y\chi\Big(\frac{x}{\varepsilon}\Big)w(x)\Big|^2 dx \leqslant C\|F\|^2_{L^\infty(Y)}\big(\|w\|^2_{L^2(\Omega)}+\varepsilon^2\|\nabla w\|^2_{L^2(\Omega)}\big) \tag{3.45}$$

and (recall that $d(Y)$ is the diameter of the cell Y)

$$\begin{aligned} &\int_{\Omega^{bl}_{\varepsilon d(Y)}} \Big|\nabla_y\chi\Big(\frac{x}{\varepsilon}\Big)w(x)\Big|^2 dx \\ &\qquad \leqslant C\|F\|^2_{L^\infty(Y)}\big(\varepsilon\|w\|^2_{L^2(\partial\Omega)} + \varepsilon^2\|\nabla w\|^2_{L^2(\Omega^{bl}_{2\varepsilon d(Y)})}\big), \end{aligned} \tag{3.46}$$

for every $w \in H^1(\Omega)$. All the constants depend only upon α β and $\partial\Omega$; they are independent of ε.

The following is a consequence of estimates (3.45) and (3.43)

Corollary 3.30. *Let χ in $H^1_{per,0}(Y)$ be the solution of (3.42). Then, for every $w \in H^1(\mathbb{R}^N)$ (resp., $H^1(\Omega)$),*

$$\chi\Big(\frac{\cdot}{\varepsilon}\Big)\,w \text{ belongs to } H^1(\mathbb{R}^N), \quad (\text{resp., } H^1(\Omega)).$$

More precisely, there is a constant C (independent of ε) such that

$$\begin{aligned} &\Big\|\chi\Big(\frac{\cdot}{\varepsilon}\Big)\,w\Big\|_{H^1(\mathbb{R}^N)} \leqslant C\|F\|_{L^\infty(Y)}\big(\varepsilon^{-1}\|w\|_{L^2(\mathbb{R}^N)} + \|\nabla w\|_{L^2(\mathbb{R}^N)}\big), \\ &\Big\|\chi\Big(\frac{\cdot}{\varepsilon}\Big)\,w\Big\|_{H^1(\Omega)} \leqslant C\|F\|_{L^\infty(Y)}\big(\varepsilon^{-1}\|w\|_{L^2(\Omega)} + \|\nabla w\|_{L^2(\Omega)}\big). \end{aligned}$$

Proof of Lemma 3.29.
Step 1. Proof of (3.43).

The Lax-Milgram theorem applied to (3.42) implies that $\|\chi\|_{H^1_{per,0}(Y)}$ is bounded above by a constant multiple of $\|F\|_{L^2(Y)}$. This gives $(3.43)_{1\text{-}2}$.

Lemma 3.28 applied to $d \doteq A(y)\nabla_y\chi(y) + F(y)$ (with $\varepsilon = 1$) shows that

$$-\text{div}\,(A(y)\nabla_y\chi(y) + F(y)) = 0 \quad \text{in } \mathcal{D}'(\mathbb{R}^N),$$

where F is extended by Y-periodicity to $\mathbb{R}^N$. Now, $(3.43)_3$ is a direct consequence of the regularity of solutions of second order elliptic equations in divergence form (Theorem of Moser-Stampacchia, see Theorem 8.18 in [114]).

Step 2. Proof of (3.44).

We start by proving (3.44) for w in $\mathcal{D}(\mathbb{R}^N)$. First apply Lemma 3.28 to

$$d = A(y)\nabla_y\chi(y) + F(y),\ f = 0, \text{ with the test function } \phi = \varepsilon\chi\Big(\frac{\cdot}{\varepsilon}\Big)w^2.$$

Note that ϕ is in $H^1(\mathbb{R}^N)$ with compact support. The result is

$$\int_{\mathbb{R}^N}\Big[A\Big(\frac{x}{\varepsilon}\Big)\nabla_y\chi\Big(\frac{x}{\varepsilon}\Big)+F\Big(\frac{x}{\varepsilon}\Big)\Big]\Big[\nabla_y\chi\Big(\frac{x}{\varepsilon}\Big)w^2(x)+2\varepsilon\chi\Big(\frac{x}{\varepsilon}\Big)w(x)\nabla w(x)\Big]dx = 0,$$

which is rewritten as

$$\begin{aligned}
&\int_{\mathbb{R}^N} A\Big(\frac{x}{\varepsilon}\Big)\Big[\nabla_y\chi\Big(\frac{x}{\varepsilon}\Big)w(x)\Big]\Big[\nabla_y\chi\Big(\frac{x}{\varepsilon}\Big)w(x)\Big]dx\\
&\qquad = -2\varepsilon\int_{\mathbb{R}^N} A\Big(\frac{x}{\varepsilon}\Big)\Big[\nabla_y\chi\Big(\frac{x}{\varepsilon}\Big)w(x)\Big]\chi\Big(\frac{x}{\varepsilon}\Big)\cdot\nabla w(x)\,dx\\
&\qquad\quad - \int_{\mathbb{R}^N} F\Big(\frac{x}{\varepsilon}\Big)w(x)\cdot\Big[\nabla_y\chi\Big(\frac{x}{\varepsilon}\Big)w(x)\Big]dx.\\
&\qquad\quad - 2\varepsilon\int_{\mathbb{R}^N} F\Big(\frac{x}{\varepsilon}\Big)\chi\Big(\frac{x}{\varepsilon}\Big)w(x)\cdot\nabla w(x)dx.
\end{aligned}\tag{3.47}$$

By the ellipticity of A, the left-hand side is bounded below as follows:

$$\alpha\int_{\mathbb{R}^N}\Big|\nabla_y\chi\Big(\frac{x}{\varepsilon}\Big)w(x)\Big|^2dx \leqslant \int_{\mathbb{R}^N} A\Big(\frac{x}{\varepsilon}\Big)\Big[\nabla_y\chi\Big(\frac{x}{\varepsilon}\Big)w(x)\Big]\Big[\nabla_y\chi\Big(\frac{x}{\varepsilon}\Big)w(x)\Big]dx.$$

We now estimate the three integrals of the right-hand side of (3.47). For the first one, use successively estimate (3.2) and Young's inequality, to get

$$\begin{aligned}
&2\varepsilon\int_{\mathbb{R}^N} A\Big(\frac{x}{\varepsilon}\Big)\Big[\nabla_y\chi\Big(\frac{x}{\varepsilon}\Big)w(x)\Big]\chi\Big(\frac{x}{\varepsilon}\Big)\cdot\nabla w(x)dx\\
&\qquad\leqslant 2\beta\int_{\mathbb{R}^N}\Big|\nabla_y\chi\Big(\frac{x}{\varepsilon}\Big)w(x)\Big|\Big|\varepsilon\chi\Big(\frac{x}{\varepsilon}\Big)\nabla w(x)\Big|dx\\
&\qquad\leqslant \frac{\alpha}{4}\int_{\mathbb{R}^N}\Big|\nabla_y\chi\Big(\frac{x}{\varepsilon}\Big)w(x)\Big|^2dx + \frac{4\beta^2}{\alpha}\int_{\mathbb{R}^N}\varepsilon^2\Big|\chi\Big(\frac{x}{\varepsilon}\Big)\nabla w(x)\Big|^2dx\\
&\qquad\leqslant \frac{\alpha}{4}\int_{\mathbb{R}^N}\Big|\nabla_y\chi\Big(\frac{x}{\varepsilon}\Big)w(x)\Big|^2dx + C\|F\|^2_{L^\infty(Y)}\int_{\mathbb{R}^N}\varepsilon^2|\nabla w(x)|^2dx,
\end{aligned}$$

where we also used estimate (3.43).

For the second integral, by Young's inequality and (3.43) again, it is easily seen that

$$\begin{aligned}
&\int_{\mathbb{R}^N} F\Big(\frac{x}{\varepsilon}\Big)w(x)\cdot\Big[\nabla_y\chi\Big(\frac{x}{\varepsilon}\Big)w(x)\Big]dx\\
&\qquad\leqslant \frac{1}{\alpha}\int_{\mathbb{R}^N}\Big|F\Big(\frac{x}{\varepsilon}\Big)w(x)\Big|^2dx+\frac{\alpha}{4}\int_{\mathbb{R}^N}\Big|\nabla_y\chi\Big(\frac{x}{\varepsilon}\Big)w(x)\Big|^2dx\\
&\qquad\leqslant C\|F\|^2_{L^\infty(Y)}\int_{\mathbb{R}^N}|w(x)|^2dx+\frac{\alpha}{4}\int_{\mathbb{R}^N}\Big|\nabla_y\chi\Big(\frac{x}{\varepsilon}\Big)w(x)\Big|^2dx.
\end{aligned}$$

Similarly, the last integral is estimated as

$$\begin{aligned}
&\varepsilon\int_{\mathbb{R}^N} F\Big(\frac{x}{\varepsilon}\Big)\chi\Big(\frac{x}{\varepsilon}\Big)w(x)\cdot\nabla w(x)dx\\
&\qquad\leqslant C\|F\|^2_{L^\infty(Y)}\Big[\int_{\mathbb{R}^N}|w(x)|^2dx+\int_{\mathbb{R}^N}\varepsilon^2|\nabla w(x)|^2dx\Big].
\end{aligned}$$

Note that all the constants in the estimates above do not depend on ε, they only depend on A (via α and β) and on Y and N.

For w in $\mathcal{D}(\mathbb{R}^N)$, the estimates above combined with (3.47) give (3.44). By density, (3.44) holds for w in $H^1(\mathbb{R}^N)$.

Step 3. Proof of (3.45) *and* (3.46).

To prove (3.45), choose w in $H^1(\Omega)$ and extend it as a function belonging to $H^1(\mathbb{R}^N)$ (using the extension operator $\mathcal{P}$ from Subsection 1.6.2, Proposition 1.74). The conclusion follows by estimates (1.80) and (3.44) since

$$\int_\Omega\Big|\nabla_y\chi\Big(\frac{x}{\varepsilon}\Big)w(x)\Big|^2dx\leqslant\int_{\mathbb{R}^N}\Big|\nabla_y\chi\Big(\frac{x}{\varepsilon}\Big)\mathcal{P}(w)(x)\Big|^2dx.$$

To prove (3.46), choose $w\in H^1(\Omega)$ and consider the map

$$x\mapsto w_\varepsilon(x)\doteq w(x)\Big[1-\inf\Big(1,\frac{\operatorname{dist}(x,\Omega^{bl}_{\varepsilon d(Y)})}{\varepsilon d(Y)}\Big)\Big].$$

The function w_ε belongs to $H^1(\Omega)$ and satisfies

$$w_\varepsilon=\begin{cases} w & \text{in } \Omega^{bl}_{\varepsilon d(Y)},\\ 0 & \text{in } \Omega\backslash\overline{\Omega}^{bl}_{2\varepsilon d(Y)}.\end{cases}$$

From (3.41) one gets

$$\begin{aligned}
\|w_\varepsilon\|^2_{L^2(\Omega)}+\varepsilon^2\|\nabla w_\varepsilon\|^2_{L^2(\Omega)}&\leqslant C\big(\|w\|^2_{L^2(\Omega^{bl}_{2\varepsilon d(Y)})}+\varepsilon^2\|\nabla w\|^2_{L^2(\Omega^{bl}_{2\varepsilon d(Y)})}\big)\\
&\leqslant C\big(\varepsilon\|w\|^2_{L^2(\partial\Omega)}+\varepsilon^2\|\nabla w\|^2_{L^2(\Omega^{bl}_{2\varepsilon d(Y)})}\big),
\end{aligned}$$

where the constant does not depend on ε. Together with (3.45), this implies estimate (3.46) and the proof is complete. ☐

Corollary 3.31. *Under the hypotheses of Theorem 3.27, if $A^\varepsilon(x) = A(x/\varepsilon)$, the strong convergence* (3.40) *can be improved as follows:*

$$u_\varepsilon - u_0 - \varepsilon \sum_{i=1}^{N} \frac{\partial u_0}{\partial x_i} \widehat{\chi}_i\Big(\frac{\cdot}{\varepsilon}\Big) \to 0 \quad \textit{strongly in } H^1_{loc}(\Omega).$$

If u_0 belongs to $H^2(\Omega)$, one has the global result,

$$u_\varepsilon - u_0 - \varepsilon \sum_{i=1}^{N} \frac{\partial u_0}{\partial x_i} \widehat{\chi}_i\Big(\frac{\cdot}{\varepsilon}\Big) \to 0 \quad \textit{strongly in } H^1(\Omega).$$

Proof. Under the assumptions of the Theorem 3.27, the solution u_0 of problem (3.8) belongs to $H^2_{loc}(\Omega)$. The correctors $\widehat{\chi}_i$ are solutions of (3.20), a problem of type (3.42). Due to the second estimate in Lemma 1.76 we get, for $\omega \Subset \Omega$ and for all $i \in \{1, \dots, N\}$,

$$\Big\| \mathcal{Q}_\varepsilon\Big(\frac{\partial u_0}{\partial x_i}\Big) - \frac{\partial u_0}{\partial x_i} \Big\|_{L^2(\omega)} \leqslant C\varepsilon \|u_0\|_{H^2(\omega)} \leqslant C\varepsilon \|f\|_{L^2(\Omega)},$$

which concludes the proof with (3.45), since the sets

$$\Big\{\nabla \mathcal{Q}_\varepsilon\Big(\frac{\partial u_0}{\partial x_i}\Big)\, \widehat{\chi}_i\Big(\frac{\cdot}{\varepsilon}\Big)\Big\}_\varepsilon \quad \text{and} \quad \Big\{\nabla \frac{\partial u_0}{\partial x_i}\, \widehat{\chi}_i\Big(\frac{\cdot}{\varepsilon}\Big)\Big\}_\varepsilon,$$

are bounded independently of ε in $L^2(\omega)$. This, together with (3.40), concludes the proof of the first statement.

If u_0 is in $H^2(\Omega)$, the previous reasoning applies directly in Ω and gives the second statement. □

Remark 3.32. *The corrector results given here are interior correctors. For correctors up to the boundary, see Chapter 14.*

3.2 Homogenization for non linear diffusion

In this section, we consider the periodic homogenization of second order non linear diffusion problems of the type introduced by Višik in [188] and [189] and further studied by Leray and Lions in [148] (the corresponding operators are often called Leray-Lions operators). Higher order problems can be studied by the same methods. We refer to [91] for a general study of the homogenization of monotone problems (see also [108]).

The starting point is a simplified form of the result of Leray and Lions [148]. The space $\mathbb{R}^N$ is given here with a euclidean scalar product denoted $\langle \cdot, \cdot \rangle_{\mathbb{R}^N}$.

Theorem 3.33. *Let p be in $(1, +\infty)$, $\mathcal{O}$ be a bounded domain in $\mathbb{R}^N$, m in $L^1(\mathcal{O})$, and f in $W^{-1,p'}(\mathcal{O})$. Let A be a map from $\mathcal{O} \times \mathbb{R}^N$ to $\mathbb{R}^N$ which satisfies the following hypotheses:*

(i) *A is of Caratheodory type (i.e., measurable with respect to* $x \in \mathcal{O}$ *and continuous with respect to* $\xi \in \mathbb{R}^N$*).*

(ii) *For a.e.* $x \in \mathcal{O}$*, the map* $\xi \mapsto A(x,\xi)$ *is strictly monotone, i.e.,*

$$\forall \xi_1 \neq \xi_2 \in \mathbb{R}^N, \quad \langle \xi_1 - \xi_2, A(x,\xi_1) - A(x,\xi_2)\rangle_{\mathbb{R}^N} > 0.$$

(iii) *There exists a strictly positive number* α *such that for all* ξ *in* $\mathbb{R}^N$*,*

$$\alpha\left(\frac{|\xi|^p}{p} + \frac{|A(\,\cdot\,,\xi)|^{p'}}{p'}\right) \leqslant \langle A(\,\cdot\,,\xi),\xi\rangle_{\mathbb{R}^N} + m \qquad \text{a.e. in } \mathcal{O}.$$

Then the following problem has a unique solution:

$$\begin{cases} -\operatorname{div} A(\,\cdot\,,\nabla u) = f & \text{in } \mathcal{D}'(\mathcal{O}),\\ u \in W_0^{1,p}(\mathcal{O}). \end{cases} \tag{3.48}$$

The variational formulation of this problem is, of course

$$\begin{cases} \displaystyle\int_\Omega A(x,\nabla u(x))\,\nabla\phi(x)\,dx = \langle f,\phi\rangle_{W^{-1,p'}(\Omega),W_0^{1,p}(\Omega)},\\ \forall \phi \in W_0^{1,p}(\Omega). \end{cases}$$

3.2.1 Maximal monotone operators

We start by recalling the definition of (multivalued) maximal monotone operators on a real reflexive Banach space X with dual denoted X'.

Consider set-valued operators $A : X \to X'$, that is, maps which take every point $\xi \in X$ to some set $A\xi \subset X'$. These applications are simply called *operators* when no confusion may occur. Similarly, when no ambiguity arises, A will also denote the *graph* of the operator A, which is the set

$$\textit{graph}\, A = \{(\xi,\eta) \in X \times X' \mid \eta \in A\xi\}.$$

The *domain* of a graph A and its range $R(A)$ are the sets

$$D(A) = \{x \in X \mid Ax \neq \varnothing\},$$
$$R(A) = \bigcup_{x\in X} Ax.$$

Definition 3.34. *The operator A is called single-valued on a set* $C \subset X$*, if*

$$\forall \xi \in C, \quad A\xi \text{ contains at most one element.}$$

The operator A is called nonexpansive or Lip_1*, if*

$$\|\eta_1 - \eta_2\|_{X'} \leqslant \|\xi_1 - \xi_2\|_X, \quad \forall\big((\xi_1,\eta_1),(\xi_2,\eta_2)\big) \in (\textit{graph}\, A)^2.$$

Every nonexpansive operator is clearly single-valued on X.

For operators A, B, we write $A \subseteq B$ whenever $A\xi \subseteq B\xi$ for every $\xi \in X$.

Definition 3.35. *The set $A \subset X \times X'$ is a* monotone graph *(or monotone operator) if for every $\big((\xi_1, \eta_1), (\xi_2, \eta_2)\big) \in (graph\, A)^2$,*

$$\langle \eta_1 - \eta_2, \xi_1 - \xi_2 \rangle_{X',X} \geqslant 0.$$

The monotone graph A is a maximal monotone graph *(or maximal monotone operator), if for every monotone graph $B \subset X \times X'$, the inclusion $A \subseteq B$ implies $A = B$.*

The following result is an easy direct consequence of maximality:

Proposition 3.36. *Let A be a maximal monotone graph from X to X'. Assume that $\{(\xi_n, \eta_n)\}_{n\in\mathbb{N}}$ converges weakly to (ξ, η) in $X \times X'$. If, furthermore,*

$$\liminf_{n\to+\infty} \langle \eta_n, \xi_n \rangle_{X',X} \leqslant \langle \eta, \xi \rangle_{X',X},$$

then (ξ, η) belongs to A and

$$\langle \eta, \xi \rangle_{X',X} = \liminf_{n\to+\infty} \langle \eta_n, \xi_n \rangle_{X',X}.$$

In particular, it is so if one of the convergences is strong (so-called demi-closure of maximal monotone operators).

Proof. For every $(\alpha, \beta) \in A$, by the monotonicity of A,

$$\langle \beta - \eta_n, \alpha - \xi_n \rangle_{X',X} \geqslant 0.$$

Therefore

$$0 \leqslant \liminf_{n\to+\infty} \langle \beta - \eta_n, \alpha - \xi_n \rangle_{X',X} \leqslant \langle \beta - \eta, \alpha - \xi \rangle_{X',X}.$$

Since A is maximal and $(\alpha, \beta) \in A$ is arbitrary, it follows that $(\xi, \eta) \in A$.

The fact that $(\xi, \eta) \in A$ and the monotonicity of A now imply

$$\begin{aligned}\liminf_{n\to+\infty} \langle \eta_n, \xi_n \rangle_{X',X} \geqslant \lim_{n\to+\infty} \langle \eta_n, \xi \rangle_{X',X} + \lim_{n\to+\infty} \langle \eta, \xi_n \rangle_{X',X} - \langle \eta, \xi \rangle_{X',X} \\ = \langle \eta, \xi \rangle_{X',X}. \quad \square\end{aligned}$$

It is known that a continuous (defined on the whole space) monotone map is maximal monotone. A famous result of Minty ([159]) in the case where X is a real Hilbert space H, characterizes the maximal monotone maps as the monotone maps A such that $Id + A$ is onto ($R(Id + A) = H$).

In order to generalize the result of Theorem 3.33 to the case of maximal monotone graphs on $\mathbb{R}^N$, including multivalued graphs, we recall below the notions of convergence of maximal monotone maps, of measurable maximal monotone-valued maps in that setting, and of the natural extension of a measurable maximal monotone-valued map to the spaces $L^p(\mathcal{O};\mathbb{R}^N)$(cf. [84] and [91] for a detailed study in the general framework of reflexive Banach spaces). To characterize these properties, since we are dealing with maximal monotone graphs on $\mathbb{R}^N$, we will use the Cayley transform (this is simpler than the resolvent families used in [84] and [91] for the general reflexive Banach space setting). More generally, this is valid on a general Hilbert space ($\mathbb{R}^N$ being a particular case of Hilbert space !).

The set of all maximal monotone operators on a Hilbert space H will be denoted $\mathfrak{M}(H)$.

Definition 3.37 (Cayley transform for maximal monotone graphs). *For A in $\mathfrak{M}(H)$ its Cayley transform, denoted $\mathcal{C}(A)$, is the single-valued map defined by its graph in $H \times H$ as follows:*

$$\mathcal{C}(A) \doteq \{(\xi+\eta, \xi-\eta) \mid (\xi,\eta) \in A\}.$$

In other words, $\mathcal{C}(A) = (Id - A) \circ (Id + A)^{-1}$, the latter being defined on the whole space precisely by Minty's theorem.

The converse map is given by

$$A = \Big\{ \Big(\frac{1}{2}(Id + \mathcal{C}(A))(x), \frac{1}{2}(Id - \mathcal{C}(A))(x) \Big) \Bigm| x \in H \Big\}. \tag{3.49}$$

The fact that A is monotone is equivalent to saying that $\mathcal{C}(A)$ is 1-Lipschitz on its domain $R(Id + A)$. The maximality of A is then equivalent to the fact that $\mathcal{C}(A)$ is defined on the whole space H (by Minty's theorem). The Cayley transform is therefore a bijection between the set of all maximal monotone operators on the Hilbert space H and the set $Lip_1(H)$ defined as

$$Lip_1(H) = \{h : H \mapsto H \mid h \text{ 1-Lipschitz } \}. \tag{3.50}$$

3.2.2 Convergence of maximal monotone graphs

Following Brezis [36] and Attouch [16], the convergence of maximal monotone graphs is defined as follows:

Definition 3.38. *Let $A^n (n \in \mathbb{N})$ and A be maximal monotone graphs from the reflexive Banach space X to its dual space X'. The sequence $\{A^n\}_{n\in\mathbb{N}}$ is said to converge to A as $n \to +\infty$, (convergence denoted $A^n \rightarrowtail A$), if for every (ξ,η) in A there exists a sequence $\{(\xi_n,\eta_n)\}_{n\in\mathbb{N}} \in A^n$ such that $(\xi_n,\eta_n) \to (\xi,\eta)$ in $X \times X'$ as $n \to +\infty$.*

In the case of a Hilbert space H, one can give an equivalent formulation in terms of the Cayley transform.

Proposition 3.39. *Let $A^n (n \in \mathbb{N})$ and A be maximal monotone graphs in H. The sequence $\{A^n\}_{n\in\mathbb{N}}$ converges to A as $n \to +\infty$, if and only if $\{\mathcal{C}(A_n)\}_{n\in\mathbb{N}}$ converges to $\mathcal{C}(A)$ pointwise (since they are in $Lip_1(H)$, by Ascoli's theorem, this is equivalent to their local uniform convergence).*

Proof. For the "if" part, it is easy to see that for given $(\xi, \eta) \in A$, the sequence $\{(\xi_n, \eta_n)\}_{n \in \mathbb{N}}$ defined by

$$\xi_n \doteq \frac{1}{2}(Id + \mathcal{C}(A^n))(\xi + \eta), \qquad \eta_n \doteq \frac{1}{2}(Id - \mathcal{C}(A^n))(\xi + \eta),$$

belongs to A^n and converges to (ξ, η) (see (3.49)).

To prove the converse implication, for $x \in H$, let $\{(\xi_n, \eta_n)\}_{n \in \mathbb{N}} \in A_n$ be such that $\xi_n + \eta_n = x$. For each $n \in \mathbb{N}$, this pair is unique (by monotonicity). Similarly, let $(\xi, \eta) \in A$ be such that $\xi + \eta = x$. The convergence of $\mathcal{C}(A^n)(x) = \xi_n - \eta_n$ to $\mathcal{C}(A)(x) = \xi - \eta$ follows from the convergence of ξ_n to ξ which we will now show.

Let $\{(\widetilde{\xi}_n, \widetilde{\eta}_n)\}_{n \in \mathbb{N}} \in A^n$ be a sequence converging to (ξ, η) given by the hypothesis $A^n \rightarrowtail A$. By the monotonicity of A^n,

$$\langle \xi_n - \widetilde{\xi}_n, \eta_n - \widetilde{\eta}_n \rangle_{X', X} \geqslant 0.$$

Setting $z_n \doteq \xi_n - \widetilde{\xi}_n$ and replacing η_n by its value $x - \xi_n$, gives

$$\langle z_n, x - \widetilde{\xi}_n - \widetilde{\eta}_n - z_n \rangle_{X', X} \geqslant 0.$$

From this inequality it follows that $\|z_n\|_{X'} \leqslant \|x - \widetilde{\xi}_n - \widetilde{\eta}_n\|_X$. But by definition, $\widetilde{\xi}_n - \widetilde{\eta}_n$ converges to $\xi - \eta = x$, so that $z_n \to 0$. This in turn implies the claim. □

Remark 3.40. *Note that if the sequence $\{\mathcal{C}(A_n)\}_{n \in \mathbb{N}}$ converges pointwise, its limit automatically belongs to $Lip_1(H)$ (see (3.50)) and is therefore associated with some maximal monotone operator A such that $A^n \rightarrowtail A$. This is why the Cayley transform is a very useful tool.*

It is now easy to extend Proposition 3.36 to the case of a converging sequence of maximal monotone operators (the proof is similar).

Proposition 3.41. *Let $A^n (n \in \mathbb{N})$, $A \subset X \times X'$ be maximal monotone graphs, and let $(\xi_n, \eta_n) \in A^n$ and $(\xi, \eta) \in X \times X'$. Assume that, as $n \to +\infty$,*

$$A^n \rightarrowtail A, \quad \xi_n \rightharpoonup \xi \quad \text{weakly in } X, \quad \eta_n \rightharpoonup \eta \quad \text{weakly in } X',$$

and

$$\liminf_{n \to +\infty} \langle \eta_n, \xi_n \rangle_{X', X} \leqslant \langle \eta, \xi \rangle_{X', X}.$$

Then,

$$(\xi, \eta) \in A \quad \text{and} \quad \liminf_{n \to +\infty} \langle \eta_n, \xi_n \rangle_{X', X} = \langle \eta, \xi \rangle_{X', X}.$$

3.2.3 Measurable families of maximal monotone graphs

In this section we introduce the definition of measurable maps from a measure space $\{\mathcal{O}, \mu\}$ to the set of maximal monotone operators on a separable Hilbert space H with inner product $(\cdot,\cdot)_H$, and give an equivalent definition via the Cayley transform.

Definition 3.42. *The map $A : \mathcal{O} \to \mathfrak{M}(H)$ is* measurable *if and only if for every open set $U \subset H \times H$ (respectively, closed set, Borel set, open ball, closed ball), the set*

$$\{t \in \mathcal{O} \mid A(t) \cap U \neq \varnothing\},$$

is measurable in $\mathcal{O}$.

The following gives an equivalent formulation which we take as an alternate definition.

Definition 3.43. *The map $A : \mathcal{O} \to \mathfrak{M}(H)$ is* measurable *if and only if its Cayley transform $\mathcal{C}(A)$ is a Caratheodory function on $\Omega \times H$ with values in $Lip_1(H)$, i.e.,*

(i) *for every $x \in H$, the map $t \mapsto \mathcal{C}(A)(t,x)$ is measurable on Ω,*

(ii) *for a.e. $t \in \mathcal{O}$, the map $x \mapsto \mathcal{C}(A)(t,x)$ belongs to $Lip_1(H)$.*

Given a measurable map $A : \mathcal{O} \to \mathfrak{M}(H)$ and p in $(1,+\infty)$, one can define a monotone graph from $L^p(\mathcal{O};H)$ to $L^{p'}(\mathcal{O};H)$ which is the *canonical extension* of A as a map on $L^p(\mathcal{O};H)$ with values in $L^{p'}(\mathcal{O};H)$.

Definition 3.44. *Let $A : \mathcal{O} \to \mathfrak{M}(H)$, the canonical extension of A from $L^p(\mathcal{O};H)$ to $L^{p'}(\mathcal{O};H)$ is defined by*

$$\mathcal{A} = \big\{(u,v) \in L^p(\mathcal{O};H) \times L^{p'}(\mathcal{O};H) \mid \big(u(t),v(t)\big) \in A(t) \text{ for a.e. } t \in \mathcal{O}\big\}.$$

One readily checks that $\mathcal{A}$ is monotone.

Proposition 3.45. *Let $A : \mathcal{O} \to \mathfrak{M}(H)$ be measurable. If $\mathcal{A} \neq \varnothing$, then $\mathcal{A}$ is maximal monotone.*

Proof. Since $\mathcal{A} \neq \varnothing$, there exists $(\alpha,\beta) \in \mathcal{A}$. Suppose that (u,v) belongs to $L^p(\mathcal{O};H) \times L^{p'}(\mathcal{O};H)$ and is such that, for every $(u',v') \in \mathcal{A}$,

$$\int_{\mathcal{O}} \big(v'(t) - v(t), u'(t) - u(t)\big)_H \, d\mu \geqslant 0.$$

For almost every t in $\mathcal{O}$, set

$$z(t) = u(t) + v(t),$$

and

$$u_*(t) = \frac{1}{2}(Id + \mathcal{C}(A(t))(z(t))$$
$$v_*(t) = \frac{1}{2}(Id - \mathcal{C}(A(t))(z(t)).$$

By construction, $(u_*(t), v_*(t))$ belongs to $A(t)$ for almost every $t \in \mathcal{O}$. Furthermore, both maps are measurable, since $t \mapsto z(t)$ is measurable and $(t, z) \mapsto \mathcal{C}(A(t))(z)$ is Caratheodory.

For $k > 0$, introduce the set $\mathcal{O}_k$,

$$\mathcal{O}_k = \{t \in \mathcal{O} \mid \|u_*(t)\|_H \leqslant k\|u(t)\|_H \quad \text{and} \quad \|v_*(t)\|_H \leqslant k\|v(t)\|_H\},$$

and set

$$(u_k, v_k) = \chi_{\mathcal{O}_k}(u_*, v_*) + \chi_{\mathcal{O}\setminus\mathcal{O}_k}(\alpha, \beta).$$

Clearly, (u_k, v_k) belongs to $\mathcal{A}$, so that

$$\int_{\mathcal{O}} \langle v_k(t) - v(t), u_k(t) - u(t)\rangle_H d\mu \geqslant 0,$$

which reads

$$\int_{\mathcal{O}_k} \langle v_*(t) - v(t), u_*(t) - u(t)\rangle_H \, d\mu \\ + \int_{\mathcal{O}\setminus\mathcal{O}_k} \langle \beta(t) - v(t), \alpha(t) - u(t)\rangle_H \, d\mu \geqslant 0.$$

Since by construction,

$$u_*(t) + v_*(t) = z(t) = u(t) + v(t),$$

the last inequality implies the following one:

$$\int_{\mathcal{O}_k} \|u(t) - u_*(t)\|_H^2 d\mu = \int_{\mathcal{O}_k} \|v(t) - v_*(t)\|_H^2 d\mu \\ \leqslant \int_{\mathcal{O}\setminus\mathcal{O}_k} \big(\beta(t) - v(t)\big)\big(\alpha(t) - u(t)\big) d\mu.$$

Lebesgue's monotone convergence theorem applied to the two left terms, and Lebesgue's dominated convergence theorem applied to the third term, together imply

$$\int_{\mathcal{O}} \|u(t) - u_*(t)\|_H^2 d\mu = \int_{\mathcal{O}} \|v(t) - v_*(t)\|_H^2 d\mu = 0.$$

Consequently $(u, v) = (u_*, v_*)$ belongs to $\mathcal{A}$. □

Remark 3.46. *The maximality of $A(t)$ for almost every $t \in \mathcal{O}$ is not sufficient to ensure the maximality of $\mathcal{A}$ because the latter can be empty. Here is such a trivial example,*

$$\mathcal{O} = (0, 1) \text{ and } A(t) = \{(\xi, \eta) \in \mathbb{R} \times \mathbb{R} \mid \eta = t^{-1/p'}\}.$$

We conclude this section with a convergence for the canonical extensions of measurable maximal monotone maps (see Definition 3.44).
For measurable maps $A^n (n \in \mathbb{N}), A : \mathcal{O} \to \mathfrak{M}(H)$ and their canonical extensions $\mathcal{A}^n$ and $\mathcal{A}$, this result gives conditions under which the pointwise convergence $A^n(t) \rightarrowtail A(t)$ for almost every $t \in \mathcal{O}$ implies the convergence of the extension graphs $\mathcal{A}^n \rightarrowtail \mathcal{A}$.

Theorem 3.47. *Let $A^n(n \in \mathbb{N}), A : \mathcal{O} \to \mathfrak{M}(H)$ be measurable. Assume that*

(i) *for almost every $t \in \mathcal{O}$, $A^n(t) \rightarrowtail A(t)$,*

(ii) *$\mathcal{A}$ and $\mathcal{A}^n$ are maximal monotone,*

(iii) *there exists $(\alpha_n, \beta_n) \in \mathcal{A}^n$, $(\alpha, \beta) \in L^p(\mathcal{O}; H) \times L^{p'}(\mathcal{O}; H)$ such that*

$$(\alpha_n, \beta_n) \to (\alpha, \beta) \quad \text{strongly in } L^p(\mathcal{O}; H) \times L^{p'}(\mathcal{O}; H).$$

Then $\mathcal{A}^n \rightarrowtail \mathcal{A}$.

Remark 3.48. *Assumption (iii) cannot be dropped. Indeed, one can have $A^n(t) \rightarrowtail A(t)$ for every $t \in \mathcal{O}$, while $\mathcal{A}^n \not\rightarrowtail \mathcal{A}$. Here is such an elementary example. For $H = \mathbb{R}$, $\mathcal{O} = (0,1)$ and $n \in \mathbb{N}$, define the operators*

$$A^n(t) = \{(x,y) \in \mathbb{R}^2 : y = n^{1/p'} \chi_{(0,1/n)}(t)\} \text{ and } A(t) = \{(x,y) \in \mathbb{R}^2 : y = 0\}.$$

Clearly, for for every $t \in (0,1)$, $A^n(t) \rightarrowtail A(t)$. On the other hand, for every $(u_n, v_n) \in \mathcal{A}^n$ and $(u,v) \in \mathcal{A}$,

$$\|v_n\|_{L^{p'}} = 1 \text{ and } \|v\|_{L^{p'}} = 0 \Longrightarrow v_n \nrightarrow v \text{ strongly in } L^{p'}(\mathcal{O}).$$

Hence, $\mathcal{A}^n \rightarrowtail \mathcal{A}$ is impossible!

Proof of Theorem 3.47. Let $(u,v) \in \mathcal{A}$. For a.e. $t \in \mathcal{O}$, let $z(t) = u(t) + v(t)$ so that

$$(u(t), v(t)) = \Big(\frac{1}{2}(Id + \mathcal{C}(A(t)))z(t)\,,\, \frac{1}{2}(Id - \mathcal{C}(A(t)))z(t)\Big).$$

For $n \in \mathbb{N}$, define

$$(u_n(t), v_n(t)) \doteq \Big(\frac{1}{2}(Id + \mathcal{C}(A^n(t)))z(t)\,,\, \frac{1}{2}(Id - \mathcal{C}(A^n(t)))z(t)\Big).$$

By hypothesis (i), the sequence $\{(u_n, d_n)\}_n$ converges a.e. to (u,v). Set

$$\mathcal{O}' = \{t \in \mathcal{O} \mid \|u_n(t)\|_H \leqslant 2\|u(t)\|_H \quad \text{and} \quad \|v_n(t)\|_H \leqslant 2\|v(t)\|_H\},$$

and for $n \geqslant 1$,

$$(u_n^*(t), v_n^*(t)) = \begin{cases} (u_n(t), v_n(t)) & \text{if } t \in \mathcal{O}', \\ (\alpha_n(t), \beta_n(t)) & \text{if } t \in \mathcal{O} \backslash \mathcal{O}'. \end{cases}$$

By construction, (u_n^*, v_n^*) belongs to $\mathcal{A}^n$. Then, the Lebesgue dominated convergence theorem implies the strong convergence,

$$(u_n^*, v_n^*) \to (u,v) \quad \text{strongly in } L^p(\mathcal{O}; H) \times L^{p'}(\mathcal{O}; H),$$

and this concludes the proof. ☐

3.2.4 Convergence results

We return to the nonlinear diffusion problems generalizing problem (3.48).

Theorem 3.49. *Let $\mathcal{O}$ be a bounded domain in $\mathbb{R}^N$ and $p \in (1, +\infty)$. Let $A, A^n : \mathcal{O} \to \mathfrak{M}(\mathbb{R}^N)$ be measurable with associated canonical extensions $\mathcal{A}^n$ and $\mathcal{A}$ and assume that the hypotheses of Theorem 3.47 are satisfied. Let $\{f_n\}_{n\in\mathbb{N}}$ be a sequence converging strongly to f in $W^{-1,p'}(\mathcal{O})$.*

Suppose that for each $n \in \mathbb{N}$, u_n is a solution of

$$\begin{cases} -\operatorname{div} d_n = f_n & \text{in } \mathcal{D}'(\mathcal{O}), \\ (\nabla u_n(x), d_n(x)) \in A^n(x) & \text{for a.e. } x \in \mathcal{O}, \\ u_n \in W_0^{1,p}(\mathcal{O}). \end{cases}$$

Then every weak limit point (u, d) (if any) of the sequence $\{(u_n, d_n)\}_n$ in the space $W_0^{1,p}(\mathcal{O}) \times L^{p'}(O)$, is a solution of

$$\begin{cases} -\operatorname{div} d = f & \text{in } \mathcal{D}'(\mathcal{O}), \\ (\nabla u(x), d(x)) \in A(x) & \text{for a.e. } x \in \mathcal{O}, \\ u \in W_0^{1,p}(\mathcal{O}). \end{cases}$$

Proof. Without loss of generality, we may assume that $\{(u_n, d_n)\}_n$ converges weakly to (u, d).

The only difficult point is to show that $\big(\nabla u(x), d(x)\big) \in A(x)$ for a.e. x in $\mathcal{O}$. In view of the definition of $\mathcal{A}$, it suffices to show that $\big(\nabla u, d\big) \in \mathcal{A}$. This follows from the convergence

$$\begin{aligned} \lim_{n\to\infty} \langle d_n, \nabla u_n \rangle_{L^{p'}(\mathcal{O}), L^p(\mathcal{O})} &= \lim_{n\to\infty} \langle f_n, u_n \rangle_{W^{-1,p'}(\mathcal{O}), W_0^{1,p}(\mathcal{O})} \\ &= \langle f, u \rangle_{W^{-1,p'}(\mathcal{O}), W_0^{1,p}(\mathcal{O})} = \langle d, \nabla u \rangle_{L^{p'}(\mathcal{O}), L^p(\mathcal{O})}, \end{aligned}$$

and from Proposition 3.41. □

3.2.5 A general existence result

Definition 3.50. *For $\mathcal{O}$ a bounded domain in $\mathbb{R}^N$, $p \in (1, \infty)$, $q \in [1, p)$, $\alpha > 0$ and m in $L^1(\mathcal{O})$, let $\mathcal{VLL}(\mathcal{O}, p, q, \alpha, m)$(12) denote the set of measurable maps $A(x, v)$ from $\mathcal{O} \times \mathbb{R}$ to $\mathfrak{M}(\mathbb{R}^N)$ which are continuous with respect to their second argument v(13) and satisfy the inequality*

$$\alpha\Big(\frac{|\xi|^p}{p} + \frac{|\eta|^{p'}}{p'}\Big) \leqslant \langle \eta, \xi \rangle_{\mathbb{R}^N} + m + |v|^q, \tag{3.51}$$

a.e. in $\mathcal{O}$, for every $v \in \mathbb{R}$, and for all $(\xi, \eta) \in A(x, v)$.

(12) $\mathcal{VLL}$ stands for Višik - Leray - Lions.

(13) In terms of the Cayley transform, this is equivalent to the fact that $\mathcal{C}(A(x, v))(\xi)$ is Caratheodory measurable with respect to x, continuous with respect to v and 1-Lipschitz with respect to ξ.

Definition 3.51. *A sequence* $\{A^n\}_{n\in\mathbb{N}}$ *in* $\mathcal{VLL}(\mathcal{O},p,q,\alpha,m)$ *is said to converges uniformly to some* A*, convergence denoted*

$$A^n \rightarrowtail A, A^n \twoheadrightarrow A$$

whenever for a.e. $x \in \mathcal{O}$ *and every* $\xi \in \mathbb{R}^N$, $\mathcal{C}(A^n(x,v))(\xi)$ *converges to* $\mathcal{C}(A(x,v))(\xi)$ *locally uniformly with respect to* $v \in \mathbb{R}$.

This is equivalent to the following convergence:

for a.e. $x \in \mathcal{O}$, for every convergent sequence $v_n \to v$ in $\mathbb{R}$,

$A^n(x,v_n)$ converges to $A(x,v)$ in $\mathfrak{M}(\mathbb{R}^N)$.

Clearly, if $A^n \rightarrowtail A$, the limit A belongs to $\mathcal{VLL}(\mathcal{O},p,q,\alpha,m)$.

Theorem 3.52. *Let* $\{A^n\}_{n\in\mathbb{N}}$ *be a sequence in* $\mathcal{VLL}(\mathcal{O},p,q,\alpha,m)$ *as defined above,* $\{f_n\}_{n\in\mathbb{N}}$ *a sequence converging to* f *in* $W^{-1,p'}(\mathcal{O})$ *and suppose that for each* $n \in \mathbb{N}$, u_n *is a solution of*

$$\begin{cases} -\text{div } d_n = f & \text{in } \mathcal{D}'(\mathcal{O}), \\ (\nabla u_n(x), d_n(x)) \in A^n(x,u_n(x)), \\ u_n \in W_0^{1,p}(\mathcal{O}). \end{cases} \tag{3.52}$$

Suppose furthermore that the sequence $\{A^n\}_{n\in\mathbb{N}}$ *converges to some* A *in the sense of the convergence defined above,* $A^n \rightarrowtail A$.

Then the sequence $\{(u_n,d_n)\}_n$ *is bounded in the space* $W_0^{1,p}(\mathcal{O}) \times L^{p'}(\mathcal{O})$*, and each of its weak limit points, say* (u,d)*, is a solution of*

$$\begin{cases} -\text{div } d = f & \text{in } \mathcal{D}'(\mathcal{O}), \\ (\nabla u(x), d(x)) \in A(x,u(x)), \\ u \in W_0^{1,p}(\mathcal{O}). \end{cases} \tag{3.53}$$

Proof. The fact that $\{(u_n,d_n)\}_{n\in\mathbb{N}}$ is bounded follows from inequality (3.51). Indeed, it implies

$$\alpha\left(\frac{\|\nabla u_n\|^p_{L^p(\mathcal{O})}}{p} + \frac{\|d_n\|^{p'}_{L^{p'}(\mathcal{O})}}{p'}\right) \leqslant \langle f_n, u_n\rangle_{W^{-1,p'}(\mathcal{O}),W_0^{1,p}(\mathcal{O})} + \int_{\mathcal{O}} m(x)\,dx + c\|u_n\|^q_{L^p(\mathcal{O})}.$$

One then can conclude using the Poincaré inequality for $W_0^{1,p}(\mathcal{O})$.

At this point, without loss of generality (by going to a subsequence), one can assume that the sequence $\{(u_n,d_n)\}_n$ converges to (u,d) weakly in $W_0^{1,p}(\mathcal{O}) \times L^{p'}(\mathcal{O})$. By the compact Sobolev embedding, one can also assume that u_n converges to u for a.e. $x \in \mathcal{O}$. The remainder of the proof then follows by applying Theorem 3.49 for the sequence $B_n(x) \doteq A^n(x,u_n(x))$ which, for a.e. $x \in \mathcal{O}$ satisfies $B_n \rightarrowtail A(x,u(x))$ in $\mathfrak{M}(\mathbb{R}^N)$. □

We finally give a proof of existence for solutions of (3.53).

Proposition 3.53. *Under the hypotheses of Theorem 3.33, problem* (3.53) *has at least one solution.*

By Theorem 3.52, it is enough to construct an approximating sequence $\{A^n(x,v)\}_n$ of A for which problem (3.52) has a solution. There is a *standard* way to achieve this. It involves the Yoshida approximations of a maximal monotone graph (see Lemma 4.3 of [91] for details).

In the following, $F_p(\xi) \doteq |\xi|^{p-2}\xi$ denote the duality map of $\mathbb{R}^N$ (associated to the gauge $|\cdot|^p$).

Lemma 3.54.
(i) Let B be a maximal monotone graph on $\mathbb{R}^N$. For $\lambda > 0$, the Yosida approximation B_λ of B is defined as

$$B_\lambda \doteq \frac{1}{\lambda}F_p\big(Id - (F_p + \lambda B)^{-1}\big).$$

It is a single valued, everywhere defined, continuous and maximal monotone map. Furthermore, it converges to B in the sense of maximal monotone graphs as $\lambda \to 0^+$.
(ii) Let $\alpha > 0$, $\lambda_0 > 0$ and $m \in \mathbb{R}$ be such that for every $(\xi,\eta) \in B$,

$$\alpha\left(\frac{|\xi|^p}{p} + \frac{|\eta|^{p'}}{p'}\right) \leqslant \langle \eta, \xi\rangle_{\mathbb{R}^N} + m.$$

Then for every $\delta > 0$, there is $\lambda_0 > 0$ such that for every $\xi \in \mathbb{R}^N$, $0 < \lambda < \lambda_0$,

$$(\alpha - \delta)\left(\frac{|\xi|^p}{p} + \frac{|B_\lambda\xi + \lambda F_p(\xi)|^{pp'}}{p'}\right) \leqslant \big\langle B_\lambda\xi + \lambda F_p(\xi),\, \xi\big\rangle_{\mathbb{R}^N} + m. \quad (3.54)$$

Proof of Proposition 3.53. Following Lemma 3.54 above, choose

$$A^n(x,\cdot) = A(x,\cdot)_{1/n} + \frac{1}{n}F_p(\cdot).$$

It clearly satisfies all the hypotheses of Theorem 3.52.

It remains to show that for each fixed $n \in \mathbb{N}$, problem (3.52) has a solution. This can be done by using a Galerkin method, as in [148]. Instead, we shall use the Schauder's fixed point theorem for the map $\mathcal{S}$ which to v given in $L^p(\mathcal{O})$, assigns the solution u^* (given by Theorem 3.33) of the problem

$$\begin{cases} -\text{div}\, d^* = f & \text{in } \mathcal{D}'(\mathcal{O}), \\ (\nabla u^*(x), d^*(x)) \in A^n(x, v(x)), \\ u^* \in W_0^{1,p}(\mathcal{O}). \end{cases}$$

From inequality (3.54), it is easy to check that there is a ball in $L^p(\mathcal{O})$ which is left invariant by $\mathcal{S}$. By Theorem 3.49, $\mathcal{S}$ is continuous for the $L^p(\mathcal{O})$ topology. It is also compact by the compactness of the embedding of $W_0^{1,p}(\mathcal{O})$ into $L^p(\mathcal{O})$. □

3.2.6 Homogenization

In this section we state the main homogenization result in the context of nonlinear diffusion. First we give the result and the proof of the convergence of solutions. Then we study the properties of the homogenized operator. Finally we give some energy convergence and a corrector result.

Theorem 3.55. *Assume p in $(1,+\infty)$, $\alpha > 0$, Ω a bounded domain of $\mathbb{R}^N$, and m in $L^1(\Omega)$. Let A_ε be in $\mathcal{VLL}(\Omega,p,q,\alpha,m)$ for some $q \in [1,p)$.*

Suppose that there exists a cell $Y \subset \mathbb{R}^N$ such that $\mathcal{T}_\varepsilon(A_\varepsilon)$ converges in $\mathcal{VLL}(\Omega \times Y, p, q, \alpha, \widetilde{m})$ to some B and for some $\widetilde{m}$ in $L^1(\Omega \times Y)$. Suppose furthermore, that this convergence is uniform with respect to the second argument v. Let $f_\varepsilon \in W^{-1,p'}(\Omega)$ such that

$$f_\varepsilon \to f_0 \quad \textit{strongly in } W^{-1,p'}(\Omega).$$

Consider a (not necessarily unique) solution $u_\varepsilon \in W_0^{1,p}(\Omega)$ of the problem

$$\begin{cases} -\text{div}\, d_\varepsilon = f_\varepsilon & \textit{in } \mathcal{D}'(\Omega), \\ (\nabla u_\varepsilon(x), d_\varepsilon(x)) \in A_\varepsilon(x, u_\varepsilon(x)) \\ u_\varepsilon \in W_0^{1,p}(\Omega). \end{cases} \tag{3.55}$$

Then the sequence $\{(u_\varepsilon, d_\varepsilon)\}_\varepsilon$ is weakly compact in $W_0^{1,p}(\Omega) \times L^{p'}(\Omega)^N$.

Furthermore, if (u_0, d_0) is one of its limit points, i.e., there is a subsequence $\{\varepsilon_n\}$ of ε converging to 0 and such that

$$\begin{aligned} u_{\varepsilon_n} &\rightharpoonup u_0 \quad \textit{weakly in } W_0^{1,p}(\Omega), \\ d_{\varepsilon_n} &\rightharpoonup d_0 \quad \textit{weakly in } L^{p'}(\Omega)^N, \end{aligned}$$

it follows that

$$\begin{cases} -\text{div}\, d_0 = f_0 & \textit{in } \mathcal{D}'(\Omega) \\ (\nabla u_0(x), d_0(x)) \in A^{hom}(x, u_0(x)), \\ u_0 \in W_0^{1,p}(\Omega) \end{cases} \tag{3.56}$$

where, for almost every $x \in \Omega$ and every $v \in \mathbb{R}$,

$$\begin{aligned} A^{hom}(x,v) = \Big\{ (\xi,\eta) \in \mathbb{R}^N \times \mathbb{R}^N \;\Big|\; & \exists (\widehat{u}, \widehat{d}) \in W^{1,p}_{\text{per},0}(Y) \times L^{p'}(Y)^N, \\ & \mathcal{M}_Y(\widehat{d}) = 0,\ -\text{div}_y\, \widehat{d} = 0 \textit{ in } (\mathcal{C}^\infty_{\text{per}})'(Y) \textit{ such that} \\ & (\xi + \nabla \widehat{u}(y), \eta + \widehat{d}(y)) \in B(x,y,v) \textit{ for a.e. } y \in Y \Big\}. \end{aligned} \tag{3.57}$$

Moreover, A^{hom} belongs to $\mathcal{VLL}(\Omega,p,q,\alpha,\overline{m})$, where $\overline{m} = \mathcal{M}_Y(\widetilde{m})$.

A short proof. The fact that $\{(u_\varepsilon, d_\varepsilon)\}_\varepsilon$ is bounded in $W_0^{1,p}(\Omega) \times L^{p'}(\Omega)^N$ follows as before from the hypothesis that A_ε belongs to $\mathcal{VLL}(\Omega,p,q,\alpha,m)$ and from the Poincaré inequality for $W_0^{1,p}(\Omega)$.

So, one can assume that, up to a subsequence,

$$u_\varepsilon \rightharpoonup u_0 \quad \text{weakly in } W_0^{1,p}(\Omega) \quad \text{and } d_\varepsilon \rightharpoonup d_0 \quad \text{weakly in } L^{p'}(\Omega)^N.$$

Up to a further subsequence, one can therefore suppose that $u_\varepsilon(x) \to u_0(x)$ for a.e. $x \in \Omega$.

By Theorem 1.41, there exists $\widehat{u}$ in $L^p(\Omega; W^{1,p}_{per,0}(Y))$ such that

$$\mathcal{T}_\varepsilon(\nabla u_\varepsilon) \rightharpoonup \nabla u_0 + \nabla_y \widehat{u} \quad \text{weakly in } L^p(\Omega \times Y)^N.$$

Similarly, by Proposition 1.9 (iii), there exists $\widehat{d} \in L^{p'}(\Omega \times Y)^N$ such that

$$\mathcal{T}_\varepsilon(d_\varepsilon) \rightharpoonup d_0 + \widehat{d} \quad \text{weakly in } L^{p'}(\Omega \times Y)^N.$$

Consider the variational formulation of problem (3.55),

$$\int_\Omega (d_\varepsilon, \nabla u_\varepsilon)\, dx = \langle f_\varepsilon, u_\varepsilon \rangle_{W^{-1,p'}(\Omega), W_0^{1,p}(\Omega)}. \tag{3.58}$$

Unfolding its left-hand side, gives

$$\frac{1}{|Y|}\int_{\Omega\times Y} \mathcal{T}_\varepsilon(d_\varepsilon)\cdot \mathcal{T}_\varepsilon(\nabla u_\varepsilon)\, dxdy + \int_{\Lambda_\varepsilon} (d_\varepsilon, \nabla u_\varepsilon)\, dx = \langle f_\varepsilon, u_\varepsilon \rangle_{W^{-1,p'}(\Omega), W_0^{1,p}(\Omega)}.$$

For the second term on the left-hand side, recalling (3.51), one has the estimate

$$-\Big(\int_{\Lambda_\varepsilon} (m(x) + |u_\varepsilon(x)|^q)\, dx\Big) \leqslant \int_{\Lambda_\varepsilon} (d_\varepsilon, \nabla u_\varepsilon)\, dx.$$

Note that the left-hand side integral goes to 0 since $|\Lambda_\varepsilon|$ goes to 0. Consequently,

$$\liminf_{\varepsilon\to 0} \int_{\Lambda_\varepsilon} (d_\varepsilon, \nabla u_\varepsilon)\, dx \geqslant 0, \tag{3.59}$$

and

$$\limsup \frac{1}{|Y|}\int_{\Omega\times Y} \mathcal{T}_\varepsilon(d_\varepsilon)\cdot \mathcal{T}_\varepsilon(\nabla u_\varepsilon)\, dxdy \leqslant \langle f_0, u_0 \rangle_{W^{-1,p'}(\Omega), W_0^{1,p}(\Omega)}. \tag{3.60}$$

Unfolding (3.58) and going to the limit along the considered sequences as in Section 3.1, leads to

$$\begin{cases} \displaystyle\frac{1}{|Y|}\int_{\Omega\times Y} \big(d_0(x) + \widehat{d}(x,y), \nabla\Psi(x) + \nabla_y \Phi(x,y)\big)\, dxdy \\ \qquad\qquad\qquad = \langle f, \Psi \rangle_{W^{-1,p'}(\Omega), W_0^{1,p}(\Omega)} \\ \forall \Psi \in W_0^{1,p}(\Omega), \end{cases} \tag{3.61}$$

which is the unfolded limit problem (3.56).

Since $\mathcal{M}_Y(\widehat{d}) = 0$ and $\mathcal{M}_Y(\nabla_y \widehat{u}) = 0$ (by the periodicity of $\widehat{u}$), from equality (3.61) it follows in particular that

$$\int_\Omega \frac{1}{|Y|}\int_{\Omega\times Y} \big(d_0(x) + \widehat{d}(x,y), \nabla u_0(x) + \nabla_y \widehat{u}(x,y)\big)\, dxdy \\ = \langle f, u_0\rangle_{W^{-1,p'}(\Omega), W_0^{1,p}(\Omega)}. \tag{3.62}$$

On the other hand, unfolding the condition $\big(\nabla u_\varepsilon(x), d_\varepsilon(x)\big) \in A_\varepsilon(x, u_\varepsilon(x))$, yields

$$\big(\mathcal{T}_\varepsilon(\nabla u_\varepsilon)(x,y), \mathcal{T}_\varepsilon(d_\varepsilon)(x,y)\big) \in \mathcal{T}_\varepsilon(A_\varepsilon)\big(x, y, \mathcal{T}_\varepsilon(u_\varepsilon)(x,y)\big).$$

Recalling that $\mathcal{T}_\varepsilon(u_\varepsilon)$ converges strongly to u_0 in $L^p(\Omega\times T)$, therefore a.e. in $\Omega \times Y$ (at least for a subsequence), the hypothesis on $\mathcal{T}_\varepsilon(A_\varepsilon)$ implies

$$\mathcal{T}_\varepsilon(A_\varepsilon)(x, y, \mathcal{T}_\varepsilon(u_\varepsilon)(x,y)) \rightarrowtail B(x, y, u_0(x)).\mathcal{T}_\varepsilon(A_\varepsilon)(x, y, \mathcal{T}_\varepsilon(u_\varepsilon)(x,y)).$$

Then, in view of Proposition 3.41, inequalities (3.59), (3.60) and (3.62),

$$\Big(d_0(x) + \widehat{d}(x,y), \nabla u_0(x) + \nabla_y \widehat{u}(x,y)\Big) \in B(x, y, u_0(x) \quad \text{for a.e. } (x,y) \in \Omega \times Y,$$

and

$$\lim_{\varepsilon\to 0} \int_{\Lambda_\varepsilon} (d_\varepsilon, \nabla u_\varepsilon)\, dx = 0.$$

Finally, formula (3.57) of A^{hom} follows from (3.61) .

It remains to show that A^{hom} belongs to $\mathcal{VLL}(\Omega, p, q, \alpha, \widetilde{m})$. To do so, first check the monotonicity of $A^{hom}(x, v)$ for fixed $v \in \mathbb{R}$ and for a.e. $x \in \Omega$. For $i = 1, 2$, let $(\xi_i, \eta_i) \in A^{hom}(x, v)$, and let $\widehat{u}_i$ and $\widehat{d}_i$ be associated with (ξ_i, η_i) by definition (3.57). By the periodicity of $\widehat{u}_i$, $\displaystyle\int_Y \nabla_y \widehat{u}_i(\cdot, y)\, dy = 0$ so that

$$\langle \eta_1 - \eta_2, \xi_1 - \xi_2\rangle_{\mathbb{R}^N} \\ = \frac{1}{|Y|}\int_Y \langle \eta_1 + \widehat{d}_1(y) - \eta_2 - \widehat{d}_2(y),\ \xi_1 + \nabla\widehat{u}_1(y) - \xi_2 - \nabla\widehat{u}_2(y)\rangle_{\mathbb{R}^N}\, dy.$$

The monotonicity of $B(x, v)$ implies that the term in the right-hand side is nonnegative. Therefore $A^{hom}(x, v)$ is monotone.

To prove that A^{hom} is in $\mathcal{VLL}(\Omega, p, q, \alpha, \widetilde{m})$, it is enough to show that it is the limit of a sequence of single valued elements of $\mathcal{VLL}(\Omega, p, q, \alpha, \widetilde{m})$. This is achieved by using the approximation given by Lemma 3.54. More precisely, for fixed $v \in \mathbb{R}$, consider

$$\widetilde{B}_n(x, y, v) \doteq B(x, y, v)_{1/n} + \frac{1}{n} F_p,$$

which is single valued, uniformly monotone, of Caratheodory type with respect to $((x, y), v)$ in $\mathcal{VLL}(\Omega \times Y, p, q, \alpha - \delta_n, \widetilde{m})$ (for some $\delta_n \geqslant 0$ which goes to 0 as $n \to +\infty$) and converges to B in the sense $\mathcal{VLL}(\Omega \times Y, p, q, \alpha', \widetilde{m})$, uniformly with respect to $v \in \mathbb{R}$ (for $0 < \alpha' < \alpha$ and n large enough).

Then set

$$B_n^{hom}(x,v) = \Big\{(\xi,\eta) \in \mathbb{R}^N \times \mathbb{R}^N; \Big| \, \exists(\widehat{u},\widehat{d}) \in W^{1,p}_{\text{per}}(Y) \times L^{p'}(Y;\mathbb{R}^N),$$
$$\mathcal{M}_Y(\widehat{u}) = \mathcal{M}_Y(\widehat{d}) = 0, \, -\text{div}_y \, \widehat{d} = 0 \text{ in } (\mathcal{C}^\infty_{\text{per}})'(Y),$$
$$\text{such that } \big(\xi + \nabla\widehat{u}(y), \eta + \widehat{d}(y)\big) \in \widetilde{B}_n(x,y,v) \text{a.e. } y \in Y\Big\}.$$

From the properties of $\widetilde{B}_n$, it is clear that $B_n^{hom}(x,v)$ is single valued and of Caratheodory type with respect to (x,v), and in $\mathcal{VLL}(\Omega, p, q, \alpha - \delta_n, \widetilde{m})$. Also, $\widetilde{B}_n$ converges to A^{hom} in $\mathcal{VLL}(\Omega, p, q, \alpha', \widetilde{m})$, uniformly with respect to $v \in \mathbb{R}$. Hence, A^{hom} in $\mathcal{VLL}(\Omega, p, q, \alpha', \widetilde{m})$ for every $\alpha' < \alpha$, which is enough for concluding the proof. □

Remark 3.56 (Convergence of the energy). *As in Section 3.1, one can establish that (at least for a subsequence),*

$$\int_{\Omega \times Y} \mathcal{T}_\varepsilon(d_\varepsilon) \cdot \mathcal{T}_\varepsilon(\nabla u_\varepsilon) \, dxdy \to \int_{\Omega \times Y} (d_0(x) + \widehat{d}(x,y), \nabla u_0(x) + \nabla_y \widehat{u}(x,y)) \, dxdy.$$

However, in general, this does not imply a corrector result. Only in the case of uniformly monotone graphs (in the sense of L^p),(14) *can one give a corrector in the same fashion as in Section 3.1.*

3.3 Unfolding of integral functionals acting on gradients

By considering the subdifferentials of convex functions, the results of the previous section can be applied in connection with the Mosco convergence of the associated functionals (see [92]). In this section, we show how the unfolding method applies to the study of the Γ-convergence of sequences of convex functionals.(15) The result presented here was published in its original form in [64] (see also [65] and [66] for some generalizations).

Recall the definition of Γ-convergence of a sequence which will be used in this chapter.

(14) In this setting, this means there is a constant $\underline{\alpha} > 0$ such that for every $v_j \in \mathbb{R}$, $j = 1, 2$ and every $(\xi_j, \eta_j) \in A_\varepsilon(v_j)$, the following inequality holds a.e. in Ω and for every ε:

$$\langle \eta_1 - \eta_2, \xi_1 - \xi_2 \rangle_{\mathbb{R}^N} + |v_1 - v_2|^p \geqslant \underline{\alpha} \, |\xi_1 - \xi_2|^p_{\mathbb{R}^N}.$$

(15) Γ-convergence, introduced by De Giorgi and his school, is often more general than Mosco-convergence (see [95, 96]).

Definition 3.57 (cf.[20, 21, 34, 45, 82, 96]). *Let X be a topological space, $\{f_n\}_{n\in\mathbb{N}}$ a sequence of functions defined on X with values in $\overline{\mathbb{R}}$. The function $f : X \to \overline{\mathbb{R}}$ is the Γ-limit (technically the Γ-limit inf) of the sequence $\{f_n\}_{n\in\mathbb{N}}$, whenever*

(*i*) *For every $x \in X$, there is a sequence $\{x_n\}_{n\in\mathbb{N}}$ converging to x, with*
$$\liminf_{n\to+\infty} f_n(x_n) = f(x).$$

(*ii*) *For every converging sequence $\{y_n\}_{n\in\mathbb{N}}$ in X,*
$$\liminf_{n\to+\infty} f_n(y_n) \geqslant f(\lim_{n\to+\infty} y_n).$$

Denote by $\mathcal{A}_0$ the class of bounded open subsets of $\mathbb{R}^N$ having a Lipschitz boundary, and by Y the unit reference cell $(0,1)^N$ (16).

Consider a Caratheodory energy density f satisfying

$$\begin{cases} f : (x,z) \in \mathbb{R}^N \times \mathbb{R}^N \mapsto f(x,z) \in [0,+\infty), \\ f(\cdot, z) \text{ Lebesgue measurable, } Y\text{-periodic } \forall z \in \mathbb{R}^N, \\ f(x,\cdot) \text{ convex for a .e. } x \in \mathbb{R}^N. \end{cases} \tag{3.63}$$

For $p \in [1,+\infty)$, and $M > 0$, introduce the two following growth conditions:

$$\begin{aligned} &\exists\, a \in L^1(Y)^+ \text{ and } Y\text{-periodic, such that } f(x,z) \leqslant a(x) + M|z|^p \\ &\qquad\qquad\qquad\qquad \text{for a.e. } x \in \mathbb{R}^N \text{ and every } z \in \mathbb{R}^N, \end{aligned} \tag{3.64}$$

and

$$\begin{aligned} &\exists\, h \in L^p(Y)^+ \text{ and } Y\text{-periodic, such that } |z|^p \leqslant M f(x,z) + h(x) \\ &\qquad\qquad\qquad\qquad \text{for a.e. } x \in \mathbb{R}^N \text{ and every } z \in \mathbb{R}^N. \end{aligned} \tag{3.65}$$

Theorem 3.58. *Let f satisfy* (3.63) *and assume that* (3.64) *holds for some $p \in (1,+\infty)$. Let Ω be in $\mathcal{A}_0$. For $v \in W^{1,p}(\Omega)$, set*

$$F_\varepsilon(v) \doteq \int_\Omega f\Big(\frac{x}{\varepsilon}, \nabla v(x)\Big)\, dx.$$

Then the Γ-limit of the sequence $\{F_\varepsilon\}_\varepsilon$ for the weak topology of $W^{1,p}(\Omega)$, is given by

$$F(u) \doteq \inf\Big\{\int_{\Omega\times Y} f(y, \nabla u(x) + \nabla_y V(x,y))\, dx\, dy \;\Big|\; V \in L^p(\Omega; W^{1,p}_{per}(Y))\Big\}.$$

In addition, if (3.65) *holds (with the same p as for* (3.64)*), this Γ-limit is represented as an integral,*

$$F(u) = \int_\Omega f^p_{hom}(\nabla u(x))\, dx,$$

(16) The results hold for any parallelotope Y; this just simplifies the notations. In particular, $|Y| = 1$.

with density defined by

$$f^p_{hom} : z \in \mathbb{R}^N \mapsto \min\Big\{ \int_Y f(y, z + \nabla v(y))\, dy \mid v \in W^{1,p}_{per}(Y) \Big\}, \tag{3.66}$$

where $W^{1,p}_{per}(Y)$ was defined in (1.32).

Furthermore, F is also the Γ-limit of the same sequence for the strong topology of $L^p(\Omega)$.

The original proof, technical and delicate, of this theorem can be found in [155] and [46], where an abstract measure-theoretical Γ-convergence method is used.

We shall give here a proof making use of the unfolding method which as for the other applications in this chapter, is short and rather easy. But before proceeding, we first recall a standard result of convex analysis and give afterwards a lemma that will be needed in the sequel.

Proposition 3.59. *For a bounded open set Ω in $\mathbb{R}^N$, let $\mathcal{F}_\Omega$ denote the functional*

$$\mathcal{F}_\Omega : V \longmapsto \int_{\Omega \times Y} f\big(y, V(x, y)\big)\, dxdy, \tag{3.67}$$

defined on $L^p(\Omega \times Y)^N$ with values in $\mathbb{R}^+$.

Then, for every bounded open set Ω,

(i) Under hypothesis (3.63), *$\mathcal{F}_\Omega$ is convex lower semicontinuous for the strong topology of $L^p(\Omega \times Y)^N$. Hence, it is also weakly sequentially lower semicontinuous on $L^p(\Omega \times Y)^N$.*

(ii) Under hypotheses (3.63) *and* (3.64), *$\mathcal{F}_\Omega$ is continuous for the strong topology of $L^p(\Omega \times Y)^N$.*

(iii) If hypothesis (3.65) *is also satisfied, $\mathcal{F}_\Omega$ is coercive on $L^p(\Omega \times Y)^N$.*

Proof. The proof of the first point is a well-known consequence of Fatou's lemma (for the strong lower semi-continuity) and the fact that for convex functions in Banach spaces, the notions of weak and strong lower semi-continuity are equivalent. The second point follows from the Lebesgue convergence theorem. The last assertion is straightforward. □

From Theorem (3.58) it follows that actually,

$$F(u) = \inf\{\mathcal{F}_\Omega(\nabla u + \nabla_y V) \mid V \in L^p(\Omega; W^{1,p}_{per}(Y))\}. \tag{3.68}$$

Technically, F depends on both p and Ω. We suppressed this dependence for simplicity, as no ambiguity occurs.

Lemma 3.60. *Let $\Omega \in \mathcal{A}_0$. Assume that f satisfies (3.63) and that (3.64) hold for some p in $(1, +\infty)$. Let u be in $W^{1,p}(\Omega)$ and $\mathcal{U}$ in $C^1(\mathbb{R}^N \times \mathbb{R}^N)$ with $\mathcal{U}(x, \cdot)$ Y-periodic for $x \in \Omega$. For every ε and for a.e. $x \in \Omega$, set*

$$u_\varepsilon(x) = u(x) + \varepsilon \mathcal{U}\Big(x, \frac{x}{\varepsilon}\Big).$$

Then, the following convergence holds:

$$\int_{\Lambda_\varepsilon} f\Big(\frac{x}{\varepsilon}, \nabla u_\varepsilon(x)\Big)\, dx \to 0.$$

Proof. By (3.63) and that (3.64), it is enough to show that

$$\int_{\Lambda_\varepsilon} a\Big(\frac{x}{\varepsilon}\Big)\, dx \to 0 \quad \text{and} \quad \int_{\Lambda_\varepsilon} |\nabla u_\varepsilon(x)|^p\, dx \to 0. \tag{3.69}$$

For the convergence of the first integral above, see (1.7).

For the second integral in (3.69), the convergence to 0 follows from the Lebesgue dominated convergence theorem. Indeed, on the one hand, $|\Lambda_\varepsilon| \to 0$ as observed in Remark 1.1. On the other hand, the elements of the sequence $\{|\nabla u_\varepsilon(x)|^p\}_\varepsilon$ are bounded above by a fixed function in $L^p(\Omega)$ (for example, $|\nabla u(\cdot)| + 2\|\mathcal{U}\|_{C^1(\overline{\Omega \times Y})}$ for $\varepsilon \leqslant 1$). □

Remark 3.61. *By density, the result of this lemma holds true for $\mathcal{U}$ in $L^p(\Omega; W^{1,p}_{per}(Y)$.*

We are now in position to give the proof of homogenization Theorem 3.58. It is contained in the three propositions given below.

Proposition 3.62. *Let Ω be in $\mathcal{A}_0$ and p in $[1, +\infty)$. Assume that f satisfies (3.63). Then, for every $u \in W^{1,p}(\Omega)$ and every sequence $\{u_\varepsilon\}_\varepsilon$ converging weakly to u in $W^{1,p}(\Omega)$,*

$$\liminf_{\varepsilon \to 0} \int_\Omega f\Big(\frac{x}{\varepsilon}, \nabla u_\varepsilon\Big)\, dx \geqslant F(u).$$

Proof. It is enough to assume that, up to a subsequence,

$$\lim \int_\Omega f\Big(\frac{x}{\varepsilon}, \nabla u_\varepsilon(x)\Big)\, dx \quad \text{exists and is finite.}$$

Up to a further subsequence, Theorem 1.41(ii) implies that there exists U in $L^p(\Omega; W^{1,p}_{per}(Y))$ with

$$\mathcal{T}_\varepsilon(\nabla u_\varepsilon) \rightharpoonup \nabla u + \nabla_y U \quad \text{weakly in } L^p(\Omega \times Y)^N. \tag{3.70}$$

According to Proposition 1.8 (i) and using definition (3.67), we have successively,

$$
\begin{aligned}
&\int_{\Omega\times Y} f\big(y,\mathcal{T}_\varepsilon(\nabla u_\varepsilon)(x,y)\big)\,dxdy = \mathcal{F}_\Omega(\mathcal{T}_\varepsilon(\nabla u_\varepsilon))\\
&= \int_{\widehat{\Omega}_\varepsilon\times Y} f\big(y,\mathcal{T}_\varepsilon(\nabla u_\varepsilon)(x,y)\big)\,dx + \int_{\Lambda_\varepsilon\times Y} f\big(y,\mathcal{T}_\varepsilon(\nabla u_\varepsilon)(x,y)\big)\,dxdy\\
&= \int_{\widehat{\Omega}_\varepsilon\times Y} \mathcal{T}_\varepsilon\Big[f\Big(\frac{x}{\varepsilon},\nabla u_\varepsilon(x)\Big)\Big](x,y)\big)\,dxdy + \int_{\Lambda_\varepsilon\times Y} f\big(y,0)(x,y)\,dxdy\\
&= \int_{\widehat{\Omega}_\varepsilon} f\Big(\frac{x}{\varepsilon},\nabla u_\varepsilon(x)\Big)\,dx + \int_{\Lambda_\varepsilon\times Y} f\big(y,0)(x,y)\,dxdy\\
&= \int_{\Omega} f\Big(\frac{x}{\varepsilon},\nabla u_\varepsilon(x)\Big)\,dx - \int_{\Lambda_\varepsilon} f\Big(\frac{x}{\varepsilon},\nabla u_\varepsilon(x)\Big)\,dx\\
&\qquad\qquad\qquad + \int_{\Lambda_\varepsilon\times Y} f\big(y,0)(x,y)\,dxdy.
\end{aligned}
\tag{3.71}
$$

Consequently,

$$
\begin{aligned}
\int_{\Omega} f\Big(\frac{x}{\varepsilon},\nabla u_\varepsilon(x)\Big)\,dx = \mathcal{F}_\Omega(\mathcal{T}_\varepsilon(\nabla u_\varepsilon)) &+ \int_{\Lambda_\varepsilon} f\Big(\frac{x}{\varepsilon},\nabla u_\varepsilon(x)\Big)\,dx\\
&- \int_{\Lambda_\varepsilon\times Y} f\big(y,0)(x,y)\,dxdy.
\end{aligned}
\tag{3.72}
$$

The second integral in the right-hand side converges to 0 by Lemma 3.60 (see also Remark 3.61). Since the Lebesgue measure of Λ_ε converges to 0, by hypothesis (3.64) the last integral also converges to 0.

Therefore, using convergence (3.70) together with Proposition 3.59 in equality (3.72), we get

$$
\liminf_{\varepsilon\to 0}\int_{\Omega} f\Big(\frac{x}{\varepsilon},\nabla u_\varepsilon\Big)\,dx \geqslant \mathcal{F}_\Omega\big(y,\nabla u+\nabla_y U\big) \geqslant F(u),
$$

which concludes the proof. □

Proposition 3.63. *Let* Ω *be in* $\mathcal{A}_0$. *Assume that* f *satisfies* (3.63) *and that* (3.64) *holds for some* $p \in (1,+\infty)$. *Then, for every* $u \in W^{1,p}(\Omega)$, *there is a sequence* $\{w_\varepsilon\}_\varepsilon$ *converging weakly to* u *in* $W^{1,p}(\Omega)$, *such that*

$$
\lim_{\varepsilon\to 0}\int_{\Omega} f\Big(\frac{x}{\varepsilon},\nabla w_\varepsilon(x)\Big)\,dx = F(u).
$$

Proof. Let u in $W^{1,p}(\Omega)$ and $\mathcal{U}$ in $C^1(\mathbb{R}^N\times\mathbb{R}^N)$ with $\mathcal{U}(x,\cdot)$ Y-periodic for $x\in\Omega$. For every ε and for a.e. $x\in\Omega$, set

$$
u_\varepsilon(x) = u(x) + \varepsilon\,\mathcal{U}\Big(x,\frac{x}{\varepsilon}\Big).
$$

Then

$$\nabla u_\varepsilon(x) = \nabla u(x) + \varepsilon \nabla \mathcal{U}\Big(x, \frac{x}{\varepsilon}\Big) + \nabla_y \mathcal{U}\Big(x, \frac{x}{\varepsilon}\Big) \quad \text{for a.e. } x \in \Omega. \tag{3.73}$$

As before (see the preceding proof), going back to formulas (3.71)-(3.72) and using (3.73), implies

$$\begin{aligned}
&\int_\Omega f\Big(\frac{x}{\varepsilon}, \nabla u_\varepsilon(x)\Big)\, dx \\
&= \int_{\Omega\times Y} f\Big(y, \mathcal{T}_\varepsilon(\nabla u)(x,y) + \varepsilon\nabla\mathcal{U}\Big(\varepsilon\Big[\frac{x}{\varepsilon}\Big]_Y + \varepsilon y, y\Big) + \nabla_y\mathcal{U}\Big(\varepsilon\Big[\frac{x}{\varepsilon}\Big]_Y + \varepsilon y, y\Big)\Big)\, dxdy \\
&\qquad - \int_{\Lambda_\varepsilon} f\Big(\frac{x}{\varepsilon}, \nabla u_\varepsilon(x)\Big)\, dx + \int_{\Lambda_\varepsilon\times Y} f(y, 0)(x,y)\, dxdy.
\end{aligned}$$

We have seen at the end of the proof of the preceding proposition that the last two integrals converge to 0 (the first one by Lemma 3.60, the latter by the properties of f and by the fact that $|\Lambda_\varepsilon| \to 0$).

It remains to pass to the limit in the first integral in the right-hand side. Due to continuity properties of $\nabla_x U$ and to the periodicity of $\nabla_y U$ (cf. Proposition 1.9(iii)), the following convergences hold:

$$\begin{aligned}
&\varepsilon\nabla_x\mathcal{U}\Big(\cdot, \frac{\cdot}{\varepsilon}\Big) \to 0 \quad \text{uniformly in } \mathbb{R}^N, \\
&\mathcal{T}_\varepsilon\Big(\varepsilon\nabla_x\mathcal{U}\Big(\cdot, \frac{\cdot}{\varepsilon}\Big)\Big) \to 0 \quad \text{uniformly in } \Omega\times Y, \\
&\nabla_y\mathcal{U}\Big(\cdot, \frac{\cdot}{\varepsilon}\Big) \rightharpoonup \int_Y \nabla_y\mathcal{U}(\cdot, y)\, dy = 0 \quad \text{weak-}* \text{ in } L^\infty(\mathbb{R}^N), \\
&\mathcal{T}_\varepsilon\Big(\nabla_y\mathcal{U}\Big(\cdot, \frac{\cdot}{\varepsilon}\Big)\Big) \rightharpoonup \nabla_y\mathcal{U} \quad \text{weak-}* \text{ in } L^\infty(\Omega\times\mathbb{R}^N).
\end{aligned}$$

This implies that

$$u_\varepsilon \rightharpoonup u \quad \text{weakly in } W^{1,p}(\Omega),$$

as well as

$$\mathcal{T}_\varepsilon(\nabla u_\varepsilon) \to \nabla u + \nabla_y\mathcal{U} \quad \text{strongly in } L^p(\Omega\times Y)^N,$$

since $\mathcal{T}_\varepsilon(\nabla u)$ converges strongly to ∇u in $L^p(\Omega\times Y)$. Therefore, due to the continuity of $\mathcal{F}_\Omega$,

$$\int_\Omega f\Big(\frac{x}{\varepsilon}, \nabla u_\varepsilon(x)\Big)\, dx \to \mathcal{F}_\Omega(\nabla u + \nabla_y\mathcal{U}).$$

Because of the density of the smooth functions in $L^p(\Omega; W^{1,p}_{per}(Y)$, and of the continuity of $\mathcal{F}_\Omega$ again, given some positive δ, it is easy to find U in $C^1(\mathbb{R}^N\times\mathbb{R}^N)$ with $U(x,\cdot)$ Y-periodic for every $x\in\Omega$, such that

$$\mathcal{F}_\Omega(\nabla u + \nabla_y U) \leqslant F(u) + \delta.$$

The corresponding sequence $\{u_\varepsilon^\delta\}_\varepsilon$ almost satisfies the required condition. It remains to apply a diagonal procedure to construct a sequence with the required property, by setting $w_\varepsilon \doteq u_\varepsilon^\varepsilon$. □

Proposition 3.64. *Let $\Omega \in \mathcal{A}_0$. Assume that f satisfies* (3.63) *and both conditions* (3.64) *and* (3.65) *for the same p in $(1, +\infty)$. Then, for every u in $W^{1,p}(\Omega)$,*

$$F(u) = \int_\Omega f^p_{hom}(\nabla u(x))\, dx,$$

where the function f^p_{hom} is defined in (3.66).

Here one uses a non trivial result due to Castaing on measurable selections. Let Ω, X be sets, and Γ a multifunction from Ω to X. A function $\sigma : \Omega \to X$ is said to be a selection of Γ if $\sigma(x) \in \Gamma(x)$ for every $x \in \Omega$. The measurable selection result below is proved in [51] (Theorem III. 6 and Proposition III. 11).

Theorem 3.65. *[51]. Let X be a separable metric space, $(\Omega; \mathcal{M})$ a measurable space, and Γ a multifunction from Ω to X. Assume that for every $x \in \Omega$, $\Gamma(x)$ is nonempty and complete in X. Assume moreover, that for every closed subset F of X, the subset $\{x \in \Omega \mid \Gamma(x) \cap F \neq \varnothing\}$ belongs to $\mathcal{M}$. Then Γ admits a $\mathcal{M}$-measurable selection.* □

Proof of Proposition 3.64. From definition (3.66), for u in $W^{1,p}(\Omega)$ and V in $L^p(\Omega; W^{1,p}_{per}(Y))$, the following inequality holds for a.e. $x \in \Omega$:

$$\begin{aligned}&\int_Y f(y, \nabla u(x) + \nabla_y V(x,y))\, dy \\ &\qquad \geqslant \inf\Big\{\int_Y f(y, \nabla u(x) + \nabla v(y))\, dy \mid v \in W^{1,p}_{per}(Y)\Big\} = f^p_{hom}(\nabla u(x)).\end{aligned}$$

Integrating over Ω, taking the infimum of the left-hand side, recalling definition (3.67) and equality (3.68), give

$$F(u) \geqslant \int_\Omega f^p_{hom}(\nabla u(x))\, dx. \tag{3.74}$$

The reverse inequality is obvious if $\displaystyle\int_\Omega f^p_{hom}(\nabla u(x))\, dx = +\infty$.

To prove it in the case where $\displaystyle\int_\Omega f^p_{hom}(\nabla u(x))\, dx < +\infty$, we make use of Castaing's selection theorem above.

Note first, that due to (3.63) and (3.64), f^p_{hom} is convex and continuous on $\mathbb{R}^N$. By (3.65) and the Poincaré-Wirtinger inequality, the infimum defining f^p_{hom} in (3.66) is achieved for every $z \in \mathbb{R}^N$. This implies that the set where this infimum is reached, denoted $\Gamma(z)$, is nonempty, bounded, convex and closed. We claim that the multi-application

$$\Gamma : z \in \mathbb{R}^N \mapsto \Big\{v \in W^{1,p}_{per}(Y) \mid \mathcal{M}_Y(v) = 0, \int_Y f(y, z + \nabla v(y))dy = f^p_{hom}(z)\Big\}$$

has a $\mathcal{B}(\mathbb{R}^N)$-measurable selection, where $\mathcal{B}(\mathbb{R}^N)$ denotes the Borel σ-algebra of $\mathbb{R}^N$.

To do so, it is enough to prove the same for the restriction of Γ to every ball in $\mathbb{R}^N$. The image of a ball by Γ is bounded, thus included in a closed ball of $W^{1,p}_{per}(Y)$. These balls are metrizable for the weak topology (since the dual space of $W^{1,p}_{per}(Y)$ is separable). Therefore, by Theorem 3.65, in order to establish the existence of a measurable selection of Γ, it is enough to show that for every bounded and weakly closed subset F of the space $W^{1,p}_{per}(Y)$,

$$\Gamma^-(F) \doteq \{\zeta \in \mathbb{R}^N \mid \Gamma(\zeta) \cap F \neq \varnothing\},$$

belongs to $\mathcal{B}(\mathbb{R}^N)$.

We now claim that $\Gamma^-(F)$ is actually closed. To do so, let $\{z_n\} \subset \Gamma^-(F)$, $z \in \mathbb{R}^N$ with $z_n \to z$. For every $n \in \mathbb{N}$, let $v_n \in \Gamma(z_n) \cap F$. The continuity and the finiteness of f^p_{hom} imply that

$$\begin{aligned}\limsup_{n\to+\infty} \int_Y |z_n + \nabla v_n(y)|^p \, dy &\leqslant \lim_{n\to+\infty} \int_Y f(y, z_n + \nabla v_n(y)) \, dy \\ &= \lim_{n\to+\infty} f^p_{hom}(z_n) = f^p_{hom}(z) < +\infty,\end{aligned}$$

so that, by the Poincaré-Wirtinger inequality for $W^{1,p}(Y)$, the sequence $\{v_n\}_n$ is bounded in $W^{1,p}_{per}(Y)$. Hence, there is a subsequence $\{v_{n_k}\}_k$ and some v_∞ such that

$$v_{n_k} \rightharpoonup v_\infty \quad \text{weakly in } W^{1,p}(Y).$$

Since F is weakly closed, the limit v_∞ is in F. Moreover, $\mathcal{M}_Y(v_\infty) = 0$.

From the continuity of f^p_{hom} and the lower-semicontinuity of the mapping

$$w \mapsto \int_Y f(y, w(y)) \, dy,$$

we get

$$\begin{aligned}f^p_{hom}(z) \leqslant \int_Y f(y, z + \nabla v_\infty(y)) \, dy &\leqslant \liminf_{k\to+\infty} \int_Y f(y, z_{h_k} + \nabla v_{h_k}(y)) \, dy \\ &= \lim_{k\to+\infty} f^p_{hom}(z_{h_k}) = f^p_{hom}(z).\end{aligned}$$

Thus v_∞ belongs to $\Gamma(z) \cap F$ and z belongs to $\Gamma^-(F)$, proving that the latter is closed. Consequently, Γ admits a measurable selection σ.

For a given u such that

$$\int_\Omega f^p_{hom}(\nabla u(x)) \, dx < +\infty,$$

let us define the function U as,

$$U(x) = \sigma(\nabla u(x)) \quad \text{for a.e. } x \in \Omega.$$

Then U is Lebesgue measurable, with values in $W^{1,p}_{per}(Y)$ and, by the definition of Γ,

$$f^p_{hom}(\nabla u(x)) = \int_Y f(y, \nabla u(x) + \nabla_y U(x)(y))\, dy \qquad \text{for a.e. } x \in \Omega. \qquad (3.75)$$

Integrating (3.75) over Ω and taking into account (3.68) yields

$$\begin{aligned}\mathcal{F}_\Omega(\nabla u + \nabla_y U) &= \int_{\Omega\times Y} f(y, \nabla u(x) + \nabla_y U(x)(y))\, dxdy \\ &= \int_\Omega f^p_{hom}(\nabla u(x))\, dx < +\infty.\end{aligned}$$

Therefore, due to (3.65), $\nabla_y U$ is in $L^p(\Omega \times Y)^N$. By the Poincaré-Wirtinger inequality, U actually belongs to $L^p(\Omega; W^{1,p}_{per}(Y))$. Consequently,

$$\int_\Omega f^p_{hom}(\nabla u(x))\, dx \geqslant \inf\left\{\mathcal{F}_\Omega(\nabla u + \nabla_y V) \mid V \in L^p(\Omega; W^{1,p}_{per}(Y))\right\} = F(u).$$

This, together with (3.74), completes the proof of Proposition 3.64 and concludes that of Theorem 3.58. □

Part II

Unfolding in Perforated Domains

Part II Unfolding in Perforated Domains

In this second part, the unfolding method is adapted to the case of periodically perforated domains, where the holes are of the same size ε as the period.

The unfolding method is very well suited for these kind of problems because although the solution for each ε-problem is in a functional space which depends wildly upon ε, the unfolded sequence is in a fixed functional space. There is no need to introduce extension operators into the holes (which require more regularity on the boundaries of the holes).

The convergence results can then be expressed in fixed spaces and in a simple form for the unfolded solutions and then translated in convenient forms for the original solutions (for example by extending them by 0 into the holes, or better yet by extending them by their local average into these holes; see Chapter 6 for such an example). This reason, in itself, justifies the method, which was originally presented in a preliminary form in [59].

Chapter 4 introduces the variants of all the unfolding operators adapted to the case of perforated domains. Most of the results are equivalent to those in Chapter 1. However, one technical hypothesis, which is automatically satisfied in the case of fixed domains has to be introduced (see Definition 4.30). The scale-splitting operators play a more central role (see Section 4.3.2). The $\mathbf{Q}_1$-interpolate has the extra property of being defined on the whole domain including the holes. In order to take into account boundary terms on the boundaries of the perforations, one also introduces a boundary unfolding operator. Chapters 5 and 6 present several applications.

The methods of Section 1.5 (unfolding with parameters and multiscales) are easily generalized to the perforated case.

In this part, due to connectedness requirements, the domains are in $\mathbb{R}^N$ *for* $N \geqslant 2$.

Chapter 4

Unfolding operators in perforated domains

In the case of periodic problems with holes (of the same size as the period), the first difficulty arises from the fact the functions (e.g. the solutions of elliptic problems) are defined on domains Ω_ε^* varying with ε (see (4.2)).

What kind of convergence can be expected for a sequence $\{u_\varepsilon\}_\varepsilon$ of such functions ? One approach for sufficiently smooth holes not intersecting the boundary of Ω is to use uniformly bounded extension operators P_ε from $H^1(\Omega_\varepsilon^*)$ to $H^1(\Omega)$ (see [20, 38, 52, 77, 78, 89, 135]). The weak convergence of $\{P_\varepsilon(u_\varepsilon)\}_\varepsilon$ in the fixed space can then be proved.

The choice of the cell Y can be critical. For example, Figure 4.1 shows two possible choices of unit cell, which differ only by the position of the cell with respect to the origin of $\mathbb{R}^N$ (the hole S is the same in both cases). The problems are therefore identical. But for the one on the left, provided the hole is with Lipschitz boundary, one can construct such an extension operator. For the choice on the right no such extension operator can be constructed on a single cell!

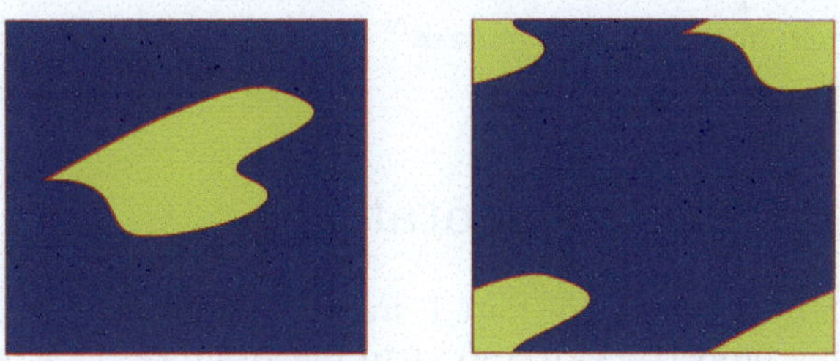

Figure 4.1: Two versions of $Y^* = Y \backslash S$ (in blue) for the same perforations

In absence of extension operators, one cannot speak about "convergence" of the sequence $\{u_\varepsilon\}_\varepsilon$ in a fixed space. As in the case without holes, an unfolding operator $\mathcal{T}_\varepsilon^*$ is introduced (with the notations from Section 4.1),

D. Cioranescu et al., *The Periodic Unfolding Method*, Series in Contemporary Mathematics 3, https://doi.org/10.1007/978-981-13-3032-2_4

for perforated domains in Section 4.2. This operator maps functions defined on Ω_ε^* to functions on the fixed domain $\Omega \times Y^*$ (see Definition 4.1). Therefore neither extension operator nor regularity hypothesis on the geometry of the holes are needed.

Subsections 4.2.2 and 4.2.3 introduce the corresponding local and averaging operators for perforated domains. Section 4.3 considers the case of sequences in $W^{1,p}(\Omega_\varepsilon^*)$. In particular, in Subsection 4.3.2 the case where $\{\|u_\varepsilon\|_{W^{1,p}(\Omega_\varepsilon^*)}\}_\varepsilon$ bounded is treated.

The two sequences $\{\mathcal{T}_\varepsilon^*(u_\varepsilon)\}_\varepsilon$ and $\{\mathcal{T}_\varepsilon^*(\nabla u_\varepsilon)\}_\varepsilon$ are bounded in the fixed spaces $L^p(\Omega; W^{1,p}(Y^*))$ and $L^p(\Omega \times Y^*)^N$, thereby allowing the use of standard convergences. In order to identify the possible limits of these two sequences and their relationships, a Poincaré-Wirtinger hypothesis in Y^* is needed (in a way similar to that of [10]). When S is not compact in Y (the case of non-isolated holes), an extra condition in terms of a Poincaré-Wirtinger inequality is required for the union of the reference cell and its translates by a period (Hypothesis $(\mathbf{H_p})$, see Definition 4.30).

In order to treat non-homogeneous Neumann conditions on the boundaries of the holes, a boundary unfolding operator is defined in Section 4.4.

Section 4.5 considers the case of cracks instead of holes, since the Poincaré-Wirtinger hypothesis does not necessarily require that Y^* be only on one side of its boundary.

The method also applies for situations where no choice of the basis of periods gives a parallelotope Y with $Y^* \doteq Y \backslash S$ connected, a condition which is necessary for the validity of the Poincaré-Wirtinger inequality in Y^*. In such situations (see Section 4.6), what is required is the existence of a reference cell (not necessarily a parallelotope) having the paving property with respect to the period basis, and in which the part occupied by the material is connected. An example is given in Figure 4.4 below. For an analogous geometry but treated differently, we refer the reader [1].

Results similar to those of Section 2.3 (for the case $p = 1$ and for measures) can be obtained for periodically perforated domains.

In this chapter the domains are in $\mathbb{R}^N$ for $N \geqslant 2$ because of the connectedness condition.

4.1 Definitions and notations

We use the general notations of Part I, in particular Definition 2.1 of a set having the paving property with respect to the group $\mathbf{G}$.

From now on, we reserve the notation Y (the reference cell) for a bounded open set having the paving property with respect to the group $\mathbf{G}$.

We also use the notation $\mathcal{P}$ for the open parallelotope generated by the basis $\mathbf{B}$, which was denoted Y in Section 1.1, and defined by (2.1), i.e.,

$$\mathcal{P} \doteq \Big\{ y \in \mathbb{R}^N \;\Big|\; y = \sum_{i=1}^N y_i \mathbf{b}_i,\ (y_1, \dots, y_n) \in (0,1)^N \Big\}. \tag{4.1}$$

It plays an important role in Section 4.6 for the definition of the macro-micro operators $\mathcal{Q}_\varepsilon^*$ and $\mathcal{R}_\varepsilon^*$ (the analogues of $\mathcal{Q}_\varepsilon$ and $\mathcal{R}_\varepsilon$ of Section 1.6).

Let S be a closed strict subset of $\overline{Y}$ and denote by Y^* the part occupied by the material,

$$Y^* = Y \backslash S.$$

Along this chapter, the sets S and Y^* will be called the reference hole and the perforated cell, respectively.

Let Ω be a given domain in $\mathbb{R}^N$. The perforated domain Ω_ε^* is obtained by removing from Ω the set of holes S_ε, i.e.,

$$\Omega_\varepsilon^* = \Omega \backslash S_\varepsilon \quad \text{where} \quad S_\varepsilon = \bigcup_{\xi \in \mathbf{G}} \varepsilon(\xi + S). \tag{4.2}$$

The following notations will be used (see Figure 4.2):

$$\begin{aligned} \widehat{\Omega}_\varepsilon^* &= \widehat{\Omega}_\varepsilon \backslash S_\varepsilon, \\ \Lambda_\varepsilon^* &= \Omega_\varepsilon^* \backslash \widehat{\Omega}_\varepsilon^* = \Lambda_\varepsilon \backslash S_\varepsilon, \\ \widehat{S}_\varepsilon &= \widehat{\Omega}_\varepsilon \cap S_\varepsilon, \end{aligned} \tag{4.3}$$

where the sets $\widehat{\Omega}_\varepsilon$ and Λ_ε are defined, as before, by (1.3). The holes in the set $\widehat{S}_\varepsilon$ do not intersect $\partial\Omega$. The boundary of the set of holes in Ω is $\partial S_\varepsilon \cap \Omega$, while $\partial \widehat{S}_\varepsilon$ is the boundary of the holes that are included in $\widehat{\Omega}_\varepsilon$.

Figure 4.2: The sets Ω_ε^*, $\widehat{\Omega}_\varepsilon^*$ (in dark blue) and Λ_ε^* (in light green)

We will also use similar notations when applied to the whole of $\mathbb{R}^N$,

$$(\mathbb{R}^N)^* = \mathbb{R}^N \backslash \bigcup_{\xi \in \mathbf{G}} (\xi + S), \quad (\mathbb{R}^N)_\varepsilon^* = \mathbb{R}^N \backslash S_\varepsilon, \tag{4.4}$$

so that $(\mathbb{R}^N)^*$ is nothing else than $\mathbb{R}^N$ perforated **G**-periodically by S, while $(\mathbb{R}^N)_\varepsilon^*$ is the $\mathbb{R}^N$ perforated by $\varepsilon\mathbf{G}$-periodically by εS. Consequently, another equivalent definition for Ω_ε^* is the following one:

$$\Omega_\varepsilon^* = (\mathbb{R}^N)_\varepsilon^* \cap \Omega.$$

In Figure 4.2 above, the cell Y is a parallelotope, this is the simplest classical situation. In general periodically perforated domains, the situation can be more complicated for problems involving derivatives. In Section 4.3 the Poincaré-Wirtinger inequality is used for Y^* which requires it to be connected (see Definition 4.30 for Hypothesis $\mathbf{H}_p$). In the example of Figure 4.3, there is no choice of a parallelotope Y leading to a connected Y^*. Nevertheless, in this case, it happens that there are many possible cells Y with Y^* connected.

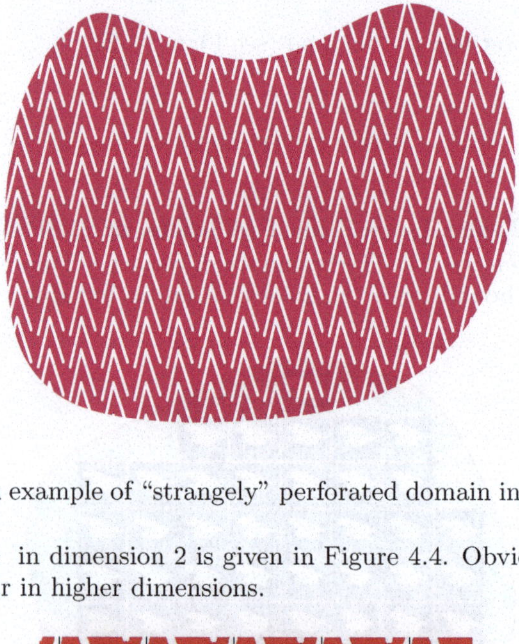

Figure 4.3: An example of "strangely" perforated domain in two dimensions

An example in dimension 2 is given in Figure 4.4. Obviously, such situations can occur in higher dimensions.

Figure 4.4: An example of Y^* (in dark blue); $\mathcal{P}$ in light beige

In dimension 2, the fact that Y^* is connected implies that S is included in Y (not only in $\overline{Y}$). In dimensions higher than 2, this is not the case. This can be seen in the example in dimension 3 from Figure 4.5 below,

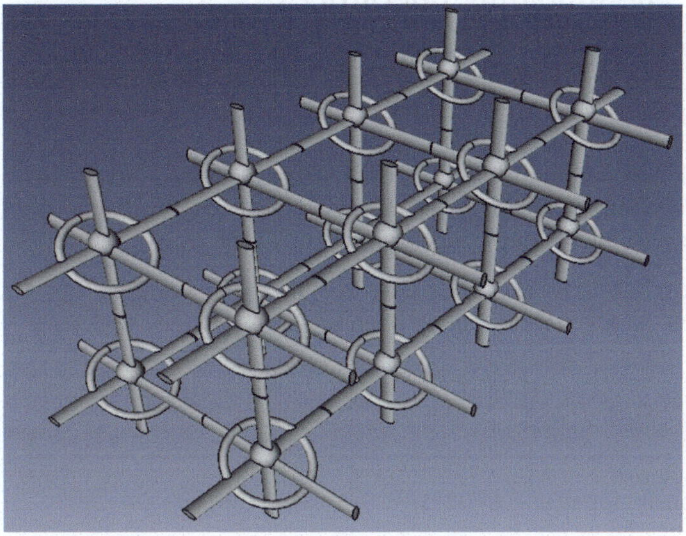

Figure 4.5: An example of set S_ε in $\mathbb{R}^3$

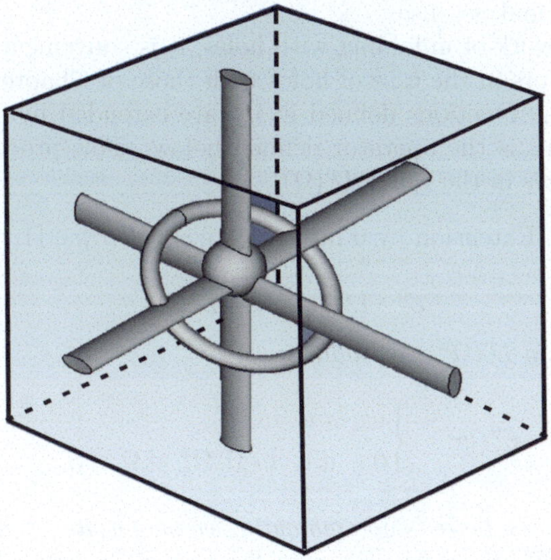

Figure 4.6: The corresponding unit cell.

Actually, in this situation (see Figure 4.6), the roles of Y^* and S can be exchanged (since both are connected) !

4.2 The operators for unfolding in periodically perforated domains

4.2.1 The unfolding operator

In this subsection, we define an unfolding operator $\mathcal{T}_\varepsilon^*$ specific to perforated domains. The first characteristic of this operator is that it maps functions defined on the oscillating domain Ω_ε^* to functions defined on the fixed domain $\Omega \times Y^*$.

The definitions used here differ slightly from those introduced originally in [72]-[73]. They follow the usage for fixed domains (see Part 1, Section 1.1). This allows to treat more general situations (such as in Section 4.6).

Definition 4.1 (The unfolding operator $\mathcal{T}_\varepsilon^*$). *For any function ϕ Lebesgue-measurable on Ω_ε^*, the unfolding operator $\mathcal{T}_\varepsilon^*$ is defined by*

$$\mathcal{T}_\varepsilon^*(\phi)(x,y) = \begin{cases} \phi\Big(\varepsilon\Big[\dfrac{x}{\varepsilon}\Big]_Y + \varepsilon y\Big) & \textit{for a.e.} \quad (x,y) \in \widehat{\Omega}_\varepsilon \times Y^*, \\ 0 & \textit{for a.e.} \quad (x,y) \in \Lambda_\varepsilon \times Y^*. \end{cases}$$

Obviously, for $v, w \in L^p(\Omega_\varepsilon^*)$,

$$\mathcal{T}_\varepsilon^*(v\,w) = \mathcal{T}_\varepsilon^*(v)\,\mathcal{T}_\varepsilon^*(w).$$

For ϕ Lebesgue-measurable on $\widehat{\Omega}_\varepsilon^*$, we extend it by zero in $\Omega_\varepsilon^* \backslash \widehat{\Omega}_\varepsilon^*$, so the above definition makes sense.

In the framework of unfolding with holes, it is convenient to connect the unfolding operators in the case of holes with those of Chapter 1. In order to do so, in [59], the functions defined in Ω_ε^* are extended by 0 into the holes (i.e. $\Omega \backslash \Omega_\varepsilon^*$). This is the operator defined below. This procedure works for weak convergences in the space $L^p(\Omega)$

Definition 4.2 (Extension by 0 into the holes). *For $p \in [1, +\infty]$, the extension operator*

$$\mathcal{E}_\varepsilon^0 : L^p(\Omega_\varepsilon^*) \longrightarrow L^p(\Omega),$$

is defined for w in $L^p(\Omega_\varepsilon^*)$, as follows:

$$\mathcal{E}_\varepsilon^0(w) = \begin{cases} w & \textit{a.e. in } \Omega_\varepsilon^*, \\ 0 & \textit{a.e. in } \Omega \backslash \Omega_\varepsilon^* \equiv \Omega \cap S_\varepsilon. \end{cases}$$

For simplicity, when there is no ambiguity, we may write $\widetilde{\cdot}^{\,\varepsilon}$ in place of $\mathcal{E}_\varepsilon^0(\cdot)$. At the level of Y^*, the extension into the hole S is denoted $\mathcal{E}_S^0$. For Ψ in $L^p(Y^*)$, this means that

$$\mathcal{E}_S^0(\Psi) = \begin{cases} \Psi & \textit{in } Y^*, \\ 0 & \textit{in } S, \end{cases}$$

while for Ψ *in* $L^p(\Omega \times Y^*)$,

$$\mathcal{E}_S^0(\Psi) = \begin{cases} \Psi & \text{in } \Omega \times Y^*, \\ 0 & \text{in } \Omega \times S. \end{cases}$$

Remark 4.3. *The relationship between* $\mathcal{T}_\varepsilon$ *and* $\mathcal{T}_\varepsilon^*$ *is*

$$\mathcal{T}_\varepsilon \circ \mathcal{E}_\varepsilon^0 = \mathcal{E}_S^0 \circ \mathcal{T}_\varepsilon^*$$

For w *defined on* Ω_ε^*,

$$\mathcal{T}_\varepsilon^*(w) = \mathcal{T}_\varepsilon \circ \mathcal{E}_\varepsilon^0(w)_{|_{\Omega \times Y^*}} = \mathcal{T}_\varepsilon(\widetilde{w}^\varepsilon)_{|_{\Omega \times Y^*}}. \tag{4.5}$$

Actually, these last equalities still hold with every extension of w *from* Ω_ε^* *into* Ω. *In particular, for* w *defined on* Ω,

$$\mathcal{T}_\varepsilon^*(w_{|_{\Omega_\varepsilon^*}}) = \mathcal{T}_\varepsilon(w)_{|_{\Omega \times Y^*}}.$$

Because of relationship (4.5), the operator $\mathcal{T}_\varepsilon^*$ enjoys properties which follow directly from those of $\mathcal{T}_\varepsilon$ listed in Proposition 1.8.

Proposition 4.4. *Suppose* $p \in [1, +\infty]$. *The operator* $\mathcal{T}_\varepsilon^*$ *is linear and continuous from* $L^p(\Omega_\varepsilon^*)$ *to* $L^p(\Omega \times Y^*)$. *For* ϕ *in* $L^1(\Omega_\varepsilon^*)$ *and* w *in* $L^p(\Omega_\varepsilon^*)$,

$$\begin{aligned} &\|\mathcal{T}_\varepsilon^*(w)\|_{L^p(\Omega \times Y^*)} = |Y|^{1/p} \, \|w \, \mathbf{1}_{\widehat{\Omega}_\varepsilon^*}\|_{L^p(\Omega_\varepsilon^*)} \leqslant |Y|^{1/p} \, \|w\|_{L^p(\Omega_\varepsilon^*)}, \\ &\frac{1}{|Y|} \int_{\Omega \times Y^*} \mathcal{T}_\varepsilon^*(\phi)(x, y) \, dx \, dy = \int_{\widehat{\Omega}_\varepsilon^*} \phi(x) \, dx = \int_{\Omega_\varepsilon^*} \phi(x) \, dx - \int_{\Lambda_\varepsilon^*} \phi(x) \, dx. \end{aligned} \tag{4.6}$$

Suppose $p \in [1, +\infty)$. *Let* $\{w_\varepsilon\}_\varepsilon$ *be a sequence in* $L^p(\Omega)$ *such that*

$$w_\varepsilon \to w \quad \text{strongly in } L^p(\Omega).$$

Then

$$\mathcal{T}_\varepsilon^*(w_\varepsilon) \to w \quad \text{strongly in } L^p(\Omega \times Y^*). \tag{4.7}$$

In particular, if $w_\varepsilon = w$, *then*

$$\mathcal{T}_\varepsilon^*(w) \to w \quad \text{strongly in } L^p(\Omega \times Y^*).$$

Corollary 4.5. *Let* $\{\phi_\varepsilon\}_\varepsilon$ *be a sequence with* $\phi_\varepsilon \in L^1(\Omega_\varepsilon^*)$ *satisfying*

$$\int_{\Lambda_\varepsilon^*} |\phi_\varepsilon| \, dx \to 0.$$

Then

$$\int_{\Omega_\varepsilon^*} \phi_\varepsilon \, dx - \frac{1}{|Y|} \int_{\Omega \times Y^*} \mathcal{T}_\varepsilon^*(\phi_\varepsilon) \, dx \, dy \to 0.$$

Proof. This convergence follows from (4.6). □

Definition 4.6 (The mean value operator $\mathcal{M}_{Y*}$). *The mean value operator for* $p \in [1, +\infty]$,

$$\mathcal{M}_{Y*} : L^p(\Omega \times Y^*) \longrightarrow L^p(\Omega),$$

is defined as follows:

$$\mathcal{M}_{Y*}(\Phi)(x) = \frac{1}{|Y^*|} \int_{Y*} \Phi(x, y)\, dy \quad \text{for a.e. } x \in \Omega.$$

The proof of the next proposition is the same as in the case without holes (see Proposition 1.11).

Proposition 4.7. *Suppose* $p \in [1, +\infty]$. *For every* $\Phi \in L^p(\Omega \times Y^*)$,

$$\|\mathcal{M}_{Y*}(\Phi)\|_{L^p(\Omega)} \leqslant |Y^*|^{-1/p} \|\Phi\|_{L^p(\Omega \times Y^*)}.$$

As a consequence of Definition 4.6, Remark 4.3 and Proposition 1.8, the following results hold:

Proposition 4.8. *Suppose* $p \in [1, +\infty)$.
(i) For every w *in* $L^p(\Omega)$,

$$\mathcal{T}^*_\varepsilon(w) \to w \quad \text{strongly in } L^p(\Omega \times Y^*).$$

(ii) Let $\{w_\varepsilon\}_\varepsilon$ *be a sequence with* w_ε *in* $L^p(\Omega^*_\varepsilon)$ *such that* $\|w_\varepsilon\|_{L^p(\Omega^*_\varepsilon)} \leqslant C$. *If*

$$\mathcal{T}^*_\varepsilon(w_\varepsilon) \rightharpoonup \widehat{w} \quad \text{weakly in } L^p(\Omega \times Y^*),$$

then

$$\mathcal{E}^0_\varepsilon(w_\varepsilon) \rightharpoonup \mathcal{M}_Y \circ \mathcal{E}^0_S(\widehat{w}) = \frac{|Y^*|}{|Y|} \mathcal{M}_{Y*}(\widehat{w}) \quad \text{weakly in } L^p(\Omega).$$

The last convergence is still valid for $p = +\infty$ *for weak-** *convergences.*

Remark 4.9. *The statement (ii) above, implies that for a sequence* $\{w_\varepsilon\}_\varepsilon$ *with* w_ε *in* $L^p(\Omega^*_\varepsilon)$ *and such that* $\|w_\varepsilon\|_{L^p(\Omega^*_\varepsilon)}$ *is uniformly bounded, the following assertions are equivalent:*
1. There is w *in* $L^p(\Omega)$ *satisfying*

$$\mathcal{E}^0_\varepsilon(w_\varepsilon) \rightharpoonup \frac{|Y^*|}{|Y|} w \quad \text{weakly in } L^p(\Omega).$$

2. All the weak limit points W *in* $L^p(\Omega \times Y^*)$ *of the sequence* $\{\mathcal{T}^*_\varepsilon(w_\varepsilon)\}_\varepsilon$ *(which is bounded) have the same average over* Y^* *(this average* $\mathcal{M}_{Y*}(w)$ *simply being* w*).*

Corollary 4.10. *Suppose* $p \in [1, +\infty]$. *Both sequences* $\{\mathbf{1}_{\Omega^*_\varepsilon}\}_\varepsilon$ *and* $\{\mathbf{1}_{\widehat{\Omega}^*_\varepsilon}\}_\varepsilon$ *converge weakly-** *in* $L^\infty(\Omega)$ *to the constant* $\frac{|Y^*|}{|Y|}$. *Consequently, if* Ω *has finite measure,*

$$\lim_{\varepsilon \to 0} |\Omega^*_\varepsilon| = \lim_{\varepsilon \to 0} |\widehat{\Omega}^*_\varepsilon| = \frac{|Y^*|}{|Y|} |\Omega|. \tag{4.8}$$

Proof. Applying Proposition 4.4 and convergence (4.7) to a function φ in $\mathcal{D}(\Omega)$ for ε small enough, gives

$$\int_{\Omega_\varepsilon^*} \varphi\, dx = \frac{1}{|Y|}\int_{\Omega\times Y*} \mathcal{T}_\varepsilon^*(\varphi)\, dxdy$$
$$\to \frac{1}{|Y|}\int_{\Omega\times Y*} \varphi\, dxdy = \frac{|Y*|}{|Y|}\int_\Omega \varphi\, dx,$$

which is precisely convergence (4.8). □

4.2.2 The local average operator

As in the case without holes, the unfolding operator $\mathcal{T}_\varepsilon^*$ allows to define the local average operator specific to perforated domains;

Definition 4.11. *For p in $[1,+\infty]$, the local average operator $\mathcal{M}_\varepsilon^*$ is defined from $L^p(\Omega_\varepsilon^*)$ to $L^p(\Omega)$ as*

$$\mathcal{M}_\varepsilon^*(\phi) = \frac{1}{|Y*|}\int_{Y*} \mathcal{T}_\varepsilon^*(\phi)(\cdot, y)\, dy = \mathcal{M}_{Y*}\circ \mathcal{T}_\varepsilon^*(\phi).$$

This operator will be denoted $\widetilde{\mathcal{M}}_\varepsilon^*$ when Ω is the whole space $\mathbb{R}^N$.

Remark 4.12. By convention the value of $\mathcal{M}_\varepsilon^*(\phi)$ on the cell $\varepsilon(\xi+Y)$ is simply denoted $\mathcal{M}_\varepsilon^*(\phi)(\varepsilon\xi)$.

It is easily seen that

$$\mathcal{M}_\varepsilon^*\circ\mathcal{M}_\varepsilon^* \equiv \mathcal{M}_\varepsilon^* \quad \text{and} \quad \mathcal{T}_\varepsilon^*\circ\mathcal{M}_\varepsilon^* \equiv \mathcal{M}_\varepsilon^*.$$

The following statements, given without proofs, are direct consequences of the definition of $\mathcal{M}_\varepsilon^*$.

Proposition 4.13. *Suppose $p\in[1,+\infty]$.*
(i) For any ϕ in $L^p(\Omega_\varepsilon^*)$,

$$\|\mathcal{M}_\varepsilon^*(\phi)\|_{L^p(\Omega)} \leqslant \Big(\frac{|Y|}{|Y*|}\Big)^{1/p}\|\phi\|_{L^p(\Omega_\varepsilon^*)}.$$

(ii) For any $\phi\in L^p(\Omega_\varepsilon^*)$ and $\psi\in L^{p'}(\Omega_\varepsilon^*)$,

$$\int_{\Omega_\varepsilon^*} \mathcal{M}_\varepsilon^*(\phi)\,\psi\, dx = \frac{|Y*|}{|Y|}\int_\Omega \mathcal{M}_\varepsilon^*(\phi)\,\mathcal{M}_\varepsilon^*(\psi)\, dx = \int_{\Omega_\varepsilon^*} \phi\,\mathcal{M}_\varepsilon^*(\psi)\, dx. \tag{4.9}$$

(iii) Suppose $p\in(1,+\infty)$. Let $\{w_\varepsilon\}$ be a sequence with w_ε in $L^p(\Omega_\varepsilon^*\times Y^*)$ and $\widehat{w}$ in $L^p(\Omega)$ such that

$$\mathcal{T}_\varepsilon^*(w_\varepsilon) \rightharpoonup \widehat{w} \quad \text{weakly in } L^p(\Omega\times Y^*).$$

Then

$$\mathcal{M}_\varepsilon^*(w_\varepsilon) \rightharpoonup \mathcal{M}_{Y*}(\widehat{w}) \quad \text{weakly in } L^p(\Omega).$$

The same holds true for the weak-$*$ convergence in $L^\infty(\Omega)$.

Remark 4.14. *Equalities (4.9) show that $\mathcal{M}^*_\varepsilon$ is formally self adjoint.*

Proposition 4.15. *Suppose $p \in [1, +\infty)$. Let $\{v_\varepsilon\}_\varepsilon$ be a sequence converging strongly to v in $L^p(\Omega)$. Then*

$$\mathcal{M}^*_\varepsilon(v_\varepsilon) \to v \quad \text{strongly in } L^p(\Omega).$$

In particular, for every $\phi \in L^p(\Omega)$,

$$\mathcal{M}^*_\varepsilon(\phi) \to \phi \quad \text{strongly in } L^p(\Omega).$$

The proof is the same as in Proposition 1.25.

Remark 4.16. *Contrary to Proposition 1.25(ii), if the sequence $\{v_\varepsilon\}_\varepsilon$ converges weakly to v in $L^p(\Omega)$, in general, the sequence $\{\mathcal{M}^*_\varepsilon(v_\varepsilon)\}_\varepsilon$ does not converge to v. For example,*

$$\mathbf{1}_{\Omega^*_\varepsilon} \rightharpoonup \frac{|Y^*|}{|Y|},$$

whereas $\mathcal{M}^*_\varepsilon(\mathbf{1}_{\Omega^*_\varepsilon}) \equiv 1$ and $\mathcal{M}^*_\varepsilon(1 - \mathbf{1}_{\Omega^*_\varepsilon}) \equiv 0$.

4.2.3 The averaging operator

We now determine the adjoint of $\mathcal{T}^*_\varepsilon$. To do so, let v be in $L^p(\Omega \times Y^*)$ and u in $L^{p'}(\Omega^*_\varepsilon)$. The computation (1.22) can be repeated replacing the set Y by Y^*. It gives

$$\int_{\Omega \times Y^*} \mathcal{T}^*_\varepsilon(u)(x,y)\, v(x,y)\, dx\, dy = \int_{\Omega^*_\varepsilon} u(x) \Big(\int_{Y^*} v\Big(\varepsilon\Big[\frac{x}{\varepsilon}\Big]_Y + \varepsilon z, \Big\{\frac{x}{\varepsilon}\Big\}_Y\Big) dz\Big) dx,$$

from which the definition of the averaging operator $\mathcal{U}^*_\varepsilon$ is hinted clearly as follows:

Definition 4.17. *For $p \in [1, +\infty]$, the averaging operator*

$$\mathcal{U}^*_\varepsilon \ : \ L^p(\Omega \times Y^*) \longrightarrow L^p(\Omega^*_\varepsilon),$$

is defined as

$$\mathcal{U}^*_\varepsilon(\Phi)(x) = \begin{cases} \dfrac{1}{|Y|} \displaystyle\int_Y \Phi\Big(\varepsilon\Big[\frac{x}{\varepsilon}\Big]_Y + \varepsilon z, \Big\{\frac{x}{\varepsilon}\Big\}_Y\Big)\, dz & \text{for a.e. } x \in \widehat{\Omega}^*_\varepsilon, \\ 0 & \text{for a.e. } x \in \Lambda^*_\varepsilon. \end{cases}$$

For $\Omega = \mathbb{R}^N$, the averaging operator is denoted $\widetilde{\mathcal{U}}^*_\varepsilon$.

Note that if Φ belongs to $L^p(\Omega \times Y)$, then

$$\mathcal{U}^*_\varepsilon(\Phi_{|\Omega \times Y^*}) = \mathcal{U}_\varepsilon(\Phi)_{|\Omega^*_\varepsilon},$$

and consequently,

$$\mathcal{U}_\varepsilon^*(v) = \mathcal{U}_\varepsilon(\overline{v})_{|_{\Omega_\varepsilon^*}}, \tag{4.10}$$

where $\overline{v}$ is any extension of v to $\Omega \times Y$.

Also, for Ψ independent of x,

$$\mathcal{U}_\varepsilon^*(\Psi) \equiv \Psi\Big(\Big\{\frac{\cdot}{\varepsilon}\Big\}_Y\Big)\,\mathbf{1}_{\widehat{\Omega}_\varepsilon^*}, \tag{4.11}$$

and for φ independent of y,

$$\mathcal{U}_\varepsilon^*(\varphi) = \mathcal{M}_\varepsilon^*(\varphi). \tag{4.12}$$

A consequence of the duality (Hölder's inequality) and of estimate (4.6), is the next result.

Proposition 4.18. *Suppose $p \in [1, +\infty]$. The averaging operator is linear and continuous from $L^p(\Omega \times Y^*)$ to $L^p(\Omega_\varepsilon^*)$ and*

$$\|\mathcal{U}_\varepsilon^*(\Phi)\|_{L^p(\Omega_\varepsilon^*)} \leqslant |Y|^{-1/p}\, \|\Phi\|_{L^p(\Omega\times Y^*)}. \qquad \square$$

For a sequence independent of y, Proposition 4.15(i) and (4.12) obviously imply the following:

Proposition 4.19. *Suppose $p \in [1, +\infty)$. For every strongly converging sequence $\{\varphi_\varepsilon\}_\varepsilon$ in $L^p(\Omega)$ with limit φ,*

$$\big\|\varphi - \mathcal{U}_\varepsilon^*(\varphi_\varepsilon)\big\|_{L^p(\Omega_\varepsilon^*)} \to 0.$$

As in the case without holes, $\mathcal{U}_\varepsilon^*$ is almost a left-inverse of $\mathcal{T}_\varepsilon^*$. On the one hand, for every ϕ in $L^p(\Omega_\varepsilon^*)$,

$$\mathcal{U}_\varepsilon^* \circ \mathcal{T}_\varepsilon^*(\phi)(x) = \begin{cases} \phi(x) & \text{for a.e. } x \in \widehat{\Omega}_\varepsilon^*, \\ 0 & \text{for a.e. } x \in \Lambda_\varepsilon^*. \end{cases}$$

On the other hand, for every Φ in $L^p(\Omega \times Y^*)$,

$$\mathcal{T}_\varepsilon^* \circ \mathcal{U}_\varepsilon^*(\Phi)(x,y) = \begin{cases} \dfrac{1}{|Y|}\displaystyle\int_Y \Phi\Big(\varepsilon\Big[\frac{x}{\varepsilon}\Big]_Y + \varepsilon z, y\Big)\,dz & \\ & \text{for a.e.} (x,y) \in \widehat{\Omega}_\varepsilon \times Y^*, \\ 0 & \text{for a.e.} (x,y) \in \Lambda_\varepsilon \times Y^*. \end{cases}$$

Proposition 4.20 (Properties of $\mathcal{U}_\varepsilon^*$ in connection with $\mathcal{T}_\varepsilon^*$, $\mathcal{M}_\varepsilon^*$ and $\mathcal{E}_\varepsilon^0$). *Suppose $p \in (1, +\infty)$.*

(i) Let $\{\Phi_\varepsilon\}_\varepsilon$ be a sequence such that $\Phi_\varepsilon \rightharpoonup \Phi$ weakly in $L^p(\Omega \times Y^*)$. Then

$$\mathcal{T}_\varepsilon^* \circ \mathcal{U}_\varepsilon^*(\Phi_\varepsilon) \rightharpoonup \Phi \quad \text{weakly in } L^p(\Omega \times Y^*),$$
$$\mathcal{M}_\varepsilon^* \circ \mathcal{U}_\varepsilon^*(\Phi_\varepsilon) \rightharpoonup \mathcal{M}_{Y^*}(\Phi) \quad \text{weakly in } L^p(\Omega).$$

The same holds true for strong convergences. For $p = +\infty$, the corresponding convergences hold for the weak-$*$ topologies.
(ii) Let $\{\Phi_\varepsilon\}_\varepsilon$ be a bounded sequence in $L^p(\Omega \times Y^*)$ which converges weakly to Φ in $L^p(\Omega \times Y^*)$. Then

$$\mathcal{E}^0_\varepsilon \circ \mathcal{U}^*_\varepsilon(\Phi_\varepsilon) \rightharpoonup \frac{|Y^*|}{|Y|}\mathcal{M}_{Y^*}(\Phi) \quad \textit{weakly in } L^p(\Omega).$$

In particular, for every $\Phi \in L^p(\Omega \times Y^*)$,

$$\mathcal{E}^0_\varepsilon \circ \mathcal{U}^*_\varepsilon(\Phi) \rightharpoonup \frac{|Y^*|}{|Y|}\mathcal{M}_{Y^*}(\Phi) \quad \textit{weakly in } L^p(\Omega).$$

Let $\{w_\varepsilon\}_\varepsilon$ be a sequence with $w_\varepsilon \in L^p(\Omega^*_\varepsilon)$ and $\widehat{w}$ in $L^p(\Omega \times Y^*)$. Then,
(iii) The following assertions are equivalent:

$$\begin{aligned}&(a) \quad \mathcal{T}^*_\varepsilon(w_\varepsilon) \to \widehat{w} \quad \textit{strongly in } L^p(\Omega \times Y^*),\\ &(b) \quad \|w_\varepsilon - \mathcal{U}^*_\varepsilon(\widehat{w})\|_{L^p(\widehat{\Omega}^*_\varepsilon)} \to 0.\end{aligned}$$

(iv) The following assertions are equivalent:

$$(c) \quad \mathcal{T}^*_\varepsilon(w_\varepsilon) \to \widehat{w} \quad \textit{strongly in } L^p(\Omega \times Y^*) \quad \textit{and} \quad \int_{\Lambda^*_\varepsilon} |w_\varepsilon|^p\, dx \to 0,$$

$$(d) \quad \|w_\varepsilon - \mathcal{U}^*_\varepsilon(\widehat{w})\|_{L^p(\Omega^*_\varepsilon)} \to 0.$$

(v) If in (iii) and (iv) resp., $\widehat{w}$ is independent of y, then (b) and (d) resp., can be replaced by

$$(b') \quad \|w_\varepsilon - \widehat{w}\|_{L^p(\widehat{\Omega}^*_\varepsilon)} \to 0,$$

and

$$(d') \quad \|w_\varepsilon - \widehat{w}\|_{L^p(\Omega^*_\varepsilon)} \to 0, \quad \textit{respectively.}$$

Proof. (i) The proof is the same as for Propositions 1.29(i) and 1.31(ii).
(ii) Applying (i) and using the equivalence of Proposition 4.8(ii), gives the result.
(iii) It is immediate that

$$\mathcal{E}^0_\varepsilon(w_\varepsilon \mathbf{1}_{\widehat{\Omega}^*_\varepsilon}) = \mathcal{E}^0_\varepsilon(w_\varepsilon)\mathbf{1}_{\widehat{\Omega}^*_\varepsilon} \quad \text{and} \quad \mathcal{E}^0_S \circ \mathcal{T}^*_\varepsilon(w_\varepsilon \mathbf{1}_{\widehat{\Omega}^*_\varepsilon}) = \mathcal{T}_\varepsilon \circ \mathcal{E}^0_\varepsilon(w_\varepsilon \mathbf{1}_{\widehat{\Omega}^*_\varepsilon}),$$

so that (a) is equivalent to

$$\mathcal{T}_\varepsilon \circ \mathcal{E}^0_\varepsilon(w_\varepsilon \mathbf{1}_{\widehat{\Omega}^*_\varepsilon}) \to \mathcal{E}^0_S(\widehat{w}) \quad \text{strongly in } L^p(\Omega \times Y).$$

This, by Proposition 1.29(ii), is equivalent to

$$\|\mathcal{E}^0_\varepsilon(w_\varepsilon \mathbf{1}_{\widehat{\Omega}^*_\varepsilon}) - \mathcal{U}_\varepsilon \circ \mathcal{E}^0_S(\widehat{w})\|_{L^p(\Omega)} \to 0 \quad \text{strongly in } L^p(\Omega),$$

which is another way of writing (b).

(iv) The proof is a simple variant of (iii), taking into account the fact that

$$\|w_\varepsilon - \mathcal{U}^*_\varepsilon(\widehat{w})\|^p_{L^p(\Omega^*_\varepsilon)} = \|w_\varepsilon - \mathcal{U}^*_\varepsilon(\widehat{w})\|^p_{L^p(\widehat{\Omega}^*_\varepsilon)} + \int_{\Lambda^*_\varepsilon} |w_\varepsilon|^p \, dx.$$

(v) In the case $\widehat{w}$ does not depend upon y, by virtue of Remark 4.19, (b) is equivalent to (b'), and (d) to (d'). □

Remark 4.21. *Statements (iii)(b) and (iv)(d) are corrector–type results.*

4.2.4 Extension by local average

There is another extension procedure into the holes, the extension by local average $\mathcal{E}_\varepsilon$. This extension was introduced in [152] for the specific case of the pressure in the homogenization of the Stokes equation (see Chapter 6). It can be used as an alternate to $\mathcal{E}^0_\varepsilon$ to prove the results of this section, by relating them to the corresponding results of Chapter 1. In particular, the use of $\mathcal{E}_\varepsilon$ simplifies the proofs of the properties of the macro-micro scale operators (see Subsection 4.3.2.2 and Remark 4.33).

Definition 4.22 (Extension by local average). *For $p \in [1, +\infty]$, the extension operator*

$$\mathcal{E}_\varepsilon : L^p(\Omega^*_\varepsilon) \longrightarrow L^p(\Omega),$$

is defined for w in $L^p(\Omega^*_\varepsilon)$, as follows:

$$\mathcal{E}_\varepsilon(w) = \begin{cases} w & a.e. \ in \ \Omega^*_\varepsilon, \\ \mathcal{M}^*_\varepsilon(w) & a.e. \ in \ \Omega \cap S_\varepsilon. \end{cases}$$

Note that by definition,

$$\mathcal{E}_\varepsilon(w) = w \quad \text{on } \Lambda^*_\varepsilon,$$

and, since $\mathcal{M}^*_\varepsilon(w) = 0$ on Λ_ε,

$$\mathcal{E}_\varepsilon(w) = 0 \quad \text{on } \Lambda_\varepsilon \backslash \Lambda^*_\varepsilon.$$

A simple computation shows that for $p \in [1, +\infty]$ and w in $L^p(\Omega^*_\varepsilon)$,

$$\|\mathcal{E}_\varepsilon(w)\|_{L^p(\Omega)} \leqslant \Big(\frac{|Y|}{|Y^*|}\Big)^{1/p} \|w\|_{L^p(\Omega^*_\varepsilon)}. \tag{4.13}$$

Remark 4.23 (Connections between $\mathcal{T}^*_\varepsilon, \mathcal{M}^*_\varepsilon, \mathcal{E}^0_\varepsilon, \mathcal{T}_\varepsilon, \mathcal{M}_\varepsilon$ and $\mathcal{E}_\varepsilon$). *The following formulas are straightforward consequences of the definitions:*

$$\mathcal{M}^*_\varepsilon \circ \mathcal{M}_\varepsilon = \mathcal{M}_\varepsilon, \quad \mathcal{M}^*_\varepsilon \circ \mathcal{M}^*_\varepsilon = \mathcal{M}^*_\varepsilon, \quad \mathcal{M}_\varepsilon \circ \mathcal{E}_\varepsilon = \mathcal{M}^*_\varepsilon, \quad \mathcal{E}_\varepsilon \circ \mathcal{M}^*_\varepsilon = \mathcal{M}^*_\varepsilon.$$

For every w in $L^p(\Omega_\varepsilon^*)$,

$$\mathcal{T}_\varepsilon \circ \mathcal{E}_\varepsilon(w) = \begin{cases} \mathcal{T}_\varepsilon^*(w) & a.e.\ in\ \Omega \times Y^*, \\ \mathcal{M}_\varepsilon^*(w) & a.e.\ in\ \Omega \times S, \end{cases}$$

while

$$\mathcal{T}_\varepsilon \circ \mathcal{E}_\varepsilon^0(w) = \begin{cases} \mathcal{T}_\varepsilon^*(w) & a.e.\ in\ \Omega \times Y^*, \\ 0 & a.e.\ in\ \Omega \times S. \end{cases}$$

Finally,

$$\mathcal{M}_\varepsilon \circ \mathcal{E}_\varepsilon^0 = \frac{|Y^*|}{|Y|}\mathcal{M}_\varepsilon^* = \frac{|Y^*|}{|Y|}\mathcal{M}_\varepsilon \circ \mathcal{E}_\varepsilon{}^{(1)}.$$

The next result is a corollary of Proposition 1.25 and of the the third formula above.

Corollary 4.24.
(i) Suppose $p \in [1, +\infty)$. Let $\{w_\varepsilon\}$ be a sequence with w_ε in $L^p(\Omega_\varepsilon^*)$, and w in $L^p(\Omega)$ such that,

$$\mathcal{E}_\varepsilon(w_\varepsilon) \to w \quad strongly\ in\ L^p(\Omega).$$

Then

$$\mathcal{M}_\varepsilon^*(w_\varepsilon) \to w \quad strongly\ in\ L^p(\Omega).$$

(ii) Suppose $p \in (1, +\infty)$. Let $\{w_\varepsilon\}$ be a sequence with w_ε in $L^p(\Omega_\varepsilon^*)$, and such that $\|w_\varepsilon\|_{L^p(\Omega_\varepsilon^*)}$ is bounded, and let w be in $L^p(\Omega)$. The following convergences are equivalent:

$$\mathcal{E}_\varepsilon(w_\varepsilon) \rightharpoonup w \quad weakly\ in\ L^p(\Omega),$$
$$\mathcal{M}_\varepsilon^*(w_\varepsilon) \rightharpoonup w \quad weakly\ in\ L^p(\Omega).$$

The same holds true for weak-$*$ convergences in $L^\infty(\Omega)$.

The previous result (i) cannot hold with $\mathcal{E}_\varepsilon^0$ in place of $\mathcal{E}_\varepsilon$, indeed the identity $\mathcal{E}_\varepsilon^0(w_\varepsilon)(1 - \mathbf{1}_{\Omega_\varepsilon^*}) \equiv 0$ implies that the limit of a strongly convergent sequence $\{\mathcal{E}_\varepsilon^0(w_\varepsilon)\}_\varepsilon$ in $L^p(\Omega)$ is 0. However, a version of (ii) holds for $\mathcal{E}_\varepsilon^0$, due to the following proposition:

Proposition 4.25 (Weak convergences with $\mathcal{E}_\varepsilon$ and $\mathcal{E}_\varepsilon^0$). *Suppose $p \in (1, +\infty)$. Let $\{w_\varepsilon\}_\varepsilon$ be a sequence with w_ε in $L^p(\Omega_\varepsilon^*)$, and w be in $L^p(\Omega)$. The following convergences are equivalent:*

$$\mathcal{E}_\varepsilon(w_\varepsilon) \rightharpoonup w \quad weakly\ in\ L^p(\Omega),$$
$$\mathcal{E}_\varepsilon^0(w_\varepsilon) \rightharpoonup \frac{|Y^*|}{|Y|} w \quad weakly\ in\ L^p(\Omega).$$

The same holds true for weak-$*$ convergences in $L^\infty(\Omega)$.

(1) This follows from the equalities

$$\mathcal{M}_\varepsilon \circ \mathcal{E}_\varepsilon^0(w) = \mathcal{M}_Y \circ \mathcal{T}_\varepsilon \circ \mathcal{E}_\varepsilon^0(w) = \frac{1}{|Y|}\int_{Y^*} \mathcal{T}_\varepsilon^*(w)dy = \frac{|Y^*|}{|Y|}\mathcal{M}_\varepsilon^*(w).$$

Proof. For φ in $L^{p'}(\Omega)$, by definition of the extension operators and using (4.9) and (1.20), and the equalities of Remark 4.23, one obtains

$$\begin{aligned}\int_\Omega \mathcal{E}^0_\varepsilon(w_\varepsilon)\mathcal{M}_\varepsilon(\varphi)\,dx &= \int_\Omega \mathcal{M}_\varepsilon \circ \mathcal{E}^0_\varepsilon(w_\varepsilon)\,\mathcal{M}_\varepsilon(\varphi)\,dx \\ &= \frac{|Y^*|}{|Y|}\int_\Omega \mathcal{M}_\varepsilon \circ \mathcal{E}_\varepsilon(w_\varepsilon)\,\mathcal{M}_\varepsilon(\varphi)\,dx \\ &= \frac{|Y^*|}{|Y|}\int_\Omega \mathcal{E}_\varepsilon(w_\varepsilon)\,\mathcal{M}_\varepsilon(\varphi)\,dx.\end{aligned}$$

Then convergence (1.21) allows to conclude the proof of the equivalence. □

The next proposition clarifies the relationship of $\mathcal{E}_\varepsilon$ with $\mathcal{T}^*_\varepsilon$ for strong convergence (for the reason stated above, there is no equivalent for $\mathcal{E}^0_\varepsilon$).

Proposition 4.26 (Strong convergences for $\mathcal{T}^*_\varepsilon$). *Suppose $p \in [1, +\infty)$. Let $\{w_\varepsilon\}_\varepsilon$ be a sequence with $w_\varepsilon \in L^p(\Omega^*_\varepsilon)$, and let $\widehat{w}$ be in $L^p(\Omega \times Y^*)$. Then, the following convergences are equivalent:*

$$\begin{aligned}&(a) \quad \mathcal{T}^*_\varepsilon(w_\varepsilon) \to \widehat{w} \quad \textit{strongly in } L^p(\Omega \times Y^*),\\ &(b) \quad \mathcal{T}_\varepsilon \circ \mathcal{E}_\varepsilon(w_\varepsilon) \to \mathcal{E}_S(\widehat{w}) \quad \textit{strongly in } L^p(\Omega \times Y),\end{aligned}$$

where $\mathcal{E}_S$ is the extension by average from Y^* to S, defined for φ in $L^p(Y^*)$ by

$$\mathcal{E}_S(\varphi) \doteq \begin{cases}\varphi & \textit{in } Y^*,\\ \mathcal{M}_{Y^*}(\varphi) & \textit{in } S.\end{cases}$$

Finally, the strong convergence involving $\mathcal{E}_\varepsilon$ is explained by making use of Propositions 4.20(v) and 4.8.

Proposition 4.27 (Strong convergences for $\mathcal{E}_\varepsilon$). *Suppose $p \in [1, +\infty)$. Let $\{w_\varepsilon\}$ be a sequence with $w_\varepsilon \in L^p(\Omega^*_\varepsilon)$, and let w be in $L^p(\Omega)$.*
(i) The following assertions are equivalent:

$$\begin{aligned}&(a) \quad \|w_\varepsilon - w\|_{L^p(\widehat{\Omega}^*_\varepsilon)} \to 0,\\ &(b) \quad \mathcal{E}_\varepsilon(w_\varepsilon)\mathbf{1}_{\widehat{\Omega}_\varepsilon} \to w \quad \textit{strongly in } L^p(\Omega),\\ &(c) \quad \mathcal{T}_\varepsilon \circ \mathcal{E}_\varepsilon(w_\varepsilon) \to w \quad \textit{strongly in } L^p(\Omega \times Y).\end{aligned}$$

(ii) The following assertions are equivalent:

$$\begin{aligned}&(d) \quad \|w_\varepsilon - w\|_{L^p(\Omega^*_\varepsilon)} \to 0,\\ &(e) \quad \mathcal{E}_\varepsilon(w_\varepsilon) \to w \quad \textit{strongly in } L^p(\Omega),\\ &(f) \quad \mathcal{T}_\varepsilon \circ \mathcal{E}_\varepsilon(w_\varepsilon) \to w \quad \textit{strongly in } L^p(\Omega \times Y) \textit{ and } \int_{\Lambda^*_\varepsilon} |w_\varepsilon|^p\,dx \to 0.\end{aligned}$$

4.3 Unfolding and gradients

In this section, we use the following notations (similar to those from(1.32)) :

$$\begin{aligned} W^{1,p}_{per}(Y^*) &\doteq \{\varphi \in W^{1,p}_{loc}((\mathbb{R}^N)^*) \mid \varphi \text{ is } \mathbf{G}\text{-periodic}\},\\ W^{1,p}_{per,0}(Y^*) &\doteq \{\varphi \in W^{1,p}_{per}(Y^*) \mid \mathcal{M}_{Y^*}(\varphi) = 0\}. \end{aligned}$$

Both of these spaces are isometric to closed subspaces of $W^{1,p}(Y^*)$ when endowed with the corresponding norm. For $p = 2$, these spaces are denoted $H^1_{per}(Y^*)$ and $H^1_{per,0}(Y^*)$, respectively[(2)].

For w in $W^{1,p}(\Omega^*_\varepsilon)$, as in the case without holes, the equivalent of Proposition 1.35 holds, that is,

$$\nabla_y(\mathcal{T}^*_\varepsilon(w)) = \varepsilon\mathcal{T}^*_\varepsilon(\nabla w) \quad \text{a.e. in } \quad \Omega \times Y^*. \tag{4.14}$$

This implies that $\mathcal{T}^*_\varepsilon$ maps $W^{1,p}(\Omega^*_\varepsilon)$ into $L^p(\Omega; W^{1,p}(Y^*))$.

We consider sequences $\{w_\varepsilon\}_\varepsilon$ such that for each ε, w_ε belongs to $W^{1,p}(\Omega^*_\varepsilon)$. As in Chapter 1, two cases are considered.

4.3.1 First case: $\|w_\varepsilon\|_{L^p(\Omega^*_\varepsilon)} + \varepsilon\|\nabla w_\varepsilon\|_{L^p(\Omega^*_\varepsilon)}$ bounded

The main result of this subsection is the following theorem:

Theorem 4.28. *Suppose $p \in (1, +\infty)$. Let $\{w_\varepsilon\}_\varepsilon$ be a sequence with w_ε in $W^{1,p}(\Omega^*_\varepsilon)$, satisfying*

$$\|w_\varepsilon\|_{L^p(\Omega^*_\varepsilon)} + \varepsilon\|\nabla w_\varepsilon\|_{L^p(\Omega^*_\varepsilon)} \leqslant C. \tag{4.15}$$

Then, there exists some $\widehat{w} \in L^p(\Omega; W^{1,p}_{per}(Y^*))$, such that, up to a subsequence,

$$\begin{aligned} &\mathcal{T}^*_\varepsilon(w_\varepsilon) \rightharpoonup \widehat{w} \quad \text{weakly in } L^p(\Omega; W^{1,p}(Y^*)),\\ &\varepsilon\mathcal{T}^*_\varepsilon(\nabla w_\varepsilon) \rightharpoonup \nabla_y\widehat{w} \quad \text{weakly in } L^p(\Omega \times Y^*)^N. \end{aligned} \tag{4.16}$$

Similar results hold for $p = +\infty$ replacing the weak topology by the weak-$*$ topology on the corresponding spaces.

The proof is very close to that of Theorem 1.36. The delicate point is the Y-periodicity of $\widehat{w}$. If the reference cell Y is a parallelotope and $S \subset Y$, one can simply argue as in the case without holes comparing the traces on opposite faces of Y (see [62]).

However, in the general case, Y is not a parallelotope. If $\overline{Y^*}$ is included in Y, i.e., $\partial Y \cap \partial Y^*$ is empty, there is no periodicity condition to verify.

If $\partial Y \cap \partial Y^*$ is not empty, the boundary of S may not be Lipschitz or the way it intersects ∂Y can be such that traces are not even meaningful.

(2) In the case where $\overline{Y^*}$ is included in Y, these spaces do not differ from $W^{1,p}(Y^*)$ and $W^{1,p}_0(Y^*)$.

To circumvent this difficulty, we use an approach which avoids the use of traces. We introduce an auxiliary bigger cell $\mathcal{Y}$ (the union of Y and the 2^N-1 contiguous copies of Y) and use a new operator $\mathcal{T}_\varepsilon^{\mathcal{Y}*}$ (see an alternative approach for a similar result in $\mathbb{R}^N$ in Subsection 2.1.1).

We stress the fact that this operator is not the unfolding operator corresponding to unit cell $\mathcal{Y}$, but is a "natural" way to extend the operator $\mathcal{T}_\varepsilon^*$ with values in $\Omega \times Y^*$, to an operator with values in the set $\Omega \times \mathcal{Y}^*$.

The formal definition of $\mathcal{Y}$ uses the notations

$$\begin{aligned}\mathcal{K} &\doteq \text{the set of vertices of } \mathcal{P}\\ &\equiv \Big\{\ell \in \mathbb{R}^N \mid \ell = \sum_{i=1}^{N} k_i \mathbf{b}_i,\ (k_1,\dots,k_n) \in \{0,1\}^N\Big\},\\ \mathcal{Y} &= \text{interior}\Big\{\bigcup_{\ell\in\mathcal{K}} (\ell + \overline{Y})\Big\}.\end{aligned}$$

The corresponding sets $\widehat{\Omega}_\varepsilon^{\mathcal{Y}}$ and $\Lambda_\varepsilon^{\mathcal{Y}}$ are defined as in Part I, that is

$$\Xi_\varepsilon^{\mathcal{Y}} = \{\xi \in \Xi_\varepsilon \mid \varepsilon(\xi + \overline{\mathcal{Y}}) \subset \Omega\}. \tag{4.17}$$

$$\widehat{\Omega}_\varepsilon^{\mathcal{Y}} = \text{interior}\Big\{\bigcup_{\xi\in\Xi_\varepsilon^{\mathcal{Y}}} \varepsilon(\xi + \overline{Y})\Big\}, \qquad \Lambda_\varepsilon^{\mathcal{Y}} = \Omega \backslash \overline{\widehat{\Omega}_\varepsilon^{\mathcal{Y}}}. \tag{4.18}$$

This definition implies that if ξ is in $\Xi_\varepsilon^{\mathcal{Y}}$, then for all ℓ in $\mathcal{K}$, $\xi+\ell$ belongs to Ξ_ε. Consequently, if a cell $\varepsilon(\xi+Y)$ is included in $\widehat{\Omega}_\varepsilon^{\mathcal{Y}}$, then all its translates $\varepsilon(\xi+\ell+Y)$ with ℓ in $\mathcal{K}$, are in $\widehat{\Omega}_\varepsilon$ (see Figure 4.7). Also note that

$$\widehat{\Omega}_\varepsilon^{\mathcal{Y}} \subset \widehat{\Omega}_\varepsilon^{\mathcal{Y}}. \tag{4.19}$$

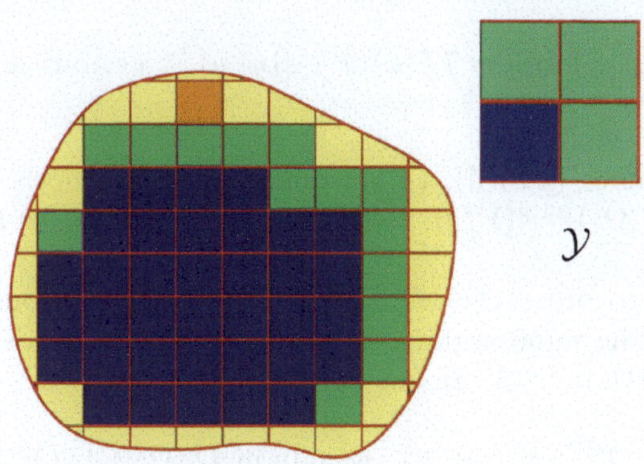

Figure 4.7: The set $\widehat{\Omega}_\varepsilon^{\mathcal{Y}}$ (in dark blue) and $\mathcal{Y}$ (in dark blue and green)

We also introduce the notation

$$\mathcal{Y}^* = \mathcal{Y} \cap (\mathbb{R}^N)^*. \tag{4.20}$$

For ϕ Lebesgue-measurable on Ω_ε^*, the unfolding operator $\mathcal{T}_\varepsilon^{\mathcal{Y}*}(\phi)$ corresponding to $\mathcal{Y}$ is defined as follows:

$$\mathcal{T}_\varepsilon^{\mathcal{Y}*}(\phi)(x,y) = \begin{cases} \phi\Big(\varepsilon\Big[\dfrac{x}{\varepsilon}\Big]_Y + \varepsilon y\Big) & \text{for a.e.} \quad (x,y) \in \widehat{\Omega}_\varepsilon^{\mathcal{Y}} \times \mathcal{Y}^*, \\ 0 & \text{for a.e.} \quad (x,y) \in \Lambda_\varepsilon^{\mathcal{Y}} \times \mathcal{Y}^*. \end{cases} \tag{4.21}$$

Due to the definition (4.18) of $\widehat{\Omega}_\varepsilon^{\mathcal{Y}}$, formula (4.21) makes sense. Note here the use of $\Big[\dfrac{x}{\varepsilon}\Big]_Y$ and not of $\Big[\dfrac{x}{\varepsilon}\Big]_{\mathcal{Y}}$.

From (4.21), it is easy to check that if w belongs to $L^p(\Omega_\varepsilon^*)$, then $\mathcal{T}_\varepsilon^{\mathcal{Y}*}(w)$ is in the space $L^p(\Omega \times \mathcal{Y}^*)$ and

$$\|\mathcal{T}_\varepsilon^{\mathcal{Y}*}(w)\|_{L^p(\Omega\times\mathcal{Y}^*)} \leqslant (2^N|Y|)^{1/p}\, \|w\|_{L^p(\Omega_\varepsilon^*)}.$$

If w belongs to $W^{1,p}(\Omega_\varepsilon^*)$, formula (4.14) holds also in $\mathcal{Y}^*$

$$\nabla_y(\mathcal{T}_\varepsilon^{\mathcal{Y}*}(w)) = \varepsilon \mathcal{T}_\varepsilon^{\mathcal{Y}*}(\nabla w) \quad \text{a.e. in} \quad \Omega \times \mathcal{Y}^*.$$

The main properties relating $\mathcal{T}_\varepsilon^{\mathcal{Y}*}$ with $\mathcal{T}_\varepsilon^*$ are, for $x \in \widehat{\Omega}_\varepsilon^{\mathcal{Y}}$ and every vector of the basis $\mathbf{B} = (\mathbf{b}_1, \ldots, \mathbf{b}_n)$,

$$\begin{aligned} &\mathcal{T}_\varepsilon^{\mathcal{Y}*}(\phi)(x,y) = \mathcal{T}_\varepsilon^*(\phi)(x,y) \quad \text{for a.e.} \quad y \in Y^*, \\ &\mathcal{T}_\varepsilon^{\mathcal{Y}*}(\phi)(x,y+\mathbf{b}_k) = \mathcal{T}_\varepsilon^*(\phi)(x+\varepsilon\mathbf{b}_k,y) \quad \text{for a.e.} \quad y \in Y^*. \end{aligned} \tag{4.22}$$

In particular, if ω is a relatively compact open subset of Ω, then for ε sufficiently small,

$$\mathcal{T}_\varepsilon^{\mathcal{Y}*}(\phi)(x,y+\mathbf{b}_k) = \mathcal{T}_\varepsilon^*(\phi)(x+\varepsilon\mathbf{b}_k,y) \quad \text{a.e. on} \quad \omega \times Y^*. \tag{4.23}$$

Proof of Theorem 4.28.

Using estimate (4.6)(ii), (4.14) and (4.15), it follows that $\{\mathcal{T}_\varepsilon^*(w_\varepsilon)\}$ is bounded in $L^p(\Omega; W^{1,p}(Y^*))$, so that convergences (4.16) hold (up to a subsequence).

It remains to prove that $\widehat{w}$ is periodic. To do so, consider the function $\mathcal{T}_\varepsilon^{\mathcal{Y}*}(w_\varepsilon)$. By the same argument as just above, up to a subsequence, there exists $\overline{w}$ in $L^p(\Omega; W^{1,p}(\mathcal{Y}^*))$, such that

$$\mathcal{T}_\varepsilon^{\mathcal{Y}*}(w_\varepsilon) \rightharpoonup \overline{w} \quad \text{weakly in } L^p(\Omega; W^{1,p}(\mathcal{Y}^*)), \tag{4.24}$$

or weak-$*$ for $p = +\infty$.

Let ω be an open bounded set whose closure is included in Ω. From (4.22) and (4.16) it follows that

$$\overline{w}(x,y) = \widehat{w}(x,y) \qquad \text{for a.e. } (x,y) \in \omega \times Y^*.$$

Now, let Φ be in $\mathcal{D}(\omega\times Y^*)$. From (4.23) for ε small enough and for every k in $\{1,\dots,N\}$, one has

$$\begin{aligned}\int_{\omega\times Y^*} & \mathcal{T}_\varepsilon^{\mathcal{Y}^*}(w_\varepsilon)(x,y+\mathbf{b}_k)\Phi(x,y)\,dx\,dy\\ &= \int_{\omega\times Y^*} \mathcal{T}_\varepsilon^*(w_\varepsilon)(x+\varepsilon\mathbf{b}_k,y)\Phi(x,y)\,dx\,dy\\ &= \int_{\omega\times Y^*} \mathcal{T}_\varepsilon^*(w_\varepsilon)(x,y)\Phi(x-\varepsilon\mathbf{b}_k,y)\,dx\,dy.\end{aligned}$$

Passing to the limit and using (4.16) and (4.24) gives

$$\int_{\omega\times Y^*} \overline{w}(x,y+\mathbf{b}_k)\Phi(x,y)\,dx\,dy = \int_{\omega\times Y^*} \widehat{w}(x,y)\Phi(x,y)\,dx\,dy,$$

hence

$$\overline{w}(x,y+\mathbf{b}_k) = \widehat{w}(x,y) \qquad \text{for a.e. } (x,y)\in\omega\times Y^*.$$

Since this holds for every $\omega \Subset \Omega$, it is also true in $\Omega\times Y^*$.

Therefore, $\overline{w}$ is actually $\mathbf{B}$-periodic and can be extended by periodicity to the whole of $\left(\mathbb{R}^N\right)^*$. Furthermore, $\widehat{w}$ is the restriction of $\overline{w}$ to Y^*, proving that it belongs to $L^p(\Omega; W^{1,p}_{per}(Y^*))$. □

Proposition 4.29. *Suppose $p\in(1,+\infty]$. Let $\{w_\varepsilon\}_\varepsilon$ be a sequence with w_ε in $L^p(\Omega_\varepsilon^*)$, such that w_ε and its gradient in the direction of a vector* $\mathbf{a}$ *satisfy*

$$\|w_\varepsilon\|_{L^p(\Omega_\varepsilon^*)} + \varepsilon\,\|\partial_{\mathbf{a}} w_\varepsilon\|_{L^p(\Omega_\varepsilon^*)} \le C.$$

Then, up to a subsequence, there exists $\widehat{w}$ in $L^p(\mathbb{R}^N; W^{1,p}_{\mathbf{a},\,\mathbf{G}\text{-}per}(Y^*))$, such that

$$\mathcal{T}_\varepsilon^*(w_\varepsilon) \rightharpoonup \widehat{w} \quad \textit{weakly in } L^p(\Omega\times Y^*),$$

$$\varepsilon\mathcal{T}_\varepsilon^*(\partial_{\mathbf{a}} w_\varepsilon) = \partial_{\mathbf{a},y}\mathcal{T}_\varepsilon^*(w_\varepsilon) \rightharpoonup \partial_{\mathbf{a},y}\widehat{w} \ \textit{ weakly in } L^p(\Omega\times Y^*),$$

or weak-$*$ for $p=+\infty$.

Here the space $W^{1,p}_{\mathbf{a},\,\mathbf{G}\text{-}per}(Y^*)$ consists of all functions w in $L^p(Y^*)$ whose $\mathbf{G}$*-periodic extension $\widehat{w}$ to $\left(\mathbb{R}^N\right)^*$ as well as the* $\mathbf{a}$*-derivative $\partial_{\mathbf{a},y}\widehat{w}$ of this extension, both belong to $L^p_{loc}\big(\left(\mathbb{R}^N\right)^*\big)$.*

Proof. The proof is done by arguments similar to those of the proof of Proposition 2.17. □

4.3.2 Second case: $\|w_\varepsilon\|_{W^{1,p}(\Omega_\varepsilon^*)}$ or $\|\nabla w_\varepsilon\|_{L^p(\Omega_\varepsilon^*)}$ bounded

We consider the space $W^{1,p}(\Omega_\varepsilon^*)$ or, for Γ_0 non empty open subset of $\partial\Omega$, its subspace denoted $W_0^{1,p}(\Omega_\varepsilon^*;\Gamma_0\cap\partial\Omega_\varepsilon^*)$ of functions vanishing on $\Gamma_0\cap\partial\Omega_\varepsilon^*$.

For the latter case, we suppose that there is an Ω' open subset of $\mathbb{R}^N$ such that $\Omega \subset \Omega'$ and $\Gamma_0 = \partial\Omega \cap \Omega'$ (the black line in Figure 4.8) and then

$$\begin{aligned} W_0^{1,p}(\Omega;\Gamma_0) = \big\{\phi \in W^{1,p}(\Omega) \mid \exists \phi' \in W^{1,p}(\Omega'),& \\ \phi = \phi'_{|\Omega},\ \ \phi' = 0 \text{ in } \Omega' \backslash \overline{\Omega}\, \big\}.& \end{aligned} \tag{4.25}$$

This holds for example, if Γ_0 has a Lipschitz neighborhood in $\partial\Omega$.

In the context of definition (4.25), the space $W_0^{1,p}(\Omega_\varepsilon^*;\Gamma_0 \cap \partial\Omega_\varepsilon^*)$ is described as follows:

$$\begin{aligned} W_0^{1,p}(\Omega_\varepsilon^*;\Gamma_0 \cap \partial\Omega_\varepsilon^*) = \big\{\psi \in W^{1,p}(\Omega_\varepsilon^*) \mid \exists \psi' \in W^{1,p}((\Omega')_\varepsilon^*),& \\ \psi = \psi'_{|\Omega_\varepsilon^*},\ \ \psi' = 0 \text{ in } (\Omega')_\varepsilon^* \backslash \overline{\Omega_\varepsilon^*}\, \big\},& \end{aligned} \tag{4.26}$$

where, recalling definition (4.2) of S_ε,

$$(\Omega')_\varepsilon^* = \Omega' \backslash S_\varepsilon. \tag{4.27}$$

By this construction obviously (see Figure 4.8),

$$\Omega_\varepsilon^* = (\Omega')_\varepsilon^* \cap \Omega.$$

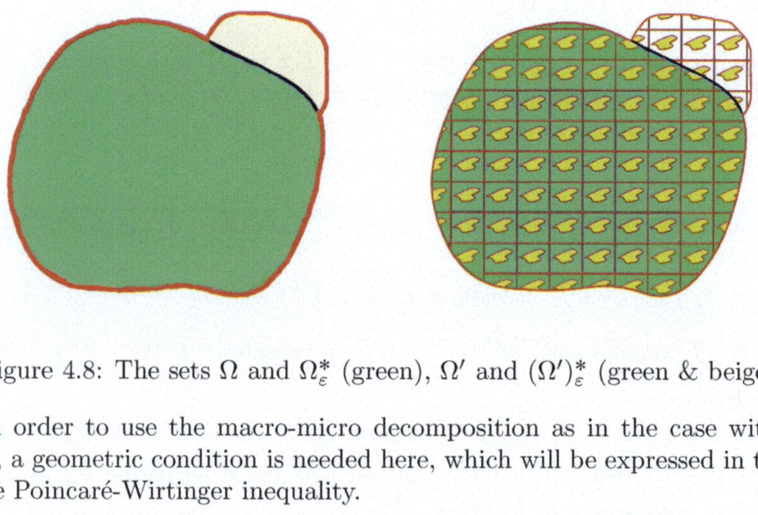

Figure 4.8: The sets Ω and Ω_ε^* (green), Ω' and $(\Omega')_\varepsilon^*$ (green & beige)

In order to use the macro-micro decomposition as in the case without holes, a geometric condition is needed here, which will be expressed in terms of the Poincaré-Wirtinger inequality.

4.3.2.1 The Poincaré-Wirtinger inequality and Hypothesis ($\mathbf{H_p}$)

We now recall Definition 2.25 (see also (1.33)).

Definition. *A bounded open set $\mathcal{O}$ satisfies the Poincaré-Wirtinger inequality for the exponent $p \in [1, +\infty]$, if there exists a constant C_p such that*

$$\|u - \mathcal{M}_{\mathcal{O}}(u)\|_{L^p(\mathcal{O})} \leqslant C_p \|\nabla u\|_{L^p(\mathcal{O})}, \quad \forall u \in W^{1,p}(\mathcal{O}).$$

This requires that $\mathcal{O}$ be connected. Conversely, there are extra conditions for this property to hold (e.g. for John domains, see [88]). The simplest one is that the boundary of $\mathcal{O}$ be Lipschitz (by the compactness of the Rellich theorem), but it is far from being necessary (see subsection 2.3.1 for more details on the Poincaré-Wirtinger inequality).

Let now state the geometric condition.

Definition 4.30 (Geometrical Hypothesis ($\mathbf{H_p}$)). *One says that* ($\mathbf{H_p}$) *is verified if the open set* Y^* *has the following properties:*

(i) the Poincaré-Wirtinger inequality holds for the exponent p ($p \in [1, +\infty]$),

(ii) there is a basis $\mathbf{B}$ *for the group* $\mathbf{G}$ *such that for every vector* $\mathbf{b}_i$ *of* $\mathbf{B}$, ($i \in \{1, \dots, N\}$), *the interior of* $\overline{Y^* \cup (\mathbf{b}_i + Y^*)}$ *is connected*[(3)].

Note that under Hypothesis ($\mathbf{H_p}$) and owing to Remark 2.27, the open set $\mathcal{Y}^*$ defined by (4.20) satisfies the Poincaré-Wirtinger inequality for the same exponent p as Y^*.

Should $\partial Y^* \cap \partial Y$ be Lipschitz – so traces exist – one could use the approach of Section 3 of [62] with the relative compactness of the sequence of local averages obtained via the Kolmogorov criterion (as in the proof of Lemma 2.3 of [10]), and the Y-periodicity of the function $\widehat{w}$ by comparing traces on opposite faces of Y^*. Here, we do not assume such regularity, allowing in particular, ∂S and ∂Y to intersect in an arbitrary fashion. We therefore use the approach of Section 1.6, i.e. the macro-micro decomposition *but without any extension outside of* Ω^*_ε.

To simplify the presentation, we consider the case where Y is a parallelotope. The general case, where Y cannot be chosen as the parallelotope $\mathcal{P}$, exhibits extra geometric complexities and is presented in Section 4.6.

4.3.2.2 The macro-micro operators $\mathcal{Q}^*_\varepsilon$ and $\mathcal{R}^*_\varepsilon$ when the reference cell Y is a parallelotope

As in the case without holes, the macro approximation will be defined at most of the points of Ξ_ε by an average, and then extended to the set $\widehat{\Omega}^{\mathcal{Y}}_\varepsilon$ by Q_1-interpolation (continuous and piece-wise polynomials of degree $\leqslant 1$ with respect to each coordinate). The notations $\mathcal{Y}, \mathcal{K}, \dots$, are those of Subsection 4.3.1. The local average operator $\mathcal{M}^*_\varepsilon$ was given by Definition 4.11.

Definition 4.31. *The operator*

$$\mathcal{Q}^*_\varepsilon \; : \; L^p(\Omega^*_\varepsilon) \longrightarrow W^{1,\infty}(\widehat{\Omega}^{\mathcal{Y}}_\varepsilon),$$

for $p \in [1, +\infty]$, *is defined as*

$$\mathcal{Q}^*_\varepsilon(\phi)(\varepsilon\xi) = \frac{1}{|Y^*|}\int_{Y^*} \phi(\varepsilon\xi + \varepsilon z)\, dz = \mathcal{M}^*_\varepsilon(\phi)(\varepsilon\xi)^{(4)} \quad \textit{for all } \xi \textit{ in } \Xi^{\mathcal{Y}}_\varepsilon + \mathcal{K},$$

(3) The latter condition easily implies that $(\mathbb{R}^N)^*$ is connected.

(4) With the convention of Remark 4.12.

and for every $x \in \widehat{\Omega}^{\mathcal{Y}}_\varepsilon$,

$$\begin{cases} \mathcal{Q}^*_\varepsilon(\phi)(x) \text{ is the } Q_1\text{-interpolate of the values of } \mathcal{Q}^*_\varepsilon(\phi) \text{ at the vertices} \\ \text{of the cell } \varepsilon\Big[\dfrac{x}{\varepsilon}\Big]_Y + \varepsilon Y. \end{cases}$$

Remark 4.32. *Using other averages, one can extend* $\mathcal{Q}^*_\varepsilon$ *to the whole of* $\widehat{\Omega}_\varepsilon$ *without changing its properties.*

Remark 4.33. *Due to Remark 4.23, note that for every* ϕ *in* $L^p(\Omega^*_\varepsilon)$

$$\mathcal{Q}^*_\varepsilon(\phi) = \mathcal{Q}_\varepsilon \circ \mathcal{E}_\varepsilon(\phi) \qquad \text{in } \widehat{\Omega}^{\mathcal{Y}}_\varepsilon. \tag{4.28}$$

As a consequence, one can transpose all the results in Section 1.6 involving functions in $L^p(\Omega)$, $p \in [1, +\infty]$ *to functions of* $L^p(\Omega^*_\varepsilon)$ *(but not of* $W^{1,p}(\Omega^*_\varepsilon)$*). An example is the next proposition.*

Proposition 4.34 (Properties of $\mathcal{Q}^*_\varepsilon$). *Suppose* $p \in [1, +\infty]$. *For* ϕ *in* $L^p(\Omega^*_\varepsilon)$, *the following estimates hold:*

$$\begin{aligned} \|\mathcal{Q}^*_\varepsilon(\phi)\|_{L^p(\widehat{\Omega}^{\mathcal{Y}}_\varepsilon)} &\leqslant \Big(\frac{2^N|Y|}{|Y^*|}\Big)^{1/p} \|\phi\|_{L^p(\Omega^*_\varepsilon)}, \\ \|\mathcal{Q}^*_\varepsilon(\phi)\|_{L^\infty(\widehat{\Omega}^{\mathcal{Y}}_\varepsilon)} &\leqslant \frac{1}{\varepsilon^{N/p}|Y^*|^{1/p}} \|\phi\|_{L^p(\Omega^*_\varepsilon)} \end{aligned} \tag{4.29}$$

As in Section 1.6, explicit formulas for ϕ in $L^p(\Omega^*_\varepsilon)$ hold. For instance,

$$\mathcal{Q}^*_\varepsilon(\phi)(x) = \sum_{\kappa \in \{0,1\}^N} \mathcal{M}^*_\varepsilon(\phi)\Big(\varepsilon\Big[\frac{x}{\varepsilon}\Big]_Y + \varepsilon\kappa\Big) \mathbf{q}_\kappa\Big(\Big\{\frac{x}{\varepsilon}\Big\}_Y\Big), \quad \forall x \in \widehat{\Omega}^{\mathcal{Y}}_\varepsilon. \tag{4.30}$$

Set $\mathbf{K}_j = \{\kappa \in \{0,1\}^N \mid \kappa_j = 1\}$, $1 \leqslant j \leqslant N$. Then another example is

$$\begin{aligned} &\frac{\partial \mathcal{Q}^*_\varepsilon(\phi)}{\partial x_j}(x) \\ &= \sum_{\kappa \in \mathbf{K}_j} \frac{\mathcal{M}^*_\varepsilon(\phi)\Big(\varepsilon\Big[\frac{x}{\varepsilon}\Big]_Y + \varepsilon(\kappa + \mathbf{b}_j)\Big) - \mathcal{M}^*_\varepsilon(\phi)\Big(\varepsilon\Big[\frac{x}{\varepsilon}\Big]_Y + \varepsilon\kappa\Big)}{\varepsilon} \mathbf{q}^j_\kappa\Big(\Big\{\frac{x}{\varepsilon}\Big\}_Y\Big), \end{aligned} \tag{4.31}$$

where the functions $\mathbf{q}_\kappa$ and $\mathbf{q}^j_\kappa$ were defined at the beginning of Section 1.6.

In the remainder of this subsection we assume that Hypothesis ($\mathbf{H_p}$) holds for some $p \in [1, +\infty]$.

Proposition 4.35. *There is a constant* C *independent of* ε *such that, for every* $\phi \in W^{1,p}(\Omega^*_\varepsilon)$,

$$(i) \quad \begin{cases} \|\phi - \mathcal{Q}^*_\varepsilon(\phi)\|_{L^p(\widehat{\Omega}^{\mathcal{Y}}_\varepsilon)} \leqslant C\varepsilon \|\nabla\phi\|_{L^p(\Omega^*_\varepsilon)}, \\ \|\mathcal{M}^*_\varepsilon(\phi) - \mathcal{Q}^*_\varepsilon(\phi)\|_{L^p(\widehat{\Omega}^{\mathcal{Y}}_\varepsilon)} \leqslant C\varepsilon \|\nabla\phi\|_{L^p(\Omega^*_\varepsilon)}, \end{cases}$$

$$(ii) \quad \|\nabla \mathcal{Q}^*_\varepsilon(\phi)\|_{L^p(\widehat{\Omega}^{\mathcal{Y}}_\varepsilon)} \leqslant C \|\nabla\phi\|_{L^p(\Omega^*_\varepsilon)}.$$

Proof. *(i)* For $\psi \in W^{1,p}(Y^*)$, by the Poincaré-Wirtinger inequality in Y^* one has

$$\|\psi - \mathcal{M}_{Y^*}(\psi)\|_{L^p(Y^*)} \leqslant C\|\nabla\psi\|_{L^p(Y^*)}.$$

For ϕ in $W^{1,p}(\Omega^*_\varepsilon)$ and $\xi \in \Xi^{\mathcal{Y}}_\varepsilon$, from this inequality by a scaling argument, it follows that

$$\|\phi - \mathcal{Q}^*_\varepsilon(\phi)(\varepsilon\xi)\|_{L^p(\varepsilon\xi+\varepsilon Y^*)} \leqslant C\varepsilon\|\nabla\phi\|_{L^p(\varepsilon\xi+\varepsilon Y^*)}. \tag{4.32}$$

Together with (4.34), this implies for every ℓ in $\mathcal{K}$,

$$\begin{aligned}
&\|\phi - \mathcal{Q}^*_\varepsilon(\phi)(\varepsilon\xi + \varepsilon\ell)\|_{L^p(\varepsilon\xi+\varepsilon Y^*)} \leqslant C\varepsilon\|\nabla\phi\|_{L^p(\varepsilon\xi+\varepsilon\mathcal{Y}^*)},\\
&\|\mathcal{M}^*_\varepsilon(\phi)(\varepsilon\xi) - \mathcal{Q}^*_\varepsilon(\phi)(\varepsilon\xi + \varepsilon\ell)\|_{L^p(\varepsilon\xi+\varepsilon Y)} \leqslant C\varepsilon\|\nabla\phi\|_{L^p(\varepsilon\xi+\varepsilon\mathcal{Y}^*)}.
\end{aligned}$$

These estimates and equality (4.30) finally give

$$\begin{aligned}
&\|\phi - \mathcal{Q}^*_\varepsilon(\phi)\|_{L^p(\varepsilon\xi+\varepsilon Y^*)} \leqslant C\varepsilon\|\nabla\phi\|_{L^p(\varepsilon\xi+\varepsilon\mathcal{Y}^*)},\\
&\|\mathcal{M}^*_\varepsilon(\phi) - \mathcal{Q}^*_\varepsilon(\phi)\|_{L^p(\varepsilon\xi+\varepsilon Y)} \leqslant C\varepsilon\|\nabla\phi\|_{L^p(\varepsilon\xi+\varepsilon\mathcal{Y}^*)},
\end{aligned}$$

from which inequalities (i) follow by summation over $\Xi^{\mathcal{Y}}_\varepsilon$.
(ii) For $\psi \in W^{1,p}(\mathcal{Y}^*)$, apply the Poincaré-Wirtinger inequality in the domain $\mathcal{Y}^*$ to obtain

$$\|\psi - \mathcal{M}_{\mathcal{Y}^*}(\psi)\|_{L^p(\mathcal{Y}^*)} \leqslant C\|\nabla\psi\|_{L^p(\mathcal{Y}^*)}. \tag{4.33}$$

Integrating $\psi - \mathcal{M}_{\mathcal{Y}^*}(\psi)$ over Y^* and $Y^* + \ell$ for ℓ in $\mathcal{K}$, and using the above inequality give, respectively,

$$|\mathcal{M}_{Y^*}(\psi) - \mathcal{M}_{\mathcal{Y}^*}(\psi)| \leqslant \frac{1}{|Y^*|^{1/p}}\|\psi - \mathcal{M}_{\mathcal{Y}^*}(\psi)\|_{L^p(\mathcal{Y}^*)} \leqslant C\|\nabla\psi\|_{L^p(\mathcal{Y}^*)},$$

and

$$|\mathcal{M}_{Y^*+\ell}(\psi) - \mathcal{M}_{\mathcal{Y}^*}(\psi)| \leqslant \frac{1}{|Y^*|^{1/p}}\|\psi - \mathcal{M}_{\mathcal{Y}^*}(\psi)\|_{L^p(\mathcal{Y}^*)} \leqslant C\|\nabla\psi\|_{L^p(\mathcal{Y}^*)}.$$

The constant C depends on the Poincaré-Wirtinger constant in (4.33) and on $|Y^*|$. Therefore, for every ℓ in $\mathcal{K}$,

$$|\mathcal{M}_{Y^*+\ell}(\psi) - \mathcal{M}_{Y^*}(\psi)| \leqslant C\|\nabla\psi\|_{L^p(\mathcal{Y}^*)}.$$

For ϕ in $W^{1,p}(\Omega^*_\varepsilon)$ and $\xi \in \Xi^{\mathcal{Y}}_\varepsilon$, this inequality applied to the function $\psi(y) \doteq \phi(\varepsilon\xi + \varepsilon y)$ gives

$$\Big|\mathcal{Q}^*_\varepsilon(\phi)(\varepsilon\xi + \varepsilon\ell) - \mathcal{Q}^*_\varepsilon(\phi)(\varepsilon\xi)\Big|^p \leqslant \varepsilon^{p-N} C^p\|\nabla\phi\|^p_{L^p(\varepsilon\xi+\varepsilon\mathcal{Y}^*)}, \tag{4.34}$$

which, in combination with (4.31), yields

$$\|\nabla\mathcal{Q}^*_\varepsilon(\phi)\|^p_{L^p(\varepsilon\xi+\varepsilon Y)} \leqslant C^p\|\nabla\phi\|^p_{L^p(\varepsilon\xi+\varepsilon\mathcal{Y}^*)}.$$

Estimate (ii) then follows by summation over $\Xi^{\mathcal{Y}}_\varepsilon$. ◻

Corollary 4.36. *There is a constant C independent of ε such that, for every ϕ in $W^{1,p}(\Omega^*_\varepsilon)$,*

$$\|\mathcal{E}_\varepsilon(\phi) - \mathcal{Q}^*_\varepsilon(\phi)\|_{L^p(\widehat{\Omega}^{\mathcal{Y}}_\varepsilon)} \leqslant C\varepsilon\,\|\nabla\phi\|_{L^p(\Omega^*_\varepsilon)}. \tag{4.35}$$

Proof. This is a simple consequence of Proposition 4.35(i). □

With the properties of the operator $\mathcal{Q}^*_\varepsilon$ in mind, every function ϕ in $L^p(\Omega^*_\varepsilon)$ is decomposed on the set

$$\widehat{\Omega}^{**}_\varepsilon \doteq \Omega^*_\varepsilon \cap \widehat{\Omega}^{\mathcal{Y}}_\varepsilon,$$

into the sum

$$\phi = \mathcal{Q}^*_\varepsilon(\phi) + \mathcal{R}^*_\varepsilon(\phi) \quad \text{a.e. in } \widehat{\Omega}^{**}_\varepsilon, \tag{4.36}$$

which defines the remainder operator $\mathcal{R}^*_\varepsilon$. Its properties are described in the next result.

Proposition 4.37. *There is a constant C independent of ε such that, for every ϕ in $W^{1,p}(\Omega^*_\varepsilon)$,*

$$\begin{aligned} &(i) \qquad \|\mathcal{R}^*_\varepsilon(\phi)\|_{L^p(\widehat{\Omega}^{**}_\varepsilon)} \leqslant C\varepsilon\,\|\nabla\phi\|_{L^p(\Omega^*_\varepsilon)},\\ &(ii) \qquad \|\nabla\mathcal{R}^*_\varepsilon(\phi)\|_{L^p(\widehat{\Omega}^{**}_\varepsilon)} \leqslant C\,\|\nabla\phi\|_{L^p(\Omega^*_\varepsilon)}. \end{aligned}$$

Proof. Estimates (i) and (ii) are simple rewrites of Proposition 4.35. □

A consequence of Proposition 4.37, is a uniform Poincaré inequality which is useful for applications. It is based on the micro-macro decomposition (carried out here on a neighborhood of Ω, which we may take to be the whole of $\mathbb{R}^N$). Let us point out that it extends to the case of non-smooth holes the result of Lemma A4 of [10].

Theorem 4.38 (Uniform Poincaré inequality)**.** *Assume that Ω is bounded in one direction and with Lipschitz boundary. Then, there exists a constant C independent of ε such that,*

$$\|\phi\|_{L^p(\Omega^*_\varepsilon)} \leqslant C\,\|\nabla\phi\|_{L^p(\Omega^*_\varepsilon)}, \quad \forall\,\phi \in W^{1,p}_0(\Omega^*_\varepsilon; \partial\Omega \cap \partial\Omega^*_\varepsilon). \tag{4.37}$$

Proof. We extend $\phi \in W^{1,p}_0(\Omega^*_\varepsilon; \partial\Omega \cap \partial\Omega^*_\varepsilon)$ by zero to the whole of $(\mathbb{R}^N)^*_\varepsilon$ (see (4.4)), extension still denoted ϕ. Then, for the macro and micro operators $\widetilde{\mathcal{Q}}^*_\varepsilon$ and $\widetilde{\mathcal{R}}^*_\varepsilon$ associated with $(\mathbb{R}^N)^*_\varepsilon$, Propositions 4.35 and 4.37 give

$$\|\nabla\widetilde{\mathcal{Q}}^*_\varepsilon(\phi)\|_{L^p(\mathbb{R}^N)} + \frac{1}{\varepsilon}\|\widetilde{\mathcal{R}}^*_\varepsilon(\phi)\|_{L^p((\mathbb{R}^N)^*_\varepsilon)} \leqslant C\,\|\nabla\phi\|_{L^p(\Omega^*_\varepsilon)}.$$

The domain Ω being bounded in one direction, for ε small enough, the support of $\widetilde{\mathcal{Q}}^*_\varepsilon(\phi)$ is contained in the neighborhood $\mathcal{V}_1(\Omega)$ (see Notation 1.63) and bounded in the same direction. Hence, by the Poincaré inequality in $\mathcal{V}_1(\Omega)$,

$$\|\widetilde{\mathcal{Q}}^*_\varepsilon(\phi)\|_{L^p(\mathbb{R}^N)} \leqslant C\|\nabla\widetilde{\mathcal{Q}}^*_\varepsilon(\phi)\|_{L^p(\mathbb{R}^N)} \leqslant C\,\|\nabla\phi\|_{L^p(\Omega^*_\varepsilon)},$$

from which (4.37) follows, since $\phi = \widetilde{\mathcal{Q}}^*_\varepsilon(\phi) + \widetilde{\mathcal{R}}^*_\varepsilon(\phi)$, a.e. in $(\mathbb{R}^N)^*_\varepsilon$. □

A similar (but more involved) computation leads to the existence of a uniform Sobolev-Poincaré inequality in Ω^*_ε with a constant independent of ε.

Remark 4.39 (Uniform Sobolev-Poincaré inequality). *Assume that Ω is bounded and with Lipschitz boundary. Suppose also that the following Sobolev-Poincaré-Wirtinger inequality holds for the set Y^*, i.e.,*

$$\|\psi - \mathcal{M}_{Y^*}(\psi)\|_{L^{p^*}(Y^*)} \leqslant C\|\nabla\psi\|_{L^p(Y^*)}, \quad \forall\psi \in W^{1,p}(Y^*),$$

where $1/p^* \doteq 1/p - 1/N$ (this assumes $N > p$) and C does not depend on ε.
Then, there exists a constant $\overline{C}$ independent of ε, such that

$$\|\phi\|_{L^{p^*}(\Omega^*_\varepsilon)} \leqslant \overline{C}\|\nabla\phi\|_{L^p(\Omega^*_\varepsilon)}, \quad \forall\, \phi \in W^{1,p}_0(\Omega^*_\varepsilon; \partial\Omega \cap \partial\Omega^*_\varepsilon)^{(5)}.$$

4.3.2.3 Convergence results

In this subsection, it is assumed that p is in $(1, +\infty]$ and that Hypothesis $(\mathbf{H_p})$ holds. For $p = +\infty$, all the stated weak convergences are to be understood as weak-$*$ convergences.

Consider sequences $\{w_\varepsilon\}_\varepsilon$ with $w_\varepsilon \in W^{1,p}(\Omega^*_\varepsilon)$, such that

$$\|w_\varepsilon\|_{W^{1,p}(\Omega^*_\varepsilon)} \leqslant C \qquad \text{or} \qquad \|\nabla w_\varepsilon\|_{L^p(\Omega^*_\varepsilon)} \leqslant C.$$

The main results in this subsection are Theorems 4.43 and 4.44 below. To state them, we need some notations and preliminary lemmas.

Notation 4.40 (Extension by zero).
To express some convergence results, it will very often be convenient to consider the extension by 0 to Ω of functions defined on a subset of Ω, which differ from it by a boundary layer of order ε (such as $\widehat{\Omega}^{\mathcal{Y}}_\varepsilon$, $\widehat{\Omega}^{**}_\varepsilon = \Omega^*_\varepsilon \cap \widehat{\Omega}^{\mathcal{Y}}_\varepsilon$ and $\Omega^{int}_{k\varepsilon\, d(Y)}$). For the extension by zero of a function Φ defined on such a subset D of Ω, we will use the (somewhat incorrect but simple) notation $\Phi\mathbf{1}_D$. Actually, $\Phi\mathbf{1}_D$ can even be considered as defined on the whole space $\mathbb{R}^N$.

Recall that

$$\Omega^{int}_{3\varepsilon\, d(Y)} \subset \widehat{\Omega}^{\mathcal{Y}}_\varepsilon \subset \Omega \qquad \text{and} \qquad \text{dist}(\Omega^{int}_{3\varepsilon\, d(Y)}, \partial\Omega) = 3\varepsilon\, d(Y).$$

Lemma 4.41. *Let $\{w_\varepsilon\}_\varepsilon$ be a sequence in $W^{1,p}(\Omega^*_\varepsilon)$, satisfying*

$$\|w_\varepsilon\|_{W^{1,p}(\Omega^*_\varepsilon)} \leqslant C. \tag{4.38}$$

Then, there exists w in $W^{1,p}(\Omega)$ such that, up to a subsequence,

$$\begin{aligned} &\mathcal{E}_\varepsilon(w_\varepsilon) \rightharpoonup w \quad \text{weakly in } L^p(\Omega), \ (\text{strongly in } L^p_{loc}(\Omega)),\\ &\mathcal{Q}^*_\varepsilon(w_\varepsilon)\mathbf{1}_{\Omega^{int}_{3\varepsilon\, d(Y)}} \rightharpoonup w \quad \text{weakly in } L^p(\Omega), \ (\text{strongly in } L^p_{loc}(\Omega)),\\ &(\nabla\mathcal{Q}^*_\varepsilon(w_\varepsilon))\mathbf{1}_{\Omega^{int}_{3\varepsilon\, d(Y)}} \rightharpoonup \nabla w \quad \text{weakly in } L^p(\Omega)^N. \end{aligned} \tag{4.39}$$

(5) In the proof, one uses the well-known fact that the Sobolev-Poincaré-Wirtinger inequality is scale-invariant, i.e. the constant is the same for Y^* and for εY^* when $p < N$.

Moreover

$$\mathcal{T}_\varepsilon\big([\nabla \mathcal{Q}^*_\varepsilon(w_\varepsilon)]\mathbf{1}_{\Omega^{int}_{3\varepsilon\, d(Y)}}\big) \rightharpoonup \nabla w \quad \textit{weakly in } L^p(\Omega\times Y)^N. \tag{4.40}$$

Furthermore, assuming that Γ_0 is a non empty open subset of $\partial\Omega$, if for every ε the function w_ε belongs to $W^{1,p}_0(\Omega^*_\varepsilon;\Gamma_0\cap\partial\Omega^*_\varepsilon)$ (see (4.26)), then w belongs to $W^{1,p}_0(\Omega;\Gamma_0)$ (i.e., the trace of w vanishes on Γ_0).

Proof. All the convergences of this proof are up to a subsequence. First, from estimates (4.13) and (4.29), both sequences $\{\mathcal{E}_\varepsilon(w_\varepsilon)\}_\varepsilon$ and $\{\mathcal{Q}^*_\varepsilon(w_\varepsilon)\mathbf{1}_{\widehat{\Omega}^{\mathcal{Y}}_\varepsilon}\}_\varepsilon$ are bounded in $L^p(\Omega)$. Hence, due to estimate $(4.35)_2$, there exists w in $L^p(\Omega)$ such that

$$\begin{aligned}&\mathcal{Q}^*_\varepsilon(w_\varepsilon)\mathbf{1}_{\widehat{\Omega}^{\mathcal{Y}}_\varepsilon} \rightharpoonup w \quad \text{weakly in } L^p(\Omega),\\ &\mathcal{E}_\varepsilon(w_\varepsilon) \rightharpoonup w \quad \text{weakly in } L^p(\Omega).\end{aligned}$$

Now, from Proposition 4.35(ii), the sequence $\{(\nabla\mathcal{Q}^*_\varepsilon(w_\varepsilon))\mathbf{1}_{\widehat{\Omega}^{\mathcal{Y}}_\varepsilon}\}_\varepsilon$ is bounded in $L^p(\Omega)^N$. Hence, there exists $F\in L^p(\Omega)^N$ such that

$$(\nabla\mathcal{Q}^*_\varepsilon(w_\varepsilon))\mathbf{1}_{\widehat{\Omega}^{\mathcal{Y}}_\varepsilon} \rightharpoonup F \quad \text{weakly in } L^p(\Omega)^N.$$

Let ω be a relatively compact open subset in Ω and ε be small enough so that ω is included in $\widehat{\Omega}^{\mathcal{Y}}_\varepsilon$. Then, by (4.28), convergence (ii) from Proposition 1.78 gives

$$\begin{aligned}&\mathcal{Q}^*_\varepsilon(w_\varepsilon)_{|\omega} = \big(\mathcal{Q}_\varepsilon\circ\mathcal{E}_\varepsilon(w_\varepsilon)\big)_{|\omega} \to w \quad \text{strongly in } L^p(\omega),\\ &\big(\nabla\mathcal{Q}^*_\varepsilon(w_\varepsilon)\big)_{|\omega} = \big(\nabla\mathcal{Q}_\varepsilon\circ\mathcal{E}_\varepsilon(w_\varepsilon)\big)_{|\omega} \rightharpoonup \nabla w \quad \text{weakly in } L^p(\omega)^N.\end{aligned}$$

Consequently $F=\nabla w$, and w belongs to $W^{1,p}(\Omega)$.

Convergence (iii) from Proposition 1.78 and the ones above allow to obtain convergence (4.40).

It remains to prove the last assertion of the theorem. Let w_ε be in $W^{1,p}_0(\Omega^*_\varepsilon;\Gamma_0\cap\partial\Omega^*_\varepsilon)$ in the sense of (4.26). We extend it by zero in $(\Omega')^*_\varepsilon\setminus\Omega^*_\varepsilon$ (see (4.27) and Figure 4.8). This extension is still denoted w_ε. It is clear from Notation 4.40 and the definition of $\mathcal{Q}^*_\varepsilon(w_\varepsilon)\mathbf{1}_{\widehat{\Omega'}^{\mathcal{Y}}_\varepsilon}$, that the above convergences hold in $L^p(\Omega')$. It follows that $w=0$ on Γ_0, since it is the restriction to Ω of a function defined in Ω' and vanishing in $\Omega'\setminus\overline{\Omega}$). □

We now prove a convergence result concerning the sequence $\{\mathcal{R}^*_\varepsilon(w_\varepsilon)\}_\varepsilon$.

Lemma 4.42. *Let $\{w_\varepsilon\}_\varepsilon$ be a sequence with $w_\varepsilon\in W^{1,p}(\Omega^*_\varepsilon)$ satisfying*

$$\|\nabla w_\varepsilon\|_{L^p(\Omega^*_\varepsilon)} \leqslant C.$$

Then, there exists $\widehat{w}'$ in $L^p(\Omega;W^{1,p}_{per}(Y^*))$ such that, up to a subsequence,

$$\mathcal{T}^*_\varepsilon\big([\nabla\mathcal{R}^*_\varepsilon(w_\varepsilon)]\mathbf{1}_{\Omega^{int}_{3\varepsilon\, d(Y)}}\big) \rightharpoonup \nabla_y\widehat{w}' \quad \textit{weakly in } L^p(\Omega\times Y^*)^N. \tag{4.41}$$

Proof. Due to Proposition 4.37, the sequence $\{\frac{1}{\varepsilon}\mathcal{T}^*_\varepsilon(\mathcal{R}^*_\varepsilon(w_\varepsilon)\mathbf{1}_{\widehat{\Omega}^{**}_\varepsilon})\}_\varepsilon$ is bounded in $L^p(\Omega; W^{1,p}(Y^*))$. So there exits $\widehat{w}'$ in $L^p(\Omega; W^{1,p}(Y^*))$ such that the following convergences hold:

$$\begin{aligned}
&\frac{1}{\varepsilon}\mathcal{T}^*_\varepsilon(\mathcal{R}^*_\varepsilon(w_\varepsilon)\mathbf{1}_{\widehat{\Omega}^{**}_\varepsilon}) \rightharpoonup \widehat{w}' \quad \text{weakly in } L^p(\Omega; W^{1,p}(Y^*)),\\
&\mathcal{T}^*_\varepsilon(\mathcal{R}^*_\varepsilon(w_\varepsilon)\mathbf{1}_{\widehat{\Omega}^{**}_\varepsilon}) \to 0 \quad \text{strongly in } L^p(\Omega; W^{1,p}(Y^*)),\\
&\mathcal{T}^*_\varepsilon([\nabla\mathcal{R}^*_\varepsilon(w_\varepsilon)]\mathbf{1}_{\widehat{\Omega}^{**}_\varepsilon}) \rightharpoonup \nabla_y\widehat{w}' \quad \text{weakly in } L^p(\Omega\times Y^*)^N.
\end{aligned}$$

To show that $\widehat{w}'$ is actually in $L^p(\Omega; W^{1,p}_{per}(Y^*))$, let ω be any relatively compact open subset of Ω. The restriction of $\frac{1}{\varepsilon}\mathcal{R}^*_\varepsilon(\phi)$ to ω^*_ε belongs to $W^{1,p}(\omega^*_\varepsilon)$. Applying Theorem 4.28 to this restriction shows that $\widehat{w}'_{|\omega\times Y^*}$ belongs to $L^p(\omega; W^{1,p}_{per}(Y^*))$. The conclusion follows due to the arbitrariness of the open subset ω. □

We are now in a position to state the main results of this section.

Theorem 4.43. *Suppose that the sequence $\{w_\varepsilon\}_\varepsilon$, $w_\varepsilon \in W^{1,p}(\Omega^*_\varepsilon)$, satisfies*

$$\|w_\varepsilon\|_{W^{1,p}(\Omega^*_\varepsilon)} \leqslant C.$$

Then, there exist w in $W^{1,p}(\Omega)$ and $\widehat{w}$ in $L^p(\Omega; W^{1,p}_{per,0}(Y^*))$ such that, up to a subsequence,

$$\begin{aligned}
&(i)\quad \begin{cases} \mathcal{T}^*_\varepsilon(w_\varepsilon) \to w & \text{strongly in } L^p_{loc}(\Omega; W^{1,p}(Y^*)),\\ \mathcal{T}^*_\varepsilon(w_\varepsilon) \rightharpoonup w & \text{weakly in } L^p(\Omega; W^{1,p}(Y^*)),\end{cases}\\
&(ii)\quad \mathcal{T}^*_\varepsilon(\nabla w_\varepsilon) \rightharpoonup \nabla w + \nabla_y\widehat{w} \quad \text{weakly in } L^p(\Omega\times Y^*)^N,\\
&(iii)\quad \begin{cases} \mathcal{E}_\varepsilon(w_\varepsilon) \to w & \text{strongly in } L^p_{loc}(\Omega),\\ \forall\omega \Subset \Omega, & \|w_\varepsilon - w\|_{L^p(\Omega^*_\varepsilon\cap\omega)} \to 0.\end{cases}
\end{aligned} \tag{4.42}$$

Moreover, assuming that Γ_0 is non empty open subset of $\partial\Omega$, if for every ε, the function w_ε belongs to $W^{1,p}_0(\Omega^*_\varepsilon; \Gamma_0 \cap \partial\Omega^*_\varepsilon)$ (with Hypothesis (4.25)), then w belongs to $W^{1,p}_0(\Omega; \Gamma_0)$ (i.e., the trace of w vanishes on Γ_0).

Proof. Convergences (4.42)(i) are consequences of $(4.39)_1$. Convergence (ii) is derived from (4.40) and (4.41) with $\widehat{w} \doteq \widehat{w}' - \mathcal{M}_{Y^*}(\widehat{w}')$. As for (iii), one uses first (i) and Proposition 4.26, and then the equivalence of Proposition 4.27(i) applied in $\omega \Subset \Omega$). □

Theorem 4.44. *Let Ω be bounded. Suppose that the sequence $\{w_\varepsilon\}_\varepsilon$ with $w_\varepsilon \in W^{1,p}_0(\Omega^*_\varepsilon; \partial\Omega \cap \partial\Omega^*_\varepsilon)$, satisfies*

$$\|\nabla w_\varepsilon\|_{L^p(\Omega^*_\varepsilon)} \leqslant C.$$

Then, there exist w in $W_0^{1,p}(\Omega)$ and $\widehat{w}$ in $L^p(\Omega; W^{1,p}_{per,0}(Y^*))$ such that, up to a subsequence,

$$\begin{aligned} &(i)\ \ \mathcal{T}^*_\varepsilon(\nabla w_\varepsilon) \rightharpoonup \nabla w + \nabla_y \widehat{w} \quad \textit{weakly in } L^p(\Omega \times Y^*)^N,\\ &(ii)\ \ \mathcal{T}^*_\varepsilon(w_\varepsilon) \to w \ \ \textit{strongly in } L^p(\Omega; W^{1,p}(Y^*)),\\ &(iii) \begin{cases} \mathcal{E}_\varepsilon(w_\varepsilon) \to w & \textit{strongly in } L^p(\Omega),\\ \|w_\varepsilon - w\|_{L^p(\Omega^*_\varepsilon)} \to 0. \end{cases} \end{aligned} \tag{4.43}$$

Proof. By Theorem 4.38 (Uniform Poincaré inequality), estimate (4.38) is satisfied, that is

$$\|w_\varepsilon\|_{W^{1,p}(\Omega^*_\varepsilon)} \leqslant C.$$

Then convergence (i) of (4.43) is obtained from Theorem 4.43(ii).

In order to obtain (ii) and (iii), without changing notations, we extend w_ε by zero to the whole of $\big(\mathbb{R}^N\big)^*_\varepsilon$ (see (4.4) for its definition). Using the unfolding operator $\widetilde{\mathcal{T}}^*_\varepsilon$ associated with $\mathbb{R}^N$, the convergence result (4.42)(i) of Theorem 4.43 implies that $\widetilde{\mathcal{T}}^*_\varepsilon(w_\varepsilon)$ converges strongly in $L^p_{loc}(\mathbb{R}^N; W^{1,p}(Y^*))$. Note that $\mathcal{T}^*_\varepsilon(w_\varepsilon) = \widetilde{\mathcal{T}}^*_\varepsilon(w_\varepsilon)_{|\widehat{\Omega}_\varepsilon \times Y^*}$. This implies (4.43)(ii).

Convergences (iii) follow first from (i) and Proposition 4.26, and then by the equivalence of Proposition 4.27(i) (applied in $\mathbb{R}^N$). ☐

We complete this subsection with a convergence result concerning the unfolding of the difference between the general term of a sequence and its local average, in the spirit of Theorem 1.41.

Proposition 4.45. *Let $\{w_\varepsilon\}_\varepsilon$ be a bounded (independently of ε) sequence in $W^{1,p}(\Omega^*_\varepsilon)$. Assume that there exist w in $W^{1,p}(\Omega)$ and $\widehat{w}$ in $L^p(\Omega; W^{1,p}_{per,0}(Y^*))$ such that,*

$$\begin{aligned} &\mathcal{T}^*_\varepsilon(w_\varepsilon) \rightharpoonup w \quad \textit{weakly in } L^p(\Omega; W^{1,p}(Y^*)),\\ &\mathcal{T}^*_\varepsilon(\nabla w_\varepsilon) \rightharpoonup \nabla w + \nabla_y \widehat{w} \quad \textit{weakly in } L^p(\Omega \times Y^*)^N. \end{aligned}$$

Then

$$\frac{1}{\varepsilon}\Big(\mathcal{T}^*_\varepsilon(w_\varepsilon) - \mathcal{M}^*_\varepsilon(w_\varepsilon)\Big) \rightharpoonup y^{c*} \cdot \nabla w + \widehat{w} \ \ \textit{weakly in } L^p(\Omega; W^{1,p}(Y^*)),$$

where $y^{c*} = y - \mathcal{M}_{Y^*}(y)$.

Proof. Let

$$Z_\varepsilon = \frac{1}{\varepsilon}\Big(\mathcal{T}^*_\varepsilon(w_\varepsilon) - \mathcal{M}^*_\varepsilon(w_\varepsilon)\Big).$$

By construction, $\mathcal{M}_{Y^*}(Z_\varepsilon) = 0$. On the other hand,

$$\nabla_y(Z_\varepsilon) = \frac{1}{\varepsilon}\nabla_y \mathcal{T}^*_\varepsilon(w_\varepsilon) = \mathcal{T}^*_\varepsilon(\nabla w_\varepsilon) \rightharpoonup \nabla w + \nabla_y \widehat{w} \ \text{ weakly in } L^p(\Omega \times Y^*)^N.$$

Applying to Z_ε the Poincaré-Wirtinger inequality in $W^{1,p}(Y^*)$, it follows that the sequence $\{Z_\varepsilon\}$ is bounded in $L^p(\Omega; W^{1,p}(Y^*))$. Since Y^* is connected, there is a unique element in this space with zero average on Y^*, whose gradient with respect to y is $\nabla w + \nabla_y \widehat{w}$. This element is nothing else than $y^{c*} \cdot \nabla w + \widehat{w}$. □

4.4 The boundary unfolding operator

In this section, we suppose that p is in $[1, +\infty]$. *As before, weak convergences in the case of* $p = +\infty$ *are to be replaced by weak-∗ convergences.*

Assume that ∂Y^* and ∂S are Lipschitz (so they have a finite number of connected components). In this setting, Hypothesis $(\mathbf{H_p})$ (see Definition 4.30) is automatically satisfied for every p if and only if for every vector $\mathbf{b}_i$, $i \in \{1, \ldots, N\}$, of the basis $\mathbf{B}$, the interior of $\overline{Y^* \cup (\mathbf{b}_i + Y^*)}$ is connected.

The boundary of the set of holes in Ω is $\partial S_\varepsilon \cap \Omega$. Recall that the boundary of the holes which are included in $\widehat{\Omega}_\varepsilon$ is $\partial \widehat{S}_\varepsilon$ as defined by (4.3).

Since each component of ∂S has a Lipschitz boundary, there is a canonical surface measure on ∂S denoted σ (similarly, we use σ_ε for the superficial measure on ∂S_ε). A well-defined and continuous trace operator also exists from $W^{1,p}(Y^*)$ to $W^{1-1/p,p}(\partial S)$. Consequently, a well-defined and continuous trace operator exists from $W^{1,p}(\widehat{\Omega}^*_\varepsilon)$ to $W^{1-1/p,p}(\partial \widehat{S}_\varepsilon)$.

The aim here is to give a meaning to the unfolding operator for such traces, and to obtain estimates and convergence results for sequences of functions in $W^{1,p}$-type spaces. To do so, we define a boundary unfolding operator which was introduced in a slightly different form in [72] and [73]. It extends the results obtained via two-scale convergence in [9] and [165].

Definition 4.46. *For* φ *Lebesgue-measurable on* $\partial \widehat{S}_\varepsilon$, *the boundary unfolding operator*

$$\mathcal{T}^b_\varepsilon : L^p(\partial \widehat{S}_\varepsilon) \longrightarrow L^p(\Omega \times \partial S),$$

is defined as follows:

$$\mathcal{T}^b_\varepsilon(\phi)(x,y) = \begin{cases} \phi\Big(\varepsilon\Big[\dfrac{x}{\varepsilon}\Big]_Y + \varepsilon y\Big) & \text{for a.e.} \quad (x,y) \in \widehat{\Omega}_\varepsilon \times \partial S, \\ 0 & \text{for a.e.} \quad (x,y) \in \Lambda_\varepsilon \times \partial S. \end{cases}$$

Like $\mathcal{T}^*_\varepsilon$, the operator $\mathcal{T}^b_\varepsilon$ transforms an integral on the rapidly oscillating set $\partial \widehat{S}_\varepsilon$ into an integral on a fixed set $\Omega \times \partial S$ as shown below.

Proposition 4.47. *For* φ *in* $L^1(\partial \widehat{S}_\varepsilon)$, *the integration formula reads*

$$\int_{\partial \widehat{S}_\varepsilon} \varphi(x)\, d\sigma_\varepsilon(x) = \frac{1}{\varepsilon |Y|} \int_{\Omega \times \partial S} \mathcal{T}^b_\varepsilon(\varphi)(x,y)\, dx\, d\sigma(y). \tag{4.44}$$

For φ *in* $L^p(\partial \widehat{S}_\varepsilon)$,

$$\|\mathcal{T}^b_\varepsilon(\varphi)\|_{L^p(\Omega \times \partial S)} = \varepsilon^{1/p} |Y|^{1/p} \|\varphi\|_{L^p(\partial \widehat{S}_\varepsilon)}. \tag{4.45}$$

If φ belongs to $W^{1,p}(\Omega_\varepsilon^*)$, then

$$\mathcal{T}_\varepsilon^b(\varphi) = \mathcal{T}_\varepsilon^*(\varphi)_{|\Omega\times\partial S}, \tag{4.46}$$

and there is a constant C such that

$$\|\varphi\|_{L^p(\partial\widehat{S}_\varepsilon)} \leqslant \varepsilon^{-1/p} C\big(\|\varphi\|_{L^p(\Omega_\varepsilon^*)} + \varepsilon\|\nabla\varphi\|_{L^p(\Omega_\varepsilon^*)}\big). \tag{4.47}$$

The presence of a power of ε in (4.44) and (4.45) requires a normalization for boundary terms, this differs from that in the bulk. Therefore, the convergence of such boundary integrals are significantly different significantly from the convergence of integrals in the bulk (see Propositions 4.53 and 4.57).

Proof of Proposition 4.47. The proof of (4.44) is straightforward and it implies (4.45). For φ in $\mathrm{W}^{1,p}(\Omega_\varepsilon^*)$, by the standard trace theorem in Y^* applied to the map $y \mapsto \mathcal{T}_\varepsilon^*(\varphi)(\cdot, y)$, one has

$$\|\mathcal{T}_\varepsilon^b(\varphi)\|_{L^p(\Omega;W^{1-1/p,p}(\partial S))} \leqslant C\big(\|\mathcal{T}_\varepsilon^*(\varphi)\|_{L^p(\Omega\times Y^*)} + \|\nabla_y\mathcal{T}_\varepsilon^*(\varphi)\|_{L^p(\Omega\times Y^*)}\big).$$

With equality (4.45), estimate (4.6)(ii) and (4.14), this implies (4.47). □

Definition 4.48 (Local average of $g \in L^p(\partial\widehat{S}_\varepsilon)$). *Let p be in $[1, +\infty]$. For φ in $L^p(\partial\widehat{S}_\varepsilon)$, the local average is defined as*

$$\mathcal{M}_\varepsilon^b(g) \doteq \frac{1}{|\partial S|}\int_{\partial S} \mathcal{T}_\varepsilon^b(g)(\cdot, y)\, d\sigma(y).$$

It belongs to $L^p(\Omega)$ and satisfies

$$\|\mathcal{M}_\varepsilon^b(g)\|_{L^p(\Omega)} \leqslant \varepsilon^{1/p}\Big(\frac{|Y|}{|\partial S|}\Big)^{1/p}\|g\|_{L^p(\partial\widehat{S}_\varepsilon)}.$$

4.4.1 The boundary averaging operator

Definition 4.49. *For Φ in $L^p(\Omega \times \partial S)$, $p \in [1, +\infty]$, the boundary unfolding operator $\mathcal{U}_\varepsilon^b$ is defined as follows:*

$$\mathcal{U}_\varepsilon^b(\Phi)(x) \doteq \begin{cases} \dfrac{1}{|Y|}\displaystyle\int_Y \Phi\Big(\varepsilon\Big[\frac{x}{\varepsilon}\Big]_Y + \varepsilon z, \Big\{\frac{x}{\varepsilon}\Big\}_Y\Big)\, dz & \text{for a.e. } x \in \partial\widehat{S}_\varepsilon, \\ 0 & \text{for a.e. } x \in \partial S_\varepsilon \cap \Lambda_\varepsilon^*. \end{cases}$$

The operator $\mathcal{U}_\varepsilon^b$ is linear and continuous from $L^p(\Omega \times \partial S)$ into $L^p(\partial\widehat{S}_\varepsilon)$. For $\Phi \in L^p(\Omega \times \partial S)$ one has

$$\|\mathcal{U}_\varepsilon^b(\Phi)\|_{L^p(\partial\widehat{S}_\varepsilon)} \leqslant \varepsilon^{-1/p}|Y|^{-1/p}\|\Phi\|_{L^p(\Omega\times\partial S)}. \tag{4.48}$$

This operator is the adjoint of $\mathcal{T}_\varepsilon^b$. Indeed, it is easily seen that for Φ in $L^p(\Omega \times \partial S)$ and ψ in $L^{p'}(\partial\widehat{S}_\varepsilon)$,

$$\int_{\partial\widehat{S}_\varepsilon} \mathcal{U}_\varepsilon^b(\Phi)(x)\,\psi(x)\, d\sigma_\varepsilon(x) = \frac{1}{\varepsilon|Y|}\int_{\Omega\times\partial S} \Phi(x, y)\, \mathcal{T}_\varepsilon^b(\psi)(x, y)\, dx d\sigma(y).$$

Then from its definition, it follows that $\mathcal{U}_\varepsilon^b$ is a left-inverse of $\mathcal{T}_\varepsilon^b$, since

$$\mathcal{U}_\varepsilon^b \circ \mathcal{T}_\varepsilon^b(\phi) = \phi, \tag{4.49}$$

for every ϕ in $L^p(\partial\widehat{S}_\varepsilon)$, whereas

$$\mathcal{T}_\varepsilon^b \circ \mathcal{U}_\varepsilon^b(\Phi)(x,y) = \begin{cases} \dfrac{1}{|Y|}\displaystyle\int_Y \Phi\Big(\varepsilon\Big[\frac{x}{\varepsilon}\Big]_Y + \varepsilon z, y\Big)\,dz & \text{for a.e. } (x,y) \in \widehat{\Omega}_\varepsilon \times \partial S, \\ 0 & \text{for a.e. } (x,y) \in \Lambda_\varepsilon \times \partial S. \end{cases}$$

4.4.2 Convergence for the boundary unfolding

Proposition 4.50. *Suppose that v belongs to $W^{1,p}(\Omega_\varepsilon^*)$ and that g is in $L^{p'}(\partial\widehat{S}_\varepsilon)$. Then,*

$$\text{(i)}\quad \Big|\int_{\partial\widehat{S}_\varepsilon} gv\,d\sigma_\varepsilon(x)\Big| \leqslant \frac{C}{\varepsilon^{1/p}}\|g\|_{L^{p'}(\partial\widehat{S}_\varepsilon)}\Big(\|v\|_{L^p(\Omega_\varepsilon^*)} + \varepsilon\|\nabla v\|_{L^p(\Omega_\varepsilon^*)}\Big),$$

$$\begin{aligned}\text{(ii)}\quad \Big|\int_{\partial\widehat{S}_\varepsilon} gv\,d\sigma_\varepsilon(x)\Big| \leqslant C\Big(&\|\mathcal{T}_\varepsilon^b(g)\|_{L^{p'}(\Omega\times\partial S)}\|\nabla v\|_{L^p(\Omega_\varepsilon^*)} \\ &+ \frac{1}{\varepsilon}\big\|\mathcal{M}_\varepsilon^b(g)\big\|_{L^{p'}(\Omega)}\|v\|_{L^p(\Omega_\varepsilon^*)}\Big).\end{aligned}$$

Proof. By Hölder's inequality, one has

$$\Big|\int_{\partial\widehat{S}_\varepsilon} gv\,d\sigma_\varepsilon(x)\Big| \leqslant \|g\|_{L^{p'}(\partial\widehat{S}_\varepsilon)}\|v\|_{L^p(\partial\widehat{S}_\varepsilon)}.$$

Applying (4.47) gives inequality (i).

Now, in view of (4.44) and (4.46), one can write

$$\int_{\partial\widehat{S}_\varepsilon} g\,v\,d\sigma_\varepsilon(x) = \frac{1}{\varepsilon|Y|}\int_{\Omega\times\partial S} \mathcal{T}_\varepsilon^b(g)(x,y)\,\mathcal{T}_\varepsilon^*(v)(x,y)dx\,d\sigma(y). \tag{4.50}$$

Thanks to Definition 4.48, this becomes

$$\begin{aligned}\int_{\partial\widehat{S}_\varepsilon} g\,v\,d\sigma_\varepsilon(x) = &\frac{1}{\varepsilon|Y|}\int_{\Omega\times\partial S} \mathcal{T}_\varepsilon^b(g)\big(\mathcal{T}_\varepsilon^*(v) - \mathcal{M}_\varepsilon^*(v)\big)\,dx\,d\sigma(y) \\ &+ \frac{|\partial S|}{\varepsilon|Y|}\int_\Omega \mathcal{M}_\varepsilon^b(g)\mathcal{M}_\varepsilon^*(v)\,dx.\end{aligned} \tag{4.51}$$

Again, by Hölder's inequality,

$$\begin{aligned}\Big|\int_{\partial\widehat{S}_\varepsilon} g\,v\,d\sigma_\varepsilon(x)\Big| \leqslant &\frac{C}{\varepsilon}\|\mathcal{T}_\varepsilon^b(g)\|_{L^{p'}(\Omega\times\partial S)}\|\mathcal{T}_\varepsilon^*(v) - \mathcal{M}_\varepsilon^*(v)\|_{L^p(\Omega\times\partial S)} \\ &+ \frac{C}{\varepsilon}\big\|\mathcal{M}_\varepsilon^b(g)\big\|_{L^{p'}(\Omega)}\|\mathcal{M}_\varepsilon^*(v)\|_{L^p(\Omega)}.\end{aligned}$$

Due to the Poincaré-Wirtinger inequality in Y^*,

$$\|\mathcal{T}_\varepsilon^*(v) - \mathcal{M}_\varepsilon^*(v)\|_{L^p(\Omega\times\partial S)} \leqslant C\|\nabla_y \mathcal{T}_\varepsilon^*(v)\|_{L^p(\Omega\times Y^*)} \leqslant C\varepsilon\|\nabla v\|_{L^p(\Omega_\varepsilon^*)}.$$

Then, thanks to the obvious estimates

$$\|\mathcal{M}_\varepsilon^*(v)\|_{L^p(\Omega)} \leqslant C\|\mathcal{T}_\varepsilon^*(v)\|_{L^p(\Omega\times Y^*)} \leqslant C\|v\|_{L^p(\Omega_\varepsilon^*)},$$

inequality (ii) follows. □

A simple consequence of formula (4.50) is the next convergence result.

Proposition 4.51. *Let $\{w_\varepsilon\}_\varepsilon$ be a sequence in $W^{1,p}(\Omega_\varepsilon^*)$ and $\{g_\varepsilon\}_\varepsilon$ a sequence in $L^{p'}(\partial\widehat{S}_\varepsilon)$, such that*

$$\begin{aligned} &(i) \quad \mathcal{T}_\varepsilon^*(w_\varepsilon) \rightharpoonup w \quad \text{weakly in } L^p(\Omega; W^{1,p}(Y^*)),\\ &(ii) \quad \mathcal{T}_\varepsilon^b(g_\varepsilon) \to g \quad \text{strongly in } L^{p'}(\Omega\times\partial S). \end{aligned} \tag{4.52}$$

Then

$$\varepsilon\int_{\partial\widehat{S}_\varepsilon} g_\varepsilon(x) w_\varepsilon(x)\, d\sigma_\varepsilon(x) \to \frac{1}{|Y|}\int_{\Omega\times\partial S} g(x,y)\, w(x,y)\, dx d\sigma(y).$$

Remark 4.52.
1. It is obvious that strong and weak convergences in Proposition 4.51 can be interchanged.
2. For every g in $L^{p'}(\Omega\times\partial S)$, there is always a sequence $\{g_\varepsilon\}_\varepsilon$ satisfying (4.52)*(ii)(see also Remark 4.55 below for a more general statement).*
3. In Proposition 4.51, if w is independent of y and $\mathcal{M}_{\partial S}(g) = 0$[6]*, then*

$$\varepsilon\int_{\partial\widehat{S}_\varepsilon} g_\varepsilon w_\varepsilon\, d\sigma_\varepsilon \to 0.$$

4. The convergence of the boundary unfolding of a sequence gives another proof of the boundary two-scale convergence (see [9, 165])

The last result of Remark 4.52, for w independent of y and $\mathcal{M}_{\partial S}(g) = 0$, was already observed in [72].

At this point, one may ask if the integral $\displaystyle\int_{\partial\widehat{S}_\varepsilon} g_\varepsilon w_\varepsilon\, d\sigma_\varepsilon$, itself could converge. Under some additional assumptions the answer is positive, as shown in the next result.

Proposition 4.53. *Let $\{w_\varepsilon\}_\varepsilon$ be a sequence with w_ε in $W^{1,p}(\Omega_\varepsilon^*)$. Suppose there exist w in $W^{1,p}(\Omega)$ and $\widehat{w}$ in $L^p(\Omega; W^{1,p}_{per,0}(Y^*))$ such that,*

$$\begin{aligned} &\mathcal{T}_\varepsilon^*(w_\varepsilon) \rightharpoonup w \quad \text{weakly in } L^p(\Omega; W^{1,p}(Y^*)),\\ &\mathcal{T}_\varepsilon^*(\nabla w_\varepsilon) \rightharpoonup \nabla w + \nabla_y\widehat{w} \quad \text{weakly in } L^p(\Omega\times Y^*)^N. \end{aligned} \tag{4.53}$$

[6] Recall that $\mathcal{M}_{\partial S}$ is the average operator on ∂S.

Suppose also that the following two convergences hold for the sequence $\{g_\varepsilon\}_\varepsilon$ *with* g_ε *in* $L^{p'}(\partial\widehat{S}_\varepsilon)$*:*

$$\begin{aligned} &\frac{1}{\varepsilon}\mathcal{M}^b_\varepsilon(g_\varepsilon) \to G \quad \text{strongly in } L^{p'}(\Omega), \\ &\mathcal{T}^b_\varepsilon(g_\varepsilon) \to g \quad \text{strongly in } L^{p'}(\Omega \times \partial S). \end{aligned} \tag{4.54}$$

Then, recalling that $y^{c*} = y - \mathcal{M}_{Y*}(y)$*, one has*

$$\begin{aligned} \int_{\partial\widehat{S}_\varepsilon} g_\varepsilon\, w_\varepsilon\, d\sigma_\varepsilon \to \frac{|\partial S|}{|Y|}\int_\Omega G\, w\, dx + \frac{|\partial S|}{|Y|}\int_\Omega \mathcal{M}_{\partial S}(y^{c*}g)\cdot\nabla w\, dx \\ + \frac{1}{|Y|}\int_{\Omega\times\partial S} g\,\widehat{w}\, dx d\sigma. \end{aligned} \tag{4.55}$$

Proof. Formula (4.51) yields

$$\begin{aligned} \int_{\partial\widehat{S}_\varepsilon} g\, v\, d\sigma_\varepsilon(x) = \frac{1}{\varepsilon|Y|}\int_{\Omega\times\partial S} \mathcal{T}^b_\varepsilon(g)\big(\mathcal{T}^*_\varepsilon(v) - \mathcal{M}^*_\varepsilon(v)\big)\, dx\, d\sigma(y) \\ + \frac{|\partial S|}{\varepsilon|Y|}\int_\Omega \mathcal{M}^b_\varepsilon(g)\mathcal{M}^*_\varepsilon(v)\, dx, \end{aligned}$$

as well as

$$\begin{aligned} \int_{\partial\widehat{S}_\varepsilon} g_\varepsilon(x)w_\varepsilon(x)\, d\sigma_\varepsilon = \frac{|\partial S|}{|Y|}\int_\Omega \frac{1}{\varepsilon}\mathcal{M}^b_\varepsilon(g_\varepsilon)(x)\mathcal{M}^*_\varepsilon(w_\varepsilon)(x)\, dx \\ + \frac{1}{|Y|}\int_{\Omega\times\partial S} \frac{1}{\varepsilon}\Big(\mathcal{T}^*_\varepsilon(w_\varepsilon) - \mathcal{M}^*_\varepsilon(w_\varepsilon)\Big)(x,y)\, \mathcal{T}^b_\varepsilon(g_\varepsilon)(x,y)\, dx\, d\sigma(y). \end{aligned}$$

Then convergence (4.55) is obtained due to the fact that $\mathcal{M}^*_\varepsilon(w_\varepsilon)$ converges weakly in $L^p(\Omega)$ to w (since $\mathcal{T}^*_\varepsilon(w_\varepsilon)$ itself converges to w which is independent of y) for the first term, and from Proposition 4.45 for the second term (which itself gives two terms in the limit). □

Remark 4.54. *Actually, Proposition 4.53 can be stated with the same conclusion with four different sets of hypotheses by exchanging weak and strong convergences in* (4.53) *and* (4.54)*. We will use the same reference for these alternate versions. An interesting case is when convergence* $(4.53)_1$ *is strong. Then convergence* (4.55) *holds assuming only a weak convergence in* $(4.54)_1$*.*

Remark 4.55. *Note that if the sequence* $\{g_\varepsilon\}_\varepsilon$, $g_\varepsilon \in L^{p'}(\partial\widehat{S}_\varepsilon)$*, is such that*

$$\begin{aligned} &\mathcal{T}^b_\varepsilon(g_\varepsilon) \rightharpoonup g \quad \text{weakly in } L^{p'}(\Omega\times\partial S), \\ &\frac{1}{\varepsilon}\mathcal{M}^b_\varepsilon(g_\varepsilon) \rightharpoonup G \quad \text{weakly in } L^{p'}(\Omega), \end{aligned}$$

then one has

$$\mathcal{M}_{\partial S}(g) = 0 \quad \text{for a.e. } x \in \Omega.$$

The above convergences also hold for $p' = +\infty$ with the weak convergences replaced by weak-$*$ convergences.

For p' finite, for every g in $L^{p'}(\Omega \times \partial S)$ with $\mathcal{M}_{\partial S}(g)(x) = 0$ for a.e. $x \in \Omega$, and G in $L^{p'}(\Omega)$, there exists a sequence $\{g_\varepsilon\}_\varepsilon$ satisfying (4.54). An example of such a sequence is given by

$$g_\varepsilon = \mathcal{U}^b_\varepsilon(g) + \varepsilon\, \mathcal{M}_\varepsilon(G)_{|\partial \widehat{S}_\varepsilon},$$

where $\mathcal{U}^b_\varepsilon$ is the boundary averaging operator (see Definition 4.49).

In the case where w_ε belongs to $W^{1,p}_0(\Omega^*_\varepsilon; \partial\Omega \cap \partial\Omega^*_\varepsilon)$ and g_ε to $L^{p'}_{loc}(\partial S_\varepsilon)$, the statements of Propositions 4.50 and 4.53 are modified (with the notations $\widetilde{\mathcal{T}}^b_\varepsilon$ and $\widetilde{\mathcal{M}}^b_\varepsilon$ corresponding to the boundary unfolding operator and the local average for $(\mathbb{R}^N)^*_\varepsilon$), as follows:

Proposition 4.56. *Assume that Ω is bounded with Lipschitz boundary. Let w belong to the space $W^{1,p}_0(\Omega^*_\varepsilon; \partial\Omega \cap \partial\Omega^*_\varepsilon)$ and g to $p'_{loc}(\partial S_\varepsilon)$. Then, for ε small enough,*

$$\Big| \int_{\partial S_\varepsilon \cap \Omega} g\, w\, d\sigma_\varepsilon \Big| \leqslant \frac{C}{\varepsilon^{1/p}} \|g\|_{L^{p'}(\partial S_\varepsilon \cap \Omega)} \big(\|w\|_{L^p(\Omega^*_\varepsilon)} + \varepsilon \|\nabla w\|_{L^p(\Omega^*_\varepsilon)} \big),$$

$$\Big| \int_{\partial S_\varepsilon \cap \Omega} g\, w\, d\sigma_\varepsilon \Big| \leqslant C \|\widetilde{\mathcal{T}}^b_\varepsilon(g)\|_{L^{p'}(\mathcal{V}_{\varepsilon d(Y)}(\Omega) \times \partial S)} \|\nabla w\|_{L^p(\Omega^*_\varepsilon)} + \frac{C}{\varepsilon} \big\|\widetilde{\mathcal{M}}^b_\varepsilon(g)\big\|_{L^{p'}(\mathcal{V}_{\varepsilon d(Y)}(\Omega))} \|w\|_{L^p(\Omega^*_\varepsilon)},$$

where $\mathcal{V}_{\varepsilon d(Y)}(\Omega)$ denotes the $\varepsilon d(Y)$-neighborhood of Ω (see Notation 1.63).

Proposition 4.57. *Let Ω be a bounded domain with Lipschitz boundary and let $\{w_\varepsilon\}_\varepsilon$ be a sequence in $W^{1,p}_0(\Omega^*_\varepsilon; \partial\Omega \cap \partial\Omega^*_\varepsilon)$ satisfying*

$$\|\nabla w_\varepsilon\|_{L^p(\Omega^*_\varepsilon)} \leqslant C.$$

Suppose that there exist w in $W^{1,p}_0(\Omega)$ and $\widehat{w}$ in $L^p(\Omega; W^{1,p}_{per,0}(Y^*))$ such that,

$$\begin{aligned} &\mathcal{T}^*_\varepsilon(w_\varepsilon) \to w \quad \text{strongly in } L^p(\Omega; W^{1,p}(Y^*)),\\ &\mathcal{T}^*_\varepsilon(\nabla w_\varepsilon) \rightharpoonup \nabla w + \nabla_y \widehat{w} \quad \text{weakly in } L^p(\Omega \times Y^*)^N. \end{aligned} \tag{4.56}$$

Let $\{g_\varepsilon\}_\varepsilon$ be a sequence in $L^{p'}_{loc}(\partial S_\varepsilon)$ and suppose furthermore that

$$\begin{aligned} &\frac{1}{\varepsilon} \widetilde{\mathcal{M}}^b_\varepsilon(g_\varepsilon) \rightharpoonup G \quad \text{weakly in } L^{p'}_{loc}(\mathbb{R}^N),\\ &\widetilde{\mathcal{T}}^b_\varepsilon(g_\varepsilon) \to g \quad \text{strongly in } L^{p'}_{loc}(\mathbb{R}^N \times \partial S). \end{aligned} \tag{4.57}$$

Then

$$\int_{\partial S_\varepsilon \cap \Omega} g_\varepsilon\, w_\varepsilon\, d\sigma_\varepsilon \to \frac{|\partial S|}{|Y|}\int_\Omega \mathcal{M}_{\partial S}(y^c * g)\cdot \nabla w\, dx \\ + \frac{|\partial S|}{|Y|}\int_\Omega G\, w\, dx + \frac{1}{|Y|}\int_{\Omega\times\partial S} \widehat{w}\; g\; dx d\sigma. \tag{4.58}$$

For $p = +\infty$, the weak convergence in (4.56) *is to be replaced by a weak-$*$ convergence.*

Proof of Propositions 4.56 and 4.57. Extend w and w_ε by 0 to the whole $(\mathbb{R}^N)^*_\varepsilon$, to get

$$\int_{\partial S_\varepsilon \cap \Omega} g_\varepsilon\, w_\varepsilon\, d\sigma_\varepsilon = \int_{(\partial \widehat{S}_\varepsilon)_1} g_\varepsilon\, w_\varepsilon\, d\sigma_\varepsilon,$$

where $(\partial \widehat{S}_\varepsilon)_1$ is the $\partial \widehat{S}_\varepsilon$ associated with $\mathcal{V}_1(\Omega)$ instead of Ω (with a similar relation for w). The results then follow from Propositions 4.50 and 4.53 applied in the domain $\mathcal{V}_1(\Omega)$. □

Remark 4.58. *In the literature, there are two standard examples of periodic functions g_ε, deriving from a function* $\mathbf{g}$ *in $L^{p'}(\partial S)$.*

For Hypothesis (4.57), *g_ε is defined as*

1. $g_\varepsilon = \varepsilon\, \mathbf{g}(\{\frac{\cdot}{\varepsilon}\}_Y)$ *if* $\mathcal{M}_{\partial S}(\mathbf{g}) \neq 0$,
2. $g_\varepsilon = \mathbf{g}(\{\frac{\cdot}{\varepsilon}\}_Y)$ *if* $\mathcal{M}_{\partial S}(\mathbf{g}) = 0$.

For Hypothesis (4.54), *since the functions g_ε have to vanish outside of $\partial \widehat{S}_\varepsilon$, the formulas are*

3. $g_\varepsilon = \varepsilon\, \mathbf{g}(\{\frac{\cdot}{\varepsilon}\}_Y)\, \mathbf{1}_{\widehat{S}_\varepsilon}$ *if* $\mathcal{M}_{\partial S}(\mathbf{g}) \neq 0$,
4. $g_\varepsilon = \mathbf{g}(\{\frac{\cdot}{\varepsilon}\}_Y)\, \mathbf{1}_{\widehat{S}_\varepsilon}$ *if* $\mathcal{M}_{\partial S}(\mathbf{g}) = 0$.

At the limit, in cases 1 and 3, $g = 0$, $G = \mathcal{M}_{\partial S}(\mathbf{g})$, whereas in cases 2 and 4, $g = \mathbf{g}$ and $G = 0$. Note that there is no way to have both g and G non zero by simply using the periodic approach. One can compare with the statement in Remark 4.55.

4.4.3 Unfolding of the space of traces

In this section we assume that S is strictly included in Y.

We recall that the space $H^{1/2}(\partial S)$ is the subspace of $L^2(\partial S)$ containing the functions satisfying

$$\int_{\partial S}\int_{\partial S} \frac{|v(y') - v(y)|^2}{|y' - y|^N} d\sigma(y') d\sigma(y) < +\infty.$$

This space is equipped with the semi-norm

$$|||v|||_{H^{1/2}(\partial S)} = \left(\int_{\partial S} \int_{\partial S} \frac{|v(y') - v(y)|^2}{|y' - y|^N} d\sigma(y') d\sigma(y) \right)^{1/2},$$

and the norm

$$\|v\|_{H^{1/2}(\partial S)} = \left(\|v\|^2_{L^2(\partial S)} + |||v|||^2_{H^{1/2}(\partial S)} \right)^{1/2}.$$

We define $H^{1/2}(\partial \widehat{S}_\varepsilon)$ as the subspace of $L^2(\partial \widehat{S}_\varepsilon)$ containing the functions whose restrictions to every connected component of $\partial \widehat{S}_\varepsilon$ belong to the space $H^{1/2}$ of this connected component. This space is equipped with the semi-norm

$$|||v|||_{H^{1/2}(\partial \widehat{S}_\varepsilon)} = \left(\sum_{\xi \in \Xi_\varepsilon} \int_{\varepsilon(\xi + \partial S)} \int_{\varepsilon(\xi + \partial S)} \frac{|v(x) - v(x')|^2}{|x - x'|^N} d\sigma_\varepsilon(x') d\sigma_\varepsilon(x) \right)^{1/2},$$

and the norm

$$\|v\|_{H^{1/2}(\partial \widehat{S}_\varepsilon)} = \left(\|v\|^2_{L^2(\partial \widehat{S}_\varepsilon)} + \varepsilon |||v|||^2_{H^{1/2}(\partial \widehat{S}_\varepsilon)} \right)^{1/2}. \tag{4.59}$$

Using the above definitions, by a change of variable

$$\|\mathcal{T}^b_\varepsilon(v)\|_{L^2(\Omega; H^{1/2}(\partial S))} = \varepsilon^{1/2} |Y|^{1/2} \|v\|_{H^{1/2}(\partial \widehat{S}_\varepsilon)}, \quad \forall v \in H^{1/2}(\partial \widehat{S}_\varepsilon). \tag{4.60}$$

This equality justifies the presence of ε in definition (4.59) of the $H^{1/2}$-norm.

Denote $H^{-1/2}(\partial \widehat{S}_\varepsilon)$ the dual space of $H^{1/2}(\partial \widehat{S}_\varepsilon)$ equipped with the dual norm. Since S is strictly included in Y, for $\Phi \in L^2(\Omega; H^{1/2}(\partial S))$, the function $\mathcal{U}^b_\varepsilon(\Phi)$ belongs to $H^{1/2}(S_\varepsilon)$ (it is not always the case if $S \cap \partial Y$ has a non null measure). This remark allows to make the following definition:

Definition 4.59. *For $g \in H^{-1/2}(\partial \widehat{S}_\varepsilon)$, the operator*

$$\mathcal{T}^b_\varepsilon : H^{-1/2}(\partial \widehat{S}_\varepsilon) \longrightarrow L^2(\Omega; H^{-1/2}(\partial S)),$$

is defined by duality for $\Phi \in L^2(\Omega; H^{1/2}(\partial S))$ as follows:

$$\langle \mathcal{T}^b_\varepsilon(g), \Phi \rangle_{L^2(\Omega; H^{-1/2}(\partial S)), L^2(\Omega; H^{1/2}(\partial S))} \doteq \varepsilon |Y| \langle g, \mathcal{U}^b_\varepsilon(\Phi) \rangle_{H^{-1/2}(\partial \widehat{S}_\varepsilon), H^{1/2}(\partial \widehat{S}_\varepsilon)}.$$

Observe that this definition makes sense because $\mathcal{U}^b_\varepsilon(\Phi)$ belongs to $H^{1/2}(\partial \widehat{S}_\varepsilon)$. Also, due to (4.49), for every $\phi \in H^{1/2}(\partial \widehat{S}_\varepsilon)$, one has

$$\begin{aligned} &\langle \mathcal{T}^b_\varepsilon(g), \mathcal{T}^b_\varepsilon(\phi) \rangle_{L^2(\Omega; H^{-1/2}(\partial S)), L^2(\Omega; H^{1/2}(\partial S))} \\ &\qquad = \varepsilon |Y| \langle g, \phi \rangle_{H^{-1/2}(\partial \widehat{S}_\varepsilon), H^{1/2}(\partial \widehat{S}_\varepsilon)}. \end{aligned} \tag{4.61}$$

By construction, if g belongs to $L^2(\partial \widetilde{S}_\varepsilon)$, then Definition 4.59 coincides with Definition 4.46.

Remark 4.60. *As for the other unfolding operators, this unfolding operator vanishes on $\Lambda_\varepsilon \times \partial S$ and is constant on the cells of $\widehat{\Omega}_\varepsilon \times \partial S$. Indeed, let (g, φ, ψ) be in $H^{-1/2}(\partial \widehat{S}_\varepsilon) \times L^2(\Omega) \times H^{1/2}(\partial S)$. Choose Φ in $L^2(\Omega; H^{1/2}(\partial S))$ defined by $\Phi(x,y) = \varphi(x)\psi(y)$ for a.e. $(x,y) \in \Omega \times \partial S$, and set*

$$\psi_\varepsilon(x) = \psi\Big(\Big\{\frac{x}{\varepsilon}\Big\}_Y\Big) \quad \text{for a.e. } x \in \partial S_\varepsilon.$$

Then one has successively,

$$\begin{aligned}
\int_\Omega \varphi(x) \langle \mathcal{T}^b_\varepsilon(g)(x,\cdot), \psi\rangle_{H^{-1/2}(\partial S), H^{1/2}(\partial S)}\, dx
&= \langle \mathcal{T}^b_\varepsilon(g), \Phi\rangle_{L^2(\Omega;H^{-1/2}(\partial S)), L^2(\Omega;H^{1/2}(\partial S))}\\
&= \varepsilon|Y|\langle g, \mathcal{U}^b_\varepsilon(\Phi)\rangle_{H^{-1/2}(\partial \widehat{S}_\varepsilon), H^{1/2}(\partial \widehat{S}_\varepsilon)}\\
&= \varepsilon|Y|\langle g, \mathcal{M}_\varepsilon(\varphi)\psi_\varepsilon\rangle_{H^{-1/2}(\partial \widehat{S}_\varepsilon), H^{1/2}(\partial \widehat{S}_\varepsilon)}\\
&= \langle \mathcal{T}^b_\varepsilon(g), \mathcal{M}_\varepsilon(\varphi)\psi\rangle_{L^2(\Omega;H^{-1/2}(\partial S)), L^2(\Omega;H^{1/2}(\partial S))}\\
&= \int_\Omega \mathcal{M}_\varepsilon(\varphi)(x) \langle \mathcal{T}^b_\varepsilon(g)(x,\cdot), \psi\rangle_{H^{-1/2}(\partial S), H^{1/2}(\partial S)}\, dx\\
&= \int_\Omega \varphi(x)\, \mathcal{M}_\varepsilon(\langle \mathcal{T}^b_\varepsilon(g)(x,\cdot), \psi\rangle_{H^{-1/2}(\partial S), H^{1/2}(\partial S)})\, dx.
\end{aligned}$$

Hence, one has the following equality:

$$\langle \mathcal{T}^b_\varepsilon(g)(x,\cdot), \psi\rangle_{H^{-1/2}(\partial S), H^{1/2}(\partial S)} = \mathcal{M}_\varepsilon(\langle \mathcal{T}^b_\varepsilon(g)(x,\cdot), \psi\rangle_{H^{-1/2}(\partial S), H^{1/2}(\partial S)},$$

and consequently, vanishes on Λ_ε and is constant on the ε-cells of $\widehat{\Omega}_\varepsilon$ for every ψ in $H^{1/2}(\partial S)$.

Lemma 4.61. *The operator $\mathcal{T}^b_\varepsilon$ is linear and continuous from $H^{-1/2}(\partial \widehat{S}_\varepsilon)$ into $L^2(\Omega; H^{-1/2}(\partial S))$. For any g in $H^{-1/2}(\partial \widehat{S}_\varepsilon)$*

$$\|\mathcal{T}^b_\varepsilon(g)\|_{L^2(\Omega;H^{-1/2}(\partial S))} = \varepsilon^{1/2}|Y|^{1/2}\|g\|_{H^{-1/2}(\partial \widehat{S}_\varepsilon)}.$$

Proof. For Φ in $L^2(\Omega; H^{1/2}(\partial S))$ one has

$$\begin{aligned}
\langle \mathcal{T}^b_\varepsilon(g), \Phi\rangle_{L^2(\Omega;H^{-1/2}(\partial S)), L^2(\Omega;H^{1/2}(\partial S))}
&\leqslant \varepsilon|Y|\, \|g\|_{H^{-1/2}(\partial \widehat{S}_\varepsilon)} \|\mathcal{U}^b_\varepsilon(\Phi)\|_{H^{1/2}(\partial \widehat{S}_\varepsilon)}.
\end{aligned}$$

On the other hand, (4.48) leads to

$$\|\mathcal{U}^b_\varepsilon(\Phi)\|_{L^2(\partial \widehat{S}_\varepsilon)} = \varepsilon^{-1/2}|Y|^{-1/2}\, \|\Phi\|_{L^2(\Omega\times\partial S)}.$$

By unfolding we get

$$\begin{aligned}
\big\|\mathcal{U}^b_\varepsilon(\Phi)\big\|_{H^{1/2}(\partial \widehat{S}_\varepsilon)} &= |Y|^{-1/2}\big\|\mathcal{T}^b_\varepsilon \circ \mathcal{U}^b_\varepsilon(\Phi)\big\|_{L^2(\Omega;H^{1/2}(\partial S))}\\
&\leqslant |Y|^{-1/2}\|\Phi\|_{L^2(\Omega;H^{1/2}(\partial S))}.
\end{aligned}$$

This inequality implies that

$$\|\mathcal{T}_\varepsilon^b(g)\|_{L^2(\Omega;H^{-1/2}(\partial S))} \leqslant \varepsilon^{1/2}|Y|^{1/2}\|g\|_{H^{-1/2}(\partial\widehat{S}_\varepsilon)}.$$

Now, due to (4.61), for ϕ in $H^{1/2}(\partial\widehat{S}_\varepsilon)$,

$$\langle g,\phi\rangle_{H^{-1/2}(\partial\widehat{S}_\varepsilon),H^{1/2}(\partial\widehat{S}_\varepsilon)} = \frac{1}{\varepsilon|Y|}\langle\mathcal{T}_\varepsilon^b(g),\mathcal{T}_\varepsilon^b(\phi)\rangle_{L^2(\Omega;H^{-1/2}(\partial S)),L^2(\Omega;H^{1/2}(\partial S))}.$$

Therefore,

$$\langle g,\phi\rangle_{H^{-1/2}(\partial\widehat{S}_\varepsilon),H^{1/2}(\partial\widehat{S}_\varepsilon)} \leqslant \frac{1}{\varepsilon|Y|}\|\mathcal{T}_\varepsilon^b(g)\|_{L^2(\Omega;H^{-1/2}(\partial S))}\|\mathcal{T}_\varepsilon^b(\phi)\|_{L^2(\Omega;H^{1/2}(\partial S))},$$

which, together with (4.60), yields

$$\|g\|_{H^{-1/2}(\partial\widehat{S}_\varepsilon)} \leqslant \varepsilon^{-1/2}|Y|^{-1/2}\|\mathcal{T}_\varepsilon^b(g)\|_{L^2(\Omega;H^{-1/2}(\partial S))},$$

and this concludes the proof of the lemma. □

Definition 4.62 (Local average of g in $H^{-1/2}(\partial\widehat{S}_\varepsilon)$). *By Lemma 4.61, for g in $H^{-1/2}(\partial\widehat{S}_\varepsilon)$, the map*

$$\begin{aligned}\varphi\in L^2(\Omega)\mapsto&\frac{1}{|\partial S|}\langle\mathcal{T}_\varepsilon^b(g),\varphi\rangle_{L^2(\Omega;H^{-1/2}(\partial S)),L^2(\Omega;H^{1/2}(\partial S))}\\ &=\frac{1}{|\partial S|}\int_\Omega\varphi(x)\langle\mathcal{T}_\varepsilon^b(g)(x,\cdot),1\rangle_{H^{-1/2}(\partial S),H^{1/2}(\partial S)}\,dx,\end{aligned}$$

defines an element of $L^2(\Omega)$ which is the local average, denoted $\mathcal{M}_\varepsilon^b(g)$, of g over $\partial\widehat{S}_\varepsilon$. It coincides with this notion for g smooth as

$$\mathcal{M}_\varepsilon^b(g) \doteq \frac{1}{|\partial S|}\langle\mathcal{T}_\varepsilon^b(g),1\rangle_{H^{-1/2}(\partial S),H^{1/2}(\partial S)},$$

and satisfies

$$\|\mathcal{M}_\varepsilon^b(g)\|_{L^2(\Omega)} \leqslant \varepsilon^{1/2}\Big(\frac{|Y|}{|\partial S|}\Big)^{1/2}\|g\|_{H^{-1/2}(\partial\widehat{S}_\varepsilon)}.$$

The next proposition is then the generalization of Proposition 4.53 to this framework and allows for g_ε in $H^{-1/2}(\partial\widehat{S}_\varepsilon)$.

Proposition 4.63. *Let $\{w_\varepsilon\}_\varepsilon$ be a sequence in $H^1(\Omega_\varepsilon^*)$. Suppose there exist w in $H^1(\Omega)$ and $\widehat{w}$ in $L^2(\Omega;H^1_{per,0}(Y^*))$ such that,*

$$\begin{aligned}&\mathcal{T}_\varepsilon^*(w_\varepsilon)\rightharpoonup w\quad\text{weakly in } L^2(\Omega;H^1(Y^*)),\\ &\mathcal{T}_\varepsilon^*(\nabla w_\varepsilon)\rightharpoonup\nabla w+\nabla_y\widehat{w}\quad\text{weakly in } L^2(\Omega\times Y^*)^N.\end{aligned}\tag{4.62}$$

Moreover, assume that the following two convergences hold for the sequence $\{g_\varepsilon\}_\varepsilon$ *with* g_ε *in* $H^{-1/2}(\partial\widehat{S}_\varepsilon)$*:*

$$\mathcal{T}^b_\varepsilon(g_\varepsilon) \to g \quad \textit{strongly in} \;\; L^2(\Omega; H^{-1/2}(\partial S)),$$

and

$$\frac{1}{\varepsilon}\mathcal{M}^b_\varepsilon(g_\varepsilon) \to G \quad \textit{strongly in} \;\; L^2(\Omega)^{(7)}.$$

Then,

$$\begin{aligned}\langle g_\varepsilon, w_\varepsilon\rangle_{H^{-1/2}(\partial\widehat{S}_\varepsilon),H^{1/2}(\partial\widehat{S}_\varepsilon)} \to \frac{|\partial S|}{|Y|}\int_\Omega G\, w\, dx \\ + \frac{1}{|Y|}\langle\, g,\, \widehat{w}\,\rangle_{L^2(\Omega;H^{-1/2}(\partial S)),L^2(\Omega;H^{1/2}(\partial S))} \\ + \frac{1}{|Y|}\int_\Omega \langle\, g,\, y^{c*}\,\rangle_{H^{-1/2}(\partial S),H^{1/2}(\partial S)} \cdot \nabla w\, dx.\end{aligned}$$

Proof. Proposition 4.15(ii), convergences (4.62) and Proposition 4.45, imply

$$\begin{aligned}&\mathcal{M}^*_\varepsilon(w_\varepsilon) \rightharpoonup w \quad \text{weakly in } L^2(\Omega),\\ &\frac{1}{\varepsilon}\mathcal{T}^*_\varepsilon\big(w_\varepsilon - \mathcal{M}^*_\varepsilon(w_\varepsilon)\big) \rightharpoonup \widehat{w} + y^{c*}\cdot\nabla w \quad \text{weakly in } L^2(\Omega; H^1(Y^*)).\end{aligned}$$

Hence,

$$\frac{1}{\varepsilon}\mathcal{T}^*_\varepsilon\big(w_\varepsilon - \mathcal{M}^*_\varepsilon(w_\varepsilon)\big) \rightharpoonup \widehat{w} + y^{c*}\cdot\nabla w \quad \text{weakly in } L^2(\Omega; H^{1/2}(\partial S)).$$

On the other hand, we can write

$$\begin{aligned}\langle g_\varepsilon, w_\varepsilon\rangle_{H^{-1/2}(\partial\widehat{S}_\varepsilon),H^{1/2}(\partial\widehat{S}_\varepsilon)} &= \langle g_\varepsilon, \mathcal{M}^*_\varepsilon(w_\varepsilon)\rangle_{H^{-1/2}(\partial\widehat{S}_\varepsilon),H^{1/2}(\partial\widehat{S}_\varepsilon)}\\ &\quad + \langle g_\varepsilon, w_\varepsilon - \mathcal{M}^*_\varepsilon(w_\varepsilon)\rangle_{H^{-1/2}(\partial\widehat{S}_\varepsilon),H^{1/2}(\partial\widehat{S}_\varepsilon)}.\end{aligned}$$

We unfold the right-hand side terms and pass to the limit. We obtain

$$\begin{aligned}&\langle g_\varepsilon, \mathcal{M}^*_\varepsilon(w_\varepsilon)\rangle_{H^{-1/2}(\partial\widehat{S}_\varepsilon),H^{1/2}(\partial\widehat{S}_\varepsilon)}\\ &\qquad = \frac{1}{\varepsilon|Y|}\,\langle\mathcal{T}^b_\varepsilon(g_\varepsilon), \mathcal{M}^*_\varepsilon(w_\varepsilon)\rangle_{L^2(\Omega;H^{-1/2}(\partial S)),L^2(\Omega;H^{1/2}(\partial S))}\\ &\qquad = \frac{|\partial S|}{|Y|}\int_\Omega \frac{1}{\varepsilon}\mathcal{M}^b_\varepsilon(g_\varepsilon)\,\mathcal{M}^*_\varepsilon(w_\varepsilon)\,dx \to \frac{|\partial S|}{|Y|}\int_\Omega G\,w\,dx.\end{aligned}$$

and also

$$\begin{aligned}&\langle g_\varepsilon, w_\varepsilon - \mathcal{M}^*_\varepsilon(w_\varepsilon)\rangle_{H^{-1/2}(\partial\widehat{S}_\varepsilon),H^{1/2}(\partial\widehat{S}_\varepsilon)}\\ &\qquad = \frac{1}{|Y|}\langle\mathcal{T}^b_\varepsilon(g_\varepsilon), \frac{1}{\varepsilon}\mathcal{T}^*_\varepsilon\big(w_\varepsilon - \mathcal{M}^*_\varepsilon(w_\varepsilon)\big)\rangle_{L^2(\Omega;H^{-1/2}(\partial S)),L^2(\Omega;H^{1/2}(\partial S))}\\ &\qquad \to \frac{1}{|Y|}\,\langle g, \widehat{w} + y^{c*}\cdot\nabla w\rangle_{L^2(\Omega;H^{-1/2}(\partial S)),L^2(\Omega;H^{1/2}(\partial S))},\end{aligned}$$

and this ends the proof of the last statement of the proposition. □

(7) This implies that $\langle g(x,\cdot), 1\rangle_{H^{-1/2}(\partial S),H^{1/2}(\partial S)} = 0$ for a.e. $x \in \Omega$.

4.5 Unfolding for cracks

In this section, we briefly consider the case of cracks instead of holes, indicating the slight changes in the theory. A combination of holes and cracks can also be considered.

The framework is the same as in the beginning of this chapter up to and including Paragraph 4.4 with the singular difference that the reference cell Y^* is not Lipschitz. It is assumed that S is a finite union of compact pieces of Lipschitz hyper-surfaces (of codimension 1) in Y which belong to the boundary of Lipschitz subdomain of Y (hence S is orientable and has two sides). Because of this property, the space $W^{1,p}(Y^*)$ is well defined in the usual way, namely

$$W^{1,p}(Y^*) \doteq \{w \in L^p(Y^*) \,|\, \nabla w \in L^p(Y^*)^N\},$$

and so is the subspace of Y-periodic functions $W^{1,p}_{per}(Y^*)$. Here the gradient is taken in Y^*, not in Y. Under these conditions, the Poincaré-Wirtinger inequality is satisfied in $W^{1,p}(Y^*)$. Note that since S is a null set, one can identify $L^p(Y^*)$ to $L^p(Y)$. This identification does not hold for $W^{1,p}(Y^*)$.

In order to define the jump across the cracks, consider each connected component S_j of cracks of S as part of the Lipschitz boundary $\widetilde{S}_j$ of a subdomain of Y. There are then two traces (from each side) in $W^{1-1/p,p}(\widetilde{S}_j)$ and their restrictions are in $W^{1-1/p,p}(S_j)$. Furthermore, the two traces coincide on the complement of $\widetilde{S}_j \backslash S_j$ so that their difference, which represents the jump across S_j of the element of $W^{1,p}(Y^*)$ and is denoted $[\cdot]_{S_j}$ (and is well-defined once a unit normal is selected), belongs to the subspace sometimes denoted $W^{1-1/p,p}_{00}(S_j)$ (see [151]). For our purpose, however, these traces being in $L^p(S_j)$ will be enough.

The same principle applies for the following spaces:

$$W^{1,p}(\Omega^*_\varepsilon) \doteq \{w \in L^p(\Omega^*_\varepsilon) \,|\, \nabla w \in L^p(\Omega^*_\varepsilon)^N\},$$
$$W^{1,p}_0(\Omega^*_\varepsilon; \partial\Omega) \doteq \{w \in W^{1,p}(\Omega^*_\varepsilon) \,|\, w \equiv 0 \text{ a.e. on } \partial\Omega\},$$

and their trace spaces[(8)].

Under these hypotheses, every integral on a crack is to be considered as the sum of the two integrals obtained on each side of the crack (the usual notation uses a subscript + or −, once a unit normal is chosen). Furthermore, the data on the crack can also be given on each side $(g_+, g_-, g_{\varepsilon+} \ldots)$.

All this is in accordance with results obtained by passing to the limit via an approximation of each crack by a "fattened crack", which is then the boundary of a regular open subset, as in the previous section (see [17]). *Consequently, Proposition 4.29, Theorems 4.38, 4.43 and 4.45 hold with the obvious changes of notation.*

[(8)] As above, $M(\alpha, \beta, \Omega^*_\varepsilon)$ can be identified to $M(\alpha, \beta, \Omega)$ and $L^2(\Omega^*_\varepsilon)$ to $L^2(\Omega)$.

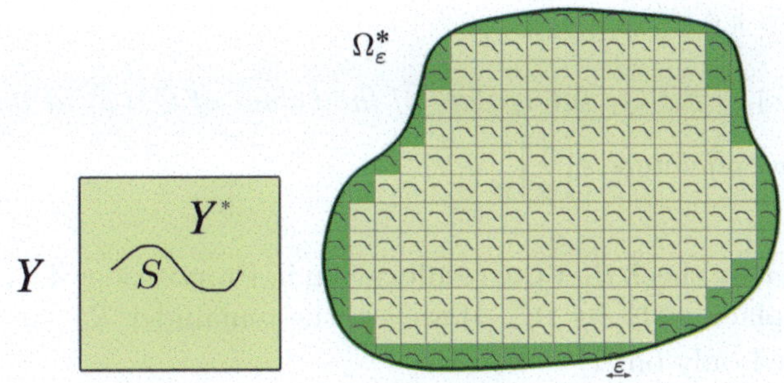

Figure 4.9: The setting of the unilateral problem for cracks

Figure 4.9 presents the simple setting with a single crack in Y in the case of a homogeneous Dirichlet condition on the outer boundary $\partial\Omega$ (for a Neumann boundary condition on $\partial\Omega$, the cracks are suppressed near $\partial\Omega$). More general situations can easily be considered with several holes and cracks in Y. For a specific application to cracks in the scalar case, see Section 5.6.

4.6 What if the periodicity cell is not a parallelotope

In this section, the set Y is not a parallelotope but has the paving property with respect to $\mathbf{G}$. We use the notations $\mathcal{P}$ (introduced in (4.1)) for a reference parallelotope and $\mathcal{K}$ is the set of all vertices of $\mathcal{P}$. Recall the definition of $\Xi_\varepsilon^{\mathcal{Y}}$, introduced in (4.17):

$$\Xi_\varepsilon^{\mathcal{Y}} = \Big\{\xi \in \Xi_\varepsilon \mid \varepsilon(\xi + \overline{\mathcal{Y}}) \subset \Omega\Big\}.$$

Just as in 4.3.2.2, the macro approximation of a function at the points of $\Xi_\varepsilon^{\mathcal{Y}}$ is constructed by an average and is then extended by Q_1-interpolation in the parallelotopes $\varepsilon(\xi + \mathcal{P})$, $\xi \in \Xi_\varepsilon^{\mathcal{Y}}$. This macro-approximation is naturally defined on the set

$$\widehat{\Omega}_\varepsilon^{\mathcal{P}} = \text{interior}\Big\{ \bigcup_{\xi \in \Xi_\varepsilon^{\mathcal{Y}}} \varepsilon(\xi + \overline{\mathcal{P}})\Big\}.$$

Note that $\widehat{\Omega}_\varepsilon^{\mathcal{P}}$ is *not necessarily included in* Ω (for example, this can occur if Y is offset with respect to $\mathcal{P}$).

Definition 4.64. *The operator $\mathcal{Q}_\varepsilon^* : L^p(\Omega_\varepsilon^*) \mapsto W^{1,\infty}(\widehat{\Omega}_\varepsilon^{\mathcal{P}})$, for $p \in [1, +\infty]$, is defined as follows*

$$\mathcal{Q}_\varepsilon^*(\phi)(\varepsilon\xi) = \frac{1}{|Y^*|}\int_{Y*} \phi(\varepsilon\xi + \varepsilon z)\, dz = \mathcal{M}_{\varepsilon\xi+\varepsilon Y*}(\phi) \quad \textit{for all } \xi \in \Xi_\varepsilon^{\mathcal{Y}} + \mathcal{K},$$

and for every $x \in \widehat{\Omega}_\varepsilon^{\mathcal{P}}$,

$$\begin{cases} \mathcal{Q}_\varepsilon^*(\phi)(x) \text{ is the } Q_1 \text{ interpolate of the values of } \mathcal{Q}_\varepsilon^*(\phi) \text{ at the vertices} \\ \text{of the parallelotope } \varepsilon\Big[\dfrac{x}{\varepsilon}\Big]_Y + \varepsilon\mathcal{P}. \end{cases}$$

We can easily check that the results given in Propositions 4.34 and 4.35 are still valid replacing $\widehat{\Omega}_\varepsilon^{\mathcal{Y}}$ by $\widehat{\Omega}_\varepsilon^{\mathcal{P}}$. However, the remainder $\mathcal{R}_\varepsilon^*(\phi) = \phi - \mathcal{Q}_\varepsilon^*(\phi)$ is now defined only on $\Omega_\varepsilon^* \cap \widehat{\Omega}_\varepsilon^{\mathcal{P}}$.

To go further, we need to estimate the L^p-norm of $\mathcal{R}_\varepsilon^*(\phi)$ only in terms of the gradient of ϕ. But this is not always possible on $\Omega_\varepsilon^* \cap \widehat{\Omega}_\varepsilon^{\mathcal{P}}$ (since this set is not always a union of cells of the type $\varepsilon(\xi + Y^*)$. We are therefore led to consider a subset of $\Omega_\varepsilon^* \cap \widehat{\Omega}_\varepsilon^{\mathcal{P}}$ defined as a union of cells of the type $\varepsilon(\xi + Y^*)$ included in Ω_ε^*. Since this subset will play the same role as $\widehat{\Omega}_\varepsilon^{**}$ in the previous case (see (4.36)), we will still denote it $\widehat{\Omega}_\varepsilon^{**}$. To give its new definition, we use the facts that the parallelotope $\mathcal{P}$ satisfies the paving property (2.1) and that Y is a bounded domain. Thus, the latter can be covered by a finite union of $\mathbf{G}$-translates of $\mathcal{P}$. More precisely, let $\{\mathbf{b}'_1, \mathbf{b}'_2, \dots, \mathbf{b}'_k\}$ be the subset of $\mathbf{G}$ defined as follows:

$$\{\mathbf{b} \in \mathbf{G} \mid (\mathbf{b} + \overline{\mathcal{P}}) \cap Y \neq \varnothing\}.$$

Then (see Figure 4.10), the set $\widehat{\mathcal{P}}$ defined as

$$\widehat{\mathcal{P}} \doteq \text{interior}\Big(\bigcup_{i=1}^{k} (\mathbf{b}'_i + \overline{\mathcal{P}})\Big) \quad \text{is connected and } Y \subset \widehat{\mathcal{P}}. \tag{4.63}$$

$\mathcal{P}$ $\widehat{\mathcal{P}}$ Y $\mathcal{Y}$

Figure 4.10: The sets $\mathcal{P}, \widehat{\mathcal{P}}, Y$ and $\mathcal{Y}$

Without loss of generality, we assume that

$$\mathcal{P} \cap Y \neq \varnothing. \tag{4.64}$$

Hence, one of the vectors $\mathbf{b}'_i$, $i = 1, \dots, k$, is equal to zero. As a consequence,

$$Y \subset \mathcal{Y} \subset \mathcal{Y}' \quad \text{where } \mathcal{Y}' \doteq \text{interior}\Big\{\bigcup_{i=1}^{k} \varepsilon(\mathbf{b}'_i + \overline{\mathcal{Y}})\Big\}.$$

Now set

$$\Xi_\varepsilon^{\mathcal{Y}'} \doteq \{\xi \in \Xi_\varepsilon \mid \varepsilon(\xi + \overline{\mathcal{Y}'}) \subset \Omega\} \subset \Xi_\varepsilon^{\mathcal{Y}}.$$

From the definition of $\Xi_\varepsilon^{\mathcal{Y}'}$, for each $\xi \in \Xi_\varepsilon^{\mathcal{Y}'}$ and every $i = 1, \dots, k$,

$$\varepsilon(\xi + \mathcal{Y}') = \varepsilon(\xi + \mathbf{b}'_i + \mathcal{Y}) \subset \Omega,$$

so that $\mathcal{Q}_\varepsilon^*(\phi)$ can be defined on $\varepsilon(\xi + \mathbf{b}'_i + \overline{\mathcal{P}})$. Thus, by (4.63), it is defined on $\varepsilon(\xi + \overline{Y})$.

In conclusion, $\mathcal{Q}_\varepsilon^*(\phi)$ is defined on the set

$$\widehat{\Omega}'_\varepsilon \doteq \text{interior}\Big\{\bigcup_{\xi \in \Xi_\varepsilon^{\mathcal{Y}'}} \varepsilon(\xi + \overline{Y})\Big\} \subset \widehat{\Omega}_\varepsilon.$$

Due to assumption (4.64), the open set $\widehat{\Omega}'_\varepsilon$ is included in $\widehat{\Omega}_\varepsilon$.

Introduce the set

$$\widehat{\Omega}_\varepsilon^{**} \doteq \Omega_\varepsilon^* \cap \widehat{\Omega}'_\varepsilon, \tag{4.65}$$

and observe that actually,

$$\widehat{\Omega}_\varepsilon^{**} = \widehat{\Omega}'_\varepsilon \backslash S_\varepsilon = \text{interior}\Big\{\bigcup_{\xi \in \Xi_\varepsilon^{\mathcal{Y}'}} \varepsilon(\xi + \overline{Y^*})\Big\}.$$

This is a generalization of the situation of Subsection 4.3.2 since for $Y = \mathcal{P}$, the sets $\mathcal{Y}$ and $\mathcal{Y}'$ (as well as $\widehat{\Omega}'_\varepsilon$ and $\widehat{\Omega}_\varepsilon^{\mathcal{Y}}$) coincide.

Note that

$$\Omega_{\varepsilon c(Y)}^{int} \subset \widehat{\Omega}'_\varepsilon \subset \Omega,$$

where $c(Y) = d(\widehat{\mathcal{P}})$ depends only on Y.

As in (4.36), for ϕ in $L^p(\Omega_\varepsilon^*)$ and p in $[1, +\infty]$, write

$$\phi = \mathcal{Q}_\varepsilon^*(\phi) + \mathcal{R}_\varepsilon^*(\phi) \quad \text{a.e. in } \widehat{\Omega}_\varepsilon^{**},$$

with $\widehat{\Omega}_\varepsilon^{**}$ defined here by (4.65). On this set, one can get the L^p-estimate for $\mathcal{R}_\varepsilon^*(\phi)$. The result below is similar to Proposition 4.37.

Proposition 4.65. *There is a constant C independent of ε such that for every ϕ in $W^{1,p}(\Omega_\varepsilon^*)$,*

$$(i) \qquad \|\mathcal{R}_\varepsilon^*(\phi)\|_{L^p(\widehat{\Omega}_\varepsilon^{**})} \leqslant \varepsilon C \|\nabla \phi\|_{L^p(\Omega_\varepsilon^*)},$$

$$(ii) \qquad \|\nabla \mathcal{R}_\varepsilon^*(\phi)\|_{L^p(\widehat{\Omega}_\varepsilon^{**})} \leqslant C\, \|\nabla \phi\|_{L^p(\Omega_\varepsilon^*)}.$$

Proof. Inequality (ii) is immediate from Proposition 4.35.

To prove inequality (i), let ϕ be in $W^{1,p}(\Omega_\varepsilon^*)$. In this setting, inequality (4.32) still holds for every $\xi \in \Xi_\varepsilon^{\mathcal{Y}'}$ and implies, for every $i = 1, \ldots, k$,

$$\|\phi - \mathcal{Q}_\varepsilon^*(\phi)(\varepsilon\xi)\|_{L^p(\varepsilon(\xi+Y*))} \leqslant C\varepsilon\|\nabla\phi\|_{L^p(\varepsilon(\xi+Y*))}. \tag{4.66}$$

Now, using inequality (4.34) and definition (4.30) of $\mathcal{Q}_\varepsilon^*(\phi)$, we get, for every ξ in $\Xi_\varepsilon^{\mathcal{Y}'}$,

$$\begin{aligned}\|\mathcal{Q}_\varepsilon^*(\phi) - \mathcal{Q}_\varepsilon^*(\phi)(\varepsilon\xi)\|_{L^p(\varepsilon(\xi+Y*))} &\leqslant \|\mathcal{Q}_\varepsilon^*(\phi) - \mathcal{Q}_\varepsilon^*(\phi)(\varepsilon\xi)\|_{L^p(\varepsilon(\xi+Y))} \\ &\leqslant \|\mathcal{Q}_\varepsilon^*(\phi) - \mathcal{Q}_\varepsilon^*(\phi)(\varepsilon\xi)\|_{L^p(\varepsilon(\xi+\widehat{\mathcal{P}}))} \\ &\leqslant C\varepsilon\|\nabla\mathcal{Q}_\varepsilon^*(\phi)\|_{L^p(\varepsilon(\xi+\widehat{\mathcal{P}}))} \\ &\leqslant C\varepsilon\sum_{i=1}^{k}\|\nabla\mathcal{Q}_\varepsilon^*(\phi)\|_{L^p(\varepsilon(\xi+\mathbf{b}_i'+\mathcal{P}))} \\ &\leqslant C\varepsilon\sum_{i=1}^{k}\|\nabla\phi\|_{L^p(\varepsilon(\xi+\mathbf{b}_i'+\mathcal{Y}*))},\end{aligned} \tag{4.67}$$

Combining (4.66) and (4.67) gives

$$\begin{aligned}\|\phi - \mathcal{Q}_\varepsilon^*(\phi)\|_{L^p(\varepsilon(\xi+Y*))} &= \|\mathcal{R}_\varepsilon^*(\phi)\|_{L^p(\varepsilon(\xi+Y*))} \\ &\leqslant C\varepsilon\sum_{i=1}^{k}\|\nabla\phi\|_{L^p(\varepsilon(\xi+\mathbf{b}_i'+\mathcal{Y}*))}.\end{aligned}$$

Finally, using definition (4.65) of $\widehat{\Omega}_\varepsilon^{**}$, inequality (i) follows by summation over $\Xi_\varepsilon^{\mathcal{Y}'}$, and this ends the proof of Proposition 4.65. ☐

Note that in the computation (4.67) we used the rescaled version of the following result applied to $\mathcal{Q}_\varepsilon^*(\phi)$:

Lemma 4.66. *There exists a constant C such that, for every $p \in [1, +\infty]$ and every continuous function ψ in $\widehat{\mathcal{P}}$ which is Q_1 in each parallelotope $\mathbf{b}_i' + \mathcal{P}$ for $i = 1, \ldots, N$,*

$$\|\psi - \psi(0)\|_{L^p(\widehat{\mathcal{P}})} \leqslant C\|\nabla\psi\|_{L^p(\widehat{\mathcal{P}})}.$$

Proof. This can be obtained by a standard contradiction argument, making use of the fact that $\widehat{\mathcal{P}}$ is connected. An alternate proof is available by finite induction connecting adjacent parallelotopes. ☐

Chapter 5

Homogenization in perforated domains

Using the notations of Chapter 4, this chapter presents the applications of the periodic unfolding method in the case of periodically (or frame-periodically) perforated domains. The various examples show the flexibility of the method as well as its ease of use. The main difficulty arising in these problems is the fact that each ε-problem is posed in a different space. Therefore, convergence of the sequence of solutions has to be interpreted in a specific way. The usual extension by zero into the holes is not very useful (as the holes oscillate wildly), and the usual method (assuming the existence of bounded extensions operators in H^1 may be too restrictive). The method has the advantage of having the unfoldings of all the solutions (for all ε) belong to a fixed space where their convergence is natural (see the discussion at the beginning of Section 5.1).

Sections 5.1 and 5.2 give the full treatment for second order linear problems in variational form. Section 5.1 concerns the Dirichlet-Neumann case (Dirichlet on the outer boundary, Neumann on the boundary of the small holes). The original treatment of this problem was via the method of oscillating test functions (see for example [69, 77]). Section 5.2 presents the case of a full Neumann problem. The limit unfolded problems are established, the homogenized limits as well and the correctors are given (making use of the convergence of the energy). The boundary unfolding operator is an essential tool in the treatment of the Neumann condition on the boundary of the holes.

Section 5.3 present the variant for frame-periodic distribution of the holes.

In Section 5.4, the geometrically more complex case of the Fredholm alternative for the Neumann problem (without zero order term) is presented in details. A uniform Poincaré-Wirtinger inequality needed for the sequence of perforated domains is established.

Finally, Section 5.6 treats a problem with periodically distributed cracks or fissures with a non-linear boundary condition on the cracks. The homo-

D. Cioranescu et al., *The Periodic Unfolding Method*, Series in Contemporary Mathematics 3, https://doi.org/10.1007/978-981-13-3032-2_5

genized problem is shown to be a Leray-Lions type problem.

As in Chapter 3, we recall that second order variational problems exhibit invariance under bi-lipschitz diffeomorphisms. Therefore, every result stated in the periodic case can be extended to the case of bi-lipschitz equivalent to periodic. This can also be combined with the local bi-lipschitz changes of variables of Section 5.3; an example where the latter combined with a global bi-lipschitz diffeomorphism can be used is the second part of [38] (where the treatment is different since it predates the unfolding method).

The method can be applied in many situations such as, among others, the heat equation (see for example [104]), the wave equation ([103]) or quasilinear elliptic problems (see [41, 100]).

In this chapter Ω is a bounded domain in $\mathbb{R}^N$ with Lipschitz boundary and Hypothesis $(\mathbf{H_2})$ is assumed (see Definition 4.30). The dimension N is greater or equal to 2 (because of the connectedness condition).

5.1 The Dirichlet-Neumann problem

Let f be in $L^2(\Omega)$ and $A^\varepsilon = (a^\varepsilon_{ij})_{1\leqslant i,j\leqslant N}$ be a matrix field in the set $M(\alpha, \beta, \Omega)$ (see Definition 3.1)

The Dirichlet-Neumann problem is

$$\begin{cases} -\operatorname{div}(A^\varepsilon \nabla u_\varepsilon) = f & \text{in } \Omega^*_\varepsilon, \\ u_\varepsilon = 0 & \text{on } \partial\Omega^*_\varepsilon \cap \partial\Omega, \\ A^\varepsilon \nabla u_\varepsilon \cdot n_\varepsilon = g_\varepsilon & \text{on } \partial S_\varepsilon \cap \Omega, \end{cases} \tag{5.1}$$

where g_ε is given in $L^2(\partial S_\varepsilon \cap \Omega)$.

The variational formulation of (5.1) is

$$\begin{cases} \text{Find } u_\varepsilon \in H^1_0(\Omega^*_\varepsilon; \partial\Omega \cap \partial\Omega^*_\varepsilon) \text{ such that} \\ \displaystyle\int_{\Omega^*_\varepsilon} A^\varepsilon \nabla u_\varepsilon \, \nabla v \, dx = \int_{\Omega^*_\varepsilon} f\, v \, dx + \int_{\partial S_\varepsilon \cap \Omega} g_\varepsilon v \, d\sigma_\varepsilon, \\ \forall v \in H^1_0(\Omega^*_\varepsilon; \partial\Omega \cap \partial\Omega^*_\varepsilon). \end{cases} \tag{5.2}$$

5.1.1 Homogenization

When studying the asymptotic behavior of (5.1), the first point is to obtain a uniform bound for u_ε solution of (5.2). To do so, first choose an extension of g_ε to $\partial S_\varepsilon \cap \mathcal{V}_1(\Omega)$ (still using the notation g_ε for this extension)(1).

(1) Recall that $\mathcal{V}_1(\Omega)$ is the 1-neighborhood of Ω (see Notation 1.63). Any extension in $L^2(\partial S_\varepsilon \cap \mathcal{V}_1(\Omega))$ will do. We usually choose either the periodic extension (if g_ε is periodic), or its extension by 0.

Then choose u_ε as a test function in (5.2). From the resulting formula, making use of Proposition 4.56 and Theorem 4.38 (for $p = 2$), one can establish the uniform bound

$$\|u_\varepsilon\|_{H^1(\Omega_\varepsilon^*)} \leqslant C\Big(\|f\|_{L^2(\Omega_\varepsilon^*)} + \varepsilon^{1/2}\|g_\varepsilon\|_{L^2(\partial S_\varepsilon \cap \mathcal{V}_1(\Omega))} + \frac{1}{\varepsilon}\big\|\mathcal{M}_\varepsilon^b(g_\varepsilon)\big\|_{L^2(\mathcal{V}_1(\Omega))}\Big). \tag{5.3}$$

In view of this estimate, the natural assumption on the function g_ε is

$$\varepsilon^{1/2}\|g_\varepsilon\|_{L^2(\partial S_\varepsilon \cap \mathcal{V}_1(\Omega))} + \frac{1}{\varepsilon}\big\|\mathcal{M}_\varepsilon^b(g_\varepsilon)\big\|_{L^2(\mathcal{V}_1(\Omega))} \leqslant C \tag{5.4}$$

where the constant C does not depend on ε. Condition (5.4) leads to a uniform bound for $\|u_\varepsilon\|_{H^1(\Omega_\varepsilon^*)}$. We can now state the homogenization result.

Theorem 5.1. *Let u_ε be the solution of Problem* (5.1). *Suppose that*

$$B^\varepsilon \doteq \mathcal{T}_\varepsilon^*(A^\varepsilon) \to B \quad a.e.\ in\ \Omega \times Y^{*(2)}. \tag{5.5}$$

Suppose furthermore that the sequence $\{g_\varepsilon\}_\varepsilon$ satisfies (5.4) *and that there exist g in $L^2(\Omega \times \partial S)$ and G in $L^2(\Omega)$ satisfying*

$$\begin{aligned} &\mathcal{T}_\varepsilon^b(g_\varepsilon) \rightharpoonup g \quad \text{weakly in } L^2(\Omega \times \partial S),\\ &\frac{1}{\varepsilon}\mathcal{M}_\varepsilon^b(g_\varepsilon) \rightharpoonup G \quad \text{weakly in } L^2(\Omega). \end{aligned} \tag{5.6}$$

Then there exist $u_0 \in H_0^1(\Omega)$ and $\widehat{u}_0 \in L^2(\Omega; H^1_{per,0}(Y^*))$ such that,

$$\begin{aligned} &(i) \quad \begin{cases} \mathcal{E}_\varepsilon(u_\varepsilon) \to u_0 & \text{strongly in } L^2(\Omega),\\ \|u_\varepsilon - u_0\|_{L^2(\Omega_\varepsilon^*)} \to 0, \end{cases}\\ &(ii) \quad \mathcal{T}_\varepsilon^*(u_\varepsilon) \to u_0 \quad \text{strongly in } L^2(\Omega; H^1(Y^*)),\\ &(iii) \quad \mathcal{T}_\varepsilon^*(\nabla u_\varepsilon) \rightharpoonup \nabla u_0 + \nabla_y \widehat{u}_0 \quad \text{weakly in } L^2(\Omega \times Y^*)^N, \end{aligned} \tag{5.7}$$

and the pair $(u_0, \widehat{u}_0)$ in $H_0^1(\Omega) \times L^2(\Omega; H^1_{per,0}(Y^*))$ is the unique solution of the homogenized problem

$$\begin{cases} \dfrac{1}{|Y|}\displaystyle\int_{\Omega\times Y^*} B(x,y)\big[\nabla u_0(x) + \nabla_y \widehat{u}_0(x,y)\big]\big[\nabla\Psi(x) + \nabla_y\Phi(x,y)\big]dxdy\\ \qquad = \dfrac{|Y^*|}{|Y|}\displaystyle\int_\Omega f\,\Psi\,dx + \dfrac{|\partial S|}{|Y|}\int_\Omega \mathcal{M}_{\partial S}(y_{\mathcal{M}}g)\cdot\nabla\Psi\,dx\\ \qquad\quad + \dfrac{|\partial S|}{|Y|}\displaystyle\int_\Omega G\,\Psi\,dx + \dfrac{1}{|Y|}\int_{\Omega\times\partial S} g\,\Phi\,dx\,d\sigma(y),\\ \forall\Psi \in H_0^1(\Omega), \quad \forall\Phi \in L^2(\Omega; H^1_{per}(Y^*)). \end{cases} \tag{5.8}$$

(2) or more generally, in measure in $\Omega \times Y^*$.

Remark 5.2. *As in the case of fixed domains (see Remark 3.6), every matrix field B belonging to $M(\alpha,\beta,\Omega\times Y^*)$ can be approached (in the sense of (5.5)) by the sequence of matrices A^ε in $M(\alpha,\beta,\Omega^*_\varepsilon)$ with A^ε defined as follows:*

$$A^\varepsilon = \begin{cases} \mathcal{U}_\varepsilon(B) & \text{in } \widehat{\Omega}^*_\varepsilon \\ \alpha I_n & \text{in } \Lambda^*_\varepsilon. \end{cases}$$

Remark 5.3. *Making use of the results of Subsection 4.4.3, one can assume that g_ε belongs to $H^{-1/2}(\partial S_\varepsilon)$ and that the first convergence of (5.6) holds in $L^2(\Omega; H^{-1/2}(\partial S))$. This remark also applies for Proposition 5.9, Theorem 5.14 and Proposition 5.18.*

The following is the equivalent of (3.14)in the case of a perforated Y^*.[3]

Lemma 5.4. *There exists a strictly positive constant c depending only upon Y^* such that for every $(v,\widehat{v})$ in $H^1(\Omega)\times L^2(\Omega; H^1_{per,0}(Y^*))$,*

$$\begin{aligned} c\big(\|\nabla v\|^2_{L^2(\Omega)} + \|\nabla_y \widehat{v}\|^2_{L^2(\Omega\times Y^*)}\big) &\leqslant \|\nabla v + \nabla_y\widehat{v}\|^2_{L^2(\Omega\times Y^*)} \\ &\leqslant \big(\|\nabla v\|_{L^2(\Omega)} + \|\nabla_y\widehat{v}\|_{L^2(\Omega\times Y^*)}\big)^2. \end{aligned}$$

Proof. The second inequality is obvious. As for the first, it is enough to show for that every $(\zeta,\widehat{v})$ in $\mathbb{R}^N\times H^1_{per,0}(Y^*)$, the following inequality holds for some strictly positive c:

$$c\big(|\zeta|^2 + \|\nabla_y\widehat{v}\|^2_{L^2(\mathcal{Y}^*)}\big) \leqslant \|\zeta + \nabla_y\widehat{v}\|^2_{L^2(\mathcal{Y}^*)}. \tag{5.9}$$

Set $w(y) \doteq \widehat{v}(y) + \zeta\cdot\big(y - \mathcal{M}_{\mathcal{Y}^*}(y)\big)$.
First, apply the Poincaré-Wirtinger inequality to w in $H^1(\mathcal{Y}^*)$. This gives

$$\|w\|_{H^1(\mathcal{Y}^*)} \leqslant C_{PW}\|\zeta + \nabla_y\widehat{v}\|_{L^2(\mathcal{Y}^*)}.$$

Since $\zeta_j(=\zeta\cdot\mathbf{b}_j) = w(y+\mathbf{b}_j) - w(y)$ for a.e. y in Y^* and for every $j = 1,\dots,N$, it follows that $|\zeta| \leqslant C\|w\|_{L^2(\mathcal{Y}^*)}$, so

$$|\zeta| \leqslant C\|\zeta + \nabla_y\widehat{v}\|_{L^2(Y^*)} \text{ hence } \|\nabla_y\widehat{v}\|_{L^2(Y^*)} \leqslant (1+C)\|\zeta + \nabla_y\widehat{v}\|_{L^2(Y^*)}.$$

Then, (5.9) easily follows. □

Proof of Theorem 5.1. First, in view of Lemma 5.4 and the ellipticity of the matrix field B (arising from A^ε in $M(\alpha,\beta,\Omega)$), problem (5.8) has a solution which is unique by direct application of the Lax-Milgram theorem in the space $H^1_0(\Omega)\times L^2(\Omega; H^1_{per,0}(Y^*))$.

Estimates (5.3) and (5.4) imply that $\|u_\varepsilon\|_{H^1(\Omega^*_\varepsilon)}$ is uniformly bounded. Then, Theorems 4.43 and 4.44 imply convergences (5.7), at least for a subsequence. The uniqueness of the solution of the limit problem will eventually imply the convergence of the whole sequence.

(3) The same proof applies to the case of $W^{1,p}$ spaces both for Y and Y^*.

Let Ψ and φ be in $\mathcal{C}_c^\infty(\Omega)$ and ψ in $H^1_{per,0}(Y^*)$. We choose in (5.2) the test function

$$v_\varepsilon = \Psi + \varepsilon\varphi\,\psi_\varepsilon \quad \text{and} \quad \psi_\varepsilon(x) = \psi\Big(\frac{x}{\varepsilon}\Big).$$

Since

$$\nabla v_\varepsilon = \nabla\Psi + \varepsilon\psi_\varepsilon\nabla\varphi + \varphi\,\nabla_y\psi\Big(\frac{\cdot}{\varepsilon}\Big),$$

by Proposition 4.8 (i),

$$\begin{aligned}
&\mathcal{T}^*_\varepsilon(v_\varepsilon) \to \Psi \quad \text{strongly in } L^2(\Omega\times Y^*),\\
&\mathcal{T}^*_\varepsilon(\varphi\,\psi_\varepsilon) \to \varphi\,\psi \ \text{ strongly in } L^2(\Omega\times Y^*),\\
&\mathcal{T}^*_\varepsilon(\nabla v_\varepsilon) \to \nabla\Psi + \varphi\nabla_y\psi \quad \text{strongly in } L^2(\Omega\times Y^*)^N.
\end{aligned} \tag{5.10}$$

Then, for ε small enough, in view of the fact that the support of v_ε remains in a fixed compact subset of Ω, Proposition 4.4 (i) and (5.7) allow passing to the limit to get

$$\begin{aligned}
\int_{\Omega^*_\varepsilon} A^\varepsilon\nabla u_\varepsilon\,\nabla v_\varepsilon\,dx &= \frac{1}{|Y|}\int_{\Omega\times Y^*} \mathcal{T}^*_\varepsilon(A^\varepsilon)\mathcal{T}^*_\varepsilon(\nabla u_\varepsilon)\,\mathcal{T}^*_\varepsilon(\nabla v_\varepsilon)\,dxdy\\
&\to \frac{1}{|Y|}\int_{\Omega\times Y^*} B\big[\nabla u_0 + \nabla_y\widehat{u}_0\big]\big[\nabla\Psi + \varphi\nabla_y\psi\big]\,dxdy,
\end{aligned} \tag{5.11}$$

as well as

$$\int_{\Omega^*_\varepsilon} f\,v_\varepsilon\,dx = \frac{1}{|Y|}\int_{\Omega\times Y^*} \mathcal{T}^*_\varepsilon(f)\mathcal{T}^*_\varepsilon(v_\varepsilon)\,dxdy \to \frac{|Y^*|}{|Y|}\int_\Omega f\,\Psi\,dx. \tag{5.12}$$

Similarly, for ε small enough, the surface integral in (5.2) takes the form

$$\int_{\partial S_\varepsilon\cap\Omega} g_\varepsilon\,v_\varepsilon\,d\sigma_\varepsilon = \int_{\widehat{\partial S}_\varepsilon} g_\varepsilon\,\Psi\,d\sigma_\varepsilon + \varepsilon\int_{\widehat{\partial S}_\varepsilon} g_\varepsilon\,\varphi\,\psi_\varepsilon\,d\sigma_\varepsilon.$$

To pass to the limit here, we use the results of Section 4.4. In view of Hypotheses (5.6) and convergences (5.10), we can apply Proposition 4.53 in the form of Remark 4.54 to the first integral in the right-hand side to obtain

$$\int_{\widehat{\partial S}_\varepsilon} g_\varepsilon\,\Psi\,d\sigma_\varepsilon \to \frac{|\partial S|}{|Y|}\int_\Omega \mathcal{M}_{\partial S}(y_{\mathcal{M}}\,g)\cdot\nabla\Psi\,dx + \frac{|\partial S|}{|Y|}\int_\Omega G\,\Psi\,dx. \tag{5.13}$$

For the second integral, Proposition 4.51 and (5.10) again give

$$\varepsilon\int_{\widehat{\partial S}_\varepsilon} g_\varepsilon\,\varphi\psi_\varepsilon\,d\sigma_\varepsilon \to \frac{1}{|Y|}\int_{\Omega\times\partial S} g(x)\,\varphi(x)\psi(y)\,dxd\sigma(y). \tag{5.14}$$

Collecting convergences (5.11), (5.12), (5.13) and (5.14), and due to the density of $\mathcal{C}_c^\infty(\Omega)$ in $H^1_0(\Omega)$ and that of the tensor product $\mathcal{C}_c^\infty(\Omega)\otimes H^1_{per}(Y^*)$ in $L^2(\Omega; H^1_{per}(Y^*))$, we get the unfolded limit formulation (5.8). □

We now obtain the homogenized limit problem in terms of u_0 alone. We begin by expressing $\widehat{u}_0$ in terms of u_0.

Proposition 5.5. *The function $\widehat{u}_0$ in Theorem 5.1 is given in terms of u_0 by the expression*

$$\widehat{u}_0(x,y) = \sum_{i=1}^{N} \frac{\partial u_0}{\partial x_i}(x)\, \widehat{\chi}_i^*(x,y) + \widehat{\chi}_0^*(x,y), \tag{5.15}$$

where the corrector functions $\widehat{\chi}_j^* \in L^\infty(\Omega\,; H^1_{per,0}(Y^*))$, $j = 1,\dots,N$ are the solutions of the cell problems (a.e. in Ω),

$$\begin{cases} \displaystyle\int_{Y^*} \sum_{i,k=1}^{N} b_{ik}(\,\cdot\,,y)\Big(\frac{\partial \widehat{\chi}_j^*}{\partial y_k}(\,\cdot\,,y) + \delta_{jk}\Big)\frac{\partial \varphi}{\partial y_i}(y)\,dy = 0, \\ \forall \varphi \in H^1_{per,0}(Y^*), \end{cases} \tag{5.16}$$

and the corrector function $\widehat{\chi}_0^* \in L^2(\Omega; H^1_{per,0}(Y^*))$ is the solution of the cell problem (a.e. in Ω),

$$\begin{cases} \displaystyle\int_{Y^*} \sum_{i,k=1}^{N} b_{ik}(\,\cdot\,,y)\frac{\partial \widehat{\chi}_0^*}{\partial y_k}(\,\cdot\,,y)\frac{\partial \varphi}{\partial y_i}(y)\,dy = \int_{\partial S} g(\,\cdot\,,y)\,\varphi(y)\,d\sigma(y), \\ \forall \varphi \in H^1_{per,0}(Y^*). \end{cases} \tag{5.17}$$

Proof. The proof is straightforward once the existence and uniqueness of the "correctors" $\widehat{\chi}_i^*$, $i = 1,\dots,N$, and $\widehat{\chi}_0^*$, is shown. For all of them, this follows from the Lax-Milgram theorem.

The case of system (5.17), which takes care of the nonhomogeneous Neumann boundary condition, requires special attention, since to apply the Lax-Milgram theorem a compatibility condition is needed. This condition is

$$\int_{\partial S} g(x,y)\,d\sigma(y) = 0 \quad \text{for a.e. } x \in \Omega,$$

and it is satisfied due to convergences (5.6) and Remark 4.55. □

The classical homogenization result for problems with holes is recovered here without the condition that the holes do not intersect the outer boundary and without the use of extension operators into the holes.

Theorem 5.6. *The homogenized formulation associated to system* (5.8) *from Theorem 5.1 is*

$$\begin{cases} -\mathrm{div}\,(A^{hom}\nabla u_0) = f + \dfrac{|\partial S|}{|Y^*|}G - \dfrac{|\partial S|}{|Y^*|}\mathrm{div}\,\big(\mathcal{M}_{\partial S}(y_{\mathcal{M}}g)\big) \\ \qquad\qquad + \mathrm{div}\,\mathcal{M}_{Y^*}\big(B(x,\cdot)\nabla_y \widehat{\chi}_0^*(x,\cdot)\big) \quad \text{in } \Omega, \\ u_0 = 0 \quad \text{on } \partial\Omega, \end{cases} \tag{5.18}$$

where $\widehat{\chi}_0^*$ *is given by* (5.17).

The homogenized matrix(4) $A^{hom} = (a_{ij}^{hom})_{1\leqslant i,j\leqslant N}$ *is elliptic and defined by*

$$a_{ij}^{hom} = \mathcal{M}_{Y*}\Big(b_{ij} + \sum_{k=1}^{N} b_{ik}\frac{\partial \widehat{\chi}_j^*}{\partial y_k}\Big) = \mathcal{M}_{Y*}(b_{ij}) + \mathcal{M}_{Y*}\Big(\sum_{k=1}^{N} b_{ik}\frac{\partial \widehat{\chi}_j^*}{\partial y_k}\Big).$$

where the $\widehat{\chi}_j^*$ *are given by* (5.16) *for* $j = 1,\dots,N$.

Remark 5.7. *For* $g \equiv 0$, *this is the standard result (where, with the different normalization, there is an extra factor* $|Y^*|/|Y|$*).*

Proof of Theorem 5.6. Inserting formula (5.15) into (5.8), and taking $\Phi = 0$ gives the result. □

In the case of a periodic g_ε (see Remark 4.58 for the notations), arising from $\mathbf{g}$ in $L^2(\partial S)$, the two cases give different results. In case 1, $g = 0$, $G = \mathcal{M}_{\partial S}(\mathbf{g})$, so that $\widehat{\chi}_0^* = 0$ and the limit problem is

$$\begin{cases} -\operatorname{div}(A^{hom}\nabla u_0) = f + \dfrac{|\partial S|}{|Y^*|}\mathcal{M}_{\partial S}(\mathbf{g}) & \text{in } \Omega,\\ u_0 = 0 \quad \text{on } \partial\Omega. \end{cases}$$

In case 2, $g = \mathbf{g}$ and $G = 0$, so the limit problem is

$$\begin{cases} -\operatorname{div}(A^{hom}\nabla u_0) = f + \operatorname{div}\big(\mathcal{M}_{Y*}(B(x,\cdot)\nabla_y\widehat{\chi}_0^*(x,\cdot))\big) & \text{in } \Omega,\\ u_0 = 0 \quad \text{on } \partial\Omega, \end{cases}$$

with $\widehat{\chi}_0^*$ given by (5.17) associated with $\mathbf{g}$.

One can remark that in the latter case, if B does not depend on x (this holds in the pure periodic case), $\widehat{\chi}_0^*$ itself is independent of x. Then, the second term in the right-hand side of the homogenized problem vanishes and $\mathbf{g}$ does not contribute to the limit problem, i.e. the limit equation is the same as that of the homogeneous Dirichlet-Neumann case.

Remark 5.8. *In problem* (5.1) *one can easily consider a sequence* $\{f_\varepsilon\}_\varepsilon$ *which converges weakly to* f *in* $L^2(\Omega)$.

However, one cannot choose a fixed f *in* $H^{-1}(\Omega)$, *simply because it cannot be restricted to* Ω_ε^* *in a meaningful way. A more general condition can be*

$$\|\varepsilon f_\varepsilon\|_{L^2(\Omega_\varepsilon^*)} \text{ and } \|\mathcal{M}_\varepsilon^*(f_\varepsilon)\|_{L^2(\Omega)} \text{ both uniformly bounded.}$$

Using arguments similar to those of Example 3.26 and assuming that $\mathcal{T}_\varepsilon^*(\varepsilon f_\varepsilon)$ *converges weakly to some* $\widehat{f}$ *in* $L^2(\Omega\times Y^*)$ *and* $\mathcal{M}_\varepsilon^*(f_\varepsilon)$ *converges to* F *in*

(4) The definition of the homogenized matrix used here differs from the "classical" one by a factor of $\dfrac{|Y^*|}{|Y|}$, and simplifies the right-hand side by the same factor.

$L^2(\Omega)$, *the first integral* $\dfrac{|Y^*|}{|Y|}\displaystyle\int_\Omega f\,\Psi\,dx$ *in the right-hand side of* (5.8) *is replaced by*

$$\frac{|Y^*|}{|Y|}\int_\Omega \Big(\mathcal{M}_{Y^*}(\widehat{f}(\cdot,y)\,y^c)\cdot\nabla\Psi + F\,\Psi\Big)\,dx + \frac{1}{|Y|}\int_{\Omega\times Y^*}\widehat{f}\,\Phi\,dxdy.$$

As in Theorem 3.23, an extra term appears in the definition (5.17) *of* $\widehat{\chi}_0^*$ *(due to* $\widehat{f}$*) and the term* f *in the right-hand side of the homogenized problem* (5.18) *is replaced by*

$$F - \operatorname{div}\mathcal{M}_{Y^*}\big(\widehat{f}(\cdot,y)\,y^c\big).$$

5.1.2 Convergence of the energy and correctors

Under stronger assumptions on the data, a convergence result stronger than (iii) of (5.7) holds for the sequences of solutions $\{u_\varepsilon\}_\varepsilon$ of Problem (5.1). It is based on the convergence of the energy for this problem.

Proposition 5.9. *Assume that the hypotheses of Theorem 5.1 are satisfied. Moreover, assume that the convergences in* (5.6) *are strong, i.e.,*

$$\begin{aligned} &\mathcal{T}_\varepsilon^b(g_\varepsilon)\to g \quad \textit{strongly in } L^2(\Omega\times\partial S),\\ &\frac{1}{\varepsilon}\mathcal{M}_\varepsilon^b(g_\varepsilon)\to G \quad \textit{strongly in } L^2(\mathcal{V}_1(\Omega)). \end{aligned}\tag{5.19}$$

Then

$$\lim_{\varepsilon\to 0}\int_{\Omega_\varepsilon^*} A^\varepsilon\nabla u_\varepsilon\nabla u_\varepsilon\,dx = \frac{1}{|Y|}\int_{\Omega\times Y^*} B\big[\nabla u_0+\nabla_y\widehat{u}_0\big]\big[\nabla u_0+\nabla_y\widehat{u}_0\big]dxdy \tag{5.20}$$

and the following strong convergence holds:

$$\mathcal{T}_\varepsilon^*(\nabla u_\varepsilon)\to\nabla u_0+\nabla_y\widehat{u}_0 \quad \textit{strongly in } L^2(\Omega\times Y^*)^N. \tag{5.21}$$

Moreover,

$$\lim_{\varepsilon\to 0}\int_{\Lambda_\varepsilon^*}|\nabla u_\varepsilon|^2\,dx = 0. \tag{5.22}$$

Proof. By Proposition 4.4 (i) and the ellipticity of A^ε,

$$\begin{aligned}\int_{\Omega\times Y^*} &\mathcal{T}_\varepsilon^*(A^\varepsilon)\,\mathcal{T}_\varepsilon^*(\nabla u_\varepsilon)\mathcal{T}_\varepsilon^*(\nabla u_\varepsilon)\,dxdy\\ &= \int_{\Omega_\varepsilon^*} A^\varepsilon\,\nabla u_\varepsilon\nabla u_\varepsilon\,dx - \int_{\Lambda_\varepsilon^*} A^\varepsilon\,\nabla u_\varepsilon\nabla u_\varepsilon\,dx\\ &\leqslant \int_{\Omega_\varepsilon^*} A^\varepsilon\,\nabla u_\varepsilon\nabla u_\varepsilon\,dx.\end{aligned}\tag{5.23}$$

Going back to (5.2), this gives successively

$$\begin{aligned}\limsup_{\varepsilon\to 0}\frac{1}{|Y|}\int_{\Omega\times Y^*} & \mathcal{T}^*_\varepsilon(A^\varepsilon)\,\mathcal{T}^*_\varepsilon(\nabla u_\varepsilon)\mathcal{T}^*_\varepsilon(\nabla u_\varepsilon)\,dxdy\\ &\leqslant \limsup_{\varepsilon\to 0}\int_{\Omega^*_\varepsilon} A^\varepsilon\,\nabla u_\varepsilon\nabla u_\varepsilon\,dx\\ &= \limsup_{\varepsilon\to 0}\Big(\int_{\Omega^*_\varepsilon} f\,u_\varepsilon\,dx + \int_{\partial S_\varepsilon\cap\Omega} g_\varepsilon u_\varepsilon\,d\sigma_\varepsilon\Big).\end{aligned}$$

But, by the three convergences of (5.7) together with formula (4.58) of Proposition 4.57, it follows that

$$\begin{aligned}&\int_{\Omega^*_\varepsilon} f\,u_\varepsilon\,dx \;\to\; \frac{|Y^*|}{|Y|}\int_\Omega f\,u_0\,dx,\\ &\int_{\partial S_\varepsilon\cap\Omega} g_\varepsilon u_\varepsilon\,d\sigma_\varepsilon \;\to\; \frac{|\partial S|}{|Y|}\int_\Omega \mathcal{M}_{\partial S}(y_{\mathcal{M}}\,g)\cdot\nabla u_0\,dx + \frac{|\partial S|}{|Y|}\int_\Omega G\,u_0\,dx\\ &\qquad\qquad + \frac{1}{|Y|}\int_{\Omega\times\partial S}\widehat{u}_0\,g\,dxd\sigma(y).\end{aligned}$$

Confronting this with (5.8) where $\Psi = u_0$ and $\Phi = \widehat{u}_0$, proves (5.20). Applying Lemma 3.11 with $D_\varepsilon = \mathcal{T}^*_\varepsilon(A^\varepsilon)$, $\zeta_\varepsilon = \mathcal{T}^*_\varepsilon(\nabla u_\varepsilon)$, yields convergences (5.20) and (5.21). Now, using (5.23) again, gives

$$\lim_{\varepsilon\to 0}\int_{\Lambda^*_\varepsilon} A^\varepsilon\,\nabla u_\varepsilon\nabla u_\varepsilon\,dx = 0,$$

hence (5.22) by ellipticity. □

We can now address the question of correctors. When $A^\varepsilon(x) = A(x/\varepsilon)$, by using extension operators and under the hypothesis that the holes do not intersect $\partial\Omega$, the following corrector result was proved in [99] (Corollary 2.2):

$$\Big\|\nabla u_\varepsilon - \nabla u_0 - \sum_{i=1}^N \frac{\partial u_0}{\partial x_i}\nabla_y\widehat{\chi}^*_i\Big(\frac{\cdot}{\varepsilon}\Big)\Big\|_{L^1(\Omega^*_\varepsilon)} \to 0. \tag{5.24}$$

Using the unfolding method, we now give a general corrector result, of which convergence (5.24) is a simple corollary. As in the case of fixed domains (see Theorem 3.27), this corrector result is a direct consequence of the strong convergence (5.21).

Theorem 5.10. *Under the hypotheses of Proposition 5.9, the following strong convergence holds:*

$$\Big\|\nabla u_\varepsilon - \nabla u_0 - \sum_{i=1}^N \mathcal{M}_\varepsilon\Big(\frac{\partial u_0}{\partial x_i}\Big)\,\mathcal{U}^*_\varepsilon(\nabla_y\widehat{\chi}^*_i) - \mathcal{U}^*_\varepsilon(\nabla_y\widehat{\chi}^*_0)\Big\|_{L^2(\Omega^*_\varepsilon)} \to 0. \tag{5.25}$$

Proof. By construction, the function $\widehat{\chi}_i^*$ belongs to $L^\infty(\Omega; H^1_{per,0}(Y^*))$, for $i = 1, \dots, N$. Due to convergences (5.21) and (5.22), Proposition 4.20 (iv) gives

$$\|\nabla u_\varepsilon - \mathcal{U}^*_\varepsilon(\nabla u_0 + \nabla_y \widehat{u}_0)\|_{L^2(\Omega^*_\varepsilon)} \to 0.$$

By Proposition 4.19 and (5.15) this implies

$$\Big\|\nabla u_\varepsilon - \nabla u_0 - \sum_{i=1}^N \mathcal{U}^*_\varepsilon\Big(\frac{\partial u_0}{\partial x_i}\nabla_y \widehat{\chi}_i^*\Big) - \mathcal{U}^*_\varepsilon(\nabla_y \widehat{\chi}_0^*)\Big\|_{L^2(\Omega^*_\varepsilon)} \to 0. \tag{5.26}$$

From formula (4.10) and Proposition 1.34 (in the latter, $\nabla_y \widehat{\chi}_i^*$ is extended by 0 in $\Omega \times S$), it follows that

$$\begin{aligned}&\Big\|\mathcal{U}^*_\varepsilon\Big(\frac{\partial u_0}{\partial x_i}\nabla_y \widehat{\chi}_i^*\Big) - \mathcal{U}_\varepsilon\Big(\frac{\partial u_0}{\partial x_i}\Big)\mathcal{U}^*_\varepsilon(\nabla_y \widehat{\chi}_i^*)\Big\|_{L^2(\Omega^*_\varepsilon)}\\ &\quad = \Big\|\mathcal{U}^*_\varepsilon\Big(\frac{\partial u_0}{\partial x_i}\nabla_y \widehat{\chi}_i^*\Big) - \mathcal{M}_\varepsilon\Big(\frac{\partial u_0}{\partial x_i}\Big)\mathcal{U}^*_\varepsilon(\nabla_y \widehat{\chi}_i^*)\Big\|_{L^2(\Omega^*_\varepsilon)} \to 0,\end{aligned}$$

which used in (5.26), concludes the proof. □

Theorem 5.11. *We assume the hypotheses of Proposition 5.9. Suppose furthermore, that the matrix field B given by (5.5) does not depend on x, and that the limit function g defined by (5.19) is of the form*

$$g(x,y) \doteq g_1(x) g_0(y) \quad \text{with} \quad g_1 \in L^2(\Omega) \text{ and } g_0 \in L^2(\partial S).$$

Then the following corrector result holds:

$$\Big\|u_\varepsilon - u_0 - \varepsilon\sum_{i=1}^N \mathcal{Q}_\varepsilon\Big(\frac{\partial u_0}{\partial x_i}\Big)\widehat{\chi}_i^*\Big(\frac{\cdot}{\varepsilon}\Big) - \varepsilon\mathcal{Q}_\varepsilon(g_1)\widetilde{\chi}_0^*\Big(\frac{\cdot}{\varepsilon}\Big)\Big\|_{H^1(\Omega^*_\varepsilon)} \to 0, \tag{5.27}$$

where $\widetilde{\chi}_0^* \in H^1_{per,0}(Y^*)$ is the solution of Problem (5.17) with g replaced by g_0, i.e.,

$$\begin{cases}\displaystyle\int_{Y^*}\sum_{i,k=1}^N b_{ik}(y)\frac{\partial \widetilde{\chi}_0^*}{\partial y_k}(y)\frac{\partial \varphi}{\partial y_i}(y)\,dy = \int_{\partial S} g_0(y)\varphi(y)\,d\sigma(y),\\ \forall \varphi \in H^1_{per,0}(Y^*).\end{cases} \tag{5.28}$$

Moreover, if the $\widehat{\chi}_i^*$'s are in $L^\infty(Y^*)$[(5)] *and if g_1 belongs to $H^1(\Omega)$ the following convergence also holds for every open $\omega \Subset \Omega$:*

$$\Big\|u_\varepsilon - u_0 - \varepsilon\sum_{i=1}^N \frac{\partial u_0}{\partial x_i}\widehat{\chi}_i^*\Big(\frac{\cdot}{\varepsilon}\Big) - \varepsilon\mathcal{Q}_\varepsilon(g_1)\widetilde{\chi}_0^*\Big(\frac{\cdot}{\varepsilon}\Big)\Big\|_{H^1(\omega\cap\Omega^*_\varepsilon)} \to 0. \tag{5.29}$$

---

(5) This condition is satisfied for example when the b_{ij}'s belong to $W^{1,\infty}_{per}(Y^*)$ and the boundary of the hole S is $C^{1,1}$ (cf. e.g. [114] Chapter 9, adapted to the periodic case).

If u_0 is in $H^2(\Omega)$ (e.g. if $\partial\Omega$ is smooth), then

$$\Big\| u_\varepsilon - u_0 - \varepsilon \sum_{i=1}^{N} \frac{\partial u_0}{\partial x_i} \widehat{\chi}_i^*\Big(\frac{\cdot}{\varepsilon}\Big) - \varepsilon \mathcal{Q}_\varepsilon(g_1)\widetilde{\chi}_0^*\Big(\frac{\cdot}{\varepsilon}\Big)\Big\|_{H^1(\Omega_\varepsilon^*)} \to 0. \tag{5.30}$$

Proof. Since B does not depend upon x, $\widehat{\chi}_j^*$ for $j = 1, \dots, N$ are independent of x. As for $\widehat{\chi}_0^*$, it is given by

$$\widehat{\chi}_0^*(x, y) = g_1(x)\widetilde{\chi}_0^*(y). \tag{5.31}$$

By (5.7)(i) and Proposition 1.71 (extending the $\widehat{\chi}_i^*$ and $\widetilde{\chi}_0^*$ by 0 in S), it follows that

$$\Big\| u_\varepsilon - u_0 - \varepsilon \sum_{i=1}^{N} \mathcal{Q}_\varepsilon\Big(\frac{\partial u_0}{\partial x_i}\Big) \widehat{\chi}_i^*\Big(\frac{\cdot}{\varepsilon}\Big) - \varepsilon \mathcal{Q}_\varepsilon(g_1)\widetilde{\chi}_0^*\Big(\frac{\cdot}{\varepsilon}\Big)\Big\|_{L^2(\Omega_\varepsilon^*)} \to 0. \tag{5.32}$$

To go further, we shall use convergence (5.25). To do so, notice that, since $\widehat{\chi}_j^*$ for $j = 1, \dots, N$ are independent of x, by (4.11)

$$\mathcal{U}_\varepsilon^*(\nabla_y \widehat{\chi}_i^*) = \nabla_y \widehat{\chi}_i^*\Big(\frac{\cdot}{\varepsilon}\Big) \quad \text{in} \quad \widehat{\Omega}_\varepsilon^*.$$

On the other hand, recalling (5.31), we get

$$\mathcal{U}_\varepsilon^*(\nabla_y \widehat{\chi}_0^*) = \nabla_y \widetilde{\chi}_0^*\Big(\frac{\cdot}{\varepsilon}\Big)\mathcal{M}_\varepsilon(g_1).$$

These last relations, replaced in (5.25), give the convergence

$$\begin{aligned} \Big\| \nabla u_\varepsilon - \nabla u_0 - \sum_{i=1}^{N} \mathcal{M}_\varepsilon\Big(\frac{\partial u_0}{\partial x_i}\Big) \nabla_y \widehat{\chi}_i^*)\Big(\frac{\cdot}{\varepsilon}\Big) & \\ - \mathcal{M}_\varepsilon(g_1)\nabla_y \widetilde{\chi}_0^*\Big(\frac{\cdot}{\varepsilon}\Big)\Big\|_{L^2(\Omega_\varepsilon^*)} & \to 0. \end{aligned} \tag{5.33}$$

Now, apply Lemma 5.12 below twice for the choices

$$\alpha = \frac{\partial u_0}{\partial x_i}, \quad \beta = \nabla_y \widehat{\chi}_i^* \quad \text{and} \quad \alpha = g_1, \quad \beta = \nabla_y \widetilde{\chi}_0^*.$$

After extending both gradients by 0 in S, this gives the estimates

$$\begin{aligned} &\Big\| \Big[\mathcal{Q}_\varepsilon\Big(\frac{\partial u_0}{\partial x_i}\Big) - \mathcal{M}_\varepsilon\Big(\frac{\partial u_0}{\partial x_i}\Big)\Big] \nabla_y \widehat{\chi}_i^*\Big(\frac{\cdot}{\varepsilon}\Big)\Big\|_{L^2(\Omega)} \\ &\quad \leqslant C\Big(\Big\|\mathcal{Q}_\varepsilon\Big(\frac{\partial u_0}{\partial x_i}\Big) - \mathcal{M}_\varepsilon\Big(\frac{\partial u_0}{\partial x_i}\Big)\Big\|_{L^2(\Omega)} + \Big\|\frac{\partial u_0}{\partial x_i}\Big\|_{L^2(\Lambda_\varepsilon)} \|\nabla_y \widehat{\chi}_i^*\|_{L^2(Y)}\Big), \end{aligned}$$

and

$$\begin{aligned}&\Big\|[\mathcal{Q}_\varepsilon(g_1)-\mathcal{M}_\varepsilon(g_1)]\nabla_y\widetilde{\chi}_0^*\Big(\frac{\cdot}{\varepsilon}\Big)\Big\|_{L^2(\Omega)}\\&\qquad\leqslant C\Big(\big\|\mathcal{Q}_\varepsilon(g_1)-\mathcal{M}_\varepsilon(g_1)\big\|_{L^2(\Omega)}+\|g_1\|_{L^2(\Lambda_\varepsilon)}\|\nabla_y\widetilde{\chi}_0^*\|_{L^2(Y)}\Big).\end{aligned}$$

Observe that the right-hand side in these estimates go to 0 (see Proposition 1.64(iv)). Therefore, from (5.33) we deduce

$$\Big\|\nabla u_\varepsilon-\nabla u_0-\sum_{i=1}^N\mathcal{Q}_\varepsilon\Big(\frac{\partial u_0}{\partial x_i}\Big)\nabla_y\widehat{\chi}_i^*\Big(\frac{\cdot}{\varepsilon}\Big)-\mathcal{Q}_\varepsilon(g_1)\nabla_y\widetilde{\chi}_0^*\Big(\frac{\cdot}{\varepsilon}\Big)\Big\|_{L^2(\Omega_\varepsilon^*)}\to 0.$$

By (1.73)(ii), $\varepsilon\nabla\mathcal{Q}_\varepsilon\Big(\dfrac{\partial u_0}{\partial x_i}\Big)$ and $\varepsilon\nabla\mathcal{Q}_\varepsilon(g_1)$ converge to 0 strongly in $L^2(\Omega)$. Consequently,

$$\Big\|\nabla\Big(u_\varepsilon-u_0-\varepsilon\sum_{i=1}^N\mathcal{Q}_\varepsilon\Big(\frac{\partial u_0}{\partial x_i}\Big)\widehat{\chi}_i^*\Big(\frac{\cdot}{\varepsilon}\Big)-\varepsilon\mathcal{Q}_\varepsilon(g_1)\widetilde{\chi}_0^*\Big(\frac{\cdot}{\varepsilon}\Big)\Big)\Big\|_{L^2(\Omega_\varepsilon^*)}\to 0.$$

This, together with convergence (5.32), proves convergence (5.27).

Assuming further that g_1 is in $H^1(\Omega)$ implies that u_0 belongs to $H^2_{loc}(\Omega)$, since ∇u_0 belongs to $H^1_{loc}(\Omega)$.

In view of (5.16), the extension by 0 to Y of the field $B(y)\nabla(\widehat{\chi}_j^*(y)+y_j)$ is divergence-free. Reasoning as in Corollary 3.31, yields convergence (5.29). Similarly, in case u_0 belongs to $H^2(\Omega)$, convergence (5.30) holds. □

In the above proof we used the following lemma:

Lemma 5.12. *There is a constant C such that, for $\alpha\in L^2(\Omega)$ and $\beta\in L^2(Y)$,*

$$\begin{aligned}&\Big\|\big(\mathcal{Q}_\varepsilon(\alpha)-\mathcal{M}_\varepsilon(\alpha)\big)\beta\Big(\frac{\cdot}{\varepsilon}\Big)\Big\|_{L^2(\Omega)}\\&\qquad\leqslant C\big(\big\|\mathcal{Q}_\varepsilon(\alpha)-\mathcal{M}_\varepsilon(\alpha)\big\|_{L^2(\Omega)}+\|\alpha\|_{L^2(\Lambda_\varepsilon)}\big)\|\beta\|_{L^2(Y)},\end{aligned}\tag{5.34}$$

and moreover,

$$\Big\|\mathcal{M}_\varepsilon(\alpha)\beta\Big(\frac{\cdot}{\varepsilon}\Big)-\mathcal{Q}_\varepsilon(\alpha)\beta\Big(\frac{\cdot}{\varepsilon}\Big)\Big\|_{L^2(\Omega)}\to 0.\tag{5.35}$$

Proof. Extending α by 0 to the whole of $\mathbb{R}^N$, it is enough to give the proof of inequality (5.34) in the case of $\Omega=\mathbb{R}^N$, noting that

$$\|\widetilde{\mathcal{M}_\varepsilon}(\alpha)\|_{L^2(\mathbb{R}^N)}\leqslant\|\mathcal{M}_\varepsilon(\alpha)\|_{L^2(\Omega)}+\|\alpha\|_{L^2(\Lambda_\varepsilon)}.$$

For $\mathbb{R}^N$, estimate (5.34) is a consequence of Proposition 1.71. Then, convergence (5.35) follows from the strong convergence of $\mathcal{Q}_\varepsilon(\alpha)$ and $\mathcal{M}_\varepsilon(\alpha)$ to α in $L^2(\Omega)$ (see Proposition 1.25 and (1.82)(i)). □

5.2 The Neumann problem

As in the previous section, let f be in $L^2(\Omega)$ and $A^\varepsilon = (a^\varepsilon_{ij})_{1\leqslant i,j\leqslant N}$ be a matrix field in the set $M(\alpha,\beta,\Omega)$.

The Neumann problem is

$$\begin{cases} -\mathrm{div}\,(A^\varepsilon \nabla u_\varepsilon) + b_\varepsilon u_\varepsilon = f & \text{in } \Omega^*_\varepsilon, \\ A^\varepsilon \nabla u_\varepsilon \cdot n_\varepsilon = 0 & \text{on } \partial\Omega^*_\varepsilon \backslash \partial\widehat{S}_\varepsilon, \\ A^\varepsilon \nabla u_\varepsilon \cdot n_\varepsilon = g_\varepsilon & \text{on } \partial\widehat{S}_\varepsilon, \end{cases} \tag{5.36}$$

where the function b_ε is measurable in Ω and g_ε is given in $L^2(\partial\widehat{S}_\varepsilon)$ (see Notation (4.3)).

The variational formulation of (5.36) is

$$\begin{cases} \text{Find } u_\varepsilon \in H^1(\Omega^*_\varepsilon) \text{ such that} \\ \displaystyle\int_{\Omega^*_\varepsilon} A^\varepsilon \nabla u_\varepsilon \nabla v \, dx + \int_{\Omega^*_\varepsilon} b_\varepsilon u_\varepsilon \, v \, dx = \int_{\Omega^*_\varepsilon} f \, v \, dx + \int_{\partial\widehat{S}_\varepsilon} g_\varepsilon v \, d\sigma_\varepsilon, \\ \forall v \in H^1(\Omega^*_\varepsilon). \end{cases} \tag{5.37}$$

Remark 5.13. *As far as we know, there is no homogenization result for Problem 5.36 if g_ε is defined on the whole of ∂S_ε and does not vanish outside of $\partial\widehat{S}_\varepsilon$. This is mainly due to the lack of uniform bounds for solutions. In several papers, the holes in the zone $\partial S_\varepsilon \backslash \partial\widehat{S}_\varepsilon$ (or in a similar boundary layer) are completely suppressed, in which case the unfolding approach works as easily (since all unfolded functions always vanish in this layer).*

5.2.1 Homogenization

To homogenize the Neumann problem, the method of the preceding section applies. We state the results without detailing the proofs. The condition on b_ε is the following:

$$\begin{cases} \text{There are two positive constants } c_0 \text{ and } C_0 \text{ and } b \in L^\infty(\Omega \times Y^*) \\ \text{such that, } c_0 \leqslant b_\varepsilon(x) \leqslant C_0 \text{ for a.e. } x \in \Omega^*_\varepsilon \\ \text{and } \mathcal{T}^*_\varepsilon(b_\varepsilon) \to b \text{ a.e. in } \Omega \times Y^{*\,(6)}. \end{cases} \tag{5.38}$$

Theorem 5.14. *Let u_ε be the solution of the Problem* (5.37). *Assume that* (5.5), (5.6) *and* (5.38) *hold and that g_ε vanishes outside of $\partial\widehat{S}_\varepsilon$. Then, there exist u in $H^1(\Omega)$ and $\widehat{u}$ in $L^2(\Omega; H^1_{per,0}(Y^*))$, such that*

(i) $\mathcal{T}^*_\varepsilon(u_\varepsilon) \rightharpoonup u$ *weakly in* $L^2(\Omega; H^1(Y^*))$ *and strongly in* $L^2_{loc}(\Omega; H^1(Y^*))$

(ii) $\begin{cases} \mathcal{E}_\varepsilon(u_\varepsilon) \to u_0 & \textit{strongly in } L^2_{loc}(\Omega), \\ \forall \omega \Subset \Omega, & \|u_\varepsilon - u_0\|_{L^2(\Omega^*_\varepsilon \cap \omega)} \to 0, \end{cases}$

(iii) $\mathcal{T}^*_\varepsilon(\nabla u_\varepsilon) \rightharpoonup \nabla u + \nabla_y \widehat{u}$ *weakly in* $L^2(\Omega \times Y^*)^N$,

(6) or more generally, in measure in $\Omega \times Y^*$.

and the pair $(u, \widehat{u})$ is the unique solution of the problem

$$\begin{cases} u \in H^1(\Omega),\ \widehat{u} \in L^2(\Omega;\ H^1_{per,0}(Y^*)), \\ \dfrac{1}{|Y|}\displaystyle\int_{\Omega\times Y^*} B(x,y)\big[\nabla u(x) + \nabla_y \widehat{u}(x,y)\big]\big[\nabla\Psi(x) + \nabla_y \Phi(x,y)\big]\,dxdy \\ \qquad\qquad + \dfrac{|Y^*|}{|Y|}\displaystyle\int_\Omega \mathcal{M}_{Y*}(b)(x)\,u(x)\,\Psi(x)\,dx \\ \quad = \dfrac{|Y^*|}{|Y|}\displaystyle\int_\Omega f(x)\,\Psi(x)\,dx + \dfrac{|\partial S|}{|Y|}\displaystyle\int_\Omega \mathcal{M}_{\partial S}(y_{\mathcal{M}} g)(x)\cdot\nabla\Psi(x)\,dx \\ \quad + \dfrac{|\partial S|}{|Y|}\displaystyle\int_\Omega G(x)\,\Psi(x)\,dx + \dfrac{1}{|Y|}\displaystyle\int_{\Omega\times\partial S} g(x,y)\Phi(x,y)\,dx\,d\sigma(y), \\ \forall\Psi \in H^1(\Omega),\ \forall\Phi \in L^2(\Omega;\ H^1_{per}(Y^*)). \end{cases}$$

Proof. The a priori estimate

$$\|u_\varepsilon\|_{H^1(\Omega^*_\varepsilon)} \leqslant C,$$

follows directly from the variational formulation (5.37) and Proposition 4.50. The remainder of the proof is the exact analog of the proof of Theorem 5.1, making use of Theorem 4.43 instead of Theorem 4.44. □

The next results are the equivalent of Proposition 5.5 and Theorem 5.6 (with obvious modifications in the proofs).

Proposition 5.15. *The function $\widehat{u}$ in Theorem 5.14 is given in terms of u by the expression,*

$$\widehat{u}(x,y) = \sum_{i=1}^N \frac{\partial u}{\partial x_i}(x)\widehat{\chi}^*_i(x,y) + \widehat{\chi}^*_0(x,y),$$

where the corrector functions $\widehat{\chi}^*_j$, $(j = 0,\dots,N)$ are given, as before, by (5.16) *and* (5.17).

Theorem 5.16. *(Standard homogenization for Neumann problem).*
Let u_ε be the solution of Problem (5.37) *and suppose that the hypotheses of Theorem 5.14 are satisfied. Then u is the solution in $H^1(\Omega)$ of the homogenized problem*

$$\begin{cases} -\mathrm{div}\,(A^{hom}\nabla u) + \mathcal{M}_{Y*}(b)\,u = f - \mathrm{div}\,\mathcal{G} \quad \text{in } \Omega, \\ A^{hom}\nabla u\cdot n = \mathcal{G}\cdot n \quad \text{on } \partial\Omega, \end{cases}$$

where

$$\mathcal{G}(x) \doteq \frac{|\partial S|}{|Y^*|}\big(G(x) + \mathcal{M}_{\partial S}(y_{\mathcal{M}} g)(x)\big) - \mathcal{M}_{Y*}\big(B(x,\cdot)\,\nabla_y\widehat{\chi}^*_0(x,\cdot)\big) \quad \text{in } \Omega.$$

The matrix field A^{hom} is the same as that defined in Theorem 5.6.

Remark 5.17. *A strange phenomenon appears in this result: the nonhomogeneous Neumann conditions on the boundary of the holes inside Ω (actually supported inside $\widehat{\Omega}_\varepsilon$) contribute to a nonhomogeneous Neumann condition on the outer boundary $\partial\Omega$ in the limit problem through the term $\mathcal{G}$. This phenomenon was also observed in the context of Γ-convergence in [55].*

When g_ε is derived from $\mathbf{g}$ in $L^2(\partial S)$, the two cases of Remark 4.58 give different results.

In case 3, the limit problem is (since $g = 0$, $\widehat{\chi}_0^* = 0$ so $\mathcal{G}$=0),

$$\begin{cases} -\text{div}\, A^{hom}\nabla u) + \mathcal{M}_{Y^*}(b)\, u = f + \dfrac{|\partial S|}{|Y^*|}\mathcal{M}_{\partial S}(\mathbf{g}) \quad \text{in } \Omega, \\ A^{hom}\nabla u \cdot n = 0 \quad \text{on } \partial\Omega. \end{cases}$$

In case 4 of Remark 4.58, the limit problem is

$$\begin{cases} -\text{div}\,(A^{hom}\nabla u) + \mathcal{M}_{Y^*}(b)\, u = f - \text{div}\,\mathcal{G} \quad \text{in } \Omega, \\ A^{hom}\nabla u \cdot n = \mathcal{G}\cdot n \quad \text{on } \partial\Omega, \end{cases}$$

with $\mathcal{G}(x) \doteq \dfrac{|\partial S|}{|Y^*|}\mathcal{M}_{\partial S}(y_{\mathcal{M}}\mathbf{g})(x) - \mathcal{M}_{Y^*}\big(B(x,\cdot)\nabla_y\widehat{\chi}_0^*(x,\cdot)\big)$ in Ω.

In this last case, the holes induce a nonhomogeneous Neumann condition at the limit.

5.2.2 Convergence of the energy and correctors

The next result is the equivalent of Proposition 5.9.

Proposition 5.18 (Convergence of the energy for Problem (5.37))**.** *Assume also that*

$$\mathcal{T}_\varepsilon^b(g_\varepsilon) \to g \quad \textit{strongly in } L^2(\Omega\times\partial S),$$
$$\frac{1}{\varepsilon}\mathcal{M}_\varepsilon^b(g_\varepsilon) \to G \quad \textit{strongly in } L^2(\Omega).$$

Then

$$\begin{aligned} &\lim_{\varepsilon\to 0}\int_{\Omega_\varepsilon^*}\Big(A^\varepsilon\nabla u_\varepsilon\nabla u_\varepsilon + b_\varepsilon\, u_\varepsilon^2\Big)dx \\ &= \frac{1}{|Y|}\int_{\Omega\times Y^*} B[\nabla u + \nabla_y\widehat{u}][\nabla u + \nabla_y\widehat{u}]dxdy + \frac{|Y^*|}{|Y|}\int_\Omega \mathcal{M}_{Y^*}(b)\, u^2\, dx, \end{aligned} \tag{5.39}$$

and

$$\lim_{\varepsilon\to 0}\Big(\int_{\Lambda_\varepsilon^*}|\nabla u_\varepsilon|^2\, dx + \int_{\Lambda_\varepsilon^*}u_\varepsilon^2\, dx\Big) = 0. \tag{5.40}$$

Moreover, one has the following convergences:

$$\begin{aligned} &\mathcal{T}_\varepsilon^*(u_\varepsilon) \to u \quad \textit{strongly in } L^2(\Omega\times Y^*), \\ &\mathcal{T}_\varepsilon^*(\nabla u_\varepsilon) \to \nabla u + \nabla_y\widehat{u} \quad \textit{strongly in } L^2(\Omega\times Y^*)^N. \end{aligned} \tag{5.41}$$

Proof. The proof of (5.39) is similar to that of Proposition 5.9. In order to prove (5.40) and (5.41), Lemma 3.11 is applied to the $(N+1)\times(N+1)$ matrix D_ε and the $N+1$ vector ξ_ε

$$D_\varepsilon = \begin{pmatrix} & \mathcal{T}_\varepsilon^*(A^\varepsilon) & & 0 \\ & & & \vdots \\ & & & 0 \\ 0 & \dots & 0 & \mathcal{T}_\varepsilon^*(b_\varepsilon) \end{pmatrix}, \quad \xi_\varepsilon = \begin{pmatrix} \mathcal{T}_\varepsilon^*\Big(\dfrac{\partial u_\varepsilon}{\partial x_1}\Big) \\ \vdots \\ \mathcal{T}_\varepsilon^*\Big(\dfrac{\partial u_\varepsilon}{\partial x_N}\Big) \\ \mathcal{T}_\varepsilon^*(u_\varepsilon) \end{pmatrix}.$$

□

As a consequence of this proposition, we have

Theorem 5.19. *As* $\varepsilon \to 0$,

$$(i)\ \begin{cases} \mathcal{E}_\varepsilon(u_\varepsilon) \to u \quad \text{strongly in } L^2(\Omega), \\ \|u_\varepsilon - u\|_{L^2(\Omega_\varepsilon^*)} \to 0, \end{cases}$$

$$(ii)\ \Big\|\nabla u_\varepsilon - \nabla u - \sum_{i=1}^N \mathcal{M}_\varepsilon\Big(\frac{\partial u}{\partial x_i}\Big)\mathcal{U}_\varepsilon^*(\nabla_y\widehat{\chi}_i^*) - \mathcal{U}_\varepsilon^*(\nabla\widehat{\chi}_0^*)\Big\|_{L^2(\Omega_\varepsilon^*)} \to 0.$$

In the case where the matrix field B does not depend on x, and where $g(x,y) \doteq g_1(x)g_0(y)$ with $g_1 \in L^2(\Omega)$ and $g_0 \in L^2(\partial S)$, the following corrector result holds:

$$\Big\|u_\varepsilon - u - \varepsilon\sum_{i=1}^N \mathcal{Q}_\varepsilon\Big(\frac{\partial u}{\partial x_i}\Big)\widehat{\chi}_i^*\Big(\frac{\cdot}{\varepsilon}\Big) - \varepsilon\mathcal{Q}_\varepsilon(g_1)\widetilde{\chi}_0^*\Big(\frac{\cdot}{\varepsilon}\Big)\Big\|_{H^1(\Omega_\varepsilon^*)} \to 0,$$

where $\widetilde{\chi}_0^* \in H^1_{per}(Y^*)$ is the solution of Problem (5.28).

Proof. Convergences (i) follow from Proposition 4.27, and from convergences (5.40)-(5.41) of Proposition 5.39. The proof of the other convergences is the same as in that of Theorem 5.10. □

5.3 The case of frame-periodic distribution of holes

The aim of this section, is to answer the question how to apply the method of Subsection 3.1.3 to the case of perforated domains.

We refer to that section for the definition of ε frame-periodic deformations (Definition 3.14) and of the set $Diff_{Lip}(Y)$ and its subset $Diff_{Lip}(\overline{Y};\partial Y)$. As noted there, an ε frame-periodic deformation defined with elements of $Diff_{Lip}(\overline{Y};\partial Y)$ is actually a global bi-Lipschitz diffeomorphism of Ω.

We consider the situation of a perforated domain as presented at the beginning of Chapter 4, the definitions of the sets Ω, Y, Y^* and Ω_ε^* being given in Section 4.1.

Following the method of Section 3.14, we introduce the subset $\mathfrak{M}^*_\varepsilon$ of $\mathfrak{M}_\varepsilon$ consisting of the elements of the latter with values in $Diff_{Lip}(\overline{Y};\partial Y)$, and we denote by $\mathfrak{S}^*$ the subset of $\mathfrak{S}$ with elements in $\mathfrak{M}^*_\varepsilon$ for each ε.

For Θ_ε in $\mathfrak{M}^*_\varepsilon$, denote by Ω^{**}_ε the image of Ω^*_ε under the map $\mathbf{apd}(\Theta_\varepsilon)$. This set is perforated by holes which are distributed in an approximately periodic fashion.

Given an element A^ε of $M(\alpha,\beta,\Omega^{**}_\varepsilon)$ (extended by αI to Ω), denote by $A^{\varepsilon *}$ its inverse image in $M(\alpha,\beta,\Omega^*_\varepsilon)$ (which is extended similarly to Ω). By Proposition 3.15, the convergence in measure of the unfolding of $\mathcal{T}_\varepsilon(A^{\varepsilon *})$ (or equivalently of $\mathcal{T}^*_\varepsilon(A^{\varepsilon *})$) is equivalent to that of $\mathcal{T}_\varepsilon(A^\varepsilon)$.

Making use of the same changes of variable as in Subsection 3.1.3 to rewrite the problem on Ω^*_ε, one can now easily prove homogenization results concerning problems in Ω^{**}_ε.

Under unfolding, Ω^{**}_ε is transformed into the open set

$$\Sigma_\varepsilon \doteq \{(x,y)\in\Omega\times Y \mid y\in Y^{\varepsilon *}(x)\}\subset\Omega\times Y,$$

where $Y^{\varepsilon *}(x)=\Theta_\varepsilon(x,Y^*)$.

We state the result for the problem with homogenous Neumann conditions on the boundaries of approximately periodic holes (more general problems can be treated in the same way),

$$\begin{cases}\text{Find } u_\varepsilon\in H^1(\Omega^{**}_\varepsilon) \text{ such that}\\ \displaystyle\int_{\Omega^*_\varepsilon} A^\varepsilon\nabla u_\varepsilon\,\nabla v\,dx+\int_{\Omega^*_\varepsilon} b_\varepsilon\,u_\varepsilon\,v\,dx=\int_{\Omega^*_\varepsilon} f\,v\,dx,\\ \forall v\in H^1(\Omega^{**}_\varepsilon).\end{cases}\tag{5.42}$$

Theorem 5.20. *Suppose there exists a map Θ_0 from Ω to $Diff_{Lip}(\overline{Y};\partial Y)$ such that the sequence $\{\Theta_\varepsilon\in\mathfrak{S}^*\}$ satisfies the hypotheses of Proposition 3.15 (see Remark 3.16 for an example of such situation).*

For every $x\in\Omega$, let $Y^*(x)$ denote $\Theta_0(x,Y^*)$, and Σ be the set defined by

$$\Sigma=\{(x,y)\in\Omega\times Y \mid y\in Y^*(x)\}.$$

Set also

$$\begin{aligned}&\mathcal{X}(\Sigma)\doteq\{w\in L^2(\Sigma)\mid\nabla_y w\in L^2(\Sigma)\},\\ &\mathcal{X}_{per}(\Sigma)\doteq\{w\in\mathcal{X}(\Sigma)\mid w(x,\cdot)\ \text{is } Y\text{-periodic for a.e. } x\in\Omega\}.\end{aligned}$$

Let u_ε be the solution of problem (5.42). Suppose that

$$B^\varepsilon\doteq\mathcal{T}_\varepsilon(A^{\varepsilon *})\to B\quad\text{in measure. in } \Omega\times Y.$$

Then, there exist $u_0\in H^1_0(\Omega)$ and $\widehat{u}_0\in\mathcal{X}_{per}(\Sigma)$ satisfying

$$\|u_\varepsilon-u_0\|_{L^2(\Omega^{**}_\varepsilon)}\to 0,$$

and such that the following convergences hold:

$$(i)\quad \begin{cases} \mathcal{T}_\varepsilon(u_\varepsilon)\mathbf{1}_{\Sigma_\varepsilon} \to u_0\mathbf{1}_\Sigma \text{ strongly in } L^2(\Omega\times Y), \\ \nabla_y\mathcal{T}_\varepsilon(u_\varepsilon)\mathbf{1}_{\Sigma_\varepsilon} \to 0 \text{ strongly in } L^2(\Omega\times Y), \end{cases}$$

$$(ii)\quad \mathcal{T}_\varepsilon(\nabla u_\varepsilon)\mathbf{1}_{\Sigma_\varepsilon} \rightharpoonup (\nabla u_0 + \nabla_y\widehat{u}_0)\mathbf{1}_\Sigma \quad \text{weakly in } L^2(\Omega\times Y^*)^N,$$

where $(u_0,\widehat{u}_0)$ *in* $H^1_0(\Omega)\times\mathcal{X}(\Sigma)$ *is the unique solution of the problem*

$$\begin{cases} \dfrac{1}{|Y|}\displaystyle\int_\Sigma B(x,y)\big[\nabla u_0(x)+\nabla_y\widehat{u}_0(x,y)\big]\big[\nabla\Psi(x)+\nabla_y\Phi(x,y)\big]dxdy \\ \qquad\qquad = \dfrac{1}{|Y|}\displaystyle\int_\Omega |Y^*(x)|\, f(x)\,\Psi(x)\,dx, \\ \mathcal{M}_{Y^*(x)}(\widehat{u}_0(x,\cdot)) = 0 \quad \text{for a.e. } x\in\Omega, \\ \forall\Psi\in H^1_0(\Omega),\quad \forall\Phi\in\mathcal{X}(\Sigma). \end{cases}$$

The proof of this theorem is obtained by applying the bi-Lipschitz transformation $\mathbf{apd}(\Theta_\varepsilon)^{-1}$ to the original problem, now posed on Ω^*_ε, applying afterwards the unfolding result for periodically perforated domains and finally, translating the result on $\Omega\times Y^*$ back to Σ via $\mathbf{apd}(\Theta)$. The homogenization result is then easily obtained.

The same method applies to many other boundary conditions. Similar results can be found in [177].

5.4 A Fredholm alternative

In this section, we follow the general notations of Chapter 4, in particular for Hypothesis $(\mathbf{H_p})$, the sets $\mathcal{P}$, $\widehat{\Omega}_\varepsilon$, Ω^*_ε, $\widehat{\Omega}^*_\varepsilon$ and the operators $\mathcal{T}^*_\varepsilon$, $\mathcal{U}^*_\varepsilon$.

Consider a Neumann problem with the notations of Subsection 3.1.2 but set in the perforated domain Ω^*_ε defined in (4.2). However, there are at least two reasons why this problem is not well-posed. First, it is not true in general that Ω^*_ε is connected. More important, even if Ω^*_ε were connected, its boundary could be so irregular that the Poincaré-Wirtinger inequality would not hold, making the definition of the space $V(\Omega^*_\varepsilon)$ invalid (see (3.31) for the definition of the latter space). Thus the original variational formulation does not make sense.

To avoid these difficulties, we consider a subset $\mathcal{C}^*_\varepsilon$ of Ω^*_ε which is connected and a Hilbert space $W(\mathcal{C}^*_\varepsilon)$ (see (5.48)) on which the approximate problem is well-posed. The sets obtained by "filling the holes" of every $\mathcal{C}^*_\varepsilon$ converge to Ω in an appropriate sense (see Theorem 5.25 and Remark 5.27). Furthermore, all the spaces $W(\mathcal{C}^*_\varepsilon)$ satisfy a Poincaré-Wirtinger inequality **with a uniform constant**, the key to obtaining the uniform estimates given in Theorem 5.25.

5.4.1 A uniform cone property for approximating domains

In this definition, a cone is the convex envelope of the union of the origin and a closed ball which does not contain the origin.

Definition 5.21. *A bounded open set $\mathcal{O}$ in $\mathbb{R}^N$ has the uniform (interior) cone property whenever there exist a finite open cover $\{\mathcal{U}_j\}_{1\leqslant j\leqslant m}$ of $\partial\mathcal{O}$, and a corresponding family $\{\mathcal{C}_j\}$ of cones isometric to a fixed cone $\mathcal{C}$, such that (i) for some strictly positive constant δ_1,*

$$\{x\in\mathcal{O}\mid dist(x,\partial\mathcal{O})<\delta_1\}\subset\bigcup_{j=1}^{m}\mathcal{U}_j,$$

(ii) for every j, $\displaystyle\bigcup_{x\in\Omega\cap\mathcal{U}_j}(x+\mathcal{C}_j)\subset\mathcal{O}$.

It is well-known (see [54]) that this property is equivalent to the fact that $\partial\mathcal{O}$ is Lipschitz.

Recall the notation (1.49), $\mathcal{O}_\delta^{int}\doteq\{x\in\mathcal{O}\mid \mathrm{dist}(x,\partial\mathcal{O})>\delta\}$.

Lemma 5.22. *Assume that the bounded open set Ω has a Lipschitz boundary. Then, there exists $\delta_0>0$ such that the sets $\partial\Omega_\delta^{int}$ for $\delta\in(0,\delta_0]$, are uniformly Lipschitz.*

Proof. We prove that the sets Ω_δ^{int} satisfy a common uniform cone condition for δ small enough. Since Ω has the uniform cone property, there exist δ_1 and a finite open cover $\{\mathcal{U}_j\}_{1\leqslant j\leqslant m}$ of $\partial\Omega$ satisfying (i) and (ii) of Definition 5.21. Set

$$\delta_2=d(\mathcal{C}),\quad \delta_3\doteq\min_{x\in\partial\Omega}\max_j dist(x,\partial\mathcal{U}_j),\quad \delta_0\doteq\frac{1}{3}\min\{\delta_1,\delta_2,\delta_3\}.$$

Since $\partial\Omega$ is compact, δ_3 is strictly positive. In the following, we use the notations of Figure 5.1.

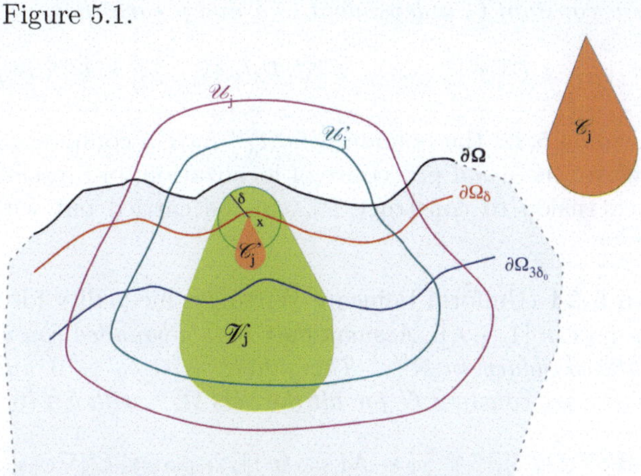

Figure 5.1: Uniform cone property

Let now $\mathcal{C}' = \frac{1}{3}\mathcal{C}$ and $\mathcal{C}_j' = \frac{1}{3}\mathcal{C}_j$ (which is isometric to $\mathcal{C}'$) and consider, for $j \in \{1, \ldots, m\}$, the following open sets:

$$\mathcal{U}_j' = \{x \in \mathcal{U}_j \mid dist(x, \partial\mathcal{U}_j) > \delta_0\}.$$

By the definition of δ_3, the family $\{\mathcal{U}_j'\}_{1 \leqslant j \leqslant m}$ still covers $\partial\Omega$. Clearly,

$$\{x \in \Omega_\delta^{int} \mid dist(x, \partial\Omega_\delta^{int}) < \delta_0\} \subset \bigcup_{j=1}^{m} \mathcal{U}_j' \quad \text{for } \delta \in (0, \delta_0].$$

This is condition (i) of Definition 5.21.

To prove condition (ii), for j in $\{1, \ldots, m\}$ and x in $\Omega_\delta^{int} \cap \mathcal{U}_j'$, one has

$$\mathcal{V}_j \doteq \{z \in \mathbb{R}^N \mid \mathrm{dist}(z, x + \mathcal{C}_j') < \delta\} \subset \bigcup_{y \in B(x;\delta)} (y + \mathcal{C}_j).$$

Furthermore, due to the uniform cone property the latter is included in Ω. Thus the set $x + \mathcal{C}_j'$ is included in Ω_δ^{int}. This proves the inclusion

$$\bigcup_{x \in \Omega_\delta^{int} \cap \mathcal{U}_j'} (x + \mathcal{C}_j') \subset \Omega_\delta^{int},$$

and this concludes the proof of Lemma 5.22. □

Corollary 5.23. *Suppose that p is in $[1, +\infty]$. For every $\delta \in (0, \delta_0]$, there exists a continuous extension operator $\mathcal{P}_\delta$ from $W^{1,p}(\Omega_\delta^{int})$ into $W^{1,p}(\Omega)$, i.e., such that for all $\phi \in W^{1,p}(\Omega_\delta^{int})$,*

$$\mathcal{P}_\delta(\phi) \in W^{1,p}(\Omega), \qquad \mathcal{P}_\delta(\phi)_{|\Omega_\delta^{int}} = \phi.$$

There exists a constant C independent of δ and p such that

$$\|\mathcal{P}_\delta(\phi)\|_{L^p(\Omega)} \leqslant C\|\phi\|_{L^p(\Omega_\delta^{int})}, \qquad \|\nabla(\mathcal{P}_\delta(\phi))\|_{L^p(\Omega)} \leqslant C\|\nabla\phi\|_{L^p(\Omega_\delta^{int})}.$$

Proof. By Lemma 5.22, the boundaries $\partial\Omega_\delta^{int}$ have a common Lipschitz constant. Therefore, the usual procedure of localization, bi-Lipschitz diffeomorphism and extension to construct $\mathcal{P}_\delta$, can be carried out with a uniform bound. □

Proposition 5.24 (Uniform Poincaré-Wirtinger inequality for Ω_δ^{int}). *Suppose that p is in $[1, +\infty]$. Assume that Ω is a bounded connected domain with Lipschitz boundary in $\mathbb{R}^N$. Then there exist $\delta_0 > 0$ and a common Poincaré-Wirtinger constant C for all the sets Ω_δ^{int} with $\delta \in (0, \delta_0]$, i.e.,*

$$\forall \phi \in W^{1,p}(\Omega_\delta^{int}), \qquad \|\phi - \mathcal{M}_{\Omega_\delta^{int}}(\phi)\|_{L^p(\Omega_\delta^{int})} \leqslant C\|\nabla\phi\|_{L^p(\Omega_\delta^{int})},$$

where C is independent of δ and p. All the Ω_δ^{int}'s are connected for $\delta \in (0, \delta_0]$.

Proof. Since Ω is bounded and connected with Lipschitz boundary, it has a Poincaré-Wirtinger constant C_0. Let ϕ be in $W^{1,p}(\Omega_\delta^{int})$. We apply the Poincaré-Wirtinger inequality to the extension $\mathcal{P}_\delta(\phi)$ of ϕ to get

$$\|\mathcal{P}_\delta(\phi) - \mathcal{M}_\Omega\big(\mathcal{P}_\delta(\phi)\big)\|_{L^p(\Omega)} \leqslant C_0\|\nabla\mathcal{P}_\delta(\phi)\|_{L^p(\Omega)} \leqslant CC_0\|\nabla\phi\|_{L^p(\Omega_\delta^{int})}.$$

Here $\mathcal{M}_\Omega\big(\mathcal{P}_\delta(\phi)\big)$ is the mean value of $\mathcal{P}_\delta(\phi)$ over Ω. Then,

$$\begin{aligned}\|\phi - \mathcal{M}_\Omega\big(\mathcal{P}_\delta(\phi)\big)\|_{L^p(\Omega_\delta^{int})} &\leqslant \|\mathcal{P}_\delta(\phi) - \mathcal{M}_\Omega\big(\mathcal{P}_\delta(\phi)\big)\|_{L^p(\Omega)}\\ &\leqslant CC_0\|\nabla\phi\|_{L^p(\Omega_\delta^{int})}.\end{aligned}$$

Taking the mean value of $\phi - \mathcal{M}_\Omega\big(\mathcal{P}_\delta(\phi)\big)$ over Ω_δ^{int} gives

$$\big|\mathcal{M}_{\Omega_\delta^{int}}(\phi) - \mathcal{M}_\Omega\big(\mathcal{P}_\delta(\phi)\big)\big| \leqslant CC_0\|\nabla\phi\|_{L^p(\Omega_\delta^{int})},$$

which used in the previous inequality concludes the proof. □

5.4.2 A uniform Poincaré-Wirtinger inequality for perforated domains

In this section, Hypothesis $(\mathbf{H_p})$ *is assumed.*

We use the notations (1.47), those of Section 4.6 and also the following ones (see Figure 5.2) to insure that for every φ in $W^{1,p}(\Omega_\varepsilon^*)$, $\mathcal{Q}_\varepsilon^*(\varphi)$ is defined on the set $\widetilde{\Omega}_\varepsilon$ (in fact in its superset $\widetilde{\Omega}_\varepsilon^Y$ which is a union of cells):

$$\begin{aligned}\widetilde{\Omega}_\varepsilon &\doteq \big\{x \in \Omega \mid \rho(x) > 2\varepsilon\, d(\mathcal{Y}')\big\},\\ \widetilde{\Xi}_\varepsilon &\doteq \big\{\xi \in \mathbf{G} \mid \varepsilon(\xi + Y) \cap \widetilde{\Omega}_\varepsilon \neq \varnothing\big\}.\end{aligned}$$

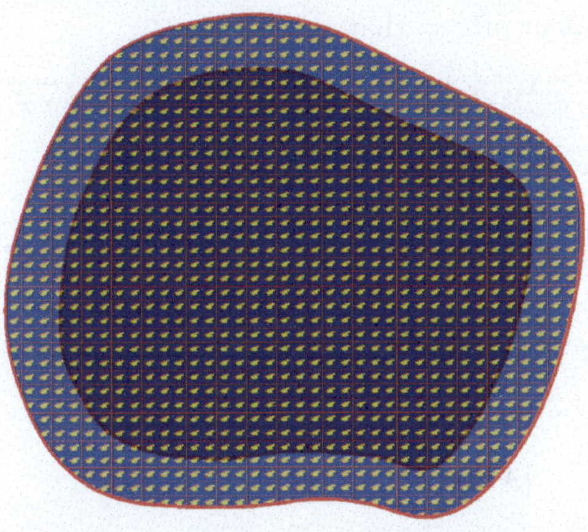

Figure 5.2: The set $\widetilde{\Omega}_\varepsilon^*$ (in dark blue)

Now define the sets

$$\widetilde{\Omega}_\varepsilon^{\mathcal{Y}'} \doteq \text{interior}\Big(\bigcup_{\xi\in\widetilde{\Xi}_\varepsilon} \varepsilon(\xi+\overline{\mathcal{Y}'})\Big), \quad \widetilde{\Omega}_\varepsilon^{Y} \doteq \text{interior}\Big(\bigcup_{\xi\in\widetilde{\Xi}_\varepsilon} \varepsilon(\xi+\overline{Y})\Big),$$

and also

$$\widetilde{\Omega}_\varepsilon^{*} \doteq \Omega_\varepsilon^* \cap \widetilde{\Omega}_\varepsilon, \qquad \widetilde{\Omega}_\varepsilon^{**} \doteq \Omega_\varepsilon^* \cap \widetilde{\Omega}_\varepsilon^{\mathcal{Y}'}.$$

The following inclusions hold:

$$\widetilde{\Xi}_\varepsilon \subset \Xi_\varepsilon^{\mathcal{Y}'} \subset \Xi_\varepsilon^{\mathcal{Y}}, \qquad \widetilde{\Omega}_\varepsilon \subset \widetilde{\Omega}_\varepsilon^{Y} \subset \widetilde{\Omega}_\varepsilon^{\mathcal{Y}'} \subset \Omega, \qquad \widetilde{\Omega}_\varepsilon^{*} \subset \widetilde{\Omega}_\varepsilon^{**} \subset \Omega_\varepsilon^{*}. \tag{5.43}$$

Theorem 5.25. *Suppose 2ε diameter$(\mathcal{Y}') \in (0,\delta_0)$. For every function φ in $W^{1,p}(\widetilde{\Omega}_\varepsilon^{**})$, there exist φ^1 in $W^{1,p}(\widetilde{\Omega}_\varepsilon)$ and φ^2 in $W^{1,p}(\widetilde{\Omega}_\varepsilon^{*})$, such that*

$$\begin{aligned}
&(i) \quad \varphi = \varphi^1 + \varphi^2, \quad \text{a.e. in } \widetilde{\Omega}_\varepsilon^{*},\\
&(ii) \quad \|\varphi^1 - \mathcal{M}_{\widetilde{\Omega}_\varepsilon^{*}}(\varphi)\|_{W^{1,p}(\widetilde{\Omega}_\varepsilon)} \leqslant C_p\|\nabla\varphi\|_{L^p(\widetilde{\Omega}_\varepsilon^{**})},\\
&(iii) \quad \|\nabla\varphi^2\|_{L^p(\widetilde{\Omega}_\varepsilon^{*})} \leqslant C_p\|\nabla\varphi\|_{L^p(\widetilde{\Omega}_\varepsilon^{**})},\\
&(iv) \quad \|\varphi - \mathcal{M}_{\widetilde{\Omega}_\varepsilon^{*}}(\varphi)\|_{L^p(\widetilde{\Omega}_\varepsilon^{*})} \leqslant C_p\|\nabla\varphi\|_{L^p(\widetilde{\Omega}_\varepsilon^{**})},
\end{aligned} \tag{5.44}$$

with a constant C_p independent of ε.

Proof. This corresponds to the decomposition $w = \mathcal{Q}_\varepsilon^*(w) + \mathcal{R}_\varepsilon^*(w)$ for the case of Ω replaced by $\widetilde{\Omega}_\varepsilon^{\mathcal{Y}'}$. Inequality (5.44)(iii) corresponds to the estimates of Proposition 4.37 for the case Y a parallelotope, and Proposition 4.65 for the general case, taking into account the inclusions from (5.43). Similarly, Proposition 4.35 or Proposition 4.37 imply

$$\|\nabla\varphi^1\|_{L^p(\widetilde{\Omega}_\varepsilon)} \leqslant C_p\|\nabla\varphi\|_{L^p(\widetilde{\Omega}_\varepsilon^{**})}.$$

By Proposition 5.24, $\widetilde{\Omega}_\varepsilon$ satisfies the Poincaré-Wirtinger inequality with a constant independent of ε so that

$$\|\varphi^1 - \mathcal{M}_{\widetilde{\Omega}_\varepsilon}(\varphi^1)\|_{L^p(\widetilde{\Omega}_\varepsilon)} \leqslant C_p\|\nabla\varphi^1\|_{L^p(\widetilde{\Omega}_\varepsilon)} \leqslant C_p\|\nabla\varphi\|_{L^p(\widetilde{\Omega}_\varepsilon^{**})}.$$

Taking the average over $\widetilde{\Omega}_\varepsilon^{*}$ gives

$$\|\mathcal{M}_{\widetilde{\Omega}_\varepsilon^{*}}(\varphi^1) - \mathcal{M}_{\widetilde{\Omega}_\varepsilon}(\varphi^1)\|_{L^p(\widetilde{\Omega}_\varepsilon)} \leqslant C_p\|\nabla\varphi\|_{L^p(\widetilde{\Omega}_\varepsilon^{**})},$$

hence,

$$\|\varphi^1 - \mathcal{M}_{\widetilde{\Omega}_\varepsilon^{*}}(\varphi^1)\|_{L^p(\widetilde{\Omega}_\varepsilon)} \leqslant C_p\|\nabla\varphi\|_{L^p(\widetilde{\Omega}_\varepsilon^{**})}.$$

Also by (5.44)(iii),

$$|\mathcal{M}_{\widetilde{\Omega}_\varepsilon^{*}}(\varphi^2)| \leqslant \varepsilon C_p\|\nabla\varphi\|_{L^p(\widetilde{\Omega}_\varepsilon^{**})},$$

from which (5.44)(ii) follows. Finally, (5.44)(iv) is an immediate consequence of estimates (5.44)(ii)-(iii). □

Definition 5.26. *The connected component of the set Ω_ε^* containing $\widetilde{\Omega}_\varepsilon^{**}$ is denoted $\mathcal{C}_\varepsilon^*$.*

Remark 5.27. *Since $\widetilde{\Omega}_\varepsilon^{**}$ contains $\widetilde{\Omega}_\varepsilon^*$, all the other connected components of Ω_ε^*, lie near the boundary of Ω. Therefore, in some sense, $\mathcal{C}_\varepsilon^*$ is the "main" connected component of Ω_ε^*. Furthermore, for w in $W^{1,p}_{loc}(\mathcal{C}_\varepsilon^*)$ with ∇w in $L^p(\mathcal{C}_\varepsilon^*)$, its restriction $w\,\mathbf{1}_{\widetilde{\Omega}_\varepsilon^*}$ belongs to $L^p(\widetilde{\Omega}_\varepsilon^*)$ (and not only to $L^p_{loc}(\widetilde{\Omega}_\varepsilon^*)$).*

As a consequence, we obtain the convergence for unfoldings which is adapted to the case of the Fredholm alternative.

Theorem 5.28. *Suppose $2\varepsilon\,\text{diameter}(\mathcal{Y}') \in (0, \delta_0)$. Let p be in $(1, +\infty]$ and $\{w_\varepsilon\}_\varepsilon$ a sequence in $W^{1,p}_{loc}(\Omega_\varepsilon^*)$ satisfying*

$$\|\nabla w_\varepsilon\|_{L^p(\mathcal{C}_\varepsilon^*)} \leqslant C,$$

and such that $\mathcal{M}_{\widetilde{\Omega}_\varepsilon^*}(w_\varepsilon) = 0$. Then, up to a subsequence, there are two functions w in $W^{1,p}(\Omega)$ and $\widehat{w}$ in $L^p(\Omega; W^{1,p}_{per,0}(Y^*))$, such that

$$\begin{aligned}
&\mathcal{T}_\varepsilon^*(w_\varepsilon \mathbf{1}_{\widetilde{\Omega}_\varepsilon^*}) \to w \quad \text{strongly in } L^p(\Omega; W^{1,p}(Y^*)),\\
&\mathcal{T}_\varepsilon^*(\nabla w_\varepsilon\, \mathbf{1}_{\mathcal{C}_\varepsilon^*}) \rightharpoonup \nabla w + \nabla_y \widehat{w} \quad \text{weakly in } L^p(\Omega \times Y^*)^N.
\end{aligned} \tag{5.45}$$

Proof. Note first that according to Theorem 5.25, $w_\varepsilon \mathbf{1}_{\widetilde{\Omega}_\varepsilon^*}$ belongs to $L^p(\widetilde{\Omega}_\varepsilon^*)$ so that the condition $\mathcal{M}_{\widetilde{\Omega}_\varepsilon^*}(w_\varepsilon) = 0$ makes sense. Then, using Theorem 5.25, we deduce that $\|w_\varepsilon\|_{W^{1,p}(\widetilde{\Omega}_\varepsilon^*)}$ is bounded (since $\widetilde{\Omega}_\varepsilon^{**}$ is included in $\mathcal{C}_\varepsilon^*$). It follows that, up to a subsequence, there exist W in $L^p(\Omega \times Y^*)$ and F in $L^p(\Omega \times Y^*)^N$ such that

$$\begin{aligned}
&\mathcal{T}_\varepsilon^*(w_\varepsilon \mathbf{1}_{\widetilde{\Omega}_\varepsilon^*}) \rightharpoonup W \quad \text{weakly in } L^p(\Omega \times Y^*),\\
&\mathcal{T}_\varepsilon^*(\nabla w_\varepsilon\, \mathbf{1}_{\mathcal{C}_\varepsilon^*}) \rightharpoonup F \quad \text{weakly in } L^p(\Omega \times Y^*)^N.
\end{aligned}$$

But, Theorem 4.43 applied in every relatively compact open subset ω of Ω implies that, again up to a subsequence, there exist w in $W^{1,p}_{loc}(\Omega)$ and $\widehat{w}$ in $L^p_{loc}(\Omega; W^{1,p}_{per,0}(Y^*))$, with

$$\begin{aligned}
&\mathcal{T}_\varepsilon^*(w_\varepsilon \mathbf{1}_{\widetilde{\Omega}_\varepsilon^*}) \rightharpoonup w \quad \text{weakly in } L^p(\omega; W^{1,p}(Y^*)),\\
&\mathcal{T}_\varepsilon^*(\nabla w_\varepsilon\, \mathbf{1}_{\mathcal{C}_\varepsilon^*}) \rightharpoonup \nabla w + \nabla_y \widehat{w} \quad \text{weakly in } L^p(\omega; L^p(Y^*))^N,
\end{aligned}$$

for all ω. These convergences give

$$w(x) = W(x, \cdot) \quad \text{and} \quad \nabla w = \mathcal{M}_{Y^*}(F) \ \text{ in } L^p(\Omega).$$

Consequently, w belongs to $W^{1,p}(\Omega)$ and $\widehat{w}$ to $L^p(\Omega; W^{1,p}_{per,0}(Y^*))$.

To obtain the strong convergence in the first line of (5.45), use the decomposition of w_ε from Theorem 5.25

$$w_\varepsilon = w_\varepsilon^1 + w_\varepsilon^2.$$

The same theorem implies first that $\|w_\varepsilon^2\|_{L^p(\widetilde{\Omega}_\varepsilon^*)} \to 0$, and then the estimate

$$\|w_\varepsilon^1\|_{W^{1,p}(\widetilde{\Omega}_\varepsilon)} \leqslant C_p \|\nabla w_\varepsilon\|_{L^p(\widetilde{\Omega}_\varepsilon^{**})},$$

where C_p is independent of ε.

Using now the extension operator $\mathcal{P}_\delta$ with $\delta = 2\varepsilon\,\text{diameter}(\mathcal{Y}') \in (0, \delta_0)$, as in the proof of Proposition 5.24, one concludes that $w_\varepsilon^1 \mathbf{1}_{\widetilde{\Omega}_\varepsilon}$ is in a compact subset of $L^p(\Omega)$. Therefore $\mathcal{T}_\varepsilon(w_\varepsilon^1 \mathbf{1}_{\widetilde{\Omega}_\varepsilon})$ is in a compact subset of $L^p(\Omega \times Y)$, so that $\mathcal{T}_\varepsilon^*(w_\varepsilon^1 \mathbf{1}_{\widetilde{\Omega}_\varepsilon^*})$ is in a compact subset of $L^p(\Omega \times Y^*)$. The proof of (5.45) follows since $\mathcal{T}_\varepsilon(w_\varepsilon^2 \mathbf{1}_{\widetilde{\Omega}_\varepsilon})$ converges strongly to 0 in the same space. □

5.4.3 A Neumann problem with Fredholm alternative

In this section, we would like to consider the homogenization of the Neumann problem whose variational formulation is

$$\begin{cases} \text{Find } u_\varepsilon \text{ in } V(\Omega_\varepsilon^*) \text{ such that} \\ \displaystyle\int_{\Omega_\varepsilon^*} A^\varepsilon \nabla u_\varepsilon \, \nabla v \, dx = \int_{\Omega_\varepsilon^*} f_\varepsilon v \, dx, \quad \forall v \in V(\Omega_\varepsilon^*), \end{cases} \tag{5.46}$$

where f_ε is given in $L^2(\Omega_\varepsilon^*)$ and the matrix field $A^\varepsilon = (a_{ij}^\varepsilon)_{1 \leqslant i,j \leqslant N}$ belongs to $M(\alpha, \beta, \Omega_\varepsilon^*)$ (see Definition 3.1).

As explained in the introduction of Section 5.4, this problem is not well-posed in general. Following Remark 5.27, the "natural" candidate for a domain on which to consider the approximate problem for (5.46) is the $\mathcal{C}_\varepsilon^*$. Assuming that the right-hand side f_ε belongs to $L^2(\mathcal{C}_\varepsilon^*)$ and satisfies the extra hypothesis

$$f_\varepsilon \textbf{ vanishes outside } \widetilde{\Omega}_\varepsilon^*, \tag{5.47}$$

there is a unique solution to the approximate problem in the space

$$W(\mathcal{C}_\varepsilon^*) \doteq \{\varphi \in H^1_{loc}(\mathcal{C}_\varepsilon^*) \mid \nabla\varphi \in L^2(\mathcal{C}_\varepsilon^*) \text{ and } \mathcal{M}_{\widetilde{\Omega}_\varepsilon^*}(\varphi) = 0\}. \tag{5.48}$$

Indeed, due to (iv) of Theorem 5.25, there exists a constant C (independent of ε) such that for all φ in $W(\mathcal{C}_\varepsilon^*)$,

$$\|\varphi\|_{L^2(\widetilde{\Omega}_\varepsilon^*)} \leqslant C \|\nabla\varphi\|_{L^2(\mathcal{C}_\varepsilon^*)}, \tag{5.49}$$

so that $W(\mathcal{C}_\varepsilon^*)$ is a Hilbert space when endowed with the norm

$$\|\varphi\|_{W(\mathcal{C}_\varepsilon^*)} \doteq \|\nabla\varphi\|_{L^2(\mathcal{C}_\varepsilon^*)}.$$

By Theorem 5.28, uniform estimates can now be obtained.

Theorem 5.29 (Existence and priori estimates for a proper approximation). *Suppose that Hypotheses* $\mathbf{H_2}$ *and* (5.47) *are satisfied.*

Then, there is a unique function u_ε in $W(\mathcal{C}^*_\varepsilon)$, solution of

$$\int_{\mathcal{C}^*_\varepsilon} A^\varepsilon(x)\nabla u_\varepsilon(x)\nabla\varphi(x)\,dx, = \int_{\mathcal{C}^*_\varepsilon} f_\varepsilon(x)\varphi(x)\,dx, \quad \forall\varphi\in W(\mathcal{C}^*_\varepsilon). \tag{5.50}$$

It satisfies the following estimates with a constant C independent of ε:

$$\begin{aligned} &\|\nabla u_\varepsilon\|_{L^2(\mathcal{C}^*_\varepsilon)} \leqslant C\|f_\varepsilon\|_{L^2(\Omega^*_\varepsilon)},\\ &\|u_\varepsilon\|_{L^2(\widetilde{\Omega}^*_\varepsilon)} \leqslant C\|f_\varepsilon\|_{L^2(\Omega^*_\varepsilon)}. \end{aligned} \tag{5.51}$$

Proof. Existence and uniqueness of the solution u_ε follow immediately from the Lax-Milgram theorem but this does not give a uniform estimate, unless C_ε in (5.49) is bounded.

To obtain a uniform estimate, use condition (5.47) to obtain the inequality

$$\alpha\|\nabla u_\varepsilon\|^2_{L^2(\mathcal{C}^*_\varepsilon)} \leqslant \|f_\varepsilon\|_{L^2(\widetilde{\Omega}^*_\varepsilon)}\|u_\varepsilon\|_{L^2(\widetilde{\Omega}^*_\varepsilon)}.$$

Since $\mathcal{M}_{\widetilde{\Omega}^*_\varepsilon}(u_\varepsilon)=0$, by Theorem 5.25,

$$\|u_\varepsilon\|_{L^2(\widetilde{\Omega}^*_\varepsilon)} \leqslant C\|\nabla u_\varepsilon\|_{L^2(\mathcal{C}^*_\varepsilon)},$$

from which (5.51) are derived. □

5.4.4 Homogenization

Theorem 5.30. *Suppose that Hypotheses $\mathbf{H_2}$ and (5.47) hold. Assume that the sequence $\{f_\varepsilon\}_\varepsilon$ satisfies*

$$\mathcal{E}^0_\varepsilon(f_\varepsilon)(=\widetilde{f}_\varepsilon) \rightharpoonup \frac{|Y^*|}{|Y|}f \quad \text{weakly in } L^2(\Omega)^{(7)}.$$

Assume furthermore, that for some matrix B,

$$\mathcal{T}^*_\varepsilon(A^\varepsilon)\to B \quad \text{a.e. in } \Omega\times Y^{*(8)}.$$

Let u_ε be the solution of Problem (5.50). Then, there exist u in $V(\Omega)$ and $\widehat{u}$ in $L^2(\Omega;H^1_{per,0}(Y^*))$, satisfying

$$\begin{aligned} &(i)\quad \mathcal{T}^*_\varepsilon(u_\varepsilon\mathbf{1}_{\widetilde{\Omega}^*_\varepsilon})\to u \quad \text{strongly in } L^2(\Omega;H^1(Y^*)),\\ &(ii)\quad \mathcal{T}^*_\varepsilon(\nabla u_\varepsilon\mathbf{1}_{\mathcal{C}^*_\varepsilon})\rightharpoonup \nabla u+\nabla_y\widehat{u} \quad \text{weakly in } L^2(\Omega\times Y^*)^N, \end{aligned} \tag{5.52}$$

and the pair $(u,\widehat{u})$ is the unique solution in of the problem

$$\begin{cases} \dfrac{1}{|Y|}\displaystyle\int_{\Omega\times Y^*} B(x,y)\big[\nabla u(x)+\nabla_y\widehat{u}(x,y)\big]\big[\nabla\Psi(x)+\nabla_y\Phi(x,y)\big]\,dxdy\\ \qquad\qquad = \dfrac{|Y^*|}{|Y|}\displaystyle\int_\Omega f(x)\,\Psi(x)\,dx,\\ \mathcal{M}_\Omega(u)=0,\\ \forall\Psi\in V(\Omega),\ \forall\Phi\in L^2(\Omega;H^1_{per}(Y^*)). \end{cases} \tag{5.53}$$

(7) See Proposition 4.25 for an equivalent statement involving $\mathcal{E}_\varepsilon(f_\varepsilon)$.

(8) or more generally, in measure in $\Omega\times Y^*$.

Proof. First, in view of Lemma 5.4 and the ellipticity of the matrix field B (arising from A^ε in $M(\alpha,\beta,\Omega)$), problem (5.53) has a solution which is unique by application of the Lax-Milgram theorem in $V(\Omega)\times L^2(\Omega;H^1_{per,0}(Y^*))$. Using estimates (5.51), convergences (5.52) follow from Theorem 5.28. This is up to a subsequence, but as usual, the uniqueness of the solution of the limit problem implies that the whole sequence converges.

At this point, the proof is somewhat standard when using unfolding.

First, unfold the left-hand side of (5.46) using as test function

$$v_\varepsilon(x)\doteq \Psi(x)+\varepsilon\,\varphi(x)\,\psi\Big(\frac{x}{\varepsilon}\Big),$$

where Ψ is in $H^1(\Omega)$, φ in $\mathcal{D}(\Omega)$ and $\psi=\psi(y)$ in $H^1_{per}(Y^*)$. Then, by Proposition 4.8(i) that

$$\begin{aligned}&\mathcal{T}^*_\varepsilon(v_\varepsilon)\to\Psi \quad \text{strongly in } L^2(\Omega\times Y^*),\\ &\mathcal{T}^*_\varepsilon\Big(\varphi\psi\Big(\frac{\cdot}{\varepsilon}\Big)\Big)\to\Phi \quad \text{strongly in } L^2(\Omega\times Y^*)\quad \text{with } \Phi(x,y)=\varphi(x)\,\psi(y).\end{aligned}$$

Since

$$\nabla v_\varepsilon(x)=\nabla\Psi(x)+\varphi(x)(\nabla_y\psi)\Big(\frac{x}{\varepsilon}\Big)+\varepsilon\nabla\varphi(x)\psi\Big(\frac{x}{\varepsilon}\Big),$$

by Proposition 4.8 (i),

$$\mathcal{T}^*_\varepsilon(\nabla v_\varepsilon)\to\nabla\Psi+\nabla_y\Phi \quad \text{strongly in } L^2(\Omega\times Y^*)^N.$$

Consequently,

$$\begin{aligned}\int_{C^*_\varepsilon}A^\varepsilon\nabla u_\varepsilon\,\nabla v_\varepsilon\,dx&=\frac{1}{|Y|}\int_{\Omega\times Y*}\mathcal{T}^*_\varepsilon(A^\varepsilon)\mathcal{T}^*_\varepsilon(\nabla u_\varepsilon\mathbf{1}_{C^*_\varepsilon})\,\mathcal{T}^*_\varepsilon(\nabla v_\varepsilon)\,dxdy\\ \to\frac{1}{|Y|}\int_{\Omega\times Y*}&B(x,y)\big[\nabla u_0(x)+\nabla_y\widehat{u}_0(x,y)\big]\big[\nabla\Psi(x)+\phi(x)\nabla\psi(y)\big]\,dx\,dy.\end{aligned}$$

Concerning the right-hand side of (5.46), as v_ε strongly converges to Ψ in $L^2(\Omega)$,

$$\lim_{\varepsilon\to 0}\int_{C^*_\varepsilon}f_\varepsilon\,v_\varepsilon\,dx=\frac{|Y^*|}{|Y|}\int_\Omega f\,\Psi\,dx.$$

Equation (5.53) is then obtained in view of the density of the tensor product $\mathcal{D}(\Omega)\otimes H^1_{per}(Y^*)$ in $L^2(\Omega;H^1_{per}(Y^*))$.

Finally, by convergence (5.52)(i),

$$\begin{aligned}\mathcal{M}_\Omega(u)=\mathcal{M}_{\Omega\times Y*}(u)&=\lim_{\varepsilon\to 0}\mathcal{M}_{\Omega\times Y*}(\mathcal{T}^*_\varepsilon(u_\varepsilon\mathbf{1}_{\widetilde{\Omega}^*_\varepsilon}))\\ &=\lim_{\varepsilon\to 0}\frac{|\widetilde{\Omega}^*_\varepsilon|}{|\Omega|}\mathcal{M}_{\widetilde{\Omega}^*_\varepsilon}(u_\varepsilon)=0,\end{aligned}$$

from which the uniqueness for the solution of (5.53) is straightforward. □

The homogenized limit problem can now be made explicit.

Theorem 5.31. *Under the hypotheses of Theorem 5.30, the homogenized formulation associated with Theorem 5.30 is*

$$\begin{cases} -\text{div}\,(A^{hom}\nabla u) = \dfrac{|Y^*|}{|Y|} f & \text{in } \Omega, \\ \mathcal{M}_\Omega(u) = 0. \end{cases}$$

The homogenized matrix $A^{hom} = (a_{ij}^{hom})_{1\leqslant i,j\leqslant N}$ is the same as the one given in Theorem 5.6.

Proof. Taking $\Psi = 0$ in Problem (5.53), yields

$$\widehat{u}(x,y) = \sum_{i=1}^{N} \frac{\partial u}{\partial x_i}(x)\widehat{\chi}_i^*(x,y),$$

where the $\widehat{\chi}_i^*$'s are the solutions of problems (5.16). Inserting into (5.53) with $\Phi = 0$ gives the result. □

Under no additional hypothesis, one has the strong convergence of the energy.

Proposition 5.32. *Under the hypotheses of Theorem 5.30, the following convergence holds:*

$$\lim_{\varepsilon\to 0} \int_{C_\varepsilon^*} A^\varepsilon \nabla u_\varepsilon \nabla u_\varepsilon \, dx = \frac{1}{|Y|}\int_{\Omega\times Y^*} B\big[\nabla u + \nabla_y \widehat{u}\big]\big[\nabla u + \nabla_y \widehat{u}\big]\, dxdy, \tag{5.54}$$

so that,

$$\mathcal{T}_\varepsilon^*(\nabla u_\varepsilon \mathbf{1}_{C_\varepsilon^*}) \to \nabla u + \nabla_y \widehat{u} \quad \textit{strongly in } L^2(\Omega\times Y^*)^N. \tag{5.55}$$

Proof. Then, by (5.45)(i) applied to u_ε, $\mathcal{T}_\varepsilon^*(u_\varepsilon \mathbf{1}_{\widetilde{\Omega}_\varepsilon^*})$ converges strongly to u in $L^2(\Omega\times Y^*)$. Due to Remark 4.9, for every weak limit point F of $\mathcal{T}_\varepsilon^*(f_\varepsilon)$,

$$\mathcal{M}_{Y^*}(F) = \frac{|Y^*|}{|Y|} f.$$

By Proposition 4.4 (i), as $\Lambda_\varepsilon \cap \widetilde{\Omega}_\varepsilon^* = \varnothing$, it follows that

$$\int_{\widetilde{\Omega}_\varepsilon^*} f_\varepsilon \, u_\varepsilon \, dx = \frac{1}{|Y|}\int_{\Omega\times Y^*} \mathcal{T}_\varepsilon^*(f_\varepsilon)\mathcal{T}_\varepsilon^*(u_\varepsilon \mathbf{1}_{\widetilde{\Omega}_\varepsilon^*})\, dxdy.$$

Consequently,

$$\lim_{\varepsilon\to 0}\int_{\widetilde{\Omega}_\varepsilon^*} f_\varepsilon \, u_\varepsilon \, dx = \frac{1}{|Y|}\int_{\Omega\times Y^*} F(x,y)u(x)\, dxdy = \frac{|Y^*|}{|Y|}\int_\Omega f\, u\, dx.$$

The proof of (5.54) is now complete, since by (5.50),

$$\int_{C_\varepsilon^*} A^\varepsilon \nabla u_\varepsilon \nabla u_\varepsilon \, dx = \int_{\tilde{\Omega}_\varepsilon^*} f_\varepsilon \, u_\varepsilon \, dx,$$

and by (5.53),

$$\frac{1}{|Y|}\int_{\Omega\times Y*} B\big[\nabla u + \nabla_y \widehat{u}\big]\,\big[\nabla u + \nabla_y \widehat{u}\big]\, dxdy = \frac{|Y^*|}{|Y|}\int_\Omega f\, u\, dx.$$

Convergence (5.55) is now obtained as in the proof of Proposition 3.9. □

Corollary 5.33 (Corrector results). *As* $\varepsilon \to 0$,

$$\Big\| \nabla u_\varepsilon - \nabla u - \sum_{i=1}^N \mathcal{M}_\varepsilon\Big(\frac{\partial u}{\partial x_i}\Big)\,\mathcal{U}_\varepsilon^*(\nabla_y \widehat{\chi}_i^*)\Big\|_{L^2(\tilde{\Omega}_\varepsilon^*)} \to 0.$$

In the case where the matrix field A does not depend on x, the following corrector result holds:

$$\Big\| u_\varepsilon - u - \varepsilon\sum_{i=1}^N \mathcal{Q}_\varepsilon\Big(\frac{\partial u}{\partial x_i}\Big)\,\widehat{\chi}_i^*\Big(\frac{\cdot}{\varepsilon}\Big)\Big\|_{H^1(\tilde{\Omega}_\varepsilon^*)} \to 0.$$

The proof is similar to those of Theorems 5.10 and 5.11. □

5.5 How to treat perforated solid structures

In this section, for simplicity one assumes $Y = (0,1)^N$. *Recall also that* $\{\mathbf{b}_1, \ldots, \mathbf{b}_N\}$ *is the usual orthonormal basis of* $\mathbb{R}^N$.

In many physical problems with holes, one cannot consider perforated open sets containing only parts of perforated cells near the boundary (e.g. in elasticity if we consider periodic truss-like structures made of thin beams or plates). In these situations, one is required to consider domains containing only whole perforated cells up to its boundary.

When one starts with a bounded domain Ω, the approximate problems have to be posed in a union of perforated cells and a first choice is given by $\widehat{\Omega}_\varepsilon^*$. But it may very well happen that this open set is not connected for every values of ε even when Ω is (see Figure 5.3). The way this difficulty is handled in the previous section (see Definition 5.26) does not necessarily yield an exact union of perforated cells. If connectedness of the approximate domain and the fact that these are unions of ε-cells are both essential, an alternative is to use the set $\widehat{\Omega}_\varepsilon^{*ext}$ defined as follows (see Figure 5.4):

$$\begin{aligned}
\Xi_\varepsilon^{ext} &= \{\xi \in \mathbf{G} \mid \varepsilon(\xi + Y) \cap \Omega \neq \varnothing\},\\
\widehat{\Omega}_\varepsilon^{ext} &= \text{interior}\Big(\bigcup_{\xi\in\Xi_\varepsilon^{ext}} \varepsilon(\xi + \overline{Y})\Big),\\
\widehat{\Omega}_\varepsilon^{*ext} &= \text{interior}\Big(\bigcup_{\xi\in\Xi_\varepsilon^{ext}} \varepsilon(\xi + \overline{Y}^*)\Big).
\end{aligned}$$

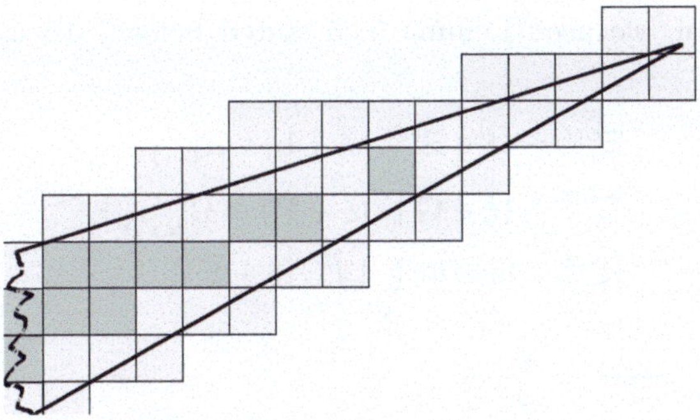

Figure 5.3: $\widehat{\Omega}_\varepsilon$ (in grey) is not an open connected set

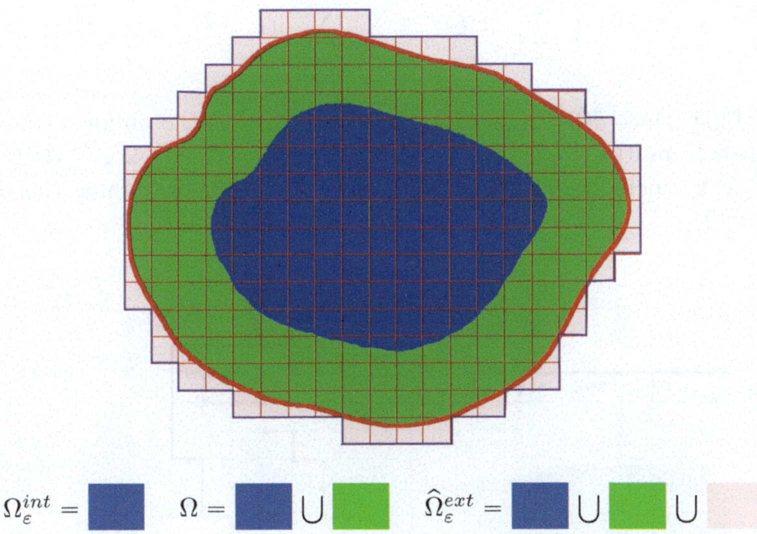

Figure 5.4: The open sets Ω_ε^{int}, Ω and $\widehat{\Omega}_\varepsilon^{ext}$.
The set $\widehat{\Omega}_\varepsilon^{*ext}$ is obtained from the latter by perforating every ε-cell.

It is easy to see that $\widehat{\Omega}_\varepsilon^{*ext}$ is connected whenever Ω is. Furthermore, Poincaré-Wirtinger and Korn inequalities hold with constants independent of ε (for ε small) as is shown in the next proposition.

Proposition 5.34 (Poincaré-Wirtinger inequality). *Assume Ω is a bounded connected domain with Lipschitz boundary. Then the open set $\widehat{\Omega}_\varepsilon^{*ext}$ satisfies a uniform Poincaré-Wirtinger inequality for every $p \in [1, +\infty)$, i.e., there exists a constant C independent of ε (it only depends on p and $\partial\Omega$) such that*

$$\|\varphi - \mathcal{M}_{\widehat{\Omega}_\varepsilon^{*ext}}(\varphi)\|_{L^p(\widehat{\Omega}_\varepsilon^{*ext})} \leqslant C\|\nabla\varphi\|_{L^p(\widehat{\Omega}_\varepsilon^{*ext})}, \quad \forall\varphi \in W^{1,p}(\widehat{\Omega}_\varepsilon^{*ext}). \quad (5.56)$$

For its proof, we need Lemma 5.35 stated below. To use it, set ($i = 1, \ldots, N$)

$$\Xi_{\varepsilon,i}^{ext} \doteq \{\xi \in \Xi_\varepsilon^{ext} \mid \xi + \mathbf{b}_i \in \Xi_\varepsilon^{ext}\},$$
$$\Xi_\varepsilon^{int} \doteq \{\xi \in \mathbf{G} \mid \varepsilon(\xi + Y) \subset \Omega_{2\varepsilon\sqrt{N}}^{int}\},$$
$$\widehat{\Omega}_\varepsilon^{int} \doteq \text{interior}\Big(\bigcup_{\xi\in\Xi_\varepsilon^{int}} \varepsilon(\xi + \overline{Y})\Big),$$

and observe that $\Xi_\varepsilon^{ext} = \bigcup_{i=1}^N \Xi_{\varepsilon,i}^{ext}$.

Lemma 5.35. *Suppose $p \in [1, +\infty)$. There exists a constant C which only depends on p and $\partial\Omega$ such that for every function ℓ defined on Ξ_ε^{ext},*

$$\sum_{\xi\in\Xi_\varepsilon^{ext}} |\ell(\xi)|^p \leqslant C\Big(\sum_{\xi\in\Xi_\varepsilon^{int}} |\ell(\xi)|^p + \sum_{i=1}^N \sum_{\xi\in\Xi_{\varepsilon,i}^{ext}} |\ell(\xi + \mathbf{b}_i) - \ell(\xi)|^p\Big). \tag{5.57}$$

Proof. First, since $\partial\Omega$ is Lipschitz, the domain Ω has the uniform cone property (see Definition 5.21). Note that every ball of radius $4\varepsilon\sqrt{N}$ contains a cell $\varepsilon(\xi + Y)$ included in $\widehat{\Omega}_\varepsilon^{int}$ (for example, the ε-cell containing the center of this ball).

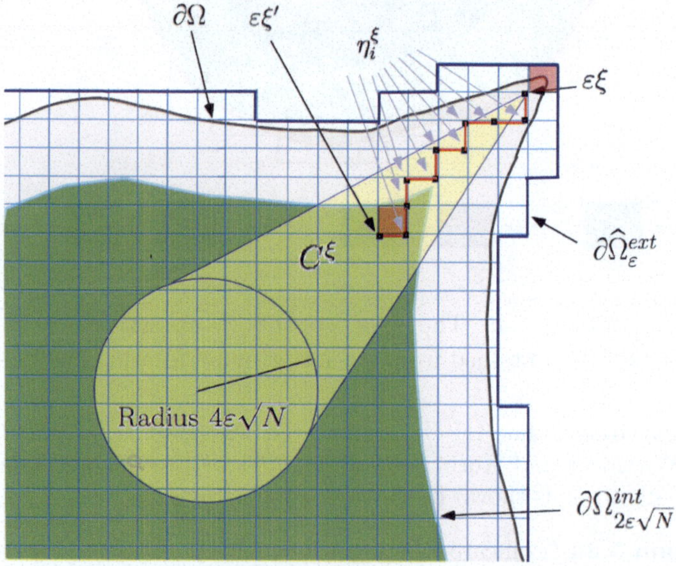

Figure 5.5: The chain from ξ to ξ'

Let ξ be in Ξ_ε^{ext}. There exists a cone of the family $\{\mathcal{C}_j\}$ (e.g., $\mathcal{C}_1$ see Definition 5.21) and a point $x_\xi \in \varepsilon(\xi + Y) \cap \Omega$ such that $x_\xi + \mathcal{C}_1 \subset \Omega$.

Consider the cone C^ξ obtained from the cone $\{x_\xi + \mathcal{C}_1\}$ (see Figure 5.5) by truncation so that it contains a tangent ball of radius $4\varepsilon\sqrt{N}$.
The domain $\{z \in \mathbb{R}^N \mid \text{dist}(z, C^\xi) < 2\varepsilon\sqrt{N}\}$ is covered by a finite number of ε-cells included in $\widehat{\Omega}_\varepsilon^{ext}$ and their number is bounded by some fixed integer κ independent of ε (depending only upon the geometry of the cone $\mathcal{C}_1$).

From such a covering, for every $\xi \in \Xi_\varepsilon^{ext}$ one constructs a finite chain $\{\eta^\xi{}_i\}_{i\in\{1,\dots k\}}$ of elements in Ξ_ε^{ext} connecting ξ to some $\xi' \in \Xi_\varepsilon^{int}$ as follows:

$$\xi' = \eta_1^\xi, \quad \xi = \eta_k^\xi, \quad \text{and} \quad \eta_{i+1}^\xi - \eta_i^\xi \in \{\pm\mathbf{b}_1, \dots, \pm\mathbf{b}_N\}, \quad i = 1, \dots, k-1.$$

Each cell $\varepsilon(\eta_i^\xi + Y)$ intersects the cone C^ξ. Since the cone C^ξ is covered by at most κ ε-cells, k is bounded by κ. Also observe that the distance (in $\mathbb{Z}^N$) between ξ and an element of the chain is less than $\kappa\sqrt{N}$.
By finite induction, one has

$$|\ell(\xi)|^p \leqslant 2^{\kappa(p-1)}\Big(|\ell(\xi')|^p + \sum_{i=1}^{k-1} |\ell(\eta_{i+1}^\xi) - \ell(\eta_i^\xi)|^p\Big). \tag{5.58}$$

Now consider the sum $\sum_{\xi\in\Xi_\varepsilon^{ext}\setminus\Xi_\varepsilon^{int}} |\ell(\xi)|^p$. Every ξ'' in Ξ_ε^{ext} appears in the right-hand side of a sum like (5.58) a finite number of times which is bounded by $(2\kappa\sqrt{N})^N$ (since the distance between ξ'' and the beginning of a chain $\xi$is less than $\kappa\sqrt{N}$).

As a consequence, one obtains

$$\sum_{\xi\in\Xi_\varepsilon^{ext}\setminus\Xi_\varepsilon^{int}} |\ell(\xi)|^p \leqslant C\Big(\sum_{\xi\in\Xi_\varepsilon^{int}} |\ell(\xi)|^p + \sum_{i=1}^{N}\sum_{\xi\in\Xi_{\varepsilon,i}^{ext}} |\ell(\xi+\mathbf{b}_i) - \ell(\xi)|^p\Big).$$

where C does not depend on ε. This proves (5.57). □

Proof of Proposition 5.34. For simplicity, one extends φ by 0 in $\mathbb{R}^N\setminus\overline{\widehat{\Omega}_\varepsilon^{*ext}}$. Set

$$\ell(\xi) = \widetilde{\mathcal{M}}_\varepsilon(\varphi)(\varepsilon\xi) - \mathcal{M}_{\Omega^{int}_{2\varepsilon\sqrt{N}}} \circ \mathcal{Q}_\varepsilon^*(\varphi),$$

and note that $\ell(\xi)$ is defined for every $\xi \in \Xi_\varepsilon^{ext}$. From Proposition 5.24, one has, for every φ in $W^{1,p}(\widehat{\Omega}_\varepsilon^{*ext})$ and for a constant depending on p and $\partial\Omega$,

$$\begin{aligned}\|\mathcal{Q}_\varepsilon^*(\varphi) - \mathcal{M}_{\Omega^{int}_{2\varepsilon\sqrt{N}}} \circ \mathcal{Q}_\varepsilon^*(\varphi)\|_{L^p(\Omega^{int}_{2\varepsilon\sqrt{N}})} &\leqslant C\|\nabla\mathcal{Q}_\varepsilon^*(\varphi)\|_{L^p(\Omega^{int}_{2\varepsilon\sqrt{N}})}\\ &\leqslant C\|\nabla\varphi\|_{L^p(\widehat{\Omega}_\varepsilon^{*ext})}.\end{aligned}$$

This implies

$$\sum_{\xi\in\Xi_\varepsilon^{int}} |\ell(\xi)|^p\varepsilon^N \leqslant C\|\mathcal{Q}_\varepsilon^*(\varphi) - \mathcal{M}_{\Omega^{int}_{2\varepsilon\sqrt{N}}} \circ \mathcal{Q}_\varepsilon^*(\varphi)\|^p_{L^p(\Omega^{int}_{2\varepsilon\sqrt{N}})} \leqslant C\|\nabla\varphi\|^p_{L^p(\widehat{\Omega}_\varepsilon^{*ext})}.$$

Moreover, by (4.34)), one has

$$\sum_{i=1}^{N}\sum_{\xi\in\Xi_{\varepsilon,i}^{ext}}\left|\frac{\ell(\xi+\mathbf{b}_i)-\ell(\xi)}{\varepsilon}\right|^p\varepsilon^N\leqslant C\|\nabla\varphi\|^p_{L^p(\widehat{\Omega}_\varepsilon^{*ext})}.$$

By Lemma 5.35 one concludes that there is a constant C independent of ε such that

$$\sum_{\xi\in\Xi_\varepsilon^{ext}}|\ell(\xi)|^p\varepsilon^N\leqslant C\|\nabla\varphi\|^p_{L^p(\widehat{\Omega}_\varepsilon^{*ext})}.$$

Recall that

$$\sum_{\xi\in\Xi_\varepsilon^{ext}}\|\varphi-\widetilde{\mathcal{M}}_\varepsilon(\varphi)(\varepsilon\xi)\|^p_{L^p(\varepsilon(\xi+Y^*))}\leqslant C\|\nabla\varphi\|^p_{L^p(\widehat{\Omega}_\varepsilon^{*ext})}.$$

From the last two inequalities one obtains

$$\|\varphi-\mathcal{M}_{\Omega^{int}_{2\varepsilon\sqrt{N}}}\circ\mathcal{Q}^*_\varepsilon(\varphi)\|_{L^p(\widehat{\Omega}_\varepsilon^{*ext})}\leqslant C\|\nabla\varphi\|_{L^p(\widehat{\Omega}_\varepsilon^{*ext})}. \tag{5.59}$$

By the Hölder inequality, this implies

$$\left\|\mathcal{M}_{\widehat{\Omega}_\varepsilon^{*ext}}(\varphi)-\mathcal{M}_{\Omega^{int}_{2\varepsilon\sqrt{N}}}\circ\mathcal{Q}^*_\varepsilon(\varphi)\right\|_{L^p(\widehat{\Omega}_\varepsilon^{*ext})}\leqslant\|\varphi-\mathcal{M}_{\Omega^{int}_{2\varepsilon\sqrt{N}}}\circ\mathcal{Q}^*_\varepsilon(\varphi)\|_{L^p(\widehat{\Omega}_\varepsilon^{*ext})}.$$

This estimate together with (5.59) imply (5.56). □

In the case of functions vanishing (at the limit as $\varepsilon\to 0$) on a part Γ_0 of $\partial\Omega$, as in Subsection 4.3.2, assume that there exists an open set Ω' with a Lipschitz boundary such that $\Omega'\cap\partial\Omega=\Gamma_0$. Set

$$\begin{aligned}W^{1,p}_{\Gamma_0}(\widehat{\Omega}_\varepsilon^{*ext})=\{\psi\in W^{1,p}(\widehat{\Omega}_\varepsilon^{*ext})\mid\exists\psi'\in W^{1,p}(\widehat{\Omega'}_\varepsilon^{*ext}),\ \psi=\psi'_{|\widehat{\Omega}_\varepsilon^{*ext}},\\ \psi'=0\text{ in }\widehat{\Omega'}_\varepsilon^{*ext}\setminus\overline{\widehat{\Omega}_\varepsilon^{*ext}}\}.\end{aligned} \tag{5.60}$$

For these spaces, there is a uniform Poincaré inequality.

Proposition 5.36 (Poincaré inequality). *For every $p\in[1,+\infty)$, there exists a constant C independent of ε (it only depends on p and $\partial\Omega$) such that*

$$\|\varphi\|_{L^p(\widehat{\Omega}_\varepsilon^{*ext})}\leqslant C\|\nabla\varphi\|_{L^p(\widehat{\Omega}_\varepsilon^{*ext})},\quad\forall\varphi\in W^{1,p}_{\Gamma_0}(\widehat{\Omega}_\varepsilon^{*ext}).$$

Proof. First we use Proposition 5.34 in $\widehat{\Omega'}_\varepsilon^{*ext}$, then we estimate $\mathcal{M}_{\widehat{\Omega}_\varepsilon^{*ext}}(\varphi)$ in $\widehat{\Omega'}_\varepsilon^{*ext}\setminus\overline{\widehat{\Omega}_\varepsilon^{*ext}}$. □

The same proof can be carried out for the case of a uniform Korn inequality for the space $W^{1,p}(\widehat{\Omega}_\varepsilon^{*ext})^N$. Here, one uses the Korn inequalities (2.12) for Y^* and $Y_j^*=\text{interior}\big(\overline{Y^*}\cup(\mathbf{b}_j+\overline{Y^*})\big)$, $j=1,\dots,N$. One also takes into account the uniform Korn inequality for the domains $\Omega^{int}_{2\varepsilon\sqrt{N}}$ (due to their uniform Lipschitz boundary, see Lemma 5.22).

Proposition 5.37 (Korn inequality). *For every displacement u belonging to $W^{1,p}(\widehat{\Omega}_\varepsilon^{*ext})^N$ there exists a rigid displacement* $\mathbf{R}$ *such that*

$$\|u-\mathbf{R}\|_{W^{1,p}(\widehat{\Omega}_\varepsilon^{*ext})} \leqslant C\|e(u)\|_{L^p(\Omega_\varepsilon^*)}.$$

Furthermore, if $u \in W^{1,p}_{\Gamma_0}(\widehat{\Omega}_\varepsilon^{*ext})^N$ (see (5.60)), then one has

$$\|u\|_{W^{1,p}(\widehat{\Omega}_\varepsilon^{*ext})} \leqslant C\|e(u)\|_{L^p(\Omega_\varepsilon^*)}.$$

The constants do not depend on ε.

5.6 Cracks: from a linear operator to a Leray-Lions operator.

In the case of fissured domains, there is a type of boundary condition which is specific to cracks and makes the problem non-linear. The non-linearity arises from a constraint on the jumps of the solution across the cracks (the most common is that the jump be with a given sign, but this can be extended to an inhomogeneous one-sided or two sided inequality satisfied by the jump). We briefly present this model problem in the scalar case (where the crack has to be orientable). The jump condition is more natural in elasticity as it represents a non-penetration condition on the displacement near the crack and is independent of the choice of direction of the unit normal. We refer to [68] for the homogenization of a linearized elastic body with cracks and to [126] for the homogenization of a diffusion-deformation problem in a fissured domain. A first version of the scalar problem was presented in [87].

In the following theorem, the general hypotheses are the same as in the preceding sections. For simplicity, we consider the case of a single crack $\mathcal{S}$ and no holes in the unit cell Y (see Figure 4.9), but one can easily consider more general situations as well). See Section 4.5 for the notations.

The ε-problem is posed in the fissured domain Ω_ε^* where the small cracks are denoted $\mathcal{S}^\varepsilon$. Since, $\mathcal{S}$ is a null set, there is no need to distinguish between $\mathcal{T}_\varepsilon$ and $\mathcal{T}_\varepsilon^*$, except when applied to elements in $H^1(\Omega_\varepsilon^*)$.

Let $\mathbf{g}$ be given in $L^1(\mathcal{S})$ and set $g_\varepsilon(x) = \varepsilon\,\mathbf{g}(\{x/\varepsilon\}_Y)$. Introduce the closed convex subset W^ε of $H^1(\Omega_\varepsilon^*)$ defined by

$$W^\varepsilon \doteq \{w \in H^1(\Omega_\varepsilon^*) \mid w \equiv 0 \text{ a.e. on } \partial\Omega \quad \text{and } [w]_{\mathcal{S}^\varepsilon} \leqslant g_\varepsilon \text{ a.e. on } \mathcal{S}^\varepsilon\},$$

where $[\,\cdot\,]_{\mathcal{S}_\varepsilon}$ is the jump along the defined normal across the crack $\mathcal{S}^\varepsilon$. The function $\mathbf{g}$ is assumed to be non-negative so that W^ε is not empty (it then contains $H_0^1(\Omega)$).

Finally, let A^ε belong to $M(\alpha,\beta,\Omega)$ and f^ε in $L^2(\Omega)$. Then the problem is: *to find u_ε in W^ε such that*

$$\int_{\Omega_\varepsilon^*} A^\varepsilon \nabla u_\varepsilon\, \nabla(w-u_\varepsilon)\,dx \geqslant \int_\Omega f(w-u_\varepsilon)\,dx, \qquad \forall w \text{ in } W^\varepsilon. \tag{5.61}$$

This variational inequality has a unique solution from standard results (see for example, [141]).

Note that here the operator is linear, the non-linearity arises solely from the constraint (W^ε is not a subspace).

In the case where A^ε is a symmetric matrix field, the problem can be reformulated as the minimization over W^ε of the functional

$$\mathbf{E}_\varepsilon(w) \doteq \frac{1}{2}\int_{\Omega_\varepsilon^*} A^\varepsilon \nabla w\, \nabla w\, dx - \int_\Omega f w\, dx,$$

which is continuous, uniformly convex and coercive.

The homogenization result is stated in the next theorems.

Theorem 5.38 (Unfolded limit). *Let $\{A^\varepsilon\}_\varepsilon$ be in $M(\alpha,\beta,\Omega)$. Assume that*

$$\mathcal{T}_\varepsilon(A^\varepsilon) \to B \text{ a.e. (or in measure) in } \Omega \times Y$$

and that f^ε converges to f weakly in $L^2(\Omega)$. Then there exists u in $H_0^1(\Omega)$, such that

$$\begin{aligned} &u_\varepsilon \to u \quad \text{strongly in } L^2(\Omega),\\ &\mathcal{T}_\varepsilon^*(u_\varepsilon) \to u \quad \text{strongly in } L^2(\Omega; H^1(Y^*)). \end{aligned} \tag{5.62}$$

There exists also $\widehat{u}$ in $L^2(\Omega; H^1_{per}(Y^*))$, such that

$$\mathcal{T}_\varepsilon^*(\nabla u_\varepsilon) \rightharpoonup \nabla u + \nabla_y \widehat{u} \quad \text{weakly in } L^2(\Omega \times Y)^N.$$

The pair $(u,\widehat{u})$ in $H_0^1(\Omega) \times L^2(\Omega;\mathcal{W})$ is the unique solution of the unfolded limit problem,

$$\begin{cases} \dfrac{1}{|Y|}\displaystyle\int_{\Omega\times Y^*} B(\nabla u + \nabla_y \widehat{u})\,(\nabla w + \nabla_y \widehat{w} - \nabla u - \nabla_y \widehat{u})\,dxdy \\ \qquad\qquad \geqslant \displaystyle\int_\Omega f(w-u)\,dx, \\ \forall (w,\widehat{w}) \in H_0^1(\Omega) \times L^2(\Omega;\mathcal{W}), \end{cases} \tag{5.63}$$

where $\mathcal{W}$ is the closed convex set defined as

$$\mathcal{W} \doteq \{w \in H^1_{per}(Y^*) \mid [w]_{\mathcal{S}} \leqslant \mathbf{g} \text{ a.e. on } \mathcal{S}, \text{ and } \mathcal{M}_Y(w) = 0\},$$

and

$$L^2(\Omega;\mathcal{W}) \doteq \{v \in L^2(\Omega; H^1_{per}(Y^*)) \mid v(x,\cdot) \in \mathcal{W} \text{ for a.e. } x \text{ in } \Omega\}.$$

Brief idea of the proof. The zero function is in $L^2(\Omega;\mathcal{W})$. Selecting it as test function in (5.61) implies that $\|u_\varepsilon\|_{H_0^1(\Omega_\varepsilon^*)}$ is bounded with respect to ε so that, up to a subsequence, convergence $(5.62)_1$ holds. Also, $\mathcal{T}_\varepsilon^*(u_\varepsilon)$ is bounded in $L^2(\Omega; H^1(Y^*))$ and $\mathcal{T}_\varepsilon(\nabla u_\varepsilon)$ is bounded in $L^2(\Omega \times Y^*)^N$. Up to a subsequence, $\mathcal{T}_\varepsilon^*(u_\varepsilon)$ converges weakly to some u in $L^2(\Omega; H^1(Y^*))$, and by $(5.62)_1$ and the fact that $\|u_\varepsilon\|_{H_0^1(\Omega_\varepsilon^*)}$ is bounded with respect to ε, the convergence is strong, which is $(5.62)_2$. Similarly, $\mathcal{T}_\varepsilon^*(\nabla u_\varepsilon)$ converges to $\nabla u + \nabla_y \widehat{u}$ for some $\widehat{u}$ in $L^2(\Omega; H^1_{per}(Y^*))$ satisfying $\mathcal{M}_Y(\widehat{u}) \equiv 0$ a.e. in Ω.

Since by Definition 4.11, $\mathcal{M}_\varepsilon^*(\phi) = \mathcal{M}_{Y*} \circ \mathcal{T}_\varepsilon^*(\phi)$ for ϕ in $L^p(\Omega_\varepsilon^*)$, Proposition 4.45(ii) gives the convergence

$$\frac{1}{\varepsilon}\Big(\mathcal{T}_\varepsilon^*(u_\varepsilon) - \mathcal{M}_{Y*} \circ \mathcal{T}_\varepsilon^*(u_\varepsilon)\Big) - y^{c*} \cdot \nabla u \rightharpoonup \widehat{u} \text{ weakly in } L^p(\Omega; W^{1,p}(Y^*)).$$

By the definition of W^ε, the jump across $\mathcal{S}$ of the left-hand side is bounded above by $\mathbf{g}$, and therefore, so is the jump of $\widehat{u}$ across $\mathcal{S}$. Hence $\widehat{u}$ belongs to $L^2(\Omega; \mathcal{W})$.

Let $w \in H_0^1(\Omega)$ and $\widehat{w} \in \mathcal{D}(\Omega; H^1_{per}(Y^*)) \cap L^2(\Omega; \mathcal{W})$ (the latter makes sense since $\mathbf{g} \geqslant \mathbf{0}$). Take the function w_ε defined by

$$w_\varepsilon(x) \doteq w(x) + \widehat{w}\Big(x, \Big\{\frac{x}{\varepsilon}\Big\}_Y\Big)$$

as test function in (5.61). Unfolding and passing to the limit gives (5.63). Here, one can use Lemma 3.11 to pass to the limit in the quadratic term in $\mathcal{T}_\varepsilon^*(\nabla u_\varepsilon)$. Finally, the density of $\mathcal{D}(\Omega; H^1_{per}(Y^*)) \cap L^2(\Omega; \mathcal{W})$ in $L^2(\Omega; \mathcal{W})$ (the standard procedure of truncation and regularization in Ω works in this case because $s\mathcal{W} \subset \mathcal{W}$ for $s \in [0, 1]$) completes the proof of (5.63).

Uniqueness of the solution of (5.63) implies the convergence of the full sequence. While existence of a solution for the unfolded problem is a consequence of the above convergence, it also follows from standard results for variational inequalities on the product space $H_0^1(\Omega) \times H^1_{per}(Y^*)$. □

In order to describe the homogenized problem satisfied by u, we introduce the cell-problem and the associated corrector corresponding to (5.63). Since the problem is nonlinear, one has to define the corrector for every vector ζ in $\mathbb{R}^N$.

But first, we write an equivalent formulation for (5.63) by replacing the test function w by $u + w$:

$$\begin{cases} \text{Find } (u, \widehat{u}) \in H_0^1(\Omega) \times L^2(\Omega; \mathcal{W}) \text{ such that} \\ \dfrac{1}{|Y|} \displaystyle\int_{\Omega \times Y*} B(\nabla u + \nabla_y \widehat{u})\,(\nabla w + \nabla_y \widehat{w} - \nabla_y \widehat{u})\,dxdy \geqslant \int_\Omega f\,w\,dx, \\ \forall (w, \widehat{w}) \in H_0^1(\Omega) \times L^2(\Omega; \mathcal{W}). \end{cases}$$

Choosing $\widehat{w} = \widehat{u}$ and then replacing w by $-w$, implies

$$\frac{1}{|Y|} \int_{\Omega \times Y*} B(\nabla u + \nabla_y \widehat{u})\,\nabla w\,dxdy = \int_\Omega f\,w\,dx, \quad \forall w \in H_0^1(\Omega). \tag{5.64}$$

Clearly, now, (5.63) is equivalent to (5.64) together with

$$\frac{1}{|Y|} \int_{\Omega \times Y*} B(\nabla u + \nabla_y \widehat{u})\,(\nabla_y \widehat{w} - \nabla_y \widehat{u})\,dxdy \geqslant 0, \quad \forall \widehat{w} \in L^2(\Omega; \mathcal{W}). \tag{5.65}$$

Definition 5.39 (The cell problems and the homogenized operator). *For ζ in $\mathbb{R}^N$ and a.e. $x \in \Omega$, let χ_ζ be the solution of the following problem:*

$$\begin{cases} \text{Find } \chi_\zeta(x,\cdot) \text{ in } \mathcal{W} \text{ such that} \\ \dfrac{1}{|Y|}\displaystyle\int_{Y*} B(\zeta + \nabla_y \chi_\zeta)(\nabla_y z - \nabla_y \chi_\zeta)\, dy \geqslant 0, \quad \forall z \in \mathcal{W}. \end{cases}$$

This is a standard variational inequality on $H^1_{per}(Y^*)$. Its unique solution for each ζ, actually belongs to $L^\infty(\Omega; H^1_{per}(Y^*))$ because

$$\|\nabla_y \chi_\zeta\|_{L^2(Y*)} \leqslant \frac{\beta}{\alpha}|\zeta||Y|^{1/2}.$$

The solution $\hat{u}$ can now be expressed from (5.65) as

$$\hat{u}(x,y) \equiv \chi_{\nabla u(x)}(x,y).$$

Going back to (5.64) one is lead to set

$$A^{hom}(x,\zeta) \doteq \frac{1}{|Y|}\int_{Y*} B(x,y)\big(\zeta + \nabla_y \chi_\zeta(x,y)\big)\, dy.$$

Then the homogenized problem satisfied by u can be written in the form

$$\int_\Omega A^{hom}(x, \nabla u(x))\, \nabla w(x)\, dx = \int_\Omega f\, w\, dx, \quad \forall w \in H^1_0(\Omega), \tag{5.66}$$

or equivalently, as

$$-\text{div}\, A^{hom}(x, \nabla u(x)) = f \quad \text{ in } \Omega.$$

Proposition 5.40. *The map A^{hom} is of Caratheodory type and strongly monotone. Hence problem* (5.66) *is of Leray-Lions type.*

Proof. It is easy to prove that the map $\zeta \mapsto \chi_\zeta(x)$ is continuous from $\mathbb{R}^N$ to $\mathcal{W}$ for a.e. $x \in \Omega$. Furthermore, using Lusin's criterion, one shows that the map A^{hom} is of Caratheodory type.

One also has

$$|A^{hom}(x,\zeta)| \leqslant \Big(\alpha + \frac{\beta}{\alpha}\Big)|\zeta| \quad \text{for a.e. } x \in \Omega,$$

and moreover, for ζ and η in $\mathbb{R}^N$,

$$\begin{aligned} &(A^{hom}(\cdot,\zeta) - A^{hom}(\cdot,\eta)) \cdot (\zeta - \eta) \\ &= \frac{1}{|Y|}\int_{Y*} \Big(B(\zeta + \nabla_y \chi_\zeta) - B(\eta + \nabla_y \chi_\eta)\Big) \cdot (\zeta - \eta)\, dy \\ &= \frac{1}{|Y|}\int_{Y*} \Big(B(\zeta + \nabla_y \chi_\zeta) - B(\eta + \nabla_y \chi_\eta)\Big) \cdot (\zeta + \nabla_y \chi_\zeta - \eta - \nabla_y \chi_\eta)\, dy \\ &\quad + \frac{1}{|Y|}\int_{Y*} \Big(B(\zeta + \nabla_y \chi_\zeta) - B(\eta + \nabla_y \chi_\eta)\Big) \cdot (\nabla_y \chi_\eta - \nabla_y \chi_\zeta)\, dy. \end{aligned}$$

The last integral above is non negative due to Definition 5.39. The next to last integral is bounded below due to the coercivity of B, by

$$\frac{\alpha}{|Y|}\int_{Y*}|\zeta-\eta+\nabla_y\chi_\zeta-\nabla_y\chi_\eta|^2\,dy.$$

Expanding and using the fact that the gradient of a Y-periodic function is of zero average, one gets

$$\big(A^{hom}(\cdot,\zeta)-A^{hom}(\cdot,\eta)\big)\cdot(\zeta-\eta)\geqslant\alpha|\zeta-\eta|^2,$$

which concludes the proof. □

Remark 5.41.
1. One can obtain a convergence of the energy and a corrector result as in the linear case.
2. In the case where A^ε is a symmetric field, one can prove that the limit problem is the minimization of a limit convex continuous energy whose derivative is A^{hom}.
3. One can also consider the more general jump condition involving a g_ε of the form

$$g_\varepsilon(x)=\varepsilon\mathbf{g}\Big(x,\Big\{\frac{x}{\varepsilon}\Big\}_Y\Big).$$

The same proof applies provided $\mathbf{g}$ is assumed to be in $C(\overline{\Omega}\times\mathbf{S})^+$.

For the first two points, we refer to [68] for a detailed presentation in the case of linear elasticity with holes and cracks.

Chapter 6

A Stokes problem in a partially porous medium

The scope of this chapter is the homogenization of the Stokes problem in a partially porous medium via the unfolding method. As in the preceding sections, Ω is a bounded open set in $\mathbb{R}^N$ with a Lipschitz boundary. Assume that it is composed by two parts, one of which is a porous medium with ε-periodically distributed pores of the size of order ε. We consider the flow of a newtonian fluid in Ω which adheres to the boundary of the pores. In mathematical terms, this signifies that the pores behave as holes with an homogeneous Dirichlet condition on their boundary. The problem is therefore a partial differential equation with unknown velocity and pressure of the fluid, stated in a partially perforated domain.

Darcy's law, proposed by H. Darcy in 1856 (originally to describe the flow of water through sand) is widely used for porous media computations. Consisting of an explicit relationship between the velocity of the fluid and the gradient of the pressure, it was at the time purely empirical.

In the framework of a completely perforated media, a remarkable mathematical result is [107], where Darcy's law was established in the homogenized limit by applying the multiple-scale asymptotic method combined with sharp evaluations of the physical quantities in the problem. To our knowledge, it is the first mathematical justification of Darcy's law.

More recently, similar mathematical results were obtained in [179] and in [185] by the method of oscillating test functions of Tartar. Using the same method, the problems associated with a family of nonhomogeneous Neumann boundary conditions on the boundary of the pores (depending on a parameter $\gamma \in (-\infty, +\infty)$) were considered in [71]. Following the values of γ, various limit problems were obtained beyond Darcy's law, such as Brinkmann's equation (also originally introduced empirically). Later, the same problems were studied in [190] using the unfolding method.

On the other hand, in [149] it was shown (still applying the multiple-scale

D. Cioranescu et al., *The Periodic Unfolding Method*, Series in Contemporary Mathematics 3, https://doi.org/10.1007/978-981-13-3032-2_6

asymptotic method) that two distinct pressures appear in the limit problem (from which Darcy's law is extracted). Contrary to that, there appeared only a single pressure in the references indicated above. In this chapter concerning a fluid flow in a partially perforated domain, the unfolding method will clarify the role of the two pressures appearing in [149], and make precise their contribution in the limit problem.

The case of partially porous media was studied in [134] and [136], where the condition on the interface was investigated. In the particular case when the latter is a hyper-plane, this paper establishes the "Beavers-Joseph-Saffmann" interface condition (originally obtained empirically).

In [61], the case of a vanishing viscosity (of order ε^2) in the porous part of the domain was considered. But this is not really natural, since it means that there are distinct fluids in the two parts of the domain. Here, we consider the more physical situation of a single fluid in both domains and a scaled external force in the perforated part. As a consequence, the limit problems in the two subdomains are decoupled since they occur at the scale 1 in the regular part and at the scale ε in the porous part (see Theorems 6.5 and 6.7 below). A corrector result is also obtained.

6.1 Setting of the problem

We first recall the definition of the Stokes problem. Since one is dealing with vector functions, introduce

Notation 6.1. *As a rule, any vector function is written in bold characters, as for example, $\boldsymbol{w}$ which means that*

$$\boldsymbol{w} = (w_1, \ldots, w_N).$$

Let Ω and Ω_1 be two bounded open connected sets with Lipschitz boundaries and such that $\Omega_1 \subset \Omega \subset \mathbb{R}^N$ $(N \geqslant 2)$. Set

$$\Omega_2 = \Omega \backslash \overline{\Omega}_1, \quad \Gamma = \partial\Omega_1 \cap \partial\Omega_2.$$

Γ is the interface between the two subdomains Ω_1 and Ω_2. For simplicity, we assume Ω_2 connected also and with Lipschitz boundary.

We use the notations introduced in Chapters 1 and 4 for the group $\mathbf{G}$, the unit parallelotope Y, the sets $\Xi_\varepsilon, \widehat{\Omega}_\varepsilon, ..,$ the operators associated with unfolding $(\mathcal{T}_\varepsilon, \mathcal{M}_\varepsilon, \mathcal{Q}_\varepsilon, ...)$.

The porous medium $\Omega_{1,\varepsilon}$ is restricted to Ω_1 and is modeled with a closed hole $S \subset \overline{Y}$, such that all the hypotheses and definitions of Section 4.1 are satisfied (in particular, $Y^* \doteq Y \backslash S$ is connected and Hypothesis $(\mathbf{H_2})$ holds).

We introduce the new notations for the cells in Ω_1

$$(\Xi_1)'_\varepsilon = \{\xi \in \mathbf{G} \mid \varepsilon\{\xi + Y\} \cap \Omega_1 \neq \varnothing,\ \varepsilon\{\xi + Y\} \cap \Gamma = \varnothing\}.$$

In Ω_1,

$$S_\varepsilon \doteq \bigcup_{\xi \in (\Xi_1)'_\varepsilon} \varepsilon\{\xi + S\}, \quad \Omega^*_{1,\varepsilon} = \Omega_1 \backslash S_\varepsilon,$$

are the pores and the part occupied by the fluid, respectively. By the choice of $(\Xi_1)'_\varepsilon$, S_ε does not intersect Γ.

The problem is posed in the open domain

$$\Omega_\varepsilon \doteq \Omega^*_{1,\varepsilon} \cup \Gamma \cup \Omega_2.$$

Figure 6.1 below depicts three possible configurations (observe that the last one corresponds to $\Omega_2 = \varnothing$, the whole of Ω being a perforated domain). In the three configurations, the holes can intersect the exterior boundary $\partial\Omega$ (but, as mentioned above, not the interface Γ which is present in the first two pictures).

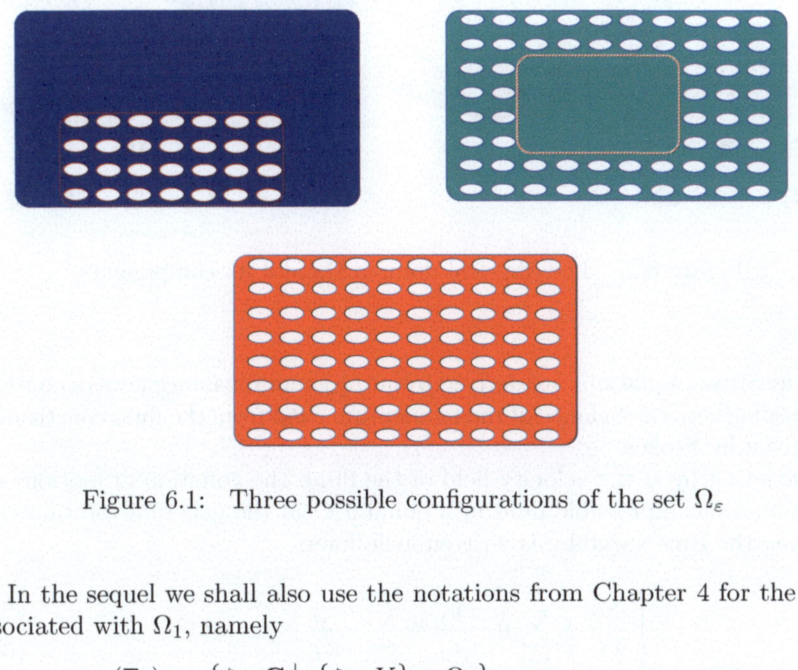

Figure 6.1: Three possible configurations of the set Ω_ε

In the sequel we shall also use the notations from Chapter 4 for the sets associated with Ω_1, namely

$$(\Xi_1)_\varepsilon = \{\xi \in \mathbf{G} \,|\, \varepsilon\{\xi + Y\} \subset \Omega_1\},$$

$$\widehat{\Omega}_{1,\varepsilon} = \text{interior}\,\Big(\bigcup_{\xi \in (\Xi_1)_\varepsilon} \varepsilon\{\xi + \overline{Y}\}\Big), \quad \Lambda_{1,\varepsilon} = \Omega_1 \backslash \widehat{\Omega}_{1,\varepsilon}$$

$$\widehat{\Omega}^*_{1,\varepsilon} = \text{interior}\,\Big(\bigcup_{\xi \in (\Xi_1)_\varepsilon} \varepsilon\{\xi + \overline{Y}^*\}\Big).$$

One can see in Figure 6.2 a general physical domain Ω_ε and the sets introduced so far. The hypothesis that the holes do not intersect the interface Γ does not in this case, exclude the holes intersecting the exterior boundary of $\Omega^*_{1,\varepsilon}$.

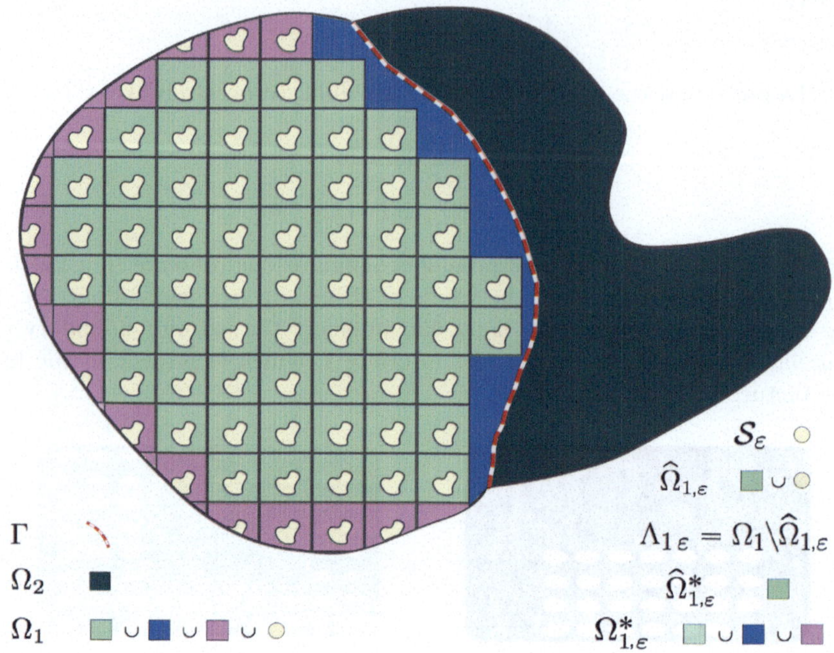

Figure 6.2: The physical domain Ω_ε and its components

The Stokes equations are derived from the general balance laws of continuous media (here the balance of the momentum) and from the fluid constitutive law given by Stokes.

Denoting by $\boldsymbol{u}$ the velocity field of the fluid, the equation of motion of a newtonian incompressible fluid in a domain $\mathcal{O}$ in the general evolution case (t being the time variable) is written as follows:

$$\begin{cases} \dfrac{\partial u_i}{\partial t} + \displaystyle\sum_{j=1}^{N} u_j \dfrac{\partial u_i}{\partial x_j} = \displaystyle\sum_{j=1}^{N} \dfrac{\partial \tau_{ij}}{\partial x_j} + f_i \quad \text{in } \mathcal{O}, \\ \operatorname{div} \boldsymbol{u} = 0 \quad \text{in } \mathcal{O}, \end{cases} \tag{6.1}$$

for $i = 1, \dots, N$, where $\boldsymbol{f}$ is an exterior body force and $\tau = (\tau_{ij})_{1 \leqslant i,j \leqslant N}$ is the stress tensor.

The Stokes constitutive law for a fluid is given by

$$\tau = -p\mathbf{I} + 2\mu D, \tag{6.2}$$

where p is the (scalar) pressure (a Lagrange multiplier for the divergence-free condition), $\mathbf{I}$ is the identity tensor, $\mu > 0$ is the viscosity constant, and

$D = (D_{ij})_{1\leqslant i,j\leqslant N}$ (known as the rate of deformation tensor) is the symmetric part of the velocity gradient, with components

$$D_{ij} = \frac{1}{2}\Big(\frac{\partial u_i}{\partial x_j} + \frac{\partial u_j}{\partial x_i}\Big).$$

Taking into account this definition and the fact that $\operatorname{div} \boldsymbol{u} = 0$, replacing τ_{ij} by its value from (6.2) in the right-hand side of (6.1), we get

$$\begin{cases} \dfrac{\partial u_i}{\partial t} + \displaystyle\sum_{j=1}^{N} u_j \frac{\partial u_i}{\partial x_j} = -\frac{\partial p}{\partial x_i} + \mu\,\Delta u_i + f_i \quad \text{in } \mathcal{O}, \\ \operatorname{div} \boldsymbol{u} = 0 \quad \text{in } \mathcal{O}. \end{cases} \tag{6.3}$$

Here, we shall only consider the steady state (no dependence in time). Moreover, for reasonably slow motions (for high viscosity for example), one can neglect the non linear terms in the left-hand side of (6.3), to end up with what is known as the Stokes equations in $\mathcal{O}$ for incompressible fluids,

$$\begin{cases} -\mu\,\Delta \boldsymbol{u} + \nabla p = \boldsymbol{f} \quad \text{in } \mathcal{O}, \\ \quad \operatorname{div} \boldsymbol{u} = 0 \quad \text{in } \mathcal{O}, \end{cases} \tag{6.4}$$

to which, to be consistent, one adds a boundary conditions on $\partial\mathcal{O}$. We consider in this section the adhesion condition, i.e., $u = 0$ on $\partial\mathcal{O}$.

Now, considering the flow of an incompressible fluid in Ω_ε, introduce the space

$$\mathbf{V}^\varepsilon = \big\{ \boldsymbol{w} \in H_0^1(\Omega_\varepsilon)^N \mid \operatorname{div} \boldsymbol{w} = 0 \big\}.$$

The variational formulation of the Stokes problem (6.4) in the domain Ω_ε is formulated as follows: find $\boldsymbol{u}_\varepsilon$ in $\mathbf{V}^\varepsilon$ such that

$$\begin{cases} \mu \displaystyle\int_{\Omega_\varepsilon} \nabla \boldsymbol{u}_\varepsilon \nabla \boldsymbol{v}\, dx = \int_{\Omega_\varepsilon} \boldsymbol{f}_\varepsilon\, \boldsymbol{v}\, dx, \\ \forall \boldsymbol{v} \in \mathbf{V}^\varepsilon, \end{cases} \tag{6.5}$$

where $\boldsymbol{f}_\varepsilon$ is given in $L^2(\Omega)^N$. It has a unique solution $\boldsymbol{u}_\varepsilon$ by the Lax-Milgram theorem applied in $\mathbf{V}^\varepsilon$. The pressure p_ε is recovered (up to constants) by the Stokes formula in $L^2(\Omega_\varepsilon)$ via its gradient.

We shall also use the following formulation which is equivalent and has a unique solution $(\boldsymbol{u}_\varepsilon, \nabla p_\varepsilon)$ in $\mathbf{V}^\varepsilon \times H^{-1}(\Omega_\varepsilon)^N$ (the pressure is normalized by the second equation to guarantee its uniqueness in the connected component of Ω_2 in Ω_ε, see for instance [187]):

$$\begin{cases} \mu \displaystyle\int_{\Omega_\varepsilon} \nabla \boldsymbol{u}_\varepsilon \nabla \boldsymbol{v}\, dx - \int_{\Omega_\varepsilon} p_\varepsilon \operatorname{div} \boldsymbol{v}\, dx = \int_{\Omega_\varepsilon} \boldsymbol{f}_\varepsilon\, \boldsymbol{v}\, dx, \\ \displaystyle\int_{\Omega_2} p_\varepsilon\, dx = 0, \\ \forall \boldsymbol{v} \in H_0^1(\Omega_\varepsilon)^N. \end{cases} \tag{6.6}$$

Our aim is to study the asymptotic behavior of u_ε, p_ε as $\varepsilon \to 0$. To do so, the first task is to obtain uniform (with respect to ε) estimates of $(u_\varepsilon, p_\varepsilon)$.

The estimates given in the next subsection will involve the right-hand side of system (6.6) through the following quantity:

$$\mathfrak{N}_\varepsilon \doteq \left(\varepsilon^2 \|\boldsymbol{f}_\varepsilon\|^2_{L^2(\Omega_1)} + \|\boldsymbol{f}_\varepsilon\|^2_{L^2(\Omega_2)}\right)^{1/2}. \tag{6.7}$$

6.2 A priori estimates

Set

$$\mathbf{Y} = \text{interior}\Big(\bigcup\{\xi + \overline{Y} \mid \xi \in \mathbf{G},\ \overline{Y} \cap \{\xi + \overline{Y}\} \neq \varnothing\}\Big),$$

which is the union of Y and the $3^N - 1$ cells surrounding it.

The basic ingredients to establish a priori estimates are the following inequalities:

1. *The Poincaré inequality in* $\mathbf{Y}$. For every $w \in H^1(\mathbf{Y})$ with $v_{|S} = 0$,

$$\|w\|_{L^2(\mathbf{Y})} \leqslant C\|\nabla w\|_{L^2(\mathbf{Y})}. \tag{6.8}$$

2. *The Nečas inequality.* For every bounded domain $\mathcal{O}$ with Lipschitz boundary, there is a constant C depending on $\mathcal{O}$ such that, for every Φ in $L^2(\mathcal{O})^N$,

$$\begin{aligned} \|\Phi - \mathcal{M}_{\mathcal{O}}(\Phi)\|_{L^2(\mathcal{O})} &\leqslant C\|\nabla\Phi\|_{H^{-1}(\mathcal{O})} \\ &= C \sup\Big\{\int_{\mathcal{O}} \Phi(y)\, \text{div}\, \Psi(y)\, dy \,\Big|\, \|\Psi\|_{H^1_0(\mathcal{O})} \leqslant 1\Big\}. \end{aligned} \tag{6.9}$$

Lemma 6.2. *For all ε satisfying $\varepsilon\, d(Y) < 1$ and for every $v \in H^1(\Omega_1)$ which vanishes in S_ε one has*

$$\|v\|_{L^2(\Omega_1)} \leqslant \varepsilon C\|\nabla v\|_{L^2(\Omega_1)}. \tag{6.10}$$

The constant is independent of ε.

Proof. Using Proposition 1.74, we extend v in a function belonging to the space $H^1(\mathbb{R}^N)$, still denoted v, and satisfying estimate $(1.80)_3$.

Now, observe that (see Notation 1.63 for $\mathcal{V}_{\varepsilon d(Y)}(\Omega_1)$),

$$\Omega_1 \subset \bigcup_{\xi \in (\Xi_1)_\varepsilon} \varepsilon(\xi + \mathbf{Y}) \subset \mathcal{V}_{\varepsilon d(Y)}(\Omega_1).$$

By a standard ε-scaling argument, inequality (6.8) implies for every cell $\varepsilon(\xi + \mathbf{Y})$ such that $\xi \in (\Xi_1)_\varepsilon$ (in this case $v = 0$ on $\varepsilon(\xi + S)$),

$$\|v\|^2_{L^2(\varepsilon(\xi+\mathbf{Y}))} \leqslant \varepsilon^2\, C\, \|\nabla v\|^2_{L^2(\varepsilon(\xi+\mathbf{Y}))}.$$

Summing this inequality over ξ in $(\Xi_1)_\varepsilon$, it follows that for all v in $H^1(\Omega_1)$ with $v_{|S_\varepsilon} = 0$,

$$\|v\|_{L^2(\Omega_1)} \leqslant \varepsilon\, C\|\nabla v\|_{L^2(\mathcal{V}_{\varepsilon d(Y)}(\Omega_1))} \leqslant \varepsilon\, C\|\nabla v\|_{L^2(\Omega_1)}.$$

$\square$

To estimate the pressure p_ε, one uses the "extension by average" operator $\mathcal{E}_\varepsilon$ introduced in Definition 4.22. In the same context, this operator was used in [152, 185].

Proposition 6.3. *Let* $(u_\varepsilon, p_\varepsilon)$ *be the solution of the Stokes problem* (6.6) *with* $\boldsymbol{f}_\varepsilon$ *given in* $L^2(\Omega)^N$. *Then the following estimates hold true:*

$$\begin{aligned}
&(i) && \|\boldsymbol{u}_\varepsilon\|_{H_0^1(\Omega)} + \|\varepsilon^{-1}\boldsymbol{u}_\varepsilon\|_{L^2(\Omega_1)} \leqslant C\,\mathfrak{N}_\varepsilon,\\
&(ii) && \|p_\varepsilon\|_{L^2(\Omega_2)} \leqslant C\,\mathfrak{N}_\varepsilon,\\
&(iii) && \|p_\varepsilon - \mathcal{M}_\varepsilon^*(p_\varepsilon)\|_{L^2(\widehat{\Omega}^*_{1,\varepsilon})} \leqslant C\,\mathfrak{N}_\varepsilon,\\
&(iv) && \big\|\mathcal{E}_\varepsilon(p_\varepsilon) - \mathcal{Q}_\varepsilon^*(p_\varepsilon)\big\|_{L^2(\Omega^{int}_{3\varepsilon d(Y)})} \leqslant C\,\mathfrak{N}_\varepsilon,\\
&(v) && \|\varepsilon\mathcal{Q}_\varepsilon^*(p_\varepsilon)\|_{H^1(\Omega^{int}_{3\varepsilon d(Y)})} \leqslant C\,\mathfrak{N}_\varepsilon,\\
&(vi) && \big\|\varepsilon\,\mathcal{E}_\varepsilon(p_\varepsilon)\mathbf{1}_{\Omega^{int}_{3\varepsilon d(Y)}}\big\|_{L^2(\Omega)} \leqslant C\,\mathfrak{N}_\varepsilon,
\end{aligned}\tag{6.11}$$

with $\mathfrak{N}_\varepsilon$ *defined by* (6.7). *The constants* C *are independent of* ε.

Proof. (i)-(ii) First, observe that the following inequalities are straightforward, the first by Lemma 6.2 and the second by the Poincaré inequality in $H_0^1(\Omega)$:

$$\begin{aligned}
&\int_{\Omega_1} f_\varepsilon\,\boldsymbol{u}_\varepsilon\,dx \leqslant \varepsilon C\,\|\boldsymbol{f}_\varepsilon\|_{L^2(\Omega_1)}\|\nabla\boldsymbol{u}_\varepsilon\|_{L^2(\Omega_1)},\\
&\int_{\Omega_2} f_\varepsilon\,\boldsymbol{u}_\varepsilon\,dx \leqslant C\,\|\boldsymbol{f}_\varepsilon\|_{L^2(\Omega_2)}\|\nabla\boldsymbol{u}_\varepsilon\|_{L^2(\Omega)}.
\end{aligned}\tag{6.12}$$

Now, take $\boldsymbol{u}_\varepsilon$ as test function in the Stokes problem (6.5) to get

$$\mu\,\|\nabla\boldsymbol{u}_\varepsilon\|^2_{L^2(\Omega)} \leqslant \int_\Omega \boldsymbol{f}_\varepsilon\,\boldsymbol{u}_\varepsilon\,dx.$$

Using (6.12), we obtain

$$\|\nabla\boldsymbol{u}_\varepsilon\|_{L^2(\Omega)} \leqslant C\,\mathfrak{N}_\varepsilon,$$

from which, by the Poincaré inequality and taking into account (6.10), estimate (6.11)(i) follows.

Furthermore, (6.11)(i) combined with estimate (6.9) written for $\mathcal{O} \doteq \Omega_2$, give (6.11)(ii) by taking an arbitrary test function $\boldsymbol{v}$ of $H_0^1(\Omega_2)^N$ in (6.6) and using the fact that $\displaystyle\int_{\Omega_2} p_\varepsilon(x)dx = 0$.

(iii) Let $\boldsymbol{\psi}$ be in $H_0^1(Y^\star)^N$ and ξ in Ξ_ε^1. Choosing as test function in (6.6)

$$\boldsymbol{v}_\xi^\varepsilon(x) = \varepsilon\boldsymbol{\psi}\Big(\Big\{\frac{x}{\varepsilon}\Big\}_Y\Big)\,\mathbf{1}_{\varepsilon\{\xi+Y*\}},$$

which is an element of $H_0^1(\Omega_\varepsilon)^N$, gives

$$\int_{\varepsilon\{\xi+Y^\star\}} p_\varepsilon\,\mathrm{div}\,\boldsymbol{v}_\xi^\varepsilon\,dx = \int_{\varepsilon\{\xi+Y^\star\}} \boldsymbol{f}_\varepsilon\,\boldsymbol{v}_\xi^\varepsilon\,dx + 2\mu\int_{\varepsilon\{\xi+Y^\star\}} \nabla\boldsymbol{u}_\varepsilon\,\nabla\boldsymbol{v}_\xi^\varepsilon\,dx.$$

Since

$$\|\nabla \boldsymbol{v}_\xi^\varepsilon\|_{L^2(\Omega_\varepsilon)} = \varepsilon^{\frac{N}{2}} \|\nabla_y(\boldsymbol{\psi})\|_{L^2(Y^\star)},$$
$$\|\boldsymbol{v}_\xi^\varepsilon\|_{L^2(\Omega_\varepsilon)} = \varepsilon^{1+\frac{N}{2}} \|\boldsymbol{\psi}\|_{L^2(Y^\star)} \leqslant C\varepsilon^{1+\frac{N}{2}} \|\nabla_y \boldsymbol{\psi}\|_{L^2(Y^\star)},$$

it follows that

$$\int_{\varepsilon\{\xi+Y^\star\}} p_\varepsilon \operatorname{div} \boldsymbol{v}_\xi^\varepsilon \, dx \leqslant C\varepsilon^{\frac{N}{2}} \,\mathfrak{M}_\varepsilon(\xi)\, \|\nabla_y \boldsymbol{\psi}\|_{L^2(Y^\star)}. \tag{6.13}$$

where

$$\mathfrak{M}_\varepsilon(\xi)^2 = \varepsilon^2 \|\boldsymbol{f}_\varepsilon\|^2_{L^2(\varepsilon\{\xi+Y^\star\})} + \|\nabla \boldsymbol{u}_\varepsilon\|^2_{L^2(\varepsilon\{\xi+Y^\star\})}.$$

On the other hand, as $(\operatorname{div} \boldsymbol{v}_\xi^\varepsilon)(\varepsilon y) = \operatorname{div}_y \boldsymbol{\psi}(y)$,

$$\varepsilon^N \int_{Y^\star} p_\varepsilon(\varepsilon(\xi+y)) \operatorname{div}_y \boldsymbol{\psi}(y)\, dy = \int_{\varepsilon\{\xi+Y^\star\}} p_\varepsilon \operatorname{div} \boldsymbol{v}_\xi^\varepsilon \, dx.$$

Consequently, by estimate (6.9) with $\mathcal{O} \doteq Y^*$, from inequality (6.13) we get

$$\varepsilon^N \|p_\varepsilon(\varepsilon(\xi+\cdot)) - \mathcal{M}_\varepsilon^*(p_\varepsilon)(\varepsilon\xi)\|^2_{L^2(Y^\star)} \leqslant C\,\mathfrak{M}_\varepsilon(\xi)^2,$$

or equivalently,

$$\varepsilon^N \|\mathcal{E}_\varepsilon(p_\varepsilon)(\varepsilon(\xi+\cdot)) - \mathcal{M}_\varepsilon^*(p_\varepsilon)(\varepsilon\xi)\|^2_{L^2(Y)} \leqslant C\,\mathfrak{M}_\varepsilon(\xi)^2.$$

Summing over $\xi \in \Xi_\varepsilon^1$ and going back to the variable x, yields

$$\|\mathcal{E}_\varepsilon(p_\varepsilon) - \mathcal{M}_\varepsilon^*(p_\varepsilon)\|^2_{L^2(\widehat{\Omega}_{1,\varepsilon})} \leqslant C \sum_{\xi\in\Xi_\varepsilon^1} \mathfrak{M}_\varepsilon(\xi)^2 \leqslant C\,\mathfrak{N}_\varepsilon^2. \tag{6.14}$$

One concludes since

$$\mathcal{E}_\varepsilon(p_\varepsilon) - \mathcal{M}_\varepsilon^*(p_\varepsilon) = \begin{cases} p_\varepsilon - \mathcal{M}_\varepsilon^*(p_\varepsilon) & \text{a.e. in } \widehat{\Omega}^*_{1,\varepsilon}, \\ 0 & \text{a.e. in } \widehat{\Omega}_{1,\varepsilon} \cap S_\varepsilon. \end{cases}$$

(iv)-(v) We use here the notations of Paragraph 4.3.1, more precisely those of (4.3.1), (4.17)-(4.20). Observe that when ξ belongs to $\Xi_\varepsilon^{\mathcal{Y}}$, one can perform the same computation as in the previous step on the set $\varepsilon(\xi+\mathcal{Y}^*)$, to get

$$\varepsilon^N \|p_\varepsilon(\varepsilon(\xi+\cdot)) - \mathcal{M}_{\mathcal{Y}^*}(p_\varepsilon)(\varepsilon\xi)\|^2_{L^2(\mathcal{Y}^*)}$$
$$\leqslant C\big(\varepsilon^2 \|\boldsymbol{f}_\varepsilon\|^2_{L^2(\varepsilon(\xi+\mathcal{Y}))} + \|\nabla \boldsymbol{u}_\varepsilon\|^2_{L^2(\varepsilon(\xi+\mathcal{Y}))}\big).$$

Note that there are 2^N cells $\varepsilon(\eta+Y^*)$ in $\varepsilon(\xi+\mathcal{Y}^*)$.

Taking averages on the contiguous cells $\varepsilon(\xi+\mathbf{b_k}+Y^*)$ and $\varepsilon(\xi+Y^*)$ and subtracting the results, yields

$$\varepsilon^N |\mathcal{M}_\varepsilon^*(p_\varepsilon)(\varepsilon(\xi+\mathbf{b}_k)) - \mathcal{M}_\varepsilon^*(p_\varepsilon)(\varepsilon\xi)|^2$$
$$\leqslant C\big(\varepsilon^2 \|\boldsymbol{f}_\varepsilon\|^2_{L^2(\varepsilon\{\xi+\mathcal{Y}\})} + \|\nabla \boldsymbol{u}_\varepsilon\|^2_{L^2(\varepsilon\{\xi+\mathcal{Y}\})}\big).$$

In view of (4.31), by summation, this inequality and (6.11)(i), imply

$$\|\varepsilon\nabla\mathcal{Q}^*_\varepsilon(p_\varepsilon)\|_{L^2(\widehat{\Omega}^{\mathcal{Y}}_\varepsilon)} \leqslant C\,\mathfrak{N}_\varepsilon. \tag{6.15}$$

Recall that $\mathcal{Q}^*_\varepsilon(p_\varepsilon)(\varepsilon\xi)$ is the value of $\mathcal{M}^*_\varepsilon(p_\varepsilon)$ over $\varepsilon(\xi+Y)$ (for all cells $\varepsilon(\xi+Y)$ included in $\widehat{\Omega}^{\mathcal{Y}}_\varepsilon$). By the estimate of Remark 1.65, since $\mathcal{Q}^*_\varepsilon = \mathcal{Q}_\varepsilon\circ\mathcal{E}_\varepsilon$ in $\widehat{\Omega}^{\mathcal{Y}}_\varepsilon$ (using the inclusion (4.19)), one obtains

$$\big\|\mathcal{Q}^*_\varepsilon(p_\varepsilon)-\mathcal{M}^*_\varepsilon(p_\varepsilon)\big\|_{L^2(\Omega^{int}_{3\varepsilon d(Y)})} \leqslant \varepsilon\,C\,\|\nabla\mathcal{Q}^*_\varepsilon(p_\varepsilon)\|_{L^2(\widehat{\Omega}^{\mathcal{Y}}_\varepsilon)}.$$

Together with (6.14), this gives (6.11)(iv).

Now, in view of (6.11)(ii), we get

$$\|\varepsilon\mathcal{Q}^*_\varepsilon(p_\varepsilon)\|_{L^2(\Omega^{int}_{3\varepsilon d(Y)}\cap\Omega_2)} \leqslant C\,\mathfrak{N}_\varepsilon,$$

which, together with (6.15), imply (6.11)(v) (here we also used Lemma 5.22 for the domains $\Omega^{int}_{3\varepsilon d(Y)}$).

(vi) It is a direct consequence of (6.11)(iv)-(v). □

6.3 Homogenization

Introduce now the following spaces:

$$\begin{aligned}
&\mathcal{H} \doteq H^1_0(\Omega_1;\Gamma) \doteq \big\{\phi\in H^1(\Omega_1)\mid \phi\equiv 0 \text{ on } \Gamma\big\},\\
&\mathbf{V} \doteq \big\{\boldsymbol{v}\in H^1_0(\Omega)^N \mid \boldsymbol{v}_{|\Omega_1}\equiv 0,\ \operatorname{div}\boldsymbol{v}\equiv 0\big\},\\
&\mathbf{H}(Y) \doteq \big\{\boldsymbol{v}\in H^1_{per}(Y)^N \mid \boldsymbol{v}\equiv 0 \text{ in } S\big\},\\
&\mathbf{V}(Y) \doteq \big\{\boldsymbol{v}\in \mathbf{H}(Y)\mid \operatorname{div}_y\boldsymbol{v}\equiv 0\big\}.
\end{aligned}$$

Note that the space $\mathbf{V}$ can equivalently be viewed as

$$\big\{\boldsymbol{v}\in H^1_0(\Omega_2)^N\mid \operatorname{div}\boldsymbol{v}\equiv 0\big\}.$$

Similarly, the space $\mathcal{H}$ can be defined as

$$H^1_0(\Omega;\Omega_2) \doteq \Big\{\phi\in H^1(\Omega)\mid \phi\equiv 0 \text{ on } \Omega_2\Big\}.$$

Theorem 6.4. *Let $(\boldsymbol{u}_\varepsilon,p_\varepsilon)$ be the solution of the Stokes problem (6.6) with $\boldsymbol{f}_\varepsilon$ given in $L^2(\Omega)^N$. Then, up to a subsequence, there are*

$$\begin{aligned}
&\boldsymbol{u}\in\mathbf{V},\\
&\overline{\boldsymbol{u}}\in L^2(\Omega;\mathbf{V}(Y)) \quad \textit{with} \quad \overline{\boldsymbol{u}}\equiv 0 \quad \textit{in} \quad \Omega_2\times Y,
\end{aligned}$$

such that the following convergences hold:

$$\begin{aligned}
&\boldsymbol{u}_\varepsilon\rightharpoonup\boldsymbol{u} \quad \textit{weakly in } H^1_0(\Omega)^N,\\
&\mathcal{T}_\varepsilon(\nabla\boldsymbol{u}_\varepsilon)\rightharpoonup\nabla\boldsymbol{u}+\nabla_y\overline{\boldsymbol{u}} \quad \textit{weakly in } L^2(\Omega\times Y)^{N\times N},\\
&\mathcal{T}_\varepsilon(\varepsilon^{-1}\boldsymbol{u}_\varepsilon\mathbf{1}_{\Omega_1})\rightharpoonup\overline{\boldsymbol{u}} \quad \textit{weakly in } L^2(\Omega;H^1(Y))^N,\\
&\varepsilon^{-1}\boldsymbol{u}_{\varepsilon|\Omega_1}\rightharpoonup\mathcal{M}_Y(\overline{\boldsymbol{u}}) \quad \textit{weakly in } L^2(\Omega_1)^N,\\
&\mathcal{T}_\varepsilon(\nabla\boldsymbol{u}_\varepsilon)\mathbf{1}_{\Omega_1}\rightharpoonup\nabla_y\overline{\boldsymbol{u}} \quad \textit{weakly in } L^2(\Omega_1\times Y)^{N\times N}.
\end{aligned} \tag{6.16}$$

There are also

$$\widehat{p} \in L^2(\Omega_1 \times Y^*), \quad p^1 \in \mathcal{H}, \quad p^2 \in L^2(\Omega_2),$$

satisfying

$$\begin{aligned} &(i) \quad \int_{\Omega_2} p^2(x)\,dx = 0, \\ &(ii) \quad \widehat{p} \equiv 0 \quad in \quad \Omega_1 \times S \quad with \quad \mathcal{M}_Y(\widehat{p}) \equiv 0 \quad in \quad \Omega_1, \end{aligned} \tag{6.17}$$

and such that the following convergences hold:

$$\begin{aligned} &p_{\varepsilon|\Omega_2} \rightharpoonup p^2 \quad weakly\ in\ L^2(\Omega_2), \\ &\varepsilon\,\mathcal{E}_\varepsilon(p_\varepsilon)\,\mathbf{1}_{\Omega^{int}_{3\varepsilon d(Y)}} \to p^1 \quad strongly\ in\ L^2(\Omega), \\ &\mathcal{T}^*_\varepsilon\big(p_\varepsilon - \mathcal{M}^*_\varepsilon(p_\varepsilon)\big)\mathbf{1}_{\Omega_1\times Y^*} \rightharpoonup \widehat{p} \quad weakly\ in\ L^2(\Omega_1 \times Y^*). \end{aligned} \tag{6.18}$$

Proof. By estimate (6.11)(i) and Theorem 1.41, up to a subsequence, there exist $(\boldsymbol{u}, \widehat{\boldsymbol{u}})$ in the space $\mathbf{V} \times L^2(\Omega; H^1_{per,0}(Y))^N$ and $\overline{\boldsymbol{u}}$ in $L^2(\Omega_1; H^1_{per}(Y)^N)$ such that convergences $(6.16)_{1,3,4,5}$ hold as well as

$$\mathcal{T}_\varepsilon(\nabla \boldsymbol{u}_\varepsilon) \rightharpoonup \nabla \boldsymbol{u} + \nabla_y \widehat{\boldsymbol{u}} \quad \text{weakly in } L^2(\Omega \times Y)^{N\times N}.$$

Since $\operatorname{div} \boldsymbol{u}_\varepsilon = 0$ in Ω, Proposition 2.8 implies $\operatorname{div}_y \widehat{\boldsymbol{u}} = 0$ in $\Omega \times Y$.

It is easy to see that for every open subset $\omega \Subset \Omega_1$, the restriction to ω of $\nabla_y \mathcal{T}_\varepsilon(\varepsilon^{-1}\boldsymbol{u}_\varepsilon \mathbf{1}_{\Omega_1})$ coincides with that of $\mathcal{T}_\varepsilon(\nabla \boldsymbol{u}_\varepsilon)$. Therefore at the limit,

$$\nabla_y \overline{\boldsymbol{u}} = \nabla_y \widehat{\boldsymbol{u}}$$

which implies $(6.16)_2$. It also implies $\operatorname{div}_y \overline{\boldsymbol{u}} = \operatorname{div}_y \widehat{\boldsymbol{u}} = 0$ in $\Omega_1 \times Y$. Recalling that $\boldsymbol{u}_\varepsilon$ vanishes in S_ε, it follows that $\overline{\boldsymbol{u}}$ vanishes on $\Omega_1 \times S$. Hence $\overline{\boldsymbol{u}}$ belongs to $L^2(\Omega_1; \mathbf{V}(Y))$.(1)

Except for the **strong** convergence $(6.18)_2$ with p_1 in $\mathcal{H}$, all the convergences (6.18), for an appropriate subsequence, are all direct consequences of estimates (6.11).

The function p^2 is the weak limit of a sequence of functions with zero mean value on Ω_2, while function $\widehat{p}$ is the weak limit of a sequence of functions with zero mean value on Y^* and vanishing on $\Omega_1 \times S$. These properties are preserved by passing to the limit, so that (6.17)(i) and (6.17)(ii) hold true.

As for p^1, by (6.11)(iv) it is the weak limit in $L^2(\Omega)$ of the sequence $\{\varepsilon \mathcal{Q}^*_\varepsilon(p_\varepsilon)\mathbf{1}_{\Omega^{int}_{3\varepsilon d(Y)}}\}_\varepsilon$.

At this point, since the sets $\Omega^{int}_{3\varepsilon d(Y)}$ are uniformly Lipschitz, one can use the extension operators $\mathcal{P}_{3\varepsilon d(Y)}$ defined in Corollary 5.23. It is immediate that the sequence $\{\mathcal{P}_{3\varepsilon d(Y)}\big(\varepsilon \mathcal{Q}^*_\varepsilon(p_\varepsilon)\mathbf{1}_{\Omega^{int}_{3\varepsilon d(Y)}}\big)\}_\varepsilon$ is bounded in $H^1(\Omega)$. Hence, up to a subsequence, it converges weakly in that space and strongly in $L^2(\Omega)$

(1) The relationship between $\overline{\boldsymbol{u}}$ and $\widehat{\boldsymbol{u}}$ is therefore: $\widehat{\boldsymbol{u}} \equiv \overline{\boldsymbol{u}} - \mathcal{M}_Y(\overline{\boldsymbol{u}})$ in $\Omega_1 \times Y$.

to the same p^1 which therefore is in $H^1(\Omega)$. Consequently, $\{\varepsilon\mathcal{Q}^*_\varepsilon(p_\varepsilon)\mathbf{1}_{\Omega^{int}_{3\varepsilon d(Y)}}\}_\varepsilon$ converges strongly to p_1 in $L^2(\Omega)$. This, by (6.11)(iv) again, implies convergence $(6.18)_2$. Since $p_\varepsilon\mathbf{1}_{\Omega_2}$ converges weakly in $L^2(\Omega_2)$, $p^1 = 0$ in Ω_2, and this ends the proof of Theorem 6.4. □

We are now in position to state the limit unfolded problem satisfied by $\boldsymbol{u}$, p^1, p^2, $\overline{\boldsymbol{u}}$, $\widehat{p}$.

Theorem 6.5. *Let $(\boldsymbol{u}_\varepsilon, p_\varepsilon)$ be the solution of the Stokes problem (6.6) with $\boldsymbol{f}_\varepsilon$ in $L^2(\Omega)^N$. Assume that there exist $\boldsymbol{f}$ in $L^2(\Omega_2)^N$ and $\widehat{\boldsymbol{f}}$ in $L^2(\Omega_1\times Y)^N$, such that*

$$\begin{aligned} &\mathcal{T}_\varepsilon(\varepsilon\boldsymbol{f}_\varepsilon\mathbf{1}_{\Omega_1}) \rightharpoonup \widehat{\boldsymbol{f}} \quad \textit{weakly in } L^2(\Omega_1\times Y)^{N\,(2)},\\ &\boldsymbol{f}_\varepsilon\mathbf{1}_{\Omega_2} \rightharpoonup \boldsymbol{f} \quad \textit{weakly in } L^2(\Omega_2)^N. \end{aligned} \tag{6.19}$$

Then the convergences in Proposition 6.4 hold for the full sequence and the limit functions $\boldsymbol{u}$, $p^1, p^2, \overline{\boldsymbol{u}}, \widehat{p}$ satisfying conditions (6.17), are the unique solution in the space

$$\mathbf{V}\times\mathcal{H}\times L^2(\Omega_2)\times L^2(\Omega_1;\mathbf{V}(Y))\times L^2(\Omega_1\times Y)$$

of the two following decoupled problems (recall that y^c is defined in (1.41)):

$$\begin{cases} \mu\displaystyle\int_{\Omega_1\times Y}\nabla_y\overline{\boldsymbol{u}}\,\nabla_y\boldsymbol{\Psi}\,dxdy - \int_{\Omega_1\times Y}(\widehat{p}-y^c\cdot\nabla p^1)\,\mathrm{div}_y\boldsymbol{\Psi}\,dxdy\\ \qquad\qquad = \displaystyle\int_{\Omega_1\times Y}(\widehat{\boldsymbol{f}}-\nabla p^1)\boldsymbol{\Psi}\,dxdy,\\ \displaystyle\int_{\Omega_1}\mathcal{M}_Y(\overline{\boldsymbol{u}})\,\nabla\varphi\,dx = 0,\\ \forall\boldsymbol{\Psi}\in L^2(\Omega_1\,;\,\mathbf{H}(Y)),\ \ \forall\varphi\in\mathcal{H}, \end{cases} \tag{6.20}$$

and

$$\begin{cases} \mu\displaystyle\int_{\Omega_2}\nabla\boldsymbol{u}\,\nabla\boldsymbol{v} - \int_{\Omega_2}p^2\,\mathrm{div}\,\boldsymbol{v}\,dx = \int_{\Omega_2}\boldsymbol{f}\,\boldsymbol{v}\,dx,\\ \forall\boldsymbol{v}\in H^1_0(\Omega_2)^N. \end{cases} \tag{6.21}$$

Remark 6.6. *In the limit problem, the two pressures p^1 and $\widehat{p}$ which appear in Ω_1 were conjectured in [149] using the multiple scale expansion method.*

Proof of Theorem 6.5. Formulation (6.6) can be rewritten as

$$\begin{cases} \mu\displaystyle\int_\Omega\nabla\boldsymbol{u}_\varepsilon\nabla\boldsymbol{v}\,dx - \int_\Omega\mathcal{E}_\varepsilon(p_\varepsilon)\,\mathrm{div}\,\boldsymbol{v}\,dx = \int_\Omega\boldsymbol{f}_\varepsilon\,\boldsymbol{v}\,dx,\\ \forall\boldsymbol{v}\in H^1_0(\Omega)^N,\ \ \boldsymbol{v}\equiv 0 \text{ on } S_\varepsilon. \end{cases} \tag{6.22}$$

Step 1. Proof of (6.21). The proof of the theorem uses the convergences established in Proposition 6.4.

(2) Note that by its definition, $\widehat{f}$ vanishes on $\Omega_2\times Y$.

Testing in (6.6) with a function $\boldsymbol{v}$ of $H_0^1(\Omega_2)^N$, one easily obtains (6.21) since $\boldsymbol{f}_\varepsilon \mathbf{1}_{\Omega_2}$ converges to $\boldsymbol{f}$.

Step 2. Proof of $(6.20)_2$. The solution $\boldsymbol{u}_\varepsilon$ of (6.6), satisfies by definition, $\operatorname{div} \boldsymbol{u}_\varepsilon = 0$ in Ω and vanishes on $\partial\Omega$. Consequently, for ϕ in $\mathcal{H}$ (which by definition vanishes on Γ), it follows that

$$0 = -\int_{\Omega_1} \operatorname{div} \boldsymbol{u}_\varepsilon \, \phi \, dx = \int_{\Omega_1} \boldsymbol{u}_\varepsilon \, \nabla\phi \, dx.$$

Multiplying with ε^{-1} and passing to the limit, which is allowed due to convergences $(6.16)_4$, implies $(6.20)_2$.

Step 3. Proof of $(6.20)_1$. Formulation (6.22) with test function

$$\boldsymbol{v}_\varepsilon(x) \doteq \varepsilon\phi(x)\boldsymbol{\psi}\Big(\Big\{\frac{x}{\varepsilon}\Big\}_Y\Big),$$

where $\boldsymbol{\psi} \in \mathbf{H}(Y)$ and $\phi \in \mathcal{D}(\Omega_1)$, becomes under unfolding for ε small enough,

$$\begin{aligned} \mu \int_{\Omega_1\times Y} \mathcal{T}_\varepsilon(\nabla \boldsymbol{u}_\varepsilon)\, \mathcal{T}_\varepsilon(\nabla \boldsymbol{v}_\varepsilon)\, dxdy - \int_{\Omega_1\times Y} \mathcal{T}_\varepsilon \circ \mathcal{E}_\varepsilon(p_\varepsilon) \mathcal{T}_\varepsilon(\operatorname{div} \boldsymbol{v}_\varepsilon)\, dxdy \\ = \int_{\Omega_1\times Y} \mathcal{T}_\varepsilon(\varepsilon \boldsymbol{f}_\varepsilon)\, \mathcal{T}_\varepsilon(\varepsilon^{-1}\boldsymbol{v}_\varepsilon)\, dxdy. \end{aligned} \tag{6.23}$$

The second term in the last equality writes successively,

$$\begin{aligned} &\int_{\Omega_1\times Y} \mathcal{T}_\varepsilon \circ \mathcal{E}_\varepsilon(p_\varepsilon) \mathcal{T}_\varepsilon(\operatorname{div} \boldsymbol{v}_\varepsilon)\, dxdy \\ &= \int_{\Omega_1\times Y} \mathcal{T}_\varepsilon \circ \mathcal{E}_\varepsilon(p_\varepsilon)(\varepsilon\mathcal{T}_\varepsilon(\nabla\phi)\, \boldsymbol{\psi}(y) + \mathcal{T}_\varepsilon(\phi) \operatorname{div}_y \boldsymbol{\psi}(y))\, dxdy \\ &= \int_{\Omega_1\times Y} \big[\varepsilon\mathcal{T}_\varepsilon \circ \mathcal{E}_\varepsilon(p_\varepsilon)\mathcal{T}_\varepsilon(\nabla\phi)\, \boldsymbol{\psi}(y) + \mathcal{T}_\varepsilon\big(\mathcal{E}_\varepsilon(p_\varepsilon) - \mathcal{M}^*_\varepsilon(p_\varepsilon)\big)\mathcal{T}_\varepsilon(\phi) \operatorname{div}_y \boldsymbol{\psi}(y)\big] dxdy \\ &\qquad + \int_{\Omega_1\times Y} \varepsilon\mathcal{M}^*_\varepsilon(p_\varepsilon) \frac{\mathcal{T}_\varepsilon(\phi) - \mathcal{M}_\varepsilon(\phi)}{\varepsilon} \operatorname{div}_y \boldsymbol{\psi}\, dxdy \\ &\qquad + \int_{\Omega_1\times Y} \mathcal{M}^*_\varepsilon(p_\varepsilon)\, \mathcal{M}_\varepsilon(\phi) \operatorname{div}_y \boldsymbol{\psi}\, dxdy. \end{aligned}$$

Now, we can pass to the limit in the the right-hand side of this equality, due to convergences (6.18) and Proposition 1.39(ii). The last term vanishes because, by the periodicity of $\boldsymbol{\psi}$,

$$\int_Y \operatorname{div}_y \boldsymbol{\psi}(y)\, dy = 0.$$

Therefore, passing to the limit, one obtains

$$\begin{aligned} &\int_{\Omega_1\times Y} \mathcal{T}_\varepsilon \circ \mathcal{E}_\varepsilon(p_\varepsilon) \mathcal{T}_\varepsilon(\operatorname{div} \boldsymbol{v}_\varepsilon)\, dxdy \\ &\to \int_{\Omega_1\times Y} \Big[p^1(x)\nabla\phi(x)\, \big(\boldsymbol{\psi}(y) + y^c \operatorname{div}_y \boldsymbol{\psi}(y)\big) + \widehat{p}(x,y)\, \phi(x) \operatorname{div}_y \boldsymbol{\psi}(y)\Big]\, dxdy. \end{aligned}$$

This last convergence allows to pass to the limit in (6.23) (using (6.16) in the first term and hypothesis (6.19) for the right-hand side), which gives

$$\mu \int_{\Omega_1 \times Y} \nabla_y \overline{\boldsymbol{u}}\, \nabla_y \boldsymbol{\psi}\, \phi\, dxdy - \int_{\Omega_1 \times Y} p^1(x)\, \nabla\phi(x) \big(\boldsymbol{\psi}(y) + y^c \mathrm{div}_y \boldsymbol{\psi}(y)\big) dxdy$$
$$- \int_{\Omega_1 \times Y} \widehat{p}(x,y)\, \phi(x)\, \mathrm{div}_y \boldsymbol{\psi}(y)\, dxdy = \int_{\Omega_1 \times Y} \widehat{\boldsymbol{f}}(x,y) \phi(x)\, \boldsymbol{\psi}(y)\, dxdy.$$

This holds for every ϕ in $\mathcal{D}(\Omega_1)$. Therefore, since p^1 belongs to $H^1(\Omega_1)$,

$$\begin{aligned} \mu \int_{\Omega_1 \times Y} & \nabla_y \overline{\boldsymbol{u}}\, \nabla_y \boldsymbol{\psi}\, \phi\, dxdy \\ & + \int_{\Omega_1 \times Y} \nabla p^1(x) \phi(x) \big(\boldsymbol{\psi}(y) + y^c \mathrm{div}_y \boldsymbol{\psi}(y)\big)\, dxdy \\ & - \int_{\Omega_1 \times Y} \widehat{p}(x,y)\, \phi(x)\, \mathrm{div}_y \boldsymbol{\psi}(y)\, dxdy \\ & = \int_{\Omega_1 \times Y} \widehat{\boldsymbol{f}}(x,y)\, \phi(x)\, \boldsymbol{\psi}(y)\, dxdy. \end{aligned} \tag{6.24}$$

Then, the density of $\mathcal{D}(\Omega_1) \otimes \mathbf{H}(Y)$ in $L^2(\Omega_1; \mathbf{H}(Y))$ proves $(6.20)_1$.

Step 4. The uniqueness of the solution of (6.21) is known, as it is the usual Stokes problem in Ω_2. The uniqueness of the solution of (6.20) holds since it is a periodic Stokes problem in Y with parameter in Ω. □

6.4 Standard homogenized problem and correctors

In order to state the (strong) homogenization result, we first introduce for $k \in \{1, \ldots, N\}$, the unique solution

$$(\boldsymbol{\theta}^k, q^k) \text{ in } \mathbf{V}(Y) \times L^2(Y^\star),$$

of the following periodic Stokes cell problem:

$$\begin{cases} \displaystyle\int_{Y^\star} \nabla_y \boldsymbol{\theta}^k\, \nabla_y \boldsymbol{\zeta}\, dy - \int_{Y^\star} q^k\, \mathrm{div}_y \boldsymbol{\zeta}\, dy = \int_{Y^\star} \zeta_k\, dy, \\ \displaystyle\int_{Y^\star} q^k\, dy = 0, \\ \forall \boldsymbol{\zeta} \in \mathbf{H}(Y). \end{cases} \tag{6.25}$$

Let $K = (k_{ij})_{1 \leqslant i,j \leqslant N}$ be the matrix with constant coefficients given by

$$k_{ij} = \int_{Y^\star} \theta^i_j(y)\, dy. \tag{6.26}$$

For a.e. $x \in \Omega_1$, let also $(\overline{\boldsymbol{u}}^0, \widehat{p}^0) \in L^2(\Omega_1; \mathbf{V}(Y)) \times L^2(\Omega_1 \times Y^*)$ be the unique solution of the problem (with $\widehat{\boldsymbol{f}}$ defined by (6.19)),

$$\begin{cases} \displaystyle\int_Y \nabla_y \overline{\boldsymbol{u}}^0(x,y)\, \nabla_y \boldsymbol{\psi}(y)\, dy - \int_Y \widehat{p}^0(x,y)\, \mathrm{div}_y \boldsymbol{\psi}(y)\, dy \\ \qquad\qquad = \displaystyle\int_Y \big(\widehat{\boldsymbol{f}}(x,y) - \mathcal{M}_Y(\widehat{\boldsymbol{f}})(x)\big)\, \boldsymbol{\psi}(y)\, dy, \\ \displaystyle\int_{Y*} \widehat{p}^0(x,y)\, dy = 0, \\ \forall \boldsymbol{\psi} \in \mathbf{H}(Y). \end{cases} \tag{6.27}$$

This is a standard periodic Stokes problem where $x \in \Omega_1$ is a parameter. Note that by the uniqueness of the solution of (6.27), $\overline{\boldsymbol{u}}^0$ and $\widehat{p}^0$ both vanish if and only if $\widehat{\boldsymbol{f}}$ is independent of y.

Theorem 6.7. *Under the assumptions of Theorem 6.5, the function p^1 belonging to $\mathbf{H}$ is the unique solution of the following elliptic problem in Ω_1:*

$$\begin{cases} -\mathrm{div}\, K\, \nabla p^1 = -\mathrm{div}\, \mathcal{M}_Y(\overline{\boldsymbol{u}}^0) - \mathrm{div}\, K \mathcal{M}_Y(\widehat{\boldsymbol{f}}) & \text{in } \Omega_1 \\ \big(K\,(\nabla p^1 - \mathcal{M}_Y(\widehat{\boldsymbol{f}})) - \mathcal{M}_Y(\overline{\boldsymbol{u}}^0)\big) \cdot n = 0 & \text{on } \partial\Omega_1 \backslash \Gamma, \end{cases} \tag{6.28}$$

where n is the unit outward normal to $\partial\Omega$.

Furthermore, $\overline{\boldsymbol{u}}$ and $\widehat{p}$ are determined uniquely as

$$\begin{aligned} \overline{\boldsymbol{u}}(x,y) &= \frac{1}{\mu}\Big(\overline{\boldsymbol{u}}^0(x,y) + \sum_{k=1}^{N} \big(\mathcal{M}_Y(\widehat{f}_k)(x) - \frac{\partial p^1}{\partial x_k}(x)\big)\, \boldsymbol{\theta}^k(y)\Big), \\ \widehat{p}(x,y) &= y^c \cdot \nabla p^1(x) + \widehat{p}^0(x,y) + \sum_{k=1}^{N} \Big(\mathcal{M}_Y(\widehat{f}_k)(x) - \frac{\partial p^1}{\partial x_k}(x)\Big) q^k(y). \end{aligned} \tag{6.29}$$

The pair of functions $(\boldsymbol{u}, p^2)$ belonging to $\mathbf{V} \times L^2(\Omega_2)$, is the unique solution of the Stokes problem in Ω_2,

$$\begin{cases} \mu \displaystyle\int_{\Omega_2} \nabla \boldsymbol{u}\, \nabla \boldsymbol{v}\, dx - \int_{\Omega_2} p^2\, \mathrm{div}\, \boldsymbol{v}\, dx = \int_{\Omega_2} \boldsymbol{f}\, \boldsymbol{v}\, dx, \\ \forall \boldsymbol{v} \in H_0^1(\Omega_2)^N. \end{cases} \tag{6.30}$$

The problems in Ω_1 and in Ω_2 are uncoupled.

Proof. The limit problem in Ω_2 is clear; indeed, (6.30) is the same as (6.21). Equation (6.24) implies that for a.e. $x \in \Omega_1$

$$\begin{aligned} \mu \int_Y \nabla_y \overline{\boldsymbol{u}}(x,y)\, \nabla_y \boldsymbol{\psi}(y)\, dy + \int_Y \nabla p^1(x)\, \Big(\boldsymbol{\psi}(y) + y^c \mathrm{div}_y \boldsymbol{\psi}(y)\Big)\, dy \\ - \int_Y \widehat{p}(x,y)\, \mathrm{div}_y \boldsymbol{\psi}(y)\, dy = \int_Y \widehat{\boldsymbol{f}}(x,y)\, \boldsymbol{\psi}(y)\, dy, \end{aligned}$$

which in turn, gives for a.e. $x \in \Omega_1$,

$$\mu \int_Y \nabla_y \overline{\boldsymbol{u}}(x,y)\, \nabla_y \boldsymbol{\psi}(y)\, dy - \int_Y \Big(\widehat{p}(x,y) - y^c\, \nabla p^1(x)\Big)\, \mathrm{div}_y \boldsymbol{\psi}(y)\, dy$$
$$= \int_Y \big(\,\widehat{\boldsymbol{f}}(x,y) - \nabla p^1(x)\big)\, \boldsymbol{\psi}(y)\, dy.$$

Using here the solutions of the cell-problems (6.25) and (6.27), implies (6.29). In particular,

$$\mu\, \mathcal{M}_Y(\overline{\boldsymbol{u}}) = \mathcal{M}_Y(\overline{\boldsymbol{u}}^0) - K\big(\nabla p^1 - \mathcal{M}_Y(\widehat{\boldsymbol{f}})\big), \tag{6.31}$$

which, replaced in the last equation of (6.20), gives the variational formulation of (6.28). □

Remark 6.8. *The third convergence of Proposition 6.4 implies that*

$$\varepsilon^{-1} u_\varepsilon \rightharpoonup \mathcal{M}_Y(\overline{\boldsymbol{u}}) \quad \textit{weakly in } L^2_{\mathrm{div}}(\Omega_1).$$

Consequently, the following limit holds in $H^{-1/2}(\partial\Omega_1)$*:*

$$\begin{aligned}\lim_{\varepsilon\to 0} \varepsilon^{-1} \boldsymbol{u}_\varepsilon \cdot n &= \mathcal{M}_Y(\overline{\boldsymbol{u}}) \cdot n\\ &= -\frac{1}{\mu}\big(K\,(\nabla p^1 - \mathcal{M}_Y(\widehat{\boldsymbol{f}})) - \mathcal{M}_Y(\overline{\boldsymbol{u}}^0)\big) \cdot n \quad \textit{on } \partial\Omega_1,\end{aligned}$$

giving the limit of the renormalized normal velocity on the interface Γ.

The macroscopic velocity in Ω_1, $\mathcal{M}_Y(\overline{\boldsymbol{u}})$, *is given by* (6.31) *which is Darcy's law in this context. If* $\widehat{\boldsymbol{f}}$ *does not depend of* y, *it implies that* $\overline{\boldsymbol{u}}^0 = 0$ *and formula* (6.31) *is the classical Darcy's law,*

$$\mathcal{M}_Y(\overline{\boldsymbol{u}}) = -\frac{1}{\mu} K(\nabla p^1 - \widehat{\boldsymbol{f}}),$$

where K *(see* (6.26)*) is known as the permeability matrix of the limit material. For a fully perforated domain* ($\Omega_2 = \varnothing$)*, this was proved in [107] and [185].*

Finally, as in preceding applications, one can investigate the convergence of the energy and obtain strong convergences.

Proposition 6.9 (Convergence of the energy). *Assume that the hypotheses of Theorem 6.5 are fulfilled and that*

$$\begin{aligned}&\mathcal{T}_\varepsilon(\varepsilon \boldsymbol{f}_\varepsilon \mathbf{1}_{\Omega_1}) \to \widehat{f} \quad \textit{strongly in } L^2(\Omega)^N,\\ &\varepsilon \boldsymbol{f}_\varepsilon \mathbf{1}_{\Omega_1 \cap \Lambda_\varepsilon} \to 0 \quad \textit{strongly in } L^2(\Omega)^N.\end{aligned}$$

Then, the following strong convergences hold:

$$\begin{aligned}&\nabla_y \mathcal{T}_\varepsilon(\varepsilon^{-1} \boldsymbol{u}_\varepsilon)_{|\Omega_1} \to \nabla_y \overline{\boldsymbol{u}} \quad \textit{strongly in } L^2(\Omega_1 \times Y)^{N\times N},\\ &\nabla \boldsymbol{u}_{\varepsilon|\Omega_2} \to \nabla \boldsymbol{u} \quad \textit{strongly in } L^2(\Omega_2)^{N\times N},\\ &p_{\varepsilon|\Omega_2} \to p^2 \quad \textit{strongly in } L^2(\Omega_2).\end{aligned} \tag{6.32}$$

Proof. The proof is based on the convergence of the energy as in Proposition 3.9. One easily checks the following successive computations, recalling that $\boldsymbol{u} = 0$ in Ω_1 and $\overline{\boldsymbol{u}} = 0$ in Ω_2:

$$
\begin{aligned}
&\frac{\mu}{|Y|}\int_{\Omega\times Y} \mathcal{T}_\varepsilon(\nabla \boldsymbol{u}_\varepsilon)\,\mathcal{T}_\varepsilon(\nabla \boldsymbol{u}_\varepsilon)\,dxdy + \mu\int_{\Lambda_\varepsilon} \nabla \boldsymbol{u}_\varepsilon\,\nabla \boldsymbol{u}_\varepsilon\,dx\\
&\qquad = \mu\int_{\Omega} \nabla \boldsymbol{u}_\varepsilon\,\nabla \boldsymbol{u}_\varepsilon\,dx = \int_{\Omega} \boldsymbol{f}_\varepsilon\,\boldsymbol{u}_\varepsilon\,dx + \int_{\Lambda_\varepsilon} \boldsymbol{f}_\varepsilon\,\boldsymbol{u}_\varepsilon\,dx\\
&\qquad = \frac{1}{|Y|}\int_{\Omega\times Y} \mathcal{T}_\varepsilon(\varepsilon \boldsymbol{f}_\varepsilon \mathbf{1}_{\Omega_1})\,\mathcal{T}_\varepsilon(\varepsilon^{-1}\boldsymbol{u}_\varepsilon \mathbf{1}_{\Omega_1})\,dxdy\\
&\qquad\qquad + \frac{1}{|Y|}\int_{\Omega\times Y} \mathcal{T}_\varepsilon(\boldsymbol{f}_\varepsilon \mathbf{1}_{\Omega_2})\,\mathcal{T}_\varepsilon(\boldsymbol{u}_\varepsilon \mathbf{1}_{\Omega_2})\,dxdy + \int_{\Lambda_\varepsilon} \boldsymbol{f}_\varepsilon\,\boldsymbol{u}_\varepsilon\,dx.
\end{aligned}
$$

By the hypotheses

$$
\begin{aligned}
&\frac{1}{|Y|}\int_{\Omega\times Y} \mathcal{T}_\varepsilon(\varepsilon \boldsymbol{f}_\varepsilon \mathbf{1}_{\Omega_1})\,\mathcal{T}_\varepsilon(\varepsilon^{-1}\boldsymbol{u}_\varepsilon \mathbf{1}_{\Omega_1})\,dxdy \to \frac{1}{|Y|}\int_{\Omega_1\times Y} \widehat{\boldsymbol{f}}\,\overline{\boldsymbol{u}}\,dxdy,\\
&\frac{1}{|Y|}\int_{\Omega\times Y} \mathcal{T}_\varepsilon(\boldsymbol{f}_\varepsilon \mathbf{1}_{\Omega_2})\,\mathcal{T}_\varepsilon(\boldsymbol{u}_\varepsilon \mathbf{1}_{\Omega_2})\,dxdy \to \int_{\Omega_2} \boldsymbol{f}\,\boldsymbol{u}\,dx.
\end{aligned}
$$

Similarly,

$$
\int_{\Lambda_\varepsilon} \boldsymbol{f}_\varepsilon\,\boldsymbol{u}_\varepsilon\,dx \to 0.
$$

Taking $\boldsymbol{\Psi} = \overline{\boldsymbol{u}}$ in (6.20) gives for value of the first integral above, namely

$$
\int_{\Omega_1\times Y} \widehat{\boldsymbol{f}}\,\overline{\boldsymbol{u}}\,dxdy = \mu\int_{\Omega_1\times Y} \nabla_y \overline{\boldsymbol{u}}\,\nabla_y \overline{\boldsymbol{u}}\,dxdy,
$$

since, according to the second condition of (6.20),

$$
\int_{\Omega_1\times Y} \overline{\boldsymbol{u}}\,\nabla p^1\,dxdy = |Y|\int_{\Omega_1} \mathcal{M}_Y(\overline{\boldsymbol{u}})\,\nabla p_1\,dx \equiv 0.
$$

Taking $\boldsymbol{v} = \boldsymbol{u}$ in (6.21) gives

$$
\int_{\Omega_2} \boldsymbol{f}\,\boldsymbol{u}\,dx = \mu\int_{\Omega_2} \nabla \boldsymbol{u}\,\nabla \boldsymbol{u}\,dx.
$$

From these equalities and by Lemma 3.11, it follows that

$$
\begin{aligned}
&\frac{1}{|Y|}\int_{\Omega\times Y} \mathcal{T}_\varepsilon(\nabla \boldsymbol{u}_\varepsilon)\,\mathcal{T}_\varepsilon(\nabla \boldsymbol{u}_\varepsilon)\,dxdy\\
&\qquad \to \frac{1}{|Y|}\int_{\Omega\times Y} \big(\nabla \boldsymbol{u} + \nabla_y \overline{\boldsymbol{u}}\big)\,\big(\nabla \boldsymbol{u} + \nabla_y \overline{\boldsymbol{u}}\big)\,dxdy, \qquad (6.33)\\
&\lim_{\varepsilon\to 0}\int_{\Lambda_\varepsilon} \nabla \boldsymbol{u}_\varepsilon\,\nabla \boldsymbol{u}_\varepsilon\,dx = 0.
\end{aligned}
$$

These convergences, using Lemma3.11, give the strong convergence,

$$\mathcal{T}_\varepsilon(\nabla \boldsymbol{u}_\varepsilon) \to \nabla \boldsymbol{u} + \nabla_y \overline{\boldsymbol{u}} \quad \text{strongly in } \; L^2(\Omega \times Y)^{N\times N}. \tag{6.34}$$

Since $\boldsymbol{u}$ vanishes in Ω_1, (6.32)(i) follows.

In Ω_2 where $\overline{\boldsymbol{u}}$ vanishes, (6.34), (6.33) and Proposition 1.29(iii), also implies (6.32)(ii) (the strong limit of $\mathcal{T}_\varepsilon(\nabla \boldsymbol{u}_\varepsilon \mathbf{1}_{\Omega_2})$ is independent of y).

Finally, going back to the Stokes problem, we get the strong convergence of ∇p_ε to ∇p^2 in $H^{-1}(\Omega_2)$, that is the strong convergence (6.32)(iii). □

As usual with unfolding, the convergences from this proposition give correctors results. There is actually no need of a corrector in Ω_2 where $\{\boldsymbol{u}_\varepsilon\}_\varepsilon$ converges strongly to $\boldsymbol{u}$ in $H^1(\Omega_2)^N$ (by (6.32)(ii)). in Ω_1, one has

Proposition 6.10. *Under the previous hypotheses, both*

$$\nabla \boldsymbol{u}_\varepsilon - \frac{1}{\mu}\mathcal{U}_\varepsilon(\nabla_y \overline{\boldsymbol{u}}^0) + \frac{1}{\mu}\mathcal{M}_\varepsilon\Big(\frac{\partial p^1}{\partial x^k}\Big)\nabla_y \boldsymbol{\theta}^k\left(\left\{\frac{\cdot}{\varepsilon}\right\}_Y\right)\mathbf{1}_{\widehat{\Omega}_{1,\varepsilon}},$$

and

$$\nabla \boldsymbol{u}_\varepsilon - \frac{1}{\mu}\mathcal{U}_\varepsilon(\nabla_y \overline{\boldsymbol{u}}^0) + \frac{1}{\mu}\,\mathcal{Q}_\varepsilon\Big(\frac{\partial p^1}{\partial x^k}\Big)\nabla_y \boldsymbol{\theta}^k\left(\left\{\frac{\cdot}{\varepsilon}\right\}_Y\right)\mathbf{1}_{\widehat{\Omega}_{1,\varepsilon}},$$

converge strongly to 0 in $L^2(\Omega_1)^{N\times N}$.

Proof. In Ω_1, by (6.33) and (6.34), Proposition 1.29(ii) gives

$$\nabla \boldsymbol{u}_\varepsilon - \mathcal{U}_\varepsilon(\nabla_y \overline{\boldsymbol{u}}) \to 0 \quad \text{strongly in } \; L^2(\Omega_1)^{N\times N}.$$

As in Theorem 3.27, using the representations (6.29)(i), one can restate this convergence in the two forms

$$\nabla \boldsymbol{u}_\varepsilon - \frac{1}{\mu}\mathcal{U}_\varepsilon(\nabla_y \overline{\boldsymbol{u}}^0) + \frac{1}{\mu}\mathcal{M}_\varepsilon\Big(\frac{\partial p^1}{\partial x^k}\Big)\mathcal{U}_\varepsilon(\nabla_y \boldsymbol{\theta}^k) \to 0 \text{ strongly in } L^2(\Omega_1)^{N\times N},$$

and

$$\nabla \boldsymbol{u}_\varepsilon - \frac{1}{\mu}\mathcal{U}_\varepsilon(\nabla_y \overline{\boldsymbol{u}}^0) + \frac{1}{\mu}\,\mathcal{Q}_\varepsilon\Big(\frac{\partial p^1}{\partial x^k}\Big)\mathcal{U}_\varepsilon(\nabla_y \boldsymbol{\theta}^k) \to 0 \text{ strongly in } L^2(\Omega_1)^{N\times N}.$$

However, the functions $\boldsymbol{\theta}^k$ do not depend upon x, so

$$\mathcal{U}_\varepsilon(\nabla_y \boldsymbol{\theta}^k) = \nabla_y \boldsymbol{\theta}^k\left(\left\{\frac{\cdot}{\varepsilon}\right\}_Y\right)\mathbf{1}_{\widehat{\Omega}_{1,\varepsilon}},$$

and this ends the proof. □

Part III

Partial Unfolding

In this short Part III, we explain how to use the unfolding method in presence of parameters or when oscillations occur only with a subset of the variables. This generalizes the results of Section 1.5.

There are two equivalent strategies, unfolding with parameters and using artificial periods, both discussed in Chapter 7.

Chapter 8 is dedicated to the particular case of oscillating boundaries, where a particular case of partial unfolding is used.

Chapter 7

Partial unfolding: a brief primer

In this short chapter, we give a brief description of the case of partial unfolding and unfolding with parameters. The theory is presented insofar as it differs from the cases of Parts I and II. Some applications are also presented.

The situation corresponds to the case where oscillations occur (in the coefficients, the domain,...) only in some directions.

We present an example to explain the ideas.

Let Ω be a domain in $\mathbb{R}^N$, and $N_1 + N_2 = N$ (both positive). Consider a problem such as (3.3) posed in Ω where the coefficient matrix A^ε is of the form

$$A\Big(\frac{x'}{\varepsilon}, x''\Big) \quad \text{where } x \equiv (x', x'') \in \Omega \quad \text{with } x' \in \mathbb{R}^{N_1},\ x'' \in \mathbb{R}^{N_2}.$$

The map $A(y', x'')$ is defined for x'' in the projection of Ω on $\mathbb{R}^{N_2}$, for y' in $\mathbb{R}^{N_1}$. It is periodic with respect to the y' with a group of periods generated by a basis of $\mathbb{R}^{N_1}$ and a corresponding parallelotope Y'.

Two distinct strategies can be applied with the unfolding method. They will of course, give the same end result.

1. Perform a partial unfolding with respect to the first N_1 variables alone; this is unfolding with parameters, also known as partial unfolding.
2. Introduce N_2 artificial periods to obtain a basis of periods in $\mathbb{R}^N$ and apply the results of Part I. Then express the results in a way which is independent of the choice of these artificial periods.

7.1 Unfolding with parameters

This is well adapted to the situation where the domain is a product and the periods bear only on one of the factors. One has to prove the results of

D. Cioranescu et al., *The Periodic Unfolding Method*, Series in Contemporary Mathematics 3, https://doi.org/10.1007/978-981-13-3032-2_7

Chapter 1 in this specific situation.

Section 1.5 deals with an example where the parameters do not appear in partial derivatives.

A slightly different case has derivatives with respect to all variables but oscillation with respect to some only.

Considering weak convergences, this program is almost straightforward due to the basic relation between integrals (the equivalent of (i) of Proposition 1.8). This allows to obtain bounds in the appropriate L^p or $W^{1,p}$ spaces. To establish strong convergences, as in the standard case, one proves convergence for the L^p norms.

We refer to Subsection 3.1.4 where this technique is applied. See also Chapter 8 below where it is applied to the case of oscillating boundaries in a slightly more general framework (the unfolded domains depend upon ε but converge in an appropriate sense).

7.2 Unfolding with artificial periods

The second strategy of adding artificial periods is straightforward to apply, but it requires to express the final result independently of the added artificial periods. Below, we give a simple example of such a situation where this is achieved simply by averaging over the artificial periods.

7.2.1 A diffusion problem

We use the preceding notations ($N_1 + N_2 = N$, $x \equiv (x', x'')$, $x' \in \mathbb{R}^{N_1}$, $x'' \in \mathbb{R}^{N_2}$). Let a basis of the given macroscopic periods be $\{\mathbf{b}_1, \ldots, \mathbf{b}_{N_1}\}$. Complete the basis with arbitrary (linearly independent) $\{\mathbf{b}_{N_1+1}, \ldots, \mathbf{b}_N\}$. This allows to define $Y = Y' \times Y''$.

What information can be obtained via unfolding which is independent of the choice of the $\{\mathbf{b}_{N_1+1}, \ldots, \mathbf{b}_N\}$?

The answer is quite straightforward, but one needs to introduce a new notation for the partial unfolding:

$$\widehat{\mathcal{T}}_\varepsilon \doteq \mathcal{M}_{Y''} \circ \mathcal{T}_\varepsilon \quad \text{where} \quad \mathcal{M}_{Y''} \text{ is the averaging operator over } Y''.$$

Applying the operator $\mathcal{M}_{Y''}$ to all the results of Chapter 1, one obtains the statements corresponding to the new unfolding operator $\widehat{\mathcal{T}}_\varepsilon$.

For example, Theorem 1.41 can be restated as follows (compare with Theorem 1.50 (vi):

Theorem 7.1. *Suppose $p \in (1, +\infty)$. Let $\{w_\varepsilon\}_\varepsilon$ be a sequence in $W^{1,p}(\Omega)$ such that*

$$w_\varepsilon \rightharpoonup w \quad \textit{weakly in } W^{1,p}(\Omega).$$

Then, up a subsequence, there exists some $\overline{w}$ *in* $L^p(\Omega; W^{1,p}_{per,0}(Y'))$ *such that*

$$(i)\quad \frac{1}{\varepsilon}\Big(\widehat{\mathcal{T}}_\varepsilon(w_\varepsilon) - \mathcal{M}_\varepsilon(w_\varepsilon)\Big) \rightharpoonup y'^c \cdot \nabla w + \overline{w} \quad \textit{weakly in } L^p(\Omega; W^{1,p}(Y')),$$

$$(ii)\quad \widehat{\mathcal{T}}_\varepsilon(\nabla w_\varepsilon) \rightharpoonup \nabla w + \nabla_{y'}\overline{w} \quad \textit{weakly in } L^p(\Omega \times Y')^N.$$

Here $\overline{w} \doteq \mathcal{M}_{Y''}(\widehat{w})$, *where* $\widehat{w}$ *is given by Theorem 1.41.*

As an application, consider the simple situation of Subsection 3.1.1, and more precisely, Theorem 3.5. Assume that the matrix field A^ε does not depend upon the x'' and that, moreover, its unfolding $B^\varepsilon \doteq \mathcal{T}_\varepsilon(A^\varepsilon)$ is independent of $y'' \in Y''$ in $\widehat{\Omega}_\varepsilon \times Y$ (and so is its limit B). More generally, we can make this assumption only at the limit, that is

$$B^\varepsilon = \mathcal{T}_\varepsilon(A^\varepsilon) \to B \quad \text{a.e. in } \Omega \times Y \quad \text{with } B \text{ independent of } y''. \tag{7.1}$$

This happens certainly in the case where A^ε is a product of two factors, one oscillating in x' and the other one non oscillating in x''. The same holds true if A^ε is of the form

$$A^\varepsilon(x) = A\Big(\frac{x'}{\varepsilon}, x''\Big) \quad \text{in } \Omega.$$

What can be concluded for the limit problem (3.13)? The answer is given in the following variant of Theorem 3.5:

Proposition 7.2. *Under the assumptions of Theorem 3.5 as well as assumption (7.1), there exist* $u_0 \in H^1_0(\Omega)$ *and* $\overline{u} \in L^2(\Omega; H^1_{per,0}(Y'))$ *such that*

$$\begin{aligned} &u_\varepsilon \rightharpoonup u_0 \quad \textit{weakly in } H^1_0(\Omega),\\ &\widehat{\mathcal{T}}_\varepsilon(u_\varepsilon) \to u_0 \quad \textit{strongly in } L^2(\Omega; H^1(Y')),\\ &\widehat{\mathcal{T}}_\varepsilon(\nabla u_\varepsilon) \rightharpoonup \nabla u_0 + \nabla_{y'}\overline{u} \quad \textit{weakly in } L^2(\Omega \times Y')^N, \end{aligned} \tag{7.2}$$

and the pair $(u_0, \overline{u})$ *is the unique solution of problem*

$$\begin{cases} \dfrac{1}{|Y'|}\displaystyle\int_{\Omega\times Y'} B(x,y')\big[\nabla u_0(x) + \nabla_{y'}\overline{u}(x,y')\big]\big[\nabla\Psi(x) + \nabla_{y'}\Phi(x,y')\big]dxdy' \\ \qquad\qquad\qquad = \langle f, \Psi\rangle_{H^{-1}(\Omega), H^1_0(\Omega)}, \\ \forall \Psi \in H^1_0(\Omega), \;\; \forall \Phi \in L^2(\Omega;\, H^1_{per}(Y')). \end{cases}$$

Proof. Indeed, convergences (7.2) follow from Theorem 7.1. Then the homogenized problem above is easily obtained because for test functions φ independent of y'' ($\varphi(x,y) \doteq \Phi(x,y')$), all the functions in (3.13) are independent of y'' except for $\widehat{u}$, so setting $\overline{u} \doteq \mathcal{M}_{Y''}(\widehat{u})$ gives the result. □

All the statements of Subsection 3.1.1 can now be rephrased with formulas using only Y'. For example, only the first N_1 correctors $\widehat{\chi}_j$ are used (they only depend upon x and y') and the remaining N_2 correctors are not used (they simply vanish !). The formula for the homogenized matrix field is also written as an average over Y'. The convergence of the energy holds and the corrector is expressed in terms of the averaging operator defined for Y', etc.

Chapter 8

Oscillating boundaries

The point of the unfolding method is to reduce rapid oscillations (of size ε) to smooth variations. In the case of problems with small holes, the unfolding method essentially reduced the problem to a fixed domain. In this chapter, we consider a situation with rapidly oscillating boundaries. The very first results for oscillating boundaries were obtained in [39] making use of an adapted extension operator. The treatment using partial unfolding, which indeed reduced it to a fixed domain, appeared in ([93]). We shall consider much more general situations, where, after unfolding, the domains still have changing boundaries, albeit without high oscillations. Consequently, one first has to understand the situation of "converging" boundaries, which is not completely standard. This is the purpose of the first section below. The corresponding results are then applied in the next section which considers the application of the unfolding method to oscillating boundaries.

In the Propositions, Corollaries and Remarks of Section 8.1, the Lebesgue exponent p is restricted to $(1, +\infty)$.

8.1 Mosco convergences for varying domains

In this section, we consider as usual a sequence $\{\varepsilon\}$ of strictly positive numbers going to 0^+ and for each element of the sequence an open bounded domain $\mathcal{O}_\varepsilon$ in $\mathbb{R}^N$.

For $p \in (1, +\infty)$ we consider the spaces $L^p(\mathcal{O}_\varepsilon)$ (respectively, $W^{1,p}(\mathcal{O}_\varepsilon)$ or $W_0^{1,p}(\mathcal{O}_\varepsilon)$) and give simple conditions for the notion of convergence to make sense for sequences of functions v_ε which, for each ε takes a value in $L^p(\mathcal{O}_\varepsilon)$(resp. in $W^{1,p}(\mathcal{O}_\varepsilon)$, $W_0^{1,p}(\mathcal{O}_\varepsilon)$).

How can one give a notion of convergence for the sequence $\{v_\varepsilon\}_\varepsilon$?

Applications to variational problems require two properties, which have been well-known since the work of Mosco. In [162], the notion of Mosco-convergence is defined for sequences of closed convex sets (and in particular of closed subspaces) $\{\mathcal{C}_\varepsilon\}_\varepsilon$. They "converge" in the Mosco sense to a closed

D. Cioranescu et al., *The Periodic Unfolding Method*, Series in Contemporary Mathematics 3, https://doi.org/10.1007/978-981-13-3032-2_8

convex subset $\mathcal{C}$ in a given reflexive Banach space X, if the following two conditions are satisfied:

1. *Strong convergence.* For each element v in $\mathcal{C}$ there is a sequence $\{v_\varepsilon\}_\varepsilon$ converging strongly to v in X and such that for each ε, v_ε belongs to $\mathcal{C}_\varepsilon$.

2. *Weak convergence (for subsequences).* For any subsequence $\{\varepsilon'\}$ of $\{\varepsilon\}$, any sequence $\{v_{\varepsilon'}\}_{\varepsilon'}$ with $v_{\varepsilon'} \in \mathcal{C}_{\varepsilon'}$ and converging weakly to some v in X, it is true that v belongs to $\mathcal{C}$.

Definition 8.1. *When the two above conditions of weak and strong convergence are satisfied, we say that the sequence $\{\mathcal{C}_\varepsilon\}_\varepsilon$ Mosco-converges to $\mathcal{C}$.*

We introduce two notions of convergence for sequences of bounded open domains. The first is measure-theoretical, the second is metric-based.

The purpose is to establish results which mimic the basic tools of weak and strong convergence in the case of fixed domains for spaces of Lebesgue or Sobolev type (see Corollaries 8.5, 8.8 and 8.11).

For simplicity, we will assume that all the $\mathcal{O}_\varepsilon$'s are included in a fixed bounded set in $\mathbb{R}^N$.

Definition 8.2 (Convergence in measure). *The sequence of (uniformly) bounded domains $\{\mathcal{O}_\varepsilon\}_\varepsilon$ converges in the measure sense to a non empty domain $\mathcal{O}$ if the sequence of characteristic functions $\{\mathbf{1}_{\mathcal{O}_\varepsilon}\}_\varepsilon$ converges to $\mathbf{1}_{\mathcal{O}}$ almost everywhere.*
Since the sets are bounded it is equivalent that the Lebesgue measure of the symmetric difference of $\mathcal{O}_\varepsilon$ and Ω_ε goes to zero[(1)].

Definition 8.3 (Hausdorff convergence). *The Hausdorff excess of a set B to a set A in a metric space is*

$$\mathfrak{E}_{\mathcal{H}}(A\,,B) \doteq \sup_{x\in B} dist(x, A).$$

The Hausdorff distance between the sets A and B is

$$d_{\mathcal{H}}(A\,,B) \doteq \max(\mathfrak{E}_{\mathcal{H}}(A\,,B), \mathfrak{E}_{\mathcal{H}}(B\,,A)).$$

The sequence of (uniformly) bounded domains $\{\mathcal{O}_\varepsilon\}_\varepsilon$ converges in the Hausdorff sense to the domain $\mathcal{O}$ whenever the sequence $\{d_{\mathcal{H}}(\mathcal{O}_\varepsilon, \mathcal{O})\}_\varepsilon$ goes to zero.

Obviously, the second convergence implies the first one. It is also known that if the $\mathcal{O}_\varepsilon$'s are with uniformly Lipschitz boundaries, the first convergence implies the second one (a consequence of Ascoli's theorem).

We shall now discuss the notion of Mosco convergence for Lebesgue and Sobolev spaces.

---

(1) Recall the definition of the symmetric difference:

$$\mathcal{O}_\varepsilon \Delta \mathcal{O} \doteq (\mathcal{O}_\varepsilon \cap (\mathbb{R}^N \backslash \mathcal{O})) \cup (\mathcal{O} \cap (\mathbb{R}^N \backslash \mathcal{O}_\varepsilon)).$$

8.1.1 Mosco convergence for the sequence $\{L^p(\mathcal{O}_\varepsilon)\}_\varepsilon$.

The space $L^p(\mathcal{O}_\varepsilon)$ is isometric to the closed subspace of $L^p(\mathbb{R}^N)$ of its elements which vanish a.e. outside of $\mathcal{O}_\varepsilon$. Without loss of generality, we can identify these spaces and consider their Mosco convergence in the space $L^p(\mathbb{R}^N)$ (or in the L^p space of a bounded set containing all the $\mathcal{O}_\varepsilon$'s).

The following is straightforward.

Proposition 8.4. *If the sequence of $\{\mathcal{O}_\varepsilon\}_\varepsilon$ converges in measure to $\mathcal{O}$, then the sequence $\{L^p(\mathcal{O}_\varepsilon)\}_\varepsilon$ Mosco-converges to $L^p(\mathcal{O})$.*

Corollary 8.5 (Weak compactness for a sequence of L^p spaces). *Under the hypothesis of the preceding proposition, if the sequence $\{v_\varepsilon\}_\varepsilon$ is such that $\{\|v_\varepsilon\|_{L^p(\mathcal{O}_\varepsilon)}\}_\varepsilon$ is bounded, then, up to a subsequence, it converges weakly to some v in $L^p(\mathcal{O})$ in the sense of the second condition of Mosco convergence. Furthermore,*

$$\|v\|_{L^p(\mathcal{O})} \leqslant \liminf_{\varepsilon\to 0} \|v_\varepsilon\|_{L^p(\mathcal{O}_\varepsilon)}.$$

Remark 8.6. *Under the same hypotheses, if the full sequence $\{v_\varepsilon\}_\varepsilon$ converges weakly and if the converse inequality holds, namely,*

$$\limsup_{\varepsilon\to 0} \|v_\varepsilon\|_{L^p(\mathcal{O}_\varepsilon)} \leqslant \|v\|_{L^p(\mathcal{O})},$$

then v_ε converges strongly to v (in the sense of $L^p(\mathbb{R}^N)$).

8.1.2 Mosco convergence for the sequence $\{W^{1,p}(\mathcal{O}_\varepsilon)\}_\varepsilon$

For spaces $W^{1,p}(\mathcal{O}_\varepsilon)$, the situation is more complex. In order to use the Mosco-convergence framework, we can identify the space $W^{1,p}(\mathcal{O}_\varepsilon)$ to its image by the map $v \mapsto (v, \nabla v)$ to a closed subspace of $L^p(\mathbb{R}^N)^{N+1}$.

Proposition 8.7. *Suppose that the domains $\mathcal{O}_\varepsilon$ converge in the Hausdorff sense to $\mathcal{O}$ and that either all the sets $\mathcal{O}_\varepsilon$ are included in $\mathcal{O}$, or the latter admits an extension operator from $W^{1,p}(\mathcal{O})$ to $W^{1,p}(\mathbb{R}^N)$. Then the sequence $\{W^{1,p}(\mathcal{O}_\varepsilon)\}_\varepsilon$ Mosco-converges to $W^{1,p}(\mathcal{O})$ in $L^p(\mathbb{R}^N)^{N+1}$.*

Proof. Step 1. *Proof of the weak convergence condition.*

It is enough to reason with the whole sequence $\{\varepsilon\}$. The result of the previous subsection can be applied to the two weakly convergent sequences $\{v_\varepsilon\}_\varepsilon$ and $\{\nabla v_\varepsilon\}_\varepsilon$ in $L^p(\mathbb{R}^N)$ (resp. $L^p(\mathbb{R}^N)^N$). It remains to show that the weak limit W of $\{\nabla v_\varepsilon\}_\varepsilon$ in the previous sense is indeed ∇v in $\mathcal{O}$.

To this aim, let φ be an element of $\mathcal{D}(\mathcal{O})$ and j be in $\{1,\dots,N\}$. The hypothesis of the Hausdorff convergence of $\mathcal{O}_\varepsilon$ to $\mathcal{O}$ implies in particular, that the excess $\mathfrak{E}_{\mathcal{H}}(\mathcal{O}, \mathcal{O}_\varepsilon)$ goes to 0. Therefore, for ε small enough, φ belongs to $\mathcal{D}(\mathcal{O}_\varepsilon)$ and it is easy to pass to the limit in the duality

$$-\Big\langle v_\varepsilon, \frac{\partial\varphi}{\partial x_j}\Big\rangle_{\mathcal{D}'(\mathcal{O}_\varepsilon),\mathcal{D}(\mathcal{O}_\varepsilon)} = \Big\langle \frac{\partial v_\varepsilon}{\partial x_j}, \varphi\Big\rangle_{\mathcal{D}'(\mathcal{O}_\varepsilon),\mathcal{D}(\mathcal{O}_\varepsilon)} = \int_{\mathrm{supp}(\varphi)} \frac{\partial v_\varepsilon}{\partial x_j}(x)\varphi(x)\,dx$$

to obtain

$$-\left\langle v, \frac{\partial \varphi}{\partial x_j} \right\rangle_{\mathcal{D}'(\mathcal{O}_\varepsilon),\mathcal{D}(\mathcal{O}_\varepsilon)} = \int_{\mathrm{supp}(\varphi)} W_j(x)\varphi(x)\,dx$$

This shows that $W_j = \dfrac{\partial v}{\partial x_j}$.

Step 2. *Proof of the strong convergence condition.*

For the *strong convergence condition*, one needs to approximate, in the sense of $L^p(\mathbb{R}^N)^{N+1}$, a given element v of $W^{1,p}(\mathcal{O})$ by a sequence each element of which is in $W^{1,p}(\mathcal{O}_\varepsilon)$. One obvious case is when each $\mathcal{O}_\varepsilon$ is included in $\mathcal{O}$ (using the sequence $\{(v\mathbf{1}_{\mathcal{O}_\varepsilon}, \nabla v\,\mathbf{1}_{\mathcal{O}_\varepsilon})\}_\varepsilon$).

Otherwise, assume that there is an extension operator from $W^{1,p}(\mathcal{O})$ to $W^{1,p}(\mathbb{R}^N)$ (which holds true as soon as $\mathcal{O}$ has a Lipschitz boundary). For v in $W^{1,p}(\mathcal{O})$, let $\widehat{v} \in W^{1,p}(\mathbb{R}^N)$ be its extension. Then, $\{(\widehat{v}|_{\mathcal{O}_\varepsilon}, \nabla\widehat{v}|_{\mathcal{O}_\varepsilon})\}_\varepsilon$ converges to $(v, \nabla v)$ strongly in the sense of $L^p(\mathbb{R}^N)^{N+1}$. □

The following corollary (similar to Corollary 8.5) is easily proved.

Corollary 8.8 (weak compactness for a sequence of $W^{1,p}$ spaces). *Under the hypothesis of the preceding proposition, if the sequence $\{v_\varepsilon\}_\varepsilon$ is such that $\{\|v_\varepsilon\|_{W^{1,p}(\mathcal{O}_\varepsilon)}\}_\varepsilon$ is uniformly bounded, then, up to a subsequence, it converges weakly to some $v \in W^{1,p}(\mathcal{O})$ in the sense of the Mosco convergence of $W^{1,p}$ spaces* (2). □

Remark 8.9. *A similar result (for weak convergences) holds when only some partial derivatives $\partial v_\varepsilon/\partial x_j$ are bounded in the sense of L^p.*

8.1.3 Mosco convergence for the sequence $\{W_0^{1,p}(\mathcal{O}_\varepsilon)\}_\varepsilon$

The space $W_0^{1,p}(\mathcal{O}_\varepsilon)$ can be seen as a closed subspace of $W^{1,p}(\mathcal{O}_\varepsilon)$ so that the previous results can be applied. However, extending its elements by zero, one can also consider $W_0^{1,p}(\mathcal{O}_\varepsilon)$ as a closed subspace of $W^{1,p}(\mathbb{R}^N)$ and one can consider Mosco convergence in the latter. It turns out that the two settings give the same notion of convergence.

Proposition 8.10. *Suppose that the domains $\mathcal{O}_\varepsilon$ converge in the Hausdorff sense to $\mathcal{O}$ and that either all the domains $\mathcal{O}_\varepsilon$ are included in $\mathcal{O}$, or the latter has a Lipschitz boundary. Then the sequence $\{W_0^{1,p}(\mathcal{O}_\varepsilon)\}_\varepsilon$ Mosco-converges to $W_0^{1,p}(\mathcal{O})$.*

Proof. Step 1. *Proof of the weak convergence condition.*

This condition requires that the "weak limit" v of a sequence $\{v_\varepsilon\}_\varepsilon$ (with $v_\varepsilon \in W_0^{1,p}(\mathcal{O}_\varepsilon)$ for every ε) belong to the subspace $W_0^{1,p}(\mathcal{O})$ of $W^{1,p}(\mathcal{O})$. This is true if all the $\mathcal{O}_\varepsilon$'s are included in $\mathcal{O}$ (so that the spaces $W_0^{1,p}(\mathcal{O}_\varepsilon)$ are all included in $W_0^{1,p}(\mathcal{O})$).

---

(2) Assuming the Hausdorff convergence of the $\mathcal{O}_\varepsilon$ to $\mathcal{O}$ holds and $\mathcal{O}$ is with Lipschitz boundary one can show, making use of (12.3), that the same subsequence, extended by zero to $\mathbb{R}^N$, converges strongly to v in $L^p(\mathbb{R}^N)$.

It is also true if $\mathcal{O}$ is with Lipschitz boundary. Indeed, the hypothesis of the Hausdorff convergence of $\mathcal{O}_\varepsilon$ to $\mathcal{O}$ implies in particular that the excess $\mathfrak{E}_{\mathcal{H}}(\mathcal{O}_\varepsilon, \mathcal{O})$ goes to 0. Therefore each v_ε vanishes outside of the open neighborhoods $\mathcal{V}_{\varepsilon'}(\mathcal{O})$ (for $\varepsilon \leqslant \varepsilon'$), so that v vanishes outside of $\overline{\mathcal{O}}$ (which is the intersection of there neighborhoods). But the fact that $\mathcal{O}$ is with Lipschitz boundary implies that this property is equivalent to belonging to $W_0^{1,p}(\mathcal{O})$.
Step 2. Proof of the strong convergence condition. It is enough to approximate a given v in $W_0^{1,p}(\mathcal{O})$ by a sequence $\{v_\varepsilon\}_\varepsilon$ in $\mathcal{D}(\mathcal{O})$ and such that $\text{supp}(v_\varepsilon) \subset \mathcal{O} \cap \mathcal{O}_\varepsilon$. By the hypothesis, it follows that v_ε belongs to $\mathcal{D}(\mathcal{O}_\varepsilon)$ (the case of a decreasing sequence of open domains $\mathcal{O}_\varepsilon$ is actually trivial, since it is enough to extend v by zero). □

The following result (similar to that stated in Corollary 8.8) is straightforward:

Corollary 8.11 (Weak compactness for a sequence of $W_0^{1,p}$ spaces). *Under the hypotheses of the preceding proposition, if the sequence $\{v_\varepsilon\}_\varepsilon$ is such that for each ε, $\{\|v_\varepsilon\|_{W_0^{1,p}(\mathcal{O}_\varepsilon)}\}_\varepsilon$ is bounded, then, up to a subsequence, it converges weakly to some $v \in W_0^{1,p}(\mathcal{O})$ in the sense of the Mosco convergence for $W^{1,p}$ spaces. In particular, the same subsequence converges strongly to v in $L^p(\mathbb{R}^N)$.*

Remark 8.12. *Here are two examples where the results of this Subsection 8.1.2 apply. In Parts I and II, the domains $\widehat{\Omega}_\varepsilon \times Y$ converge to $\Omega \times Y$ in the Hausdorff sense (likewise $\widehat{\Omega}_\varepsilon \times Y^*$ converge to $\Omega \times Y^*$). In Section 5.3, the domains Σ_ε also converge to Σ in the Hausdorff sense.*

In these two parts, making use of the results of Subsection 8.1.2, one can consider situations with varying domains ins the place of a fixed Ω.

8.2 An example of oscillating boundaries

In this section, we give a model example of the use of partial unfolding to treat the case of oscillating boundaries in two dimensions. Problems in higher dimensions can be treated in the same way (but the figures are in dimension 4 at least !). Three dimensional models are presented at the end of this chapter. Problems of this type with tooth-like periodic oscillating boundaries were studied originally in [39], see also [81] for nonlinear operators and more recently [109] for the case of generalized solutions in a non-periodic setting (and the list of references therein).

8.2.1 Setting of the problem

Consider a sequence of two-dimensional domains $\{\Omega_\varepsilon\}_\varepsilon$ which are all included in the rectangle Ω (see Figure 8.1),

$$\Omega \doteq (0,1) \times (-1,1), \quad \Omega_A = (0,1) \times (0,1) \quad \text{and} \quad \Omega_B = (0,1) \times (-1,0).$$

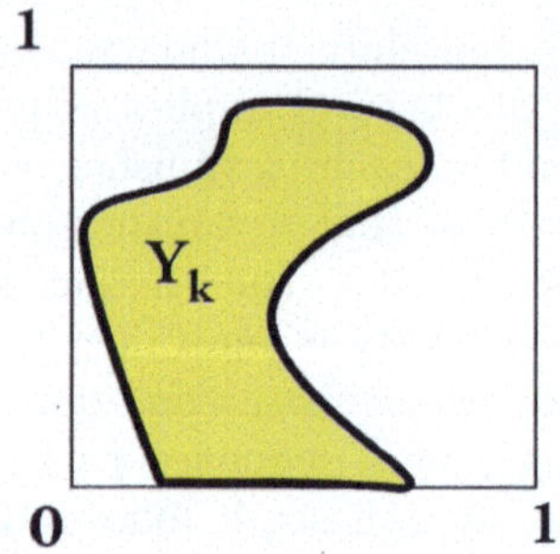

Figure 8.1: An example of a set Y_k

For a positive integer N, set $\varepsilon = 1/N$. For each $k = 0, \ldots, N-1$, let Y_k be an open Lipschitz subdomain of Ω_A (with variable $y = (y_1, y_2)$) such that the interior of $\partial Y_k \bigcap \{y_2 = 0\}$ is not empty in $(0,1)$. Set also

$$Y_k^\varepsilon \doteq \Big\{(x_1, x_2) \in (0,1)^2 \mid \Big(\frac{x_1 - k}{\varepsilon}, x_2\Big) \in Y_k\Big\}.$$

The definitions of the domains Ω_A^ε and Ω_ε are given, respectively, by

$$\Omega_A^\varepsilon \doteq \bigcup_{k=0}^{N-1} Y_k^\varepsilon, \quad \Omega^\varepsilon = \text{interior}\ \Big(\overline{\Omega}_A^\varepsilon \cup \overline{\Omega}_B\Big).$$

Note that part Ω_A^ε contained in Ω_A, exhibits an oscillating boundary. Figure 8.2 gives two examples of such domains. The second one corresponds to the case studied in [93] where the Y_k's are all the same rectangle.

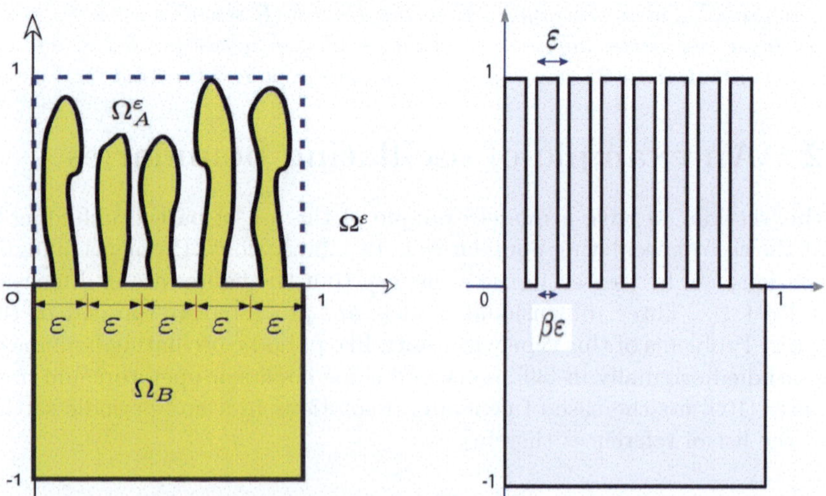

Figure 8.2: Examples of sets Ω^ε

Notation 8.13. *We set*

$$I = (0,1) \times \{0\} \text{ (a line segment in } \Omega), \quad J^\varepsilon = \text{interior}\big(\partial\Omega_a^\varepsilon \bigcup \partial\Omega_b^\varepsilon\big).$$

which is a subset of I.

In this geometrical framework, we are in a position to introduce the problem in Ω_ε. For $A^\varepsilon \doteq (a^\varepsilon_{ij})_{1\leqslant i,j\leqslant 2}$ in $M(\alpha,\beta,\Omega)$, $c>0$ a constant, and f_ε given in $L^2(\Omega)$, consider the following Neumann problem:

$$\begin{cases} -\displaystyle\sum_{i,j=1,2}\frac{\partial}{\partial x_j}\Big(a^\varepsilon_{ij}\frac{\partial u_\varepsilon}{\partial x_i}\Big)+c\,u_\varepsilon=f_\varepsilon & \text{in } \Omega_\varepsilon,\\ \displaystyle\sum_{i,j=1,2} a^\varepsilon_{ij}\frac{\partial u_\varepsilon}{\partial x_i}n_j=0 & \text{on } \partial\Omega_\varepsilon,\end{cases}$$

the weak formulation of which is

$$\int_{\Omega_\varepsilon}\Big(\sum_{i,j=1,2} a^\varepsilon_{ij}\frac{\partial u}{\partial x_i}\frac{\partial \varphi}{\partial x_j}+c\,u_\varepsilon\Big)\,dx=\int_{\Omega_\varepsilon} f_\varepsilon\varphi\,dx,\quad \forall\varphi\in H^1(\Omega_\varepsilon). \tag{8.1}$$

Note that f_ε cannot be chosen as an element of $H^{-1}(\Omega)$ since, in that case, $f_{\varepsilon|\Omega_\varepsilon}$ does not make sense.

With the above assumptions on the data, the Lax-Milgram theorem provides a unique solution u_ε in $H^1(\Omega_\varepsilon)$ of (8.1) such that

$$\|u_\varepsilon\|_{H^1(\Omega_\varepsilon)}\leqslant C, \tag{8.2}$$

where C is independent of ε. Our aim is to study what happens when $\varepsilon\to 0$.

8.2.2 The partial unfolding operator

On Ω_B we use the standard unfolding operator $\mathcal{T}_\varepsilon$ (see Chapter 1) corresponding to the reference cell $Y=(0,1)^2$. On Ω^ε_A, we introduce a partial unfolding operator which unfolds only on the first variable x_1. To do so, let Σ_ε denote the open domain defined as follows (see Figure 8.3):

$$\Sigma_\varepsilon \doteq \text{interior}\Big(\bigcup_{k=0}^{N-1}[k\varepsilon,(k+1)\varepsilon)\times Y_k\Big)\subset(0,1)^3.$$

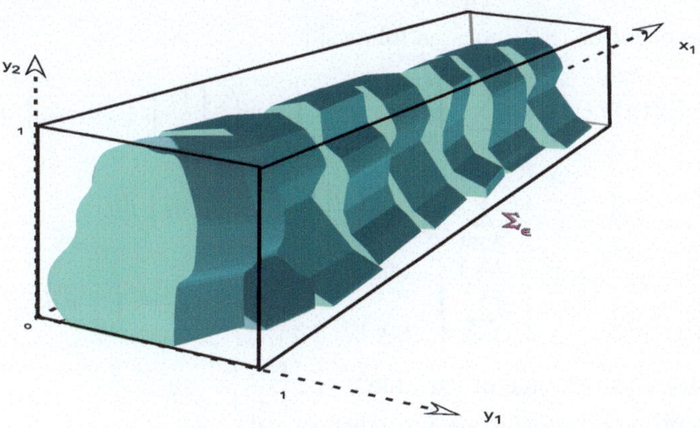

Figure 8.3: An example of a set Σ_ε

Definition 8.14 (The partial unfolding operator $\mathcal{T}_\varepsilon^A$).
For a function $u : \Omega_A^\varepsilon \to \mathbb{R}$*, its unfolding by the partial operator* $\mathcal{T}_\varepsilon^A$ *is the function* $\mathcal{T}_\varepsilon^A(u) : \Sigma_\varepsilon \to \mathbb{R}$*, defined (almost everywhere) as*

$$\mathcal{T}_\varepsilon^A(u)(x_1, y) \doteq u\left(\varepsilon\left[\frac{x_1}{\varepsilon}\right] + \varepsilon y_1, y_2\right).$$

Proposition 8.15 (Properties of $\mathcal{T}_\varepsilon^A$). *The properties of this partial unfolding operator are similar to those of previous partial unfolding operators (but with the variable domain* Σ_ε*).*

(i) *The operator* $\mathcal{T}_\varepsilon^A$ *is linear.*

(ii) *For two functions* $u, v\ \Omega_A^\varepsilon \to \mathbb{R}$, $\mathcal{T}_\varepsilon^A(uv) = \mathcal{T}_\varepsilon^A(u)\mathcal{T}_\varepsilon^A(v)$.

(iii) *Let* $u \in L^1(\Omega_A^\varepsilon)$*. Then* $\mathcal{T}_\varepsilon^A(u)$ *belongs to* $L^1(\Sigma_\varepsilon)$ *and*

$$\int_{\Sigma_\varepsilon} \mathcal{T}_\varepsilon^A(u) dx_1 dy = \int_{\Omega_A^\varepsilon} u\, dx. \tag{8.3}$$

So, $\mathcal{T}_\varepsilon^A$ *is an isometry from* $L^p(\Omega_A^\varepsilon)$ *to* $L^p(\Sigma_\varepsilon)$ *for every* $p \in [1, +\infty]$.

(iv) *Let* $u \in H^1(\Omega_A^\varepsilon)$*. Then*

$$\frac{\partial}{\partial y_2}\mathcal{T}_\varepsilon^A(u) = \mathcal{T}_\varepsilon^A\Big(\frac{\partial u}{\partial x_2}\Big), \qquad \frac{\partial}{\partial y_1}\mathcal{T}_\varepsilon^A(u) = \varepsilon\mathcal{T}_\varepsilon^A\Big(\frac{\partial u}{\partial x_1}\Big).$$

(v) *For every* $x_1 \in (0,1)$*, let* $\Sigma_\varepsilon^{\{x_1\}}$ *denote the section of* Σ_ε *defined by* x_1*, that is,*

$$\Sigma_\varepsilon^{\{x_1\}} \doteq \{(y_1, y_2) \mid (x_1, y_1, y_2) \in \Sigma_\varepsilon\}.$$

Then,

$$\mathcal{T}_\varepsilon^A(u)(x_1, \cdot, \cdot) \in H^1(\Sigma_\varepsilon^{\{x_1\}}),$$

and

$$x_1 \mapsto \|u(x_1, \cdot, \cdot)\|_{H^1\left(\Sigma_\varepsilon^{\{x_1\}}\right)} \quad \text{is in} \quad L^2(0,1),$$

and is bounded above by $\|u\|_{H^1(\Omega_A^\varepsilon)}$.

Proof. The proof of (8.3) goes as follows:

$$\begin{aligned}\int_{\Sigma_\varepsilon} \mathcal{T}_\varepsilon^A(u) dx_1 dy &= \sum_{k=0}^{N-1} \int_{k\varepsilon}^{(k+1)\varepsilon} dx_1 \int_{Y_k} u\left(\varepsilon\left[\frac{x_1}{\varepsilon}\right] + \varepsilon y_1, y_2\right) dy \\ &= \sum_{k=0}^{N-1} \varepsilon \int_{Y_k} u\left(\varepsilon\left[\frac{x_1}{\varepsilon}\right] + \varepsilon y_1, y_2\right) dy \\ &= \sum_{k=0}^{N-1} \int_{(z,y_2)\in Y_k^\varepsilon} u(z, y_2)\, dz dy_2 = \int_{\Omega_A^\varepsilon} u\, dx,\end{aligned}$$

where we used the change of variable $z = \varepsilon\left[\dfrac{x_1}{\varepsilon}\right] + \varepsilon y_1$.

All the other properties are straightforward. □

In Ω_B the standard homogenization operator $\mathcal{T}_\varepsilon$ (of Part I is used with the standard period $Y = (0,1)^2$).

8.2.3 Homogenization

8.2.3.1 Geometric assumption

We make the following assumption on the sequence of domains $\{\Sigma_\varepsilon\}_\varepsilon$: there exists some Σ with a Lipschitz boundary such that,

$$\Sigma_\varepsilon \to \Sigma \quad \text{in the Hausdorff sense,} \tag{8.4}$$

see Definition 8.3. Observe that Σ_ε is with Lipschitz boundary by construction.

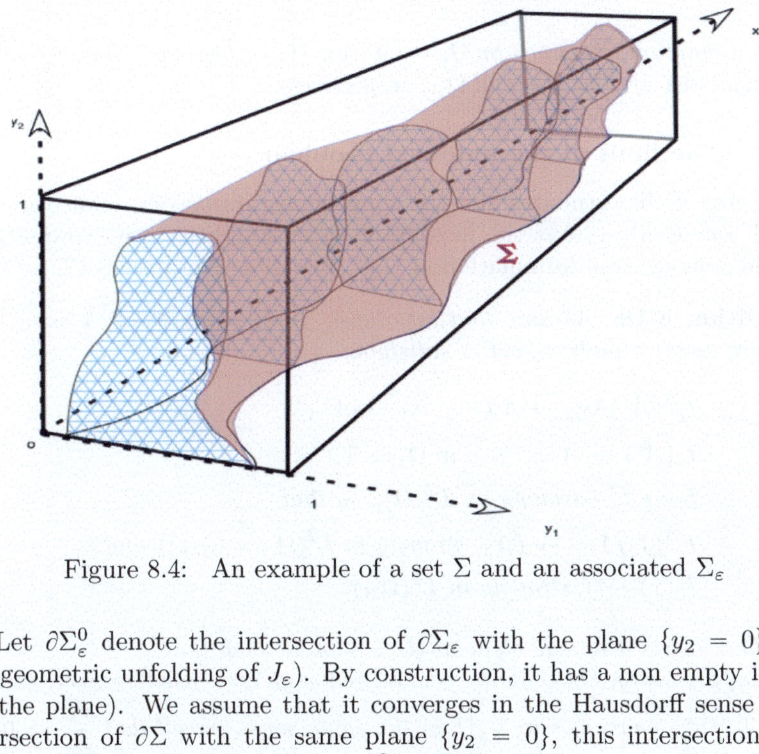

Figure 8.4: An example of a set Σ and an associated Σ_ε

Let $\partial\Sigma_\varepsilon^0$ denote the intersection of $\partial\Sigma_\varepsilon$ with the plane $\{y_2 = 0\}$ (it is the geometric unfolding of J_ε). By construction, it has a non empty interior (in the plane). We assume that it converges in the Hausdorff sense to the intersection of $\partial\Sigma$ with the same plane $\{y_2 = 0\}$, this intersection being denoted $\partial\Sigma^0$ (by this hypothesis, $\partial\Sigma^0$ is not empty).

For simplicity we write $d\sigma$ for $dx_1 dy$, the volume element of Σ.

Remark 8.16 (Examples). *In the model problem studied in [93], the previous assumption on $\{\Sigma_\varepsilon\}_\varepsilon$ is satisfied trivially since the sets Σ_ε do not depend upon ε and all coincide with Σ.*

A more general example is constructed in the following way. Let Σ be a given subset of subset of $(0,1)^3$ with Lipschitz boundary and such that $\partial\Sigma \bigcap \{(x_1, y) \mid x_1 = \ell, y_2 = 0\}$ contains a non empty interval for $\ell \in [0,1]$. For $\varepsilon > 0$, let $\xi_{\varepsilon,k}$ be any point in $[k\varepsilon, (k+1)\varepsilon)$ $(k = 0, \ldots, N-1)$ and set $Y_k \doteq \Sigma \cap \{x_1 = \xi_{\varepsilon,k}\}$. Then, the corresponding sequence $\{\Sigma_\varepsilon\}_\varepsilon$ converges in the Hausdorff sense to Σ (see Figure 8.4).

Notation 8.17.
1. In the sequel, $\mathbb{V}$ *denotes the following Hilbert space:*

$$\mathbb{V} \doteq \Big\{V \in L^2(\Sigma) \mid \frac{\partial V}{\partial y_1} \equiv 0, \ \frac{\partial V}{\partial y_2} \in L^2(\Sigma)\Big\},$$

endowed with the obvious (graph) norm. By definition, the elements of $\mathbb{V}$ *have traces on the set* $\partial\Sigma^0$*, belonging to* $L^2(\partial\Sigma^0)$.
2. Also, $\mathbb{W}$ *denotes the Hilbert space of pairs of functions,*

$$\mathbb{W} \doteq \Big\{(V, v) \in \mathbb{V} \times H^1(\Omega_B) \,\big|\, V_{|\partial\Sigma^0}(x_1, y_1) = v_{|I}(x_1) \text{ for a.e. } (x_1, y_1) \in \partial\Sigma_0 \times \Omega_B\Big\}.$$

3. For a function w *defined on* $\Omega_\varepsilon = \text{interior}\,\big(\overline{\Omega}^\varepsilon_A \cup \overline{\Omega}_B\big)$, w^a *and* w^b*, denote the restrictions of* w *to* Ω^ε_A *and* Ω_B*, respectively.*

8.2.3.2 The limit of the unfolded problem

Making use of the estimates and the convergence properties defined in Section 8.1, one easily proves the next results (for simplicity, we express them with the homogenized formulation in Ω_B). Recalling Notation 8.17, we have

Proposition 8.18. *Assume that hypothesis (8.4) holds true. Assume also that there exists a matrix field* A *satisfying*

$$\begin{aligned}
&\mathcal{T}^A_\varepsilon(A^\varepsilon)\,\mathbf{1}_{\Sigma_\varepsilon} \to A\,\mathbf{1}_\Sigma \quad \text{a.e. in } \Omega_A,\\
&\mathcal{T}_\varepsilon(A^\varepsilon) \to A \quad \text{a.e. in } \Omega_B \times Y,\\
&f_\varepsilon \to f \quad \text{strongly in } L^2(\Omega), \text{ so that}\\
&\mathcal{T}^A_\varepsilon(f_\varepsilon)\,\mathbf{1}_{\Sigma_\varepsilon} \to \widehat{f}\,\mathbf{1}_\Sigma \text{ strongly in } L^2(\Omega_A \times (0,1)) \text{ and}\\
&\mathcal{T}_\varepsilon(f) \to f \text{ strongly in } L^2(\Omega_B).
\end{aligned}$$

Then there exists a unique element (U, u^b) *in* $\mathbb{W}$ *such that*
(i) $\mathcal{T}^A_\varepsilon(u^a_\varepsilon)$ *converges to* U *in the generalized weak* L^2 *sense (see Section 8.1),*
(ii) $\nabla_y \mathcal{T}^A_\varepsilon(u^a_\varepsilon)$ *converges to* $\nabla_y U$ *in the same sense (recall that* $\dfrac{\partial U}{\partial y_1} = 0$*),*
(iii) $u^b_\varepsilon \rightharpoonup u^b$ *weakly in* $H^1(\Omega_B)$,
(iv) the pair (U, u^b) *is the unique solution in* $\mathbb{W}$ *of the following problem:*

$$\begin{cases}
\displaystyle\int_\Sigma \Big(a_{22}(x_1, y) - \frac{a_{12}(x_1, y)a_{21}(x_1, y)}{a_{11}(x_1, y)}\Big)\frac{\partial U}{\partial y_2}(x_1, y)\,\frac{\partial \Psi}{\partial y_2}(x_1, y)\,d\sigma\\
\displaystyle+ c\int_\Sigma U(x_1, y)\Psi(x_1, y)\,d\sigma + \int_{\Omega_B}\sum_{i,j=1,2} a^{hom}_{ij}(x)\,\frac{\partial u^b}{\partial x_i}(x)\,\frac{\partial \varphi}{\partial x_j}(x)\,dx\\
\displaystyle+ c\int_{\Omega_B} u^b(x)\varphi(x)\,dx = \int_\Sigma \widehat{f}(x_1, y_2)\,\Psi(x_1, y)\,d\sigma + \int_{\Omega_B} f(x)\,\varphi(x)\,dx,\\
\forall (\Psi, \varphi) \in \mathbb{W}.
\end{cases} \tag{8.5}$$

Proof. Step 1. Problem (8.5) *is well-posed.*

Note that since the matrix field A belongs to $M(\alpha, \beta, \Omega)$, it follows that the function

$$a_{22} - \frac{a_{12}a_{21}}{a_{11}} = \frac{\det A}{a_{11}},$$

takes its values in the interval $[\alpha^2/\beta, \beta^2/\alpha]$. Therefore, the bilinear form in (8.5) is coercive and bounded on the space $\mathbb{W}$.

Step 2. Existence of the pair (U, u^b) *in* $\mathbb{W}$.

Under the hypotheses of the statement, the sequences $\{\|u_a^\varepsilon\|_{H^1(\Omega_A^\varepsilon)}\}_\varepsilon$ and $\{u_b^\varepsilon\}_\varepsilon$ are bounded in $\mathbb{R}$ and in $H^1(\Omega_B)$, respectively. Therefore, the sequences $\{\|\mathcal{T}_\varepsilon^A(u_\varepsilon)\|_{L^2(\Sigma_\varepsilon)}\}_\varepsilon$ and $\{\|\mathcal{T}_\varepsilon^A(\nabla_y u_\varepsilon)\|_{L^2(\Sigma_\varepsilon)}\}_\varepsilon$ are also bounded.

The existence of a pair (U, u^b) in $L^2(\Sigma) \times H^1(\Omega_B)$ such that convergences (i)-(iii) hold, is obvious in view of hypothesis (8.4).

From the properties of the unfolding operator, it easily also follows that, at the limit,

$$\frac{\partial U}{\partial y_1} \equiv 0, \quad \text{and} \quad \frac{\partial U}{\partial y_2} \in L^2(\Sigma).$$

Step 3. (U, u^b) *belongs to* $\mathbb{W}$.

The subsequence $\{u_\varepsilon^b\}_\varepsilon$ converges weakly to u^b in $H^1(\Omega_B)$ so that by the trace theorem, the traces $u_\varepsilon^b(\cdot, 0)$ converge strongly to $u^b(\cdot, 0)$ in $L^2(0, 1)$.

Therefore, by the properties of the one dimensional unfolding operator, the unfolding of this trace defined as

$$\mathcal{T}_\varepsilon(u_\varepsilon^b)_{|(x_2=0)}(x_1, y_1) \doteq u_\varepsilon^b\Big(\varepsilon\Big[\frac{x_1}{\varepsilon}\Big] + \varepsilon y_1, 0\Big),$$

converges strongly to u^b in $L^2((0,1) \times (0,1))$.

On the other hand, by definition, the traces of u_ε^a from above and u_ε^b from below on J^ε (see Notation 8.13) coincide. Therefore, on $\partial\Sigma_\varepsilon^0$, the following equality holds:

$$\mathcal{T}_\varepsilon^A(u_\varepsilon^a)(x_1, y_1)\, \mathbf{1}_{\partial\Sigma_\varepsilon^0} = \mathcal{T}_\varepsilon(u_\varepsilon^b)_{|(x_2=0)}(x_1, y_1)\, \mathbf{1}_{\partial\Sigma_\varepsilon^0}.$$

Then, the hypotheses on Σ_ε and $\partial\Sigma_\varepsilon^0$ allow to pass to the limit as $\varepsilon \to 0$ (along the subsequence), and one concludes that

$$U(x_1, y_1, 0) = u^b(x_1) \quad \text{a.e. on} \quad \partial\Sigma^0.$$

Step 4. (U, u^b) *is a solution of* (8.5).

By (8.2)

$$\Big\|\frac{\partial u_\varepsilon}{\partial x_1}\Big\|_{L^2(\Omega_A^\varepsilon)} \leqslant C,$$

so that, up to a subsequence, its unfolding converges weakly (in the generalized sense) to some W_1 which belongs to $L^2(\Sigma)$. To determine this limit, consider in (8.1) a test functions φ^ε of the form

$$\varphi^\varepsilon(x) = \varepsilon\,\varphi(x)\,\psi\Big(\Big\{\frac{x_1}{\varepsilon}\Big\}\Big), \tag{8.6}$$

where $\varphi \in C^\infty([0,1]^2)$ and vanishes near $\{x_2 = 0\}$, and $\psi \in C^\infty([0,1])$. The function φ^ε is piecewise continuous in x_1 but smooth in Ω^ε_A (the jump possibilities are only at $x_1 = k\varepsilon$ for $k = 1, \dots, n$, outside of Ω^ε_A).

For the integral in the left hand side, we shall use the unfolding operator $\mathcal{T}^A_\varepsilon$. From its definition

$$\mathcal{T}^A_\varepsilon(\varphi^\varepsilon) \doteq \varepsilon\varphi\Big(\varepsilon\Big[\frac{x_1}{\varepsilon}\Big] + \varepsilon y_1, y_2\Big)\,\psi(y_1),$$

it follows that

$$\begin{aligned}
\mathcal{T}^A_\varepsilon\left(\frac{\partial\varphi^\varepsilon}{\partial x_1}\right) &= \frac{1}{\varepsilon}\frac{\partial}{\partial y_1}(\mathcal{T}^A_\varepsilon\varphi^\varepsilon) = \varepsilon\frac{\partial\varphi}{\partial x_1}\Big(\varepsilon\Big[\frac{x_1}{\varepsilon}\Big] + \varepsilon y_1, y_2\Big)\,\psi(y_1) \\
&\qquad + \varphi\Big(\varepsilon\Big[\frac{x_1}{\varepsilon}\Big] + \varepsilon y_1, y_2\Big)\,\psi'(y_1), \\
\mathcal{T}^A_\varepsilon\left(\frac{\partial\varphi^\varepsilon}{\partial y_2}\right) &= \frac{\partial}{\partial y_2}(\mathcal{T}^A_\varepsilon\varphi^\varepsilon) = \varepsilon\frac{\partial\varphi}{\partial x_2}\Big(\varepsilon\Big[\frac{x_1}{\varepsilon}\Big] + \varepsilon y_1, y_2\Big)\,\psi(y_1).
\end{aligned}$$

Consequently,

$$\|\mathcal{T}^A_\varepsilon(\varphi^\varepsilon)\|_{L^2(\Sigma_\varepsilon)} \to 0 \quad\text{and}\quad \left\|\mathcal{T}^A_\varepsilon\left(\frac{\partial\varphi^\varepsilon}{\partial x_2}\right)\right\|_{L^2(\Sigma_\varepsilon)} \to 0,$$

while

$$\mathcal{T}^A_\varepsilon\Big(\frac{\partial\varphi^\varepsilon}{\partial x_1}\Big)_{|\Sigma_\varepsilon} \to \varphi(x_1,y_2)\psi'(y_1)_{|\partial\Sigma^0}.$$

Therefore,

$$\begin{aligned}
&\int_{\Omega_\varepsilon}\Big(\sum_{i,j=1,2} a^\varepsilon_{ij}(x)\,\frac{\partial u^a_\varepsilon}{\partial x_i}(x)\,\frac{\partial\varphi^\varepsilon}{\partial x_j}(x) + c\,u^a_\varepsilon(x)\,\varphi^\varepsilon(x)\Big)\,dx \\
&= \int_{\Omega^\varepsilon_A}\Big(\sum_{i,j=1,2} a^\varepsilon_{ij}(x)\,\frac{\partial u^a_\varepsilon}{\partial x_i}(x)\,\frac{\partial\varphi^\varepsilon}{\partial x_j}(x) + c\,u^a_\varepsilon(x)\,\varphi^\varepsilon(x)\Big)\,dx \\
&= \int_{\Sigma_\varepsilon}\Big(\sum_{i,j=1,2} \mathcal{T}^A_\varepsilon(a^\varepsilon_{ij})\,\mathcal{T}^A_\varepsilon\Big(\frac{\partial u^a_\varepsilon}{\partial x_i}\Big)\mathcal{T}^A_\varepsilon\Big(\frac{\partial\varphi^\varepsilon}{\partial x_j}\Big) + c\,\mathcal{T}^A_\varepsilon(u^a_\varepsilon)\,\mathcal{T}^A_\varepsilon(\varphi^\varepsilon)\Big)\,d\sigma \\
&\to \int_\Sigma\Big(a_{11}(x_1,y_2)\,W_1(x_1,y) + a_{21}(x_1,y_2)\,\frac{\partial U}{\partial y_2}(x_1,y)\Big)\,\varphi(x_1,y_2)\,\psi'(y_1)\,d\sigma.
\end{aligned}$$

For the right-hand side,

$$\int_{\Omega_\varepsilon} f_\varepsilon\,\varphi^\varepsilon\,dx = \int_{\Omega^\varepsilon_A} f_\varepsilon\,\varphi^\varepsilon\,dx \to 0.$$

Therefore the limit of (8.1) with the choice (8.6) as test function is

$$\int_\Sigma\Big(a_{11}(x_1,y_2)\,W_1(x_1,y) + a_{21}(x_1,y_2)\frac{\partial U}{\partial y_2}(x_1,y)\Big)\varphi(x_1,y_2)\psi'(y_1)\,d\sigma = 0.$$

Observe that φ is an arbitrary function in $C^\infty([0,1]^2)$ as is ψ' in $C^\infty([0,1])$. Thus, the set of test functions $\varphi(x_1,y_2)\psi'(y_1)$ is dense in $L^2(\Sigma)$, from which

$$a_{11}(x_1,y_2)\,W_1(x_1,y)+a_{21}(x_1,y_2)\frac{\partial U}{\partial y_2}(x_1,y)\equiv 0 \quad \text{a.e. in } \Sigma. \tag{8.7}$$

Hence

$$W_1(x_1,y)=-\frac{a_{21}}{a_{11}}(x_1,y_2)\,\frac{\partial U}{\partial y_2}(x_1,y) \quad \text{a.e. in } \Sigma.$$

Now take $\Psi\in C^\infty(\overline{\Omega})$ as test function in (8.1). The left-hand side of (8.1) thus obtained is

$$\begin{aligned}&\int_{\Omega_A^\varepsilon}\Big[\sum_{i,j=1,2}a_{ij}^\varepsilon(x)\,\frac{\partial u_\varepsilon}{\partial x_i}(x)\,\frac{\partial\Psi}{\partial x_j}(x)+c\,u_\varepsilon(x)\,\Psi(x)\Big]\,dx\\&\qquad+\int_{\Omega_B}\Big[\sum_{i,j=1,2}a_{ij}^\varepsilon(x)\,\frac{\partial u_\varepsilon}{\partial x_i}(x)\,\frac{\partial\Psi}{\partial x_j}(x)+c\,u_\varepsilon(x)\,\Psi(x)\Big]\,dx.\end{aligned}$$

The first integral is unfolded as

$$\int_{\Sigma_\varepsilon}\Big[\sum_{i,j=1,2}\mathcal{T}_\varepsilon^A(a_{ij}^\varepsilon)\,\mathcal{T}_\varepsilon^A\Big(\frac{\partial u_\varepsilon}{\partial x_i}\Big)\,\mathcal{T}_\varepsilon^A\Big(\frac{\partial\Psi}{\partial x_j}\Big)+c\,\mathcal{T}_\varepsilon^A(u_\varepsilon)\mathcal{T}_\varepsilon^A(\Psi)\Big]\,d\sigma$$

and converges to

$$\int_\Sigma\Big[\Big(a_{11}\,W_1+a_{21}\,\frac{\partial U}{\partial y_2}\Big)\frac{\partial\Psi}{\partial x_1}+\Big(a_{12}\,W_1+a_{22}\,\frac{\partial U}{\partial y_2}\Big)\frac{\partial\Psi}{\partial x_2}+c\,U\Psi\Big]\,d\sigma,$$

which, by (8.7), reduces to

$$\int_\Sigma\Big[\Big(a_{22}-\frac{a_{21}a_{12}}{a_{11}}\Big)\,\frac{\partial U}{\partial y_2}\,\frac{\partial\Psi}{\partial x_2}+c\,U\Psi\Big]\,d\sigma. \tag{8.8}$$

For the second integral, using the standard unfolding $\mathcal{T}_\varepsilon$ in Ω_B, we get

$$\begin{aligned}&\int_{\Omega_B}\Big[\sum_{i,j=1,2}a_{ij}^\varepsilon\,\frac{\partial u_\varepsilon}{\partial x_i}\,\frac{\partial\Psi}{\partial x_j}+c\,u_\varepsilon\Psi\Big]\,dx\\&\qquad\to\int_{\Omega_B}\Big[\sum_{i,j=1,2}a_{ij}^{hom}\,\frac{\partial u^b}{\partial x_i}\,\frac{\partial\Psi}{\partial x_j}+c\,u^b\Psi\Big]\,dx,\end{aligned} \tag{8.9}$$

where A^{hom} is the homogenized matrix field obtained by homogenization in the domain Ω_B (see formulas (3.19)).

For the right-hand side of (8.1), we have

$$\begin{aligned}\int_{\Omega_\varepsilon}f_\varepsilon\,\Psi\,dx&=\int_{\Omega_A^\varepsilon}f_\varepsilon\,\Psi\,dx+\int_{\Omega_B}f_\varepsilon\,\Psi\,dx=\int_{\Sigma_\varepsilon}\mathcal{T}_\varepsilon^A(f_\varepsilon\Psi)\,d\sigma+\int_{\Omega_B}f_\varepsilon\,\Psi\,dx\\&\to\int_\Sigma\widehat{f}(x_1,y_2)\Psi(x_1,y)\,d\sigma+\int_{\Omega_B}f\,\Psi\,dx.\end{aligned}$$

This, together with (8.8) and (8.9), used in (8.1) gives problem (8.5) and concludes the proof in view of the density of $C^\infty(\overline{\Omega})$ in $\mathbb{W}$. □

8.2.3.3 Unfolded homogenization result

Although U is a better choice, traditionally, the weak limit in $L^2(\Omega_A)$ of $\widetilde{u}^a_\varepsilon$, the extension by 0 of u^a_ε, is considered as representing the limit solution. The following lemma allows to obtain it from the convergence of $\mathcal{T}^A_\varepsilon(u^a_\varepsilon)$.

Lemma 8.19. *Let p be in $(1,+\infty)$ and assume that the hypothesis (8.4) holds. Let w_ε belong to $L^p(\Sigma_\varepsilon)$ and be such that the sequence $\{\mathcal{T}^A_\varepsilon(w_\varepsilon)\}_\varepsilon$ converges weakly to some $W \in L^p(\Sigma)$ in the generalized sense. Then, $\widetilde{w}_\varepsilon$, the extension by 0 of w_ε , satisfies*

$$\widetilde{w}_\varepsilon \rightharpoonup \int_0^1 \widetilde{W}(\,\cdot\,, y_1, \cdot\,)\, dy_1 \quad \textit{weakly in } L^p((0,1)^2),$$

where $\widetilde{W}$ is the extension by 0 of W to the set $(0,1)^3$.

Proof. Similar to $\mathcal{T}^A_\varepsilon$ associated with Ω^ε_A, one can define an unfolding operator $\widetilde{\mathcal{T}_\varepsilon}^A$ for the set $(0,1)^2$ which contains Ω^ε_A. The properties of this operator are the same as those of $\mathcal{T}^A_\varepsilon$ but the domains are fixed. By testing against fixed functions in $L^{p'}((0,1)^2)$, it is easily shown that if for a sequence $\{\varphi_\varepsilon\}_\varepsilon$ in $L^p((0,1)^2)$,

$$\widetilde{\mathcal{T}_\varepsilon}^A(\varphi_\varepsilon) \rightharpoonup \Phi \quad \text{weakly in } L^p((0,1)^3),$$

then,

$$\varphi_\varepsilon \rightharpoonup \int_0^1 \widetilde{\Phi}(\,\cdot\,, y_1, \cdot\,)\, dy_1 \quad \text{weakly in } L^p((0,1)^2).$$

One concludes, since $\widetilde{\mathcal{T}_\varepsilon}^A(\widetilde{w}_\varepsilon)$ is simply the extension by 0 of $\mathcal{T}^A_\varepsilon(w_\varepsilon)$. ☐

Given $\lambda \equiv (\lambda_1, \lambda_2)$ in $(0,1)^2$, denote by $\mathfrak{s}(\lambda)$ the section of Σ by the line $\{(x_1, y_2) = \lambda\}$. Lemma 8.19 implies the weak convergence in $L^2((0,1)^2)$

$$\widetilde{u}^a_\varepsilon \rightharpoonup u^a \quad \text{weakly in } L^2((0,1)^2),$$

where

$$u^a(x_1, x_2) = \int_{\mathfrak{s}(x_1,x_2)} U(x_1, y_1, x_2)\, dy_1. \tag{8.10}$$

On $y_2 = 0$, the trace of U is independent of y_1 so that the transmission condition is expressed as

$$u^a(x_1, 0) = |\mathfrak{s}(x_1, 0)|\, u^b(x_1, 0).$$

However, in order to express the limit problem in terms of (u_a, u_b) one requires to invert (8.10), in order to express U in terms of u^a. In general, this is impossible. Below we present the only case where this seems possible.

Assume the section $\mathfrak{s}(\lambda)$ is **connected** (i.e. a possibly empty interval) for every $\lambda \in \Omega_A$, and denote by $\ell(\lambda)$ its length. For simplicity, assume also that $\ell(\lambda)$ is uniformly bounded away from zero near $\{\lambda_2 = 0\}$. Recall that $\dfrac{\partial U}{\partial y_1} = 0$, so that U does not depend upon y_1 and (8.10) becomes simply

$$u^a(x_1, y_2) = \ell(x_1, y_2)\, U(x_1, y_2),$$

and by a change of variables (to go back to the original variables (x_1, x_2)),

$$U(x) = \frac{u^a(x)}{\ell(x)} \quad \text{for every } x \in \Omega_A \text{ such that } \ell(x) > 0.$$

The effective upper part of the domain for the problem satisfied by (u^a, u^b) is then

$$\Omega_A^r \doteq \{x \in \Omega_A \mid \ell(x) > 0\},$$

which is the projection of Σ on the plane of coordinates (x_1, y_2). Finally, the reduced domain is

$$\Omega^r \doteq \text{interior } (\, \overline{\Omega_A^r} \bigcup \overline{\Omega}_B),$$

which is connected.

For convenience, extend ℓ to be just 1 on Ω_B and define the measure μ as

$$\mu(x) = \ell(x)\, dx \quad \text{on} \quad \Omega^r.$$

Let $\widehat{\mathbb{W}}$ be the space

$$\widehat{\mathbb{W}} \doteq \Big\{\phi \,\Big|\, \frac{\phi}{\ell} \in L^2(\Omega^r; \mu),\ \frac{\partial}{\partial x_2}\Big(\frac{\phi}{\ell}\Big) \in L^2(\Omega^r; \mu),$$
$$\phi(x_1, 0^+) \equiv \ell(0)\, \phi(x_1, 0^-) \quad \text{for } x_1 \in \partial\Omega_a^r \cap \Omega_b\Big\}.$$

Then the proper variational space for the homogenized problem is

$$\mathcal{V} = \{\ell\, \phi \mid \phi \in \widehat{\mathbb{W}}\}.$$

The pair (u^a, u^b) defines one element u in $\mathcal{V}$, which can be seen as the homogenized solution. This generalizes the result of [93], where the Y_k are all equal to fixed rectangle (second example of Figure 8.2).

Theorem 8.20. *Under the above hypotheses, there exists a unique element $u \in \mathcal{V}$ such that,*

$$\widetilde{u}_\varepsilon \rightharpoonup u \quad \textit{weakly in } \mathcal{V}.$$

The function u is the unique solution of the following variational problem (where y_2 is replaced by x_2 to revert to the original variables): To find $u \in \mathcal{V}$ such that

$$\int_{\Omega^r} \sum_{i,j=1,2} \widetilde{\mathrm{a}}_{ij}^{hom} \frac{\partial}{\partial x_i}\Big(\frac{u}{\ell}\Big) \frac{\partial}{\partial x_j}\Big(\frac{\varphi}{\ell}\Big)\, d\mu = \int_{\Omega^r} f\, \varphi\, d\mu, \quad \forall \varphi \in \mathcal{V},$$

where

$$\tilde{\mathbf{a}}_{ij}^{hom}(x) \doteq \begin{cases} 0 & \text{for } (i,j) \neq (2,2) \text{ and } x \text{ in } \Omega_A^r, \\ \dfrac{1}{\ell(x)} \displaystyle\int_{\mathfrak{s}(x)} \Big(a_{22}(x_1, y_1, x_2) - \dfrac{a_{12}(x_1, y_1, x_2) a_{21}(x_1, y_1, x_2)}{a_{11}(x_1, y_1, x_2)} \Big) dy_1 & \\ \qquad \text{for } i = j = 2 \quad \text{and } x \text{ in } \Omega_A^r, & \\ a_{ij}{}^{hom}(x) \quad \text{for } i, j = 1, 2 \quad \text{and } x \text{ in } \Omega_B. & \end{cases}$$

Furthermore, recalling that ℓ is a non negative element of $L^\infty(\Omega_A)$ which can take the value 0, one also has the convergences,

$$\begin{aligned} &\widetilde{u_{\varepsilon|\Omega_A^\varepsilon}} \rightharpoonup \ell\, u_{|\Omega_A} \quad \text{weakly in } L^2(\Omega_A), \\ &\widetilde{\frac{\partial u_\varepsilon}{\partial x_2}}_{|\Omega_A^\varepsilon} \rightharpoonup \ell \frac{\partial u}{\partial x_2}_{|\Omega_A} \quad \text{weakly in } L^2(\Omega_A), \\ &u_{\varepsilon|\Omega_B} \rightharpoonup u_{|\Omega_B} \quad \text{weakly in } H^1(\Omega_B). \end{aligned}$$

In Ω_A^r, this is a degenerate problem with the bounded weight (x) which is not necessarily bounded below away from zero.

Remark 8.21. *The case of variable thickness of the narrow bands (defined above as uniformly distributed as $k\varepsilon$) can be easily reduced to the previous case by a bi-Lipschitz change of variable affecting only x_1, and the results from this section can then be applied, provided the sequence of change of variables converges uniformly. Another possibility is to modulate the unfolding operator itself (see [15] for an example of such a treatment).*

Convergence of the energy and corrector In this situation one can prove the convergence of the energy, from which as usual, corrector result follows. The corrector is the standard one in Ω_B. In Ω_A the corrector is of different type: it is the extension of the solution u_ε by its local average on the lines x_2 constant. This extension then converges strongly to the limit $u_{|\Omega_A}$ in $L^2(\Omega_A)$ (see [93] for details).

Coefficients oscillating at a different scale In case the coefficients oscillate at a different scale than the upper boundary, one can use the same method as in Subsection 3.1.4. Whether the scale for the coefficients is smaller or larger than that of the boundary, the double unfolded problem is posed in a domain which is supposed to converge in the Hausdorff sense.

Remark 8.22 (Transformation via a bi-Lipschitz diffeomorphism). *As noted in the introduction of Chapter 3, the previous problems can be transformed via a global bi-Lipschitz diffeomorphism, in which case the periodicity becomes a local periodicity varying in a bi-Lipschitz way with each point.*

The next figure presents several "nasty" devices (Brushes or radiators ?) for which the homogenization method presented in this section applies.

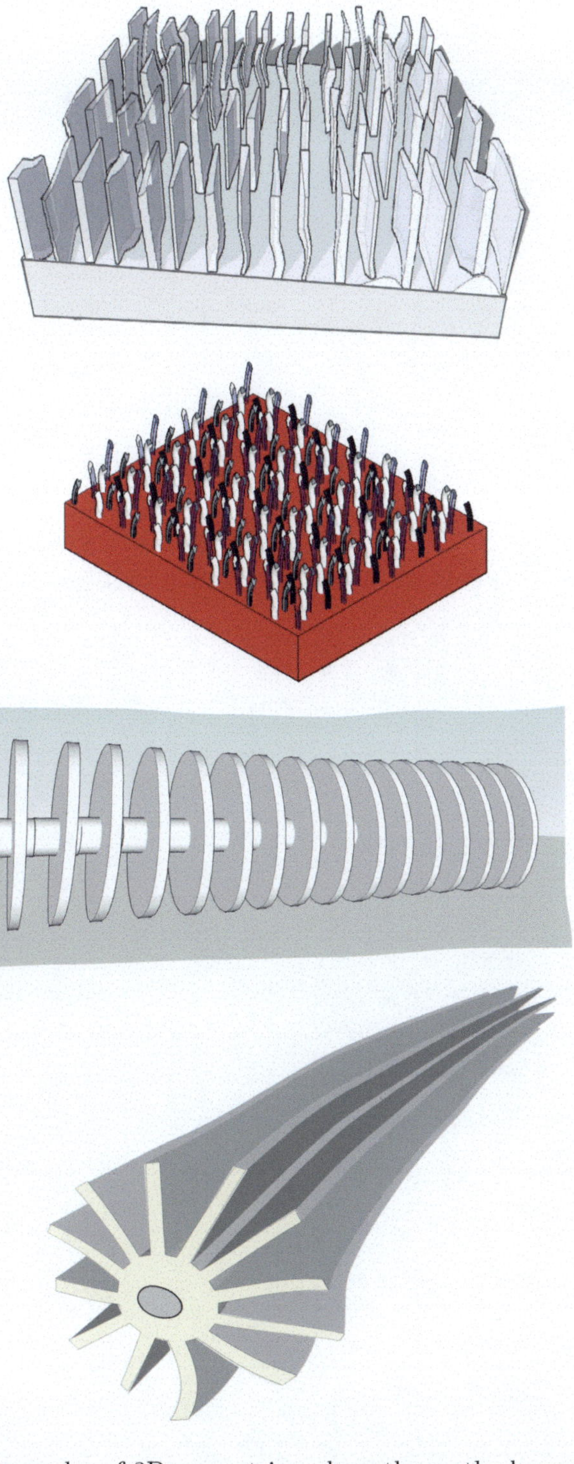

Figure 8.5: Examples of 3D geometries where the method can be applied

Part IV

Unfolding for small obstacles and strange terms

Part IV Unfolding for small obstacles and strange terms

In Part IV of this book, the unfolding method is presented for domains in $\mathbb{R}^N$, $N \geqslant 2$, with "small holes" and obstacles which, at the critical size, give rise to the so-called "strange terms". It turns out that the unfolding method is very well adapted for these situations.

Chapter 9 presents the notations, tools and main convergence results in the case of volume-distributed small holes and in the case of a single layer along a hyperplane which is well-suited for sieve-type problems (boundary layer unfolding). They depend on two small parameters, the first, ε, is the scale of periodicity, and the second, δ, defines the size $\varepsilon\delta$ of the holes. This approach never assumes the existence of an extension operator in the cells but is based on the Sobolev-Poincaré-Wirtinger inequality. The latter requires that the dimension N of the domain Ω be larger than the Lebesgue exponent p for functions in $W^{1,p}(\Omega)$.

In Chapter 10 several linear applications are presented in various settings, exemplifying the flexibility and effectiveness of the method.

It is worth noticing that the same method applies, *mutatis mutandis*, in the case combining small holes with ε-size holes ε-periodically distributed as in Part I, provided the two type of holes stay away from each other (in the macroscopic period Y).

Chapter 9

Unfolding operators: the case of "small holes"

This chapter presents the unfolding method adapted to perforated domains by holes of size $\varepsilon\delta$, distributed periodically with period ε (on whose boundary a Dirichlet condition is imposed). We consider here the case where $\delta = \delta(\varepsilon)$ is such that $\delta \to 0$ as $\varepsilon \to 0$. Such kind of holes are known in literature as "small holes" (see e.g. [74], [156]), or "tiny holes" ([3], [142]). For this new geometrical framework, we define specific unfolding operators.

The related estimates are based on the Sobolev-Poincaré-Wirtinger inequality, presented in Section 9.1 with some of its consequences.

In Section 9.2, the unfolding operator $\mathcal{T}_{\varepsilon,\delta}$ is introduced for domains in $\mathbb{R}^N, N \in \mathbb{N}\backslash\{0,1\}$ when the small holes are distributed in volume.

Section 9.3 presents the case where the "holes" are concentrated in a layer along a section of the domain in $\mathbb{R}^N$ by a hyperplane, which can be near the boundary of the domain or internal (e.g. the Dirichlet screen). The boundary-layer unfolding operator $\mathcal{T}^{bl}_{\varepsilon,\delta}$ and the corresponding local average operator $\mathcal{M}^{bl}_{\varepsilon}$ are defined.

For sieve problems, one has to distinguish between the two sides of the hyperplane, so that there are two local average operators $\mathcal{M}^{bl+}_{\varepsilon}$ and $\mathcal{M}^{bl-}_{\varepsilon}$. The case of a layer along a Lipschitz orientable hypersurface can be reduced to a hyperplane via a bi-Lipschitz diffeomorphism as in previous chapters.

Other type of transmission conditions on the boundary of ε-periodically distributed subdomains of size $\varepsilon\delta$ can be treated with the unfolding method.

9.1 Sobolev - Poincaré - Wirtinger inequality

The Sobolev-Poincaré-Wirtinger inequality combines the Sobolev embedding and the Poincaré-Wirtinger inequality for a domain D where the latter holds.

D. Cioranescu et al., *The Periodic Unfolding Method*, Series in Contemporary Mathematics 3, https://doi.org/10.1007/978-981-13-3032-2_9

Definition 9.1. *Let p be in $[1, N)$ and let*

$$p^* \doteq \frac{pN}{N-p},$$

be the associated Sobolev exponent. A bounded domain D in $\mathbb{R}^N$ satisfies the Sobolev-Poincaré-Wirtinger inequality if there exists a constant $C = C(D)$ such that,

$$\|\varphi - \mathcal{M}_D(\varphi)\|_{L^{p^*}(D)} \leqslant C\, \|\nabla\varphi\|_{L^p(D)}, \quad \forall\varphi \in W^{1,p}(D).$$

This is the case for D bounded with Lipschitz boundary. The constant C is the same for all scaled domains δD for $\delta > 0$ (as can be checked by rescaling), and also for their translates.

We recall the definition of the "homogeneous" space $\dot{W}^{1,p}(\mathbb{R}^N)$: it is the completion of $\mathcal{D}(\mathbb{R}^N)$ for the norm

$$\|\phi\|_{\dot{W}^{1,p}(\mathbb{R}^N)} = \|\nabla\phi\|_{L^p(\mathbb{R}^N)}.$$

The Sobolev imbedding theorem implies that for $p < N$,

$$\dot{W}^{1,p}(\mathbb{R}^N) = \{\phi \in L^{p^*}(\mathbb{R}^N) \,|\, \nabla\phi \in L^p(\mathbb{R}^N)^N\}.$$

Therefore, all its elements admit 0 as limit at ∞ of $\mathbb{R}^N$ in the weak sense of $L^{p^*}(\mathbb{R}^N)$, see [2].

The next proposition generalizes this result to functions having a gradient in $L^p(\mathbb{R}^N)^N$. It shows that they have a constant generalized limit at infinity of $\mathbb{R}^N$.

Proposition 9.2. *Let p be in $[1, N)$. Consider the space of functions*

$$\boldsymbol{\mathcal{W}}^{1,p}(\mathbb{R}^N) \doteq \{W^{1,p}_{loc}(\mathbb{R}^N) \,|\, \nabla\Phi \in L^p(\mathbb{R}^N)^N\}.$$

It is isomorphic to the space $\dot{W}^{1,p}(\mathbb{R}^N) \oplus \mathbb{R}$.

More precisely, for every Φ in $\boldsymbol{\mathcal{W}}^{1,p}(\mathbb{R}^N)$ there exists a real number $\Phi(\infty)$ (which is the weak limit of Φ at infinity) such that

$$\Phi - \Phi(\infty) \in \dot{W}^{1,p}(\mathbb{R}^N).$$

Moreover,

$$\|\Phi - \Phi(\infty)\|_{L^{p^*}(\mathbb{R}^N)} \leqslant C\|\nabla\Phi\|_{L^p(\mathbb{R}^N)}, \;\; \forall\,\Phi \in \boldsymbol{\mathcal{W}}^{1,p}(\mathbb{R}^N), \tag{9.1}$$

where C is the constant for the Sobolev embedding of $\dot{W}^{1,p}(\mathbb{R}^N)$ into $L^{p^*}(\mathbb{R}^N)$ (it is actually the Sobolev-Poincaré-Wirtinger constant of the balls of $\mathbb{R}^N$).

Remark 9.3. *Assume that T is a distribution on $\mathbb{R}^N$ such that ∇T is in $L^p(\mathbb{R}^N)^N$. Making use of the Poincaré-Wirtinger inequality for balls in $\mathbb{R}^N$ applied to the distributions $T_\varepsilon \doteq T * \rho_\varepsilon$ (ρ_ε here is a mollifying sequence), one can show that T_ε and then T itself actually belong to $W^{1,p}_{loc}(\mathbb{R}^N)$. So Proposition 9.2 holds under the apparently more general hypothesis that Φ is in $\mathcal{D}'(\mathbb{R}^N)$ with gradient in $L^p(\mathbb{R}^N)^N$.*

Proof of Proposition 9.2. Let B be the unit ball in $\mathbb{R}^N$ and C be the Sobolev-Poincaré-Wirtinger constant for B. By the scaling invariance of this constant, it is the constant for all balls nB with $n \in \mathbb{N}^*$.

Consequently, for Φ in $\mathcal{W}^{1,p}(\mathbb{R}^N)$,

$$\|\Phi - \mathcal{M}_{nB}(\Phi)\|_{L^{p^*}(nB)} \leqslant C\|\nabla\Phi\|_{L^p(nB)} \leqslant C\|\nabla\Phi\|_{L^p(\mathbb{R}^N)}. \tag{9.2}$$

Thus, $\Psi_n \doteq (\Phi - \mathcal{M}_{nB}(\Phi))\mathbf{1}_{nB}$ is bounded in $L^{p^*}(\mathbb{R}^N)$ and admits weak limit points in that space. Let Ψ be such a limit point. Clearly, $\nabla\Psi = \nabla\Phi$ in every nB, so also in $\mathbb{R}^N$. This implies the uniqueness of the weak limit point for the sequence $\{\Psi_n\}_n$ which therefore converges. It also implies that Ψ belongs to $\dot{W}^{1,p}(\mathbb{R}^N)$. As for $\Phi - \Psi$, it is constant (since its gradient is 0 on $\mathbb{R}^N$!) and it is also the limit of the sequence of averages $\{\mathcal{M}_{nB}(\Phi)\}_n$. Denoting it $\Phi(\infty)$ and going to the limit in inequality (9.2) gives (9.1). □

With these notations, a norm on $\mathcal{W}^{1,p}(\mathbb{R}^N)$ can be taken as

$$\|\Phi\| = \big(\|\nabla\Phi\|^p_{L^p(\mathbb{R}^N)} + |\Phi(\infty)|^p\big)^{1/p},$$

for which it is a Banach space. Then the linear form $\Phi \mapsto \Phi(\infty)$ is continuous. A similar result holds for half-spaces.

We use the notations

$$\mathbb{R}^N_+ \doteq \mathbb{R}^{N-1} \times (0, +\infty) \text{ and } \quad \mathbb{R}^N_- \doteq \mathbb{R}^{N-1} \times (-\infty, 0).$$

The equivalent of Proposition 9.2 for $\mathbb{R}^N_+$ is stated below without proof. The corresponding spaces are in this case $\dot{W}^{1,p}(\mathbb{R}^N_+)$, the completion of the space of functions in $\mathcal{C}^\infty\big(\overline{\mathbb{R}^N_+}\big)$ with bounded support for the norm $\phi \mapsto \|\nabla\phi\|_{L^p(\mathbb{R}^N_+)}$, and

$$\mathcal{W}^{1,p}(\mathbb{R}^N_+) \doteq \Big\{W^{1,p}_{loc}(\mathbb{R}^N_+) \mid \nabla\Phi \in L^p(\mathbb{R}^N_+)^N\Big\}.$$

Proposition 9.4. *Let p be in $[1, N)$. The space $\mathcal{W}^{1,p}(\mathbb{R}^N_+)$ is isomorphic to the space $\dot{W}^{1,p}(\mathbb{R}^N_+) \oplus \mathbb{R}$. More precisely, for every Φ in $\mathcal{W}^{1,p}(\mathbb{R}^N_+)$ there is a real number $\Phi(+\infty)$ such that*

$$\Phi - \Phi(+\infty) \in \dot{W}^{1,p}(\mathbb{R}^N_+),$$

and satisfying

$$\|\Phi - \Phi(+\infty)\|_{L^{p^*}(\mathbb{R}^N_+)} \leqslant C\|\nabla\Phi\|_{L^p(\mathbb{R}^N_+)},$$

where C is the constant for the Sobolev embedding of $\dot{W}^{1,p}(\mathbb{R}^N_+)$ into $L^{p^*}(\mathbb{R}^N_+)$ (it is actually the Sobolev-Poincaré-Wirtinger of every half-ball in $\mathbb{R}^N$).

The norm on $\mathcal{W}^{1,p}(\mathbb{R}^N_+)$ is given by

$$\|\Phi\| = \left(\|\nabla\Phi\|^p_{L^p(\mathbb{R}^N_+)} + |\Phi(+\infty)|^p\right)^{1/p},$$

and the linear form $\Phi \mapsto \Phi(+\infty))$ is continuous.

Of course, for functions defined on $\mathbb{R}^N_+ \bigcup \mathbb{R}^N_-$, the two versions of Proposition 9.4 can be combined. Therefore,

Corollary 9.5. *On the space of functions Φ belonging to $W^{1,p}_{loc}(\mathbb{R}^N_+ \cup \mathbb{R}^N_-)$ with $\nabla\Phi$ belonging to $L^p(\mathbb{R}^N_+ \cup \mathbb{R}^N_-)^N$, there exist two continuous linear forms $\Phi \mapsto \Phi(+\infty)$ and $\Phi \mapsto \Phi(-\infty)$, such that*

$$\Phi^+ - \Phi(+\infty) \in \dot{W}^{1,p}(\mathbb{R}^N_+) \quad \textit{and} \quad \Phi^- - \Phi(-\infty) \in \dot{W}^{1,p}(\mathbb{R}^N_-).$$

9.2 Unfolding for volume-distributed holes

Recall that $\mathbf{G}$ is the subgroup of $\mathbb{R}^N$ generated by the basis $\mathbf{B} = (\mathbf{b}_1, \dots, \mathbf{b}_N)$ of periods, see Section 1.1. In the remainder of this chapter, Ω is a bounded domain in $\mathbb{R}^N$ and the reference cell is the open parallelotope Y with coordinates relative to the basis $\mathbf{B}$ in $(-1/2, 1/2)^N$.

Below, we will consider domains with εY-periodically distributed holes of size $\varepsilon\delta$ ($1 > \delta = \delta(\varepsilon) \to 0$ as $\varepsilon \to 0$), see Figure 9.1.

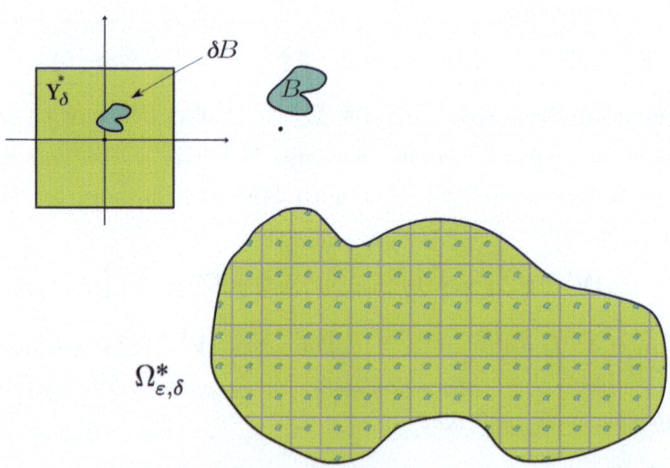

Figure 9.1: The sets B and Y^*_δ and the corresponding $\Omega^*_{\varepsilon,\delta}$

More precisely, for a given open $B \Subset Y$, set

$$Y_\delta^* \doteq Y \backslash \delta \overline{B},$$

and define the perforated domain $\Omega_{\varepsilon,\delta}^*$ as

$$\Omega_{\varepsilon,\delta}^* = \left\{ x \in \Omega \;\middle|\; \left\{ \frac{x}{\varepsilon} \right\}_Y \in Y_\delta^* \right\}. \tag{9.3}$$

Functions belonging to the space $W_0^{1,p}(\Omega_{\varepsilon,\delta}^*)$, $p \in [1, +\infty]$ vanish on the whole boundary of the perforated domain $\Omega_{\varepsilon,\delta}^*$ and their natural extensions by zero to the whole of Ω belong to $W_0^{1,p}(\Omega)$. **Consequently, in this chapter and in the next one, we will not distinguish elements of $W_0^{1,p}(\Omega_{\varepsilon,\delta}^*)$ and their extensions by 0 in $W_0^{1,p}(\Omega)$.**

This geometry of domains with small holes requires another unfolding operator $\mathcal{T}_{\varepsilon,\delta}$ depending on both parameters ε and δ. It will be defined for functions on the whole of Ω.

Definition 9.6 (The unfolding operator in domains with small holes). *For ϕ in $L^p(\Omega)$, $p \in [1, +\infty]$, the unfolding operator*

$$\mathcal{T}_{\varepsilon,\delta} : L^p(\Omega) \longrightarrow L^p(\Omega \times \mathbb{R}^N),$$

is defined by

$$\mathcal{T}_{\varepsilon,\delta}(\phi)(x, z) = \begin{cases} \mathcal{T}_\varepsilon(\phi)(x, \delta z) & \text{for a.e. } (x, z) \in \widehat{\Omega}_\varepsilon \times \dfrac{1}{\delta} Y, \\ 0 & \text{otherwise.} \end{cases}$$

The next results follow from Proposition 1.8 by using the change of variable $z = (1/\delta) y$.

Theorem 9.7. (Properties of the operator $\mathcal{T}_{\varepsilon,\delta}$).
Suppose $p \in [1, +\infty]$. The operator $\mathcal{T}_{\varepsilon,\delta}$ is linear and continuous from $L^p(\Omega)$ to $L^p(\Omega \times \mathbb{R}^N)$.

(i) For every $v, w \in L^p(\Omega)$, $\mathcal{T}_{\varepsilon,\delta}(vw) = \mathcal{T}_{\varepsilon,\delta}(v)\mathcal{T}_{\varepsilon,\delta}(w)$.

(ii) For every $u \in L^1(\Omega)$,

$$\frac{\delta^N}{|Y|} \int_{\Omega \times \mathbb{R}^N} \mathcal{T}_{\varepsilon,\delta}(u)(x, z)\, dxdz = \frac{1}{|Y|} \int_{\Omega \times Y} \mathcal{T}_\varepsilon(u)(x, y)\, dx\, dy = \int_{\widehat{\Omega}_\varepsilon} u\, dx.$$

(iii) For every $u \in L^p(\Omega)$,

$$\|\mathcal{T}_{\varepsilon,\delta}(u)\|_{L^p(\Omega \times \mathbb{R}^N)} \leqslant \left(\frac{|Y|}{\delta^N} \right)^{1/p} \|u\|_{L^p(\Omega)}.$$

(iv) For every $u \in L^1(\Omega)$,

$$\left| \int_\Omega u\, dx - \frac{\delta^N}{|Y|} \int_{\Omega \times \mathbb{R}^N} \mathcal{T}_{\varepsilon,\delta}(u)\, dxdz \right| \leqslant \int_{\Lambda_\varepsilon} |u|\, dx.$$

(v) Let $u \in W^{1,p}(\Omega)$. Then

$$\begin{aligned} \mathcal{T}_{\varepsilon,\delta}(\nabla_x u) &= \frac{1}{\varepsilon\delta}\nabla_z\big(\mathcal{T}_{\varepsilon,\delta}(u)\big) \quad \text{in} \quad \Omega \times \frac{1}{\delta}Y, \\ \|\nabla_z\big(\mathcal{T}_{\varepsilon,\delta}(u)\big)\|_{L^p(\Omega\times\frac{1}{\delta}Y)} &\leqslant \frac{\varepsilon\,|Y|^{\frac{1}{p}}}{\delta^{\frac{N}{p}-1}}\|\nabla u\|_{L^p(\Omega)}. \end{aligned} \tag{9.4}$$

(vi) Suppose $p \in [1, N)$ and let ω be a bounded open set in $\mathbb{R}^N$. For every u in $W^{1,p}(\Omega)$ the following estimates hold:

$$\begin{aligned} \|\mathcal{T}_{\varepsilon,\delta}\big(u - \mathcal{M}_\varepsilon(u)\big)\|_{L^p(\Omega;L^{p^*}(\mathbb{R}^N))} &\leqslant C\|\nabla_z\big(\mathcal{T}_{\varepsilon,\delta}(u)\big)\|_{L^p(\Omega\times\frac{1}{\delta}Y)} \\ &\leqslant C\frac{\varepsilon\,|Y|^{\frac{1}{p}}}{\delta^{\frac{N}{p}-1}}\|\nabla u\|_{L^p(\Omega)}, \end{aligned} \tag{9.5}$$

and

$$\begin{aligned} &\|\mathcal{T}_{\varepsilon,\delta}(u)\|_{L^p(\Omega\times\omega)} \\ &\quad\leqslant |\omega|^{1/N}\|\mathcal{T}_{\varepsilon,\delta}\big(u - \mathcal{M}_\varepsilon(u)\big)\|_{L^p(\Omega;L^{p^*}(\mathbb{R}^N))} + |\omega|^{1/p}||u||_{L^p(\Omega)} \\ &\quad\leqslant C\frac{\varepsilon\,|Y|^{\frac{1}{p}}}{\delta^{\frac{N}{p}-1}}|\omega|^{1/N}\|\nabla u\|_{L^p(\Omega)} + |\omega|^{1/p}\,\|u\|_{L^p(\Omega)}, \end{aligned} \tag{9.6}$$

where C denotes the Sobolev-Poincaré-Wirtinger constant for $W^{1,p}(Y)$.

(vii) Suppose $p \in (1, N)$. Let $\{w_{\varepsilon,\delta}\}_{(\varepsilon,\delta)}$ be a sequence in $W^{1,p}(\Omega)$ which is uniformly bounded when both ε and δ go to zero. Then, up to a subsequence, there is W in $L^p(\Omega; \dot{W}^{1,p}(\mathbb{R}^N))$ such that

$$\begin{aligned} &\frac{\delta^{\frac{N}{p}-1}}{\varepsilon}\mathcal{T}_{\varepsilon,\delta}\big(w_{\varepsilon,\delta} - \mathcal{M}_\varepsilon(w_{\varepsilon,\delta})\big) \rightharpoonup W \quad \text{weakly in } L^p(\Omega; L^{p^*}(\mathbb{R}^N)), \\ &\frac{\delta^{\frac{N}{p}-1}}{\varepsilon}\nabla_z\big(\mathcal{T}_{\varepsilon,\delta}(w_{\varepsilon,\delta})\big)\,\mathbf{1}_{\frac{1}{\delta}Y} \rightharpoonup \nabla_z W \quad \text{weakly in } L^p(\Omega \times \mathbb{R}^N)^N. \end{aligned}$$

Assuming furthermore, that

$$\limsup_{(\varepsilon,\delta)\to(0,0)} \frac{\delta^{\frac{N}{p}-1}}{\varepsilon} < +\infty,$$

we choose the subsequence above and some U in $L^p(\Omega; \dot{W}^{1,p}(\mathbb{R}^N) \oplus \mathbb{R})$ with

$$\frac{\delta^{\frac{N}{p}-1}}{\varepsilon}\mathcal{T}_{\varepsilon,\delta}(w_{\varepsilon,\delta}) \rightharpoonup U \quad \text{weakly in } L^p(\Omega; L^p_{loc}(\mathbb{R}^N)).$$

Remark 9.8. *Estimates (9.5)-(9.6) are derived from (9.4) by using the Sobolev-Poincaré-Wirtinger inequality (because of its scale-invariance). The standard Poincaré-Wirtinger inequality would give (see Proposition 1.38),*

$$\|\mathcal{T}_{\varepsilon,\delta}\big(u - \mathcal{M}_\varepsilon(u)\big)\|_{L^p(\Omega\times\mathbb{R}^N)} \leqslant \frac{C\varepsilon}{\delta^{N/p}}\|\nabla u\|_{L^p(\Omega)},$$

where C is the Poincaré-Wirtinger constant of Y. This estimate is not as sharp as (9.5).

Remark 9.9. *One can also define a corresponding averaging operator $\mathcal{U}_{\varepsilon,\delta}$, adjoint of $\mathcal{T}_{\varepsilon,\delta}$. It can be expressed in terms of $\mathcal{U}_\varepsilon$ together with the change of variable $z = (1/\delta)y$. The properties of $\mathcal{U}_{\varepsilon,\delta}$ follow directly from those of $\mathcal{U}_\varepsilon$.*

9.3 Boundary-layer unfolding operator

The sieve-type problems (Sections 10.2 and 10.3 in the next chapter), consider the case of holes periodically distributed in a layer Σ_ε of thickness of order ε, parallel to a hyperplane (e.g. $x_N = constant$) in the domain Ω of $\mathbb{R}^N$.

Denote $\mathbf{G}'$ the subgroup generated by the vectors $\mathbf{B}' \doteq (\mathbf{b}_1, \dots, \mathbf{b}_{N-1})$.

Notation 9.10.
1. $x' \doteq (x_1, \cdots, x_{N-1}, 0)$.
2. $\Pi \doteq \{x_N = 0\}$ *and* $\Sigma = \Pi \cap \Omega$.
3. $Y' \doteq Y \cap \{y_N = 0\}$ *the* $(N-1)$*-dimensional parallelotope associated with this system.*

The layer Σ_ε is defined as follows:

$$\Sigma_\varepsilon = \Omega \cap \Big\{ x \in \mathbb{R}^N \;\Big|\; |x_N| < \frac{\varepsilon}{2} \Big\}. \tag{9.7}$$

By analogy with (1.3), introduce the corresponding sets

$$\begin{aligned} &\Xi'_\varepsilon = \{\xi \in \mathbf{G}', \, \varepsilon(\xi + Y') \subset \Sigma\}, \\ &\widehat{\Sigma}'_\varepsilon = \text{interior}\{\bigcup_{\xi \in \Xi'_\varepsilon} \varepsilon(\xi + \overline{Y'})\}, \quad \widehat{\Sigma}_\varepsilon = \widehat{\Sigma}'_\varepsilon \times \Big(-\frac{\varepsilon}{2}, \frac{\varepsilon}{2}\Big), \\ &\Lambda'_\varepsilon = \Sigma \backslash \widehat{\Sigma}'_\varepsilon, \quad \Lambda_\varepsilon = \Lambda'_\varepsilon \times \Big(-\frac{\varepsilon}{2}, \frac{\varepsilon}{2}\Big). \end{aligned}$$

The set $\widehat{\Sigma}'_\varepsilon$ is the largest union of $\varepsilon Y'$ cells contained in Σ (see Figure 9.2).

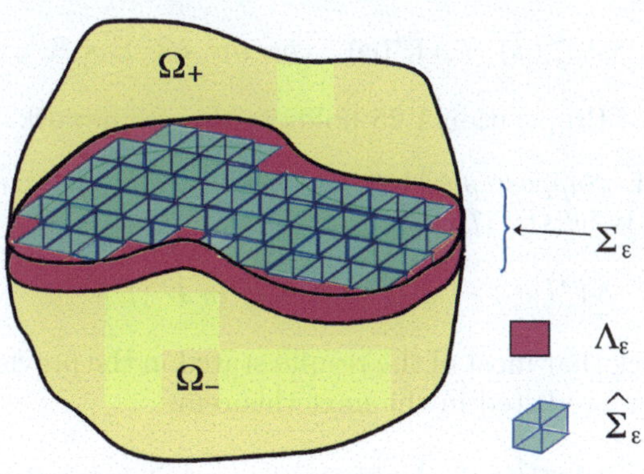

Figure 9.2: The sets $\Sigma_\varepsilon, \widehat{\Sigma}_\varepsilon$ and Λ_ε

As we did for perforated domains in Section 4.4 (see Definition 4.46), we now define a specific boundary layer operator for this particular geometry.

Definition 9.11 (The boundary layer unfolding operator).
For $\phi \in L^p(\Sigma_\varepsilon)$, $p \in [1, +\infty]$, the boundary unfolding operator

$$\mathcal{T}^{bl}_{\varepsilon,\delta} : L^p(\Sigma_\varepsilon) \longrightarrow L^p(\Sigma \times \mathbb{R}^N),$$

is defined by

$$\mathcal{T}^{bl}_{\varepsilon,\delta}(\phi)(x', z) = \begin{cases} \phi\Big(\varepsilon\Big[\dfrac{x'}{\varepsilon}\Big]_{Y'} + \varepsilon\delta z\Big) & \text{for a.e. } (x', z) \in \widehat{\Sigma}'_\varepsilon \times \dfrac{1}{\delta}Y, \\ 0 & \text{otherwise.} \end{cases}$$

This operator, designed to capture the contribution of the barriers in the limit process, was originally used in [169].

We also introduce the local average operator related to the hyperplane Σ.

Definition 9.12. *The local average*

$$\mathcal{M}^{bl}_\varepsilon : L^p(\Sigma_\varepsilon) \longrightarrow L^p(\Sigma),$$

is defined for every ϕ in $L^p(\Sigma_\varepsilon)$, $p \in [1, +\infty]$, by

$$\begin{aligned} \mathcal{M}^{bl}_\varepsilon(\phi)(x') &= \begin{cases} \dfrac{1}{\varepsilon^N|Y|}\displaystyle\int_{\varepsilon[\frac{x'}{\varepsilon}]_{Y'}+\varepsilon Y} \phi(\zeta)\, d\zeta & \text{for a.e. } x' \in \widehat{\Sigma}'_\varepsilon, \\ 0 \quad \text{for a.e. } x' \in \Lambda'_\varepsilon, \end{cases} \\ &= \frac{\delta^N}{|Y|}\int_{\frac{1}{\delta}Y} \mathcal{T}^{bl}_{\varepsilon,\delta}(\phi)(x', z)\, dz. \end{aligned}$$

Remark 9.13. *Since elements in $L^p(\Sigma)$ can be considered as functions in $L^p(\Sigma_\varepsilon)$, $\mathcal{M}^{bl}_\varepsilon$ can be applied to them. With this convention,*

$$\mathcal{T}^{bl}_{\varepsilon,\delta} \circ \mathcal{M}^{bl}_\varepsilon(\phi) = \mathcal{M}^{bl}_\varepsilon(\phi) \qquad \text{on the set } \Sigma \times \mathbb{R}^N.$$

An equivalent of Proposition 1.25 holds in this framework.

Proposition 9.14. *Suppose $p \in [1, +\infty)$. Let $\{w_\varepsilon\}_\varepsilon$ be a sequence such that $w_\varepsilon \rightharpoonup w$ weakly in $W^{1,p}(\Omega)$. Then*

$$\mathcal{M}^{bl}_\varepsilon(w_\varepsilon) \to w_{|\Sigma} \quad \text{strongly in } L^p(\Sigma).$$

It is easy to check that most of the results stated in the previous subsection extend to $\mathcal{T}^{bl}_{\varepsilon,\delta}$, they are listed in the next theorem.

Theorem 9.15. (Properties of the operator $\mathcal{T}^{bl}_{\varepsilon,\delta}$) *Suppose $p \in [1, +\infty]$.*
(i) For every $v, w \in L^p(\Sigma_\varepsilon)$,

$$\mathcal{T}^{bl}_{\varepsilon,\delta}(vw) = \mathcal{T}^{bl}_{\varepsilon,\delta}(v)\mathcal{T}^{bl}_{\varepsilon,\delta}(w).$$

(ii) *For every* $u \in L^1(\Sigma_\varepsilon)$,

$$\frac{\varepsilon\delta^N}{|Y|}\int_{\Sigma\times\mathbb{R}^N}\mathcal{T}^{bl}_{\varepsilon,\delta}(u)\,dx'\,dz = \int_{\widehat{\Sigma}_\varepsilon} u\,dx.$$

(iii) *For every* $u \in L^p(\Sigma_\varepsilon)$,

$$\big\|\mathcal{T}^{bl}_{\varepsilon,\delta}(u)\big\|_{L^p(\Sigma\times\mathbb{R}^N)} \leqslant \Big(\frac{|Y|}{\varepsilon\delta^N}\Big)^{1/p}\|u\|_{L^p(\Sigma_\varepsilon)}.$$

(iv) *For every* $u \in L^1(\Sigma_\varepsilon)$,

$$\Big|\int_{\Sigma_\varepsilon} u\,dx - \frac{\varepsilon\delta^N}{|Y|}\int_{\Sigma\times\mathbb{R}^N}\mathcal{T}^{bl}_{\varepsilon,\delta}(u)\,dx'\,dz\Big| \leqslant \int_{\Lambda_\varepsilon}|u|\,dx.$$

(v) *Let* u *be in* $W^{1,p}(\Sigma_\varepsilon)$. *Then*

$$\mathcal{T}^{bl}_{\varepsilon,\delta}(\nabla_x u) = \frac{1}{\varepsilon\delta}\nabla_z\big(\mathcal{T}^{bl}_{\varepsilon,\delta}(u)\big) \quad in \quad \Sigma\times\frac{1}{\delta}Y,$$

and

$$\big\|\nabla_z\big(\mathcal{T}^{bl}_{\varepsilon,\delta}(u)\big)\big\|_{L^p(\Sigma\times\frac{1}{\delta}Y)} \leqslant |Y|^{1/p}\,\frac{\varepsilon^{1-\frac{1}{p}}}{\delta^{\frac{N}{p}-1}}\,\|\nabla u\|_{L^p(\Sigma_\varepsilon)}.$$

(vi) *Suppose* $p \in [1, N)$ *and let* ω *be a bounded open set in* $\mathbb{R}^N$. *Then, for every* u *in* $W^{1,p}(\Sigma_\varepsilon)$, *the following estimates hold:*

$$\big\|\mathcal{T}^{bl}_{\varepsilon,\delta}\big(u - \mathcal{M}^{bl}_\varepsilon(u)\big)\big\|_{L^p(\Sigma;L^{p^*}(\mathbb{R}^N))} \leqslant C|Y|^{1/p}\,\frac{\varepsilon^{1-\frac{1}{p}}}{\delta^{\frac{N}{p}-1}}\,\|\nabla u\|_{L^p(\Sigma_\varepsilon)},$$

and

$$\big\|\mathcal{T}^{bl}_{\varepsilon,\delta}(u)\big\|_{L^p(\Sigma\times\omega)} \leqslant C|Y|^{1/p}\,\frac{\varepsilon^{1-\frac{1}{p}}}{\delta^{\frac{N}{p}-1}}\,|\omega|^{1/N}\,\|\nabla u\|_{L^p(\Sigma_\varepsilon)} + |\,\omega\,|^{1/p}\,\|u\|_{L^p(\Sigma_\varepsilon)},$$

where C *denotes the Sobolev-Poincaré-Wirtinger constant for* $W^{1,p}(Y)$.

(vii) *Suppose* $p \in (1, N)$. *Let* $\{w_{\varepsilon,\delta}\}_{(\varepsilon,\delta)}$ *be a sequence in* $W^{1,p}(\Sigma_\varepsilon)$ *such that* $\|\nabla w_{\varepsilon,\delta}\|_{L^p(\Sigma_\varepsilon)}$ *is bounded. Then, up to a subsequence, there exists* W *in* $L^p(\Sigma;\dot{W}^{1,p}(\mathbb{R}^N))$ *satisfying,*

$$\frac{\delta^{\frac{N}{p}-1}}{\varepsilon^{1-\frac{1}{p}}}\,\mathcal{T}^{bl}_{\varepsilon,\delta}\big(w_{\varepsilon,\delta} - \mathcal{M}^{bl}_\varepsilon(w_{\varepsilon,\delta})\big) \rightharpoonup W \qquad weakly\ in\ L^p(\Sigma;L^{p^*}(\mathbb{R}^N)),$$

$$\frac{\delta^{\frac{N}{p}-1}}{\varepsilon^{1-\frac{1}{p}}}\,\nabla_z\big(\mathcal{T}^{bl}_{\varepsilon,\delta}(w_{\varepsilon,\delta})\big)\,\mathbf{1}_{\frac{1}{\delta}Y'} \rightharpoonup \nabla_z W \qquad weakly\ in\ L^p(\Sigma\times\mathbb{R}^N)^N.$$

Furthermore, assuming

$$\limsup_{(\varepsilon,\delta)\to(0,0)}\frac{\delta^{\frac{N}{p}-1}}{\varepsilon^{1-\frac{1}{p}}} < +\infty,$$

one can choose the subsequence above and U *in* $L^p(\Sigma;\dot{W}^{1,p}(\mathbb{R}^N)\oplus\mathbb{R})$ *with*

$$\frac{\delta^{\frac{N}{p}-1}}{\varepsilon^{1-\frac{1}{p}}}\,\mathcal{T}^{bl}_{\varepsilon,\delta}(w_{\varepsilon,\delta}) \rightharpoonup U \qquad weakly\ in\ L^p(\Sigma;L^p_{loc}(\mathbb{R}^N)).$$

For sieve problems, there is a need to distinguish between the subdomains above (index +) and below (index -) Σ. Set

$$\begin{aligned} \Omega_+ &= \mathbb{R}^N_+ \cap \Omega, \quad \Sigma^+_\varepsilon = \mathbb{R}^N_+ \cap \Sigma_\varepsilon, \quad Y_+ = \mathbb{R}^N_+ \cap Y, \\ \Omega_- &= \mathbb{R}^N_- \cap \Omega, \quad \Sigma^-_\varepsilon = \mathbb{R}^N_- \cap \Sigma_\varepsilon, \quad Y_- = \mathbb{R}^N_- \cap Y. \end{aligned} \tag{9.8}$$

Both domains Ω_+ and Ω_- are assumed to have a Lipschitz boundary.

For every function u defined in Ω, u^+ denotes its restriction to the domain Ω_+, i.e., $u^+ \equiv u_{|\Omega_+}$; analogously, $u^- \equiv u_{|\Omega_-}$. The restrictions of the operator $\mathcal{T}^{bl}_{\varepsilon,\delta}$ to the spaces $L^p(\Sigma^+_\varepsilon)$ and $L^p(\Sigma^-_\varepsilon)$ ($p \in [1, +\infty]$), are still denoted $\mathcal{T}^{bl}_{\varepsilon,\delta}$.

In this framework, two local averages (defined below) are used.

Definition 9.16. *Let p be in $[1, +\infty]$. The local average operators,*

$$\mathcal{M}^{bl+}_\varepsilon : L^p(\Sigma^+_\varepsilon) \to L^p(\Sigma), \qquad \mathcal{M}^{bl-}_\varepsilon : L^p(\Sigma^-_\varepsilon) \to L^p(\Sigma),$$

are defined by

$$\begin{aligned} \mathcal{M}^{bl+}_\varepsilon(\phi)(x') &\doteq \frac{\delta^N}{|Y_+|} \int_{\frac{1}{\delta}Y_+} \mathcal{T}^{bl}_{\varepsilon,\delta}(\phi)(x', z)\, dz, \quad \forall \phi \in L^p(\Sigma^+_\varepsilon), \\ \mathcal{M}^{bl-}_\varepsilon(\phi)(x') &\doteq \frac{\delta^N}{|Y_-|} \int_{\frac{1}{\delta}Y_-} \mathcal{T}^{bl}_{\varepsilon,\delta}(\phi)(x', z)\, dz, \quad \forall \phi \in L^p(\Sigma^-_\varepsilon). \end{aligned}$$

The next results are obtained in the same way as at the beginning of this section and are stated without proof. Unless specified otherwise, all the results stated for Ω_+ hold with obvious modifications for Ω_-.

Proposition 9.17. *Suppose $p \in [1, +\infty)$. Let $\{w_\varepsilon\}_\varepsilon$ be a sequence such that $w_\varepsilon \rightharpoonup w^+$ weakly in $W^{1,p}(\Omega_+)$. Then*

$$\mathcal{M}^{bl+}_\varepsilon(w_\varepsilon) \to {w^+}_{|\Sigma} \quad \textit{strongly in } L^p(\Sigma).$$

Theorem 9.18.

(i) Suppose $p \in [1, +\infty]$. For every $u \in L^p(\Sigma^+_\varepsilon)$ and $v \in W^{1,p}(\Sigma^+_\varepsilon)$,

$$\big\|\mathcal{T}^{bl}_{\varepsilon,\delta}(u)\big\|_{L^p(\Sigma \times \mathbb{R}^N_+)} \leqslant \Big(\frac{|Y_+|}{\varepsilon \delta^N}\Big)^{1/p} \|u\|_{L^p(\Sigma^+_\varepsilon)},$$

$$\big\|\nabla_z\big(\mathcal{T}^{bl}_{\varepsilon,\delta}(v)\big)\big\|_{L^p(\Sigma \times \frac{1}{\delta}Y_+)} \leqslant C|Y_+|^{1/p} \frac{\varepsilon^{1-\frac{1}{p}}}{\delta^{\frac{N}{p}-1}} \|\nabla v\|_{L^p(\Sigma_{\varepsilon+})}.$$

(ii) Suppose $p \in [1, N)$ and let ω be bounded open set in $\mathbb{R}^N_+$. Then, for every u in $W^{1,p}(\Sigma^+_\varepsilon)$, the following estimates hold:

$$\big\|\mathcal{T}^{bl}_{\varepsilon,\delta}\big(u - \mathcal{M}^{bl+}_\varepsilon(u)\big)\big\|_{L^p(\Sigma; L^{p^*}(\mathbb{R}^N_+))} \leqslant C|Y_+|^{1/p} \frac{\varepsilon^{1-\frac{1}{p}}}{\delta^{\frac{N}{p}-1}} \|\nabla u\|_{L^p(\Sigma^+_\varepsilon)},$$

and

$$\|\mathcal{T}^{bl}_{\varepsilon,\delta}(u)\|_{L^p(\Sigma\times\omega)} \leqslant C|Y_+|^{1/p}\,\frac{\varepsilon^{1-\frac{1}{p}}}{\delta^{\frac{N}{p}-1}}\,|\omega|^{1/N}\|\nabla u\|_{L^p(\Sigma^+_\varepsilon)} + |\,\omega\,|^{1/p}\,\|u\|_{L^p(\Sigma^+_\varepsilon)},$$

where C denotes the Sobolev-Poincaré-Wirtinger constant for $W^{1,p}(Y)$.

(iii) Suppose $p\in(1,N)$. Let $\{w_{\varepsilon,\delta}\}_{(\varepsilon,\delta)}$ be a sequence in $W^{1,p}(\Sigma^+_\varepsilon)$ such that $\|\nabla w_{\varepsilon,\delta}\|_{L^p(\Sigma_{\varepsilon+})}$ is bounded. Then, up to a subsequence, there exists W^+ in $L^p(\Sigma;\dot{W}^{p,1}(\mathbb{R}^N_+))$ satisfying,

$$\frac{\delta^{\frac{N}{p}-1}}{\varepsilon^{1-\frac{1}{p}}}\,\mathcal{T}^{bl}_{\varepsilon,\delta}\big(w_{\varepsilon,\delta}-\mathcal{M}^{bl+}_{\varepsilon}(w_{\varepsilon,\delta})\big) \rightharpoonup W^+ \qquad \text{weakly in } L^p(\Sigma;L^{p^*}(\mathbb{R}^N_+)),$$

$$\frac{\delta^{\frac{N}{p}-1}}{\varepsilon^{1-\frac{1}{p}}}\,\nabla_z\big(\mathcal{T}^{bl}_{\varepsilon,\delta}(w_{\varepsilon,\delta})\big)\,\mathbf{1}_{\frac{1}{\delta}Y_+} \rightharpoonup \nabla_z W^+ \qquad \text{weakly in } L^p(\Sigma\times\mathbb{R}^N_+)^N.$$

Assuming furthermore, that

$$\limsup_{(\varepsilon,\delta)\to(0,0)}\frac{\delta^{\frac{N}{p}-1}}{\varepsilon^{1-\frac{1}{p}}} < +\infty,$$

one can choose the subsequence above and U^+ in $L^p(\Sigma;\dot{W}^{p,1}(\mathbb{R}^N_+)\oplus\mathbb{R})$ with

$$\frac{\delta^{\frac{N}{p}-1}}{\varepsilon^{1-\frac{1}{p}}}\,\mathcal{T}^{bl}_{\varepsilon,\delta}(w_{\varepsilon,\delta}) \rightharpoonup U^+ \qquad \text{weakly in } L^p(\Sigma;L^p_{loc}(\mathbb{R}^N_+)).$$

Remark 9.19. *As in Remark 9.9, the adjoint of operator $\mathcal{T}^{bl}_{\varepsilon,\delta}$ can be defined as an averaging operator in this setting.*

Chapter 10

Homogenization in domains with "small holes"

This chapter is devoted to examples of the use of the unfolding method for various linear problems in domains perforated with small holes and with oscillating coefficients. For simplicity, in most cases, we prescribe a homogeneous Dirichlet boundary condition on the outer boundary of the domain, but more general boundary conditions can be handled provided the outer boundary is Lipschitz and the perforations do not intersect it. In each case, we obtain both the unfolded and the classical (standard) form for the limit problem. The operator $\mathcal{T}_\varepsilon$ allows to homogenize the coefficients of the differential operators, whereas the operator $\mathcal{T}_{\varepsilon\delta}$ or $\mathcal{T}^{bl}_{\varepsilon,\delta}$, generate the "strange terms" in the limit. In each example, the convergence of the energy is investigated and used to obtain corrector results. It should be noted that for technical reasons, the method fails to apply in dimension $N = 2$ for linear problems. See [13] for the nonlinear case.

For each class of problems (e.g., problems in domains with "small holes" distributed in volume or in a layer such as sieves), specific unfolding operators are designed on the model of the unfolding operators introduced in Chapters 1 and 3. This emphasizes that unfolding is a general and powerful method. Moreover, since by definition, the unfolding separates the scales, it allows to have very general problems, combining at the same time highly oscillating coefficients, various holes of size ε and small holes of size $\varepsilon\delta$ with varying conditions on their boundaries.

Section 10.1 concerns the homogenization of elliptic problems with oscillating coefficients and ε-periodically volume-distributed small holes of size $\varepsilon\delta$ having an homogeneous Dirichlet condition on their boundary. These results are well known for the Laplace operator, with the appearance of the "strange

D. Cioranescu et al., *The Periodic Unfolding Method*, Series in Contemporary Mathematics 3, https://doi.org/10.1007/978-981-13-3032-2_10

term" (see [74], [75] and references therein). For the case of oscillating coefficients, we also refer to [83] where H-convergence is used. The similar problem in the nonlinear case (associated with Leray-Lions type operators) can be treated in the same way, with the complications due to non-linearity. We refer to [90] for a detailed presentation.

The same kind of small holes of size $\varepsilon\delta$ are considered in Section 10.2, now distributed ε-periodically in a layer of thickness ε on which the solution has to vanish (the electrostatic screen). The homogenization results are obtained by using the boundary unfolding operator $\mathcal{T}^{bl}_{\varepsilon,\delta}$. They generalize the results of [166, 180] and [75] to the case of oscillating coefficients.

Section 10.3 deals with the Neumann sieve problem with zero thickness and oscillating coefficients. For the case of constant coefficients, we refer the reader to [11, 17, 85, 163, 170] and [176]. We also refer to [12] for a different approach.

In Section 10.4, the problem of a thick sieve is treated (see [98] for the case of constant coefficients).

The unfolding method was applied for the first time for sieve problems in [169], also in the case of constant coefficients.

In Section 10.5, some variational inequalities with periodically oscillating obstacles are studied.

The last section presents some applications to the case of periodically distributed singular homogeneities near the boundary. Here again, the use of the boundary unfolding operator allows to give precise statements concerning the behavior in the boundary layer, and the limit results.

The methods presented here can be applied to evolution problems (see [42] for the heat and wave equations).

In this chapter the domains are in $\mathbb{R}^N$ *for* $N \geqslant 3$.

10.1 Volume distributed small holes

We use the notations of Section 9.2. In particular, B is a compact set in $\mathbb{R}^N$. The perforated domain $\Omega^*_{\varepsilon,\delta}$ is defined by (9.3). Assume that the matrix field $A^\varepsilon = (a^\varepsilon_{ij})_{1\leqslant i,j\leqslant N}$ belongs to $M(\alpha,\beta,\Omega)$ (see Definition 3.1). For f in $L^2(\Omega)$, consider the problem

$$\begin{cases} \text{Find } u_{\varepsilon,\delta} \in H^1_0(\Omega^*_{\varepsilon,\delta}) \text{ satisfying} \\ \displaystyle\int_{\Omega^*_{\varepsilon,\delta}} A^\varepsilon \nabla u_{\varepsilon,\delta}\, \nabla\phi\, dx = \int_{\Omega^*_{\varepsilon,\delta}} f\phi\, dx, \\ \forall \phi \in H^1_0(\Omega^*_{\varepsilon,\delta}). \end{cases} \tag{10.1}$$

In order to state the homogenization result, one introduces some functional spaces. The first are the spaces $\dot{H}^1(\mathbb{R}^N)$ and $\dot{H}^1(\mathbb{R}^N) \oplus \mathbb{R}$. They correspond to the statement of Proposition 9.2 for the case $p = 2$.

Here, the important space for the applications is the subspace $\mathbf{K}_B$ of

$\dot{H}^1(\mathbb{R}^N) \oplus \mathbb{R}$ defined as follows (recall that B is a given open set):

$$\begin{aligned}\mathbf{K}_B &\doteq \{\Phi \in \dot{H}^1(\mathbb{R}^N) \oplus \mathbb{R} \mid \Phi = 0 \text{ on } B\} \\ &= \{\Phi \in H^1_{loc}(\mathbb{R}^N) \mid \nabla\Phi \in L^2(\mathbb{R}^N)^N \text{ and } \Phi = 0 \text{ on } B\}\end{aligned} \tag{10.2}$$

with the norm $\|\Phi\|_{\mathbf{K}_B} \doteq \|\nabla\Phi\|_{L^2(\mathbb{R}^N)}$. Since the elements of $\dot{H}^1(\mathbb{R}^N)$ vanish at infinity of $\mathbb{R}^N$, the component in $\mathbb{R}$ of an element Φ of $\mathbf{K}_B$ is actually its limit at infinity, denoted $\Phi(\infty)$. Thus, $\mathbf{K}_B$ is a Hilbert space isometric to the closed subspace of $\dot{H}^1(\mathbb{R}^N)$ consisting of functions which are constant on B (the isometry is $\Phi \mapsto \Phi - \Phi(\infty)$).

Finally, consider the space

$$\begin{aligned}\mathbf{L}_B &= \{V \in L^2(\Omega; \mathbf{K}_B) \mid V(\cdot, \infty) \in H^1_0(\Omega)\} \\ &= \{V \in L^2(\Omega; \dot{H}^1(\mathbb{R}^N)) \oplus H^1_0(\Omega) \mid V = 0 \quad \text{a.e. in } \Omega \times B\},\end{aligned} \tag{10.3}$$

with the norm

$$\|V\|_{\mathbf{L}_B} = \left(\|\nabla V(\infty)\|^2_{L^2(\Omega)} + \|\nabla_y V\|^2_{L^2(\Omega\times\mathbb{R}^N)}\right)^{1/2}.$$

10.1.1 Homogenization

We now derive the unfolded formulation of the limit of problem (10.1). The result will show the contribution of the periodic oscillations of A^ε, as well as the contribution of the perforations.

Proposition 10.1. *Suppose that $\delta = \delta(\varepsilon)$ is such that*

$$k_1 \doteq \lim_{\varepsilon\to 0} \frac{\delta^{\frac{N}{2}-1}}{\varepsilon} \quad \textit{exists in } \mathbb{R}^+. \tag{10.4}$$

Suppose also that there exist a matrix A such that,

$$\mathcal{T}_\varepsilon(A^\varepsilon)(x,y) \to A(x,y) \quad \textit{for a.e. } (x,y) \textit{ in } \Omega \times Y, \tag{10.5}$$

and a matrix A_0 such that,

$$\mathcal{T}_{\varepsilon,\delta}(A^\varepsilon)(x,z) \to A_0(x,z) \quad \textit{for a.e. } (x,z) \textit{ in } \Omega \times (\mathbb{R}^N \setminus \overline{B})^{(1)}. \tag{10.6}$$

(1) It is easy to see that there is no particular connection between A and A_0. Actually, every pair (A, A_0) in $M(\alpha.\beta, \Omega\times Y) \times M(\alpha,\beta,\Omega\times\mathbb{R}^N\setminus\overline{B})$ can be obtained as the limit of the $(\mathcal{T}_\varepsilon(A^\varepsilon), \mathcal{T}_{\varepsilon,\delta}(A^\varepsilon))$ for the sequence $\{A^{\varepsilon,\delta}\}$ in $\mathrm{M}(\alpha,\beta,\Omega))$ defined by

$$A^{\varepsilon,\delta} \doteq \mathcal{U}_\varepsilon(A)\mathbf{1}_{\Omega^*_{\varepsilon,\sqrt{\delta}}} + \mathcal{U}_{\varepsilon,\delta}(A_0)\mathbf{1}_{\overline{\Omega\setminus\Omega^*_{\varepsilon,\sqrt{\delta}}}},$$

where $\mathcal{U}_{\varepsilon,\delta}$ is the adjoint of $\mathcal{T}_{\varepsilon,\delta}$ (see Remark 9.9).

Let $u_{\varepsilon,\delta}$ be the solution of problem (10.1). There exist u_0 in $H_0^1(\Omega)$ and U in $\mathbf{L}_B$ such that

$$
\begin{aligned}
&u_{\varepsilon,\delta} \rightharpoonup u_0 \quad \text{weakly in } H_0^1(\Omega),\\
&\mathcal{T}_{\varepsilon,\delta}(u_{\varepsilon,\delta}) \rightharpoonup U \quad \text{weakly in } L^2(\Omega; L^2_{loc}(\mathbb{R}^N)) \ \text{ with } \ U(\cdot,\infty) = u_0,\\
&\nabla_z\big(\mathcal{T}_{\varepsilon,\delta}(u_{\varepsilon,\delta})\big)\, \mathbf{1}_{\frac{1}{\delta}Y} \rightharpoonup \nabla_z U \quad \text{weakly in } L^2(\Omega \times \mathbb{R}^N)^N.
\end{aligned} \tag{10.7}
$$

Moreover, there exists $\widehat{u}$ in $L^2(\Omega; H^1_{per,0}(Y))$ such that

$$
\mathcal{T}_\varepsilon(\nabla u_{\varepsilon,\delta}) \rightharpoonup \nabla u_0 + \nabla_y \widehat{u} \quad \text{weakly in } L^2(\Omega \times Y)^N.
$$

The pair $(U, \widehat{u}) \in \mathbf{L}_B \times L^2(\Omega; H^1_{per,0}(Y))$ satisfies the unfolded limit problem

$$
\begin{cases}
\dfrac{1}{|Y|} \displaystyle\int_{\Omega\times Y} A\big(\nabla U(\cdot,\infty) + \nabla_y \widehat{u}\big)\big(\nabla V(\cdot,\infty) + \nabla_y \widehat{v}\big)\, dxdy \\
\qquad + \dfrac{k_1^2}{|Y|} \displaystyle\int_{\Omega\times(\mathbb{R}^N\setminus\overline{B})} A_0\, \nabla_z U\, \nabla_z V\, dxdz = \int_\Omega f\, V(\cdot,\infty)\, dx, \\
\forall V \in \mathbf{L}_B, \quad \forall \widehat{v} \in L^2(\Omega; H^1_{per}(Y)).
\end{cases} \tag{10.8}
$$

Problem (10.8) has a unique solution by the Lax-Milgram theorem applied in the space $\mathbf{L}_B \times L^2(\Omega; H^1_{per,0}(Y))$ due to the ellipticity of the matrix fields A and A_0 (arising from A^ε in $M(\alpha, \beta, \Omega)$) and identity (3.14).

The proof of Proposition 10.1 makes use of the next two elementary results. Since B is compact in $\mathbb{R}^N$, take $R > 1$ so that B is included in the ball $B(O; R)$ (the ball centered at the origin and of radius R). There exists $\delta_0 > 0$ such that

$$
B(O; 2\delta_0 R) \subset \overline{Y}.
$$

Lemma 10.2.
(i) For $p \in [1, +\infty)$, the set $\displaystyle\bigcup_{\delta\in(0\delta_0]} \big\{\varphi \in W^{1,p}(Y) \mid \varphi \text{ const. on } \delta B\big\}$ is dense in $W^{1,p}(Y)$.
(ii) For $p \in [1, N)$, the set $\displaystyle\bigcup_{\delta\in(0\delta_0]} \big\{\phi \in W^{1,p}(Y) \mid \phi = 0 \text{ on } \delta B\big\}$ is dense in $W^{1,p}(Y)$[2]*.*

Proof. It is enough to show the result in the case where B is the ball $B(O; R)$, in which situation the sets are subspaces of $W^{1,p}(Y)$. Denote by θ the function in $W^{1,\infty}(\mathbb{R}_+)$ defined by

$$
\theta(t) = \begin{cases} 0 & \text{if } t \in [0,1], \\ t-1 & \text{if } t \in (1,2), \\ 1 & \text{if } t \geqslant 2, \end{cases}
$$

[2] This density does not hold for $p > N$ since, under that hypothesis, $W^{1,p}(Y)$ is included in $\mathcal{C}(\overline{Y})$ by the Sobolev embedding theorem.

and set

$$\Theta_\delta(y) = \theta\Big(\frac{|y|}{\delta}\Big), \qquad \forall y \in Y.$$

The function Θ_δ is with values in $[0,1]$. It vanishes on $B(O;\delta R)$ and equals 1 outside of $B(O;2\delta R)$. Thus the sequence $\{\Theta_\delta\}_\delta$ converges to 1 for every $y \in Y\backslash\{0\}$ and its gradient is bounded by $1/\delta$. It is used as a cut-off.

Let φ be in $W^{1,p}(Y)$, for simplicity set

$$\mathbf{m}_\delta = \mathcal{M}_{B(O;2\delta R)}(\varphi) = \frac{1}{|B(O;2\delta R)|}\int_{B(O;2\delta R)} \varphi(y)\,dy.$$

(*i*) From the (rescaled) Poincaré-Wirtinger inequality, there exists a constant C which does not depend on δ such that

$$\|\varphi - \mathbf{m}_\delta\|_{L^p(B(O;2\delta R))} \leqslant C\delta\|\nabla\varphi\|_{L^p(B(O;2\delta R)}.$$

For $\delta \in (0,\delta_0]$, set

$$\varphi_\delta = \mathbf{m}_\delta + \Theta_\delta\big(\varphi - \mathbf{m}_\delta\big).$$

It belongs to $W^{1,p}(Y)$ by construction, and is constant on $B(O;\delta R)$. We claim that

$$\varphi_\delta \;\to \varphi \quad \text{strongly in } \; W^{1,p}(Y). \tag{10.9}$$

Indeed,

$$\varphi - \varphi_\delta = (1-\Theta_\delta)\big(\varphi - \mathbf{m}_\delta\big),$$

and from the properties of Θ_δ,

$$\begin{aligned}
\|\varphi - \varphi_\delta\|_{L^p(Y)} &\leqslant \|\varphi - \mathbf{m}_\delta\|_{L^p(B(O;2\delta R))} \leqslant C\delta\|\nabla\varphi\|_{L^p(B(O;2\delta R))},\\
\|\nabla(\varphi - \varphi_\delta)\|_{L^p(Y)} &\leqslant \|\nabla\Theta_\delta\big(\varphi - \mathbf{m}_\delta\big)\|_{L^p(B(O;2\delta R))}\\
&\qquad\qquad + \|(1-\Theta_\delta)\nabla\big(\varphi - \mathbf{m}_\delta\big)\|_{L^p(Y)}\\
&\leqslant C\|\nabla\varphi\|_{L^p(B(O;2\delta R))},
\end{aligned}$$

where the constants do not depend on δ.
By dominated convergence, $\lim_{\delta\to 0}\|\nabla\varphi\|_{L^p(B(O;2\delta R))} \to 0$, the claim (10.9) follows.

(*ii*) Assume $1 \leqslant p < N$. We claim that $\Theta_\delta\varphi$, which vanishes on $B(O;\delta R)$, converges to φ in $W^{1,p}(Y)$. The Sobolev embedding theorem applies since $p < N$. Therefore, φ belongs to $L^{p^*}(Y)$ where p^* is the Sobolev conjugate of p, defined by $\dfrac{1}{p^*} = \dfrac{1}{p} - \dfrac{1}{N}$.

By the properties of Θ_δ,

$$\|(1-\Theta_\delta)\varphi\|_{L^p(Y)} \to 0 \text{ by dominated convergence,}$$

so that $\Theta_\delta\varphi$ converges to φ in $L^p(Y)$. Similarly,

$$\begin{aligned}\|\nabla((1-\Theta_\delta)\varphi)\|_{L^p(Y)}& \\ &\leqslant \|(1-\Theta_\delta)\nabla\varphi\|_{L^p(B(O;2\delta R))} + \frac{C}{\delta}\|\varphi\|_{L^p(B(O;2\delta R))}.\end{aligned} \tag{10.10}$$

By dominated convergence, the first term in the right-hand side goes to 0. By Hölder's inequality, $\|\varphi\|_{L^p(B(O;2\delta R))}$ is bounded above by

$$|B(O;2\delta R)|^{(\frac{1}{p}-\frac{1}{p^*})}\|\varphi\|_{L^{p^*}(B(O;2\delta R))} = C\delta\|\varphi\|_{L^{p^*}(B(O;2\delta R))}.$$

Consequently, the last term in (10.10) is bounded above by $C\|\varphi\|_{L^{p^*}(B(O;2\delta R))}$. The claim is proved since the last norm goes to 0 (by dominated convergence again since φ belongs to $L^{p^*}(Y)$). □

Remark 10.3. *Lemma 10.2 holds also true in the space $W^{1,p}_{per}(Y)$ (in place of $W^{1,p}(Y)$).*

Lemma 10.4. *Let $p \in [1,+\infty)$ and v be in $W^{1,p}_{loc}(\mathbb{R}^N)$, such that $\nabla_z v$ is in $L^p(\mathbb{R}^N)^N$ and has a compact support. Set*

$$v_{\varepsilon,\delta}(x) = v\Big(\frac{1}{\delta}\Big\{\frac{x}{\varepsilon}\Big\}_Y\Big) \quad \textit{for a.e. } x \in \Omega. \tag{10.11}$$

By Proposition 9.2, it has a limit at infinity denoted $v(\infty)$. If δ is small enough, the function $v_{\varepsilon,\delta}$ belongs to $W^{1,p}(\Omega)$ and

$$v_{\varepsilon,\delta} \to v(\infty) \quad \textit{strongly in } L^p(\Omega). \tag{10.12}$$

Moreover, if

$$\frac{\delta^{\frac{N}{p}-1}}{\varepsilon} \quad \textit{is uniformly bounded,} \tag{10.13}$$

the following convergence hold:

$$v_{\varepsilon,\delta} \rightharpoonup v(\infty) \quad \textit{weakly in } W^{1,p}(\Omega). \tag{10.14}$$

Proof. Let v be in $W^{1,p}_{loc}(\mathbb{R}^N)$ satisfying the hypotheses. By construction,

$$\mathrm{supp}(v_{\varepsilon,\delta} - v(\infty)) \subset \Omega \bigcap \bigcup_{\xi\in G} \varepsilon(\xi + \delta\,\mathrm{supp}(\nabla_z v)).$$

Consequently, for δ small enough, $v_{\varepsilon,\delta} - v(\infty)$ belongs to $W^{1,p}(\Omega)$ since it vanishes on the boundary of the ε-cells (so that there is no jump discontinuity going from a cell to a neighboring one). Moreover, the measure of its support is of order δ^N. Applying the Poincaré inequality, we have[3]

$$\|v - v(\infty)\|_{L^p(\mathbb{R}^N)} \leqslant C_v\|\nabla_z v\|_{L^p(\mathbb{R}^N)}.$$

(3) Here, the constant C_v depends only upon the support of $\nabla_z v$.

Rescaling with (10.11), yields

$$\begin{aligned}\|v_{\varepsilon,\delta} - v(\infty)\|_{L^p(\Omega)} &\leqslant C_v \delta^{N/p} \|v - v(\infty)\|_{L^p(\mathbb{R}^N)} \\ &\leqslant C_v \delta^{N/p} \|\nabla_z v\|_{L^p(\mathbb{R}^N)},\end{aligned}$$

which implies convergence (10.12).

Similarly rescaling for the gradient gives

$$\|\nabla v_{\varepsilon,\delta}\|_{L^p(\Omega)} \leqslant C \frac{\delta^{\frac{N}{p}-1}}{\varepsilon} \|\nabla_z v\|_{L^p(\mathbb{R}^N)},$$

with a constant C depending only upon Ω (related to $|\Omega|$). In turn, due to (10.13), this implies the boundedness of the sequence $\{v_{\varepsilon,\delta}\}_{(\varepsilon,\delta)}$ in $W^{1,p}(\Omega)$. This boundedness together with convergence (10.12) gives (10.14). □

Proof of Proposition 10.1 for the case $k_1 > 0$[4].

Step 1. Observe first that by the Lax-Milgram theorem, there exists a unique solution $u_{\varepsilon,\delta}$ of (10.1) satisfying

$$\|u_{\varepsilon,\delta}\|_{H_0^1(\Omega^*_{\varepsilon,\delta})} \leqslant C \|f\|_{L^2(\Omega)},$$

which implies the first convergence in (10.7), up to a subsequence. Next, by Theorem 1.41, there exists $\widehat{u}$ in $L^2(\Omega; H^1_{per,0}(Y))$, such that

$$\mathcal{T}_\varepsilon(\nabla u_{\varepsilon,\delta}) \rightharpoonup \nabla u_0 + \nabla_y \widehat{u} \quad \text{weakly in } L^2(\Omega \times Y)^N. \tag{10.15}$$

For ψ in $\mathcal{D}(\Omega)$ and ϕ in $H^1_{per}(Y)$ vanishing in a neighborhood of the origin, set

$$\Phi_\varepsilon(x) = \varepsilon \psi(x) \phi\Big(\frac{x}{\varepsilon}\Big).$$

By construction, Φ_ε is an admissible test function in (10.1), and this gives

$$\begin{aligned}\varepsilon \int_{\Omega^*_{\varepsilon,\delta}} A^\varepsilon \nabla u_{\varepsilon,\delta}\, \nabla\psi\, \phi\Big(\frac{x}{\varepsilon}\Big)\, dx + \int_{\Omega^*_{\varepsilon,\delta}} A^\varepsilon \nabla u_{\varepsilon,\delta}\, \psi\, \nabla\phi\Big(\frac{x}{\varepsilon}\Big)\, dx \\ = \varepsilon \int_{\Omega^*_{\varepsilon,\delta}} f\, \psi\, \phi\Big(\frac{x}{\varepsilon}\Big)\, dx.\end{aligned}$$

It is easy to see that the first integral, as well as the right-hand side of the above equality, converge to zero. Unfolding the second integral with $\mathcal{T}_\varepsilon$, and taking into account the fact that

$$\mathcal{T}_\varepsilon\Big(\nabla\phi\Big(\frac{x}{\varepsilon}\Big)\Big)(x,y) = \nabla_y \phi(y),$$

[4] See Remark 10.36 for the cases $k_1 = 0$ and $k_1 = +\infty$.

we get,

$$\int_{\Omega^*_{\varepsilon,\delta}} A^\varepsilon \nabla u_{\varepsilon,\delta}\, \psi\, \nabla\phi\Big(\frac{x}{\varepsilon}\Big)\, dx$$
$$= \frac{1}{|Y|}\int_{\Omega\times Y} \mathcal{T}_\varepsilon(A^\varepsilon)(x,y)\, \mathcal{T}_\varepsilon(\nabla u_{\varepsilon,\delta})(x,y)\, \nabla_y\phi(y)\, \mathcal{T}_\varepsilon(\psi)(x,y)\, dxdy.$$

Passing to the limit in this identity using convergences (10.5) and (10.15), yields

$$\lim_{\varepsilon\to 0}\int_{\Omega^*_{\varepsilon,\delta}} A^\varepsilon \nabla u_{\varepsilon,\delta}\, \psi\, \nabla\phi\Big(\frac{\cdot}{\varepsilon}\Big)\, dx = \int_{\Omega\times Y} A\big(\nabla u_0 + \nabla_y \widehat{u}\big)\, \nabla_y\phi\; \psi\, dxdy = 0.$$

Then, by Lemma 10.2(ii),

$$\int_{\Omega\times Y} A\big(\nabla u_0 + \nabla_y \widehat{u}\big)\, \nabla_y \widehat{v}\, dxdy = 0, \quad \forall \widehat{v} \in L^2(\Omega; H^1_{per}(Y)). \tag{10.16}$$

This equation describes the effect of the periodic oscillations of the coefficients in problem (10.1).

Step 2. By Theorem 9.7(vii), there exists U in $L^2(\Omega; L^2_{loc}(\mathbb{R}^N))$ such that, up to a subsequence,

$$\mathcal{T}_{\varepsilon,\delta}(u_{\varepsilon,\delta}) \rightharpoonup U \quad \text{weakly in } L^2(\Omega; L^2_{loc}(\mathbb{R}^N)). \tag{10.17}$$

By Proposition 1.25, one has

$$\mathcal{M}_\varepsilon(u_{\varepsilon,\delta}) \to u_0 \quad \text{strongly in } L^2(\Omega),$$

so that,

$$\mathcal{M}_\varepsilon(u_{\varepsilon,\delta})\mathbf{1}_{\frac{1}{\delta}Y} \to u_0 \quad \text{strongly in } L^2(\Omega; L^2_{loc}(\mathbb{R}^N)). \tag{10.18}$$

Again by Theorem 9.7(vii), there exists W in $L^2(\Omega; L^{2^*}(\mathbb{R}^N))$ with $\nabla_z W$ in $L^2(\Omega\times\mathbb{R}^N)$, such that

$$\mathcal{T}_{\varepsilon,\delta}\big(u_{\varepsilon,\delta} - \mathcal{M}_\varepsilon(u_{\varepsilon,\delta})\big) \rightharpoonup W \quad \text{weakly in } L^2(\Omega; L^{2^*}(\mathbb{R}^N)). \tag{10.19}$$

From (10.17), (10.18) and (10.19), one draws the conclusion

$$U = W + u_0 \quad \text{and} \quad \nabla_z U = \nabla_z W,$$

and, by Theorem 9.7(vii) once more,

$$\nabla_z\big(\mathcal{T}_{\varepsilon,\delta}(u_{\varepsilon,\delta})\big)\, \mathbf{1}_{\frac{1}{\delta}Y} \rightharpoonup \nabla_z U \quad \text{weakly in } L^2(\Omega\times\mathbb{R}^N)^N.$$

From Definition 9.6, $\mathcal{T}_{\varepsilon,\delta}(u_{\varepsilon,\delta}) = 0$ in $\Omega\times B$, and thanks to (10.17),

$$U = 0 \text{ on } \Omega\times B. \tag{10.20}$$

Due to (10.19)-(10.20), U belongs to $\mathbf{L}_B$ and $U(\cdot,\infty) = u_0$.

Step 3. Let ψ be in $\mathcal{D}(\Omega)$ and $v \in \mathbf{K}_B$ such that $\nabla_z v$ has a compact support. We take $v_{\varepsilon,\delta}\,\psi$ as test function in problem (10.1) where $v_{\varepsilon,\delta}$ is defined by (10.11). The result is

$$\int_{\Omega^*_{\varepsilon,\delta}} A^\varepsilon \nabla u_{\varepsilon,\delta}\,\nabla v_{\varepsilon,\delta}\,\psi\,dx + \int_{\Omega^*_{\varepsilon,\delta}} A^\varepsilon \nabla u_{\varepsilon,\delta}\,\nabla\psi\; v_{\varepsilon,\delta}\,dx = \int_{\Omega^*_{\varepsilon,\delta}} f\, v_{\varepsilon,\delta}\,\psi\,dx. \tag{10.21}$$

The first term in this equation is unfolded with $\mathcal{T}_{\varepsilon,\delta}$. From the obvious inequality

$$\|\mathcal{T}_{\varepsilon,\delta}(\psi) - \psi\|_{L^\infty(\widehat{\Omega}_\varepsilon \times \frac{1}{\delta}Y)} \leqslant C\,\varepsilon \|\nabla\psi\|_{L^\infty(\Omega)}, \tag{10.22}$$

it follows that

$$\mathcal{T}_{\varepsilon,\delta}(\psi)\nabla_z v \to \psi\nabla_z v \quad \text{strongly in } L^2(\Omega\times\mathbb{R}^N)^N. \tag{10.23}$$

Convergences (10.1.1), (10.23) together with the strong convergence in $L^2(\Omega)$ of $v_{\varepsilon,\delta}$ (given by (10.12)), as well as hypothesis (10.6), allow to pass to the limit in the first integral in (10.21). Recalling that $\mathcal{T}_{\varepsilon,\delta}(\nabla v_{\varepsilon,\delta}) = \dfrac{1}{\varepsilon\delta}\nabla_z v$, one has successively,

$$\begin{aligned}
&\lim_{\varepsilon\to 0} \int_{\Omega^*_{\varepsilon,\delta}} A^\varepsilon \nabla u_{\varepsilon,\delta}\,\nabla v_{\varepsilon,\delta}\,\psi\,dx \\
&\quad = \lim_{\varepsilon\to 0} \frac{\delta^{N-2}}{\varepsilon^2|Y|} \int_{\Omega\times\frac{1}{\delta}Y} \mathcal{T}_{\varepsilon,\delta}(A^\varepsilon)\,\nabla_z \mathcal{T}_{\varepsilon,\delta}(u_{\varepsilon,\delta})\,\nabla_z v(z)\,\mathcal{T}_{\varepsilon,\delta}(\psi)\,dxdz \qquad (10.24)\\
&\quad = \frac{k_1^2}{|Y|}\int_{\Omega\times(\mathbb{R}^N\setminus\overline{B})} A_0(x,z)\,\nabla_z U(x,z)\,\nabla_z v(z)\,\psi(x)\,dxdz,
\end{aligned}$$

which by density, is true for every $v \in \mathbf{K}_B$.

The second term in (10.21) is unfolded with $\mathcal{T}_\varepsilon$. Due to Proposition 1.9 and to convergence (10.15), using again the strong convergence of $v_{\varepsilon,\delta}$ as well as hypothesis (10.5), we pass to the limit, to get

$$\begin{aligned}
&\lim_{\varepsilon\to 0} \int_{\Omega^*_{\varepsilon,\delta}} A^\varepsilon \nabla u_{\varepsilon,\delta}\; v_{\varepsilon,\delta}\,\nabla\psi\,dx \\
&\quad = \lim_{\varepsilon\to 0} \frac{1}{|Y|}\int_{\Omega\times Y} \mathcal{T}_\varepsilon(A^\varepsilon)\,\mathcal{T}_\varepsilon(\nabla u_{\varepsilon,\delta})\mathcal{T}_\varepsilon(v_{\varepsilon,\delta})\,\mathcal{T}_\varepsilon(\nabla\psi)\,dxdy \qquad (10.25)\\
&\quad = \frac{v(\infty)}{|Y|}\int_{\Omega\times Y} A(\nabla u_0 + \nabla_y \widehat{u})\,\nabla\psi\,dxdy.
\end{aligned}$$

Passing to the limit in (10.21) and using (10.24)-(10.25), give

$$\frac{v(\infty)}{|Y|}\int_{\Omega\times Y} A(\nabla u_0+\nabla_y\widehat{u})\,\nabla\psi\,dxdy \\ +\frac{k_1^2}{|Y|}\int_{\Omega\times(\mathbb{R}^N\setminus\overline{B})} A_0\,\nabla_z U\,\nabla_z v\,\psi\,dxdz = v(\infty)\int_\Omega f\,\psi\,dx,$$

from which, by density,

$$\frac{1}{|Y|}\int_{\Omega\times Y} A(\nabla u_0+\nabla_y\widehat{u})\,\nabla V(\cdot,\infty)\,dxdy \\ +\frac{k_1^2}{|Y|}\int_{\Omega\times(\mathbb{R}^N\setminus\overline{B})} A_0\,\nabla_z U\,\nabla_z V\,dxdz = \int_\Omega f\,V(\cdot,\infty)\,dx$$

for every $V\in\mathbf{L}_B$. Then, taking into account (10.16), the unfolded problem (10.8) satisfied by the pair $(U,\widehat{u})$ follows. The Lax-Milgram theorem gives the existence of a unique solution for (10.8). □

10.1.2 Standard form of the limit problem

Here we show that the unfolded problem is well-posed and give the formulation in terms of the macroscopic solution u_0 alone.

First express $\widehat{u}$ in terms of the correctors $\widehat{\chi}_j, j=1,\dots,N$ defined by the cell problems (3.10). That gives

$$\widehat{u}(x,y)=\sum_{j=1}^N\frac{\partial u_0}{\partial x_j}(x)\,\widehat{\chi}_j(x,y)=\sum_{j=1}^N\frac{\partial U}{\partial x_j}(x,\infty)\,\widehat{\chi}_j(x,y),$$

for a.e. $(x,y)\in\Omega\times Y$. Using it in equation (10.8) from Proposition 10.1, yields

$$\int_\Omega A^{hom}\nabla U(\cdot,\infty)\,\nabla V(\cdot,\infty)\,dxdy+\frac{k_1^2}{|Y|}\int_{\Omega\times\mathbb{R}^N} A_0\,\nabla_z U\,\nabla_z V\,dxdz \\ =\int_\Omega f\,V(\cdot,\infty)\,dx,\quad \forall V\in\mathbf{L}_B, \tag{10.26}$$

where (cf. (3.22)), for a.e. $x\in\Omega$,

$$A^{hom}(x)\nabla U(x,\infty)\doteq\frac{1}{|Y|}\int_Y A(x,y)\big[\nabla U(x,\infty)+\nabla_y\widehat{u}(x,y)\big]\,dy. \tag{10.27}$$

Let $\theta\in L^\infty(\Omega;\mathbf{K}_B)$, $\theta(x,\infty)\equiv 1$, be the solution of the "cell problem"[(5)]

$$\begin{cases}\displaystyle\int_{\mathbb{R}^N\setminus\overline{B}} A_0(x,z)\,\nabla_z\theta(x,z)\,\nabla_z\Psi(z)\,dz=0\quad\text{for a.e. } x\in\Omega,\\ \forall\Psi\in\dot{H}^1(\mathbb{R}^N)\quad\text{with } \Psi=0\quad\text{on } B.\end{cases} \tag{10.28}$$

(5) The cell problem has a unique solution by the Lax-Milgram theorem applied in the affine subspace of $\mathbf{K}_B$ consisting of functions which are constant equal to 1 at infinity.

In view of (10.26) and since $U(\cdot, \infty) = u_0$, it follows that for a.e. $x \in \Omega$, the two functions $U(x, \cdot)$ and $u_0(x)\theta(x, \cdot)$ coincide since they both belong to $\mathbf{K}_B$ and are solutions of the same problem, (10.28) multiplied by $u_0(x)$, which has a unique solution.

Going back to equation (10.26) with $V(x, z) = \psi(x)\theta(z)$ for a.e. (x, z) in $\Omega \times \mathbb{R}^N$, where ψ is arbitrary in $H_0^1(\Omega)$, gives

$$\int_\Omega A^{hom} \nabla u_0 \, \nabla \psi \, dxdy + k_1^2 \int_\Omega \boldsymbol{\Theta} \, u_0 \, \psi \, dxdz = \int_\Omega f \, \psi \, dx,$$

where, for a.e. $x \in \Omega$,

$$\boldsymbol{\Theta}(x) = \frac{1}{|Y|} \int_{\mathbb{R}^N \setminus \overline{B}} A_0(x, z) \, \nabla_z \theta(x, z) \, \nabla_z \theta(x, z) \, dz.$$

This integral is non-negative, so that $\boldsymbol{\Theta}$ can be interpreted as the local capacity of the set B.

In summary, we have proved the following result:

Theorem 10.5. *Under the assumptions of Proposition 10.1 with $k_1 > 0$, the limit function $u_0 \in H_0^1(\Omega$ is the unique solution of the homogenized equation,*

$$\begin{cases} \displaystyle\int_\Omega A^{hom} \nabla u_0 \, \nabla \psi \, dx + k_1^2 \int_\Omega \boldsymbol{\Theta} \, u_0 \, \psi \, dx = \int_\Omega f \, \psi \, dx, \\ \forall \psi \in H_0^1(\Omega). \end{cases} \tag{10.29}$$

Remark 10.6. *The contribution of the oscillations of the matrix A^ε in the homogenized problem is reflected by the first term of the left-hand side in* (10.29)*, that is the presence of the operator A^{hom}. The contribution of the perforations is the appearance in the limit equation of the zero order "strange term" $k_1^2 \, \boldsymbol{\Theta} \, u_0$.*

Remark 10.7.

1. The proof for the case $k_1 = 0$ (k_1 defined by (10.4)*) is actually simpler, but requires a slight modification, namely setting in place of* $(10.7)_2$*, the following convergence:*

$$\frac{\delta^{\frac{N}{2}-1}}{\varepsilon} \mathcal{T}_{\varepsilon,\delta}(u_{\varepsilon,\delta}) \rightharpoonup U \quad \textit{weakly in } L^2(\Omega; L^2_{loc}(\mathbb{R}^N)) \quad \textit{with } U(\cdot, \infty) = u_0,$$

Then the corresponding statement is Theorem 10.5 with $k_1 = 0$; the holes are too small to have any influence at the limit.

2. The case of $k_1 = +\infty$ is easy to analyze with the help of Theorem 9.7(vi). Indeed, by (9.5)*,*

$$\mathcal{T}_{\varepsilon,\delta}(u_{\varepsilon,\delta}) \rightharpoonup u_0 \quad \textit{weakly in } L^2(\Omega; L^2_{loc}(\mathbb{R}^N)).$$

On the other hand, since

$$\mathcal{T}_{\varepsilon,\delta}(u_{\varepsilon,\delta}) = 0 \quad in\ \Omega \times B,$$

this implies that

$$u_0 = 0.$$

This means that in this case, the holes are too big and force the limit to vanish on the whole of Ω.

3. In [75], the small holes in the perforated domain which led to a "strange term" were balls of the critical radius

$$r_\varepsilon^c = C\varepsilon^{N/N-2} \quad for\ N > 2 \quad and \quad C\ a\ constant.$$

Like in the items 1 and 2 above, for a radius $r_\varepsilon \lambda r_\varepsilon^c$ *the holes do not contribute at the limit, whereas for* $r_\varepsilon^c \lambda r_\varepsilon$, *the limit solution is 0. It is easily seen that looking for* δ *of the form* $C\varepsilon^\alpha$ *with* $\alpha > 0$, *and satisfying* (10.4), *one finds the critical size* r_ε^c *above.*

10.1.3 Convergence of the energy

Proposition 10.8 (Convergence of the energy). *Under the hypotheses of Theorem 10.5, one has*

$$\begin{aligned}
(i)\quad \lim_{\varepsilon\to 0}\int_{\Omega^*_{\varepsilon,\delta}} A^\varepsilon \nabla u_{\varepsilon,\delta}\, \nabla u_{\varepsilon,\delta}\, dx &= \int_\Omega A^{hom}\nabla u_0\, \nabla u_0\, dx + k_1^2 \int_\Omega \boldsymbol{\Theta}\, u_0^2\, dx \\
&= \frac{1}{|Y|}\int_{\Omega\times Y} A\big(\nabla U(\cdot,\infty) + \nabla_y \widehat{u}\big)\big(\nabla U(\cdot,\infty) + \nabla_y \widehat{u}\big)\, dxdy \\
&\quad + \frac{k_1^2}{|Y|}\int_{\Omega\times(\mathbb{R}^N\setminus\overline{B})} A_0\, \nabla_z U\, \nabla_z U\, dxdz,
\end{aligned}$$

$$(ii)\quad \lim_{\varepsilon\to 0}\int_{\Lambda_\varepsilon} |\nabla u_{\varepsilon,\delta}|^2\, dx = 0.$$

Proof. Taking $u_{\varepsilon,\delta}$ as test-function in (10.1) and unfolding with $\mathcal{T}_\varepsilon$ gives

$$\frac{1}{|Y|}\int_{\Omega\times Y} \mathcal{T}_\varepsilon(A^\varepsilon)\mathcal{T}_\varepsilon(\nabla u_{\varepsilon,\delta})\mathcal{T}_\varepsilon(\nabla u_{\varepsilon,\delta})dxdy + \int_{\Lambda_\varepsilon} A^\varepsilon \nabla u_{\varepsilon,\delta}\nabla u_{\varepsilon,\delta}\, dx = \int_{\Omega^*_{\varepsilon,\delta}} f\, u_{\varepsilon,\delta}\, dx.$$

Set

$$\widetilde{Y}_\delta = Y\setminus\sqrt{\delta}\,\overline{B}.$$

Transforming the first integral on the left-hand side by using the change of variable $y = \delta z$ in $Y \setminus \widetilde{Y}_\delta$, we get

$$\begin{aligned}
&\frac{1}{|Y|}\int_{\Omega\times\widetilde{Y}_\delta} \mathcal{T}_\varepsilon(A^\varepsilon)\,\mathcal{T}_\varepsilon(\nabla u_{\varepsilon,\delta})\,\mathcal{T}_\varepsilon(\nabla u_{\varepsilon,\delta})\,dxdy\\
&\qquad + \frac{\delta^{N-2}}{\varepsilon^2|Y|}\int_{\Omega\times\frac{1}{\sqrt{\delta}}B} \mathcal{T}_{\varepsilon,\delta}(A^\varepsilon)\mathcal{T}_{\varepsilon,\delta}(\nabla u_{\varepsilon,\delta})\,\mathcal{T}_{\varepsilon,\delta}(\nabla u_{\varepsilon,\delta})\,dxdz\\
&\qquad + \int_{\Lambda_\varepsilon} A^\varepsilon\nabla u_{\varepsilon,\delta}\,\nabla u_{\varepsilon,\delta}\,dx = \int_{\Omega^*_{\varepsilon,\delta}} f\,u_{\varepsilon,\delta}\,dx.
\end{aligned}$$

For simplicity, set

$$\begin{aligned}
\mathcal{A}_\varepsilon &= \frac{1}{|Y|}\int_{\Omega\times\widetilde{Y}_\delta} \mathcal{T}_\varepsilon(A^\varepsilon)\mathcal{T}_\varepsilon(\nabla u_{\varepsilon,\delta})\,\mathcal{T}_\varepsilon(\nabla u_{\varepsilon,\delta})\,dx,\\
\mathcal{B}_\varepsilon &= \frac{\delta^{N-2}}{\varepsilon^2|Y|}\int_{\Omega\times\frac{1}{\sqrt{\delta}}B} \mathcal{T}_{\varepsilon,\delta}(A^\varepsilon)\mathcal{T}_{\varepsilon,\delta}(\nabla u_{\varepsilon,\delta})\,\mathcal{T}_{\varepsilon,\delta}(\nabla u_{\varepsilon,\delta})\,dxdz, \qquad (10.30)\\
\mathcal{C}_\varepsilon &= \int_{\Lambda_\varepsilon} A^\varepsilon\nabla u_{\varepsilon,\delta}\,\nabla u_{\varepsilon,\delta}\,dx.
\end{aligned}$$

From (10.15) and (10.1.1) we have the convergences

$$\begin{aligned}
&\mathcal{T}_\varepsilon(\nabla u_{\varepsilon,\delta})\mathbf{1}_{\Omega\times\widetilde{Y}_\delta} \rightharpoonup \nabla u_0 + \nabla_y\widehat{u} = \nabla U(\cdot,\infty) + \nabla_y\widehat{u} \quad \text{weakly in } L^2(\Omega\times Y)^N,\\
&\nabla_z\big(\mathcal{T}_{\varepsilon,\delta}(u_{\varepsilon,\delta})\big)\,\mathbf{1}_{\frac{1}{\sqrt{\delta}}B} \rightharpoonup \nabla_z U \quad \text{weakly in } L^2(\Omega\times\mathbb{R}^N)^N.
\end{aligned}$$

Consequently,

$$\begin{aligned}
&\frac{1}{|Y|}\int_{\Omega\times Y} A\big(\nabla U(\cdot,\infty) + \nabla_y\widehat{u}\big)\,\big(\nabla U(\cdot,\infty) + \nabla_y\widehat{u}\big)\,dxdy\\
&\qquad\qquad + \frac{k_1^2}{|Y|}\int_{\Omega\times(\mathbb{R}^N\setminus\overline{B})} A_0\,\nabla_z U\,\nabla_z U\,dxdz\\
&\leqslant \liminf_{\varepsilon\to 0}\mathcal{A}_\varepsilon + \liminf_{\varepsilon\to 0}\mathcal{B}_\varepsilon \leqslant \liminf_{\varepsilon\to 0}\Big(\int_{\Omega^*_{\varepsilon,\delta}} A^\varepsilon\nabla u_{\varepsilon,\delta}\,\nabla u_{\varepsilon,\delta}\,dx - \mathcal{C}_\varepsilon\Big)\\
&\leqslant \limsup_{\varepsilon\to 0}\Big(\int_{\Omega^*_{\varepsilon,\delta}} A^\varepsilon\nabla u_{\varepsilon,\delta}\,\nabla u_{\varepsilon,\delta}\,dx - \mathcal{C}_\varepsilon\Big) \leqslant \limsup_{\varepsilon\to 0}\int_{\Omega^*_{\varepsilon,\delta}} A^\varepsilon\nabla u_{\varepsilon,\delta}\,\nabla u_{\varepsilon,\delta}\,dx\\
&= \lim_{\varepsilon\to 0}\int_{\Omega^*_{\varepsilon,\delta}} f\,u_{\varepsilon,\delta}\,dx = \int_\Omega f\,U(\cdot,\infty)\,dx.
\end{aligned}$$

Then, from (10.8), all inequalities above are actually equalities and this implies convergences (i) and (ii) from the statement of the proposition. □

The standard method applies then to obtain a corrector result.

Corollary 10.9. *Under the hypotheses of Theorem 10.5, the following strong convergences hold:*

$$\begin{aligned}
&\mathcal{T}_\varepsilon(\nabla u_{\varepsilon,\delta})\mathbf{1}_{\Omega\times\widetilde{Y}_\delta} \to \nabla u_0 + \nabla_y\widehat{u} \quad \textit{strongly in } L^2(\Omega\times Y)^N,\\
&\nabla_z\big(\mathcal{T}_{\varepsilon,\delta}(u_{\varepsilon,\delta})\big)\,\mathbf{1}_{\frac{1}{\sqrt{\delta}}B} \to \nabla_z U \quad \textit{strongly in } L^2(\Omega\times\mathbb{R}^N)^N.
\end{aligned} \qquad (10.31)$$

The first convergence implies the corrector result

$$\nabla u_{\varepsilon,\delta}\mathbf{1}_{\Omega^*_{\varepsilon,\sqrt{\delta}}} - \nabla u_0 - \mathcal{U}_\varepsilon(\nabla_y \widehat{u}) \to 0 \quad \textit{strongly in } L^2(\Omega)^N,$$

while the second convergence implies that

$$\|\nabla u_{\varepsilon,\delta}\|_{L^2(\Omega^*_{\varepsilon,\delta}\setminus\Omega^*_{\varepsilon,\sqrt{\delta}})} \to k_1\|\nabla_z U\|_{L^2(\Omega\times\mathbb{R}^N)}.$$

Furthermore,

$$\mathcal{T}_{\varepsilon,\delta}(u_{\varepsilon,\delta}) \to U \quad \textit{strongly in } \; L^2(\Omega; L^2_{loc}(\mathbb{R}^N)). \tag{10.32}$$

Proof. Using again notations (10.30), the previous proposition proved that

$$\liminf_{\varepsilon\to 0} \mathcal{A}_\varepsilon + \liminf_{\varepsilon\to 0} \mathcal{B}_\varepsilon = \lim_{\varepsilon\to 0} \big(\mathcal{A}_\varepsilon + \mathcal{B}_\varepsilon\big).$$

On the other hand, it is always true that

$$\begin{aligned}\lim_{\varepsilon\to 0} \big(\mathcal{A}_\varepsilon + \mathcal{B}_\varepsilon\big) &\geqslant \limsup_{\varepsilon\to 0} \mathcal{A}_\varepsilon + \liminf_{\varepsilon\to 0} \mathcal{B}_\varepsilon\\ \lim_{\varepsilon\to 0} \big(\mathcal{A}_\varepsilon + \mathcal{B}_\varepsilon\big) &\geqslant \liminf_{\varepsilon\to 0} \mathcal{A}_\varepsilon + \limsup_{\varepsilon\to 0} \mathcal{B}_\varepsilon.\end{aligned}$$

Hence,

$$\liminf_{\varepsilon\to 0} \mathcal{A}_\varepsilon = \limsup_{\varepsilon\to 0} \mathcal{A}_\varepsilon, \qquad \liminf_{\varepsilon\to 0} \mathcal{B}_\varepsilon = \limsup_{\varepsilon\to 0} \mathcal{B}_\varepsilon,$$

which concludes the proof of (10.31).

To prove (10.32), let ω be an open bounded set and $R > 0$ such that $\omega \cup B \subset B(O;R)$ (where $B(O;R)$ is the ball of radius R and center O in $\mathbb{R}^N$). Since $\mathcal{T}_{\varepsilon,\delta}(u_{\varepsilon,\delta})(x,\cdot)$ and $U(x,\cdot)$ vanishes on B for a.e. $x \in \Omega$, one has

$$\begin{aligned}&\|\mathcal{T}_{\varepsilon,\delta}(u_{\varepsilon,\delta})(x,\cdot) - U(x,\cdot)\|^2_{L^2(B(O;R))}\\ &\qquad\qquad \leqslant C\|\nabla z\big(\mathcal{T}_{\varepsilon,\delta}(u_{\varepsilon,\delta})(x,\cdot) - U(x,\cdot)\big)\|^2_{L^2(B(O;R))},\end{aligned}$$

for a.e. x in $\widehat{\Omega}_\varepsilon$. The constant depends on R and B. Then, integrating over $\widehat{\Omega}_\varepsilon$, gives

$$\|\mathcal{T}_{\varepsilon,\delta}(u_{\varepsilon,\delta}) - U\|^2_{L^2(\widehat{\Omega}_\varepsilon\times B(O;R))} \leqslant C\big(\|\nabla z\big(\mathcal{T}_{\varepsilon,\delta}(u_{\varepsilon,\delta}) - U\big)\|^2_{L^2(\widehat{\Omega}_\varepsilon\times B(O;R))}.$$

As a consequence,

$$\begin{aligned}\|\mathcal{T}_{\varepsilon,\delta}(u_{\varepsilon,\delta}) - U\|^2_{L^2(\Omega\times B(O;R))} &\leqslant C\big(\|\nabla z(\mathcal{T}_{\varepsilon,\delta}(u_{\varepsilon,\delta}) - U)\|^2_{L^2(\Omega\times B(O;R))}\\ &\qquad + \|\nabla_z U\|^2_{L^2(\Lambda_\varepsilon\times B(O;R))} + \|U\|^2_{L^2(\Lambda_\varepsilon\times B(O;R))}\big).\end{aligned}$$

For δ small enough the ball $B(O;R)$ is included in $\dfrac{1}{\sqrt{\delta}}B$. Then (10.31) and the fact that $\omega \subset B(O;R) \subset \dfrac{1}{\sqrt{\delta}}B$, imply

$$\mathcal{T}_{\varepsilon,\delta}(u_{\varepsilon,\delta}) \to U \quad \text{strongly in } \; L^2(\Omega\times\omega),$$

which ends the proof of (10.32). □

10.2 Electrostatic screens

This section considers the homogenization of an electrostatic screen in a domain, namely when small sets on which the solution is to vanish, are periodically distributed in a layer. The notations of Sections 9.3 and 10.1 are used. In particular, $Y_\delta^* = Y \setminus \delta\overline{B}$.

Here, the perforated layer $\Sigma_{\varepsilon,\delta}$ is given by (recall Notations (9.7) and 9.10)

$$\Sigma_{\varepsilon,\delta} = \Big\{ x \in \Sigma_\varepsilon \mid \Big\{\frac{x}{\varepsilon}\Big\}_Y \in Y_\delta^* \Big\} = \Omega \cap \Big\{ x \in \mathbb{R}^N \mid |x_N| < \frac{\varepsilon}{2} \text{ and } \Big\{\frac{x}{\varepsilon}\Big\}_Y \in Y_\delta^* \Big\},$$

and the perforated domain is now (see Figure 10.1 for an example)

$$\Omega_{\varepsilon,\delta} = \Omega \setminus \overline{\Big\{ x \in \Sigma_\varepsilon \mid \Big\{\frac{x}{\varepsilon}\Big\}_Y \in \delta B \Big\}}.$$

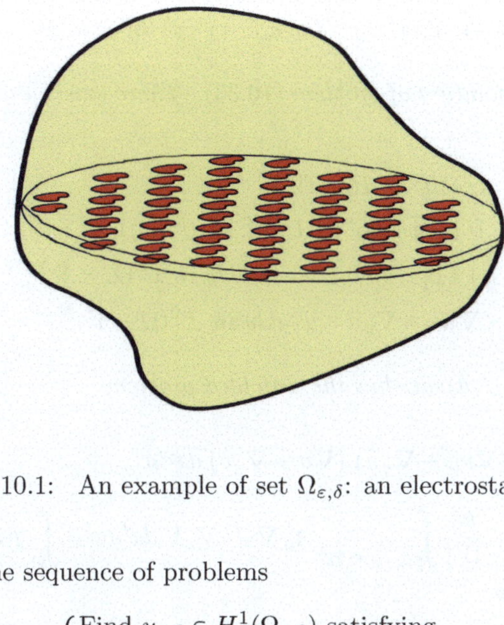

Figure 10.1: An example of set $\Omega_{\varepsilon,\delta}$: an electrostatic screen

Consider the sequence of problems

$$\begin{cases} \text{Find } u_{\varepsilon,\delta} \in H_0^1(\Omega_{\varepsilon,\delta}) \text{ satisfying} \\ \displaystyle\int_{\Omega_{\varepsilon,\delta}} A^\varepsilon \nabla u_{\varepsilon,\delta} \nabla \phi \, dx = \int_{\Omega_{\varepsilon,\delta}} f \phi \, dx, \\ \forall \phi \in H_0^1(\Omega_{\varepsilon,\delta}), \end{cases} \tag{10.33}$$

where f is given in $L^2(\Omega)$.

The following space will play an important role ($\mathbf{K}_B$ was defined in (10.2)):

$$\mathbf{L}'_B = \big\{ V \in L^2(\Sigma; \mathbf{K}_B) \mid V(\cdot, \infty) \in H^{1/2}(\Sigma) \big\},$$

with the norm $\|V\|_{\mathbf{L}_B} = \big(\|V(\infty)\|^2_{H^{1/2}(\Sigma)} + \|\nabla_y V\|^2_{L^2(\Sigma \times \mathbb{R}^N)} \big)^{1/2}$.

10.2.1 Homogenization

Set

$$\mathbb{W} \doteq \{(v, V, \widehat{v}) \in H_0^1(\Omega) \times \mathbf{L}'_B \times L^2(\Omega; H^1_{per}(Y)) \,|\, V(x', \infty) = v(x') \quad \text{for a.e. } x' \in \Sigma\}.$$

Proposition 10.10. *Suppose that $\delta = \delta(\varepsilon)$ is such that*

$$\lim_{\varepsilon\to 0} \frac{\delta^{\frac{N}{2}-1}}{\sqrt{\varepsilon}} \doteq k_2 \geqslant 0. \tag{10.34}$$

Let A^ε belong to $M(\alpha, \beta, \Omega)$. Suppose that, as $\varepsilon \to 0$, there exist a matrix A such that (10.5) is satisfied and a matrix A'_0 such that

$$\mathcal{T}^{bl}_{\varepsilon,\delta}(A^\varepsilon)(x', z) \to A'_0(x', z) \quad \text{for a.e. } (x', z) \text{ in } \Sigma \times (\mathbb{R}^N \backslash \overline{B}). \tag{10.35}$$

Let $u_{\varepsilon,\delta}$ be the solution of problem (10.33). There exists a unique $(u_0, U, \widehat{u})$ in $\mathbb{W}$ such that,

$$\begin{aligned} &u_{\varepsilon,\delta} \rightharpoonup u_0 \quad \text{weakly in } H^1_0(\Omega),\\ &\mathcal{T}^{bl}_{\varepsilon,\delta}(u_{\varepsilon,\delta}) \rightharpoonup U \quad \text{weakly in } L^2(\Sigma; L^2_{loc}(\mathbb{R}^N)),\\ &\nabla_z\big(\mathcal{T}^{bl}_{\varepsilon,\delta}(u_{\varepsilon,\delta})\big)\, \mathbf{1}_{\frac{1}{\delta}Y} \rightharpoonup \nabla_z U \quad \text{weakly in } L^2(\Sigma \times \mathbb{R}^N)^N,\\ &\mathcal{T}_\varepsilon(\nabla u_{\varepsilon,\delta}) \rightharpoonup \nabla u_0 + \nabla_y \widehat{u} \quad \text{weakly in } L^2(\Omega \times Y)^N. \end{aligned} \tag{10.36}$$

The triple $(u_0, U, \widehat{u})$ satisfies the unfolded problem

$$\begin{cases} \dfrac{1}{|Y|}\displaystyle\int_{\Omega\times Y} A(\nabla u_0 + \nabla_y \widehat{u})\,(\nabla v + \nabla_y \widehat{v})\, dx dy \\ \qquad + \dfrac{k_2^2}{|Y|}\displaystyle\int_{\Sigma\times(\mathbb{R}^N\backslash\overline{B})} A'_0\, \nabla_z U\, \nabla_z V\, dx' dz = \int_\Omega f\, v\, dx, \\ \forall (v, V, \widehat{v}) \in \mathbb{W}. \end{cases} \tag{10.37}$$

Problem (10.37) has a unique solution by the Lax-Milgram theorem applied in the space $\mathbb{W}$ due to the ellipticity of the matrix fields A and A'_0 (arising from A^ε in $M(\alpha, \beta, \Omega)$) and identity (3.14).

Proof of Proposition 10.10. We assume $k_2 > 0$ and denote $u_{\varepsilon,\delta}$ the extension by zero to the whole of Ω of the solution of (10.33). The reasoning is similar to that of the previous section.
Step 1. Choosing $u_{\varepsilon,\delta}$ itself as test function in (10.33), gives the a priori estimate

$$\|u_{\varepsilon,\delta}\|_{H^1_0(\Omega)} \leqslant C\|f\|_{L^2(\Omega)},$$

so, up to a subsequence, there exist u_0 in $H_0^1(\Omega)$ and $\widehat{u}$ in $L^2(\Omega; H^1_{per,0}(Y))$ such that

$$\begin{aligned} &u_{\varepsilon,\delta} \rightharpoonup u_0 \quad \text{weakly in } H_0^1(\Omega),\\ &\mathcal{T}_\varepsilon(\nabla u_{\varepsilon,\delta}) \rightharpoonup \nabla u_0 + \nabla_y \widehat{u} \quad \text{weakly in } L^2(\Omega \times Y)^N. \end{aligned} \tag{10.38}$$

As in Step 1 of the proof of Proposition 10.1, we obtain

$$\int_{\Omega\times Y} A\big(\nabla u_0 + \nabla_y \widehat{u}\big)\, \nabla_y \widehat{v}\, dxdy = 0, \quad \forall \widehat{v} \in L^2(\Omega; H^1_{per}(Y)). \tag{10.39}$$

Step 2. By Theorem 9.15(vii), there exists U in $L^2(\Sigma; L^2_{loc}(\mathbb{R}^N))$ such that, up to a subsequence,

$$\mathcal{T}^{bl}_{\varepsilon,\delta}(u_{\varepsilon,\delta}) \rightharpoonup U \quad \text{weakly in } L^2(\Sigma; L^2_{loc}(\mathbb{R}^N)). \tag{10.40}$$

Thanks to Proposition 9.14,

$$\mathcal{M}^{bl}_\varepsilon(u_{\varepsilon,\delta}) \to u_0 \quad \text{strongly in } L^2(\Sigma),$$

so that

$$\mathcal{M}^{bl}_\varepsilon(u_{\varepsilon,\delta})\mathbf{1}_{\frac{1}{\delta}Y} \to u_0 \quad \text{strongly in } L^2(\Sigma; L^2_{loc}(\mathbb{R}^N)). \tag{10.41}$$

On the other hand, again by Theorem 9.15(vii), there exists W in the space $L^2(\Sigma; L^{2^*}(\mathbb{R}^N))$ with $\nabla_z W \in L^2(\Sigma \times \mathbb{R}^N)$, and such that

$$\mathcal{T}^{bl}_{\varepsilon,\delta}\big(u_{\varepsilon,\delta} - \mathcal{M}^{bl}_\varepsilon(u_{\varepsilon,\delta})\big) \rightharpoonup W \quad \text{weakly in } L^2(\Sigma; L^{2^*}(\mathbb{R}^N)). \tag{10.42}$$

From (10.40), (10.41) and (10.42), one concludes

$$U = W + u_0 \quad \text{and} \quad \nabla_z U = \nabla_z W,$$

and, by Theorem 9.15(vii) once more

$$\nabla_z\big(\mathcal{T}^{bl}_{\varepsilon,\delta}(u_{\varepsilon,\delta})\big)\,\mathbf{1}_{\frac{1}{\delta}Y} \rightharpoonup \nabla_z U \quad \text{weakly in } L^2(\Sigma \times \mathbb{R}^N)^N. \tag{10.43}$$

Since $\mathcal{T}^{bl}_{\varepsilon,\delta}(u_{\varepsilon,\delta}) = 0$ on $\Sigma \times B$ by Definition 9.6, and thanks to (10.40),

$$U = 0 \quad \text{on } \Sigma \times B. \tag{10.44}$$

Due to (10.42)-(10.44), U belongs to $\mathbf{L_B}$ and moreover, $U(\cdot, \infty) = u_0$ on Σ.

Step 3. Now, let ψ be in $\mathcal{D}(\Omega)$ and $v \in \mathbf{K}_B$ such that $\nabla_z v$ has a compact support. Take $v^{bl}_{\varepsilon,\delta}\psi$ as test function in problem (10.33), where

$$v^{bl}_{\varepsilon,\delta}(x) = v\Big(\frac{1}{\delta}\Big\{\frac{x}{\varepsilon}\Big\}_Y\Big).$$

Note that for δ small enough, the gradient of $v^{bl}_{\varepsilon,\delta}$ vanishes outside of $\Sigma_{\varepsilon,\delta}$. Therefore,

$$\int_{\Sigma_{\varepsilon,\delta}} A^\varepsilon \nabla u_{\varepsilon,\delta}\, \nabla v^{bl}_{\varepsilon,\delta}\, \psi\, dx + \int_{\Omega_{\varepsilon,\delta}} A^\varepsilon \nabla u_{\varepsilon,\delta}\, \nabla \psi\, v^{bl}_{\varepsilon,\delta}\, dx = \int_{\Omega_{\varepsilon,\delta}} f\, v^{bl}_{\varepsilon,\delta}\, \psi\, dx. \quad (10.45)$$

We will pass to limit in this identity. To do so, we will treat separately each of its terms.

For the first one, observe that since $v^{bl}_{\varepsilon,\delta}$ vanishes in the holes,

$$\int_{\Sigma_{\varepsilon,\delta}} A^\varepsilon \nabla u_{\varepsilon,\delta}\, \nabla v^{bl}_{\varepsilon,\delta}\, \psi\, dx = \int_{\Sigma_\varepsilon} A^\varepsilon \nabla u_{\varepsilon,\delta}\, \nabla v^{bl}_{\varepsilon,\delta}\, \psi\, dx,$$

which unfolded with $\mathcal{T}^{bl}_{\varepsilon,\delta}$ gives (for ε small enough),

$$\begin{aligned}&\int_{\Sigma_\varepsilon} A^\varepsilon \nabla u_{\varepsilon,\delta}\, \nabla v^{bl}_{\varepsilon,\delta}\, \psi\, dx\\ &\qquad = \frac{\varepsilon\delta^N}{|Y|} \int_{\Sigma\times\mathbb{R}^N} \mathcal{T}^{bl}_{\varepsilon,\delta}(A^\varepsilon)\mathcal{T}^{bl}_{\varepsilon,\delta}(\nabla u_{\varepsilon,\delta})\, \mathcal{T}^{bl}_{\varepsilon,\delta}(\nabla v^{bl}_{\varepsilon,\delta})\mathcal{T}^{bl}_{\varepsilon,\delta}(\psi)\, dx'dz.\end{aligned}$$

Since from Theorem 9.15(v),

$$\mathcal{T}^{bl}_{\varepsilon,\delta}(\nabla v^{bl}_{\varepsilon,\delta}) = \frac{1}{\varepsilon\delta}\nabla_z v,$$

the former identity writes

$$\begin{aligned}&\int_{\Sigma_{\varepsilon,\delta}} A^\varepsilon \nabla u_{\varepsilon,\delta}\, \nabla v^{bl}_{\varepsilon,\delta}\, \psi\, dx\\ &= \Big(\frac{\delta^{\frac{N}{2}-1}}{\sqrt{\varepsilon}}\Big)^2 \frac{1}{|Y|} \int_{\Sigma\times\mathbb{R}^N} \mathcal{T}^{bl}_{\varepsilon,\delta}(A^\varepsilon)\, \nabla_z\big(\mathcal{T}^{bl}_{\varepsilon,\delta}(u_{\varepsilon,\delta})\big)\, \nabla_z v\, \mathcal{T}^{bl}_{\varepsilon,\delta}(\psi)\, dx'dz.\end{aligned} \quad (10.46)$$

On the other hand, thanks to (10.22) one has the convergence,

$$\mathcal{T}^{bl}_{\varepsilon,\delta}(\psi)\nabla_z v \to \psi\nabla_z v \quad \text{strongly in } \; L^2(\Sigma\times\mathbb{R}^N)^N.$$

This, together with convergences (10.34) and (10.43), as well as hypothesis (10.35), allow to pass to the limit in (10.46) to get,

$$\begin{aligned}&\lim_{\varepsilon\to 0}\int_{\Sigma_{\varepsilon,\delta}} A^\varepsilon \nabla u_{\varepsilon,\delta}\, \nabla v^{bl}_{\varepsilon,\delta}\, \psi\, dx\\ &\qquad = \frac{k_2^2}{|Y|} \int_{\Sigma\times\mathbb{R}^N} A_0'(x',z)\, \nabla_z U(x',z)\, \nabla_z v(z)\, \psi(x',0)\, dx'dz.\end{aligned} \quad (10.47)$$

By a density argument, this is true for every v in $\mathbf{K}_B$.

The second term in (10.45) is unfolded with $\mathcal{T}_\varepsilon$. By convergence $(10.38)_2$,

$$\lim_{\varepsilon\to 0}\int_{\Omega_{\varepsilon,\delta}} A^\varepsilon \nabla u_{\varepsilon,\delta}\, v^{bl}_{\varepsilon,\delta}\, \nabla\psi\, dx = \frac{v(\infty)}{|Y|}\int_{\Omega\times Y} A(\nabla u_0 + \nabla_y \widehat{u})\,\nabla\psi\, dxdy,$$

which with (10.47), gives

$$\frac{v(\infty)}{|Y|}\int_{\Omega\times Y} A(\nabla u_0 + \nabla_y \widehat{u})\,\nabla\psi\, dxdy + \frac{k_2^2}{|Y|}\int_{\Sigma\times\mathbb{R}^N} A_0'\,\nabla_z U\,\nabla_z v\,\psi_{|\Sigma}\, dx'dz$$
$$= v(\infty)\int_\Omega f\,\psi\, dx.$$

By density, we deduce

$$\frac{1}{|Y|}\int_{\Omega\times Y} A(\nabla u_0 + \nabla_y \widehat{u})\,\nabla v\, dxdy + \frac{k_2^2}{|Y|}\int_{\Sigma\times\mathbb{R}^N} A_0'\,\nabla_z U\,\nabla_z V\, dx'dz = \int_\Omega f\, v\, dx,$$

for every $v \in H^1_0(\Omega)$ and $V \in \mathbf{L}'_B$ such that $V(\cdot,\infty) = v$ on Σ.

Taking into account (10.39) and (10.36), we obtain the unfolded problem (10.37) satisfied by the triple $(u_0, U, \widehat{u})$. The uniqueness of the solution for (10.37) implies the convergence of the whole sequence. □

10.2.2 Standard form of the limit problem

Like in Subsection 10.1.2, one can rewrite system (10.37) in the standard form. The result is stated in the next theorem, the proof of which follows along the lines that of Theorem 10.5.

First, express $\widehat{u}$ in terms of the correctors $\widehat{\chi}_j$, $j = 1,\dots,N$ defined by the cell problems (3.10),

$$\widehat{u}(x,y) = \sum_{j=1}^N \frac{\partial u_0}{\partial x_j}(x)\,\widehat{\chi}_j(x,y) \quad \text{for a.e. } (x,y)\in\Omega\times Y.$$

This, used in equation (10.37), gives

$$\begin{cases} \displaystyle\int_\Omega A^{hom}\nabla u_0\,\nabla v\, dxdy + \frac{k_2^2}{|Y|}\int_{\Sigma\times\mathbb{R}^N} A_0'\,\nabla_z U\,\nabla_z V\, dx'dz = \int_\Omega f\, v\, dx, \\ \forall v\in H^1_0(\Omega),\ \ \forall V\in \mathbf{L}'_B \text{ such that } V(\cdot,\infty) = v \ \text{ a.e. on } \Sigma. \end{cases} \tag{10.48}$$

Here, as before, for a.e. $x\in\Omega$,

$$A^{hom}(x)\nabla u_0(x) \doteq \frac{1}{|Y|}\int_Y A(x,y)\big[\nabla u_0(x) + \nabla_y \widehat{u}(x,y)\big]\, dy.$$

Let θ' be the solution of the corresponding "cell problem"[(6)]

$$\begin{cases} \theta' \in L^\infty(\Sigma; \mathbf{K}_B), \; \theta'(x', \infty) \equiv 1, \\ \displaystyle\int_{\mathbb{R}^N \setminus \overline{B}} A_0'(x', z) \nabla_z \theta(x', z) \nabla_z \Psi(z) \, dz = 0 \quad \text{for a.e. } x' \in \Sigma, \\ \forall \Psi \in \dot{H}^1(\mathbb{R}^N) \quad \text{with } \Psi = 0 \quad \text{on } B. \end{cases}$$

For a.e. $x' \in \Sigma$, set

$$\boldsymbol{\Theta}'(x') = \frac{1}{|Y|} \int_{\mathbb{R}^N \setminus \overline{B}} A_0'(x', z) \nabla_z \theta'(x', z) \nabla_z \theta'(x', z) \, dz, \tag{10.49}$$

which is non-negative and can be interpreted as the local capacity of the set B. In view of (10.37),

$$U(x', z) = u_0(x') \, \theta'(x', z) \quad \text{for a.e. } (x', z) \in \Sigma \times \mathbb{R}^N.$$

To summarize, one has the following statement:

Theorem 10.11. *Under the assumptions of Proposition 10.10 with $k_2 > 0$, the limit function u_0 in $H_0^1(\Omega)$ is the solution of the homogenized problem*

$$\begin{cases} \displaystyle\int_\Omega A^{hom} \nabla u_0 \nabla \psi \, dx + k_2^2 \int_\Sigma \boldsymbol{\Theta}' \, u_0 \, \psi \, dx' = \int_\Omega f \, \psi \, dx \\ \forall \psi \in H_0^1(\Omega), \end{cases} \tag{10.50}$$

where $\boldsymbol{\Theta}'$ is defined by (10.49).

Proof. Let ψ be in $H_0^1(\Omega)$, we choose v and $V(x', z) = \psi(x')\theta'(z)$ for a.e. $(x', z) \in \Sigma \times \mathbb{R}^N$ as a test function in (10.48); that gives (10.50). □

Remark 10.12. *The strong formulation for* (10.50) *is the following:*

$$\begin{cases} -\text{div } A^{hom} \nabla u_0 = f \quad \textit{in } \Omega \setminus \Sigma, \\ -\left[A^{hom} \nabla u_0\right]_\Sigma = k_2^2 \boldsymbol{\Theta}' \, u_0 \quad \textit{on } \Sigma, \\ u_0 = 0 \quad \textit{on } \partial\Omega, \end{cases}$$

where $\left[A^{hom} \nabla u_0\right]_\Sigma$ denotes the jump across Σ, that is

$$\left[A^{hom} \nabla u_0\right]_\Sigma \doteq A^{hom} \nabla u_0^- \, n^- + A^{hom} \nabla u_0^+ \, n^+ \quad \textit{on } \Sigma,$$

and n^+ and n^- are the respective (opposite) exterior unit normal to Ω_+ and Ω_- on Σ.

(6) The cell problem has a unique solution by the Lax-Milgram theorem applied in the affine subspace of $\mathbf{K}_B$ consisting of functions which are constant equal to 1 at infinity.

Remark 10.13.
1. The proof for the case $k_2 = 0$ (k_2 defined by (10.34)), is actually simpler but requires a slight modification, namely setting in $(10.7)_2$,

$$\frac{\delta^{\frac{N}{2}-1}}{\sqrt{\varepsilon}}\mathcal{T}_{\varepsilon,\delta}(u_{\varepsilon,\delta}) \rightharpoonup U.$$

The corresponding statement is that of Theorem 10.11 for $k_2 = 0$. The small holes have no influence at the limit, i.e. the equation $\mathrm{div}\,\big(A^{hom}\nabla u_0\big) = f$ is satisfied in the whole of Ω.
2. As in Remark 10.7, for the case of $k_2 = +\infty$, Theorem 9.15(vi) implies the convergence

$$\mathcal{T}^{bl}_{\varepsilon,\delta}(u_{\varepsilon,\delta}) \rightharpoonup u_{0|\Sigma} \quad \text{weakly in } L^2(\Sigma; L^2_{loc}(\mathbb{R}^N)).$$

On the other hand, $\mathcal{T}^{bl}_{\varepsilon,\delta}(u_{\varepsilon,\delta}) = 0$ in $\Sigma \times B$ implies that $u_{0|\Sigma} = 0$. Thus the limit problem splits into two separate homogeneous Dirichlet problems in Ω_+ and Ω_-. In Ω_+, the problem is

$$\begin{cases} -\mathrm{div}\,\big(A^{hom}\nabla u_0\big) = f & \text{in } \Omega_+,\\ u_0 = 0 & \text{on } \partial\Omega_+,\end{cases}$$

and a similar one in Ω_-.

10.2.3 Convergence of the energy

The proofs of the next propositions follow the same lines as that of Proposition 10.8 and its corollary.

Proposition 10.14 (Convergence of the energy). *Under the hypotheses of Theorem 10.11, one has*

$$\begin{aligned}\lim_{\varepsilon\to 0}\int_{\Omega^*_{\varepsilon,\delta}} A^\varepsilon\nabla u_{\varepsilon,\delta}\,\nabla u_{\varepsilon,\delta}\,dx &= \int_\Omega A^{hom}\nabla u_0\,\nabla u_0\,dx + k_2^2\int_\Sigma \Theta'\,u_0^2\,dx'\\ &= \frac{1}{|Y|}\int_{\Omega\times Y} A\big(\nabla u_0 + \nabla_y\widehat{u}\big)\big(\nabla u_0 + \nabla_y\widehat{u}\big)\,dxdy\\ &\quad + \frac{k_2^2}{|Y|}\int_{\Sigma\times(\mathbb{R}^N\setminus\overline{B})} A_0'\,\nabla_z U\,\nabla_z U\,dx'dz,\end{aligned}$$

$$\lim_{\varepsilon\to 0}\int_{\Lambda_\varepsilon} |\nabla u_{\varepsilon,\delta}|^2\,dx = 0.$$

Corollary 10.15. *Under the hypotheses of Theorem 10.11, the following convergences hold:*

$$\begin{aligned}&\mathcal{T}_\varepsilon(\nabla u_{\varepsilon,\delta})\mathbf{1}_{(\Omega\setminus\overline{\Sigma}_\varepsilon)\times Y} \to \nabla u_0 + \nabla_y\widehat{u} \quad \text{strongly in } L^2(\Omega\times Y)^N,\\ &\nabla_z\big(\mathcal{T}^{bl}_{\varepsilon,\delta}(u_{\varepsilon,\delta})\big)\,\mathbf{1}_{\frac{1}{\delta}Y} \to \nabla_z U \quad \text{strongly in } L^2(\Sigma\times\mathbb{R}^N)^N.\end{aligned}$$

The first convergence implies the corrector result

$$\nabla u_{\varepsilon,\delta}\mathbf{1}_{(\Omega\setminus\overline{\Sigma}_\varepsilon)} - \nabla u_0 - \mathcal{U}_\varepsilon(\nabla_y \widehat{u}) \to 0 \quad \textit{strongly in } L^2(\Omega)^N.$$

The second one implies that

$$\|\nabla u_{\varepsilon,\delta}\|_{L^2(\Sigma_\varepsilon)} \to k_2 \|\nabla_z U\|_{L^2(\Sigma\times\mathbb{R}^N)}.$$

This shows how a non negligible part of the energy concentrates near Σ, *in general.*

10.3 The thin Neumann sieve

We use the framework of Sections 9.3 and 10.2, in particular Notation 9.10 and (9.7)-(9.8). For an open subset S of $Y \cap \Pi$ such that $\overline{S} \subset Y \cap \Pi$, set

$$Y_\delta = Y_+ \cup Y_- \cup \delta S,$$

and

$$S_{\varepsilon,\delta} = \Big\{ x' \in \Sigma \;\Big|\; \Big\{\frac{x'}{\varepsilon}\Big\}_{Y'} \in \delta S \Big\}.$$

For Ω bounded domain in $\mathbb{R}^N$, define

$$\Omega^{bl}_{\varepsilon\delta} = \Omega_+ \cup \Omega_- \cup S_{\varepsilon,\delta} \quad \text{and} \quad \Sigma^{bl}_{\varepsilon,\delta} \doteq \Sigma_\varepsilon \cap \Omega^{bl}_{\varepsilon\delta}.$$

The connection between Ω_+ and Ω_- occurs through the "sieve" consisting of the set $S_{\varepsilon,\delta}$ (see Figure 10.2).

Consider the space

$$\mathbb{V} = \{ v \in H^1(\Omega_+ \cup \Omega_-) \mid v = 0 \quad \text{on} \quad \partial\Omega \},$$

which is a Hilbert space for the scalar product

$$\langle u, v \rangle_{\mathbb{V}} = \int_{\Omega_+ \cup \Omega_-} \nabla u \, \nabla v \, dx \quad \text{for all} \quad (u,v) \in \mathbb{V}^2.$$

For simplicity, when v belongs to $\mathbb{V}$, we denote ∇v the $L^2(\Omega)$-function which equals the gradient of v in $\Omega_+ \cup \Omega_-$ (this is not the distributional gradient of v in Ω but its restriction to $\Omega_+ \cup \Omega_-$). Every element v of $\mathbb{V}$ has two traces in $H^{1/2}(\Sigma)$, one from above, $v^+_{|\Sigma}$ and one from below $v^-_{|\Sigma}$. The jump of v across Σ is given by

$$[v]_\Sigma \doteq v^+_{|\Sigma} - v^-_{|\Sigma} \in H^{1/2}(\Sigma)^{(7)}.$$

Finally set

$$\mathbb{V}_{\varepsilon,\delta} = \{ v \in \mathbb{V} \mid [v]_\Sigma = 0 \quad \text{on} \quad S_{\varepsilon,\delta} \}.$$

(7) With these notations, the distributional gradient of v in whole of Ω is $\nabla v + [v]_\Sigma \mathbf{e}_N d\sigma$.

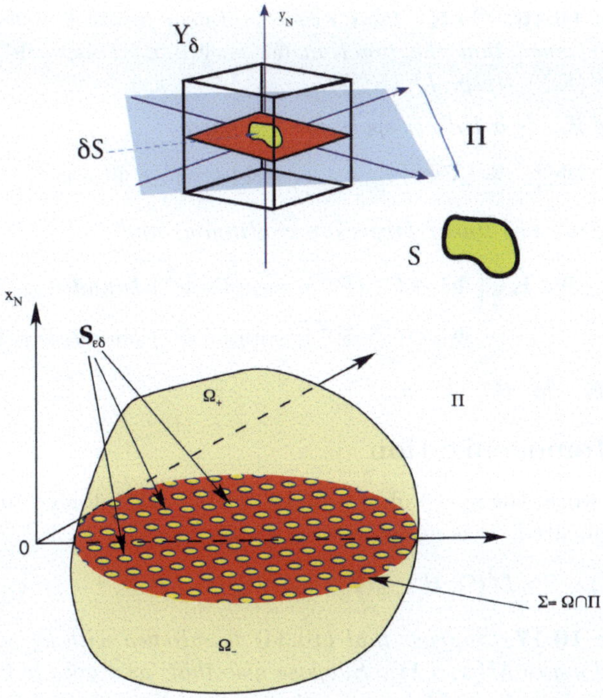

Figure 10.2: The set Y_δ and the thin sieve $\Omega^{bl}_{\varepsilon\delta}$

The thin Neumann sieve model is

$$\begin{cases} \text{Find } u_{\varepsilon,\delta} \in \mathbb{V}_{\varepsilon,\delta} \text{ satisfying} \\ \displaystyle\int_{\Omega^{bl}_{\varepsilon,\delta}} A^\varepsilon \nabla u_{\varepsilon,\delta} \nabla \phi \, dx = \int_{\Omega^{bl}_{\varepsilon,\delta}} f\phi \, dx, \quad f \in L^2(\Omega), \\ \forall \phi \in \mathbb{V}_{\varepsilon,\delta}. \end{cases} \tag{10.51}$$

Let $\mathbb{R}^N_+$ be the set $\mathbb{R}^{N-1} \times (0, +\infty)$ and similarly $\mathbb{R}^N_- \doteq \mathbb{R}^{N-1} \times (-\infty, 0)$. In this problem, the equivalent of the space $\mathbf{K}_B$ (see (10.2)), is

$$\widehat{\mathbf{K}}_S = \Big\{ \Phi \in H^1_{loc}(\mathbb{R}^N_+ \cup \mathbb{R}^N_-) \;\Big|\; \nabla\Phi \in L^2(\mathbb{R}^N_+ \cup \mathbb{R}^N_-),\ [\Phi]_\Sigma = 0 \text{ on } S \Big\}. \tag{10.52}$$

For $\Phi \in \widehat{\mathbf{K}}_S$, we denote

$$\Phi_{|\mathbb{R}^N_+} = \Phi^+, \qquad \Phi_{|\mathbb{R}^N_-} = \Phi^-.$$

The following proposition is an easy consequence of Corollary 9.5.

Proposition 10.16. *On $\widehat{\mathbf{K}}_S$ there exist two linear forms $\Phi \mapsto \Phi(+\infty)$ (resp. $\Phi \mapsto \Phi(-\infty)$) such that the function $\Phi^+ - \Phi(+\infty)$ (resp. $\Phi^- - \Phi(-\infty)$) belongs to $L^{2^*}(\mathbb{R}^N_+)$ (resp. $L^{2^*}(\mathbb{R}^N_-)$).*

The space $\widehat{\mathbf{K}}_S$ is a Hilbert space for the norm

$$\|\Phi\|^2_{\widehat{\mathbf{K}}_S} \doteq \|\nabla\Phi\|^2_{L^2(\mathbb{R}^N_+ \cup \mathbb{R}^N_-)} + \Phi(+\infty)^2 + \Phi(-\infty)^2.$$

Furthermore, the two linear forms are continuous and

$$\widehat{\mathbf{K}}_S^\infty \doteq \Big\{\Phi \in \widehat{\mathbf{K}}_S \;\Big|\; \Phi^+ \in C^\infty(\mathbb{R}^N_+),\ \mathrm{supp}(\nabla\Phi^+) \text{ bounded in } \mathbb{R}^N_+,$$
$$\Phi^- \in C^\infty(\mathbb{R}^N_-),\ \mathrm{supp}(\nabla\Phi^-) \text{ bounded in } \mathbb{R}^N_- \Big\},$$

is dense in $\widehat{\mathbf{K}}_S$ for this norm.

10.3.1 Homogenization

In this framework, the space of test functions for the unfolded limit problem is more complicated. It is given by

$$\mathbb{W} \doteq \big\{(v, \widehat{v}, V) \in \mathbb{V} \times L^2(\Omega; H^1_{per,0}(Y)) \times L^2(\Sigma; \widehat{\mathbf{K}}_S) \,|\, V(\cdot, \pm\infty) = v^\pm_{|\Sigma} \text{ a.e. on } \Sigma\big\}.$$

Proposition 10.17. *Suppose that (10.34) is satisfied with $k_2 > 0$.*

Let A^ε belong to $M(\alpha, \beta, \Omega)$. Suppose also that, as ε goes to 0, there exist a matrix A such that (10.5) holds, and a matrix A'_0 such that

$$\mathcal{T}^{bl}_{\varepsilon,\delta}(A^\varepsilon)(x', z) \to A'_0(x', z) \quad \text{for a.e. } (x', z) \text{ in } \Sigma \times \mathbb{R}^N. \tag{10.53}$$

Let $u_{\varepsilon,\delta}$ be the solution of problem (10.51). Then, there exists $(u_0, \widehat{u}, U)$ in $\mathbb{W}$ such that

$$\begin{aligned}
&u_{\varepsilon,\delta} \rightharpoonup u_0 \quad \text{weakly in } \mathbb{V},\\
&\mathcal{T}^{bl}_{\varepsilon,\delta}(u^\pm_{\varepsilon,\delta}) \rightharpoonup U^\pm \quad \text{weakly in } L^2(\Sigma; L^2_{loc}(\mathbb{R}^N_\pm)),\\
&\nabla_z(\mathcal{T}^{bl}_{\varepsilon,\delta}(u^\pm_{\varepsilon,\delta}))\,\mathbf{1}_{\frac{1}{\delta}Y_\pm} \rightharpoonup \nabla_z U^\pm \quad \text{weakly in } L^2(\Sigma \times \mathbb{R}^N_\pm)^N,\\
&\mathcal{T}_\varepsilon(\nabla u^\pm_{\varepsilon,\delta}) \rightharpoonup \nabla u^\pm_0 + \nabla_y \widehat{u} \quad \text{weakly in } L^2(\Omega_\pm \times Y)^N.
\end{aligned}$$

The triple $(u_0, \widehat{u}, U) \in \mathbb{W}$ satisfies the following unfolded limit problem:

$$\begin{cases}
\dfrac{1}{|Y|}\displaystyle\int_{\Omega_+ \times Y} A(\nabla u^+_0 + \nabla_y \widehat{u})\,(\nabla v^+ + \nabla_y \widehat{v})\,dx\,dy\\
\qquad + \dfrac{1}{|Y|}\displaystyle\int_{\Omega_- \times Y} A(\nabla u^-_0 + \nabla_y \widehat{u})\,(\nabla v^- + \nabla_y \widehat{v})\,dx\,dy\\
\qquad + \dfrac{k_2^2}{|Y|}\displaystyle\int_{\Sigma \times \mathbb{R}^N_+} A'_0\,\nabla_z U^+\,\nabla_z V^+\,dx'\,dz\\
\qquad + \dfrac{k_2^2}{|Y|}\displaystyle\int_{\Sigma \times \mathbb{R}^N_-} A'_0\,\nabla_z U^-\,\nabla_z V^-\,dx'\,dz = \displaystyle\int_\Omega f v\,dx,\\
\forall (v, \widehat{v}, V) \in \mathbb{W}.
\end{cases} \tag{10.54}$$

Problem (10.54) has a unique solution by the Lax-Milgram theorem applied in the space $\mathbb{W}$ due to the ellipticity of the matrix fields A and A_0' (arising from A^ε in $M(\alpha,\beta,\Omega)$) and identity (3.14).

Proof of Proposition 10.17. Use $u_{\varepsilon,\delta}$ as test function in (10.51). The corresponding equality, together with the Poincaré inequality on Ω_+ and Ω_-, imply that there is a constant C (independent of (ε,δ)) such that,

$$||u_{\varepsilon,\delta}||_{\mathbb{V}} \leqslant C||f||_{L^2(\Omega)}.$$

Step 1. Making use of Theorem 1.41(ii), separately in Ω_+ and Ω_-, there exist (up to a subsequence) $u_0 \in \mathbb{V}$ and $\widehat{u} \in L^2(\Omega; H^1_{per}(Y))$ such that,

$$\begin{aligned}
&u_{\varepsilon,\delta} \rightharpoonup u_0 \quad \text{weakly in } \mathbb{V},\\
&\mathcal{T}_\varepsilon(\nabla u^+_{\varepsilon,\delta}) \rightharpoonup \nabla u_0^+ + \nabla_y \widehat{u} \ \text{ weakly in } L^2(\Omega_+ \times Y)^N,\\
&\mathcal{T}_\varepsilon(\nabla u^-_{\varepsilon,\delta}) \rightharpoonup \nabla u_0^- + \nabla_y \widehat{u} \ \text{ weakly in } L^2(\Omega_- \times Y)^N.
\end{aligned}$$

Now, let ϕ be in $H^1_{per}(Y)$ and ψ in $\mathcal{D}(\Omega)$ such that $\text{supp}(\psi) \subset \Omega_+ \cup \Omega_-$. Using

$$\Phi(x) = \varepsilon\psi(x)\phi\Big(\frac{x}{\varepsilon}\Big),$$

as a test function in (10.51), yields

$$\int_{\Omega_+\times Y} A\big(\nabla u_0^+ + \nabla_y\widehat{u}\big)\,\psi\,\nabla_y\phi\,dxdy + \int_{\Omega_-\times Y} A\big(\nabla u_0^- + \nabla_y\widehat{u}\big)\,\psi\,\nabla_y\phi\,dxdy = 0.$$

By density it follows that for every $\widehat{v} \in L^2(\Omega; H^1_{per}(Y))$,

$$\begin{aligned}
&\int_{\Omega_+\times Y} A\big(\nabla u_0^+ + \nabla_y\widehat{u}\big)\,\nabla_y\widehat{v}\,dxdy\\
&\qquad + \int_{\Omega_-\times Y} A\big(\nabla u_0^- + \nabla_y\widehat{u}\big)\,\nabla_y\widehat{v}\,dxdy = 0.
\end{aligned} \tag{10.55}$$

Step 2. By Theorem 9.18(iii), there exist U^+ and U^- in $L^2(\Sigma; L^2_{loc}(\mathbb{R}^N_+))$ and $L^2(\Sigma; L^2_{loc}(\mathbb{R}^N_-))$ respectively, such that (up to a subsequence),

$$\mathcal{T}^{bl}_{\varepsilon,\delta}(u^\pm_{\varepsilon,\delta}) \rightharpoonup U^\pm \quad \text{weakly in } L^2(\Sigma; L^2_{loc}(\mathbb{R}^N_\pm)). \tag{10.56}$$

By construction

$$\mathcal{T}^{bl}_{\varepsilon,\delta}\big(\mathcal{M}^{bl+}_\varepsilon(u^+_{\varepsilon,\delta})\big) = \mathcal{M}^{bl+}_\varepsilon(u^+_{\varepsilon,\delta})\mathbf{1}_{\frac{1}{\delta}Y_+},$$

and by Proposition 9.17,

$$\mathcal{M}^{bl+}_\varepsilon(u^+_{\varepsilon,\delta})\mathbf{1}_{\frac{1}{\delta}Y_+} \to u^+_{0|\Sigma} \quad \text{strongly in } L^2(\Sigma; L^2_{loc}(\mathbb{R}^N_+)). \tag{10.57}$$

By Theorem 9.18(iii) there exists W^+ in $L^2(\Sigma; L^{2^*}(\mathbb{R}^N_+))$ with $\nabla_z W^+$ in $L^2(\Sigma \times \mathbb{R}^N_+)$, such that

$$\mathcal{T}^{bl}_{\varepsilon,\delta}\big(u^+_{\varepsilon,\delta} - \mathcal{M}^{bl+}_{\varepsilon}(u^+_{\varepsilon,\delta})\big) \rightharpoonup W^+ \quad \text{weakly in } L^2(\Sigma; L^{2^*}(\mathbb{R}^N_+)). \tag{10.58}$$

From (10.56), (10.57) and (10.58), one concludes

$$U^+ = W^+ + u^+_{0|\Sigma}, \quad \text{and} \quad \nabla_z U^+ = \nabla_z W^+. \tag{10.59}$$

Similarly, there exists $W^- \in L^2(\Sigma; L^{2^*}(\mathbb{R}^N_-))$ with $\nabla_z W^- \in L^2(\Sigma \times \mathbb{R}^N_-)$ and

$$U^- = W^- + u^-_{0|\Sigma}, \quad \text{and} \quad \nabla_z U^- = \nabla_z W^-. \tag{10.60}$$

Again by Theorem 9.18(iii), one has the convergences

$$\nabla_z\big(\mathcal{T}^{bl}_{\varepsilon,\delta}(u^\pm_{\varepsilon,\delta})\big) \rightharpoonup \nabla_z U^\pm \quad \text{weakly in } L^2(\Sigma \times \mathbb{R}^N_\pm)^N. \tag{10.61}$$

From Definition 9.11, $\mathcal{T}^{bl}_{\varepsilon,\delta}(u^+_{\varepsilon,\delta}) = \mathcal{T}^{bl}_{\varepsilon,\delta}(u^-_{\varepsilon,\delta})$ on $\Sigma \times S$, so that by convergences (10.56) and (10.61),

$$[U(x', \cdot)]_\Sigma = 0 \quad \text{on } S \quad \text{for a.e. } x' \in \Sigma.$$

Therefore, $U \in L^2(\Sigma; \widehat{\mathbf{K}}_S)$, and (10.59)-(10.60) provide the compatibility condition guaranteeing that $(u_0, \widehat{u}, U)$ belongs to $\mathbb{W}$.

Now, choose a function v in $\widehat{\mathbf{K}}^\infty_S$ and set

$$v_{\varepsilon,\delta}(x', x_N) = v\Big(\frac{1}{\delta}\Big\{\frac{x'}{\varepsilon}\Big\}_{Y'}, \frac{x_N}{\varepsilon\delta}\Big) \qquad \text{for a.e. } x = (x', x_N) \in \mathbb{R}^N.$$

Clearly, $[v_{\varepsilon,\delta}] = 0$ on $S_{\varepsilon,\delta}$ and $\nabla v^+_{\varepsilon,\delta}$, $\nabla v^-_{\varepsilon,\delta}$ vanish outside $\Sigma^{bl}_{\varepsilon,\delta}$ for δ small enough. As in Step 2 of the proof of Proposition 10.1, we get

$$\begin{aligned} &\mathcal{T}_\varepsilon(v^\pm_{\varepsilon,\delta}) \to v(\pm\infty) \quad \text{strongly in } L^2(\Omega_\pm \times Y), \\ &v^\pm_{\varepsilon,\delta} \rightharpoonup v(\pm\infty) \quad \text{weakly in } H^1(\Omega_\pm). \end{aligned} \tag{10.62}$$

For $\psi \in \mathcal{D}(\Omega)$, using $\psi\, v_{\varepsilon,\delta}$ as a test function in problem (10.51), gives the following formula:

$$\begin{aligned} \int_{\Omega^{bl}_{\varepsilon,\delta}} A^\varepsilon \nabla u_{\varepsilon,\delta}\, \nabla\psi\, v_{\varepsilon,\delta}\, dx &+ \int_{\Sigma_{\varepsilon,\delta}} A^\varepsilon \nabla u_{\varepsilon,\delta}\, \nabla v_{\varepsilon,\delta}\, \psi\, dx \\ &= \int_{\Omega^{bl}_{\varepsilon,\delta}} f\, v_{\varepsilon,\delta}\, \psi\, dx, \end{aligned} \tag{10.63}$$

The first term in (10.63) is unfolded with $\mathcal{T}_\varepsilon$ as usual. This yields

$$\int_{\Omega^{bl}_{\varepsilon,\delta}} A^\varepsilon \nabla u_{\varepsilon,\delta}\, \nabla\psi\, v_{\varepsilon,\delta}\, dx = \frac{1}{|Y|}\int_{\Omega\times Y} \mathcal{T}_\varepsilon(A^\varepsilon)\mathcal{T}_\varepsilon(\nabla u_{\varepsilon,\delta})\, \mathcal{T}_\varepsilon(\nabla\psi)\mathcal{T}_\varepsilon(w_{\varepsilon,\delta})\, dxdy.$$

Applying (10.62) and passing to the limit gives

$$\lim_{\varepsilon\to 0}\int_{\Omega^{bl}_{\varepsilon,\delta}} A^\varepsilon \nabla u_{\varepsilon,\delta}\,\nabla\psi\, v_{\varepsilon,\delta}\,dx = \frac{v(+\infty)}{|Y|}\int_{\Omega_+\times Y} A(\nabla u_0^+ + \nabla_y \widehat{u})\,\nabla\psi\,dxdy + \frac{v(-\infty)}{|Y|}\int_{\Omega_-\times Y} A(\nabla u_0^- + \nabla_y \widehat{u})\,\nabla\psi\,dxdy.$$

The second term in (10.63) is unfolded with $\mathcal{T}^{bl}_{\varepsilon,\delta}$. To do so, first recall that

$$\mathcal{T}^{bl}_{\varepsilon,\delta}(\nabla v_{\varepsilon,\delta}) = \frac{1}{\varepsilon\delta}\nabla_z v.$$

As a consequence,

$$\int_{\Sigma_{\varepsilon,\delta}} A^\varepsilon \nabla u_{\varepsilon,\delta}\nabla v_{\varepsilon,\delta}\,\psi\,dx = \Big(\frac{\delta^{\frac{N}{2}-1}}{\sqrt{\varepsilon}}\Big)^2\frac{1}{|Y|}\int_{\Sigma\times\mathbb{R}^N} \mathcal{T}^{bl}_{\varepsilon,\delta}(A^\varepsilon)\,\nabla_z \mathcal{T}^{bl}_{\varepsilon,\delta}(u_{\varepsilon,\delta})\,\nabla_z v\,\mathcal{T}^{bl}_{\varepsilon,\delta}(\psi)\,dx'dz.$$

Convergences (10.53) and (10.61) allow to pass to the limit in this equality to obtain

$$\lim_{\varepsilon\to 0}\int_{\Sigma_{\varepsilon,\delta}} A^\varepsilon \nabla u_{\varepsilon,\delta}\,\nabla v_{\varepsilon,\delta}\,\psi\,dx = \frac{k_2^2}{|Y|}\int_{\Sigma\times\mathbb{R}^N_+} A_0'\,\nabla_z U^+(x',z)\,\nabla_z v^+\,\psi\,dx'dz + \frac{k_2^2}{|Y|}\int_{\Sigma\times\mathbb{R}^N_-} A_0'\,\nabla_z U^-(x',z)\,\nabla_z v^-\,\psi\,dx'dz.$$

Then, gathering the above convergences in the limit of (10.63), yields

$$\begin{aligned}\frac{v(+\infty)}{|Y|}&\int_{\Omega_+\times Y} A(\nabla u_0^+ + \nabla_y\widehat{u})\,\nabla\psi\,dxdy\\ &+\frac{v(-\infty)}{|Y|}\int_{\Omega_-\times Y} A(\nabla u_0^- + \nabla_y\widehat{u})\,\nabla\psi\,dxdy\\ &+\frac{k_2}{|Y|}\int_{\Sigma\times\mathbb{R}^N_+} A_0'\,\nabla_z U^+\,\nabla_z v^+\,\psi\,dx'dz\\ &+\frac{k_2}{|Y|}\int_{\Sigma\times\mathbb{R}^N_-} A_0'\,\nabla_z U^-\,\nabla_z v^-\,\psi\,dx'dz\\ &= v(+\infty)\int_{\Omega_+} f\psi\,dx + v(-\infty)\int_{\Omega_-} f\psi\,dx.\end{aligned}$$

Then by density, we get

$$
\begin{aligned}
\frac{1}{|Y|}\int_{\Omega_+\times Y} & A\big(\nabla u_0^+ + \nabla_y \widehat{u}\big)\,\nabla v^+ dxdy \\
&+ \frac{1}{|Y|}\int_{\Omega_-\times Y} A\big(\nabla u_0^- + \nabla_y \widehat{u}\big)\,\nabla v^- dxdy \\
&+ \frac{k_2^2}{|Y|}\int_{\Sigma\times \mathbb{R}^N_+} A_0'\,\nabla_z U^+\,\nabla_z V^+\,dx'dz \\
&+ \frac{k_2^2}{|Y|}\int_{\Sigma\times \mathbb{R}^N_-} A_0'\,\nabla_z U^-\,\nabla_z V^-\,dx'dz = \int_\Omega f\,v\,dx,
\end{aligned}
$$

for every $(v, V) \in V \times L^2(\Sigma; \widehat{\mathbf{K}}_S)$ with $V(\cdot, \pm\infty) = v^{\pm}_{|\Sigma}$ a.e. on Σ. Together with (10.55), this gives (10.54). □

10.3.2 Standard form of the limit problem

As in Subsections 10.1.2 - 10.2.2, one can write system (10.54) in a standard form, with only u_0^+ and u_0^- as unknowns.

Using the correctors $\widehat{\chi}_j$ $(j = 1, \dots, N)$ defined by the cell problems (3.10), we can express $\widehat{u}$ in terms of these correctors to get the formulas,

$$
\widehat{u}(x,y) = \sum_{j=1}^{N} \frac{\partial u_0^+}{\partial x_j}(x)\,\widehat{\chi}_j(x,y) \qquad \text{for a.e. } (x,y) \in \Omega_+ \times Y,
$$

$$
\widehat{u}(x,y) = \sum_{j=1}^{N} \frac{\partial u_0^-}{\partial x_j}(x)\,\widehat{\chi}_j(x,y) \qquad \text{for a.e. } (x,y) \in \Omega_- \times Y,
$$

which used in equation (10.37), yields

$$
\begin{aligned}
\int_{\Omega_+} A^{hom} \nabla u_0^+\,\nabla v^+\,dx &+ \int_{\Omega_-} A^{hom} \nabla u_0^-\,\nabla v^-\,dx \\
&+ \frac{k_2^2}{|Y|}\int_{\Sigma\times \mathbb{R}^N_+} A_0'\,\nabla_z U^+\,\nabla_z V^+ dx'dz \\
&+ \frac{k_2^2}{|Y|}\int_{\Sigma\times \mathbb{R}^N_-} A_0'\,\nabla_z U^-\,\nabla_z V^- dx'dz = \int_\Omega f\,v\,dx,
\end{aligned}
\tag{10.64}
$$

for every $v \in \mathbb{V}$ and for every $V \in L^2(\Sigma; \widehat{\mathbf{K}}_S)$, such that $V(\cdot, +\infty) = v^+_{|\Sigma}$, $V(\cdot, -\infty) = v^-_{|\Sigma}$ a.e. on Σ, and where for a.e. $x \in \Omega$,

$$
A^{hom}(x)\nabla u_0^+(x) \doteq \frac{1}{|Y|}\int_Y A(x,y)\big[\nabla u_0^+(x) + \nabla_y \widehat{u}(x,y)\big]\,dy \quad \text{in } \Omega_+,
$$

$$
A^{hom}(x)\nabla u_0^-(x) \doteq \frac{1}{|Y|}\int_Y A(x,y)\big[\nabla u_0^-(x) + \nabla_y \widehat{u}(x,y)\big]\,dy \quad \text{in } \Omega_-.
$$

Let $\theta \in L^\infty(\Sigma; \widehat{\mathbf{K}}_S)$ be the solution of the following cell problem:

$$\begin{cases} \displaystyle\int_{\mathbb{R}^N_+} A'_0(x',z)\,\nabla_z\theta^+(x',z)\,\nabla_z\Psi^+(z)\,dz \\ \qquad + \displaystyle\int_{\mathbb{R}^N_-} A'_0(x',z)\,\nabla_z\theta^-(x',z)\,\nabla_z\Psi^-(z)\,dz = 0 \quad \text{for a.e. } x' \in \Sigma, \\ \theta(x',+\infty) \equiv 1, \quad \theta(x',-\infty) \equiv 0, \\ \forall \Psi \in \widehat{\mathbf{K}}_S \ \text{ with } \Psi(+\infty) = \Psi(-\infty) = 0. \end{cases}$$

From (10.54) it follows that

$$\begin{aligned} U^+(\cdot,z) &= \theta^+(\cdot,z)u^+_{0|\Sigma} + \big(1-\theta^+(\cdot,z)\big)u^-_{0|\Sigma} \qquad &&\text{in } \Sigma \times \mathbb{R}^N_+, \\ U^-(\cdot,z) &= \theta^-(\cdot,z)u^+_{0|\Sigma} + \big(1-\theta^-(\cdot,z)\big)u^-_{0|\Sigma} \qquad &&\text{in } \Sigma \times \mathbb{R}^N_-. \end{aligned}$$

Let v be in $\mathbb{V}$ and V in $L^2(\Sigma; \widehat{\mathbf{K}}_S)$, such that

$$\begin{aligned} V^+(\cdot,z) &= \theta^+(\cdot,z)v^+_{|\Sigma} + \big(1-\theta^+(\cdot,z)\big)v^-_{|\Sigma} \qquad &&\text{in } \Sigma \times \mathbb{R}^N_+, \\ V^-(\cdot,z) &= \theta^-(\cdot,z)v^+_{|\Sigma} + \big(1-\theta^-(\cdot,z)\big)v^-_{|\Sigma} \qquad &&\text{in } \Sigma \times \mathbb{R}^N_-. \end{aligned}$$

Going back to (10.64) with this test function gives

$$\begin{aligned} &\int_{\Omega_+} A^{hom}\nabla u_0^+\,\nabla v^+\,dx + \int_{\Omega_-} A^{hom}\nabla u_0^-\,\nabla v^-\,dx \\ &\quad + \frac{k_2^2}{|Y|}\int_{\Sigma\times\mathbb{R}^N_+} A'_0\,\nabla_z\theta^+\,\nabla_z\theta^+\,(u^+_{0|\Sigma} - u^-_{0|\Sigma})\,(v^+_{|\Sigma} - v^-_{|\Sigma})dx'dz \\ &\quad + \frac{k_2^2}{|Y|}\int_{\Sigma\times\mathbb{R}^N_-} A'_0\,\nabla_z\theta^-\,\nabla_z\theta^-\,(u^+_{0|\Sigma} - u^-_{0|\Sigma})\,(v^+_{|\Sigma} - v^-_{|\Sigma})\,dx'dz \\ &\quad = \int_\Omega f\,v\,dx. \end{aligned}$$

Set

$$\begin{aligned} \mathbf{\Theta}''(x') = \frac{1}{|Y|}\int_{\mathbb{R}^N_+} A'_0\,\nabla_z\theta^+\,\nabla_z\theta^+\,dz \\ + \frac{1}{|Y|}\int_{\mathbb{R}^N_-} A'_0\,\nabla_z\theta^-\,\nabla_z\theta^-\,dz \quad \text{for a.e. } x' \in \Sigma. \end{aligned}$$

We proved the following theorem:

Theorem 10.18. *Suppose that (10.34) is satisfied with $k_2 > 0$. The limit function $u_0 \in \mathbb{V}$ given by Proposition 10.17, is the solution of the variational homogenized equation*

$$\begin{cases} \displaystyle\int_{\Omega_+} A^{hom}\nabla u_0^+\,\nabla v^+\,dx + \int_{\Omega_-} A^{hom}\nabla u_0^-\,\nabla v^-\,dx + k_2^2\int_\Sigma \mathbf{\Theta}''[u_0]_\Sigma\,[v]_\Sigma dx' \\ \qquad = \displaystyle\int_\Omega f\,v\,dx, \quad \forall v \in \mathbb{V}. \end{cases}$$

Remark 10.19. *The strong formulation of the limit problem is,*

$$\begin{cases} -\operatorname{div} A^{hom} \nabla u_0 = f & \text{in } \Omega \backslash \Sigma, \\ A^{hom} \nabla u_0 \, n^-_{|\Sigma} = -A^{hom} \nabla u_0 \, n^+_{|\Sigma} = k_2^2 \, \boldsymbol{\Theta}'' \, [u_0]_\Sigma, \\ u_0 = 0 & \text{on } \partial\Omega. \end{cases}$$

Remark 10.20. *In the case where A'_0 is even with respect to z_N, θ vanishes on S. Then, $\boldsymbol{\Theta}''$ can be interpreted as the local capacity of the set S.*

Remark 10.21.
1. The proof for the case $k_2 = 0$ (k_2 defined by (10.34)) is actually simpler but, as in the preceding examples, requires a slight modification, namely setting in $(10.7)_2$,

$$\frac{\delta^{\frac{N}{2}-1}}{\sqrt{\varepsilon}} \mathcal{T}_{\varepsilon,\delta}(u_{\varepsilon,\delta}) \rightharpoonup U.$$

The corresponding statement is that of Theorem 10.18 with $k_2 = 0$. The holes are too small to keep any connection between Ω_+ and Ω_-. The limit problem is split into two independent problems in each of these sets, with mixed homogeneous boundary conditions,

$$\begin{cases} -\operatorname{div} A^{hom} \nabla u_0 = f & \text{in } \Omega \pm, \\ A^{hom} \nabla u_0 \cdot n^\pm_{|\Sigma} = 0 & \text{on } \Sigma, \\ u_0 = 0 & \text{on } \partial\Omega. \end{cases}$$

2. For the case of $k_2 = +\infty$, Theorem 9.18(ii) implies

$$\begin{aligned} \mathcal{T}^{bl}_{\varepsilon,\delta}(u^+_{\varepsilon,\delta}) &\rightharpoonup u^+_{0|\Sigma} \quad \text{weakly in } L^2(\Sigma; L^2_{loc}(\mathbb{R}^N_+)), \\ \mathcal{T}^{bl}_{\varepsilon,\delta}(u^-_{\varepsilon,\delta}) &\rightharpoonup u^-_{0|\Sigma} \quad \text{weakly in } L^2(\Sigma; L^2_{loc}(\mathbb{R}^N_-)). \end{aligned}$$

On the other hand, $[\mathcal{T}^{bl}_{\varepsilon,\delta}(u_{\varepsilon,\delta})]_{\Sigma\times S} = 0$ on $\Sigma \times S$ implies that $[u_0]_\Sigma = 0$. Hence, u_0 is in $H^1_0(\Omega)$ and the limit problem is satisfied in the whole of Ω.

10.3.3 Convergence of the energy

The proofs of the next proposition and corollary follow the same lines as that of Proposition 10.8 and its corollary.

Proposition 10.22 (Convergence of the energy). *Under the hypotheses of Theorem 10.18, one has*

$$\lim_{\varepsilon\to 0} \int_{\Omega^{bl}_{\varepsilon,\delta}} A^\varepsilon \nabla u_{\varepsilon,\delta} \nabla u_{\varepsilon,\delta} \, dx = \int_{\Omega\backslash\Sigma} A^{hom} \nabla u_0 \nabla u_0 \, dx + k_2^2 \int_\Sigma \boldsymbol{\Theta}'' \, [u_0]^2_\Sigma \, dx',$$

$$\lim_{\varepsilon\to 0} \int_{\Lambda_\varepsilon} |\nabla u_{\varepsilon,\delta}|^2 \, dx = 0.$$

Corollary 10.23. *Under the hypotheses of Theorem 10.18, one has*

$$\mathcal{T}_\varepsilon(\nabla u_{\varepsilon,\delta})\mathbf{1}_{(\Omega\setminus\overline{\Sigma}_\varepsilon)\times Y} \to \nabla u_0 + \nabla_y \widehat{u} \quad \text{strongly in } L^2(\Omega\times Y)^N,$$
$$\nabla_z\big(\mathcal{T}^{bl}_{\varepsilon,\delta}(u_{\varepsilon,\delta})\big)\mathbf{1}_{\frac{1}{\delta}Y} \to \nabla_z U \quad \text{strongly in } L^2(\Sigma\times\mathbb{R}^N)^N.$$

The first convergence implies the corrector result

$$\nabla u_{\varepsilon,\delta}\mathbf{1}_{(\Omega\setminus\overline{\Sigma}_\varepsilon)} - \nabla u_0 - \mathcal{U}_\varepsilon(\nabla_y\widehat{u}) \to 0 \quad \text{strongly in } L^2(\Omega)^N,$$

while the second convergence implies that

$$\|\nabla u_{\varepsilon,\delta}\|_{L^2(\Sigma^\pm_\varepsilon)} \to k_2\|\nabla_z U\|_{L^2(\Sigma\times\mathbb{R}^N_\pm)}.$$

10.4 The thick Neumann sieve

In this section we extend the results of Section 10.3 to the case of a thick Neumann sieve of thickness of order of $\varepsilon > 0$. We will use the same notations, unless specified otherwise, and we only sketch the main modifications of setting and of the proofs.

Recall Notation 9.10, as in particular, $\Pi = \{x_N = 0\}$. Then, for an open subset S such that

$$S \subset Y \cap \Pi \quad \text{satisfying } \overline{S} \subset Y \cap \Pi,$$

we introduce the class $\mathcal{F}_S$ of admissible sets, that we will use in order to describe a thick sieve with holes shaped according to S.

Definition 10.24. *The subset set F of $\mathbb{R}^N$ is in $\mathcal{F}_S$, if*
i) F is closed with connected complement in $\mathbb{R}^N$,
ii) F is symmetric with respect to all the hyperplanes of equations $z_j = 0$, $j \in \{1,\dots,N-1\}$, and $F = F_+ \cup F_- \cup (\Pi\setminus\overline{S})$,
iii) F is such that for every $0 < \delta << 1$,

$$F \cap \frac{1}{\delta}\overline{Y} \subset \Big\{|z_N| \leqslant \frac{1}{2\delta}\Big\},$$

iv) F_+ and F_- are unbounded with Lipschitz boundary,
v) there exists some positive R such that the boundaries ∂F_+ and ∂F_- outside the ball of radius R, are Lipschitz graphs over $\mathbb{R}^{N-1}$.

For $F \in \mathcal{F}_S$, set

$$F_\delta = \delta F \cap Y \quad \text{and} \quad F_{\varepsilon,\delta} = \Big\{x \in \Sigma^{bl}_\varepsilon \;\Big|\; \Big\{\frac{x}{\varepsilon}\Big\}_Y \in F_\delta\Big\}.$$

Due to the hypotheses ii) and iii), the pieces of $F_{\varepsilon,\delta}$ from neighboring ε-cells match.

Figure 10.3 presents an example of admissible set F in dimension 3.

Figure 10.3: An example of set F: the hole in the sieve

Below, Figure 10.4 is a sieve corresponding to the geometry of Figure 10.3, while Figure 10.5 is its two dimensional cross-section.

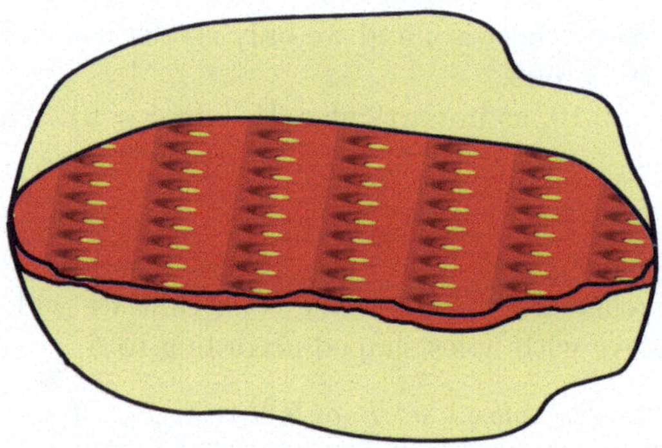

Figure 10.4: The 3D geometry of the thick Neumann sieve

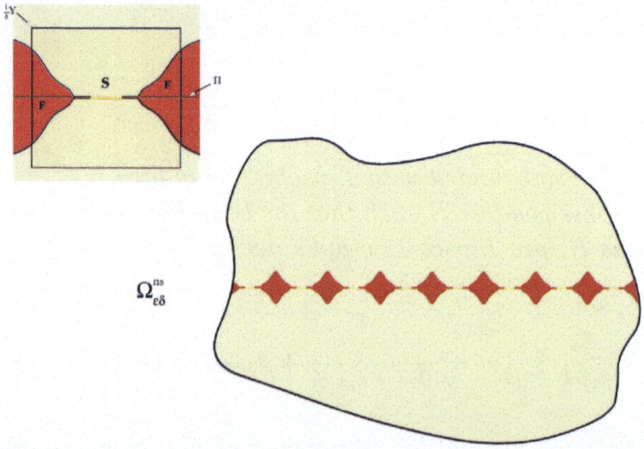

Figure 10.5: A 2D cross-section of F and of the domain $\Omega^{ns}_{\varepsilon,\delta}$

Define

$$\Omega^{ns}_{\varepsilon\delta} = \Omega \backslash \overline{F_{\varepsilon\delta}} \quad \text{and} \quad S_{\varepsilon,\delta} = \Omega^{ns}_{\varepsilon\delta} \cap \Pi.$$

We use the same space $\mathbb{V}$ as in Section 10.3, while the $\mathbb{V}_{\varepsilon,\delta}$ is now

$$\mathbb{V}_{\varepsilon,\delta} = \Big\{ v \in H^1(\Omega^{ns}_{\varepsilon\delta +} \cup \Omega^{ns}_{\varepsilon\delta -}) \mid v_{|\partial(\Omega^{ns}_{\varepsilon\delta} \backslash \overline{S_{\varepsilon,\delta}})} = 0, \; [v]_{S_{\varepsilon,\delta}} = 0 \Big\}.$$

Every function v belonging to $\mathbb{V}_{\varepsilon,\delta}$ is extended by 0 in

$$(\Omega_+ \backslash \overline{\Omega^{ns}_{\varepsilon\delta +}}) \cup (\Omega_- \backslash \overline{\Omega^{ns}_{\varepsilon\delta -}}),$$

and this extension, belonging to $\mathbb{V}$, is still denoted v.

The equivalent of the space $\widehat{\mathbf{K}}_S$ (see (10.52)) is the following:

$$\widehat{\mathbf{K}}_G = \Big\{ \Phi \in H^1_{loc}(G) \mid \nabla \Phi \in L^2(G) \Big\},$$

where G denotes the complement of F (not to be confused with the group $\mathbf{G}$).

Proposition 10.25. *There exist two linear forms $\Phi \mapsto \Phi(+\infty)$ (respectively $\Phi \mapsto \Phi(-\infty)$) on $\widehat{\mathbf{K}}_G$ such that, for every Φ in $\widehat{\mathbf{K}}_G$, the function $\Phi^+ - \Phi(+\infty)$ (resp. $\Phi^- - \Phi(-\infty)$) belongs to $L^{2^*}(\mathbb{R}^N_+ \backslash F_+)$ (resp. to $L^{2^*}(\mathbb{R}^N_- \backslash F_-)$).*

The space $\widehat{\mathbf{K}}_G$ is a Hilbert space for the norm

$$\|\Phi\|^2_{\widehat{\mathbf{K}}_G} \doteq \|\nabla \Phi\|^2_{L^2((\mathbb{R}^N_+ \cup \mathbb{R}^N_-) \backslash F)} + \Phi(+\infty)^2 + \Phi(-\infty)^2.$$

Furthermore, for this norm, the two linear forms are continuous and

$$\widehat{\mathbf{K}}^\infty_G \doteq \Big\{ \Phi \in \widehat{\mathbf{K}}_G \mid \Phi \in C^\infty(G), \; \text{supp}(\nabla\Phi) \textit{ bounded in } G \Big\},$$

is dense in $\widehat{\mathbf{K}}_G$.

Proof. The proof is the same as that of Proposition 10.16. The only modification concerns the sequence of sets on which the Sobolev-Poincaré-Wirtinger inequality (with a uniform constant) is applied. In view of Definition 10.24 (iv), this can be achieved on the set

$$\frac{1}{\delta} Y_+ \cap \{z_N > R\} \cap G \quad \text{(respectively, } \frac{1}{\delta} Y_- \cap \{z_N < -R\} \cap G),$$

making use of results from [184]. □

10.4.1 Homogenization

The thick Neumann sieve problem can be stated as follows:

$$\begin{cases} \text{Find } u_{\varepsilon,\delta} \in \mathbb{V}_{\varepsilon,\delta} \text{ satisfying} \\ \displaystyle\int_{\Omega^{ns}_{\varepsilon,\delta}} A^\varepsilon \nabla u_{\varepsilon,\delta} \nabla \phi \, dx = \int_{\Omega^{ns}_{\varepsilon\delta}} f\phi \, dx, \qquad f \in L^2(\Omega), \\ \forall \phi \in \mathbb{V}_{\varepsilon,\delta}. \end{cases}$$

The unfolded limit problem and the standard homogenized equation are given in the next two theorems. Up to the modifications of notations indicated above, theirs proofs are the same as those in Section 10.3.

Proposition 10.26. *Suppose that (10.34) is satisfied with $k_2 > 0$. Let A^ε belong to $M(\alpha, \beta, \Omega)$. Suppose also that, as $\varepsilon \to 0$, there exist a matrix A such that (10.5) is satisfied and a matrix A_0' such that*

$$\mathcal{T}^{bl}_{\varepsilon,\delta}(A^\varepsilon)(x', z) \to A_0'(x', z) \quad \textit{a.e. in } \Sigma \times (\mathbb{R}^N \backslash F).$$

Let $u_{\varepsilon,\delta}$ be the solution of the problem (10.51). Then there exists $(u_0, \widehat{u}, U)$ in $\mathbb{V} \times L^2(\Omega; H^1_{per,0}(Y)) \times L^2(\Sigma; \widehat{\mathbf{K}}_G)$ with $U(\cdot, \pm\infty) = u^{\pm}_{0|\Sigma}$ such that,

$$\begin{aligned}
&u_{\varepsilon,\delta} \rightharpoonup u_0 \quad \textit{weakly in } \mathbb{V},\\
&\mathcal{T}^{bl}_{\varepsilon,\delta}(u^{\pm}_{\varepsilon,\delta}) \rightharpoonup U^\pm \quad \textit{weakly in } L^2(\Sigma; L^2_{loc}((\mathbb{R}^N \backslash F)_\pm)),\\
&\nabla_z\big(\mathcal{T}^{bl}_{\varepsilon,\delta}(u^{\pm}_{\varepsilon,\delta})\big)\,\mathbf{1}_{\frac{1}{\delta}Y_\pm} \rightharpoonup \nabla_z U^\pm \quad \textit{weakly in } L^2(\Sigma \times (\mathbb{R}^N \backslash F)_\pm)^N,\\
&\mathcal{T}_\varepsilon(\nabla u^{\pm}_{\varepsilon,\delta}) \rightharpoonup \nabla u_0^\pm + \nabla_y \widehat{u} \quad \textit{weakly in } L^2(\Omega_\pm \times Y)^N.
\end{aligned}$$

The triple $(u_0, \widehat{u}, U)$ belongs to $\mathbb{V} \times L^2(\Omega; H^1_{per,0}(Y)) \times L^2(\Sigma; \widehat{\mathbf{K}}_G)$ with $U(\cdot, \pm\infty) = u^{\pm}_{0|\Sigma}$, and satisfies the unfolded problem

$$\begin{cases}
\dfrac{1}{|Y|}\displaystyle\int_{\Omega_+ \times Y} A(\nabla u_0^+ + \nabla_y \widehat{u})\,(\nabla v^+ + \nabla_y \widehat{v})\,dxdy\\
\qquad + \dfrac{1}{|Y|}\displaystyle\int_{\Omega_- \times Y} A(\nabla u^- + \nabla_y \widehat{u})\,(\nabla v^- + \nabla_y \widehat{v})\,dxdy\\
\qquad + \dfrac{k_2^2}{|Y|}\displaystyle\int_{\Sigma \times (\mathbb{R}^N \backslash F)_+} A_0'\,\nabla_z U^+\,\nabla_z V^+\,dx'dz\\
\qquad + \dfrac{k_2^2}{|Y|}\displaystyle\int_{\Sigma \times (\mathbb{R}^N \backslash F)_-} A_0'\,\nabla_z U^-\,\nabla_z V^-\,dx'dz = \int_\Omega fv\,dx,\\
\forall (v, \widehat{v}, V) \textit{ in } \mathbb{V} \times L^2(\Omega; H^1_{per}(Y)) \times L^2(\Sigma; \widehat{\mathbf{K}}_G) \textit{ with } V(\cdot, \pm\infty) = v^{\pm}_{|\Sigma}.
\end{cases}$$

Let $\theta \in L^\infty(\Sigma; \widehat{\mathbf{K}}_G)$ be the solution of the following cell problem, defined for a.e. $x' \in \Sigma$ as follows:

$$\begin{cases}
\displaystyle\int_{\mathbb{R}^N_+ \backslash F_+} A_0'(x', z)\,\nabla_z \theta^+(x', z)\,\nabla_z \Psi^+(z)\,dz\\
\qquad + \displaystyle\int_{\mathbb{R}^N_- \backslash F_-} A_0'(x', z)\,\nabla_z \theta^-(x', z)\,\nabla_z \Psi^-(z)\,dz = 0,\\
\theta^+(x', +\infty) \equiv 1, \quad \theta^-(x', -\infty) \equiv 0,\\
\forall \Psi \in \widehat{\mathbf{K}}_G \text{ with } \Psi^+(+\infty) = \Psi^-(-\infty) = 0.
\end{cases}$$

With this definition, it follows that

$$U(x',z) = \theta(x',z)\, u^+_{0|\Sigma}(x') + (1-\theta(x',z))\, u^-_{0|\Sigma}(x').$$

Set

$$\begin{aligned}\mathbf{\Theta_G} = \frac{1}{|Y|}\int_{\mathbb{R}^N_+\setminus F_+} & A'_0\, \nabla_z\theta^+\, \nabla_z\theta^+ dz \\ & + \frac{1}{|Y|}\int_{\mathbb{R}^N_-\setminus F_-} A'_0\, \nabla_z\theta^-\, \nabla_z\theta^- dz \quad \text{a.e. in } \Sigma.\end{aligned}$$

Theorem 10.27. *Suppose that (10.34) is satisfied with $k_2 > 0$. The limit function $u_0 \in \mathbb{V}$ given by Proposition 10.26, is the solution of the homogenized equation*

$$\begin{cases}\displaystyle\int_{\Omega_+} A^{hom}\nabla u_0^+\, \nabla v^+\, dx + \int_{\Omega_-} A^{hom}\nabla u_0^-\, \nabla v^-\, dx \\ \displaystyle\qquad\qquad + k_2^2\int_\Sigma \mathbf{\Theta_G}\, [u_0]_\Sigma\, [v]_\Sigma\, dx' = \int_\Omega f\, v\, dx, \\ \forall v \in \mathbb{V}.\end{cases}$$

Remark 10.28. *The function $\mathbf{\Theta_G}(x')$ can be interpreted as the local relative capacity (in G) of the set $\mathcal{C}(x') \doteq \{x' \in G \,|\, \theta(x',\cdot) = 1/2\}$, the capacitary potential being $(2\theta(x',\cdot)-1)$ "above $\mathcal{C}(x')$" and $(1-2\theta(x',\cdot))$ "below $\mathcal{C}(x')$".*

10.4.2 Convergence of the energy

Here also, the proofs of the next proposition and corollary follow the same lines as that of Proposition 10.8 and its corollary.

Proposition 10.29 (Convergence of the energy). *Under the hypotheses of Theorem 10.27, one has*

$$\begin{aligned}\lim_{\varepsilon\to 0}\int_{\Omega^{bl}_{\varepsilon,\delta}} A^\varepsilon\nabla u_{\varepsilon,\delta}\, \nabla\phi\, dx = & \int_{\Omega_+} A^{hom}\nabla u_0^+\, \nabla u_0^+\, dx \\ & + \int_{\Omega_-} A^{hom}\nabla u_0^-\, \nabla u_0^-\, dx + k_2^2\int_\Sigma \mathbf{\Theta_G}\, [u_0]^2_\Sigma\, dx',\end{aligned}$$

$$\lim_{\varepsilon\to 0}\int_{\Lambda_\varepsilon} |\nabla u_{\varepsilon,\delta}|^2\, dx = 0.$$

Corollary 10.30. *Under the hypotheses of Theorem 10.27, one has*

$$\begin{aligned}&\mathcal{T}_\varepsilon(\nabla u_{\varepsilon,\delta})\mathbf{1}_{(\Omega\setminus\overline{\Sigma}_\varepsilon)\times Y} \to \nabla u_0 + \nabla_y\widehat{u} \quad \textit{strongly in } L^2(\Omega\times Y)^N,\\ &\nabla_z\big(\mathcal{T}^{bl}_{\varepsilon,\delta}(u_{\varepsilon,\delta})\big) \to \nabla_z U \quad \textit{strongly in } L^2(\Sigma\times G)^N.\end{aligned}$$

The first convergence implies the corrector result

$$\nabla u_{\varepsilon,\delta}\mathbf{1}_{(\Omega\setminus\overline{\Sigma}_\varepsilon)} - \nabla u_0 - \mathcal{U}_\varepsilon(\nabla_y \widehat{u}) \to 0 \quad \textit{strongly in } L^2(\Omega)^N.$$

The second convergence implies that

$$\|\nabla u_{\varepsilon,\delta}\|_{L^2(\Sigma^\pm_\varepsilon)} \to k_2\|\nabla_z U\|_{L^2(\Sigma\times G)_\pm)}.$$

10.5 Inequalities with obstacles: fakir's carpet

The problem considered here is a variational inequality with small obstacles. The terminology "fakir's carpet" was introduced in [74] where the problem was originally studied. The unfolding method gives an alternate proof in the general case of oscillating coefficients.

Since the small obstacles are distributed in volume, this section is related to Section 10.1 the notations of which will be used, in particular for the sets $\mathbf{K}_B$ and $\mathbf{L}_B$ (see (10.2)-(10.3)).

The set $B_{\varepsilon,\delta} \doteq \Omega\setminus\Omega^*_{\varepsilon,\delta}$ is where the unilateral constraint is imposed.

Assume that the matrix field $A^\varepsilon = (a^\varepsilon_{ij})_{1\leqslant i,j\leqslant N}$ belongs to $M(\alpha,\beta,\Omega)$. Introduce the closed convex cone in $H^1_0(\Omega)$,

$$\mathbf{V}^+_{\varepsilon,\delta} = \Big\{v \in H^1_0(\Omega) \mid v \geqslant 0 \quad \text{a.e. in } B_{\varepsilon,\delta}\Big\}.$$

For $f \in L^2(\Omega)$ consider the variational inequality,

$$\begin{cases} \text{Find } u_{\varepsilon,\delta} \in \mathbf{V}^+_{\varepsilon,\delta} \text{ satisfying} \\ \displaystyle\int_\Omega A^\varepsilon \nabla u_{\varepsilon,\delta}\,\nabla(v - u_{\varepsilon,\delta})\,dx \geqslant \int_\Omega f(v - u_{\varepsilon,\delta})\,dx, \\ \forall v \in \mathbf{V}^+_{\varepsilon,\delta}. \end{cases} \tag{10.65}$$

It is well-known (see e.g. [141, Theorem 2.1, Chapter II]) that problem 10.65 admits a unique solution, which furthermore satisfies the estimate,

$$\|u_{\varepsilon,\delta}\|_{H^1(\Omega)} \leqslant C\|f\|_{L^2(\Omega)}. \tag{10.66}$$

The constant does not depend on ε and δ.

Remark 10.31. *If the matrix field A^ε is symmetric, problem* (10.65) *is equivalent to the following one:*

$$\begin{cases} \textit{Find a minimizer over } \mathbf{V}^+_{\varepsilon,\delta} \textit{ of the functional} \\ \displaystyle\mathcal{E}_{\varepsilon,\delta}(v) = \frac{1}{2}\int_\Omega A^\varepsilon \nabla v\,\nabla v\,dx - \int_\Omega f\,v\,dx. \end{cases}$$

Since $\mathcal{E}_{\varepsilon,\delta}$ is bounded from below, strictly convex and strongly continuous (hence weakly lower semi-continuous) and coercive, there exists a minimizer $u_{\varepsilon,\delta} \in \mathbf{V}^+_{\varepsilon,\delta}$ of $\mathcal{E}_{\varepsilon,\delta}$ which is unique.

10.5.1 Homogenization

We now derive the unfolded limit formulation for problem (10.65) as ε and δ tend to 0. To this end, we introduce the associated variational setting and use the notations of Proposition 9.2.

For V in $L^2(\Omega; \dot{H}^1(\mathbb{R}^N) \oplus \mathbb{R})$, it is clear that $V(\cdot, \infty)$ belongs to $L^2(\Omega)$. Introduce the following space:

$$\mathfrak{W} \doteq \{V \in L^2(\Omega; \dot{H}^1(\mathbb{R}^N) \oplus \mathbb{R}) \mid V(\cdot, \infty) \in H_0^1(\Omega)\},$$

which is a Hilbert space for the norm,

$$\|V\|_{\mathfrak{W}}^2 = \|\nabla_z V\|_{L^2(\Omega \times \mathbb{R}^N)}^2 + \|\nabla V(\cdot, \infty)\|_{L^2(\Omega)}^2.$$

Finally, introduce the closed convex cone

$$\mathbf{L}_{\mathbf{B}}^+ = \{V \in \mathfrak{W} \mid V \geqslant 0 \text{ a.e. on } \Omega \times B\}.$$

Proposition 10.32. *Suppose that $\delta = \delta(\varepsilon)$ is such that*

$$k_1 \doteq \lim_{\varepsilon \to 0} \frac{\delta^{\frac{N}{2}-1}}{\varepsilon} \quad \textit{exists in } \mathbb{R}^+.$$

Suppose also that, as ε goes to 0, there exist a matrix A such that (10.5) *is satisfied and a matrix A_0 such that* (10.6) *is satisfied.*

Let $u_{\varepsilon,\delta}$ be the solution of problem (10.65). *Then there exists $(u_0, \widehat{u}, U)$ in $H_0^1(\Omega) \times L^2(\Omega; H_{per,0}^1(Y)) \times \mathbf{L}_{\mathbf{B}}^+$ such that,*

$$\begin{aligned}
&u_{\varepsilon,\delta} \rightharpoonup u_0 \quad \textit{weakly in } H_0^1(\Omega),\\
&\mathcal{T}_{\varepsilon,\delta}(u_{\varepsilon,\delta}) \rightharpoonup U \quad \textit{weakly in } L^2(\Omega; L_{loc}^2(\mathbb{R}^N)),\\
&\nabla_z\big(\mathcal{T}_{\varepsilon,\delta}(u_{\varepsilon,\delta})\big)\, \mathbf{1}_{\frac{1}{\delta}Y} \rightharpoonup \nabla_z U \quad \textit{weakly in } L^2(\Omega \times \mathbb{R}^N)^N,\\
&U(x, \infty) = u_0(x) \quad \textit{for a.e. } x \in \Omega,\\
&\mathcal{T}_\varepsilon(\nabla u_{\varepsilon,\delta}) \rightharpoonup \nabla u_0 + \nabla_y \widehat{u} \quad \textit{weakly in } L^2(\Omega \times Y)^N.
\end{aligned} \tag{10.67}$$

The pair $(\widehat{u}, U) \in L^2(\Omega; H_{per,0}^1(Y)) \times \mathbf{L}_{\mathbf{B}}^+$ satisfies the unfolded variational inequality,

$$\begin{cases}
\dfrac{1}{|Y|} \displaystyle\int_{\Omega \times Y} A(\nabla U(\cdot, \infty) + \nabla_y \widehat{u})\, (\nabla(V - U)(\cdot, \infty) + \nabla_y(\widehat{v} - \widehat{u}))dxdy\\
\quad + \dfrac{k_1^2}{|Y|} \displaystyle\int_{\Omega \times \mathbb{R}^N} A_0\, \nabla_z U\, \nabla_z(V - U)\, dxdz \geqslant \int_\Omega f\,(V - U)(\cdot, \infty)dx,\\
\forall (\widehat{v}, V) \in L^2(\Omega; H_{per}^1(Y)) \times \mathbf{L}_{\mathbf{B}}^+.
\end{cases} \tag{10.68}$$

The weak limit u_0 of the sequence of solutions of (10.65) *is then recovered as* $u_0 = U(\cdot, \infty)$.

Problem (10.68) has a unique solution as a standard variational inequality in the closed convex set $L^2(\Omega; H^1_{per,0}(Y)) \times \mathbf{L}^+_{\mathbf{B}}$ due to the ellipticity of the matrix fields A and A_0 (arising from A^ε in $M(\alpha,\beta,\Omega)$) and identity (3.14).

Proof of Proposition 10.32 for the case $k_1 > 0$[8].

Step 1. Due to estimate (10.66), up to a subsequence, we get the first convergence of (10.67). Then, as in the proof of the Proposition 10.1,

$$\mathcal{M}_\varepsilon(u_{\varepsilon,\delta})\mathbf{1}_{\frac{1}{\delta}Y} \to u_0 \quad \text{strongly in } L^2(\Omega; L^2_{loc}(\mathbb{R}^N)). \tag{10.69}$$

By Theorem 1.41, there exists $\widehat{u} \in L^2(\Omega; H^1_{per,0}(Y))$ such that, up to a subsequence,

$$\mathcal{T}_\varepsilon(\nabla u_{\varepsilon,\delta}) \rightharpoonup \nabla u_0 + \nabla_y \widehat{u} \;\text{ weakly in } L^2(\Omega \times Y)^N.$$

By Theorem 9.7(vii), there exists U in $L^2(\Omega; L^2_{loc}(\mathbb{R}^N))$ such that, up to a subsequence

$$\mathcal{T}_{\varepsilon,\delta}(u_{\varepsilon,\delta}) \rightharpoonup U \quad \text{weakly in } L^2(\Omega; L^2_{loc}(\mathbb{R}^N)). \tag{10.70}$$

Again by Theorem 9.7(vii) there exists W in $L^2(\Omega; L^{2^*}(\mathbb{R}^N))$ with $\nabla_z W$ in $L^2(\Omega \times \mathbb{R}^N)$ such that

$$\mathcal{T}_{\varepsilon,\delta}\big(u_{\varepsilon,\delta} - \mathcal{M}_\varepsilon(u_{\varepsilon,\delta})\big) \rightharpoonup W \quad \text{weakly in } L^2(\Omega; L^{2^*}(\mathbb{R}^N)). \tag{10.71}$$

By (10.69)-(10.71), one concludes

$$U = W + u_0 \quad \text{and} \quad \nabla_z U = \nabla_z W. \tag{10.72}$$

From Definition 9.6, $\mathcal{T}_{\varepsilon,\delta}(u_{\varepsilon,\delta}) \geqslant 0$ in $\Omega \times B$, so that by (10.70),

$$U \geqslant 0 \;\text{ in }\; \Omega \times B. \tag{10.73}$$

Due to (10.72) and (10.73), $U(\cdot,\infty) = u_0$ and U belongs to $\mathbf{L}^+_{\mathbf{B}}$.

Step 2. From (10.65) and for every test-function $v \in \mathbf{V}_{\varepsilon,\delta}$, we can write

$$\int_\Omega A^\varepsilon \nabla u_{\varepsilon,\delta}\, \nabla u_{\varepsilon,\delta}\, dx \leqslant \int_\Omega A^\varepsilon \nabla u_{\varepsilon,\delta}\, \nabla v\, dx - \int_\Omega f(v - u_{\varepsilon,\delta})\, dx. \tag{10.74}$$

Then using both unfoldings, and proceeding as in the proof of Proposition 10.8, we obtain for the left hand side of the above inequality

$$\begin{aligned}
&\frac{1}{|Y|}\int_{\Omega\times Y} A\big(\nabla U(\cdot,\infty) + \nabla_y \widehat{u}\big)\big(\nabla U(\cdot,\infty) + \nabla_y \widehat{u}\big)\, dxdy \\
&+ \frac{k_1^2}{|Y|}\int_{\Omega\times\mathbb{R}^N} A_0\, \nabla_z U\, \nabla_z U\, dxdz \leqslant \liminf_{\varepsilon\to 0}\int_\Omega A^\varepsilon \nabla u_{\varepsilon,\delta}\, \nabla u_{\varepsilon,\delta}\, dx.
\end{aligned} \tag{10.75}$$

[8] See Remark 10.36 for the cases $k_1 = 0$ and $k_1 = +\infty$.

Let $(\psi_1, \psi_2) \in \mathcal{D}(\Omega) \times \mathcal{D}(\Omega)$, $\phi \in H^1_{per}(Y)$ vanishing in a neighborhood of the origin, $v \in \mathbf{K}_B$ such that $\nabla_z v$ has a compact support (we recall that v vanishes on B) and $\Psi \in \mathcal{D}(\Omega \times \mathbb{R}^N) \cap \mathbf{L}^+_{\mathbf{B}}$.

Consider in (10.65), the test function $\Phi_{\varepsilon,\delta} = \Psi^0_{\varepsilon,\delta} + \Psi_{\varepsilon,\delta}$ where

$$\Psi^0_{\varepsilon,\delta}(x) = \varepsilon\psi_1(x)\,\phi\Big(\frac{x}{\varepsilon}\Big) + \psi_2(x)\,v_{\varepsilon,\delta}(x),$$

and

$$v_{\varepsilon,\delta}(x) = v\Big(\frac{1}{\delta}\Big\{\frac{x}{\varepsilon}\Big\}_Y\Big), \quad \Psi_{\varepsilon,\delta}(x) = \Psi\Big(x, \frac{1}{\delta}\Big\{\frac{x}{\varepsilon}\Big\}_Y\Big),$$

for a.e. x in Ω.

First, observe that if ε is small enough, the support of $\Psi^0_{\varepsilon,\delta}$ is included in $\Omega^*_{\varepsilon,\delta}$. Hence $\Psi^0_{\varepsilon,\delta}$ belongs to $\mathbf{V}^+_{\varepsilon,\delta}$.

Second, since $\Psi_{\varepsilon,\delta}$ vanishes on the boundary of the ε-cells, it also belongs to $\mathbf{V}^+_{\varepsilon,\delta}$. Therefore $\Phi_{\varepsilon,\delta}$ itself is in $\mathbf{V}^+_{\varepsilon,\delta}$.

Clearly, $\varepsilon\psi_1\,\phi(\cdot/\varepsilon)$ converges to 0 and $v_{\varepsilon,\delta}$ to $v(\infty)$ strongly in $L^2(\Omega)$ (see Lemma 10.4). Moreover, $\Psi_{\varepsilon,\delta}$ also converges to 0 strongly in $L^2(\Omega)$ since the support of the function $\Psi_{\varepsilon,\delta}$ is included in a set whose measure is at most of order δ^N. Hence we get

$$\int_\Omega f(v - u_{\varepsilon,\delta})\,dx \to \int_\Omega f(x)\,\big(\psi_2(x)v(\infty) - U(x,\infty)\big)dx.$$

Now,

$$\begin{aligned}\int_\Omega A^\varepsilon \nabla u_{\varepsilon,\delta} \nabla \Phi_{\varepsilon,\delta} dx = &\int_\Omega A^\varepsilon \nabla u_{\varepsilon,\delta}\, \psi_1 \nabla\phi\Big(\frac{\cdot}{\varepsilon}\Big)dx + \varepsilon\int_\Omega A^\varepsilon \nabla u_{\varepsilon,\delta} \nabla\psi_1\, \phi\Big(\frac{\cdot}{\varepsilon}\Big)dx\\ &+ \int_\Omega A^\varepsilon \nabla u_{\varepsilon,\delta} \nabla\psi_2\, v_{\varepsilon,\delta}\,dx + \int_\Omega A^\varepsilon \nabla u_{\varepsilon,\delta} \nabla v_{\varepsilon,\delta}\,\psi_2\,dx\\ &+ \int_\Omega A^\varepsilon \nabla u_{\varepsilon,\delta} \nabla \Psi_{\varepsilon,\delta}\,dx + \frac{1}{\varepsilon\delta}\int_\Omega A^\varepsilon \nabla u_{\varepsilon,\delta} \nabla_z \Psi_{\varepsilon,\delta}\,dx.\end{aligned}$$

Following the same arguments as in steps 2 and 3 of the proof of the Proposition 10.1 gives

$$\begin{aligned}\lim_{\varepsilon\to 0}\int_\Omega A^\varepsilon \nabla u_{\varepsilon,\delta}\, \nabla\Phi_{\varepsilon,\delta} dx = &\int_{\Omega\times Y} A\big(\nabla u_0 + \nabla_y \widehat{u}\big)\, \nabla_y\phi\,\psi_1\,dxdy\\ &+ \frac{v(\infty)}{|Y|}\int_{\Omega\times Y} A\big(\nabla u_0 + \nabla_y\widehat{u}\big)\,\nabla\psi_2\,dxdy \qquad (10.76)\\ &+ \frac{k_1^2}{|Y|}\int_{\Omega\times\mathbb{R}^N} A_0\,\nabla_z U\big(\nabla_z v\,\psi_2 + \nabla_z\Psi\big)dxdz.\end{aligned}$$

Taking into account (10.75), (10.74) and (10.76), we get

$$\begin{aligned}
\frac{1}{|Y|}\int_{\Omega\times Y} A\big(\nabla U(\cdot,\infty)+\nabla_y\widehat{u}\big)\big(\nabla U(\cdot,\infty)+\nabla_y\widehat{u}\big)\,dxdy \\
+\frac{k_1^2}{|Y|}\int_{\Omega\times\mathbb{R}^N} A_0\,\nabla_z U\,\nabla_z U\,dxdz \leqslant \int_{\Omega\times Y} A\big(\nabla u_0+\nabla_y\widehat{u}\big)\,\nabla_y\phi\,\psi_1\,dxdy \\
+\frac{v(\infty)}{|Y|}\int_{\Omega\times Y} A\big(\nabla u_0+\nabla_y\widehat{u}\big)\,\nabla\psi_2\,dxdy \qquad (10.77)\\
+\frac{k_1^2}{|Y|}\int_{\Omega\times\mathbb{R}^N} A_0\,\nabla_z U\big(\nabla_z v\,\psi_2+\nabla_z\Psi\big)dxdz \\
-\int_\Omega f(x)\,\big(v(\infty)\psi_2(x)-U(x,\infty)\big)dx.
\end{aligned}$$

Then, by Lemma 10.2 and due to the density of the product $\mathcal{D}(\Omega)\otimes H^1_{per}(Y)$ in $L^2(\Omega;H^1_{per}(Y))$, one can replace in (10.77), $\nabla_y\phi\,\psi_1$ by $\nabla_y\widehat{v}$ for arbitrary $\widehat{v}$ in $L^2(\Omega;H^1_{per}(Y))$. By density again, inequality (10.77) holds true for every Ψ in $\mathbf{L}^+_{\mathbf{B}}$ such that $\Psi(\cdot,\infty)=0$ and ψ_2 in $H^1_0(\Omega)$. As a consequence, setting

$$V(x,z)\doteq v(z)\psi_2(x)+\Psi(x,z),$$

inequality (10.68) holds true for every V in $\mathbf{L}^+_{\mathbf{B}}$.

Uniqueness of the solution of (10.68) is easily verified by cross-testing. □

10.5.2 Standard form of the limit problem

Here we first give the variational inequality satisfied by U and then the one satisfied by the macroscopic term u_0 alone.

In (10.68), we take $V=U$ and $\widehat{v}=\widehat{w}+\widehat{u}$ (resp. $\widehat{v}=-\widehat{w}+\widehat{u}$), where $\widehat{w}\in L^2(\Omega;H^1_{per}(Y))$. This gives

$$\int_{\Omega\times Y} A\big(\nabla u_0+\nabla_y\widehat{u}\big)\,\nabla_y\widehat{w}\,dxdy=0,\quad \forall\widehat{w}\in L^2(\Omega;H^1_{per}(Y)).$$

Using the correctors $\widehat{\chi}_j$ ($j=1,\dots,N$) defined by the cell problems (3.10), we express $\widehat{u}$ as

$$\widehat{u}(x,y)=\sum_{j=1}^N\frac{\partial u_0}{\partial x_j}(x)\,\widehat{\chi}_j(x,y)=\sum_{j=1}^N\frac{\partial U}{\partial x_j}(x,\infty)\,\widehat{\chi}_j(x,y),$$

for a.e. $(x,y)\in\Omega\times Y$. This used in (10.68) implies

$$\begin{aligned}
\int_\Omega A^{hom}\nabla U(\cdot,\infty)\,\nabla(V-U)(\cdot,\infty)\,dx \\
+\frac{k_1^2}{|Y|}\int_{\Omega\times\mathbb{R}^N} A_0\,\nabla_z U\,\nabla_z(V-U)\,dxdz \qquad (10.78)\\
\geqslant\int_\Omega f\,(V-U)(\cdot,\infty)\,dx,\quad \forall V\in\mathbf{L}^+_{\mathbf{B}},
\end{aligned}$$

where the matrix A^{hom} is given by (10.27).

Now, if we choose $V \in \mathbf{L}^+_{\mathbf{B}}$ such that $V(\cdot,\infty) = U(\cdot,\infty)(= u_0)$, we get

$$\int_{\Omega\times\mathbb{R}^N} A_0\,\nabla_z U\,\nabla_z (V-U)\,dxdz \geqslant 0. \tag{10.79}$$

In order to solve this variational inequality, introduce the convex set

$$\mathbf{C}^+_{\mathbf{B}} = \Big\{\Psi \in \dot{H}^1(\mathbb{R}^N)\oplus\mathbb{R} \mid \Psi \geqslant 0 \text{ a.e. on } B,\quad \Psi(\infty) = -1\Big\},$$

and let $\theta_I \in L^\infty(\Omega\,;\,\mathbf{C}^+_{\mathbf{B}})$ be the solution of the variational inequality,

$$\begin{cases} \displaystyle\int_{\mathbb{R}^N} A_0(x,z)\,\nabla_z\theta_I(x,z)\,\nabla_z\big(\Psi(z)-\theta_I(x,z)\big)dz \geqslant 0 \text{ for a.e. } x\in\Omega,\\ \forall\Psi\in\mathbf{C}^+_{\mathbf{B}}. \end{cases} \tag{10.80}$$

Existence and uniqueness of the solution to this problem follows from standard results (see [141, Theorem 2.1, Chapter II]).

The unique solution of (10.79) can be now determined: multiplying the equation in (10.80) by $(u_0^-)^2$, one can see that the solution U is given by

$$U(x,z) = u_0^+(x) + u_0^-(x)\theta_I(x,z) \quad \text{for a.e. } (x,z)\in\Omega\times\mathbb{R}^N,$$

where u_0^+ is added to satisfy the condition at ∞, and the test function V is written as $u_0^+(x) + u_0^-(x)\Psi(x,z)$.

Now, for ψ arbitrary in $H_0^1(\Omega)$, choose as test-function in (10.78)

$$V(x,z) = U(x,z) - \psi(x)\vartheta(z) \qquad \text{for a.e. } (x,z)\in\Omega\times\mathbb{R}^N,$$

where ϑ belongs to $\mathbf{K}_{\mathbf{B}}\cap\mathbf{C}^+_{\mathbf{B}}$. This gives

$$\begin{aligned} \int_\Omega A^{hom}\nabla u_0\,\nabla\psi\,dx - \frac{k_1^2}{|Y|}\int_\Omega u_0^-\psi\Big(\int_{\mathbb{R}^N} A_0\,\nabla_z\theta_I\,\nabla_z\vartheta\,dz\Big)dx& \\ \geqslant \int_\Omega f\,\psi\,dx.& \end{aligned} \tag{10.81}$$

Choose successively $\Psi = \vartheta$ and $2\theta_I - \vartheta$ as test functions in inequality (10.80), to obtain

$$\int_{\mathbb{R}^N} A_0\,\nabla_z\theta_I\,\nabla_z\theta_I\,dz = \int_{\mathbb{R}^N} A_0\,\nabla_z\theta_I\,\nabla_z\vartheta\,dz \quad \text{a.e. in } \Omega.$$

Therefore, setting for a.e. x in Ω,

$$\mathbf{\Theta_I}(x) = \frac{1}{|Y|}\int_{\mathbb{R}^N} A_0(x,z)\,\nabla_z\theta_I(x,z)\,\nabla_z\theta_I(x,z)\,dz, \tag{10.82}$$

inequality (10.81) becomes

$$\int_\Omega A^{hom}\nabla u_0\,\nabla\psi\,dx - k_1^2\int_\Omega \boldsymbol{\Theta}_{\mathbf{I}}\;u_0^-\;\psi\,dx \geqslant \int_\Omega f\,\psi\,dx.$$

Since ψ is arbitrary in $H_0^1(\Omega)$, the latter inequality is actually an equality which is the limit homogenized problem,

$$\begin{cases} \displaystyle\int_\Omega A^{hom}\nabla u_0\,\nabla\psi\,dx - k_1^2\int_\Omega \boldsymbol{\Theta}_{\mathbf{I}}\;u_0^-\psi\,dx = \int_\Omega f\,\psi\,dx, \\ \forall\psi\in H_0^1(\Omega) \end{cases} \tag{10.83}$$

Theorem 10.33. *Under the assumptions of Proposition 10.32 with $k_1 > 0$, the limit function u_0 in $H_0^1(\Omega)$ is the unique solution of the homogenized problem* (10.83) *which is the variational formulation of the equation*

$$\begin{cases} -\mathrm{div}\left(A^{hom}\nabla u_0\right) - k_1^2\,\boldsymbol{\Theta}_{\mathbf{I}}\;u_0^- = f, \\ u_0 = 0 \quad on\,\partial\Omega. \end{cases} \tag{10.84}$$

This equation is well-posed since the function $\boldsymbol{\Theta}_{\mathbf{I}}$ given by (10.82) *is non negative a.e. in Ω.*

Remark 10.34. *The contribution of the oscillations of the matrix A^ε in the homogenized problem is reflected by the first term in the left-hand side of* (10.84). *The contribution of the perforations is represented by the second term of the left-hand side.*

Remark 10.35. *If the matrix field A^{hom} is symmetric, problem* (10.84) *is equivalent to the following one:*

$$\begin{cases} \text{Find a minimizer over } H_0^1(\Omega) \text{ of the functional} \\ \displaystyle\mathcal{E}(v) = \frac{1}{2}\int_\Omega A^{hom}\nabla v\,\nabla v\,dx + \frac{k_1^2}{2}\int_\Omega \boldsymbol{\Theta}_{\mathbf{I}}\,(v^-)^2\,dx - \int_\Omega f\,v\,dx. \end{cases}$$

Remark 10.36.
1. The proof for the case $k_1 = 0$ (k_1 defined by (10.4)*), is actually simpler but requires a slight modification, setting in* $(10.67)_2$

$$\frac{\delta^{\frac{N}{2}-1}}{\varepsilon}\mathcal{T}_{\varepsilon,\delta}(u_{\varepsilon,\delta}) \rightharpoonup U.$$

The corresponding statement is that of Theorem 10.5 with $k_1 = 0$; in this case, the holes are too small to have any influence at the limit.
2. The case of $k_1 = +\infty$ is easy to analyze with the help of Theorem 9.7(vi). By (9.5),

$$\mathcal{T}_{\varepsilon,\delta}(u_{\varepsilon,\delta}) \rightharpoonup u_0 \text{ weakly in } L^2(\Omega; L^2_{loc}(\mathbb{R}^N)).$$

Thus, U is independent of z and is equal to u_0. Since $\mathcal{T}_{\varepsilon,\delta}(u_{\varepsilon,\delta}) \geqslant 0$ in $\Omega \times B$, it follows that u_0 belongs to $H_0^1(\Omega)^+$. Then, is enough to test with functions which are also in $H_0^1(\Omega)^+$.

The limit problem is obtained from (10.78) *with test functions independent of z also. It reduces to the following standard variational inequality with solution u_0 in $H_0^1(\Omega)^+$:*

$$\begin{cases}\displaystyle\int_\Omega A^{hom}\nabla u_0\,\nabla(v-u_0)\,dx \geqslant \int_\Omega f\,(v-u_0)dx,\\ \forall v \in H_0^1(\Omega)^+.\end{cases}$$

10.5.3 Convergence of the energy

Proposition 10.37 (Convergence of the energy). *Under the hypotheses of Theorem 10.32,*

$$\lim_{\varepsilon\to 0}\int_\Omega A^\varepsilon\nabla u_{\varepsilon,\delta}\,\nabla u_{\varepsilon,\delta}\,dx = \int_\Omega A^{hom}\nabla u_0\,\nabla u_0\,dx + k_1^2\int_\Omega \mathbf{\Theta_I}\,(u_0^-)^2\,dx,$$

$$\lim_{\varepsilon\to 0}\int_{\Lambda_\varepsilon}|\nabla u_{\varepsilon,\delta}|^2\,dx = 0,$$

where

$$\int_\Omega A^{hom}\nabla u_0\,\nabla u_0\,dx = \frac{1}{|Y|}\int_{\Omega\times Y} A\big(\nabla U(\cdot,\infty)+\nabla_y\widehat{u}\big)\big(\nabla U(\cdot,\infty)+\nabla_y\widehat{u}\big)\,dxdy, \tag{10.85}$$

$$k_1^2\int_\Omega \mathbf{\Theta_I}\,(u_0^-)^2\,dx = \frac{k_1^2}{|Y|}\int_{\Omega\times\mathbb{R}^N} A_0\,\nabla_z U\,\nabla_z U\,dxdz.$$

Proof. First, taking into account (10.75) and (10.85), we get

$$\int_\Omega A^{hom}\nabla u_0\,\nabla u_0\,dx + k_1^2\int_\Omega \mathbf{\Theta_I}\,u_0^-\,u_0^-\,dx \leqslant \liminf_{\varepsilon\to 0}\int_\Omega A^\varepsilon\nabla u_{\varepsilon,\delta}\,\nabla u_{\varepsilon,\delta}\,dx.$$

We now prove the converse inequality. To do so, let $\{\theta_I^{(\eta)}\}_\eta$ be a sequence in $L^\infty(\Omega;\mathbf{C}_\mathbf{B}^+)\cap C^1(\Omega\times\mathbb{R}^N)$ with $\nabla_z\theta_I^{(\eta)}$ having a compact support, and such that

$$\theta_I^{(\eta)} - \theta_I \to 0 \qquad \text{strongly in } L^2(\Omega;\dot{H}(\mathbb{R}^N)).$$

Choosing

$$v_{\varepsilon,\delta}^{(\eta)}(x) = 2u_0^+(x) + 2u_0^-(x)\theta_I^{(\eta)}\Big(x,\frac{1}{\delta}\Big\{\frac{x}{\varepsilon}\Big\}_Y\Big) \qquad \text{for a.e. } x\in\Omega,$$

as test function in (10.74), and proceeding as in the proof of the Proposition 10.32, we obtain

$$\lim_{\varepsilon\to 0}\int_\Omega A^\varepsilon \nabla u_{\varepsilon,\delta}\,\nabla v^{(\eta)}_{\varepsilon,\delta}dx = \frac{2}{|Y|}\int_{\Omega\times Y} A(\nabla u_0+\nabla_y\widehat{u})\,\nabla u_0\,dxdy$$
$$+\,2\frac{k_1^2}{|Y|}\int_{\Omega\times\mathbb{R}^N} A_0\,\nabla_z U\,(\nabla_z\theta^{(\eta)}_I u_0^-)dxdz,$$

$$\lim_{\varepsilon\to 0}\int_\Omega f(v^{(\eta)}_{\varepsilon,\delta}-u_{\varepsilon,\delta})\,dx = \int_\Omega f(x)\,(2u_0(x)-U(x,\infty))dx = \int_\Omega f(x)\,u_0(x)\,dx.$$

Summarizing the above convergences gives

$$\int_\Omega A^{hom}\nabla u_0\,\nabla u_0\,dx + k_1^2\int_\Omega \mathbf{\Theta_I}\,u_0^-\,u_0^-\,dx$$
$$\leqslant \liminf_{\varepsilon\to 0}\int_\Omega A^\varepsilon\nabla u_{\varepsilon,\delta}\,\nabla u_{\varepsilon,\delta}\,dx \leqslant \limsup_{\varepsilon\to 0}\int_\Omega A^\varepsilon\nabla u_{\varepsilon,\delta}\,\nabla u_{\varepsilon,\delta}\,dx$$
$$\leqslant \lim_{\varepsilon\to 0}\int_\Omega A^\varepsilon\nabla u_{\varepsilon,\delta}\,\nabla v^{(\eta)}_{\varepsilon,\delta}dx - \lim_{\varepsilon\to 0}\int_\Omega f(v^{(\eta)}_{\varepsilon,\delta}-u_{\varepsilon,\delta})\,dx$$
$$= \int_\Omega A^{hom}\nabla u_0\nabla(2u_0)dx + 2\frac{k_1^2}{|Y|}\int_{\Omega\times\mathbb{R}^N} A_0\,\nabla_z U(\nabla_z\theta^{(\eta)}_I u_0^-)dxdz - \int_\Omega f\,u_0\,dx.$$

Passing to the limit with respect to η and using (10.83) gives

$$\lim_{\varepsilon\to 0}\int_\Omega A^\varepsilon\nabla u_{\varepsilon,\delta}\,\nabla u_{\varepsilon,\delta}\,dx = \int_\Omega A^{hom}\nabla u_0\,\nabla u_0\,dx + k_1^2\int_\Omega \mathbf{\Theta_I}\,u_0^{-2}\,dx.$$

This gives the convergence of the energy. As in the proof Proposition 10.8, one also obtains $\lim_{\varepsilon\to 0}\int_{\Lambda_\varepsilon}|\nabla u_{\varepsilon,\delta}|^2\,dx = 0.$ ☐

Corollary 10.38. *Under the hypotheses of Theorem 10.5, the following strong convergences hold:*

$$\mathcal{T}_\varepsilon(\nabla u_{\varepsilon,\delta})\mathbf{1}_{\Omega\times\tilde{Y}_\delta} \to \nabla u_0+\nabla_y\widehat{u} = \nabla U(\cdot,\infty)+\nabla_y\widehat{u}\quad \textit{strongly in } L^2(\Omega\times Y)^N,$$
$$\nabla_z\big(\mathcal{T}_{\varepsilon,\delta}(u_{\varepsilon,\delta})\big)\,\mathbf{1}_{\frac{1}{\sqrt{\delta}}B} \to \nabla_z U\quad \textit{strongly in } L^2(\Omega\times\mathbb{R}^N)^N.$$

The first convergence implies the corrector result

$$\nabla u_{\varepsilon,\delta}\mathbf{1}_{\Omega^*_{\varepsilon,\sqrt{\delta}}} - \nabla u_0 - \mathcal{U}_\varepsilon(\nabla_y\widehat{u}) \to 0\quad \textit{strongly in } L^2(\Omega)^N.$$

The second convergence implies that

$$\|\nabla u_{\varepsilon,\delta}\|_{L^2(\Omega\setminus\Omega^*_{\varepsilon,\sqrt{\delta}})} \to k_1\|\nabla_z U\|_{L^2(\Omega\times\mathbb{R}^N)}.$$

10.6 Singular inhomogeneities near boundary

In this section we give a brief presentation of the use of the unfolding method for the homogenization of a boundary value problem for the Poisson equation with singular asymptotically high contrast zero order term and right-hand side, the support of which is concentrated near a fixed subset of the domain boundary and with a periodic microstructure. It involves two small parameters (both intended to go to zero), the first one, ε, as usual characterizes the boundary microstructure period, while the second, δ, characterizes the volume fraction of the set where the source term is large, as well as the portion of the boundary where the Dirichlet condition is imposed. The boundary condition alternates rapidly between Dirichlet and Neumann on this subset.

This gives rise to a boundary layer and the unfolding method is well adapted to give actual proofs for formal results common in this theory. Depending on the ratio between ε and δ, one can obtain various boundary conditions in the limit. We follow the method of [53] where the unfolding approach was used for the first time in this kind of problems.

Problems in domains with singularly perturbed density ("concentrated masses") have been widely discussed in the literature. The behavior of solutions of a wave equation with one concentrated mass and the vibration of a body with a concentrated mass were studied in [181] and [182], respectively. We refer to [53] for a detailed list of references concerning problems with concentrated masses.

10.6.1 Setting the problem

For a given fixed $h \geqslant 0$, consider a domain Ω in $\mathbb{R}^N$, $N \geqslant 3$, which lies in the upper half space, with a piecewise smooth boundary consisting of three parts $\partial\Omega = \Gamma_1 \cup \Gamma_2 \cup \Gamma_3$ (see Figure 10.6). The part Γ_3 is the $(N-1)$-dimensional unit cube

$$\Gamma_3 \doteq \Big\{x \,\Big|\, -\frac{1}{2} < x_i < \frac{1}{2} \ \text{ for } i = 1, \dots N-1, \ \ x_N = 0\Big\}.$$

The part Γ_2 is the union of Γ_2^i for $i = 1, \dots N-1$, where

$$\Gamma_2^i \doteq \Big\{x \,\Big|\, x_i = \pm\frac{1}{2}, \ -\frac{1}{2} < x_j < \frac{1}{2} \ \text{ for } j \neq i, \ 1 \leqslant j \leqslant N-1, \ 0 \leqslant x_N \leqslant h\Big\}.$$

The remainder Γ_1 is the part of $\partial\Omega$ located in the half-space $x_N \geqslant h$. Moreover, Γ_3 (see Figure 10.7) has a periodic micro-structure associated to the small parameters $\delta < 1$ and $\varepsilon = \frac{1}{(2\mathcal{N}+1)}$ where $\mathcal{N} \in \mathbb{N}$, $\mathcal{N} \geqslant 1$ and $\delta = \delta(\varepsilon) \to 0$ as $\varepsilon \to 0$.

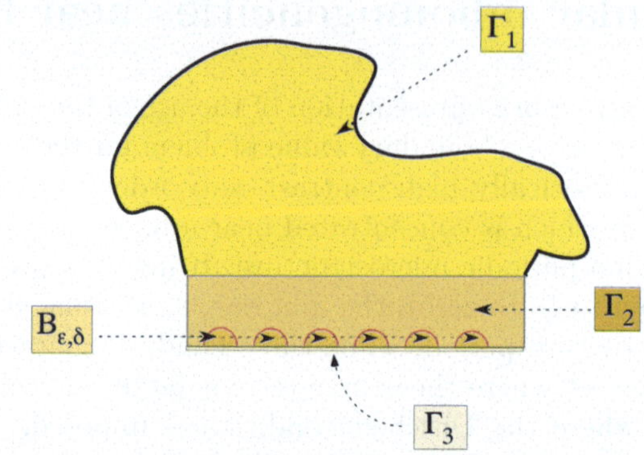

Figure 10.6: Side view

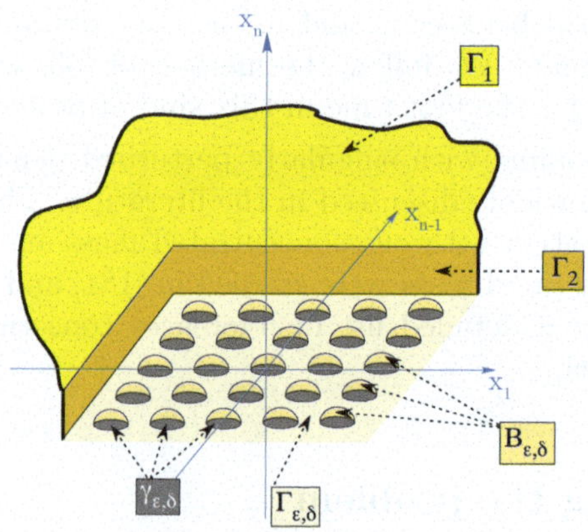

Figure 10.7: Perspective view

To describe the microstructure, let D be the hyperdisk

$$D = \{z \in \mathbb{R}^N \mid |z| < 1,\ z_N = 0\}.$$

Then

$$\Gamma_3 = \Gamma_{\varepsilon,\delta} \cup \gamma_{\varepsilon,\delta},$$

where

$$\gamma_{\varepsilon,\delta} \doteq \Gamma_3 \bigcap \Big(\bigcup\nolimits_{\xi \in \mathbb{Z}^{N-1}} \varepsilon(\xi + \delta D) \Big) \quad \text{and} \quad \Gamma_{\varepsilon,\delta} \doteq \Gamma_3 \backslash \gamma_{\varepsilon,\delta}.$$

Let B be the half-ball

$$B = \{z \mid |z| < 1,\ z_N > 0\}.$$

Then $B_{\varepsilon,\delta}$ is the set

$$B_{\varepsilon,\delta} = \Omega \bigcap \big(\bigcup\nolimits_{\xi \in \mathbb{Z}^{N-1}} \varepsilon(\xi + \delta B) \big).$$

When the small parameters go to zero, we have a homogenization process where the periodic cell Y is the cube $Y = (-1/2, 1/2)^N$, but the periodicity is restricted to the boundary Γ_3. The notations of the boundary layer unfolding from Section 9.3 will be used in the sequel.

Remark 10.39. *The set B defined above as a half-ball, can be replaced by any bounded connected open subset of the upper half-space with Lipschitz boundary. All the results of this section remain valid for such a choice of B.*

The problem is to study the asymptotic behavior as ε and δ go to 0, of the solutions of the following boundary value problem:

$$\begin{cases} -\Delta u_{\varepsilon,\delta} + (\varepsilon\delta)^{-\varkappa} g_{\varepsilon,\delta} u_{\varepsilon,\delta} = f_{\varepsilon,\delta} \quad \text{in } \Omega, \\ u_{\varepsilon,\delta} = 0 \quad \text{on } \gamma_{\varepsilon,\delta} \cup \Gamma_1, \\ \dfrac{\partial u_{\varepsilon,\delta}}{\partial n} = 0 \quad \text{on } \Gamma_{\varepsilon,\delta} \cup \Gamma_2, \end{cases} \tag{10.86}$$

where n is the outward unit normal on $\partial\Omega$, and $(\varepsilon\delta)^{-\varkappa} g_{\varepsilon,\delta}$ with $\varkappa \geqslant 0$, is a non-negative density supported in $B_{\varepsilon,\delta}$, such that $g_{\varepsilon,\delta}$ is bounded in $L^\infty(B_{\varepsilon,\delta})$. The right-hand side $f_{\varepsilon,\delta} \in L^2(\Omega)$ is of the form

$$f_{\varepsilon,\delta}(x) = \begin{cases} f & \text{in } \Omega \backslash \overline{B}_{\varepsilon,\delta}, \\ \widetilde{f}_{\varepsilon,\delta} & \text{in } B_{\varepsilon,\delta}. \end{cases}$$

where $f_{\varepsilon,\delta}$ belongs to $L^2(\Omega)$, and, for simplicity, f is fixed in $L^2(\Omega)$. The function $\widetilde{f}_{\varepsilon,\delta}$ is not necessarily bounded with respect to ε and δ, the exact scaling being specified in Proposition 10.42 below.

Let $\mathcal{V}_{\varepsilon,\delta}$ be the space

$$\mathcal{V}_{\varepsilon,\delta} = \{v \in H^1(\Omega) \,|\, v = 0 \text{ on } \Gamma_1 \cup \gamma_{\varepsilon,\delta}\}.$$

The variational formulation of problem (10.86) is now

$$\begin{cases} \text{Find } u_{\varepsilon,\delta} \in \mathcal{V}_{\varepsilon,\delta} \text{ satisfying} \\ \displaystyle\int_\Omega \nabla u_{\varepsilon,\delta} \nabla\phi \, dx + \int_\Omega (\varepsilon\delta)^{-\varkappa} \, g_{\varepsilon,\delta} \, u_\varepsilon \, \phi \, dx = \int_\Omega f_{\varepsilon,\delta} \, \phi \, dx, \\ \forall\phi \in \mathcal{V}_{\varepsilon,\delta}. \end{cases} \tag{10.87}$$

Introduce the spaces

$$\begin{aligned} \mathcal{V}_0 &= \{w \in H^1(\Omega) \,|\, w = 0 \text{ on } \Gamma_1\}, \\ \mathcal{V}_0^c &= \{w \in C^\infty(\overline{\Omega}) \,|\, w = 0 \text{ in some neighborhood of } \Gamma_1\}. \end{aligned} \tag{10.88}$$

Observe that the latter is a dense subspace of the former.

Remark 10.40. *The spaces $\mathcal{V}_{\varepsilon,\delta}$ are all closed subspaces of $H^1(\Omega)$ and more precisely, of $\mathcal{V}_0$. The Poincaré inequality holds for $\mathcal{V}_0$, hence uniformly for the spaces $\mathcal{V}_{\varepsilon,\delta}$.*

Remark 10.41. *The presence of the density $(\varepsilon\delta)^{-\varkappa} g_{\varepsilon,\delta}$ means that equation (10.86) is asymptotically singular near the boundary Γ_3 within the set $B_{\varepsilon,\delta}$.*

10.6.2 A priori estimates

The estimates are established assuming that ε and δ are independent small parameters.

Proposition 10.42. *There is a constant C independent of ε and δ such that*

$$\|u_{\varepsilon,\delta}\|_{H^1(\Omega)} + (\varepsilon\delta)^{-\varkappa/2}\|g_{\varepsilon,\delta}^{1/2}\, u_{\varepsilon,\delta}\|_{L^2(B_{\varepsilon,\delta})} \leqslant C\Big(\|f\|_{L^2(\Omega)} + \min\big\{(\varepsilon\delta)^{\varkappa/2}\|g_{\varepsilon,\delta}^{-1/2}\widetilde{f}_{\varepsilon,\delta}\|_{L^2(B_{\varepsilon,\delta})},\, \varepsilon\delta\|\widetilde{f}_{\varepsilon,\delta}\|_{L^2(B_{\varepsilon,\delta})}\big\}\Big).$$

Proof. Using $u_{\varepsilon,\delta}$ as a test function in (10.87), we obtain

$$\int_\Omega |\nabla u_{\varepsilon,\delta}|^2\, dx + (\varepsilon\delta)^{-\varkappa}\int_\Omega g_{\varepsilon,\delta}\, u_{\varepsilon,\delta}^2\, dx = \int_{B_{\varepsilon,\delta}} \widetilde{f}_{\varepsilon,\delta}\, u_{\varepsilon,\delta}\, dx + \int_{\Omega\setminus B_{\varepsilon,\delta}} f\, u_{\varepsilon,\delta}\, dx.$$

By the Poincaré inequality for $\mathcal{V}_0$ and the standard Young inequality,

$$\frac{1}{2}\int_\Omega |\nabla u_{\varepsilon,\delta}|^2 dx + (\varepsilon\delta)^{-\varkappa}\int_\Omega g_{\varepsilon,\delta}\, u_{\varepsilon,\delta}^2\, dx \leqslant \int_{B_{\varepsilon,\delta}} |\widetilde{f}_{\varepsilon,\delta}\, u_{\varepsilon,\delta}|\, dx + C\|f\|_{L^2(\Omega)}^2. \tag{10.89}$$

The Cauchy-Schwarz and the Young inequalities, give

$$\int_{B_{\varepsilon,\delta}} |\widetilde{f}_{\varepsilon,\delta}|\, |u_{\varepsilon,\delta}|\, dx \leqslant \|g_{\varepsilon,\delta}^{-1/2}\, \widetilde{f}_{\varepsilon,\delta}\|_{L^2(B_{\varepsilon,\delta})}\, \|g_{\varepsilon,\delta}^{1/2}\, u_{\varepsilon,\delta}\|_{L^2(B_{\varepsilon,\delta})},$$

and

$$\int_\Omega |\nabla u_{\varepsilon,\delta}|^2\, dx + (\varepsilon\delta)^{-\varkappa}\int_\Omega g_{\varepsilon,\delta}\, u_{\varepsilon,\delta}^2\, dx \leqslant C\Big((\varepsilon\delta)^{\varkappa}\|g_{\varepsilon,\delta}^{-1/2}\, \widetilde{f}_{\varepsilon,\delta}\|_{L^2(B_{\varepsilon,\delta})}^2 + \|f\|_{L^2(\Omega)}^2\Big). \tag{10.90}$$

On the other hand, by scaling the Poincaré inequality in the set B for functions vanishing on $D \cap \overline{B}$, it follows that

$$\|u_{\varepsilon,\delta}\|_{L^2(B_{\varepsilon,\delta})} \leqslant C\varepsilon\delta\|\nabla u_{\varepsilon,\delta}\|_{L^2(B_{\varepsilon,\delta})}.$$

Using this estimate and the Young inequality again in (10.89), we get

$$\frac{1}{4}\int_\Omega |\nabla u_{\varepsilon,\delta}|^2\, dx + (\varepsilon\delta)^{-\varkappa}\int_\Omega g_{\varepsilon,\delta}\, u_{\varepsilon,\delta}^2\, dx \leqslant C\Big(\varepsilon\delta\, \|\widetilde{f}_{\varepsilon,\delta}\|_{L^2(B_{\varepsilon,\delta})}^2 + \|f\|_{L^2(\Omega)}^2\Big). \tag{10.91}$$

The claim is then obtained by combining (10.90) and (10.91). □

Corollary 10.43. *If*

$$\min\left\{(\varepsilon\delta)^{\varkappa/2}\|g_{\varepsilon,\delta}^{-1/2}\,\widetilde{f}_{\varepsilon,\delta}\|_{L^2(B_{\varepsilon,\delta})},\ \varepsilon\delta\|\widetilde{f}_{\varepsilon,\delta}\|_{L^2(B_{\varepsilon,\delta})}\right\} \leqslant C,$$

then $u_{\varepsilon,\delta}$ is bounded in $H^1(\Omega)$. Furthermore,

$$\|g_{\varepsilon,\delta}^{1/2}\,u_{\varepsilon,\delta}\|_{L^2(B_{\varepsilon,\delta})} \leqslant C(\varepsilon\delta)^{\varkappa/2}.$$

10.6.3 The boundary-layer unfolding operator

The setting is similar to that of Section 9.3, with the modification due to the fact that the layer in on one side of the boundary only (and not an internal boundary).

Recall the notation $x' \doteq (x_1,\ldots,x_{N-1})$. We use the periodicity cells

$$Y' \doteq (-1/2,1/2)^{N-1}, \qquad Y \doteq Y' \times (0,1)^{(9)},$$

and define the layer Σ_ε as

$$\Sigma_\varepsilon = \Omega \cap \{x \mid 0 < x_N < \varepsilon\}.$$

For a.e. y' in $\mathbb{R}^{N-1}$, one writes

$$y' = [y']_{Y'} + \{y'\}_{Y'}, \qquad [y']_{Y'} \in \mathbb{Z}^{N-1},\ \ \{y'\}_{Y'} \in Y'.$$

From now on, when referring to a point $(x',0)$ in Γ_3, we often drop the last coordinate and just write x'.

For simplicity, we assume that Γ_3 is the exact union of $\varepsilon Y'$-cells, so that, with the notations of Section 9.3 (see also (1.3)),

$$\Lambda_\varepsilon = \varnothing \quad\text{and}\quad \widehat{\Sigma}_\varepsilon \equiv \Sigma_\varepsilon.$$

The general case of Γ_3 with Lipschitz boundary can actually be handled as in Section 9.3.

In this situation, Definitions 9.11-9.12 become

Definition 10.44. *For $\phi \in L^p(\Sigma_\varepsilon)$, $p \in [1,+\infty)$, the unfolding operator $\mathcal{T}^{bl}_{\varepsilon,\delta} : L^p(\Sigma_\varepsilon) \to L^p(\Gamma_3 \times \mathbb{R}^N_+)$ is defined by*

$$\mathcal{T}^{bl}_{\varepsilon,\delta}(\phi)(x',z) = \begin{cases} \phi\Big(\varepsilon\Big[\dfrac{x'}{\varepsilon}\Big]_{Y'} + \varepsilon\delta z\Big) & \text{if } (x',z) \in \Gamma_3 \times \dfrac{1}{\delta}Y, \\ 0 & \text{otherwise.}\end{cases}$$

Definition 10.45. *The local average $\mathcal{M}^{bl}_\varepsilon : L^p(\Sigma_\varepsilon) \mapsto L^p(\Gamma_3)$, is defined for every ϕ in $L^p(\Sigma_\varepsilon)$, $p \in [1,+\infty)$, by*

$$\mathcal{M}^{bl}_\varepsilon(\phi)(x') = \delta^N \int_{\frac{1}{\delta}Y} \mathcal{T}^{bl}_{\varepsilon,\delta}(\phi)(x',z)\,dz = \varepsilon^{-N}\int_{\varepsilon[\frac{x'}{\varepsilon}]+\varepsilon Y} \phi(\zeta)\,d\zeta, \quad \text{for } x' \in \Gamma_3.$$

(9) Since the measure $|Y|$ of Y is 1, it will not appear explicitly in this section.

Remark 10.46. *Since elements of $L^p(\Gamma_3)$ can be considered as functions of $L^p(\Sigma_\varepsilon)$, $\mathcal{M}^{bl}_\varepsilon$ can be applied to them. With this convention,*

$$\mathcal{T}^{bl}_{\varepsilon,\delta} \circ \mathcal{M}^{bl}_\varepsilon(\phi) = \mathcal{M}^{bl}_\varepsilon(\phi) \quad \text{on } \Gamma_3.$$

Proposition 9.14 and Theorem 9.15 hold in the present setting and they read as follows:

Proposition 10.47. *Let $\{w_\varepsilon\}_\varepsilon$ be a sequence such that $w_\varepsilon \rightharpoonup w$ weakly in $H^1(\Omega)$. Then the following convergence holds:*

$$\mathcal{M}^{bl}_\varepsilon(w_\varepsilon) \to w_{|\Gamma_3} \quad \text{strongly in } L^2(\Gamma_3).$$

Theorem 10.48. (Properties of the operator $\mathcal{T}^{bl}_{\varepsilon,\delta}$)
(i) For every v, w measurable on Σ_ε,

$$\mathcal{T}^{bl}_{\varepsilon,\delta}(vw) = \mathcal{T}^{bl}_{\varepsilon,\delta}(v)\mathcal{T}^{bl}_{\varepsilon,\delta}(w).$$

(ii) For every $u \in L^1(\Sigma_\varepsilon)$,

$$\varepsilon\delta^N \int_{\Gamma_3 \times \mathbb{R}^N_+} \mathcal{T}^{bl}_{\varepsilon,\delta}(u)\, dx'dz = \int_{\Sigma_\varepsilon} u\, dx,$$

and

$$\varepsilon\delta^N \int_{\Gamma_3 \times \mathbb{R}^N_+} |\mathcal{T}^{bl}_{\varepsilon,\delta}(u)|\, dx'dz = \int_{\Sigma_\varepsilon} |u|\, dx.$$

(iii) For every $u \in L^2(\Sigma_\varepsilon)$,

$$\|\mathcal{T}^{bl}_{\varepsilon,\delta}(u)\|^2_{L^2(\Gamma_3 \times \mathbb{R}^N_+)} = \frac{1}{\varepsilon\delta^N}\|u\|^2_{L^2(\Sigma_\varepsilon)}.$$

(iv) Let u be in $H^1(\Sigma_\varepsilon)$. Then,

$$\mathcal{T}^{bl}_{\varepsilon,\delta}(\nabla u) = \frac{1}{\varepsilon\delta}\nabla_z\big(\mathcal{T}^{bl}_{\varepsilon,\delta}(u)\big) \quad \text{in } \Gamma_3 \times \frac{1}{\delta}Y.$$

(v) Let ω be an open bounded set in $\mathbb{R}^N_+$. Then the following estimates hold:

$$\|\nabla_z(\mathcal{T}^{bl}_{\varepsilon,\delta}(u))\|^2_{L^2(\Gamma_3 \times \frac{1}{\delta}Y)} \leqslant \frac{\varepsilon}{\delta^{N-2}}\|\nabla u\|^2_{L^2(\Sigma_\varepsilon)},$$
$$\|\mathcal{T}^{bl}_{\varepsilon,\delta}(u - \mathcal{M}^{bl}_\varepsilon(u))\|^2_{L^2(\Gamma_3; L^{2^*}(\mathbb{R}^N_+))} \leqslant C\frac{\varepsilon}{\delta^{N-2}}\|\nabla u\|^2_{L^2(\Sigma_\varepsilon)},$$

and

$$\|\mathcal{T}^{bl}_{\varepsilon,\delta}(u)\|^2_{L^2(\Gamma_3; L^{2^*}(\omega))} \leqslant C\frac{\varepsilon}{\delta^{N-2}}\|\nabla u\|^2_{L^2(\Sigma_\varepsilon)} + 2|\,\omega\,|^{2/2^*}\,\|u\|^2_{L^2(\Sigma_\varepsilon)},$$

where 2^* is the Sobolev exponent defined by

$$\frac{1}{2^*} \doteq \frac{1}{2} - \frac{1}{N}.$$

(vi) Assume $\dfrac{\varepsilon}{\delta^{N-2}}$ *is bounded. Let* $w_{\varepsilon,\delta}$ *be in* $H^1(\Sigma_\varepsilon)$ *such that*

$$\|\nabla w_{\varepsilon,\delta}\|_{L^2(\Sigma_\varepsilon)} \leqslant C.$$

Then, up to a subsequence, there exist two functions $W \in L^2(\Gamma_3; L^{2^*}(\mathbb{R}^N_+))$ *and* U *in* $L^2(\Gamma_3; H^1_{\mathrm{loc}}(\mathbb{R}^N_+))$ *with* $\nabla_z W$ *and* $\nabla_z U$ *in* $L^2(\Gamma_3 \times \mathbb{R}^N_+)$*, such that*

$$\begin{aligned}
&\mathcal{T}^{bl}_{\varepsilon,\delta}\big(w_{\varepsilon,\delta} - \mathcal{M}^{bl}_\varepsilon(w_{\varepsilon,\delta})\big) \rightharpoonup W \quad \text{weakly in } L^2(\Gamma_3; L^{2^*}(\mathbb{R}^N_+)),\\
&\mathcal{T}^{bl}_{\varepsilon,\delta}(w_{\varepsilon,\delta}) \rightharpoonup U \quad \text{weakly in } L^2(\Gamma_3; L^{2^*}_{loc}(\mathbb{R}^N_+)),\\
&\nabla_z\big(\mathcal{T}^{bl}_{\varepsilon,\delta}(w_{\varepsilon,\delta})\big)\,\mathbf{1}_{\frac{1}{\delta}Y} \rightharpoonup \nabla_z U \quad \text{weakly in } L^2(\Gamma_3 \times \mathbb{R}^N_+)^N.
\end{aligned}$$

Furthermore, $\nabla_z W = \nabla_z U$*,* $U - W$ *is independent of* z *and*

$$w_{\varepsilon,\delta|\Gamma_3} \to U - W \quad \text{strongly in } L^2(\Gamma_3).$$

10.6.4 The functional setting

In the study of the limit behavior of problem (10.86) as $\varepsilon, \delta \to 0$, the functional setting is similar to that of Section 10.2 with the modification for a half space due to the geometry of the boundary layer. It can be also compared to "half" of the situation of Section 10.3.

Introduce the following functional space:

$$\begin{aligned}
\mathbf{K}_D &\doteq \{\Phi \in H^1_{loc}(\mathbb{R}^N_+) \mid \nabla\Phi \in L^2(\mathbb{R}^N_+)^N \text{ and } \Phi = 0 \text{ on } D\}\\
&= \{\Phi \in \dot{H}^1(\mathbb{R}^N_+) \oplus \mathbb{R} \mid \Phi = 0 \text{ on } D\}.
\end{aligned} \tag{10.92}$$

It is isomorphic to the subspace of the Hilbert space $\dot{H}^1(\mathbb{R}^N_+)$ consisting of functions constant on D[10] so that $\|\Phi\|_{\mathbf{K}_D} \doteq \|\nabla\Phi\|_{L^2(\mathbb{R}^N_+)}$ is a Hilbert norm on $\mathbf{K}_D$. Also, the space

$$\mathbf{K}^c_D \doteq \{\Phi \in \mathbf{K}_D \cap C^\infty(\overline{\mathbb{R}^N_+}) \mid \text{ support of } \nabla\Phi \text{ is bounded}\},$$

is dense in $\mathbf{K}_D$. Let $\mathbf{K}^0_D$ be the subspace $\mathbf{K}_D \cap \dot{H}^1(\mathbb{R}^N_+)$.

The effective (homogenized) boundary condition in problem (10.86) depends crucially on the ratio between ε and δ^{N-2}. We assume that δ is a function of ε such that there exists $k_2 \in [0, +\infty]$ satisfying hypothesis (10.34).

10.6.5 Homogenization

10.6.5.1 The unfolded limit for $0 < k_2 < +\infty$.

In this subsection, we only state the main results. They can be proved in a similar way as in the previous subsections (one can also refer to [53]).

(10) The isomorphism is the map $\Phi \mapsto \Phi - \Phi(\infty)$.

We make the assumption that (10.34) holds with $0 < k_2 < +\infty$. Note that this implies the relation $\varepsilon\delta^N \sim (k_2)^2(\varepsilon\delta)^2$.

Suppose that the following conditions on the functions $g_{\varepsilon,\delta}$ and $\widetilde{f}_{\varepsilon,\delta}$ are satisfied:

H1. The functions $g_{\varepsilon,\delta}$ satisfy the estimate $\|g_{\varepsilon,\delta}\|_{L^\infty(B_{\varepsilon,\delta})} \leqslant C$ uniformly in ε, δ, and the sequence $\{\mathcal{T}^{bl}_{\varepsilon,\delta}(g_{\varepsilon,\delta})\}_{(\varepsilon,\delta)}$ converges in measure (or almost everywhere) in $\Gamma_3 \times B$ to a function g.

H2. The sequence $\{\varepsilon\delta^N \mathcal{T}^{bl}_{\varepsilon,\delta}(\widetilde{f}_{\varepsilon,\delta})\}_{(\varepsilon,\delta)}$ converges weakly in $L^2(\Gamma_3 \times B)$ to some $\widetilde{f}$.

Hypothesis **H2** implies that $\varepsilon\delta\|\widetilde{f}\|_{L^2(B_{\varepsilon,\delta})}$ is uniformly bounded, so that Corollary 10.43 applies.

Example 10.49. *A typical example of* $g_{\varepsilon,\delta}$ *and* $\widetilde{f}_{\varepsilon,\delta}$ *satisfying Hypotheses* **H1** *and* **H2** *is the case where*

$$g_{\varepsilon,\delta}(x) = g\Big(\frac{1}{\delta}\Big\{\frac{x'}{\varepsilon}\Big\}, \frac{x_N}{\varepsilon\delta}\Big), \qquad \widetilde{f}_{\varepsilon,\delta}(x) = \frac{1}{\varepsilon\delta^N}\widetilde{f}\Big(\frac{1}{\delta}\Big\{\frac{x'}{\varepsilon}\Big\}, \frac{x_N}{\varepsilon\delta}\Big),$$

with $g \in L^\infty(B)$ *and* $\widetilde{f} \in L^2(B)$. *Observe that in these cases, one has* $\mathcal{T}^{bl}_{\varepsilon,\delta}(g_{\varepsilon,\delta}) = g$ *and* $\varepsilon\delta^N\mathcal{T}^{bl}_{\varepsilon,\delta}(\widetilde{f}_{\varepsilon,\delta}) = \widetilde{f}$.

Consider first the case $\varkappa = 2$ and recall definition (10.88) of $\mathcal{V}_0$. Then

Proposition 10.50. *Let* $u_{\varepsilon,\delta}$ *be a solution of problem* (10.86). *Assume that* $\varkappa = 2$, *and that conditions* **H1** *and* **H2** *are fulfilled. Then there exists* (u_0, U) *in* $\mathcal{V}_0 \times L^2(\Gamma_3; \mathbf{K}_D)$ *with* $U(\cdot,\infty) = u_0$ *a.e. on* Γ_3, *such that*

$$\begin{aligned} & u_{\varepsilon,\delta} \rightharpoonup u_0 \quad \text{weakly in } \mathcal{V}_0,\\ & \mathcal{T}^{bl}_{\varepsilon,\delta}(u_{\varepsilon,\delta}) \rightharpoonup U \quad \text{weakly in } L^2(\Gamma_3; L^2_{loc}(\mathbb{R}^N_+)),\\ & \varepsilon\delta\mathcal{T}^{bl}_{\varepsilon,\delta}(\nabla u_{\varepsilon,\delta}) = \nabla_z\big(\mathcal{T}^{bl}_{\varepsilon,\delta}(u_{\varepsilon,\delta})\big)\mathbf{1}_{\frac{1}{\delta}Y} \rightharpoonup \nabla_z U\\ & \qquad\qquad\qquad\qquad\qquad\qquad \text{weakly in } L^2(\Gamma_3 \times \mathbb{R}^N_+)^N, \end{aligned} \tag{10.93}$$

and such that the pair (u_0, U) *is the unique solution of the unfolded problem*

$$\begin{cases} \displaystyle\int_\Omega \nabla u_0\,\nabla v\,dx + (k_2)^2\int_{\Gamma_3\times\mathbb{R}^N_+} \nabla_z U\,\nabla_z V\,dx'dz + (k_2)^2\int_{\Gamma_3\times B} g\,U\,V\,dx'dz\\ \displaystyle\qquad\qquad = \int_\Omega f\,v\,dx + \int_{\Gamma_3\times B} \widetilde{f}\,V\,dx'dz\\ \forall (v,V) \in \mathcal{V}_0 \times L^2(\Omega; \mathbf{K}_D) \text{ such that } V(\cdot,\infty) = v \text{ a.e. on } \Gamma_3. \end{cases} \tag{10.94}$$

Let $\theta'_1 \in L^\infty(\Gamma_3; \mathbf{K}_D)$ with $\theta'_1(x',\infty) \equiv 1$ be the solution of the cell problem[11]

$$\begin{cases} \displaystyle\int_{\mathbb{R}^N_+} \nabla_z \theta'_1(x', z) \nabla_z \Psi(z)\, dz = 0 \qquad \text{for a.e. } x' \in \Gamma_3, \\ \forall \Psi \in \mathbf{K}^0_D, \end{cases} \tag{10.95}$$

and $\theta'_0 \in L^2(\Gamma_3\, ;\, \mathbf{K}^0_D)$ the solution of

$$\begin{cases} \displaystyle\int_{\mathbb{R}^N_+} \nabla_z \theta'_0(x', z) \nabla_z \Psi(z)\, dz + \int_B g(x', z)\, \theta'_0(x', z)\, \Psi(z)\, dz \\ \displaystyle\qquad\qquad = \int_B \widetilde{f}(x', z)\, \Psi(z)\, dz \quad \text{for a.e. } x' \in \Gamma_3, \\ \forall \Psi \in \mathbf{K}^0_D. \end{cases} \tag{10.96}$$

According to (10.95)-(10.96), for a.e. $(x', z) \in \Gamma_3 \times \mathbb{R}^N_+$

$$U(x', z) = u_0(x')\theta'_1(x', z) + \frac{1}{(k_2)^2}\theta'_0(x', z).$$

Set,

$$\mathbf{\Theta}'_1(x') = \int_{\mathbb{R}^N_+} \nabla_z \theta'_1(x', z) \nabla_z \theta'_1(x', z)\, dz \quad \text{for a.e. } x' \in \Gamma_3 \tag{10.97}$$

which is non-negative and can be interpreted as a local capacity. We can now state the homogenization result concerning u_0.

Theorem 10.51. *Let $\varkappa = 2$. Then, under the assumptions of Proposition 10.50, the limit function $u_0 \in \mathcal{V}_0$ is the solution of the homogenized equation,*

$$\begin{cases} \displaystyle\int_\Omega \nabla u_0 \nabla \psi\, dx + (k_2)^2 \int_{\Gamma_3} \mathbf{\Theta}'_1\, u_0\, \psi\, dx' = \int_\Omega f\, \psi\, dx + \int_{\Gamma_3} F\psi\, dx', \\ \forall \psi \in \mathcal{V}_0, \end{cases}$$

where $\mathbf{\Theta}'_1$ is defined by (10.97) *and $F \in L^2(\Gamma_3)$ is given by*

$$F(x') = \int_B \widetilde{f}(x', z)\, \theta'_1(x', z)\, dz \quad \textit{for a.e. } x' \in \Gamma_3.$$

The next statement treats the case $\varkappa < 2$.

Proposition 10.52. *Let $u_{\varepsilon,\delta}$ be a solution of problem* (10.86). *Assume that $\varkappa < 2$, and that conditions* **H1** *and* **H2** *are fulfilled.*

---

[11] These cell problems have a unique solution by the Lax-Milgram theorem.

Then there exists (u_0, U) *in* $\mathcal{V}_0 \times L^2(\Gamma_3; \mathbf{K}_D)$ *with* $U(\cdot, \infty) = u_0$ *a.e. on* Γ_3*, such that* (10.93) *holds and the pair* (u_0, U) *is the unique solution of*

$$\begin{cases} \displaystyle\int_\Omega \nabla u_0 \, \nabla v \, dx + (k_2)^2 \int_{\Gamma_3 \times \mathbb{R}^N_+} \nabla_z U \, \nabla_z V \, dx' dz \\ \displaystyle\qquad\qquad = \int_\Omega f \, v \, dx + \int_{\Gamma_3 \times B} \widetilde{f} \, V \, dx' dz, \\ \forall (v, V) \in \mathcal{V}_0 \times L^2(\Gamma_3; \mathbf{K}_D) \text{ such that } V(\cdot, \infty) = v \text{ a.e.on } \Gamma_3. \end{cases}$$

Theorem 10.53. *In this case, the homogenized equation is actually the same as that of Theorem 10.11 with* $A^{hom} \equiv Id$*, but obviously, the function spaces are not the same.*

We now consider the case $\varkappa > 2$. For simplicity we assume,
H1'. There is a subset B' of B with Lipschitz boundary such that $\mathcal{T}^{bl}_{\varepsilon,\delta}(g_{\varepsilon,\delta})$ vanishes on $\Gamma_3 \times (B \backslash B')$, and

$$\mathcal{T}^{bl}_{\varepsilon,\delta}(g_{\varepsilon,\delta}) \longrightarrow g \text{ a.e. in } \Gamma_3 \times B',$$

where $g > 0$ a.e. on $\Gamma_3 \times B'$.

We introduce the notations

$$B'' = B' \cup D, \qquad D'' = (\partial B' \cup D) \backslash (\partial B' \cap D),$$

and define in the same way as $\mathbf{K}_D$ (see (10.92)) the space

$$\mathbf{K}_{B''} \doteq \Big\{ \Phi \in \dot{H}^1(\mathbb{R}^N_+) \oplus \mathbb{R} \mid \Phi = 0 \text{ on } B'' \Big\}.$$

The spaces $\mathbf{K}^c_{B''}$ and $\mathbf{K}^0_{B''}$ are defined similarly.

Proposition 10.54. *Let* $u_{\varepsilon,\delta}$ *be a solution of problem* (10.86)*. Assume that* $\varkappa > 2$*, and that conditions* **H1'** *and* **H2** *are satisfied.*

Then there exists $(u_0, U) \in \mathcal{V}_0 \times L^2(\Gamma_3; \mathbf{K}_{B''})$ *with* $U(\cdot, \infty) = u_0$ *a.e. on* Γ_3*, such that convergences* (10.93) *hold, and the pair* (u_0, U) *is the unique solution of*

$$\begin{cases} \displaystyle\int_\Omega \nabla u_0 \, \nabla v \, dx + (k_2)^2 \int_{\Gamma_3 \times \mathbb{R}^N_+} \nabla_z U \, \nabla_z V \, dx' dz \\ \displaystyle\qquad\qquad = \int_\Omega f \, v \, dx + \int_{\Gamma_3 \times B} \widetilde{f} \, V \, dx' dz, \\ \forall (v, V) \in \mathcal{V}_0 \times L^2(\Gamma_3; \mathbf{K}_{B''}) \text{ such that } V(\cdot, \infty) = v \text{ a.e. on } \Gamma_3. \end{cases}$$

Theorem 10.55. *In this case, the homogenized formulation is actually the same as that of Theorem 10.11 with* $A^{hom} \equiv Id$.

Remark 10.56. In Proposition 10.54 assumption **H1'** can be relaxed to the case where the subset B' depends on $x' \in \Gamma_3$ in a enough regular way, in which case D'' depends on x' too.

10.6.5.2 Unfolded limit for $k_2 = 0$.

In the case $k_2 = 0$, the contributions of inhomogeneities $\gamma_{\varepsilon,\delta}$ and $B_{\varepsilon,\delta}$ are asymptotically negligible, and the limit problem includes the Neumann boundary condition on Γ_3.

We make the following hypothesis:

H3. The number $\varkappa$ and the functions $g_{\varepsilon,\delta}$ and $\widetilde{f}_{\varepsilon,\delta}$ are such that, as ε and δ tend to zero, $\min\big\{(\varepsilon\delta)^{\varkappa/2}\|g_{\varepsilon,\delta}^{-1/2}\widetilde{f}_{\varepsilon,\delta}\|_{L^2(B_{\varepsilon,\delta})}\,,\,\varepsilon\delta\|\widetilde{f}_{\varepsilon,\delta}\|_{L^2(B_{\varepsilon,\delta})}\big\}$ is bounded (so that Corollary 10.43 applies).

Theorem 10.57. *Let $u_{\varepsilon,\delta}$ be a solution of problem* (10.86). *Assume that hypothesis* **H3** *is fulfilled and $k_2 = 0$. Then*

$$u_{\varepsilon,\delta} \rightharpoonup u_0 \quad \textit{weakly in } \mathcal{V}_0,$$

and u_0 is the solution of the variational problem

$$\begin{cases} \displaystyle\int_\Omega \nabla u_0 \nabla \psi \, dx = \int_\Omega f \psi \, dx, \\ \forall \psi \in \mathcal{V}_0. \end{cases} \tag{10.98}$$

Remark 10.58. *Note that for $k_2 = 0$ the value of $\varkappa$ has no influence on the structure of the limit problem. Its formulation* (10.98) *only involves the function u_0 and thus, represents the macroscopic limit problem.*

10.6.5.3 Unfolded limit for $k_2 = +\infty$.

In this case, the "spots" $\gamma_{\varepsilon,\delta}$ are large enough to ensure the homogeneous Dirichlet boundary condition on Γ_3 in the limit problem.

Theorem 10.59. *Let $u_{\varepsilon,\delta}$ be the solution of problem* (10.86). *Assume that the hypothesis* **H3** *is fulfilled, and $k_2 = +\infty$. Then*

$$u_{\varepsilon,\delta} \rightharpoonup u_0 \quad \textit{weakly in } \mathcal{V}_0,$$

and u_0 is the solution of the variational problem

$$\begin{cases} \displaystyle\int_\Omega \nabla u_0 \nabla \psi \, dx = \int_\Omega f \psi \, dx, \\ u_0 = 0 \quad \textit{on } \Gamma_3, \\ \forall \psi \in \mathcal{V}_0 \quad \textit{with } \psi = 0 \quad \textit{on } \Gamma_3. \end{cases}$$

Remark 10.60. *The assertion of Remark 10.58 for $k_2 = 0$, is the same for $k_2 = +\infty$, the value of $\varkappa$ has no influence on the structure of the limit problem as it only involves the function u_0.*

10.6.6 Convergence of the energy

In preceding section, the sequence $\{u_{\varepsilon,\delta}\}_{(\varepsilon,\delta)}$ was shown to converge weakly to u_0 in the space $\mathcal{V}_0$. Following the same procedure as in Proposition 10.8, one can prove the next theorem.

Theorem 10.61. *Under the hypotheses of Theorem 10.10, the following strong convergences hold:*

$$u_{\varepsilon,\delta} \to u_0 \quad \textit{strongly in } \mathcal{V}_0,$$
$$\mathcal{T}^{bl}_{\varepsilon,\delta}(u_{\varepsilon,\delta}) \to U \quad \textit{strongly in } L^2(\Gamma_3; L^2_{loc}(\mathbb{R}^N_+)),$$
$$(\nabla_z \mathcal{T}^{bl}_{\varepsilon,\delta}(u_{\varepsilon,\delta}))\mathbf{1}_{\frac{1}{\delta}Y} \to \nabla_z U \quad \textit{strongly in } L^2(\Gamma_3 \times \mathbb{R}^N_+)^N.$$

The limit U of the boundary layer term is in $L^2(\Gamma_3; \mathbf{K}_D)$. Due to the discontinuity of the boundary layer term $\mathcal{T}^{bl}_{\varepsilon,\delta}(u_{\varepsilon,\delta})$ at the boundary of $\frac{1}{\delta}Y$, one cannot expect a convergence in this space.

10.6.7 Complement

All the above results can be extended for the case of a second order elliptic operator with a possibly oscillating matrix $A_{\varepsilon,\delta}$. The original problem is changed to

$$\begin{cases} \text{Find } u_{\varepsilon,\delta} \in \mathcal{V}_{\varepsilon,\delta} \text{ satisfying} \\ \displaystyle\int_\Omega A_{\varepsilon,\delta} \nabla u_{\varepsilon,\delta}\, \nabla \phi dx + \int_\Omega (\varepsilon\delta)^{-\varkappa} g_{\varepsilon,\delta}\, u_{\varepsilon,\delta}\, \phi\, dx = \int_\Omega f_{\varepsilon,\delta}\, \phi\, dx, \\ \forall \phi \in \mathcal{V}_{\varepsilon,\delta}. \end{cases}$$

The extra hypotheses for $A_{\varepsilon,\delta}$ are

A1. $A_{\varepsilon,\delta}$ is positive definite and bounded uniformly in ε and δ, a.e. in Ω.

A2. $A_{\varepsilon,\delta}$ H-converges to some limit matrix A^{hom} (see [164] for the definition of H-convergence)[(12)].

A3. $\mathcal{T}^{bl}_{\varepsilon,\delta}(A_{\varepsilon,\delta})$ converges to some A^0 a.e. on $\Gamma_3 \times \mathbb{R}^N_+$.

Let us briefly describe the limit problem in the case $0 < k_2 < +\infty$ and $\varkappa = 2$. Equation (10.94) becomes

$$\begin{cases} \displaystyle\int_\Omega A^{hom} \nabla u_0\, \nabla v\, dx + (k_2)^2 \int_{\Gamma_3 \times \mathbb{R}^N_+} A^0 \nabla_z U\, \nabla_z V\, dx'dz \\ \displaystyle\qquad\qquad + k^2 \int_{\Gamma_3 \times B} g\, U\, V\, dx'dz = \int_\Omega f\, v\, dx + \int_{\Gamma_3 \times B} \widetilde{f}\, V\, dx'dz \\ \forall (v, V) \in \mathcal{V}_0 \times L^2(\Omega; \mathbf{K}_D)) \text{ such that } V(\cdot, \infty) = v \text{ a.e. on } \Gamma_3, \end{cases}$$

(12) In particular, this holds when (with the notations of Part I) $\mathcal{T}_\varepsilon(A_{\varepsilon,\delta})$ converges to B a.e. in $\Omega \times Y$, in which case A^{hom} is given by formula (3.19).

which has a unique solution $(u_0, U) \in \mathcal{V}_0 \times L^2(\Omega; \mathbf{K}_D)$ such that $U(\cdot, \infty) = u_0$ a.e. on Γ_3.

The convergence of the energy still holds (with obvious modifications) and implies the strong convergence for the boundary layer term. The strong convergence of $u_{\varepsilon,\delta} \to u_0$ in $\mathcal{V}_0$ is replaced by the standard corrector result associated with the H-convergence of the operators $A_{\varepsilon,\delta}$.

The other cases for k_2 and $\varkappa$ are modified accordingly.

Part V

Linear Elasticity

In this part, we investigate the interaction of homogenization and domain reduction, involving two small parameters in a plate-like domain with thickness δ and periodicity ε in its large dimensions (the case of a beam-like domain is treated in [131]).

Such a study covers two aspects: the homogenization ($\varepsilon \to 0$) and the reduction of dimension ($\delta \to 0$). For such problems, usually one studies the cases δ goes to 0 before ε, and vice-versa. In general the two limit problems are different (see e.g. [44, 110]). The purpose of Chapter 11 is to explore the general situation where both parameters go to 0 independently.

The reduction of dimension is done by the standard scaling to a fixed thickness (see the pioneer papers [44, 94]). In this setting, an appropriate tool was introduced in [117, 118] and developed in [121, 123, 124], decomposing a general displacement into the sum of an elementary displacement which is affine in the transverse dimension(s), plus a residual part. Together with the unfolding operator, one proceed to the reduction of dimension and homogenization processes.

The decomposition of the displacements in thin structures has been also used in [22–24, 127–130] in particular for the homogenization of the junction of rods and $3D$ domains or plates.

In nonlinear elasticity for deformations of beams or shells and for structures made of beams, similar decompositions are considered in [25–27]. These decompositions are applied to obtain the asymptotic behavior of structures made of a plate and a beam [28, 29].

Chapter 11

Homogenization of an elastic thin plate

This chapter concerns the modeling of a thin elastic plate $\Omega_\delta = \omega \times (-\delta, \delta)$. Its thickness is 2δ, $(\delta \in (0,1))$, and its mid-surface ω is a bounded domain in $\mathbb{R}^2$ with Lipschitz boundary. The plate is made of mix of materials periodically distributed in cells isometric to the domain $(0, \varepsilon)^2 \times (-\delta, \delta)$. Both parameters ε and δ are assumed to go to zero.

Supposing that the limit

$$\theta = \lim_{(\varepsilon,\delta)\to(0,0)} \frac{\varepsilon}{\delta} \in [0, +\infty]$$

exists (which is always true, at least for a subsequence), a limit problem is obtained. Varying θ, this gives rise to a continuum of limit problems. It is also shown that the limit problem corresponding to $\theta = 0$ is the same as the one obtained when homogenization is carried out first ($\varepsilon \to 0$) followed by dimension reduction ($\delta \to 0$), while the limit problem for $\theta = +\infty$ is the same as the one obtained when dimension reduction is performed before homogenization. These two limits are usually different.

Section 11.1 introduces the plate and recalls results concerning the decomposition of plate displacements. A displacement u is decomposed into the sum of an elementary displacement U_e, affine along the fibers $\{x'\} \times (-\delta, \delta)$ $(x' \in \omega)$ and a residual part $\overline{u}$. The elementary displacement represents the displacement of the mid-surface ω and the small rotations of the fibers (see Definition 11.2). The residual part is the warping: it takes into account the deformations of the fibers. A Korn inequality adapted to the decomposition is established.

Section 11.2 presents the general hypotheses for the study of the linear elasticity problem in this setting.

Section 11.3 introduces the classical unfolding operator $\mathcal{T}_\varepsilon$ in ω (associated with the unit cell $Y' \doteq (0,1)^2$), and a rescaling operator $\mathfrak{T}_\delta$ in the direction

---

The original version of this chapter was revised: The proof for Lemma 11.11 has been included. The correction to this chapter is available at https://doi.org/10.1007/978-981-13-3032-2_15

D. Cioranescu et al., *The Periodic Unfolding Method*, Series in Contemporary Mathematics 3, https://doi.org/10.1007/978-981-13-3032-2_11

of the thickness of the plate, the latter in order to work in a fixed domain of thickness 2. Rewriting the estimates obtained in the preceding section highlights the role of the limit θ.

Section 11.4 is dedicated to an "abstract" unfolding result with two parameters which is used in the sequel.

In Section 11.5, Proposition 11.13 gives the asymptotic behavior of the rescaled strain tensor in $\Omega = \omega \times (-1,1)$ and Proposition 11.14 gives the same for the rescaled and unfolded tensor in $\omega \times Y$ with respect to θ, where $Y = Y' \times (-1,1)$.

Theorem 11.20 in Section 11.6 presents the unfolded and rescaled limit elasticity problem.

Section 11.7 gives the homogenized limit via the mid-surface displacement as the solution of a problem posed in ω and depending on θ.

Finally, Section 11.8 considers the two problems where dimension reduction is performed before homogenization and vice-versa, then proves the continuity of the limit mid-surface displacement with respect to θ in $[0,+\infty]$.

11.1 Geometry and preliminary results

11.1.1 The plate and notations

The set $\{\mathbf{e}_1, \mathbf{e}_2, \mathbf{e}_3\}$ is the usual orthonormal basis for $\mathbb{R}^3$. Let ω be a bounded domain in $\mathbb{R}^2$ with Lipschitz boundary, and let γ_0 be a subset of $\partial\omega$ with non null measure. For $\delta \in (0,1)$, the plate is the domain

$$\Omega_\delta \doteq \omega \times (-\delta, \delta).$$

It is clamped (i.e. with zero displacement) on the part $\Gamma_{0,\delta}$ of its lateral boundary

$$\Gamma_{0,\delta} \doteq \gamma_0 \times (-\delta, \delta),$$

The displacement of the clamped plate belongs to the space $\mathbb{D}_\delta$ defined below. We will also use the space $\mathbb{D}_M$:

$$\begin{aligned} \mathbb{D}_\delta &\doteq \{u \in H^1(\Omega_\delta)^3 \mid u = 0 \text{ on } \Gamma_{0,\delta}\}, \\ \mathbb{D}_M &\doteq H^1_{\gamma_0}(\omega) \times H^1_{\gamma_0}(\omega) \times H^2_{\gamma_0}(\omega). \end{aligned} \tag{11.1}$$

Here, $H^1_{\gamma_0}(\omega)$ is the same as the space $H^1_0(\omega;\gamma_0)$ with the notations of Chapter 4, and $H^2_{\gamma_0}(\omega)$ is the subspace of $H^2(\omega)$ of functions which vanish on γ_0 as well as their first derivatives (equivalently, their normal derivative).

Notation 11.1.
1. In this chapter the latin indices i, j, k, l belong to $\{1,2,3\}$ while the greek indices α, β, α', β' belong to $\{1,2\}$. We also use the Einstein summation convention on repeated indices (unless indicated otherwise).
2. The mid-surface component of every $x = (x_1, x_2, x_3)$ in Ω_δ is denoted $x' \doteq (x_1, x_2) \in \omega$.

3. The calligraphic letters (e.g. $\mathcal{U}$, $\mathcal{R}$, ...) are used for functions or fields defined on ω.
4. Fields depending on the variable x_3 (resp. y_3)[(1)] *which satisfy the three vanishing average conditions ($i = 1, 2, 3$)*

$$\int_{-\delta}^{\delta} \mathring{u}_i(\cdot, x_3)\, dx_3 = 0 \quad \Big(\textit{resp.} \int_{-1}^{1} \mathring{\psi}_i(\cdot, y_3)\, dy_3 = 0\Big) \tag{11.2}$$

are indicated with the modifier $^\circ$ (e.g., $\mathring{u}$, $\mathring{\psi}$ ...).
5. Similarly, fields satisfying (11.2) as well as the two extra vanishing moment conditions ($\alpha = 1, 2$)

$$\int_{-\delta}^{\delta} x_3\, \overline{u}_\alpha(\cdot, x_3)\, dx_3 = 0 \quad (\textit{resp.} \int_{-1}^{1} y_3\, \overline{\psi}_\alpha(\cdot, y_3)\, dy_3 = 0),$$

are indicated with an overbar (e.g. $\overline{u}$, $\overline{\psi}$ ...).
6. Recall the strain tensor $(e_{ij})_{1 \leqslant i,j \leqslant 3}$, defined for every u in $H^1(\Omega_\delta)^3$:

$$e_{ij}(u) \doteq \frac{1}{2}\Big(\frac{\partial u_j}{\partial x_i} + \frac{\partial u_i}{\partial x_j}\Big), \qquad e(u) \doteq \frac{1}{2}\big(\nabla u + (\nabla u)^T\big).$$

Its kernel consists of the set of rigid displacements (in $\mathbb{R}^3$), defined as

$$\{R : x \mapsto \mathbf{a} + \mathbf{b} \wedge x \mid \mathbf{a}, \mathbf{b} \in \mathbb{R}^3\}.$$

11.1.2 A decomposition for the plate displacements

Definition 11.2 (elementary and residual displacements). *Elementary displacement are elements u_e of $H^1(\Omega_\delta)^3$ satisfying for a.e. $x = (x', x_3) \in \Omega_\delta$ (where $x' \in \omega$)*

$$\begin{aligned} u_{e,1}(x) &= \mathcal{U}_1(x') + x_3 \mathcal{R}_1(x'), \\ u_{e,2}(x) &= \mathcal{U}_2(x') + x_3 \mathcal{R}_2(x'), \\ u_{e,3}(x) &= \mathcal{U}_3(x'). \end{aligned}$$

Here

$$\mathcal{U} = (\mathcal{U}_1, \mathcal{U}_2, \mathcal{U}_3) \in H^1(\omega)^3 \quad \textit{and} \quad \mathcal{R} = \mathcal{R}_1 \mathbf{e}_1 + \mathcal{R}_2 \mathbf{e}_2 \in H^1(\omega)^2.$$

Elementary displacements are a generalization of the notion of Kirchhoff-Love displacement[(2)]. *The first part $\mathcal{U}(x')$ of the elementary displacement is*

---

(1) Simultaneous dependence upon both x_3 and y_3 never occurs.
(2) A Kirchhoff-Love displacement is a displacement of the form

$$u_e(x) = \Big(\mathcal{U}_\alpha(x') - x_3 \frac{\partial \mathcal{U}_3}{\partial x_1}(x')\Big)\mathbf{e}_\alpha + \mathcal{U}_3(x')\mathbf{e}_3, \quad \text{for a.e. } x = (x', x_3) \in \Omega_\delta.$$

the mid-surface displacement at the point $x' \in \omega$, while $x_3\mathcal{R}(x')$ represents the small linearized rotation[(3)] *of the fiber $\{x'\} \times (-\delta, \delta)$.*

Residual displacements are elements $\overline{u}$ of $H^1(\Omega_\delta)^3$ which satisfy (see Notation 11.1-5) the five conditions

$$\int_{-\delta}^{\delta} \overline{u}_i(x', x_3)\, dx_3 = \int_{-\delta}^{\delta} x_3\, \overline{u}_\alpha(x', x_3)\, dx_3 = 0 \quad \text{for a.e. } x' \in \omega. \tag{11.3}$$

Definition 11.3 (Decomposition of a displacement). *Every displacement u in $H^1(\Omega_\delta)^3$ is* uniquely *decomposed as the sum*

$$u = u_e + \overline{u} \tag{11.4}$$

of an elementary displacement u_e and a residual displacement $\overline{u} \in H^1(\Omega_\delta)^3$. Due to the properties of the residual part, the components $\mathcal{U}$ and $\mathcal{R}$ of u_e are given for a.e. $x' \in \omega$, by

$$\begin{aligned} \mathcal{U}(x') &= \frac{1}{2\delta}\int_{-\delta}^{\delta} u(x', x_3)\, dx_3, \\ \mathcal{R}(x') &= \frac{3}{2\delta^3}\int_{-\delta}^{\delta} x_3\big(u_1(x', x_3)\mathbf{e}_1 + u_2(x', x_3)\mathbf{e}_2\big)\, dx_3. \end{aligned} \tag{11.5}$$

The sum

$$\mathcal{U}_m = \mathcal{U}_1\, \mathbf{e}_1 + \mathcal{U}_2\, \mathbf{e}_2,$$

is the membrane displacement, while $\mathcal{U}_3$ represents the bending of the mid-surface. The residual part $\overline{u}$ is the warping; it stands for the deformation of the fibers $\{x'\} \times (-\delta, \delta)$.

With the above notations, the explicit expressions of the components of the strain tensor of u are

$$\begin{aligned} e_{11}(u) &= \frac{\partial \mathcal{U}_1}{\partial x_1} + x_3\frac{\partial \mathcal{R}_1}{\partial x_1} + \frac{\partial \overline{u}_1}{\partial x_1}, \quad e_{22}(u) = \frac{\partial \mathcal{U}_2}{\partial x_2} + x_3\frac{\partial \mathcal{R}_2}{\partial x_2} + \frac{\partial \overline{u}_2}{\partial x_2}, \\ e_{12}(u) &= \frac{1}{2}\Big(\Big[\frac{\partial \mathcal{U}_1}{\partial x_2} + \frac{\partial \mathcal{U}_2}{\partial x_1}\Big] + x_3\Big[\frac{\partial \mathcal{R}_1}{\partial x_2} + \frac{\partial \mathcal{R}_2}{\partial x_1}\Big] + \Big[\frac{\partial \overline{u}_1}{\partial x_2} + \frac{\partial \overline{u}_2}{\partial x_1}\Big]\Big), \\ e_{13}(u) &= \frac{1}{2}\Big(\Big[\frac{\partial \mathcal{U}_3}{\partial x_1} + \mathcal{R}_1\Big] + \Big[\frac{\partial \overline{u}_3}{\partial x_1} + \frac{\partial \overline{u}_1}{\partial x_3}\Big]\Big), \\ e_{23}(u) &= \frac{1}{2}\Big(\Big[\frac{\partial \mathcal{U}_3}{\partial x_2} + \mathcal{R}_2\Big] + \Big[\frac{\partial \overline{u}_3}{\partial x_2} + \frac{\partial \overline{u}_2}{\partial x_3}\Big]\Big), \\ e_{33}(u) &= \frac{\partial \overline{u}_3}{\partial x_3}. \end{aligned} \tag{11.6}$$

Theorem 11.4, stated below, was proved in [124].

(3) The axis of this small rotation is directed by the vector $\mathcal{R}_2(x')\mathbf{e}_1 - \mathcal{R}_1(x')\mathbf{e}_2$ and its angle is approximately equal to the euclidean norm of this vector. Here, in the framework of small displacements, the symmetric part of the rotation is neglected.

Theorem 11.4. *Let u be a displacement in $H^1(\Omega_\delta)^3$ and $(\mathcal{U}, \mathcal{R}, \overline{u})$ be its decomposition given by Definition 11.3. Then,*

$$\begin{aligned}
&\delta \big\| e_{\alpha\beta}(\mathcal{R}) \big\|_{L^2(\omega)} + \big\| e_{\alpha\beta}(\mathcal{U}) \big\|_{L^2(\omega)} \leqslant \frac{C}{\delta^{1/2}} \| e(u) \|_{L^2(\Omega_\delta)},\\
&\Big\| \frac{\partial \mathcal{U}_3}{\partial x_\alpha} + \mathcal{R}_\alpha \Big\|_{L^2(\omega)} \leqslant \frac{C}{\delta^{1/2}} \| e(u) \|_{L^2(\Omega_\delta)},
\end{aligned} \tag{11.7}$$

$$\begin{aligned}
&\| \overline{u} \|_{L^2(\Omega_\delta)} \leqslant C\,\delta\, \| e(u) \|_{L^2(\Omega_\delta)},\\
&\| \nabla \overline{u} \|_{L^2(\Omega_\delta)} \leqslant C \| e(u) \|_{L^2(\Omega_\delta)},
\end{aligned} \tag{11.8}$$

with a constant C independent of δ.

11.1.3 Rescaling of the plate and Korn type inequalities

The reference plate is the domain

$$\Omega_1 \doteq \omega \times (-1, 1),$$

denoted Ω for simplicity.

We rescale Ω_δ to Ω in the vertical direction.

Definition 11.5 (The rescaling operator $\mathfrak{T}_\delta$)**.** *For every measurable function Ψ over Ω_δ, we define the measurable function $\mathfrak{T}_\delta(\Psi)$ over Ω by*

$$\mathfrak{T}_\delta(\Psi)(x', y_3) = \Psi(x', \delta y_3) \quad \textit{for a.e.} \quad (x', y_3) \in \Omega.$$

For every $\Psi \in L^2(\Omega_\delta)$, $\mathfrak{T}_\delta(\Psi)$ satisfies

$$\| \mathfrak{T}_\delta(\Psi) \|_{L^2(\Omega)} = \frac{1}{\sqrt{\delta}} \| \Psi \|_{L^2(\Omega_\delta)}.$$

The next proposition gives a Korn-type inequality for displacements in the space $H^1(\Omega_\delta)^3$.

Proposition 11.6. *For every u in $H^1(\Omega_\delta)^3$, there exists a rigid displacement R in $\mathbb{R}^3$ such that $v = u - R$ satisfies*

$$\begin{aligned}
&\| v_\alpha \|_{L^2(\Omega_\delta)} + \delta \| v_3 \|_{L^2(\Omega_\delta)} \leqslant C \| e(u) \|_{L^2(\Omega_\delta)},\\
&\sum_{\alpha,\beta=1}^{2} \Big\| \frac{\partial v_\alpha}{\partial x_\beta} \Big\|_{L^2(\Omega_\delta)} + \Big\| \frac{\partial v_3}{\partial x_3} \Big\|_{L^2(\Omega_\delta)} \leqslant C \| e(u) \|_{L^2(\Omega_\delta)},\\
&\sum_{\alpha=1}^{2} \Big(\Big\| \frac{\partial v_\alpha}{\partial x_3} \Big\|_{L^2(\Omega_\delta)} + \Big\| \frac{\partial v_3}{\partial x_\alpha} \Big\|_{L^2(\Omega_\delta)} \Big) \leqslant \frac{C}{\delta} \| e(u) \|_{L^2(\Omega_\delta)}.
\end{aligned} \tag{11.9}$$

The constant does not depend on δ.

Proof. The displacement u is decomposed as (11.4). Below we show that there exists a rigid displacement R such that $v_e = u_e - R$ satisfies the estimates (11.9). Combining with (11.8), this prove the proposition.

Consider the displacement $U_e = \mathfrak{T}_\delta(u_e)$ on Ω. For a.e. $(x', y_3) \in \Omega$,

$$\begin{aligned} U_{e,1}(x', y_3) &= \mathcal{U}_1(x') + y_3\, \delta\, \mathcal{R}_1(x'), \\ U_{e,2}(x', y_3) &= \mathcal{U}_2(x') + y_3\, \delta\, \mathcal{R}_2(x'), \\ U_{e,3}(x', y_3) &= \mathcal{U}_3(x'). \end{aligned}$$

The displacement $\widetilde{U}_e \doteq U_{e,\alpha}\mathbf{e}_\alpha + \delta U_{e,3}\mathbf{e}_3$ belongs to $H^1(\Omega)^3$ and its strain tensor $e_{x',y_3}(\widetilde{U}_e)(\cdot, y_3)$ is given a.e. in Ω by

$$\begin{pmatrix} e_{11}(\mathcal{U}) + y_3\, \delta e_{11}(\mathcal{R}) & e_{12}(\mathcal{U}) + y_3\, \delta e_{12}(\mathcal{R}) & \dfrac{\delta}{2}\Big(\dfrac{\partial \mathcal{U}_3}{\partial x_1} + \mathcal{R}_1\Big) \\ e_{12}(\mathcal{U}) + y_3\, \delta e_{12}(\mathcal{R}) & e_{22}(\mathcal{U}) + y_3\, \delta e_{11}(\mathcal{R}) & \dfrac{\delta}{2}\Big(\dfrac{\partial \mathcal{U}_3}{\partial x_2} + \mathcal{R}_2\Big) \\ \dfrac{\delta}{2}\Big(\dfrac{\partial \mathcal{U}_3}{\partial x_1} + \mathcal{R}_1\Big) & \dfrac{\delta}{2}\Big(\dfrac{\partial \mathcal{U}_3}{\partial x_2} + \mathcal{R}_2\Big) & 0 \end{pmatrix}.$$

Estimates (11.7) give

$$\|e_{x',y_3}(\widetilde{U}_e)\|_{L^2(\Omega)} \leqslant \frac{C}{\delta^{1/2}} \|e(u)\|_{L^2(\Omega_\delta)}.$$

Using the first Korn inequality (see (2.12) in Proposition 2.11), there exist a rigid displacement $\widetilde{R}$ and a constant C which only depends on Ω

$$\widetilde{R}(x', y_3) \doteq \begin{pmatrix} a_1 - b_3 x_2 + \delta b_2 y_3 \\ a_2 + b_3 x_1 - \delta b_1 y_3 \\ \delta a_3 + \delta b_1 x_2 - \delta b_2 x_1 \end{pmatrix} \qquad \text{for } (x', y_3) \in \Omega,$$

such that,

$$\|\widetilde{U}_e - \widetilde{R}\|_{H^1(\Omega)} \leqslant C \|e_{x',y_3}(\widetilde{U}_e)\|_{L^2(\Omega)} \leqslant \frac{C}{\delta^{1/2}} \|e(u)\|_{L^2(\Omega_\delta)}.$$

After a simple calculation, this reduces to

$$\begin{aligned} &\big\|\mathcal{U}_1 - a_1 + b_3 x_2\big\|_{H^1(\omega)} + \delta \big\|\mathcal{R}_1 - b_2\big\|_{H^1(\omega)} \leqslant \frac{C}{\delta^{1/2}} \|e(u)\|_{L^2(\Omega_\delta)}, \\ &\big\|\mathcal{U}_2 - a_2 - b_3 x_1\big\|_{H^1(\omega)} + \delta \big\|\mathcal{R}_2 + b_1\big\|_{H^1(\omega)} \leqslant \frac{C}{\delta^{1/2}} \|e(u)\|_{L^2(\Omega_\delta)}, \\ &\delta \big\|\mathcal{U}_3 - a_3 - b_1 x_2 + b_2 x_1\big\|_{H^1(\omega)} \leqslant \frac{C}{\delta^{1/2}} \|e(u)\|_{L^2(\Omega_\delta)}. \end{aligned}$$

Set $R(x) = \mathbf{a} + \mathbf{b} \wedge x$, for all $x \in \Omega_\delta$, where $\mathbf{a} = a_i \mathbf{e}_i$ and $\mathbf{b} = b_i \mathbf{e}_i$. Then, as an immediate consequence, the displacement $v_e = u_e - R$ satisfies (11.9). ☐

For u in $\mathbb{D}_\delta$ (i.e., vanishing on $\Gamma_{0,\delta}$), the formulas for its components $\mathcal{U}$, $\mathcal{R}$ and $\overline{u}$ ((11.3)-(11.5)) show that

$$\mathcal{U} = 0, \ \ \mathcal{R} = 0 \quad \text{a.e. on } \gamma_0 \qquad \text{and} \qquad \overline{u} = 0 \quad \text{a.e. on } \Gamma_{0,\delta}.$$

Proposition 11.7. *For every u in $\mathbb{D}_\delta$ decomposed as* (11.4), *one has*

$$\begin{aligned}
&\|\mathcal{U}_\alpha\|_{H^1(\omega)} + \delta\big(\|\mathcal{U}_3\|_{H^1(\omega)} + \|\mathcal{R}\|_{H^1(\omega)}\big) \leqslant \frac{C}{\delta^{1/2}}\|e(u)\|_{L^2(\Omega_\delta)},\\
&\|u_\alpha\|_{L^2(\Omega_\delta)} + \delta\|u_3\|_{L^2(\Omega_\delta)} \leqslant C\|e(u)\|_{L^2(\Omega_\delta)},\\
&\sum_{\alpha,\beta=1}^{2}\Big\|\frac{\partial u_\alpha}{\partial x_\beta}\Big\|_{L^2(\Omega_\delta)} + \Big\|\frac{\partial u_3}{\partial x_3}\Big\|_{L^2(\Omega_\delta)} \leqslant C\|e(u)\|_{L^2(\Omega_\delta)},\\
&\sum_{\alpha=1}^{2}\Big(\Big\|\frac{\partial u_\alpha}{\partial x_3}\Big\|_{L^2(\Omega_\delta)} + \Big\|\frac{\partial u_3}{\partial x_\alpha}\Big\|_{L^2(\Omega_\delta)}\Big) \leqslant \frac{C}{\delta}\|e(u)\|_{L^2(\Omega_\delta)}.
\end{aligned} \tag{11.10}$$

The constant does not depend on δ.

Proof. Since $\mathcal{U} = 0$, $\mathcal{R} = 0$ a.e. on γ_0, one has $\widetilde{U}_e = 0$ on $\gamma_0 \times (-1,1)$, where $\widetilde{U}_e$ has been introduced in the proof of Proposition 11.6. Then the second Korn inequality (see (2.13) in Proposition 2.11) gives

$$\|\widetilde{U}_e\|_{H^1(\Omega)} \leqslant C\|e_{x',y_3}(\widetilde{U}_e)\|_{L^2(\Omega)} \leqslant \frac{C}{\delta^{1/2}}\|e(u)\|_{L^2(\Omega_\delta)}.$$

The constant only depends on Ω and Γ_0. Since

$$\begin{aligned}
\|\widetilde{U}_{e,\alpha}\|^2_{H^1(\Omega)} &= 2\|\mathcal{U}_\alpha\|^2_{H^1(\omega)} + \frac{2\delta^2}{3}\|\mathcal{R}_\alpha\|_{H^1(\omega)},\\
\|\widetilde{U}_{e,3}\|^2_{H^1(\Omega)} &= 2\delta^2\|\mathcal{U}_3\|^2_{H^1(\omega)},
\end{aligned}$$

inequality $(11.10)_1$ follows. Estimates $(11.10)_{2,3,4}$ are a consequence, taking into account (11.8). □

11.2 The linear elasticity problem

11.2.1 Hypotheses on the Hook tensor field

Set

$$Y' \doteq (0,1)^2, \qquad Y \doteq Y' \times (-1,1).$$

For a.e. $x' \in \mathbb{R}^2$, one has (cf. (1.1); for simplicity, we omit the subscripts Y')

$$x' = [x'] + \{x'\}, \qquad [x'] \in \mathbb{Z}^2, \quad \{x'\} \in Y'.$$

From now on, we assume that the plate Ω_δ is made up of a mix of various linear elastic materials.

For $1 \leqslant i, j, k, l \leqslant 3$, let a_{ijkl} be in $L^\infty(Y)$ and satisfy the symmetry conditions

$$a_{ijkl}(y) = a_{jikl}(y) = a_{klij}(y) \qquad \text{for a.e. } y \in Y,$$

as well as the coercivity condition

$$a_{ijkl}(y)\tau_{ij}\tau_{kl} \geqslant c_0\, \tau_{ij}\tau_{ij} \qquad \text{for a.e. } y \in Y \tag{11.11}$$

for every symmetric matrix $\tau = (\tau_{ij})$ of order 3 (c_0 is a given strictly positive number).

The coefficients $a^{\varepsilon,\delta}_{ijkl}$ of the Hooke tensor are then given by

$$a^{\varepsilon,\delta}_{ijkl}(x) = a_{ijkl}\Big(\Big\{\frac{x'}{\varepsilon}\Big\}, \frac{x_3}{\delta}\Big) \quad \text{for a.e. } x \in \Omega_\delta.$$

The constitutive law of the materials is the classical relation between the strain tensor and the stress tensor,

$$\sigma^{\varepsilon,\delta}_{ij}(v) = a^{\varepsilon,\delta}_{ijkl}\, e_{kl}(v), \qquad \forall v \in \mathbb{D}_\delta. \tag{11.12}$$

For given applied forces f_δ (to simplify the presentation, we consider only body forces), the displacement $u_{\varepsilon,\delta} \in \mathbb{D}_\delta$ of the plate is the solution of the following elasticity problem:

$$\begin{cases} \displaystyle\int_{\Omega_\delta} \sigma^{\varepsilon,\delta}_{ij}(u_{\varepsilon,\delta})\, e_{ij}(v)\, dx = \int_{\Omega_\delta} f_\delta(x)\cdot v(x)\, dx \\ \forall v \in \mathbb{D}_\delta. \end{cases} \tag{11.13}$$

11.2.2 Assumptions on the forces and a priori estimate

We make the assumption that the applied body forces f_δ are of the form

$$f_\delta(x) = \big(\delta\, f_\alpha(x') + x_3 g_\alpha(x')\big)\mathbf{e}_\alpha + \delta^2 f_3(x')\mathbf{e}_3 \quad \text{for a.e. } x \in \Omega_\delta,$$

where $f = (f_1, f_2, f_3)$ is in $L^2(\omega)^3$ and $g = (g_1, g_2)$ in $L^2(\omega)^2$.

This allows to obtain an a priori estimate as follows. Using the decomposition (11.4) for a $u \in \mathbb{D}_\delta$, one can write

$$\frac{1}{2}\int_{\Omega_\delta} f_\delta \cdot u\, dx = \delta^2 \int_\omega f_\alpha \mathcal{U}_\alpha\, dx' + \frac{\delta^3}{3}\int_\omega g_\alpha \mathcal{R}_\alpha\, dx' + \delta^3 \int_\omega f_3\, \mathcal{U}_3\, dx',$$

where the warping $\overline{u}$ does not appear (in view of the conditions (11.3) which it satisfies).

From estimates $(11.10)_{1,2}$, one deduces an upper bound for this integral:

$$\Big|\int_{\Omega_\delta} f_\delta(x)\cdot u(x)\, dx\Big| \leqslant C\delta^{3/2}\big(\|f\|_{L^2(\omega)} + \|g\|_{L^2(\omega)}\big)\|e(u)\|_{L^2(\Omega_\delta)}.$$

Applying this estimate for $u = u_{\varepsilon,\delta}$ taken as test function in (11.13), gives the estimate

$$\|e(u_{\varepsilon,\delta})\|_{L^2(\Omega_\delta)} \leqslant C\delta^{3/2}\big(\|f\|_{L^2(\omega)} + \|g\|_{L^2(\omega)}\big).$$

Definition 11.8. *To simplify the notations, for every* $u \in H^1(\Omega_\delta)^3$, *set*

$$\mathbf{N}(u) = \mathbf{N}_\delta(u) \doteq \delta^{-3/2}\|e(u)\|_{L^2(\Omega_\delta)}.$$

The last inequality now simply reads

$$\mathbf{N}(u_{\varepsilon,\delta}) \leqslant C\big(\|f\|_{L^2(\omega)} + \|g\|_{L^2(\omega)}\big). \tag{11.14}$$

In the following sections we study the asymptotic behavior of the sequence $\{u_{\varepsilon,\delta}\}_{\varepsilon,\delta}$ when both parameters ε and δ go to 0.

11.3 Unfolding the rescaled plate

From now on, we will use the usual unfolding operator in ω, as well as in Ω (where y_3 is a parameter in the vertical direction).(4) The subset of ω included in the ε-cells intersecting its boundary $\partial\omega$ is denoted $\widetilde{\Lambda}_\varepsilon$.

At some point we may identify $\Omega \times Y'$ with $\omega \times Y$(5).

Definition 11.9 (The unfolding operators). *The unfolding* $\mathcal{T}_\varepsilon(\psi')$ *(resp.* $\mathcal{T}_\varepsilon(\psi)$*) of a measurable function* ψ' *defined on* ω, *(resp.* ψ *on* Ω*) is measurable on* $\omega \times Y'$ *(resp. on* $\Omega \times Y'$*) and given by*

$$\begin{aligned}
&\mathcal{T}_\varepsilon(\psi')(x',y') = \psi'\Big(\varepsilon\Big[\frac{x'}{\varepsilon}\Big] + \varepsilon y'\Big) && \text{for a.e.} \quad (x',y') \in \widehat{\omega}_\varepsilon \times Y',\\
&\mathcal{T}_\varepsilon(\psi')(x',y') = 0 && \text{for a.e.} \quad (x',y') \in \widetilde{\Lambda}_\varepsilon \times Y',
\end{aligned}$$

respectively,

$$\begin{aligned}
&\mathcal{T}_\varepsilon(\psi)(x',y',y_3) = \psi\Big(\varepsilon\Big[\frac{x'}{\varepsilon}\Big] + \varepsilon y', y_3\Big) && \text{for a.e.} \quad (x',y',y_3) \in \widehat{\omega}_\varepsilon \times Y,\\
&\mathcal{T}_\varepsilon(\psi)(x,y',y_3) = 0 && \text{for a.e.} \quad (x',y') \in \widetilde{\Lambda}_\varepsilon \times Y.)
\end{aligned}$$

We recall (see Section 1.1) that if ψ' is in $L^2(\omega)$ then $\mathcal{T}_\varepsilon(\psi')$ satisfies

$$\|\mathcal{T}_\varepsilon(\psi')\|_{L^2(\omega\times Y')} \leqslant \|\psi'\|_{L^2(\omega)}.$$

A similar estimate holds for ψ in $L^2(\Omega)$.

Finally, to simplify the presentation, introduce the

Notation 11.10 (The rescaling-unfolding operator for thin plates).

$$\mathfrak{T}_{\varepsilon,\delta} \doteq \mathcal{T}_\varepsilon \circ \mathfrak{T}_\delta.$$

(4) See Section 1.5 for notations and details.

(5) The variables in the former are (x', x_3, y_1, y_2), while they are (x', y_1, y_2, y_3) in the latter, so the identification consists of replacing x_3 by y_3.

Estimates (11.8) of $\overline{u}$, transposed by $\mathfrak{T}_\delta$ in the rescaled domain Ω, are now

$$\begin{aligned}
&\| \mathfrak{T}_\delta(\overline{u})\|_{L^2(\Omega)} \leqslant C\delta^2 \mathbf{N}(u)\\
&\Big\| \frac{\partial \mathfrak{T}_\delta(\overline{u})}{\partial x_\alpha}\Big\|_{L^2(\Omega)} \leqslant C\delta \mathbf{N}(u),\\
&\Big\| \frac{\partial \mathfrak{T}_\delta(\overline{u})}{\partial y_3}\Big\|_{L^2(\Omega)} \leqslant C\delta^2 \mathbf{N}(u).
\end{aligned} \tag{11.15}$$

The same estimates (11.8) of $\overline{u}$, transposed by $\mathfrak{T}_{\varepsilon,\delta}$ in $\omega \times Y$, become

$$\begin{aligned}
&\| \mathfrak{T}_{\varepsilon,\delta}(\overline{u})\|_{L^2(\omega\times Y)} \leqslant C\delta^2 \mathbf{N}(u)\\
&\Big\| \frac{\partial \mathfrak{T}_{\varepsilon,\delta}(\overline{u})}{\partial y_\alpha}\Big\|_{L^2(\omega\times Y)} \leqslant C\varepsilon\delta \mathbf{N}(u) = C\,\frac{\varepsilon}{\delta}\,\delta^2 \mathbf{N}(u),\\
&\Big\| \frac{\partial \mathfrak{T}_{\varepsilon,\delta}(\overline{u})}{\partial y_3}\Big\|_{L^2(\omega\times Y)} \leqslant C\delta^2 \mathbf{N}(u).
\end{aligned} \tag{11.16}$$

From now on, assume that

$$(\varepsilon,\delta) \to (0,0) \quad \text{and} \quad \lim \frac{\varepsilon}{\delta} = \theta \in [0,+\infty]. \tag{11.17}$$

In view of estimates (11.16), we are led to distinguish three cases, namely

$$\theta = 0, \qquad \theta \in (0,+\infty), \qquad \theta = +\infty.$$

11.4 An unfolding result with two parameters

We give here a general unfolding result involving two parameters (ε and δ, both supposed to go to 0). In this section (and only in this section), we go back to the notations of Part I, namely Ω is assumed to be a bounded domain in $\mathbb{R}^N$, the basis of $\mathbf{G}$ (for simplicity) is the usual euclidean basis $(\mathbf{e}_1,\dots,\mathbf{e}_N)$ and $Y = (0,1)^N$.

Both ε and δ are positive and converge to 0.

Lemma 11.11. *Suppose $p \in (1,+\infty)$. Assume (11.17) holds (and defines θ).*
Let $\{(u_{\varepsilon,\delta}, v_{\varepsilon,\delta})\}_{\varepsilon,\delta}$ be a sequence converging weakly to (u,v) in the space $W^{1,p}(\Omega)\times W^{1,p}(\Omega)^N$. Assume furthermore that there exist $\mathcal{X}$ in $L^p(\Omega)^N$ and $\widehat{v}$ in $L^p(\Omega; W^{1,p}_{per,0}(Y))^N$ such that

$$\begin{aligned}
&\frac{1}{\delta}(\nabla u_{\varepsilon,\delta} + v_{\varepsilon,\delta}) \rightharpoonup \mathcal{X} \quad \textit{weakly in } L^p(\Omega)^N,\\
&\mathcal{T}_\varepsilon(\nabla v_{\varepsilon,\delta}) \rightharpoonup \nabla v + \nabla_y \widehat{v} \quad \textit{weakly in } L^p(\Omega\times Y)^N.
\end{aligned} \tag{11.18}$$

Then u belongs to $W^{2,p}(\Omega)$, and one of the following assertions, depending of the values of θ, holds:

(i) if θ is finite, there exists $\widehat{\mathfrak{u}} \in L^p(\Omega; W^{1,p}_{per,0}(Y))$ such that, up to a subsequence,

$$\frac{1}{\delta}\mathcal{T}_\varepsilon(\nabla u_{\varepsilon,\delta} + v_{\varepsilon,\delta}) \rightharpoonup \mathcal{X} + \nabla_y \widehat{\mathfrak{u}} + \theta\widehat{v} \quad \textit{weakly in} \quad L^p(\Omega \times Y)^N. \tag{11.19}$$

(ii) if $\theta = +\infty$, there exists $\widehat{\mathfrak{u}}$ in $L^p(\Omega; W^{1,p}_{per,0}(Y) \cap W^{2,p}_{per}(Y))$ such that

$$\widehat{v} = \nabla_y \widehat{\mathfrak{u}}. \tag{11.20}$$

Proof. Step 1. Proof of the assertion $u \in W^{2,p}(\Omega)$.

Passing to the limit in assumption $(11.18)_1$ implies that

$$\nabla u = -v. \tag{11.21}$$

The latter belongs to $W^{1,p}(\Omega)^N$, so u is in $W^{2,p}(\Omega)$ and

$$\nabla v = -D^2(u) \text{ (the Hessian of } u). \tag{11.22}$$

By Theorem 1.41, up to a subsequence, there exists a function $\widehat{u}$ in $L^p(\Omega; W^{1,p}_{per,0}(Y))$ such that

$$\mathcal{T}_\varepsilon(\nabla u_{\varepsilon,\delta}) \rightharpoonup \nabla u + \nabla_y \widehat{u} \quad \text{weakly in} \quad L^p(\Omega \times Y)^N,$$

while by Proposition 1.9, $\mathcal{T}_\varepsilon(v_{\varepsilon,\delta})$ converges weakly to v in $L^p(\Omega \times Y)^N$.[(6)]

Consequently,

$$\nabla u + \nabla_y \widehat{u} + v = 0 \quad \text{a.e. in } \Omega \times Y.$$

Together with (11.21), this implies $\nabla_y \widehat{u} = 0$. Thus $\widehat{u} = 0$ (since $\mathcal{M}_Y(\widehat{u}) = 0$ a.e. in Ω). By Theorem 1.41, it follows that

$$\frac{1}{\varepsilon}\big(\mathcal{T}_\varepsilon(v_{\varepsilon,\delta}) - \mathcal{M}_\varepsilon(v_{\varepsilon,\delta})\big) \rightharpoonup \nabla v \cdot y^c + \widehat{v} = -D^2 u \cdot y^c + \widehat{v} \tag{11.23}$$

weakly in $L^p(\Omega; W^{1,p}(Y))^N$.

Step 2. Proof of convergence (11.19).

From (11.17) there exists a finite C (independent of ε and δ) such that

$$0 < \varepsilon \leqslant C\delta. \tag{11.24}$$

Introduce the function $Z_{\varepsilon,\delta}$ belonging to $L^p(\Omega; W^{1,p}(Y))$, defined as

$$Z_{\varepsilon,\delta} = \frac{1}{\varepsilon}\big(\mathcal{T}_\varepsilon(u_{\varepsilon,\delta}) - \mathcal{M}_\varepsilon(u_{\varepsilon,\delta})\big) - \mathcal{M}_\varepsilon(\nabla u_{\varepsilon,\delta}) \cdot y^c. \tag{11.25}$$

(6) The full sequence converges strongly to v in $L^p_{loc}(\Omega; L^p(Y))^N$ (and actually in $L^p(\Omega; L^p(Y))^N$ whenever $W^{1,p}(\Omega)$ is compactly embedded in $L^p(\Omega)$).

Its gradient with respect to y is

$$\begin{aligned}\nabla_y Z_{\varepsilon,\delta} &= \mathcal{T}_\varepsilon(\nabla u_{\varepsilon,\delta}) - \mathcal{M}_\varepsilon(\nabla u_{\varepsilon,\delta})\\ &= \mathcal{T}_\varepsilon(\nabla u_{\varepsilon,\delta} + v_{\varepsilon,\delta}) - \big(\mathcal{T}_\varepsilon(v_{\varepsilon,\delta}) - \mathcal{M}_\varepsilon(v_{\varepsilon,\delta})\big) - \mathcal{M}_\varepsilon(\nabla u_{\varepsilon,\delta} + v_{\varepsilon,\delta}).\end{aligned} \tag{11.26}$$

Convergences $(11.18)_1$ and (11.23) (together with (11.24)) give

$$\|\nabla_y Z_{\varepsilon,\delta}\|_{L^p(\Omega\times Y)} \leqslant C(\delta+\varepsilon) \leqslant C\delta.$$

Since $\mathcal{M}_Y(Z_{\varepsilon,\delta}) = 0$ a.e. in Ω, the Poincaré-Wirtinger inequality implies

$$\|Z_{\varepsilon,\delta}\|_{L^p(\Omega;W^{1,p}(Y))} \leqslant C\delta.$$

Consequently, up to a subsequence, there exists $\widehat{Z}$ in $L^p(\Omega;W^{1,p}(Y))$ such that,

$$\frac{1}{\delta} Z_{\varepsilon,\delta} \rightharpoonup \widehat{Z} \quad \text{weakly in } L^p(\Omega;W^{1,p}(Y)). \tag{11.27}$$

By (11.26)

$$\frac{1}{\delta}\mathcal{T}_\varepsilon\big(\nabla u_{\varepsilon,\delta} + v_{\varepsilon,\delta}\big) = \frac{1}{\delta}\nabla_y Z_{\varepsilon,\delta} + \frac{\varepsilon}{\delta}\,\frac{\mathcal{T}_\varepsilon(v_{\varepsilon,\delta}) - \mathcal{M}_\varepsilon(v_{\varepsilon,\delta})}{\varepsilon} + \frac{1}{\delta}\mathcal{M}_\varepsilon\big(\nabla u_{\varepsilon,\delta} + v_{\varepsilon,\delta}\big),$$

and going to the limit, using (11.27), (11.23) and $(11.18)_1$,

$$\begin{aligned}&\frac{1}{\delta}\mathcal{T}_\varepsilon\big(\nabla u_{\varepsilon,\delta} + v_{\varepsilon,\delta}\big)\\ &\qquad \rightharpoonup \nabla_y\widehat{Z} + \theta\big(\nabla v\cdot y^c + \widehat{v}\big) + \mathcal{X} \quad \text{weakly in } L^p(\Omega\times Y)^N.\end{aligned} \tag{11.28}$$

Here (with the Einstein summation convention), $\nabla v\cdot y^c$ is given componentwise by $(\nabla v\cdot y^c)_i \doteq \nabla v_i\cdot y^c = \dfrac{\partial v_i}{\partial x_j}\, y^c_j$ $(i = 1,\dots,N)$, or in terms of u, by (11.22),

$$(\nabla v\cdot y^c)_i = -\frac{\partial^2 u}{\partial x_i\partial x_j}\, y^c_j = -\frac{1}{2}\frac{\partial}{\partial y_i}\left(\frac{\partial^2 u}{\partial x_j\partial x_k}\, y^c_j y^c_k\right).$$

In other words,

$$\nabla v\cdot y^c = -D^2u\cdot y^c = -\frac{1}{2}\nabla_y\left(\frac{\partial^2 u}{\partial x_j\partial x_k}\Big(y^c_j y^c_k - \mathcal{M}_Y(y^c_j y^c_k)\Big)\right)^{(7)}.$$

Going back to (11.28), one obtains (11.19) by setting

$$\widehat{\mathfrak{u}} \doteq \widehat{Z} - \frac{\theta}{2}\left(\frac{\partial^2 u}{\partial x_j\partial x_k}\Big(y^c_j y^c_k - \mathcal{M}_Y(y^c_j y^c_k)\Big)\right) \tag{11.29}$$

(7) Note that for $j\neq k$, $\mathcal{M}_Y(y^c_j y^c_k) = \mathcal{M}_Y(y^c_j)\mathcal{M}_Y(y^c_k) = 0$.

which belongs to $L^p(\Omega; W^{1,p}(Y))$ with mean value 0. It remains only to show that $\widehat{u}$ is Y-periodic.

To show the periodicity with respect to the first variable (the computation is the same for the others variables), we proceed as in the proof of Theorem 1.36 by evaluating the difference of the traces of $Z_{\varepsilon,\delta}$ on the faces $Y_1 = \{0\} \times (0,1)^{N-1}$ and $Y_1 + \mathbf{e}_1$. For a.e. $(x, y') \in \Omega \times Y_1$, one has

$$
\begin{aligned}
&Z_{\varepsilon,\delta}(x, y' + \mathbf{e}_i) - Z_{\varepsilon,\delta}(x, y') \\
&\quad = \frac{1}{\varepsilon}\big(\mathcal{T}_\varepsilon(u_{\varepsilon,\delta})(x, y' + \mathbf{e}_1) - \mathcal{T}_\varepsilon(u_{\varepsilon,\delta})(x, y')\big) - \mathcal{M}_\varepsilon\Big(\frac{\partial u_{\varepsilon,\delta}}{\partial x_1}\Big)(x) \\
&\quad = \frac{1}{\varepsilon}\big(\mathcal{T}_\varepsilon(u_{\varepsilon,\delta})(x + \varepsilon\mathbf{e}_1, y') - \mathcal{T}_\varepsilon(u_{\varepsilon,\delta})(x, y')\big) - \mathcal{M}_\varepsilon\Big(\frac{\partial u_{\varepsilon,\delta}}{\partial x_1}\Big)(x).
\end{aligned}
$$

Let Φ be in $\mathcal{D}(\Omega \times Y_1)^N$. By standard transformations, one has

$$
\begin{aligned}
&\int_{\Omega\times Y_1} \big(Z_{\varepsilon,\delta}(x, y' + \mathbf{e}_1) - Z_{\varepsilon,\delta}(x, y')\big)\Phi(x, y')\,dxdy' \\
&= \int_{\Omega\times Y_1} \Big[\frac{1}{\varepsilon}\Big(\mathcal{T}_\varepsilon(u_{\varepsilon,\delta})(x + \varepsilon\mathbf{e}_1, y') - \mathcal{T}_\varepsilon(u_{\varepsilon,\delta})(x, y')\Big)\Big]\Phi(x, y')\,dxdy' \\
&\qquad\qquad - \int_{\Omega\times Y_1} \mathcal{M}_\varepsilon\Big(\frac{\partial u_{\varepsilon,\delta}}{\partial x_1}\Big)(x)\Phi(x, y')\,dxdy' \\
&= \int_{\Omega\times Y_1} \mathcal{T}_\varepsilon(u_{\varepsilon,\delta})(x, y')\frac{1}{\varepsilon}\Big[\Phi(x - \varepsilon\mathbf{e}_1, y') - \Phi(x, y')\Big]dxdy' \\
&\qquad\qquad - \int_{\Omega\times Y_1} \mathcal{M}_\varepsilon\Big(\frac{\partial u_{\varepsilon,\delta}}{\partial x_1}\Big)(x)\Phi(x, y')\,dxdy',
\end{aligned}
$$

which can be rewritten as

$$
\begin{aligned}
&\int_{\Omega\times Y_1} \mathcal{T}_\varepsilon(u_{\varepsilon,\delta})(x, y')\frac{1}{\varepsilon}\Big[\Phi(x - \varepsilon\mathbf{e}_1, y') - \Phi(x, y') + \varepsilon\frac{\partial\Phi}{\partial x_1}(x, y')\Big]dxdy' \\
&\qquad + \int_{\Omega\times Y_1} \big(u_{\varepsilon,\delta}(x) - \mathcal{T}_\varepsilon(u_{\varepsilon,\delta})(x, y')\big)\frac{\partial\Phi}{\partial x_1}(x, y')dxdy' \\
&\qquad + \int_{\Omega\times Y_1} \Big[\frac{\partial u_{\varepsilon,\delta}}{\partial x_1} - \mathcal{M}_\varepsilon\Big(\frac{\partial u_{\varepsilon,\delta}}{\partial x_1}\Big)\Big]\Phi(x, y')\,dxdy',
\end{aligned}
$$

or as

$$
\begin{aligned}
&\int_{\Omega\times Y_1} \mathcal{T}_\varepsilon(u_{\varepsilon,\delta})(x, y')\frac{1}{\varepsilon}\Big[\Phi(x - \varepsilon\mathbf{e}_1, y') - \Phi(x, y') + \varepsilon\frac{\partial\Phi}{\partial x_1}(x, y')\Big]dxdy' \\
&\qquad + \int_{\Omega\times Y_1} \big(\mathcal{M}_\varepsilon(u_{\varepsilon,\delta})(x) - \mathcal{T}_\varepsilon(u_{\varepsilon,\delta})(x, y')\big)\frac{\partial\Phi}{\partial x_1}(x, y')dxdy' \\
&\qquad + \int_{\Omega\times Y_1} \big(u_{\varepsilon,\delta}(x) - \mathcal{M}_\varepsilon(u_{\varepsilon,\delta})(x)\frac{\partial\Phi}{\partial x_1}(x, y')dxdy' \\
&\qquad + \int_{\Omega\times Y_1} \Big(\frac{\partial u_{\varepsilon,\delta}}{\partial x_1} - \mathcal{M}_\varepsilon\Big(\frac{\partial u_{\varepsilon,\delta}}{\partial x_1}\Big)\Big)\Phi(x, y')\,dxdy'.
\end{aligned}
$$

The last two integrals can be modified using Proposition 1.24 (ii):

$$\begin{aligned}&\int_{\Omega\times Y_1}(u_{\varepsilon,\delta}(x)-\mathcal{M}_\varepsilon(u_{\varepsilon,\delta})(x)\frac{\partial\Phi}{\partial x_1}(x,y')dxdy'\\&=\int_\Omega u_{\varepsilon,\delta}(x)\Big(\int_Y\frac{\partial\Phi}{\partial x_1}(x,y')dy'-\mathcal{M}_\varepsilon\Big(\int_Y\frac{\partial\Phi}{\partial x_1}(x,y')dy'\Big)dx,\\&\int_{\Omega\times Y_1}\Big(\frac{\partial u_{\varepsilon,\delta}}{\partial x_1}-\mathcal{M}_\varepsilon\Big(\frac{\partial u_{\varepsilon,\delta}}{\partial x_1}\Big)\Big)\Phi(x,y')\,dxdy'\\&=\int_\Omega\frac{\partial u_{\varepsilon,\delta}}{\partial x_1}(x)\Big(\int_Y\Phi(x,y')dy'-\mathcal{M}_\varepsilon\Big(\int_Y\Phi(x,y')dy'\Big)dx.\end{aligned}$$

Finally,

$$\begin{aligned}&\int_{\Omega\times Y_1}(Z_{\varepsilon,\delta}(x,y'+\mathbf{e}_i)-Z_{\varepsilon,\delta}(x,y'))\Phi(x,y')\,dxdy'\\&=\int_{\Omega\times Y_1}\varepsilon\,\mathcal{T}_\varepsilon(u_{\varepsilon,\delta})(x,y')\frac{1}{\varepsilon^2}\Big[\Phi(x-\varepsilon\mathbf{e}_1,y')-\Phi(x,y')+\varepsilon\frac{\partial\Phi}{\partial x_1}(x,y')\Big]dxdy'\\&\quad+\int_{\Omega\times Y_1}(\mathcal{M}_\varepsilon(u_{\varepsilon,\delta})(x)-\mathcal{T}_\varepsilon(u_{\varepsilon,\delta})(x,y'))\frac{\partial\Phi}{\partial x_1}(x,y')dxdy'\\&\quad+\int_\Omega u_{\varepsilon,\delta}(x)\Big(\int_Y\frac{\partial\Phi}{\partial x_1}(x,y')dy'-\mathcal{M}_\varepsilon\Big(\int_Y\frac{\partial\Phi}{\partial x_1}(x,y')dy'\Big)dx,\\&\quad+\int_\Omega\frac{\partial u_{\varepsilon,\delta}}{\partial x_1}(x)\Big(\int_Y\Phi(x,y')dy'-\mathcal{M}_\varepsilon\Big(\int_Y\Phi(x,y')dy'\Big)dx.\end{aligned}\tag{11.30}$$

Dividing (11.30) by δ and passing to the limit (using Propositions 1.38 and 1.39) gives

$$\begin{aligned}&\lim_{\varepsilon,\delta\to(0,0)}\int_{\Omega\times Y_1}\frac{1}{\delta}(Z_{\varepsilon,\delta}(x,y'+\mathbf{e}_i)-Z_{\varepsilon,\delta}(x,y'))\Phi(x,y)\,dxdy'\\&=\frac{\theta}{2}\int_{\Omega\times Y_1}u(x)\,\frac{\partial^2\Phi}{\partial x_1^2}(x,y')\,dxdy'-\theta\int_{\Omega\times Y_1}\Big(\sum_{j=1}^N\frac{\partial u}{\partial x_j}(x)\,y_j^c\Big)\,\frac{\partial\Phi}{\partial x_1}(x,y')dxdy'.\end{aligned}$$

The latter is transformed by integration by parts with respect to x_1 into

$$\begin{aligned}&\frac{\theta}{2}\int_{\Omega\times Y_1}\frac{\partial^2 u}{\partial x_1^2}(x)\;\Phi(x,y'),dxdy'+\theta\int_{\Omega\times Y_1}\Big(\sum_{j=1}^N\frac{\partial^2 u}{\partial x_1\partial x_j}(x)\,y_j^c\Big)\;\Phi(x,y')dxdy'\\&\qquad=\theta\int_{\Omega\times Y_1}\sum_{j=2}^N\frac{\partial^2 u}{\partial x_1\partial x_j}(x)y_j'^c\;\Phi(x,y')dxdy'\end{aligned}$$

since y_1^c is just $-1/2$ on Y_1. Hence, for a.e. $(x,y')\in\Omega\times Y_1$

$$\widehat{Z}(x,y'+\mathbf{e}_i)-\widehat{Z}(x,y')=\theta\sum_{j=2}^N\frac{\partial^2 u}{\partial x_1\partial x_j}(x)y_j'^c.$$

Going back to its definition (11.29), this proves that $\widehat{\mathfrak{u}}$ is indeed $\mathbf{e}_1$-periodic.

Step 3. Proof of the existence of $\widehat{\mathrm{u}}$ *when* $\theta = +\infty$.

Again we consider the function $Z_{\varepsilon,\delta}$ introduced in (11.25), which, in this case, satisfies

$$\|Z_{\varepsilon,\delta}\|_{L^p(\Omega;W^{1,p}(Y))} \leqslant C\varepsilon.$$

Hence, up to a subsequence, there exists $\widehat{Z}$ in $L^p(\Omega; W^{1,p}(Y))$ such that,

$$\frac{1}{\varepsilon} Z_{\varepsilon,\delta} \rightharpoonup \widehat{Z} \quad \text{weakly in} \quad L^p(\Omega; W^{1,p}(Y)).$$

Observe that

$$\begin{aligned}
\frac{1}{\varepsilon}\mathcal{T}_\varepsilon\big(\nabla u_{\varepsilon,\delta} + v_{\varepsilon,\delta}\big) &= \frac{\delta}{\varepsilon}\frac{1}{\delta}\mathcal{T}_\varepsilon\big(\nabla u_{\varepsilon,\delta} + v_{\varepsilon,\delta}\big) \longrightarrow 0 \quad \text{strongly in} \quad L^p(\Omega \times Y)^N,\\
\frac{1}{\varepsilon}\mathcal{M}_\varepsilon\big(\nabla u_{\varepsilon,\delta} + v_{\varepsilon,\delta}\big) &= \frac{\delta}{\varepsilon}\frac{1}{\delta}\mathcal{M}_\varepsilon\big(\nabla u_{\varepsilon,\delta} + v_{\varepsilon,\delta}\big) \longrightarrow 0 \quad \text{strongly in} \quad L^p(\Omega)^N.
\end{aligned}$$

Since

$$\frac{1}{\varepsilon}\mathcal{T}_\varepsilon\big(\nabla u_{\varepsilon,\delta} + v_{\varepsilon,\delta}\big) = \frac{1}{\varepsilon}\nabla_y(Z_{\varepsilon,\delta}) + \frac{\mathcal{T}_\varepsilon(v_{\varepsilon,\delta}) - \mathcal{M}_\varepsilon(v_{\varepsilon,\delta})}{\varepsilon} + \frac{1}{\varepsilon}\mathcal{M}_\varepsilon\big(\nabla u_{\varepsilon,\delta} + v_{\varepsilon,\delta}\big),$$

at the limit, by (11.23),

$$0 = \nabla_y \widehat{Z} + \nabla v \cdot y^c + \widehat{v} = \nabla_y \widehat{Z} - D^2 u \cdot y^c + \widehat{v}. \tag{11.31}$$

Set

$$\widehat{\mathrm{u}} \doteq \widehat{Z} - \frac{1}{2}\sum_{j,k=1}^{N} \frac{\partial^2 u}{\partial x_j \partial x_k}\Big(y_j^c y_k^c - \mathcal{M}_Y(y_j^c y_k^c)\Big).$$

Clearly, $\widehat{\mathrm{u}}$ belongs to $L^p(\Omega; W^{1,p}(Y))$, has mean value 0 on Y and by (11.31) $\widehat{v} = -\nabla_y \widehat{\mathrm{u}}$. This last equality now implies that $\widehat{\mathrm{u}}$ belongs to $L^p(\Omega; W^{2,p}(Y))$. The proof of the Y-periodicity of $\widehat{\mathrm{u}}$ is the same as in Step 2. □

The above proof is easily adapted to the case of vector-valued fields, making use of the symmetric gradient e and the Korn inequality.

Lemma 11.12. *Let* Ω *be a bounded domain in* $\mathbb{R}^N$ *with Lipschitz boundary and* $Y = (0,1)^N$ *and* p *be in* $(1, +\infty)$. *Suppose* $\{(u_{\varepsilon,\delta}, v_{\varepsilon,\delta})\}_{\varepsilon,\delta}$ *is a sequence in* $W^{1,p}(\Omega)^N \times W^{1,p}(\Omega)^{N\times N}$ *(with* $v_{\varepsilon,\delta}$ *a symmetric matrix field) converging weakly to* (u, v) *in* $W^{1,p}(\Omega)^N \times W^{1,p}(\Omega)^{N\times N}$.
Assume that there exist $\mathcal{X}$ *in* $L^p(\Omega)^{N\times N}$ *and* $\widehat{v}$ *in* $L^p(\Omega; W^{1,p}_{per,0}(Y))^{N\times N}$ *such that as* $(\varepsilon, \delta) \to (0, 0)$

$$\begin{aligned}
&\frac{1}{\delta}\big(e(u_{\varepsilon,\delta}) + v_{\varepsilon,\delta}\big) \rightharpoonup \mathcal{X} \quad \text{weakly in} \quad L^p(\Omega)^{N\times N},\\
&\mathcal{T}_{\varepsilon,\delta}(\nabla v_{\varepsilon,\delta}) \rightharpoonup \nabla v + \nabla_y \widehat{v} \quad \text{weakly in} \quad L^p(\Omega \times Y)^{N\times N\times N}.
\end{aligned} \tag{11.32}$$

Then u belongs to $W^{2,p}(\Omega)^N$ and there exists $\widehat{\mathrm{u}} \in L^p(\Omega; W^{1,p}_{per,0}(Y))^N$ such that, up to a subsequence,

$$\begin{aligned}
&\text{if } \frac{\varepsilon}{\delta} \to \theta \in [0, +\infty), \\
&\qquad \frac{1}{\delta}\mathcal{T}_{\varepsilon,\delta}\big(e(u_{\varepsilon,\delta}) + v_{\varepsilon,\delta}\big) \rightharpoonup \mathcal{X} + e_y(\widehat{\mathrm{u}}) + \theta\,\widehat{v} \text{ weakly in } L^p(\Omega \times Y)^{N\times N}; \\
&\text{if } \frac{\varepsilon}{\delta} \to \theta + \infty, \qquad \widehat{v} = e_y(\mathrm{u}).
\end{aligned}$$

11.5 Asymptotic behavior of the strain tensor

In this subsection we investigate the behavior of a sequence $\{u_{\varepsilon,\delta}\}_{(\varepsilon,\delta)}$ in $\mathbb{D}_\delta$ for which (cf. estimate (11.14)),

$$\mathbf{N}(u_{\varepsilon,\delta}) \leqslant C. \tag{11.33}$$

We denote by $\mathcal{U}_{\varepsilon,\delta}$, $\mathcal{R}_{\varepsilon,\delta}$ and $\overline{u}_{\varepsilon,\delta}$ the terms of the decomposition of $u_{\varepsilon,\delta}$ (see (11.4) in Subsection 11.1.2).

11.5.1 Limit of the rescaled strain tensor

We start by introducing the spaces

$$\begin{aligned}
H^1_{per}(Y') &= \Big\{\phi \in H^1(Y') \mid \phi(0, y_2) = \phi(1, y_2) \text{ for a.e. } y_2 \in (0,1) \\
&\qquad\qquad \phi(y_1, 0) = \phi(y_1, 1) \text{ for a.e. } y_1 \in (0,1)\Big\}, \\
\boldsymbol{\mathcal{H}}^1_{per}(Y) &\doteq \Big\{\psi \in H^1(Y) \mid \psi_{|y_1=0} = \psi_{|y_1=1},\ \ \psi_{|y_2=0} = \psi_{|y_2=1}\Big\}, \\
\mathcal{D}_w &\doteq \Big\{\phi \in L^2(\omega; H^1(-1,1))^3 \mid \int_{-1}^1 \phi_i(\cdot, y_3)\, dy_3 = \int_{-1}^1 y_3\phi_\alpha(\cdot, y_3)\, dy_3 = 0\Big\}.
\end{aligned}$$

Note that the space $\boldsymbol{\mathcal{H}}^1_{per}(Y)$ defined here (with periodicity with respect to the first two dimensions only) is different from the space $H^1_{per}(Y)$ used in previous chapters (where periodicity is in all dimensions).

Proposition 11.13. *Under hypothesis* (11.33) *and assumption* (11.17), *there exist*

$$\mathcal{U}_m = (\mathcal{U}_1, \mathcal{U}_2) \in H^1_{\gamma_0}(\omega)^2, \quad \mathcal{U}_3 \in H^2_{\gamma_0}(\omega), \quad \mathcal{Z} \in L^2(\omega)^2, \quad \overline{u} \in \mathcal{D}_w,$$

such that, up to a subsequence, the following convergences hold:

$$\begin{aligned}
&\frac{1}{\delta}\mathcal{U}_{\varepsilon,\delta,m} \rightharpoonup \mathcal{U}_m \quad \textit{weakly in } H^1_{\gamma_0}(\omega)^2,\\
&\mathcal{U}_{\varepsilon,\delta,3} \longrightarrow \mathcal{U}_3 \quad \textit{strongly in } H^1_{\gamma_0}(\omega),\\
&\mathcal{R}_{\varepsilon,\delta} \rightharpoonup -\nabla\mathcal{U}_3 \quad \textit{weakly in } H^1_{\gamma_0}(\omega)^2,\\
&\frac{1}{\delta}\big(\nabla\mathcal{U}_{\varepsilon,\delta,3} + \mathcal{R}_{\varepsilon,\delta}\big) \rightharpoonup \mathcal{Z} \quad \textit{weakly in } L^2(\omega)^2,\\
&\frac{1}{\delta^2}\mathfrak{T}_\delta(\overline{u}_{\varepsilon,\delta}) \rightharpoonup \overline{u} \quad \textit{weakly in } \mathcal{D}_w,\\
&\frac{1}{\delta}\mathfrak{T}_\delta(\overline{u}_{\varepsilon,\delta}) \rightharpoonup 0 \quad \textit{weakly in } H^1(\Omega)^3,
\end{aligned} \tag{11.34}$$

as well as,

$$\begin{aligned}
&\frac{1}{\delta}\mathfrak{T}_\delta(u_{\varepsilon,\delta,\alpha}) \rightharpoonup \mathcal{U}_\alpha - y_3\frac{\partial\mathcal{U}_3}{\partial x_\alpha} \quad \textit{weakly in } H^1(\Omega),\\
&\mathfrak{T}_\delta(u_{\varepsilon,\delta,3}) \rightharpoonup \mathcal{U}_3 \quad \textit{weakly in } H^1(\Omega).^{(8)}
\end{aligned} \tag{11.35}$$

Set

$$\mathring{u} = \overline{u} + y_3\mathcal{Z}_\alpha\mathbf{e}_\alpha.$$

Then

$$\frac{1}{\delta}\mathfrak{T}_\delta\big(e(u_{\varepsilon,\delta})\big) \rightharpoonup \begin{pmatrix} E_{11}(\mathcal{U}) & E_{12}(\mathcal{U}) & \dfrac{1}{2}\dfrac{\partial\mathring{u}_1}{\partial y_3}\\[2ex] E_{12}(\mathcal{U}) & E_{22}(\mathcal{U}) & \dfrac{1}{2}\dfrac{\partial\mathring{u}_2}{\partial y_3}\\[2ex] \dfrac{1}{2}\dfrac{\partial\mathring{u}_1}{\partial y_3} & \dfrac{1}{2}\dfrac{\partial\mathring{u}_2}{\partial y_3} & \dfrac{\partial\mathring{u}_3}{\partial y_3}\end{pmatrix} \textit{ weakly in } L^2(\Omega)^9, \tag{11.36}$$

where

$$\mathcal{U} = (\mathcal{U}_m, \mathcal{U}_3), \quad E_{\alpha\beta}(\mathcal{U}) = e_{\alpha\beta}(\mathcal{U}_m) - y_3\frac{\partial^2\mathcal{U}_3}{\partial x_\alpha\partial x_\beta}.$$

Furthermore, there exist

$$\widehat{U}_\alpha,\ \widehat{R}_\alpha \in L^2(\omega; H^1_{per,0}(Y')) \quad \textit{and} \quad \widehat{Z}_\alpha \in L^2(\omega\times Y'),$$

with $\mathcal{M}_{Y'}(\widehat{Z}_\alpha) = 0$ *a.e. in* ω*, such that*

$$\begin{aligned}
&\frac{1}{\delta}\mathcal{T}_\varepsilon(\nabla\mathcal{U}_{\varepsilon,\delta,\alpha}) \rightharpoonup \nabla\mathcal{U}_\alpha + \nabla_{y'}\widehat{U}_\alpha \quad \textit{weakly in } L^2(\omega\times Y')^2,\\
&\mathcal{T}_\varepsilon(\nabla\mathcal{R}_{\varepsilon,\delta}) \rightharpoonup -D^2\mathcal{U}_3 + \nabla_{y'}\widehat{R} \quad \textit{weakly in } L^2(\omega\times Y')^{2\times 2},\\
&\frac{1}{\delta}\mathcal{T}_\varepsilon\big(\nabla\mathcal{U}_{\varepsilon,\delta,3} + \mathcal{R}_{\varepsilon,\delta}\big) \rightharpoonup \mathcal{Z} + \widehat{Z} \quad \textit{weakly in } L^2(\omega\times Y')^2.
\end{aligned} \tag{11.37}$$

---

(8) This mean that the original displacement in the domain Ω_δ is asymptotically of Kirchhoff-Love type.

There are three cases according to the value of θ.

<u>*Case 1*</u>. *If $\theta = 0$, there exist $\widehat{\overline{u}}$ in $L^2(\Omega; H^1_{per,0}(Y'))^3$, $\widehat{U}_3$ in $L^2(\omega; H^1_{per,0}(Y'))$ satisfying*

$$\int_{-1}^{1} \widehat{\overline{u}}_i(\cdot, y_3)\, dy_3 = \int_{-1}^{1} y_3\, \widehat{\overline{u}}_\alpha(\cdot, y_3)\, dy_3 = 0 \quad \text{a.e. in } \omega \times Y', \tag{11.38}$$

and such that

$$\begin{aligned}
&\frac{1}{\delta^2} \mathfrak{T}_{\varepsilon,\delta}(\overline{u}_{\varepsilon,\delta}) \rightharpoonup \overline{u} \quad \text{weakly in } L^2(\omega; H^1(Y))^3,\\
&\frac{1}{\varepsilon\delta} \frac{\partial \mathfrak{T}_{\varepsilon,\delta}(\overline{u}_{\varepsilon,\delta})}{\partial y_\alpha} \rightharpoonup \frac{\partial \widehat{\overline{u}}}{\partial y_\alpha} \quad \text{weakly in } L^2(\omega \times Y)^3,\\
&\widehat{Z} = \nabla_{y'} \widehat{U}_3.
\end{aligned} \tag{11.39}$$

<u>*Case 2*</u>. *If $\theta \in (0, +\infty)$, there exist $\widehat{\overline{u}}$ in $L^2(\omega; \boldsymbol{H}^1_{per}(Y))^3$ satisfying (11.38) and as before $\widehat{U}_3$ in $L^2(\omega; H^1_{per,0}(Y'))$ such that*

$$\begin{aligned}
&\frac{1}{\delta^2} \mathfrak{T}_{\varepsilon,\delta}(\overline{u}_{\varepsilon,\delta}) \rightharpoonup \overline{u} + \widehat{\overline{u}} \quad \text{weakly in } L^2(\omega; H^1(Y))^3,\\
&\widehat{Z} = \nabla_{y'} \widehat{U}_3 + \theta \widehat{R}.
\end{aligned} \tag{11.40}$$

<u>*Case 3*</u>. *If $\theta = +\infty$, there exist $\widehat{\overline{u}}$ in $L^2(\omega \times Y'; H^1(-1,1))^3$ satisfying (11.38) and $\widehat{U}_3$ in $L^2(\omega; H^2_{per,0}(Y'))$ such that*

$$\begin{aligned}
&\frac{1}{\delta^2} \mathfrak{T}_{\varepsilon,\delta}(\overline{u}_{\varepsilon,\delta}) \rightharpoonup \overline{u} + \widehat{\overline{u}} \quad \text{weakly in } L^2(\omega \times Y'; H^1(-1,1))^3,\\
&\frac{1}{\varepsilon\delta} \frac{\partial \mathfrak{T}_{\varepsilon,\delta}(\overline{u}_{\varepsilon,\delta})}{\partial y_\alpha} \rightharpoonup 0 \quad \text{weakly in } L^2(\omega \times Y)^3,\\
&\widehat{R} = -\nabla_{y'} \widehat{U}_3.
\end{aligned} \tag{11.41}$$

Proof. The sequence $\{u_{\varepsilon,\delta}\}_{\varepsilon,\delta}$ satisfies $\mathbf{N}(u_{\varepsilon,\delta}) \leqslant C$. Therefore, the inequalities of $(11.10)_1$ give the convergences $(11.34)_{1,2}$ and

$$\mathcal{R}_{\varepsilon,\delta} \rightharpoonup \mathcal{R} \quad \text{weakly in} \quad H^1_{\gamma_0}(\omega)^2,$$

where $\mathcal{R} \in H^1_{\gamma_0}(\omega)^2$. Equality $\dfrac{\partial \mathcal{U}_3}{\partial x_\alpha} = -\mathcal{R}_\alpha$ and $(11.34)_4$ are consequences of $(11.32)_2$. Estimates (11.15) and equalities (11.3) yield $(11.34)_{5,6}$.

Convergences (11.35) follow from (11.34), while convergence (11.36) is also a consequence of (11.34) and of the expression (11.6) of the strain tensor. Then (11.37) is obtained by applying Theorem 1.41.

For the residual displacement $\overline{u}_{\varepsilon,\delta}$, estimates (11.16) and (11.33) imply

$$\Big\| \frac{1}{\delta^2} \mathfrak{T}_{\varepsilon,\delta}(\overline{u}_{\varepsilon,\delta}) \Big\|_{L^2(\omega \times Y)} + \Big\| \frac{1}{\delta^2} \frac{\partial \mathfrak{T}_{\varepsilon,\delta}(\overline{u}_{\varepsilon,\delta})}{\partial y_3} \Big\|_{L^2(\omega \times Y)} \leqslant C,$$

and

$$\Big\| \frac{1}{\delta^2} \frac{\partial \mathfrak{T}_{\varepsilon,\delta}(\overline{u}_{\varepsilon,\delta})}{\partial y_1} \Big\|_{L^2(\omega\times Y)} + \Big\| \frac{1}{\delta^2} \frac{\partial \mathfrak{T}_{\varepsilon,\delta}(\overline{u}_{\varepsilon,\delta})}{\partial y_2} \Big\|_{L^2(\omega\times Y)} \leqslant \frac{\varepsilon}{\delta} C.$$

Then convergences $(11.39)_1$ and $(11.39)_2$ follow by using Lemma 1.44, and convergences $(11.40)_1$ by using Theorem 1.36 and (11.41). On the other hand, convergences $(11.34)_1$, $(11.34)_3$, $(11.34)_5$ and Lemma 11.11, give $(11.39)_3$ and $(11.40)_2$.

In the case $\theta = +\infty$, observe that

$$\Big\| \frac{1}{\varepsilon\delta} \mathfrak{T}_{\varepsilon,\delta}(\overline{u}_{\varepsilon,\delta}) \Big\|_{L^2(\omega\times Y)} + \Big\| \frac{1}{\varepsilon\delta} \frac{\partial \mathfrak{T}_{\varepsilon,\delta}(\overline{u}_{\varepsilon,\delta})}{\partial y_3} \Big\|_{L^2(\omega\times Y)} \leqslant \frac{\varepsilon}{\delta} C,$$

and

$$\Big\| \frac{1}{\varepsilon\delta} \frac{\partial \mathfrak{T}_{\varepsilon,\delta}(\overline{u}_{\varepsilon,\delta})}{\partial y_1} \Big\|_{L^2(\omega\times Y)} + \Big\| \frac{1}{\varepsilon\delta} \frac{\partial \mathfrak{T}_{\varepsilon,\delta}(\overline{u}_{\varepsilon,\delta})}{\partial y_2} \Big\|_{L^2(\omega\times Y)} \leqslant C.$$

As a consequence,

$$\frac{1}{\varepsilon\delta} \mathfrak{T}_{\varepsilon,\delta}(\overline{u}_{\varepsilon,\delta}) \rightharpoonup 0 \quad \text{weakly in} \quad L^2(\omega; H^1(Y))^3,$$

from which convergence $(11.41)_2$ is straightforward. Finally, Lemma 11.11 provides the existence of $\widehat{U}_3$ in $L^2(\omega; H^2_{per}(Y'))$ satisfying $\mathcal{M}_{Y'}(\widehat{U}_3) = 0$ a.e. in ω, and such that $\widehat{R} = -\nabla_{y'}\widehat{U}_3$. □

11.5.2 Limit of the unfolded scaled strain tensor

As a consequence of Proposition 11.13, we can give the different weak limits of the unfolded and rescaled strain tensor in $L^2(\omega \times Y)^9$. Before doing so, we introduce some spaces and notations which will be needed in the sequel.

- *The plate mid-surface limit displacements*

The mid-surface limit displacements are elements of $\mathbb{D}_M$ (see (11.1)). For every $\mathcal{V}$ belonging to $\mathbb{D}_M$, the associated membrane displacement is denoted $\mathcal{V}_m = (\mathcal{V}_1, \mathcal{V}_2)$ or $\mathcal{V}_1\mathbf{e}_1 + \mathcal{V}_2\mathbf{e}_2$, while the bending is denoted $\mathcal{V}_3$. Thus, $\mathcal{V} = (\mathcal{V}_m, \mathcal{V}_3) = (\mathcal{V}_1, \mathcal{V}_2, \mathcal{V}_3)$.

For $\mathcal{V}$ in $\mathbb{D}_M$, define the symmetric matrix $E_M(\mathcal{V})$ as

$$E_M(\mathcal{V}) = \begin{pmatrix} E_{11}(\mathcal{V}) & E_{12}(\mathcal{V}) & 0 \\ E_{12}(\mathcal{V}) & E_{22}(\mathcal{V}) & 0 \\ 0 & 0 & 0 \end{pmatrix},$$

where

$$E_{\alpha\beta}(\mathcal{V}) = e_{\alpha\beta}(\mathcal{V}_m) - y_3 \frac{\partial^2 \mathcal{V}_3}{\partial x_\alpha \partial x_\beta}.$$

• *The warping displacements*

They are the relative displacements of the vertical fibers, and belong to the space $L^2(\omega; \mathcal{W})$ where (see (11.2)),

$$\mathcal{W} = \Big\{ \mathring{\psi} \in H^1(-1,1)^3 \;\Big|\; \int_{-1}^{1} \mathring{\psi}(y_3)\, dy_3 = 0 \Big\}.$$

For every $\mathring{v} \in L^2(\omega; \mathcal{W})$, define the symmetric matrix $E_w(\mathring{v})$ by

$$E_w(\mathring{v}) = \begin{pmatrix} 0 & 0 & \dfrac{1}{2}\dfrac{\partial \mathring{v}_1}{\partial y_3} \\ 0 & 0 & \dfrac{1}{2}\dfrac{\partial \mathring{v}_2}{\partial y_3} \\ \dfrac{1}{2}\dfrac{\partial \mathring{v}_1}{\partial y_3} & \dfrac{1}{2}\dfrac{\partial \mathring{v}_2}{\partial y_3} & \dfrac{\partial \mathring{v}_3}{\partial y_3} \end{pmatrix}.$$

• *The microscopic displacements or the limit periodic cell displacements*

They are defined separately in the three cases of θ.

Case 1. If $\theta = 0$, set

$$\mathcal{D}_0 = \Big\{ \widehat{\psi} \in L^2(-1,1; H^1_{per}(Y'))^3 \;\Big|\; \int_{Y'} \widehat{\psi}(y', y_3)\, dy' = 0 \ \text{ for a.e. } y_3 \in (-1,1) \Big\}$$

and define the symmetric tensor $\mathcal{E}^0_y(\widehat{\psi})$ by

$$\mathcal{E}^0_y(\widehat{\psi}) = \begin{pmatrix} e_{11,y'}(\widehat{\psi}) & e_{12,y'}(\widehat{\psi}) & \dfrac{1}{2}\dfrac{\partial \widehat{\psi}_3}{\partial y_1} \\ e_{12,y'}(\widehat{\psi}) & e_{22,y'}(\widehat{\psi}) & \dfrac{1}{2}\dfrac{\partial \widehat{\psi}_3}{\partial y_2} \\ \dfrac{1}{2}\dfrac{\partial \widehat{\psi}_3}{\partial y_1} & \dfrac{1}{2}\dfrac{\partial \widehat{\psi}_3}{\partial y_2} & 0 \end{pmatrix}.$$

Case 2. If $\theta \in (0, +\infty)$, set

$$\mathcal{D} = \Big\{ \widehat{\psi} \in \boldsymbol{\mathcal{H}}^1_{per}(Y) \;\Big|\; \int_{Y'} \widehat{\psi}(y', y_3)\, dy' = 0 \ \text{ for a.e. } y_3 \in (-1,1) \Big\}$$

In order to simplify some notations in the statements below, we will write $\mathcal{D}_\theta$ instead of $\mathcal{D}$ for $\theta \in (0, +\infty)$, even though this space does not actually depend on θ (only its norm does).

Introduce also the tensor

$$\mathcal{E}_y^\theta(\widehat{\psi}) = \begin{pmatrix} e_{11,y'}(\widehat{\psi}) & e_{12,y'}(\widehat{\psi}) & \frac{1}{2}\Big(\frac{\partial \widehat{\psi}_3}{\partial y_1} + \theta \frac{\partial \widehat{\psi}_1}{\partial y_3}\Big) \\ e_{12,y'}(\widehat{\psi}) & e_{22,y'}(\widehat{\psi}) & \frac{1}{2}\Big(\frac{\partial \widehat{\psi}_3}{\partial y_2} + \theta \frac{\partial \widehat{\psi}_2}{\partial y_3}\Big) \\ \frac{1}{2}\Big(\frac{\partial \widehat{\psi}_3}{\partial y_1} + \theta \frac{\partial \widehat{\psi}_1}{\partial y_3}\Big) & \frac{1}{2}\Big(\frac{\partial \widehat{\psi}_3}{\partial y_2} + \theta \frac{\partial \widehat{\psi}_2}{\partial y_3}\Big) & \theta \frac{\partial \widehat{\psi}_3}{\partial y_3} \end{pmatrix}.$$

This symmetric tensor $\mathcal{E}_y^\theta$ depends on θ and so does the associated semi-norm (see (11.43) below) on $\mathcal{D}_\theta$ in which it intervenes. In fact, one easily has

$$\|\mathcal{E}_y^\theta(\widehat{\psi})\|_{L^2(Y)} \leqslant (1+\theta)\|\widehat{\psi}\|_{H^1(Y)}, \quad \forall \widehat{\psi} \in \mathcal{D}_\theta \tag{11.42}$$

where C does not depend on θ.
<u>Case 3</u>. If $\theta = +\infty$, set

$$\mathcal{D}_\infty \equiv \Big\{\widehat{\psi} \in L^2(Y'; H^1(-1,1))^3 \mid \int_{Y'} \widehat{\psi}(y', y_3)\, dy' = 0 \ \text{ for a.e. } y_3 \in (-1,1),$$
$$\int_{-1}^{1} \widehat{\psi}_\alpha(\cdot, y_3)\, dy_3 \in H^1_{per}(Y'), \ \int_{-1}^{1} \widehat{\psi}_3(\cdot, y_3)\, dy_3 \in H^2_{per}(Y')\Big\}.$$

Then, every $\widehat{\psi}$ in $\mathcal{D}_\infty$ can be uniquely written as

$$\widehat{\psi} = \Big[\widehat{\Psi}_\alpha - y_3 \frac{\partial \widehat{\Psi}_3}{\partial y_\alpha}\Big]\mathbf{e}_\alpha + \widehat{\Psi}_3 \mathbf{e}_3 + \overline{\widehat{\psi}}, \qquad \widehat{\Psi} \doteq \frac{1}{2}\int_{-1}^{1} \widehat{\psi}(\cdot, y_3)\, dy_3,$$

where $(\widehat{\Psi}_m, \widehat{\Psi}_3, \overline{\widehat{\psi}})$ belongs to $H^1_{per}(Y')^2 \times H^2_{per}(Y') \times L^2(Y'; \mathcal{D}_w)$.

The symmetric tensor $\mathcal{E}_y^\infty(\widehat{\psi})$ is defined as

$$\mathcal{E}_y^\infty(\widehat{\psi}) = \begin{pmatrix} E_{11,y}(\widehat{\psi}) & E_{12,y}(\widehat{\psi}) & \frac{1}{2}\frac{\partial \overline{\widehat{\psi}}_1}{\partial y_3} \\ E_{12,y}(\widehat{\psi}) & E_{22,y}(\widehat{\psi}) & \frac{1}{2}\frac{\partial \overline{\widehat{\psi}}_2}{\partial y_3} \\ \frac{1}{2}\frac{\partial \overline{\widehat{\psi}}_1}{\partial y_3} & \frac{1}{2}\frac{\partial \overline{\widehat{\psi}}_2}{\partial y_3} & \frac{\partial \overline{\widehat{\psi}}_3}{\partial y_3} \end{pmatrix},$$

where

$$E_{\alpha\beta,y}(\widehat{\psi}) = e_{\alpha\beta,y}(\widehat{\Psi}_m) - y_3 \frac{\partial^2 \widehat{\Psi}_3}{\partial y_\alpha \partial y_\beta}.$$

Introduce the sets $\mathbb{D}_0$, $\mathbb{D}_\theta$ (for $\theta \in (0,+\infty)$) and $\mathbb{D}_\infty$ as follows:

$$\begin{aligned}\mathbb{D}_0 &\doteq \{v = (\mathcal{V}, \mathring{v}, \widehat{v}) \in \mathbb{D}_M \times L^2(\omega; \mathcal{W}) \times L^2(\omega; \mathcal{D}_0)\},\\ \mathbb{D}_\theta &\doteq \{v = (\mathcal{V}, \mathring{v}, \widehat{v}) \in \mathbb{D}_M \times L^2(\omega; \mathcal{W}) \times L^2(\omega; \mathcal{D}_\theta)\}^{(9)},\\ \mathbb{D}_\infty &\doteq \{v = (\mathcal{V}, \mathring{v}, \widehat{v}) \in \mathbb{D}_M \times L^2(\omega; \mathcal{W}) \times L^2(\omega; \mathcal{D}_\infty)\}.\end{aligned}$$

To every v in $\mathbb{D}_\theta$, $\theta \in [0,+\infty]$, we associate the symmetric tensor

$$E_M(\mathcal{V}) + E_w(\mathring{v}) + \mathcal{E}_y^\theta(\widehat{v}),$$

and the semi-norm

$$\|v\|_\theta = \|E_M(\mathcal{V}) + E_w(\mathring{v}) + \mathcal{E}_y^\theta(\widehat{v})\|_{L^2(\omega\times Y)}. \tag{11.43}$$

In Lemma 11.19 it will be shown that this semi-norm is actually a norm.

We are now in a position to give the limits of the rescaled and unfolded strain tensor. Proposition 11.14 below completes the results of Proposition 11.13. We continue to distinguish the cases $\theta = 0$, $\theta \in (0,+\infty)$, $\theta = +\infty$.

Proposition 11.14. *Under the hypotheses and with the notations of Proposition 11.13,*

$$\frac{1}{\delta}\mathfrak{T}_{\varepsilon,\delta}(e(u_{\varepsilon,\delta})) \rightharpoonup E_M(\mathcal{U}) + E_w(\mathring{u}) + \mathcal{E}_y^\theta(\widehat{u}) \quad \text{weakly in } L^2(\omega \times Y)^9, \tag{11.44}$$

where the function $\widehat{u}$ is specified below in each of the three cases.

Case 1. For $\theta = 0$,

$$\widehat{u}(\cdot, y) = \widehat{U}(\cdot, y') + y_3\widehat{R}(\cdot, y') + \widehat{\overline{u}}(\cdot, y), \qquad \widehat{u} \in L^2(\omega; \mathcal{D}_0). \tag{11.45}$$

Case 2. For $\theta \in (0,+\infty)$

$$\widehat{u}(\cdot, y) = \widehat{U}(\cdot, y') + y_3\widehat{R}(\cdot, y') + \frac{1}{\theta}\widehat{\overline{u}}(\cdot, y), \qquad \widehat{u} \in L^2(\omega; \mathcal{D}_\theta). \tag{11.46}$$

Case 3. For $\theta = +\infty$

$$\begin{aligned}\widehat{u}(\cdot, y) = &\Big[\widehat{U}_\alpha(\cdot, y') - y_3\frac{\partial \widehat{U}_3}{\partial y_\alpha}(\cdot, y')\Big]\mathbf{e}_\alpha + \widehat{U}_3(\cdot, y')\mathbf{e}_3\\ &+ \big(\widehat{\overline{u}}(\cdot, y) + y_3\widehat{Z}_\alpha(\cdot)\mathbf{e}_\alpha\big), \qquad \widehat{u} \in L^2(\omega; \mathcal{D}_\infty).\end{aligned} \tag{11.47}$$

Observe that in the third case, since the field $\widehat{\overline{u}}$ satisfies conditions (11.38), the function $\widehat{\overline{u}}(\cdot, y) + y_3\widehat{Z}_\alpha(\cdot)\mathbf{e}_\alpha$ belongs to the space

$$\Big\{v \in L^2(\omega \times Y'; H^1(-1,1))^3 \mid \int_{-1}^1 v(\cdot, y_3) = 0, \ \int_{Y'} v(\cdot, y', y_3)dy' = 0\Big\}.$$

(9) For the same reason as for the space $\mathcal{D}_\theta$, the space $\mathbb{D}_\theta$ does not depend on θ.

Proof of Proposition 11.14. In all three cases, it is enough to consider the limits of the following two terms:

$$\frac{1}{\delta}\mathfrak{T}_{\varepsilon,\delta}\big(e_{11}(u_{\varepsilon,\delta})\big) \quad\text{and}\quad \frac{1}{\delta}\mathfrak{T}_{\varepsilon,\delta}\big(e_{13}(u_{\varepsilon,\delta})\big), \tag{11.48}$$

since the limit of all the terms of the form $\frac{1}{\delta}\mathfrak{T}_{\varepsilon,\delta}\big(e_{\alpha\beta}(u_{\varepsilon,\delta})\big)$ are obtained in the same way as the former, and the limit of $\frac{1}{\delta}\mathfrak{T}_{\varepsilon,\delta}\big(e_{23}(u_{\varepsilon,\delta})\big)$ is obtained in the same way as the latter. The limit of $\frac{1}{\delta}\mathfrak{T}_{\varepsilon,\delta}\big(e_{33}(u_{\varepsilon,\delta})\big) = \frac{1}{\delta^2}\frac{\partial \mathfrak{T}_{\varepsilon,\delta}(\overline{u}_{\varepsilon,\delta,3})}{\partial y_3}$ was already obtained in Proposition 11.13 ((11.39), (11.40) and (11.41)).

<u>Case 1</u>. $\theta = 0$. In this step, in order to treat the terms from (11.48), we refer to Proposition 11.13, in particular to convergence (11.39), and use the explicit expression (11.6) of the strain tensor. The result is

$$\begin{aligned}\frac{1}{\delta}\mathfrak{T}_{\varepsilon,\delta}\big(e_{11}(u_{\varepsilon,\delta})\big) &= \frac{1}{\delta}\mathcal{T}_\varepsilon\Big(\frac{\partial \mathcal{U}_{\varepsilon,\delta,1}}{\partial x_1}\Big) + y_3\mathcal{T}_\varepsilon\Big(\frac{\partial \mathcal{R}_{\varepsilon,\delta,1}}{\partial x_1}\Big) + \frac{1}{\varepsilon\delta}\frac{\partial \mathfrak{T}_{\varepsilon,\delta}(\overline{u}_{\varepsilon,\delta,1})}{\partial y_1}\\ &\rightharpoonup \frac{\partial \mathcal{U}_1}{\partial x_1} + \frac{\partial \widehat{U}_1}{\partial y_1} + y_3\Big(\frac{\partial \mathcal{R}_1}{\partial x_1} + \frac{\partial \widehat{R}_1}{\partial y_1}\Big) + \frac{\partial \widehat{\overline{u}}_1}{\partial y_1} \quad \text{weakly in } L^2(\omega\times Y),\end{aligned}$$

and similarly,

$$\begin{aligned}&\frac{1}{\delta}\mathfrak{T}_{\varepsilon,\delta}\big(e_{13}(u_{\varepsilon,\delta})\big)\\ &\quad= \frac{1}{2}\Big[\frac{1}{\delta}\mathcal{T}_\varepsilon\Big(\frac{\partial \mathcal{U}_{\varepsilon,\delta,3}}{\partial x_1} + \mathcal{R}_{\varepsilon,\delta,1}\Big) + \frac{1}{\delta^2}\frac{\partial \mathfrak{T}_{\varepsilon,\delta}(\overline{u}_{\varepsilon,\delta,1})}{\partial y_3} + \frac{1}{\varepsilon\delta}\frac{\partial \mathfrak{T}_{\varepsilon,\delta}(\overline{u}_{\varepsilon,\delta,3})}{\partial y_1}\Big]\\ &\qquad \rightharpoonup \frac{1}{2}\Big[\mathcal{Z}_1 + \widehat{\mathcal{Z}}_1 + \frac{\partial \overline{u}_1}{\partial y_3} + \frac{\partial \widehat{\overline{u}}_3}{\partial y_1}\Big] \quad \text{weakly in } L^2(\omega\times Y).\end{aligned}$$

Recalling that

$$\mathcal{R}_\alpha = -\frac{\partial \mathcal{U}_3}{\partial x_\alpha}, \quad \widehat{\mathcal{Z}}_1 = \frac{\partial \widehat{U}_3}{\partial y_1}, \quad \frac{\partial \mathring{u}_1}{\partial y_3} = \mathcal{Z}_1 + \frac{\partial \overline{u}_1}{\partial y_3},$$

and introducing the field $\widehat{u}$ (see (11.45)), convergence (11.44) is proved.

<u>Case 2</u>. $\theta \in (0, +\infty)$. Here again, consider the terms from (11.48). In view of (11.40) one obtains

$$\begin{aligned}\frac{1}{\delta}\mathfrak{T}_{\varepsilon,\delta}\big(e_{11}(u_{\varepsilon,\delta})\big) &= \frac{1}{\delta}\mathcal{T}_\varepsilon\Big(\frac{\partial \mathcal{U}_{\varepsilon,\delta,1}}{\partial x_1}\Big) + y_3\mathcal{T}_\varepsilon\Big(\frac{\partial \mathcal{R}_{\varepsilon,\delta,1}}{\partial x_1}\Big) + \frac{1}{\varepsilon\delta}\frac{\partial \mathfrak{T}_{\varepsilon,\delta}(\overline{u}_{\varepsilon,\delta,1})}{\partial y_1}\\ &\rightharpoonup \frac{\partial \mathcal{U}_1}{\partial x_1} + \frac{\partial \widehat{U}_1}{\partial y_1} + y_3\Big(\frac{\partial \mathcal{R}_1}{\partial x_1} + \frac{\partial \widehat{R}_1}{\partial y_1}\Big) + \frac{1}{\theta}\frac{\partial \overline{u}_1}{\partial y_1} \quad \text{weakly in } L^2(\omega\times Y).\end{aligned}$$

Similarly,

$$\frac{1}{\delta}\mathfrak{T}_{\varepsilon,\delta}(e_{13}(u_{\varepsilon,\delta}))$$
$$=\frac{1}{2}\Big[\frac{1}{\delta}\mathcal{T}_\varepsilon\Big(\frac{\partial \mathcal{U}_{\varepsilon,\delta,3}}{\partial x_1}+\mathcal{R}_{\varepsilon,\delta,1}\Big)+\frac{1}{\delta^2}\frac{\partial \mathfrak{T}_{\varepsilon,\delta}(\overline{u}_{\varepsilon,\delta,1})}{\partial y_3}+\frac{1}{\varepsilon\delta}\frac{\partial \mathfrak{T}_{\varepsilon,\delta}(\overline{u}_{\varepsilon,\delta,3})}{\partial y_1}\Big]$$
$$\rightharpoonup \frac{1}{2}\Big[\mathcal{Z}_1+\widehat{\mathcal{Z}}_1+\frac{\partial \overline{u}_1}{\partial y_3}+\frac{\partial \widehat{\overline{u}}_1}{\partial y_3}+\frac{1}{\theta}\frac{\partial \widehat{\overline{u}}_3}{\partial y_1}\Big] \quad \text{weakly in } L^2(\omega\times Y).$$

In this case

$$\widehat{\mathcal{Z}}_1=\frac{\partial \widehat{U}_3}{\partial y_1}+\theta\widehat{R}_1 \quad \text{and} \quad \frac{\partial \mathring{u}_1}{\partial y_3}=\mathcal{Z}_1+\frac{\partial \overline{u}_1}{\partial y_3}.$$

As in the Case 1, with the help of the field $\widehat{u}$ (see (11.46)), we get convergence (11.44).

<u>Case 3</u>. $\theta=+\infty$. The corresponding convergences for the terms in (11.48) are now (see (11.41)),

$$\frac{1}{\delta}\mathfrak{T}_{\varepsilon,\delta}(e_{11}(u_{\varepsilon,\delta}))=\frac{1}{\delta}\mathcal{T}_\varepsilon\Big(\frac{\partial \mathcal{U}_{\varepsilon,\delta,1}}{\partial x_1}\Big)+y_3\mathcal{T}_\varepsilon\Big(\frac{\partial \mathcal{R}_{\varepsilon,\delta,1}}{\partial x_1}\Big)+\frac{1}{\varepsilon\delta}\frac{\partial \mathfrak{T}_{\varepsilon,\delta}(\overline{u}_{\varepsilon,\delta,1})}{\partial y_1}$$
$$\rightharpoonup \frac{\partial \mathcal{U}_1}{\partial x_1}+\frac{\partial \widehat{U}_1}{\partial y_1}+y_3\Big(\frac{\partial \mathcal{R}_1}{\partial x_1}-\frac{\partial^2 \widehat{U}_3}{\partial y_1^2}\Big) \quad \text{weakly in } L^2(\omega\times Y),$$

and

$$\frac{1}{\delta}\mathfrak{T}_{\varepsilon,\delta}(e_{13}(u_{\varepsilon,\delta}))$$
$$=\frac{1}{2}\Big[\frac{1}{\delta}\mathcal{T}_\varepsilon\Big(\frac{\partial \mathcal{U}_{\varepsilon,\delta,3}}{\partial x_1}+\mathcal{R}_{\varepsilon,\delta,1}\Big)+\frac{1}{\delta^2}\frac{\partial \mathfrak{T}_{\varepsilon,\delta}(\overline{u}_{\varepsilon,\delta,1})}{\partial y_3}+\frac{1}{\varepsilon\delta}\frac{\partial \mathfrak{T}_{\varepsilon,\delta}(\overline{u}_{\varepsilon,\delta,3})}{\partial y_1}\Big]$$
$$\rightharpoonup \frac{1}{2}\Big[\mathcal{Z}_1+\frac{\partial \overline{u}_1}{\partial y_3}+\widehat{\mathcal{Z}}_1+\frac{\partial \widehat{\overline{u}}_1}{\partial y_3}\Big] \quad \text{weakly in } L^2(\omega\times Y).$$

Due to (11.20), and using again the field $\widehat{u}$ (see (11.47)), convergence (11.44) follows. □

As a first consequence of Proposition 11.14, we obtain the limit of the unfolded stress tensor (11.12).

Corollary 11.15. *For the rescaled and unfolded stress tensor, one has the convergence*

$$\frac{1}{\delta}\mathfrak{T}_{\varepsilon,\delta}(\sigma^{\varepsilon,\delta})\rightharpoonup\Sigma^\theta \quad \textit{weakly in } L^2(\omega\times Y),$$

i.e., for $1\leqslant i,j,k,l\leqslant 3$,

$$\frac{1}{\delta}\mathfrak{T}_{\varepsilon,\delta}(\sigma^{\varepsilon,\delta}_{ij}(u_{\varepsilon,\delta}))\rightharpoonup\Sigma^\theta_{ij}=a_{ijkl}\,E_{kl,M}(\mathcal{U})+a_{ijkl}\,E_{kl,w}(\mathring{u})+a_{ijkl}\,\mathcal{E}^\theta_{kl,y}(\widehat{u}).$$

Remark 11.16. *As can be seen from* (11.44), *the tensor* $E_M(\mathcal{U}) + E_w(\mathring{u})$ *is the limit strain tensor of the sequence* $\{u_{\varepsilon,\delta}\}_{\varepsilon,\delta}$ *in the rescaled domain* Ω. *The third tensor appearing in* (11.44) *captures the strong oscillations.*

In the case $\theta = 0$, *the cell* $(0,\varepsilon)^2 \times (-\delta,\delta)$ *looks like a thin beam. The limit tensor* $\mathcal{E}^0_y(\widehat{u})$ *represents the behavior of the symmetric gradient of a thin periodic beam (the periods are in the directions of its smallest dimensions).*

In the case $\theta = +\infty$, *the tensor* $\mathcal{E}^\infty_y(\widehat{u})$ *corresponds to the limit tensor of a periodic thin plate. That is why this tensor looks like* $E_M(\mathcal{U}) + E_w(\mathring{u})$.

The lemma below gives a Korn inequality for fields in $\mathcal{D}_0$ (resp. $\mathcal{D}_\infty$).

Lemma 11.17. *For every* $\widehat{\psi}$ *in* $\mathcal{D}_0$ *the following estimate holds:*

$$\|\widehat{\psi}\|_{L^2(-1,1;H^1(Y'))} \leqslant C\|\mathcal{E}^0_y(\widehat{\psi})\|_{L^2(Y)}. \tag{11.49}$$

For every $\widehat{\psi}$ *in* $\mathcal{D}_\infty$, *decomposed as*

$$\widehat{\psi} = \Big[\widehat{\Psi}_\alpha - y_3\frac{\partial \widehat{\Psi}_3}{\partial y_\alpha}\Big]\mathbf{e}_\alpha + \widehat{\Psi}_3\mathbf{e}_3 + \overline{\widehat{\psi}}, \tag{11.50}$$

where $\big(\widehat{\Psi}_m, \widehat{\Psi}_3, \overline{\widehat{\psi}}\,\big)$ *belongs to* $H^1_{per}(Y')^2 \times H^2_{per}(Y') \times L^2(Y';\mathcal{D}_w)$, *the following estimates hold:*

$$\begin{aligned} &\sum_{\alpha=1}^2 \|\widehat{\Psi}_\alpha\|_{H^1(Y')} + \|\widehat{\Psi}_3\|_{H^2(Y')} \leqslant C\|\mathcal{E}^\infty_y(\widehat{\psi})\|_{L^2(Y)},\\ &\|\overline{\widehat{\psi}}\|_{L^2(Y';H^1(-1,1))} \leqslant C\|\mathcal{E}^\infty_y(\widehat{\psi})\|_{L^2(Y)}. \end{aligned} \tag{11.51}$$

Proof. Estimate (11.49) for $\widehat{\psi}_\alpha, \alpha = 1,2$, is an immediate consequence of the definitions of $\mathcal{D}_0$, $\mathcal{E}^0_y$ and the 2-dimensional Korn inequality for periodic functions (see (2.14) in Proposition 2.11), which writes

$$\sum_{\alpha=1}^2 \|\widehat{\psi}_\alpha\|_{L^2(-1,1;H^1(Y'))} \leqslant C\sum_{\alpha,\beta=1}^2 \|e_{\alpha\beta}(\widehat{\psi})\|_{L^2(Y)}.$$

Recall that

$$\forall\phi \in L^2(-1,1;H^1(Y')) \quad \text{such that} \int_{Y'} \phi(\cdot, y_3)dy' = 0,$$

the 2-dimensional Poincaré-Wirtinger inequality yields

$$\|\phi\|_{L^2(-1,1;H^1(Y'))} \leqslant C\Big(\Big\|\frac{\partial\phi}{\partial y_1}\Big\|_{L^2(Y)} + \Big\|\frac{\partial\phi}{\partial y_2}\Big\|_{L^2(Y)}\Big). \tag{11.52}$$

This, applied to $\widehat{\psi}_3$ gives (11.49) (for $\widehat{\psi}_3$).

To obtain estimates (11.51), with $\widehat{\psi}$ decomposed as in (11.50), we apply again the 2-dimensional Korn and Poincaré-Wirtinger inequalities. Taking into account the periodicity give $(11.51)_1$.

Finally, the Poincaré-Wirtinger inequality yields $(11.51)_2$. □

In $\mathcal{D}_\theta$, $\theta \in (0, +\infty)$, the semi-norm $\|\mathcal{E}_y^\theta(\cdot)\|_{L^2(Y)}$ is clearly equivalent to the usual norm of $H^1(Y)^3$, but with constants which depend upon θ. In the next lemma, we make these constants explicit.

Lemma 11.18. *For every $\psi \in \mathcal{D}_\theta$, $\theta \in (0, +\infty)$, the following estimates hold:*
(i) If $\theta \geqslant 1$, then

$$\begin{aligned}
&\|\psi_\alpha\|_{L^2(Y)} + \frac{1}{\theta}\|\psi_3\|_{L^2(Y)} \leqslant C\|\mathcal{E}_y^\theta(\psi)\|_{L^2(Y)},\\
&\sum_{\alpha,\beta=1}^{2}\Big\|\frac{\partial\psi_\alpha}{\partial y_\beta}\Big\|_{L^2(Y)} + \theta\Big\|\frac{\partial\psi_3}{\partial y_3}\Big\|_{L^2(Y)} \leqslant C\|\mathcal{E}_y^\theta(\psi)\|_{L^2(Y)}, \qquad (11.53)\\
&\sum_{\alpha=1}^{2}\Big(\Big\|\frac{\partial\psi_\alpha}{\partial y_3}\Big\|_{L^2(Y)} + \frac{1}{\theta}\Big\|\frac{\partial\psi_3}{\partial y_\alpha}\Big\|_{L^2(Y)}\Big) \leqslant C\|\mathcal{E}_y^\theta(\psi)\|_{L^2(Y)}.
\end{aligned}$$

(ii) If $\theta \leqslant 1$, then

$$\|\psi\|_{L^2(Y)} + \sum_{\alpha=1}^{2}\Big\|\frac{\partial\psi}{\partial y_\alpha}\Big\|_{L^2(Y)} + \theta\Big\|\frac{\partial\psi}{\partial y_3}\Big\|_{L^2(Y)} \leqslant C\|\mathcal{E}_y^\theta(\psi)\|_{L^2(Y)}. \qquad (11.54)$$

The constants do not depend on θ.

Proof. We show that, after a simple change of variables, the tensor $\mathcal{E}_y^\theta(\psi)$ for $\psi \in \mathcal{D}_\theta$, appears as the symmetric gradient of a displacement belonging to $H^1(Y_\theta)^3$ periodic in the directions $\mathbf{e}_\alpha$ $(\alpha = 1, 2)$ where

$$Y_\theta = Y' \times \Big(-\frac{1}{\theta}, +\frac{1}{\theta}\Big).$$

For $\psi \in \mathcal{D}_\theta$, $\theta \in (0, +\infty)$, define the displacement Ψ as

$$\Psi(z) = \psi\big(z_1, z_2, \theta z_3\big) \quad \text{for a.e. } z \in Y_\theta. \qquad (11.55)$$

For a.e. $z \in Y_\theta$, one has

$$\begin{aligned}
e_{\alpha\beta,z}(\Psi)(z) &= e_{\alpha\beta,y}(\psi)\big(z_1, z_2, \theta z_3\big),\\
e_{\alpha3,z}(\Psi)(z) &= \frac{1}{2}\Big(\theta\frac{\partial\psi_\alpha}{\partial y_3} + \frac{\partial\psi_3}{\partial y_\alpha}\Big)\big(z_1, z_2, \theta z_3\big),\\
e_{33,z}(\Psi)(z) &= \theta\frac{\partial\psi_3}{\partial y_3}\big(z_1, z_2, \theta z_3\big).
\end{aligned}$$

By (11.55), a simple computation shows that

$$\|e_z(\Psi)\|_{L^2(Y_\theta)} = \frac{1}{\sqrt{\theta}}\|\mathcal{E}_y^\theta(\psi)\|_{L^2(Y)}.$$

Case $\theta \geqslant 1$. Observe that in this case, the set Y_θ is actually, a plate of thickness $2/\theta$ and of mid-surface Y'. Hence, by applying Proposition 11.6 with $\delta = 1/\theta$, there exists a rigid displacement R such that

$$\|\Psi_\alpha - R_\alpha\|_{L^2(Y_\theta)} + \frac{1}{\theta}\|\Psi_3 - R_3\|_{L^2(Y_\theta)} \leqslant C\|e_z(\Psi)\|_{L^2(Y_\theta)},$$

$$\sum_{\alpha,\beta=1}^{2} \Big\|\frac{\partial(\Psi - R)_\alpha}{\partial z_\beta}\Big\|_{L^2(Y_\theta)} + \Big\|\frac{\partial \Psi}{\partial z_3}\Big\|_{L^2(Y_\theta)} \leqslant C\|e_z(\Psi)\|_{L^2(Y_\theta)},$$

$$\sum_{\alpha=1}^{2} \Big(\Big\|\frac{\partial(\Psi - R)_\alpha}{\partial z_3}\Big\|_{L^2(Y_\theta)} + \Big\|\frac{\partial(\Psi - R)_3}{\partial z_\alpha}\Big\|_{L^2(Y_\theta)}\Big) \leqslant C\theta\|e_z(\Psi)\|_{L^2(Y_\theta)}.$$

Using the periodicity of Ψ and the fact that

$$\int_{Y'} \Psi(z_1, z_2, z_3)\, dz_1 dz_2 = 0 \quad \text{for a.e. } z_3 \in \Big(-\frac{1}{\theta}, +\frac{1}{\theta}\Big),$$

the rigid displacement can be eliminated in the above estimates, to get

$$\|\Psi_\alpha\|_{L^2(Y_\theta)} + \frac{1}{\theta}\|\Psi_3\|_{L^2(Y_\theta)} \leqslant C\|e_z(\Psi)\|_{L^2(Y_\theta)},$$

$$\sum_{\alpha,\beta=1}^{2} \Big\|\frac{\partial \Psi_\alpha}{\partial z_\beta}\Big\|_{L^2(Y_\theta)} + \Big\|\frac{\partial \Psi}{\partial z_3}\Big\|_{L^2(Y_\theta)} \leqslant C\|e_z(\Psi)\|_{L^2(Y_\theta)},$$

$$\sum_{\alpha=1}^{2} \Big(\Big\|\frac{\partial \Psi_\alpha}{\partial z_3}\Big\|_{L^2(Y_\theta)} + \Big\|\frac{\partial \Psi_3}{\partial z_\alpha}\Big\|_{L^2(Y_\theta)}\Big) \leqslant C\theta\|e_z(\Psi)\|_{L^2(Y_\theta)}.$$

Then, going back to the variable y_3, we obtain (11.53).

Case $\theta \leqslant 1$. In this case, we cannot assimilate the domain to a thin plate. Set

$$\mathcal{N} = \Big[\frac{1}{\theta}\Big],$$

for the integer part of $1/\theta$, and consider the domain

$$\mathcal{Y}_\theta = \big(0, (\mathcal{N} + 1)\big)^2 \times \Big(-\frac{1}{\theta}, +\frac{1}{\theta}\Big).$$

Now, extend the function Ψ from (11.55) by periodicity to $\mathcal{Y}_\theta$.

The domain $\mathcal{Y}_\theta$ is star-shaped with respect to a ball of radius $1/(2\theta)$, since

$$\Big(0, \frac{1}{\theta}\Big)^2 \times \Big(-\frac{1}{\theta}, +\frac{1}{\theta}\Big) \subset \mathcal{Y}_\theta,$$

and its diameter is less than $4/\theta$. Thus, the three-dimensional Korn inequality holds in $\mathcal{Y}_\theta$ with a constant independent of θ (cf. [124, Theorem 2.3]). Here, this gives

$$\|\nabla\Psi - \mathbf{A}\|_{L^2(\mathcal{Y}_\theta)} \leqslant C\|e_z(\Psi)\|_{L^2(\mathcal{Y}_\theta)},$$

for some antisymmetric matrix $\mathbf{A}$ and C independent of θ. By periodicity,

$$\|\nabla\Psi - \mathbf{A}\|_{L^2(\mathcal{Y}_\theta)} = (\mathcal{N}+1)\|\nabla\Psi - \mathbf{A}\|_{L^2(Y_\theta)},$$
$$\|e_z(\Psi)\|_{L^2(\mathcal{Y}_\theta)} = (\mathcal{N}+1)\|e_z(\Psi)\|_{L^2(Y_\theta)}.$$

Hence,

$$\|\nabla\Psi - \mathbf{A}\|_{L^2(Y_\theta)} \leqslant C\|e_z(\Psi)\|_{L^2(Y_\theta)}.$$

This, with the periodicity of Ψ and the Cauchy-Schwarz inequality, gives

$$|Y_\theta|\,|\mathbf{A}\mathbf{e}_\alpha| = \Big|\int_{Y_\theta}\Big(\frac{\partial\Psi}{\partial y_\alpha} - \mathbf{A}\mathbf{e}_\alpha\Big)dy\Big| \leqslant |Y_\theta|^{1/2}\Big\|\frac{\partial\Psi}{\partial y_\alpha} - \mathbf{A}\mathbf{e}_\alpha\Big\|_{L^2(Y_\theta)}$$
$$\leqslant C|Y_\theta|^{1/2}\|e_z(\Psi)\|_{L^2(Y_\theta)}.$$

Since the constant matrix $\mathbf{A}$ is antisymmetric,

$$|Y_\theta|^{1/2}\,|\mathbf{A}| = \|\mathbf{A}\|_{L^2(Y_\theta)} \leqslant C\|e_z(\Psi)\|_{L^2(Y_\theta)},$$

so that

$$\|\nabla\Psi\|_{L^2(Y_\theta)} \leqslant C\|e_z(\Psi)\|_{L^2(Y_\theta)}.$$

Then estimate (11.52) yields

$$\|\Psi\|_{L^2(Y_\theta)} \leqslant C\|e_z(\Psi)\|_{L^2(Y_\theta)}.$$

Going back to Y, one gets (11.54). □

Lemma 11.19. *The semi-norm $\|\cdot\|_\theta$, $\theta \in [0,+\infty]$, is equivalent to the product-norm of the space $\mathbb{D}_\theta$.*

Proof. Step 1. Proof of the first inequality ($\leqslant$).

By the definitions of the tensor $\mathcal{E}^\theta_y$ and of the space $\mathcal{D}_\theta$, $\theta \in [0,+\infty]$, for every $\widehat{\psi} \in L^2(\omega;\mathcal{D}_\theta)$,

$$\int_{Y'}\mathcal{E}^\theta_y(\widehat{\psi})(x',y',y_3)dy' = 0 \quad \text{for a.e. } (x',y_3) \in \Omega.$$

Then, for every $v = (\mathcal{V},\mathring{v},\widehat{v}) \in \mathbb{D}_\theta$ (recall that the semi-norm $\|\cdot\|_\theta$ is defined by (11.43)), one has

$$\|v\|_\theta^2 = \|E_M(\mathcal{V})\|^2_{L^2(\Omega)} + \|E_w(\mathring{v})\|^2_{L^2(\Omega)} + \|\mathcal{E}^\theta_y(\widehat{v})\|_{L^2(\omega\times Y)}. \tag{11.56}$$

Hence, for all v in $\mathbb{D}_\theta$, $\theta \in [0,+\infty]$,

$$\|v\|_0 \leqslant C\big(\|\mathcal{V}_m\|_{H^1(\omega)} + \|\mathcal{V}_3\|_{H^2(\omega)} + \|\widehat{v}\|_{L^2(\Omega;H^1(Y'))} + \|\mathring{v}\|_{L^2(\omega;H^1(-1,1))}\big),$$
$$\|v\|_\theta \leqslant C\big(\|\mathcal{V}_m\|_{H^1(\omega)} + \|\mathcal{V}_3\|_{H^2(\omega)} + (1+\theta)\|\widehat{v}\|_{L^2(\omega;H^1(Y))}$$
$$+ \|\mathring{v}\|_{L^2(\omega;H^1(-1,1))}\big),$$
$$\|v\|_\infty \leqslant C\big(\|\mathcal{V}_m\|_{H^1(\omega)} + \|\mathcal{V}_3\|_{H^2(\omega)} + \|\widehat{\mathcal{V}}_m\|_{L^2(\omega;H^1(Y'))} + \|\widehat{\mathcal{V}}_3\|_{L^2(\omega;H^2(Y'))}$$
$$+ \|\widehat{v}\|_{L^2(\omega\times Y';H^1(-1,1))} + \|\mathring{v}\|_{L^2(\omega;H^1(-1,1))}\big),$$

which proves that the semi-norm $\|\cdot\|_\theta$ is bounded above by a multiple of the norm of the corresponding product space. The constants do not depend on θ.

Step 2. Proof of the converse inequality ($\geqslant$).

Let $v = (\mathcal{V}, \mathring{v}, \widehat{v})$ be in $\mathbb{D}_\theta$, $\theta \in [0, +\infty]$. From (11.56), on the one hand, one has

$$\sum_{\alpha,\beta=1}^{2} \Big(2\|e_{\alpha\beta}(\mathcal{V}_m)\|^2_{L^2(\omega)} + \frac{2}{3}\Big\|\frac{\partial^2 \mathcal{V}_3}{\partial x_\alpha \partial x_\beta}\Big\|^2_{L^2(\omega)}\Big) = \|E_M(\mathcal{V})\|^2_{L^2(\Omega)}$$

and on the other hand,

$$\frac{1}{4}\sum_{\alpha=1}^{2} \Big\|\frac{\partial \mathring{v}_\alpha}{\partial y_3}\Big\|^2_{L^2(\Omega)} + \Big\|\frac{\partial \mathring{v}_3}{\partial y_3}\Big\|^2_{L^2(\Omega)} = \|E_w(\mathring{v})\|^2_{L^2(\Omega)}.$$

The first equality, together with the 2-dimensional Korn and Poincaré inequalities in ω and the boundary conditions, implies

$$\|\mathcal{V}_m\|_{H^1(\omega)} + \|\mathcal{V}_3\|_{H^2(\omega)} \leqslant C\|E_M(\mathcal{V})\|_{L^2(\Omega)} \leqslant C\|v\|_\theta,$$

From the Poincaré-Wirtinger inequality applied in the second equality, one gets

$$\|\mathring{v}\|_{L^2(\omega;H^1(-1,1))} \leqslant C\Big\|\frac{\partial \mathring{v}}{\partial y_3}\Big\|_{L^2(\Omega)} \leqslant C\|E_w(\mathring{v})\|_{L^2(\Omega)} \leqslant C\|v\|_\theta.$$

The constants do not depend on θ.

To conclude the proof, the three cases are treated separately.

Case 1. $\theta = 0$. Lemma 11.17 and (11.56) give

$$\|\widehat{v}\|_{L^2(\Omega;H^1(Y'))} \leqslant C\|\mathcal{E}^0_y(\widehat{v})\|_{L^2(\omega\times Y)} \leqslant C\|v\|_0.$$

Case 2. $\theta \in (0, +\infty)$. Due to Lemma 11.18 and (11.56), there exists a constant C_θ such that

$$\|\widehat{v}\|_{L^2(\omega;H^1(Y))} \leqslant C_\theta\|\mathcal{E}^\theta_y(\widehat{v})\|_{L^2(\omega\times Y)} \leqslant C_\theta\|v\|_\theta.$$

Case 2. $\theta = +\infty$. Decompose $\widehat{v}$ as

$$\widehat{v}(\cdot, y) = \Big[\widehat{V}_\alpha(\cdot, y') - y_3\frac{\partial \widehat{V}_3}{\partial y_\alpha}(\cdot, y')\Big]\mathbf{e}_\alpha + \widehat{V}_3(\cdot, y')\mathbf{e}_3 + \overline{\widehat{v}}(\cdot, y).$$

Then, again Lemma 11.17 and (11.56) give

$$\sum_{\alpha=1}^{2} \|\widehat{V}_\alpha\|_{L^2(\omega\,;H^1(Y'))} + \|\widehat{V}_3\|_{L^2(\omega;H^2(Y'))} \leqslant C\|\mathcal{E}^\infty_y(\widehat{v})\|_{L^2(\omega\times Y)} \leqslant C\|v\|_\infty,$$

and

$$\|\overline{\widehat{v}}\|_{L^2(\omega\times Y';H^1(-1,1))} \leqslant C\|\mathcal{E}^\infty_y(\widehat{v})\|_{L^2(\omega\times Y)} \leqslant C\|v\|_\infty.$$

Finally, summarizing the results of Steps 1 and 2, the proof of Lemma 11.19 is complete. □

11.6 The unfolded limit problems

Theorem 11.20. *Let $u_{\varepsilon,\delta}$ be the solution of the elasticity problem* (11.13). *Then, for the whole sequence* (ε,δ) *and for* $\theta \in [0,+\infty]$, *the following convergence holds:*

$$\frac{1}{\delta}\mathfrak{T}_{\varepsilon,\delta}\big(e(u_{\varepsilon,\delta})\big) \to E_M(\mathcal{U})+E_w(\mathring{u})+\mathcal{E}^\theta_y(\widehat{u}) \quad \textit{strongly in } L^2(\omega\times Y)^9, \tag{11.57}$$

and $u^\theta = u = (\mathcal{U},\mathring{u},\widehat{u})$ *belonging to* $\mathbb{D}_\theta$ *is the solution of the rescaled and unfolded problem*

$$\begin{cases} \dfrac{1}{2}\displaystyle\int_{\omega\times Y} a_{ijkl}\big(E_{ij,M}(\mathcal{U}) + E_{ij,w}(\mathring{u}) + \mathcal{E}^\theta_{ij,y}(\widehat{u})\big) \\ \qquad\qquad \times\big(E_{kl,M}(\mathcal{V}) + E_{kl,w}(\mathring{v}) + \mathcal{E}^\theta_{kl,y}(\widehat{v})\big)dx'dy \\ \qquad = \displaystyle\int_\omega \Big(f\cdot\mathcal{V} - \frac{1}{3}g_\alpha\frac{\partial\mathcal{V}_3}{\partial x_\alpha}\Big)dx', \quad \forall v = (\mathcal{V},\mathring{v},\widehat{v}) \in \mathbb{D}_\theta. \end{cases} \tag{11.58}$$

Proof. Due to hypotheses of Section 11.2 and of Lemma 11.19, the Lax-Milgram theorem applies to Problem (11.58) which, therefore, has a unique solution. This uniqueness implies it is enough to prove convergence (11.57) for a subsequence, as is done now.

Case 1. $\theta \in [0,+\infty)$.

Let $v = (\mathcal{V},\mathring{v},\widehat{v})$ be given in $\mathbb{D}_\theta$ with $\mathcal{V} = (\mathcal{V}_m,\mathcal{V}_3) \in \mathbb{D}_M$ and $(\mathring{v},\widehat{v})$ in $\mathring{V}\times V'_{per}$ where

$$\mathring{V} \doteq \Big\{\Phi \in \mathcal{C}^1(\overline{\omega}\times[-1,1])^3 \mid \Phi(\cdot,y_3) = 0 \text{ on } \partial\omega \text{ for all } y_3 \in [-1,1]\Big\},$$

$$V_{per} \doteq \Big\{\Psi \in \mathcal{C}^1(\overline{\omega}\times\overline{Y})^3 \mid \Psi\ Y'\text{-periodic and } \Psi(\cdot,y) = 0 \text{ on } \partial\omega \text{ for all } y \in \overline{Y}\Big\}.$$

Consider the following test displacement:

$$w_{\varepsilon,\delta} = v_\delta + v_{\varepsilon,\delta},$$

where, for every x in Ω_δ,

$$v_\delta(x) = \Big[\delta\mathcal{V}_\alpha(x') - x_3\frac{\partial\mathcal{V}_3}{\partial x_\alpha}(x') + \delta^2\mathring{v}_\alpha\Big(x',\frac{x_3}{\delta}\Big)\Big]\mathbf{e}_\alpha + \Big[\mathcal{V}_3(x') + \delta^2\mathring{v}_3\Big(x',\frac{x_3}{\delta}\Big)\Big]\mathbf{e}_3,$$

$$v_{\varepsilon,\delta}(x) = \delta\varepsilon\widehat{v}\Big(x',\Big\{\frac{x'}{\varepsilon}\Big\},\frac{x_3}{\delta}\Big).$$

A straightforward computation gives

$$\frac{1}{\delta}\mathfrak{T}_\delta\big(e(v_\delta)\big) \to E_M(\mathcal{V}) + E_w(\mathring{v}) \quad \text{strongly in} \quad L^2(\Omega)^9.$$

Hence,

$$\frac{1}{\delta}\mathfrak{T}_{\varepsilon,\delta}\big(e(v_\delta)\big) \to E_M(\mathcal{V}) + E_w(\mathring{v}) \quad \text{strongly in} \quad L^2(\omega\times Y)^9.$$

Also,

$$
\begin{aligned}
e_{\alpha\beta}(v_{\varepsilon,\delta})(x) &= \frac{\varepsilon\delta}{2}\Big(\frac{\partial \widehat{v}_\alpha}{\partial x_\beta} + \frac{\partial \widehat{v}_\beta}{\partial x_\alpha}\Big)\Big(x', \Big\{\frac{x'}{\varepsilon}\Big\}, \frac{x_3}{\delta}\Big)\\
&\quad + \frac{\delta}{2}\Big(\frac{\partial \widehat{v}_\alpha}{\partial y_\beta} + \frac{\partial \widehat{v}_\beta}{\partial y_\alpha}\Big)\Big(x', \Big\{\frac{x'}{\varepsilon}\Big\}, \frac{x_3}{\delta}\Big),\\
e_{\alpha 3}(v_{\varepsilon,\delta})(x) &= \frac{\varepsilon\delta}{2}\Big(\frac{\partial \widehat{v}_\alpha}{\partial x_3} + \frac{\partial \widehat{v}_3}{\partial x_\alpha}\Big)\Big(x', \Big\{\frac{x'}{\varepsilon}\Big\}, \frac{x_3}{\delta}\Big)\\
&\quad + \frac{\varepsilon}{2}\frac{\partial \widehat{v}_\alpha}{\partial y_3}\Big(x', \Big\{\frac{x'}{\varepsilon}\Big\}, \frac{x_3}{\delta}\Big) + \frac{\delta}{2}\frac{\partial \widehat{v}_3}{\partial y_\alpha}\Big(x', \Big\{\frac{x'}{\varepsilon}\Big\}, \frac{x_3}{\delta}\Big)\\
e_{33}(v_{\varepsilon,\delta})(x) &= \varepsilon\delta\frac{\partial \widehat{v}_3}{\partial x_3}\Big(x', \Big\{\frac{x'}{\varepsilon}\Big\}, \frac{x_3}{\delta}\Big) + \varepsilon\frac{\partial \widehat{v}_3}{\partial y_3}\Big(x', \Big\{\frac{x'}{\varepsilon}\Big\}, \frac{x_3}{\delta}\Big).
\end{aligned}
$$

It is easily seen that

$$\frac{1}{\delta}\mathfrak{T}_{\varepsilon,\delta}\big(e(v_{\varepsilon,\delta})\big) \to \mathcal{E}^\theta_y(\widehat{v}) \quad \text{strongly in} \quad L^2(\omega\times Y)^9,$$

and then,

$$\frac{1}{\delta}\mathfrak{T}_{\varepsilon,\delta}\big(e(v_{\varepsilon,\delta})\big) \to E_M(\mathcal{V}) + E_w(\mathring{v}) + \mathcal{E}^\theta_y(\widehat{v}) \quad \text{strongly in} \quad L^2(\omega\times Y)^9.$$

Taking $v_{\varepsilon,\delta}$ as test displacement in (11.13), unfolding the equality with $\mathfrak{T}_{\varepsilon,\delta}$, dividing by $2\delta^3$, and passing to the limit, give (11.58) with v as the test function. The density of the product space $\mathring{V}\times V_{per}$ in $L^2(\omega;\mathcal{W})\times L^2(\omega;\mathcal{D}_\theta)$ give (11.58) for every $v\in \mathbb{D}_\theta$, $\theta\in[0,+\infty)$.

Case 2. $\theta = +\infty$.

Choose as test function $w_{\varepsilon,\delta} = v_\delta + v_{\varepsilon,\delta}$ with v_δ given as in Case 1 and

$$
\begin{aligned}
v_{\varepsilon,\delta}(x) = \delta\varepsilon\Big[\widehat{V}_\alpha\Big(x', \Big\{\frac{x'}{\varepsilon}\Big\}\Big) - \frac{x_3}{\delta}\frac{\partial \widehat{V}_3}{\partial y_\alpha}\Big(x', \Big\{\frac{x'}{\varepsilon}\Big\}\Big)\Big]\mathbf{e}_\alpha + \delta\varepsilon\widehat{V}_3\Big(x', \Big\{\frac{x'}{\varepsilon}\Big\}\Big)\mathbf{e}_3\\
+ \delta^2\overline{\widehat{v}}\Big(x', \Big\{\frac{x'}{\varepsilon}\Big\}, \frac{x_3}{\delta}\Big),
\end{aligned}
$$

with $\widehat{V}\in V'_{per}$ and $\overline{\widehat{v}}\in \underline{V}$, where

$$
\begin{aligned}
&V'_{per} \doteq \Big\{\Psi\in \mathcal{C}^1(\overline{\omega}\times\overline{Y}')^3 \mid \Psi(\cdot,y') = 0 \text{ on } \partial\omega\ \forall y'\in\overline{Y}';\ \Psi\ Y'-\text{periodic}\Big\},\\
&\underline{V} \doteq \Big\{\psi\in \mathcal{C}^1(\overline{\omega}\times\overline{Y}'\times[-1,1])^3 \mid \psi(\cdot,y_3) = 0 \ \text{ on } \ \partial(\omega\times Y')\ \forall\, y_3\in[-1,1]\Big\}.
\end{aligned}
$$

By a simple computation, one can check that

$$\frac{1}{\delta}\mathfrak{T}_{\varepsilon,\delta}\big(e(v_{\varepsilon,\delta})\big) \to \mathcal{E}^\infty_y(\widehat{v}) \quad \text{strongly in} \quad L^2(\omega\times Y)^9,$$

with

$$\widehat{v}(\cdot,y) = \Big[\widehat{V}_\alpha(\,\cdot\,,y') - y_3\frac{\partial \widehat{V}_3}{\partial y_\alpha}(\,\cdot\,,y')\Big]\mathbf{e}_\alpha + \widehat{V}_3(\,\cdot\,,y')\mathbf{e}_3 + \overline{\widehat{v}}(\,\cdot\,,y),$$

which means in fact, that

$$\frac{1}{\delta}\mathfrak{T}_{\varepsilon,\delta}\big(e(w_{\varepsilon,\delta})\big) \to E_M(\mathcal{V}) + E_w\big(\mathring{v}\big) + \mathcal{E}_y^{\infty}(\widehat{v}) \quad \text{strongly in} \quad L^2(\omega\times Y)^9,$$

and we conclude as in Case 1. □

11.7 Homogenization

The aim of this section is to give the expressions of the warping $\mathring{u}$ and the microscopic displacement $\widehat{u}$ in terms of the membrane displacement $\mathcal{U}_m$ and of the bending $\mathcal{U}_3$.

With the choice $\mathcal{V}=0$, Problem 11.58 becomes

$$\frac{1}{2}\int_{\omega\times Y} a_{ijkl}\big(E_{ij,M}(\mathcal{U}) + E_{ij,w}\big(\mathring{u}\big) + \mathcal{E}^{\theta}_{ij,y}(\widehat{u})\big)\big(E_w\big(\mathring{v}\big) + \mathcal{E}^{\theta}_{kl,y}(\widehat{v})\big)\,dx'dy = 0.$$

This shows that the warping $\mathring{u}$ and the microscopic displacement $\widehat{u}$ can be written in terms of the tensor $E_M(\mathcal{U})$. For $\theta\in[0,+\infty]$, define the space

$$\mathcal{WD}_\theta = \Big\{\mathring{\psi} + \widehat{\psi}\;\big|\;\mathring{\psi}\in\mathcal{W}, \widehat{\psi}\in\mathcal{D}_\theta\Big\}.^{(10)}$$

Note that for $\widetilde{\psi}$ in $\mathcal{WD}_\theta$,

$$\mathring{\psi} = \int_{Y'}\widetilde{\psi}\,dy',\qquad \widehat{\psi} = \widetilde{\psi} - \int_{Y'}\widetilde{\psi}\,dy'.$$

To every function $\widetilde{\psi}$ in $\mathcal{WD}_\theta$, associate the symmetric tensor

$$\mathbb{E}^{\theta}_y(\widetilde{\psi}) = E_w\big(\mathring{\psi}\big) + \mathcal{E}^{\theta}_y(\widehat{\psi}).$$

Due to the properties of the functions in $\mathcal{D}_\theta$, one has

$$\begin{aligned}\big\|\mathbb{E}^{\theta}_y(\widetilde{\psi})\big\|^2_{L^2(Y)} &= \frac{1}{4}\Big\|\frac{\partial\mathring{\psi}_1}{\partial y_3}\Big\|^2_{L^2(-1,1)} + \frac{1}{4}\Big\|\frac{\partial\mathring{\psi}_2}{\partial y_3}\Big\|^2_{L^2(-1,1)}\\ &\quad + \Big\|\frac{\partial\mathring{\psi}_3}{\partial y_3}\Big\|^2_{L^2(-1,1)} + \big\|\mathcal{E}^{\theta}_y(\widehat{\psi})\big\|^2_{L^2(Y)}.\end{aligned}\tag{11.59}$$

Set

$$\mathbf{M}^{11} = \begin{pmatrix}1&0&0\\0&0&0\\0&0&0\end{pmatrix},\quad \mathbf{M}^{12} = \begin{pmatrix}0&1&0\\1&0&0\\0&0&0\end{pmatrix},\quad \mathbf{M}^{22} = \begin{pmatrix}0&0&0\\0&1&0\\0&0&0\end{pmatrix}.$$

(10) As for the space $\mathcal{D}_\theta$ itself, $\mathcal{WD}_\theta$ only takes three values, $\mathcal{WD}_0$, $\mathcal{WD}$, $\mathcal{WD}_\infty$.

The tensor E_M having 6 components (due to symmetries) one introduces 6 correctors

$$\widetilde{\chi^{\alpha\beta}_{m,\theta}} \in \mathcal{WD}_\theta, \quad \widetilde{\chi^{\alpha\beta}_{b,\theta}} \in \mathcal{WD}_\theta, \qquad (\alpha,\beta) \in \{(1,1),(1,2),(2,2)\},$$

defined respectively by

$$\begin{aligned} &\int_Y a_{ijkl}(y)\,\big(\mathbf{M}^{\alpha\beta}_{ij} + \mathbb{E}^\theta_{ij,y}(\widetilde{\chi^{\alpha\beta}_{m,\theta}})(y)\big)\,\mathbb{E}^\theta_{kl,y}(\widetilde{\psi})(y)\,dy = 0,\\ &\int_Y a_{ijkl}(y)\big(y_3\mathbf{M}^{\alpha\beta}_{ij} + \mathbb{E}^\theta_{ij,y}(\widetilde{\chi^{\alpha\beta}_{b,\theta}})(y)\big)\mathbb{E}^\theta_{kl,y}(\widetilde{\psi})(y)\,dy = 0, \end{aligned} \tag{11.60}$$

for all $\widetilde{\psi} \in \mathcal{WD}_\theta$.

As a consequence, $\widetilde{u}$ can be written in the form

$$\begin{aligned} \widetilde{u}(\cdot,y) &= \mathring{u}(\cdot,y_3) + \widehat{u}(\cdot,y)\\ &= e_{\alpha\beta}\,(\mathcal{U}_m)\widetilde{\chi^{\alpha\beta}_{m,\theta}}(y) + \frac{\partial^2\mathcal{U}_3}{\partial x_\alpha\partial x_\beta}\,\widetilde{\chi^{\alpha\beta}_{b,\theta}}(y) \quad \text{for a.e. } y \in Y. \end{aligned} \tag{11.61}$$

11.7.1 The limit problems in the mid surface

From now on, we add the superscript θ, $\theta \in [0,+\infty]$, to the limit membrane fields $\mathcal{U}_m$ and to the bending $\mathcal{U}_3$.

Theorem 11.21. *The limit displacement*

$$\mathcal{U}^\theta = (\mathcal{U}^\theta_m, \mathcal{U}^\theta_3) \ \ \textit{belongs to} \ \ \mathbb{D}_M = H^1_{\gamma_0}(\omega)^2 \times H^2_{\gamma_0}(\omega).$$

It is the solution of the homogenized problem

$$\begin{cases} \displaystyle\int_\omega \Big[a^{hom,\theta}_{\alpha\beta\alpha'\beta'}\, e_{\alpha\beta}(\mathcal{U}^\theta_m)e_{\alpha\beta}(\mathcal{V}_m) + b^{hom,\theta}_{\alpha\beta\alpha'\beta'}\Big(e_{\alpha\beta}(\mathcal{U}^\theta_m)\frac{\partial^2\mathcal{V}_3}{\partial x_{\alpha'}\partial x_{\beta'}}\\ \displaystyle\qquad + e_{\alpha'\beta'}(\mathcal{V}_m)\frac{\partial^2\mathcal{U}^\theta_3}{\partial x_\alpha\partial x_\beta}\Big) + c^{hom,\theta}_{\alpha\beta\alpha'\beta'}\frac{\partial^2\mathcal{U}^\theta_3}{\partial x_\alpha\partial x_\beta}\frac{\partial^2\mathcal{V}_3}{\partial x_{\alpha'}\partial x_{\beta'}}\Big]dx'\\ \displaystyle\qquad = \int_\omega f\cdot\mathcal{V}\,dx' - \frac{1}{3}\int_\omega g_\alpha\frac{\partial\mathcal{V}_3}{\partial x_\alpha}\,dx',\\ \forall\mathcal{V} = (\mathcal{V}_m,\mathcal{V}_3) \in \mathbb{D}_M, \end{cases} \tag{11.62}$$

where

$$\begin{aligned} a^{hom,\theta}_{\alpha\beta\alpha'\beta'} &= \frac{1}{2}\int_Y a_{ijkl}(y)\big[\mathbf{M}^{\alpha\beta}_{ij} + \mathbb{E}^\theta_{ij,y}(\widetilde{\chi^{\alpha\beta}_{m,\theta}})\big]\,\mathbf{M}^{\alpha'\beta'}_{kl}\,dy,\\ b^{hom,\theta}_{\alpha\beta\alpha'\beta'} &= \frac{1}{2}\int_Y a_{ijkl}(y)\big[y_3\mathbf{M}^{\alpha\beta}_{ij} + \mathbb{E}^\theta_{ij,y}(\widetilde{\chi^{\alpha\beta}_{b,\theta}})\big]\mathbf{M}^{\alpha'\beta'}_{kl}\,dy,\\ c^{hom,\theta}_{\alpha\beta\alpha'\beta'} &= \frac{1}{2}\int_Y a_{ijkl}(y)\big[y_3\mathbf{M}^{\alpha\beta}_{ij} + \mathbb{E}^\theta_{ij,y}(\widetilde{\chi^{\alpha\beta}_{b,\theta}})\big]y_3\mathbf{M}^{\alpha'\beta'}_{kl}\,dy. \end{aligned} \tag{11.63}$$

Proof. In problem 11.58, choose as test displacement $\mathcal{V} = (\mathcal{V}_m, \mathcal{V}_3)$ in $\mathbb{D}_M$ and $\mathring{v} = 0$, $\widehat{v} = 0$. Replacing $\widetilde{u}$ by its expression (11.61), yields

$$\int_{\omega\times Y} a_{ijkl}(y)\Big[e_{\alpha\beta}(\mathcal{U}_m)(\mathbf{M}_{ij}^{\alpha\beta} + \mathbb{E}_{ij,y}^{\theta}(\widetilde{\chi_{m,\theta}^{\alpha\beta}})) + \frac{\partial^2 \mathcal{U}_3}{\partial x_\alpha \partial x_\beta}(y_3 \mathbf{M}_{ij}^{\alpha\beta} + \mathbb{E}_{ij,y}^{\theta}(\widetilde{\chi_{b,\theta}^{\alpha\beta}}))\Big]$$
$$\times \mathbf{M}_{kl}^{\alpha'\beta'}\Big[e_{\alpha'\beta'}(\mathcal{V}_m) + y_3 \frac{\partial^2 \mathcal{V}_3}{\partial x_{\alpha'} \partial x_{\beta'}}\Big] dx'\, dy$$
$$= 2\int_\omega f\cdot\mathcal{V}\, dx' - \frac{2}{3}\int_\omega g_\alpha \frac{\partial \mathcal{V}_3}{\partial x_\alpha}\, dx'.$$

Then Problem (11.62) with the homogenized coefficients given by (11.63) is obtained by a simple computation.

Taking into account the variational problems (11.60) satisfied by the correctors, it is easily seen that the homogenized coefficients are also given by the following expressions:

$$\begin{aligned}
a_{\alpha\beta\alpha'\beta'}^{hom,\theta} &= \frac{1}{2}\int_Y a_{ijkl}(y)\big[\mathbf{M}_{ij}^{\alpha\beta} + \mathbb{E}_{ij,y}^{\theta}(\widetilde{\chi_{m,\theta}^{\alpha\beta}})\big]\big[\mathbf{M}_{kl}^{\alpha'\beta'} + \mathbb{E}_{kl,y}^{\theta}(\widetilde{\chi_{m,\theta}^{\alpha'\beta'}})\big]dy,\\
b_{\alpha\beta\alpha'\beta'}^{hom,\theta} &= \frac{1}{2}\int_Y a_{ijkl}(y)\big[y_3\mathbf{M}_{ij}^{\alpha\beta} + \mathbb{E}_{ij,y}^{\theta}(\widetilde{\chi_{b,\theta}^{\alpha\beta}})\big]\big[\mathbf{M}_{kl}^{\alpha'\beta'} + \mathbb{E}_{kl,y}^{\theta}(\widetilde{\chi_{m,\theta}^{\alpha'\beta'}})\big]dy,\\
&= \frac{1}{2}\int_Y a_{ijkl}(y)\big[\mathbf{M}_{ij}^{\alpha\beta} + \mathbb{E}_{ij,y}^{\theta}(\widetilde{\chi_{m,\theta}^{\alpha\beta}})\big]\big[y_3\mathbf{M}_{kl}^{\alpha'\beta'} + \mathbb{E}_{kl,y}^{\theta}(\widetilde{\chi_{b,\theta}^{\alpha'\beta'}})\big]dy,\\
c_{\alpha\beta\alpha'\beta'}^{hom,\theta} &= \frac{1}{2}\int_Y a_{ijkl}(y)\big[y_3\mathbf{M}_{ij}^{\alpha\beta} + \mathbb{E}_{ij,y}^{\theta}(\widetilde{\chi_{b,\theta}^{\alpha\beta}})\big]\big[y_3\mathbf{M}_{kl}^{\alpha'\beta'} + \mathbb{E}_{kl,y}^{\theta}(\widetilde{\chi_{b,\theta}^{\alpha'\beta'}})\big]dy
\end{aligned} \tag{11.64}$$

The second expression of the $b_{\alpha\beta\alpha'\beta'}^{hom,\theta}$'s is a consequence of the properties of the coefficients of the Hooke tensor. □

Remark 11.22. *From* (11.60), *we deduce that the left-hand side operator in Problem* (11.62) *is uniformly bounded.*

Lemma 11.23. *The left-hand side operator in Problem* (11.62) *is uniformly elliptic.*

Proof. Using formulas (11.64) of the homogenized coefficients, one obtains

$$a_{\alpha\beta\alpha'\beta'}^{hom,\theta}\tau_{\alpha\beta}^m\tau_{\alpha'\beta'}^m + b_{\alpha\beta\alpha'\beta'}^{hom,\theta}\big(\tau_{\alpha\beta}^m\tau_{\alpha'\beta'}^b + \tau_{\alpha\beta}^b\tau_{\alpha'\beta'}^m\big) + c_{\alpha\beta\alpha'\beta'}^{hom,\theta}\tau_{\alpha\beta}^b\tau_{\alpha'\beta'}^b$$
$$= \frac{1}{2}\int_Y a_{ijkl}\big[M_{ij} + \mathbb{E}_{ij,y}^{\theta}(\Psi)\big]\big[M_{kl} + \mathbb{E}_{kl,y}^{\theta}(\Psi)\big]dy,$$

where

$$M = \big(\tau_{\alpha\beta}^m + y_3\tau_{\alpha\beta}^b\big)M^{\alpha\beta},$$

and

$$\Psi = \tau_{\alpha\beta}^m\widetilde{\chi_{m,\theta}^{\alpha\beta}} + \tau_{\alpha\beta}^b\widetilde{\chi_{b,\theta}^{\alpha\beta}}.$$

Then, in view of (11.11),

$$\int_Y a_{ijkl}\big[M_{ij} + \mathbb{E}^\theta_{ij,y}(\Psi)\big]\big[M_{kl} + \mathbb{E}^\theta_{kl,y}(\Psi)\big]dy$$
$$\geqslant c_0 \int_Y \big[M_{ij} + \mathbb{E}^\theta_{ij,y}(\Psi)\big]\big[M_{ij} + \mathbb{E}^\theta_{ij,y}(\Psi)\big]dy.$$

Following the same lines of Step 1 in Lemma 11.19, we get

$$\int_Y \big[M_{ij} + \mathbb{E}^\theta_{ij,y}(\Psi)\big]\big[M_{ij} + \mathbb{E}^\theta_{ij,y}(\Psi)\big]dy \geqslant C\big(\tau^m_{\alpha\beta}\tau^m_{\alpha\beta} + \tau^b_{\alpha\beta}\tau^b_{\alpha\beta}\big),$$

where the constant C is independent of $\theta \in [0, +\infty]$. □

11.8 Complements

Again, we consider the solution $u_{\varepsilon,\delta}$ of Problem (11.13). In this section, we will follow the lines of the previous sections, especially those of Section 11.5. In the next subsection 11.8.1 we shall pass to the limit with $\delta \to 0$ (dimension reduction) first, and with $\varepsilon \to 0$ (homogenization) afterwards. In Subsection 11.8.2, the order of passing to the limit is inverted, $\varepsilon \to 0$ first, followed by $\delta \to 0$. The proofs are similar to the ones given in the previous sections and they are only sketched.

11.8.1 Dimension reduction followed by homogenization

Proposition 11.24. *Let $u_{\varepsilon,\delta}$ be the solution of* (11.13). *If first $\delta \to 0$ (with ε fixed) and then $\varepsilon \to 0$, the limit problem is the same as that obtained when $(\varepsilon, \delta) \to (0, 0)$ with $\dfrac{\varepsilon}{\delta} \to +\infty$.*

Proof. Step 1. Reduction of dimension.

As is Section 11.5, denote by $\mathcal{U}_{\varepsilon,\delta}$, $\mathcal{R}_{\varepsilon,\delta}$ and $\overline{u}_{\varepsilon,\delta}$, the elements of the decomposition of $u_{\varepsilon,\delta}$ given by Theorems 11.4. Then, up to subsequence of $\{\delta\}$, there exist $\mathcal{U}_\varepsilon = \big(\mathcal{U}_{\varepsilon,m}, \mathcal{U}_{\varepsilon,3}\big) \in \mathbb{D}_M$ and $\overline{u}_\varepsilon \in L^2(\omega; H^1(-1,1))^3$, with

$$\int_{-1}^1 \overline{u}_{\varepsilon,i}(x', y_3)\, dy_3 = \int_{-1}^1 \overline{u}_{\varepsilon,\alpha}(x', y_3) y_3\, dy_3 = 0 \quad \text{a.e. for } x' \in \omega,$$

such that as $\delta \to 0$ for ε fixed,

$$\begin{aligned}
&\frac{1}{\delta}\mathcal{U}_{\varepsilon,\delta,m} \rightharpoonup \mathcal{U}_{\varepsilon,m} \quad \text{weakly in } H^1_{\gamma_0}(\omega)^2,\\
&\mathcal{U}_{\varepsilon,\delta,3} \longrightarrow \mathcal{U}_{\varepsilon,3} \quad \text{strongly in } H^1_{\gamma_0}(\omega),\\
&\mathcal{R}_{\varepsilon,\delta} \rightharpoonup -\nabla\mathcal{U}_{\varepsilon,3} \quad \text{weakly in } H^1_{\gamma_0}(\omega)^2,\\
&\frac{1}{\delta}\big(\nabla\mathcal{U}_{\varepsilon,\delta,3} + \mathcal{R}_{\varepsilon,\delta}\big) \rightharpoonup \mathcal{Z}_\varepsilon \quad \text{weakly in } L^2(\omega)^2,
\end{aligned}$$

as well as

$$\frac{1}{\delta^2}\mathfrak{T}_\delta(\overline{u}_{\varepsilon,\delta}) \rightharpoonup \overline{u}_\varepsilon \quad \text{weakly in } L^2(\omega; H^1(-1,1))^3,$$
$$\frac{1}{\delta}\mathfrak{T}_\delta(e(u_{\varepsilon,\delta})) \rightharpoonup E_M(\mathcal{U}_\varepsilon) + E_w(\mathring{u}_\varepsilon) \quad \text{weakly in } L^2(\Omega)^9,$$

where $\mathring{u}_\varepsilon(\cdot, y_3) = \overline{u}_\varepsilon(\cdot, y_3) + y_3\, \mathcal{Z}_{\alpha,\varepsilon}\, \mathbf{e}_\alpha$ and $\mathcal{U}_\varepsilon = (\mathcal{U}_{\varepsilon,m}, \mathcal{U}_{\varepsilon,3}) \in \mathbb{D}_M$.

Due to estimates (11.9) and (11.15),

$$\|\mathcal{U}_{\varepsilon,m}\|_{H^1(\omega)} + \|\mathcal{U}_{\varepsilon,3}\|_{H^2(\omega)} + \|\mathring{u}_\varepsilon\|_{L^2(\omega;H^1(-1,1))} \leqslant C,$$

with a constant not depending on ε.

Now choose in Problem (11.13) the test displacement v_δ introduced in the proof of Theorem 11.20. Dividing by $2\delta^3$, unfolding the equality by $\mathfrak{T}_\delta$, and then passing to the limit as $\delta \to 0$, allows to obtain

$$\begin{aligned}&\frac{1}{2}\int_\Omega a_{ijkl}\Big(\Big\{\frac{x'}{\varepsilon}\Big\}, y_3\Big)\Big(E_{ij,M}(\mathcal{U}_\varepsilon) + E_{ij,w}(\mathring{u}_\varepsilon)\Big)\\ &\times \Big(E_{kl,M}(\mathcal{V}) + E_{kl,w}(\mathring{v})\Big)dx'dy_3 = \int_\omega f\cdot\mathcal{V}\,dx' - \frac{1}{3}\int_\omega g_\alpha \frac{\partial \mathcal{V}_3}{\partial x_\alpha}dx'.\end{aligned} \tag{11.65}$$

Applying the results of Part I, in particular Theorem 1.47 and Proposition 2.13, there exist

$$\mathcal{U} = (\mathcal{U}_m, \mathcal{U}_3) \in \mathbb{D}_M,\ \mathring{u} \in L^2(\omega; \mathcal{W}),\ \widehat{U}_m \in L^2(\omega; H^1_{per,0}(Y'))^2,$$
$$\widehat{U}_3 \in L^2(\omega; H^1_{per,0}(Y') \cap H^2_{per}(Y')),\ \widehat{\mathring{u}} \in L^2(\omega\times Y'; \mathcal{W}),$$

such that

$$\widehat{u}(\cdot, y) \doteq \Big[\widehat{U}_\alpha(\cdot, y') - y_3\frac{\partial \widehat{U}_3}{\partial y_\alpha}(\cdot, y')\Big]\mathbf{e}_\alpha + \widehat{U}_3(\cdot, y')\mathbf{e}_3 + \widehat{\mathring{u}}(\cdot, y) \in L^2(\omega; \mathcal{D}_\infty).$$

Step 2. Homogenization.

Now, letting ε go to 0, the following convergences hold:

$$\begin{aligned}&\mathcal{U}_{\varepsilon,m} \rightharpoonup \mathcal{U}_m \quad \text{weakly in } H^1_{\gamma_0}(\omega)^2,\\ &\mathcal{U}_{\varepsilon,3} \rightharpoonup \mathcal{U}_3 \quad \text{weakly in } H^2_{\gamma_0}(\omega),\\ &\mathcal{T}_\varepsilon(\nabla\mathcal{U}_{\varepsilon,m}) \rightharpoonup \nabla\mathcal{U}_m + \nabla_{y'}\widehat{U}_m \quad \text{weakly in } L^2(\omega; H^1(Y'))^4,\\ &\mathcal{T}_\varepsilon(D^2\mathcal{U}_{\varepsilon,3}) \rightharpoonup D^2\mathcal{U}_3 + D^2_{y'}\widehat{U}_3 \quad \text{weakly in } L^2(\omega\times Y')^4,\\ &\mathcal{T}_\varepsilon(\mathring{u}_\varepsilon) \rightharpoonup \mathring{u} + \widehat{\mathring{u}} \quad \text{weakly in } L^2(\omega\times Y'; H^1(-1,1))^3,\end{aligned}$$

and finally,

$$\mathcal{T}_\varepsilon(E_M(\mathcal{U}_\varepsilon) + E_w(\mathring{u}_\varepsilon)) \rightharpoonup E_M(\mathcal{U}) + E_w(\mathring{u}) + \mathbb{E}^\infty_y(\widehat{u}) \text{ weakly in } L^2(\omega\times Y)^9,$$

where $(\mathcal{U}, \mathring{u}, \widehat{u}) \in \mathbb{D}_\infty$.

For $\mathcal{V} = (\mathcal{V}_1, \mathcal{V}_2, \mathcal{V}_3) \in \mathbb{D}_M$, the next step consists in choosing the test function in problem (11.65) as $(\mathcal{V}_{\varepsilon,m}, \mathcal{V}_{\varepsilon,3}, \mathring{v}_\varepsilon)$ defined for all $x \in \Omega_\delta$, by

$$\mathcal{V}_{\varepsilon,\alpha}(x) = \mathcal{V}_\alpha(x') + \varepsilon\widehat{\mathcal{V}}_\alpha\Big(x', \Big\{\frac{x'}{\varepsilon}\Big\}\Big), \qquad \widehat{\mathcal{V}}_\alpha \in \mathcal{C}_c^\infty(\omega; \mathcal{C}^1_{per}(\overline{Y'})),$$

$$\mathcal{V}_{\varepsilon,3}(x) = \mathcal{V}_3(x') + \varepsilon^2\widehat{\mathcal{V}}_3\Big(x', \Big\{\frac{x'}{\varepsilon}\Big\}\Big), \qquad \widehat{\mathcal{V}}_3 \in \mathcal{C}_c^\infty(\omega; \mathcal{C}^2_{per}(\overline{Y'})),$$

and

$$\mathring{v}_\varepsilon(x) = \mathring{v}(x', y_3) + \widehat{\mathring{v}}\Big(x', \Big\{\frac{x'}{\varepsilon}\Big\}, y_3\Big), \quad \mathring{v} \in L^2(\omega; \mathcal{W}), \ \ \widehat{\mathring{v}} \in L^2(\omega \times Y'; \mathcal{W}).$$

Unfolding the resulting identity, then passing to the limit as ε goes to 0 and using a density argument allow to obtain Problem 11.58 for $\theta = +\infty$. □

11.8.2 Homogenization followed by dimension reduction

Proposition 11.25. *Let $u_{\varepsilon,\delta}$ be the solution of* (11.13). *If first $\varepsilon \to 0$ (with δ fixed) and then $\delta \to 0$, the limit problem is the same as that obtained when $(\varepsilon, \delta) \to (0, 0)$ and $\dfrac{\varepsilon}{\delta}$ goes to 0.*

Proof. Step 1. Homogenization.

Following the homogenization procedure from Chapter 3, thanks to the a priori estimates established in the preceding subsection, there exist u_δ in $\mathbb{D}_\delta$ and $\widehat{u}_\delta$ in $L^2(\Omega_\delta; H^1_{per,0}(Y'))^3$ such that, up to a subsequence, for $\varepsilon \to 0$ and δ fixed,

$$\begin{aligned} &u_{\varepsilon,\delta} \rightharpoonup u_\delta \quad \text{weakly in } H^1(\Omega_\delta)^3,\\ &\mathcal{T}_\varepsilon(\nabla u_{\varepsilon,\delta}) \rightharpoonup \nabla u_\delta + \nabla_{y'}\widehat{u}_\delta \quad \text{weakly in } L^2(\Omega_\delta \times Y')^3. \end{aligned}$$

Therefore (see Proposition 2.13),

$$\mathcal{T}_\varepsilon\big(e(u_{\varepsilon,\delta})\big) \rightharpoonup e(u_\delta) + e_{y'}(\widehat{u}_\delta) \quad \text{weakly in } \ L^2(\Omega_\delta \times Y')^6,$$

where

$$e_{y'}(\widehat{u}_\delta) = \begin{pmatrix} e_{11,y'}(\widehat{u}_\delta) & e_{12,y'}(\widehat{u}_\delta) & \dfrac{1}{2}\dfrac{\partial \widehat{u}_{\delta,3}}{\partial y_1} \\ e_{12,y'}(\widehat{u}_\delta) & e_{22,y'}(\widehat{u}_\delta) & \dfrac{1}{2}\dfrac{\partial \widehat{u}_{\delta,3}}{\partial y_2} \\ \dfrac{1}{2}\dfrac{\partial \widehat{u}_{\delta,3}}{\partial y_1} & \dfrac{1}{2}\dfrac{\partial \widehat{u}_{\delta,3}}{\partial y_2} & 0 \end{pmatrix}.$$

Choose in Problem (11.13) as test displacement

$$v(x) + \varepsilon\widehat{v}\Big(x, \Big\{\frac{x'}{\varepsilon}\Big\}\Big) \quad \text{with } v \in \mathbb{D}_\delta, \ \widehat{v} \in \mathcal{C}_c^\infty(\Omega_\delta; \mathcal{C}^1_{per}(\overline{Y'}))^3.$$

By unfolding the result, passing to the limit and using a density argument, the following variational problem for $(u_\delta, \widehat{u}_\delta) \in \mathbb{D}_\delta \times L^2(\Omega_\delta; H^1_{per,0}(Y'))^3$ is obtained:

$$\begin{cases} \displaystyle\int_{\Omega_\delta \times Y'} a_{ijkl}\Big(y', \frac{x_3}{\delta}\Big)\big(e_{ij}(u_\delta) + e_{ij,y'}(\widehat{u}_\delta)\big)\big(e_{kl}(v) + e_{kl,y'}(\widehat{v})\big)\, dxdy' \\ \qquad = \displaystyle\int_{\Omega_\delta} f_\delta \cdot v\, dx, \qquad \forall (v, \widehat{v}_\delta) \in \mathbb{D}_\delta \times L^2(\Omega_\delta; H^1_{per,0}(Y'))^3. \end{cases} \tag{11.66}$$

Step 2. Reduction of dimension.

Estimates (11.9)-(11.33), the Poincaré-Wirtinger inequality as well as the two-dimensional Korn inequality for periodic displacements (see (2.14) in Proposition 2.11), give the following a priori estimate for the homogenized solution u_δ:

$$\|e(u_\delta)\|_{L^2(\Omega_\delta)} + \|e_{y'}(\widehat{u}_\delta)\|_{L^2(\Omega_\delta \times Y')} + \|\widehat{u}_\delta\|_{L^2(\Omega_\delta; H^1_{per}(Y'))} \leqslant C\delta^{3/2},$$

where C is independent of δ.

Now, decompose u_δ as done Subsection 11.1.2 and denote by $\mathcal{U}_\delta$, $\mathcal{R}_\delta$ and $\overline{u}_\delta$ the terms of the decomposition (11.4). Thanks to the above a priori estimate, there exist $\widehat{u}$ in $L^2(\omega; \mathcal{D}_0)$, and

$$\mathcal{U} = (\mathcal{U}_m, \mathcal{U}_3) \in \mathbb{D}_M, \quad \overline{u} \in L^2(\omega; H^1(-1,1))^3),$$

satisfying

$$\int_{-1}^{1} \overline{u}_i(x', y_3)\, dy_3 = \int_{-1}^{1} \overline{u}_\alpha(x', y_3) y_3\, dy_3 = 0 \quad \text{for a.e. } x' \in \omega,$$

such that, up to a subsequence,

$$\begin{aligned} &\frac{1}{\delta}\mathcal{U}_{\delta,m} \rightharpoonup \mathcal{U}_m \quad \text{weakly in } H^1_{\gamma_0}(\omega)^2, \\ &\mathcal{U}_{\delta,3} \rightharpoonup \mathcal{U}_3 \quad \text{weakly in } H^1_{\gamma_0}(\omega), \\ &\mathcal{R}_\delta \rightharpoonup -\nabla\mathcal{U}_3 \quad \text{weakly in } H^1_{\gamma_0}(\omega)^2, \\ &\frac{1}{\delta}\big(\nabla\mathcal{U}_{\delta,3} + \mathcal{R}_\delta\big) \rightharpoonup \mathcal{Z} \quad \text{weakly in } L^2(\omega)^2, \end{aligned}$$

and also

$$\begin{aligned} &\frac{1}{\delta}\mathfrak{T}_\delta(\overline{u}_\delta) \rightharpoonup \overline{u} \quad \text{weakly in } L^2(\omega; H^1(-1,1))^3, \\ &\frac{1}{\delta}\mathfrak{T}_\delta\big(e(u_\delta)\big) \rightharpoonup E_M(\mathcal{U}) + E_w\big(\mathring{u}\big) \quad \text{weakly in } L^2(\Omega)^9, \\ &\frac{1}{\delta}\mathfrak{T}_\delta\big(e_{y'}(\widehat{u}_\delta)\big) \rightharpoonup \mathbb{E}^0_y(\widehat{u}) \quad \text{weakly in } L^2(\omega \times Y)^9, \end{aligned}$$

where

$$\mathring{u}(\cdot, y_3) = \overline{u}(\cdot, y_3) + y_3\mathcal{Z}_\alpha \mathbf{e}_\alpha, \quad (\mathcal{U}, \mathring{u}, \widehat{u}) \in \mathbb{D}_0, \quad \mathcal{U} = (\mathcal{U}_m, \mathcal{U}_3) \in \mathbb{D}_M.$$

Take as test function in Problem (11.66),

$$\Big(v_\delta(x), \delta\,\widehat{v}\Big(x', y', \frac{x_3}{\delta}\Big)\Big),$$

where v_δ was introduced in Step 1 of the proof of Theorem 11.20, and $\widehat{v}$ is in $L^2(\omega; \mathcal{D}_0)$. Dividing by $2\delta^3$, applying the operator $\mathfrak{T}_\delta$ and passing to the limit with $\delta \to 0$, allow to obtain Problem 11.58 with $\theta = 0$. □

Remark 11.26. *As noted in [44], the two limiting processes do not commute since the limit problems vary with θ.*

11.8.3 Continuous dependence with respect to θ

We start by exhibiting some properties of the homogenized coefficients obtained in the preceding sections. Choose successively $\widetilde{\chi^{\alpha\beta}_{m,\theta}}$ and $\widetilde{\chi^{\alpha\beta}_{b,\theta}}$ as test functions in (11.60) to obtain the estimates

$$\|\mathbb{E}^\theta_y(\widetilde{\chi^{\alpha\beta}_{m,\theta}})\|_{L^2(\Omega)} \leqslant C, \qquad \|\mathbb{E}^\theta_y(\widetilde{\chi^{\alpha\beta}_{b,\theta}})\|_{L^2(\Omega)} \leqslant C, \tag{11.67}$$

where the constant C does not depend on θ. Then, from (11.42) it follows that for all $\theta \in (0, +\infty)$,

$$\|\widetilde{\chi^{\alpha\beta}_{m,\theta}}\|_{H^1(\Omega)} \leqslant C(1+\theta), \quad \|\widetilde{\chi^{\alpha\beta}_{b,\theta}}\|_{H^1(\Omega)} \leqslant C(1+\theta).$$

Lemma 11.27. *The homogenized coefficients* (11.64) *are continuous functions with respect to θ on $[0, +\infty]$.*

Proof. Step 1. We first prove the continuity of the function $\theta \mapsto (\widetilde{\chi^{\alpha\beta}_{m,\theta}}, \widetilde{\chi^{\alpha\beta}_{b,\theta}})$ on $(0, +\infty)$ for the strong topology of the space $\boldsymbol{\mathcal{H}}^1_{per}(Y)$.

Let (θ, θ') be in $(0, +\infty)^2$. Then, for every $\widetilde{\psi} = \mathring{\psi} + \widehat{\psi} \in \mathcal{WD} \subset \boldsymbol{\mathcal{H}}^1_{per}(Y)^3$,

$$\mathbb{E}^\theta_y(\widetilde{\psi}) - \mathbb{E}^{\theta'}_y(\widetilde{\psi}) = \mathcal{E}^\theta_y(\widehat{\psi}) - \mathcal{E}^{\theta'}_y(\widehat{\psi}) = \frac{\theta - \theta'}{2}\begin{pmatrix} 0 & 0 & \dfrac{\partial\widehat{\psi}_1}{\partial y_3} \\ 0 & 0 & \dfrac{\partial\widehat{\psi}_2}{\partial y_3} \\ \dfrac{\partial\widehat{\psi}_1}{\partial y_3} & \dfrac{\partial\widehat{\psi}_2}{\partial y_3} & 2\dfrac{\partial\widehat{\psi}_3}{\partial y_3}\end{pmatrix}.$$

This equality, together with equations (11.60) and (11.53)-(11.54), give the following estimates with constants independent of (θ, θ'):

$$\begin{aligned}
&\Big|\int_Y a_{ijkl}\mathbb{E}^\theta_{ij,y}(\widetilde{\chi^{\alpha\beta}_{m,\theta}} - \widetilde{\chi^{\alpha\beta}_{m,\theta'}})\,\mathbb{E}^\theta_{kl,y}(\widetilde{\psi})\,dy\Big| \\
&\qquad\leqslant C\,|\theta - \theta'|\Big(\frac{1}{\theta'} + 1\Big)\|\mathbb{E}^{\theta'}_y(\widetilde{\chi^{\alpha\beta}_{m,\theta'}})\|_{L^2(Y)}\|\mathbb{E}^\theta_y(\widetilde{\psi})\|_{L^2(Y)} \\
&\qquad\quad + C\,|\theta - \theta'|\Big(\frac{1}{\theta} + 1\Big)\Big(1 + \|\mathbb{E}^{\theta'}_y(\widetilde{\chi^{\alpha\beta}_{m,\theta'}})\|_{L^2(Y)}\Big)\|\mathbb{E}^\theta_y(\widetilde{\psi})\|_{L^2(Y)}.
\end{aligned}$$

Take $\widetilde{\chi^{\alpha\beta}_{m,\theta}} - \widetilde{\chi^{\alpha\beta}_{m,\theta'}}$ as test function and use (11.67) to get

$$\|\mathbb{E}^{\theta}_{ij,y}(\widetilde{\chi^{\alpha\beta}_{m,\theta}} - \widetilde{\chi^{\alpha\beta}_{m,\theta'}})\|_{L^2(Y)} \leqslant C\,|\theta - \theta'|\Big(\frac{1}{\theta'} + \frac{1}{\theta} + 1\Big).$$

Then

$$\|\widetilde{\chi^{\alpha\beta}_{m,\theta}} - \widetilde{\chi^{\alpha\beta}_{m,\theta'}}\|_{H^1(Y)} \leqslant C\,|\theta - \theta'|\Big(\frac{1}{\theta'} + \frac{1}{\theta} + \theta + \theta'\Big). \tag{11.68}$$

In the same way,

$$\|\widetilde{\chi^{\alpha\beta}_{b,\theta}} - \widetilde{\chi^{\alpha\beta}_{b,\theta'}}\|_{H^1(Y)} \leqslant C\,|\theta - \theta'|\Big(\frac{1}{\theta'} + \frac{1}{\theta} + \theta + \theta'\Big).$$

As a consequence, the homogenized coefficients (11.64) are locally Lipschitz continuous on $(0, +\infty)$.

Step 2. Similarly, for $\theta = 0$, we prove the continuity at 0 of the function $\theta \mapsto (\widetilde{\chi^{\alpha\beta}_{m,\theta}}, \widetilde{\chi^{\alpha\beta}_{b,\theta}})$ for the strong topology of $\mathcal{WD}_0$.

Recall that

$$\widetilde{\chi^{\alpha\beta}_{m,\theta}} = \mathring{\chi}^{\alpha\beta}_{m,\theta} + \widehat{\chi^{\alpha\beta}_{m,\theta}}, \qquad \widetilde{\chi^{\alpha\beta}_{b,\theta}} = \mathring{\chi}^{\alpha\beta}_{b,\theta} + \widehat{\chi^{\alpha\beta}_{b,\theta}}. \tag{11.69}$$

From (11.67) and (11.54), up to a subsequence, there exist $\widetilde{\Upsilon^{\alpha\beta}_m}$ and $\widetilde{\Upsilon^{\alpha\beta}_b}$ in $\mathcal{WD}_0$, such that as $\theta \to 0$,

$$\begin{aligned}
\widetilde{\chi^{\alpha\beta}_{m,\theta}} &\rightharpoonup \widetilde{\Upsilon^{\alpha\beta}_m} \quad \text{weakly in } \mathcal{WD}_0,\\
\theta\frac{\partial\widehat{\chi^{\alpha\beta}_{m,\theta}}}{\partial y_3} &\rightharpoonup 0 \quad \text{weakly in } L^2(Y)^3,\\
\widetilde{\chi^{\alpha\beta}_{b,\theta}} &\rightharpoonup \widetilde{\Upsilon^{\alpha\beta}_b} \quad \text{weakly in } \mathcal{WD}_0,\\
\theta\frac{\partial\widehat{\chi^{\alpha\beta}_{b,\theta}}}{\partial y_3} &\rightharpoonup 0 \quad \text{weakly in } L^2(Y)^3.
\end{aligned}$$

Then, taking $\widetilde{\psi} = \mathring{\psi} + \widehat{\psi}$ in $\mathcal{WD}$ as test function in (11.60), and passing to the limit, gives

$$\int_Y a_{ijkl}\big(\mathbf{M}^{\alpha\beta}_{ij} + \mathbb{E}^0_{ij,y}(\widetilde{\Upsilon^{\alpha\beta}_m})\big)\mathbb{E}^0_{ij,y}(\widetilde{\psi})\,dy = 0,$$

$$\int_Y a_{ijkl}\big(y_3\mathbf{M}^{\alpha\beta}_{ij} + \mathbb{E}^0_{ij,y}(\widetilde{\Upsilon^{\alpha\beta}_b})\big)\mathbb{E}^0_{ij,y}(\widehat{\psi})\,dy = 0.$$

By density, these equalities hold true for any $\widetilde{\psi}$ in $\mathcal{WD}_0$. Hence,

$$\widetilde{\Upsilon^{\alpha\beta}_m} = \widetilde{\chi^{\alpha\beta}_{m,0}} \quad \text{and} \quad \widetilde{\Upsilon^{\alpha\beta}_b} = \widetilde{\chi^{\alpha\beta}_{b,0}}.$$

It is then easily proved that the whole sequence strongly converges to its limit in $\mathcal{WD}_0$. Consequently, the homogenized coefficients are continuous functions at $\theta = 0$.

Step 3. The continuity of the homogenized coefficients at $+\infty$, is proved through the continuity of the function

$$\theta \mapsto \big(\mathbb{E}^\theta_y(\widetilde{\chi^{\alpha\beta}_{m,\theta}}), \mathbb{E}^\theta_y(\widetilde{\chi^{\alpha\beta}_{b,\theta}})\big)$$

for the strong topology of $L^2(Y)^6$. To do so, it is sufficient to show that the following function is continuous:

$$\theta \mapsto \mathbb{E}^\theta_y(\widetilde{\chi^{\alpha\beta}_{m,\theta}}).$$

A similar proof corresponding to $\mathbb{E}^\theta_y(\widetilde{\chi^{\alpha\beta}_{b,\theta}})$ will give the result.

Recalling (11.69), estimates (11.59) and (11.67) imply

$$\|\mathring{\chi}^{\alpha\beta}_{m,\theta}\|_{H^1(-1,1)} \leqslant C, \qquad \|\mathcal{E}^\theta_y(\widehat{\chi^{\alpha\beta}_{m,\theta}})\|_{L^2(Y)} \leqslant C. \tag{11.70}$$

In Proposition 11.14 the domain Y_θ was viewed as a thin periodic plate of thickness $2/\theta$. Now the displacement $\widehat{\chi^{\alpha\beta}_{m,\theta}}$ is decomposed in the same way as in Step 1 above. Due to estimates (11.68) and (11.70), up to a subsequence (of $\{\theta\}$), there exist $\mathring{\Upsilon}^{\alpha\beta}_m$ in $\mathcal{W}$ and $\widehat{\Upsilon^{\alpha\beta}_m}$ in $\mathcal{D}_\infty$, such that

$$\begin{aligned} &\mathring{\chi}^{\alpha\beta}_{m,\theta} \rightharpoonup \mathring{\Upsilon}^{\alpha\beta}_m \quad \text{weakly in } \mathcal{W},\\ &\mathcal{E}^\theta_y(\widehat{\chi^{\alpha\beta}_{m,\theta}}) \rightharpoonup \mathcal{E}^\infty_y(\widehat{\Upsilon^{\alpha\beta}_m}) \quad \text{weakly in } L^2(Y)^6. \end{aligned}$$

Take as test function in (11.60) $\widetilde{\psi} = \mathring{\psi} + \widehat{\psi}$ in $\mathcal{WD}$, where

$$\widehat{\psi}(y) = \Big[\widehat{\Psi}_\alpha(y') - y_3\frac{\partial\widehat{\Psi}_3}{\partial y_\alpha}(y')\Big]\mathbf{e}_\alpha + \widehat{\Psi}_3(y')\mathbf{e}_3 + \frac{1}{\theta}\overline{\widehat{\psi}}(y', y_3) \quad \text{for a.e. } y \in Y.$$

Passing to the limit gives (here we set $\widetilde{\Upsilon^{\alpha\beta}_m} = \mathring{\Upsilon}^{\alpha\beta}_m + \widehat{\Upsilon^{\alpha\beta}_m}$)

$$\int_Y a_{ijkl}\big(\mathbf{M}^{\alpha\beta}_{ij} + \mathbb{E}^\infty_{ij,y}(\widetilde{\Upsilon^{\alpha\beta}}_m)\big)\mathbb{E}^\infty_{kl,y}(\widetilde{\psi})\, dy = 0.$$

By the density of $\mathcal{WD}$ in $\mathcal{WD}_\infty$, this equality holds true for any $\widetilde{\psi}$ in $\mathcal{WD}_\infty$. Hence

$$\widetilde{\Upsilon^{\alpha\beta}_m} = \widetilde{\chi^{\alpha\beta}_{m,\infty}},$$

and as before, the whole sequence strongly converges. Consequently, the homogenized coefficients are continuous at $\theta = +\infty$. □

The required continuity is now a consequence of Lemmas 11.23, 11.27 and Remark 11.22 :

Theorem 11.28. *The function*

$$\theta \longmapsto \mathcal{U}^\theta = (\mathcal{U}_m^\theta, \mathcal{U}_3^\theta),$$

is continuous on $[0, +\infty]$ *for the strong topology of* $\mathbb{D}_M$.

Part VI

An application: sharp error estimates

The goal of the last part of this book is to establish precise error estimates and correctors for the homogenization of second order elliptic problems (Chapters 13 and 14).

Some preliminary results are obtained in Chapter 12 which starts with classical estimates for functions of $W^{1,p}(\Omega)$ in layers of small thickness δ near the boundary of Ω. These estimates are then used to control the distance between a function and its local average for the case of $[W^{1,p'}(\Omega)]'$ and $W^{-1,p}(\Omega)$ (Proposition 12.5 which extends Proposition 1.38). The latter are used in the proof of Proposition 12.10, which, for φ in $W^{1,p}(\Omega)$, controls the distance between $\nabla \mathcal{Q}_\varepsilon(\varphi)$ and its unfolding in $L^\infty(Y; (W^{1,p'}(\Omega))')$. This is an important ingredient to obtain the projection theorems of Chapter 13 and the error estimates of Chapter 14.

The boundary layer corrector for the homogenization of the Dirichlet problem is the solution of a Dirichlet problem with highly oscillating boundary conditions. These problems are of interest in their own right and Chapter 13 concerns their study and the corresponding error estimates. Here, the homogenized limit depends on the choice of the subsequence of ε and on the nature of the boundary of the domain (hyperplane or smooth hypersurface; see Sections 13.1 and 13.7). This precludes the existence of any asymptotic expansion beyond the second term for the solution of a general homogenization problem. The last section of Chapter 13 concerns projection results from $W^{1,p}(Y)$ to $W^{1,p}_{per}(Y)$ measured by the "periodicity defect" (see Definition 13.33 and Proposition 13.38).

In Chapter 14, all the previous results are combined to get the error estimates for the homogeneous Dirichlet problem (Theorem 14.3) and for the homogeneous Neumann problem (Theorem 14.10).

Chapter 12

The scale-splitting operators revisited

In Section 1.4, some estimates and convergence were obtained for functions in $W^{1,p}(\mathbb{R}^N)$, $p \in [1, +\infty]$ (cf. Proposition 1.38 to Theorem 1.41). This chapter is devoted to similar estimates and convergences for functions in $L^p(\mathbb{R}^N)$. Accordingly, the estimates are shifted by one order of derivation, e.g. (12.6). These results play a fundamental role in the following chapters on error estimates. They also use estimates involving oscillating functions which are given at the end of this chapter.

12.1 Notations and preliminary results

Let Ω be a domain in $\mathbb{R}^N$ with bounded Lipschitz boundary. Recall some notations introduced in Chapter 1 and that will be used in the sequel:

$$
\begin{aligned}
&d(Y) = \text{the diameter of } Y,\\
&\rho(x) = \operatorname{dist}(x, \partial\Omega), \quad \forall x \in \mathbb{R}^N \text{ (see (1.47))},\\
&W^{1,p}_\rho(\Omega) = \big\{\phi \in L^p(\Omega) \mid \rho\nabla\phi \in L^p(\Omega)^N\big\} \text{ (see Subsection 1.4.3)}.
\end{aligned}
$$

Set

$$
\rho_\varepsilon(x) = \inf\Big\{1, \frac{\rho(x)}{\varepsilon d(Y)}\Big\} \quad \text{for } x \in \mathbb{R}^N.
$$

For $\delta > 0$, introduce the sets

$$
\begin{aligned}
\Omega^{bl}_\delta &\doteq \big\{x \in \Omega \mid \rho(x) < \delta\big\},\\
\Omega^{b}_\delta &\doteq \big\{x \in \mathbb{R}^N \mid \rho(x) < \delta\big\}.
\end{aligned}
\tag{12.1}
$$

The next lemma and proposition give some classical estimates in a neighborhood of the boundary of an open set.

D. Cioranescu et al., *The Periodic Unfolding Method*, Series in Contemporary Mathematics 3, https://doi.org/10.1007/978-981-13-3032-2_12

Lemma 12.1. *Suppose $p \in [1,+\infty]$. Let $\mathcal{O}$ be an open set given in an orthonormal coordinate system $(O;\mathbf{e}_1,\ldots,\mathbf{e}_N)$ by a map $f : [-\boldsymbol{\alpha},\boldsymbol{\alpha}]^{N-1} \to \mathbb{R}$, Lipschitz continuous, such that*

$$\mathcal{O} = \big\{(x',x_N) \in \mathbb{R}^N \mid f(x') < x_N < f(x') + \boldsymbol{\beta},\ \ x' \in (-\boldsymbol{\alpha},\boldsymbol{\alpha})^{N-1}\big\}.$$

where $x' = (x_1,\ldots,x_{N-1})$ and $\boldsymbol{\alpha}$, $\boldsymbol{\beta}$ are strictly positive constants.

Denote by $\partial\mathcal{O}_f$ the lower part of the boundary of $\mathcal{O}$, i.e.,

$$\partial\mathcal{O}_f = \big\{(x',x_N) \in \mathbb{R}^N \mid x_N = f(x'),\ \ x' \in (-\boldsymbol{\alpha},\boldsymbol{\alpha})^{N-1}\big\}.$$

For every $\eta \in (0,\boldsymbol{\beta}]$, set

$$\mathcal{O}_\eta = \big\{(x',x_N) \in \mathbb{R}^N \mid f(x') < x_N < f(x') + \eta,\ \ x' \in (-\boldsymbol{\alpha},\boldsymbol{\alpha})^{N-1}\big\}.$$

Then, for every $\varphi \in W^{1,p}(\mathcal{O})$, $p \in [1,+\infty]$, one has

$$\begin{aligned} \|\varphi\|_{L^p(\mathcal{O}_\eta)} &\leqslant C\big(\eta^{1/p}\|\varphi\|_{L^p(\partial\mathcal{O}_f)} + \eta\|\nabla\varphi\|_{L^p(\mathcal{O}_\eta)}\big),\\ \|\varphi\|_{L^p(\partial\mathcal{O}_f)} &\leqslant C\eta^{-1/p}\big(\|\varphi\|_{L^p(\mathcal{O}_\eta)} + \eta\|\nabla\varphi\|_{L^p(\Omega)}\big). \end{aligned} \tag{12.2}$$

The constant depends on $\boldsymbol{\alpha}$ and $\boldsymbol{\beta}$ as well as the Lipschitz constant of f but not on p.

Proof. Let φ be in $\mathcal{C}^1(\overline{\mathcal{O}})$ and $p \in [1,+\infty)$. For all $x' \in (-\boldsymbol{\alpha},\boldsymbol{\alpha})^{N-1}$, one has

$$\int_{I(x')} \big|\varphi(x',t)\big|^p dt \leqslant 2^{p-1}\eta\big|\varphi(x',f(x'))\big|^p + \frac{2^{p-1}}{p}\eta^p \int_{I(x')} \Big|\frac{\partial\varphi}{\partial x_N}(x',t)\Big|^p dt,$$

where $I(x') = [f(x'), f(x') + \eta]$. Hence, for a.e. $x' \in (-\boldsymbol{\alpha},\boldsymbol{\alpha})^{N-1}$

$$\begin{aligned} \int_{I(x')} \big|\varphi(x',t)\big|^p dt \leqslant{}& 2^p\eta\big|\varphi(x',f(x'))\big|^p\sqrt{1+|\nabla f(x')|^2}\\ &+ 2^p\eta^p \int_{I(x')} \Big|\frac{\partial\varphi}{\partial x_N}(x',t)\Big|^p dt. \end{aligned}$$

Integrating with respect to $x' \in (-\boldsymbol{\alpha},\boldsymbol{\alpha})^{N-1}$ yields

$$\|\varphi\|_{L^p(\mathcal{O}_\eta)} \leqslant C\big(\eta^{1/p}\|\varphi\|_{L^p(\partial\mathcal{O}_f)} + \eta\|\nabla\varphi\|_{L^p(\mathcal{O}_\eta)}\big).$$

The constant depends on $\boldsymbol{\alpha}$ and $\boldsymbol{\beta}$ but not on p. By the density of $\mathcal{C}^1(\overline{\mathcal{O}})$ in $W^{1,p}(\mathcal{O})$, this inequality is also satisfied for every φ in $W^{1,p}(\mathcal{O})$, $p \in [1,+\infty)$. Passing to the limit with respect to p, this estimate holds true for every φ in $W^{1,\infty}(\mathcal{O})$.

Inequality $(12.2)_2$ is shown in a similar way. ☐

Proposition 12.2. *Suppose $p \in [1,+\infty]$. There exist δ_0 and C which only depend on $\partial\Omega$, such that for every ϕ in $W^{1,p}(\Omega)$ and every $\delta \in (0,\delta_0]$,*

$$\begin{aligned} \|\phi\|_{L^p(\Omega^{bl}_\delta)} &\leqslant C\big(\delta^{1/p}\|\phi\|_{L^p(\partial\Omega)} + \delta\|\nabla\phi\|_{L^p(\Omega^{bl}_\delta)}\big)\ \big(\leqslant C\delta^{1/p}\|\phi\|_{W^{1,p}(\Omega)}\big),\\ \|\phi\|_{L^p(\partial\Omega)} &\leqslant C\delta^{-1/p}\big(\|\phi\|_{L^p(\Omega^{bl}_\delta)} + \delta\|\nabla\phi\|_{L^p(\Omega)}\big). \end{aligned} \tag{12.3}$$

Remark 12.3. *As soon as* $(12.3)_1$ *holds for every* δ *in some* $(0, \delta_0]$, *it holds for every* $\delta > 0$. *Indeed, for* δ *in* $(0, \delta_0]$ *this inequality is given by* $(12.3)_1$ *with constants independent of* δ *while for* $\delta \geqslant \delta_0$

$$\begin{aligned}\|\phi\|_{L^p(\Omega^{bl}_\delta)} \leqslant \|\phi\|_{L^p(\Omega)} &\leqslant C\big(\|\phi\|_{L^p(\partial\Omega)} + \|\nabla\phi\|_{L^p(\Omega)}\big)\\ &\leqslant \frac{C}{\inf\{1, \delta_0^{1/p}\}}\big(\delta^{1/p}\|\phi\|_{L^p(\partial\Omega)} + \delta\|\nabla\phi\|_{L^p(\Omega)}\big).\end{aligned}$$

Proof of Proposition 12.2. Since the boundary of Ω is Lipschitz, there exist constants $\boldsymbol{\alpha}, \boldsymbol{\beta}, C, M$ strictly positive, a finite number $\mathcal{N}$ of orthonormal local coordinate systems in $(O_r; \mathbf{e}_{1r}, \ldots, \mathbf{e}_{Nr})$ and $\mathcal{N}$ maps $f_r : [-\boldsymbol{\alpha}, \boldsymbol{\alpha}]^{N-1} \to \mathbb{R}$, Lipschitz continuous with constant M, $1 \leqslant r \leqslant \mathcal{N}$, such that,

$$\begin{aligned}\partial\Omega &= \bigcup_{r=1}^{\mathcal{N}}\Big\{(x'_r, x_{N,r}) \mid x_{N,r} = f_r(x'_r), \quad x'_r \in (-\boldsymbol{\alpha}, \boldsymbol{\alpha})^{N-1}\Big\},\\ \Omega^{bl}_C &\subset \bigcup_{r=1}^{\mathcal{N}}\Big\{(x'_r, x_{N,r}) \mid f_r(x'_r) < x_{N,r} < f_r(x'_r) + \boldsymbol{\beta},\ x'_r \in (-\boldsymbol{\alpha}, \boldsymbol{\alpha})^{N-1}\Big\} \subset \Omega,\\ \Omega^{b}_C &\subset \bigcup_{r=1}^{\mathcal{N}}\Big\{(x'_r, x_{N,r}) \mid |x_{N,r} - f_r(x'_r)| < \boldsymbol{\beta},\ x'_r \in (-\boldsymbol{\alpha}, \boldsymbol{\alpha})^{N-1}\Big\},\end{aligned}$$

where $x'_r = (x_{1,r}, \ldots, x_{N-1,r})$.

Set

$$\delta_0 \doteq \inf\{C, \boldsymbol{\beta}/\sqrt{1+M^2} \quad \text{and} \quad \bar{\delta} = \delta\sqrt{1+M^2}.$$

Through easy geometrical arguments one shows that for every $\delta \in (0, \delta_0]$,

$$\Omega^b_\delta \subset \bigcup_{r=1}^{\mathcal{N}}\Big\{(x'_r, x_{N,r}) \mid |x_{N,r} - f_r(x'_r)| < \bar{\delta},\ x'_r \in (-\boldsymbol{\alpha}, \boldsymbol{\alpha})^{N-1}\Big\} \subset \Omega^b_{\bar{\delta}}.$$

Then, combining estimates (12.2) for every local maps, estimates (12.3) are obtained for every $\delta \in (0, \delta_0]$. □

Recalling that Λ_ε is included in $\Omega^{bl}_{\varepsilon d(Y)}$, from Remark 12.3 it follows that for every $\varepsilon > 0$,

$$\|\varphi\|_{L^p(\Lambda_\varepsilon)} \leqslant \|\varphi\|_{L^p(\Omega^{bl}_{\varepsilon d(Y)})} \leqslant C\big(\varepsilon^{1/p}\|\varphi\|_{L^p(\partial\Omega)} + \varepsilon\|\nabla\varphi\|_{L^p(\Omega)}\big), \tag{12.4}$$

where C depends only on $\partial\Omega$ and $d(Y)$, but not upon p and ε.

Corollary 12.4. *There is a constant* C *depending on* $\partial\Omega$ *and independent of* p *and* δ, *such that for every* ψ *in* $W^{1,p}_0(\Omega)$ *(resp. in* $W^{1,p}(\mathbb{R}^N)$*)*,

$$\begin{aligned}\|\psi\|_{L^p(\Omega^{bl}_\delta)} &\leqslant C\delta\|\nabla\psi\|_{L^p(\Omega)},\\ (\textit{resp. } \|\psi\|_{L^p(\Omega^b_\delta)} &\leqslant C\delta^{1/p}\|\psi\|_{W^{1,p}(\mathbb{R}^N)}).\end{aligned} \tag{12.5}$$

Proof. Estimate $(12.5)_1$ is an immediate consequence of Remark 12.3. Proceeding as in the proof of Proposition 12.2 and due to Remark 12.3, it gives

$$\|\phi\|_{L^p(\Omega^b_\delta \setminus \overline{\Omega^{bl}_\delta})} \leqslant C\big(\delta^{1/p}\|\phi\|_{L^p(\partial\Omega)} + \delta\|\nabla\phi\|_{L^p(\mathbb{R}^N)}\big) \leqslant C\delta^{1/p}\|\phi\|_{W^{1,p}(\mathbb{R}^N)}.$$

This inequality, together with that in Remark 12.3, gives $(12.5)_2$. □

12.2 Shifted estimates

Proposition 12.5 (Compare with Proposition 1.38). *Suppose $p \in [1, +\infty]$. Let Ω be a bounded domain in $\mathbb{R}^N$ with a Lipschitz boundary. Then, for every φ in $L^p(\Omega)$,*

$$\begin{aligned}
&\|\varphi \mathbf{1}_{\widehat{\Omega}_\varepsilon} - \mathcal{M}_\varepsilon(\varphi)\|_{[W^{1,p'}(\Omega)]'} \leqslant C\varepsilon\|\varphi\|_{L^p(\Omega)},\\
&\|\varphi - \mathcal{M}_\varepsilon(\varphi)\|_{[W^{1,p'}(\Omega)]'} \leqslant C\varepsilon\|\varphi\|_{L^p(\Omega)} + C\varepsilon^{1/p'}\|\varphi\|_{L^p(\Omega^{bl}_{\varepsilon d(Y)})},\\
&\|\varphi - \mathcal{M}_\varepsilon(\varphi)\|_{W^{-1,p}(\Omega)} \leqslant C\varepsilon\|\varphi\|_{L^p(\Omega)}.
\end{aligned} \tag{12.6}$$

Likewise, for every φ in $L^p(\mathbb{R}^N)$,

$$\begin{aligned}
&(i) \quad \|\varphi - \widetilde{\mathcal{M}}_\varepsilon(\varphi)\|_{W^{-1,p}(\mathbb{R}^N)} \leqslant C\varepsilon\|\varphi\|_{L^p(\mathbb{R}^N)},\\
&(ii) \quad \frac{1}{\varepsilon}\big(\varphi - \widetilde{\mathcal{M}}_\varepsilon(\varphi)\big) \rightharpoonup 0 \quad \text{weakly in } W^{-1,p}(\mathbb{R}^N).
\end{aligned} \tag{12.7}$$

For $p = +\infty$, the convergence holds in $W^{-1,\infty}(\mathbb{R}^N)$ for the weak-$*$ topology.

Proof. Let φ be in $L^p(\Omega)$ and ψ be in $W^{1,p'}(\Omega)$. One has

$$\int_\Omega \big(\varphi\mathbf{1}_{\widehat{\Omega}_\varepsilon} - \mathcal{M}_\varepsilon(\varphi)\big)\,\psi\,dx = \int_{\widehat{\Omega}_\varepsilon} \varphi\,\big(\psi - \mathcal{M}_\varepsilon(\psi)\big)\,dx.$$

Then estimate $(1.39)_1$ implies $(12.6)_1$.

Again, let φ be in $L^p(\Omega)$ and ψ be in $W^{1,p'}(\Omega)$ (resp. in $W^{1,p'}_0(\Omega)$). The above equalities and the definition of $\mathcal{M}_\varepsilon$ give

$$\int_\Omega \big(\varphi - \mathcal{M}_\varepsilon(\varphi)\big)\,\psi\,dx = \int_{\widehat{\Omega}_\varepsilon} \varphi\,\big(\psi - \mathcal{M}_\varepsilon(\psi)\big)\,dx + \int_{\Lambda_\varepsilon} \varphi\,\psi\,dx.$$

Estimates $(1.39)_1$ and (12.4) then give $(12.6)_{2,3}$.

For $\Omega = \mathbb{R}^N$, the same computation, together with inequality $(1.39)_1$, proves (12.7)(i).

It remains to prove convergence in (12.7)(ii). To do so, for ψ in $W^{1,p'}(\mathbb{R}^N)$, consider

$$\begin{aligned}
\int_{\mathbb{R}^N} \frac{\varphi - \widetilde{\mathcal{M}}_\varepsilon(\varphi)}{\varepsilon}\,\psi\,dx &= \int_{\mathbb{R}^N} \varphi\,\frac{\psi - \widetilde{\mathcal{M}}_\varepsilon(\psi)}{\varepsilon}\,dx\\
&= \frac{1}{|Y|}\int_{\mathbb{R}^N \times Y} \widetilde{\mathcal{T}}_\varepsilon(\varphi)\,\frac{\widetilde{\mathcal{T}}_\varepsilon(\psi) - \widetilde{\mathcal{M}}_\varepsilon(\psi)}{\varepsilon}\,dxdy.
\end{aligned}$$

Applying Proposition 1.39 for $\Omega = \mathbb{R}^N$, we obtain

$$\lim_{\varepsilon\to 0} \int_{\mathbb{R}^N} \frac{\varphi - \widetilde{\mathcal{M}}_\varepsilon(\varphi)}{\varepsilon} \psi\, dx = \frac{1}{|Y|} \int_{\mathbb{R}^N \times Y} \varphi(x)\, y^c \cdot \nabla\psi(x)\, dxdy = 0.$$

This ends the proof. □

Remark 12.6. *The equivalent of* (1.39) *(ii) of Proposition 1.38 with a derivative shift does not hold in general. Indeed, this would imply the existence of a uniform estimates of the form*

$$\forall\varphi \in L^p(\Omega), \quad \|\widetilde{\mathcal{T}}_\varepsilon(\varphi) - \widetilde{\mathcal{M}}_\varepsilon(\varphi)\|_{W^{-1,p}(\Omega\times Y)} \leqslant C\varepsilon\|\varphi\|_{L^p(\Omega)}, \tag{12.8}$$

but the latter is false. A counter example to this inequality is constructed with the help of the sequence $\{\varphi_\varepsilon\}_\varepsilon$ *defined by* $\varphi_\varepsilon(x) \doteq \Phi(\{x/\varepsilon\}_Y)$ *for a non constant* Φ *in* $L^p(Y)$ *satisfying* $\mathcal{M}_Y(\Phi) = 0$. *Since* $\widetilde{\mathcal{T}}_\varepsilon(\varphi_\varepsilon)(x,y) = \Phi(y)\mathbf{1}_{\widehat{\Omega}_\varepsilon}$ *a.e. in* $\Omega \times Y$, *Proposition 1.8 implies*

$$\|\varphi_\varepsilon\|_{L^p(\Omega)} \to \Big(\frac{|\Omega|}{|Y|}\Big)^{1/p} \|\Phi\|_{L^p(Y)}.$$

Also $\widetilde{\mathcal{M}}_\varepsilon(\varphi_\varepsilon) = 0$. *This contradicts* (12.8). *Further down (see* (12.15)*), we show that an estimate similar to* (12.8) *can be established when replacing* φ *by the scale-splitting operator* $\mathcal{Q}_\varepsilon(\varphi)$.

Recall the definition of $\breve{w}$ for w measurable on $\mathbb{R}^N$:

$$\breve{w}(x) = w(-x) \quad \text{for a.e. } x \in \mathbb{R}^N.$$

Lemma 12.7. *Suppose* $p \in [1, +\infty]$. *For every* φ *in* $L^p(\mathbb{R}^N)$ *and* θ *in* $L^{p'}(\mathbb{R}^N)$, *the following equalities hold:*

$$\begin{aligned}
&(i) \int_{\mathbb{R}^N} \widetilde{\mathcal{M}}_\varepsilon(\varphi)(x)\, \widetilde{\mathcal{M}}_\varepsilon(\theta)(x)\, dx = \int_{\mathbb{R}^N} \widetilde{\mathcal{M}}_\varepsilon(\breve{\varphi})(x)\, \widetilde{\mathcal{M}}_\varepsilon(\breve{\theta})(x)\, dx,\\
&(ii) \int_{\mathbb{R}^N} \widetilde{\mathcal{Q}}_\varepsilon(\varphi)(x)\, \widetilde{\mathcal{M}}_\varepsilon(\theta)(x)\, dx = \int_{\mathbb{R}^N} \widetilde{\mathcal{M}}_\varepsilon(\breve{\varphi})(x)\, \widetilde{\mathcal{Q}}_\varepsilon(\breve{\theta})(x)\, dx,\\
&(iii) \int_{\mathbb{R}^N} \widetilde{\mathcal{T}}_\varepsilon \circ \widetilde{\mathcal{Q}}_\varepsilon(\varphi)(x,y)\, \theta(x)\, dx = \int_{\mathbb{R}^N} \breve{\varphi}(x)\, \widetilde{\mathcal{T}}_\varepsilon \circ \widetilde{\mathcal{Q}}_\varepsilon(\breve{\theta})(x,y)\, dx.
\end{aligned} \tag{12.9}$$

The last equality is satisfied for every $y \in Y$.

Proof. (*i*) For a.e. x in $\mathbb{R}^N$ one can write

$$x = \varepsilon\xi + \varepsilon y, \quad \text{with } \xi = \Big[\frac{x}{\varepsilon}\Big]_Y \in \mathbf{G}, \quad \mathbf{y} = \Big\{\frac{\mathbf{x}}{\varepsilon}\Big\}_{\mathbf{Y}} \in \mathbf{Y}.$$

Hence, with $\mathbf{b} = \sum_{i=1}^N \mathbf{b}_i$, one has $-x = -\varepsilon(\xi + \mathbf{b}) + \varepsilon(\mathbf{b} - y)$. Since $-(\xi + \mathbf{b})$ belongs to $\mathbf{G}$ and $\mathbf{b} - y$ to Y, they are $\Big[\frac{-x}{\varepsilon}\Big]_Y$ and $\Big\{\frac{-x}{\varepsilon}\Big\}_{Y}$, respectively.

Then definition 1.21 and Remark 1.22 give

$$\begin{aligned}
\widetilde{\mathcal{M}}_\varepsilon(w)(-x) &= \widetilde{\mathcal{M}}_\varepsilon(w)\big(-\varepsilon(\xi+\mathbf{b})\big)\\
&= \frac{1}{|Y|}\int_Y w\big(-\varepsilon(\xi+\mathbf{b})+\varepsilon z\big)dz = \frac{1}{|Y|}\int_Y \breve{w}\big(\varepsilon\xi+\varepsilon(\mathbf{b}-z)\big)dz\\
&= \frac{1}{|Y|}\int_Y \breve{w}\big(\varepsilon\xi+\varepsilon z'\big)dz' = \widetilde{\mathcal{M}}_\varepsilon(\breve{w})(\varepsilon\xi) = \widetilde{\mathcal{M}}_\varepsilon(\breve{w})(x),
\end{aligned}$$

which implies identity (i).
(*ii*) It is immediate from the above equalities that for every $\xi\in\mathbf{G}$,

$$\widetilde{\mathcal{M}}_\varepsilon(w)(-\varepsilon\xi) = \widetilde{\mathcal{M}}_\varepsilon(\breve{w})(\varepsilon\xi-\varepsilon\mathbf{b}).$$

Now, using the explicit expression (1.63) of $\widetilde{\mathcal{Q}}_\varepsilon(\varphi)$ and the first equality in (1.56), we compute successively

$$\begin{aligned}
\int_{\mathbb{R}^N} \widetilde{\mathcal{Q}}_\varepsilon(\varphi)(x)\widetilde{\mathcal{M}}_\varepsilon(\theta)(x)\,dx &= \frac{\varepsilon^N|Y|}{2^N}\sum_{\xi\in\mathbf{G}}\sum_{\kappa\in\{0,1\}^N} \widetilde{\mathcal{M}}_\varepsilon(\varphi)\big(\varepsilon\xi+\varepsilon\kappa\big)\widetilde{\mathcal{M}}_\varepsilon(\theta)(\varepsilon\xi)\\
&= \frac{\varepsilon^N|Y|}{2^N}\sum_{\xi\in\mathbf{G}}\sum_{\kappa\in\{0,1\}^N} \widetilde{\mathcal{M}}_\varepsilon(\varphi)(\varepsilon\xi)\widetilde{\mathcal{M}}_\varepsilon(\theta)(\varepsilon\xi-\varepsilon\kappa)\\
&= \frac{\varepsilon^N|Y|}{2^N}\sum_{\xi\in\mathbf{G}}\sum_{\kappa\in\{0,1\}^N} \widetilde{\mathcal{M}}_\varepsilon(\varphi)(-\varepsilon\xi)\widetilde{\mathcal{M}}_\varepsilon(\theta)(-\varepsilon\xi-\varepsilon\kappa)\\
&= \frac{\varepsilon^N|Y|}{2^N}\sum_{\xi\in\mathbf{G}}\sum_{\kappa\in\{0,1\}^N} \widetilde{\mathcal{M}}_\varepsilon(\breve{\varphi})(\varepsilon\xi-\varepsilon\mathbf{b})\widetilde{\mathcal{M}}_\varepsilon(\breve{\theta})(\varepsilon\xi+\varepsilon\kappa-\varepsilon\mathbf{b})\\
&= \frac{\varepsilon^N|Y|}{2^N}\sum_{\xi\in\mathbf{G}}\sum_{\kappa\in\{0,1\}^N} \widetilde{\mathcal{M}}_\varepsilon(\breve{\varphi})(\varepsilon\xi)\widetilde{\mathcal{M}}_\varepsilon(\breve{\theta})(\varepsilon\xi+\varepsilon\kappa)\\
&= \int_{\mathbb{R}^N} \widetilde{\mathcal{M}}_\varepsilon(\breve{\varphi})(x)\,\widetilde{\mathcal{Q}}_\varepsilon(\breve{\theta})(x)\,dx.
\end{aligned}$$

(*iii*) Again, due to the explicit expression (1.63) of $\widetilde{\mathcal{Q}}_\varepsilon(\varphi)$ and to the first equality in (1.56), we proceed as in the proof of (12.9)(ii), to get

$$\begin{aligned}
\int_{\mathbb{R}^N} \widetilde{\mathcal{T}}_\varepsilon\circ\widetilde{\mathcal{Q}}_\varepsilon(\varphi)(\cdot,y)\theta\,dx &= \int_{\mathbb{R}^N} \widetilde{\mathcal{T}}_\varepsilon\circ\widetilde{\mathcal{Q}}_\varepsilon(\varphi)(\cdot,y)\widetilde{\mathcal{M}}_\varepsilon(\theta)\,dx\\
&= \varepsilon^N|Y|\sum_{\xi\in\mathbf{G}}\sum_{\kappa\in\{0,1\}^N} \mathbf{q}_\kappa(y)\widetilde{\mathcal{M}}_\varepsilon(\varphi)\big(\varepsilon\xi+\varepsilon\kappa\big)\widetilde{\mathcal{M}}_\varepsilon(\theta)(\varepsilon\xi)\\
&= \int_{\mathbb{R}^N} \widetilde{\mathcal{M}}_\varepsilon(\breve{\varphi})\widetilde{\mathcal{T}}_\varepsilon\circ\widetilde{\mathcal{Q}}_\varepsilon(\breve{\theta})(\cdot,y)\,dx\\
&= \int_{\mathbb{R}^N} \breve{\varphi}\,\widetilde{\mathcal{T}}_\varepsilon\circ\widetilde{\mathcal{Q}}_\varepsilon(\breve{\theta})(\cdot,y)\,dx,
\end{aligned}$$

which ends the proof of the lemma. □

Proposition 12.8 (Compare with Proposition 12.5). *Suppose* $p \in [1, +\infty]$.
(i) For every $\varphi \in L^p(\mathbb{R}^N)$

$$\|\widetilde{\mathcal{Q}}_\varepsilon(\varphi) - \varphi\|_{W^{-1,p}(\mathbb{R}^N)} = \|\widetilde{\mathcal{R}}_\varepsilon(\varphi)\|_{W^{-1,p}(\mathbb{R}^N)} \leqslant C\varepsilon\|\varphi\|_{L^p(\mathbb{R}^N)}. \tag{12.10}$$

(ii) Moreover (compare with (1.76)*)*

$$\frac{1}{\varepsilon}\widetilde{\mathcal{R}}_\varepsilon(\varphi) \rightharpoonup -\sum_{i=1}^N \mathcal{M}_Y(y_i)\frac{\partial\varphi}{\partial x_i} \quad \textit{weakly in} \quad W^{-1,p}(\mathbb{R}^N). \tag{12.11}$$

For $p = +\infty$, *the convergence holds in* $W^{-1,\infty}(\mathbb{R}^N)$ *for the weak-∗ topology.*

Proof. (i) Due to $(12.7)_1$, to prove (12.10), it is enough to estimate the quantity $\|\widetilde{\mathcal{Q}}_\varepsilon(\varphi) - \widetilde{\mathcal{M}}_\varepsilon(\varphi)\|_{W^{-1,p}(\mathbb{R}^N)}$. To do so, let ψ in $W^{1,p'}(\mathbb{R}^N)$. A simple computation shows that

$$\begin{aligned}\int_{\mathbb{R}^N} \big(\widetilde{\mathcal{M}}_\varepsilon(\varphi) - \widetilde{\mathcal{Q}}_\varepsilon(\varphi)\big)\,\psi\,dx &= \int_{\mathbb{R}^N} \big(\widetilde{\mathcal{M}}_\varepsilon(\varphi) - \widetilde{\mathcal{Q}}_\varepsilon(\varphi)\big)\,\widetilde{\mathcal{M}}_\varepsilon(\psi)\,dx\\ &\quad + \int_{\mathbb{R}^N} \big(\widetilde{\mathcal{M}}_\varepsilon(\varphi) - \widetilde{\mathcal{Q}}_\varepsilon(\varphi)\big)\big(\psi - \widetilde{\mathcal{M}}_\varepsilon(\psi)\big)\,dx.\end{aligned} \tag{12.12}$$

Propositions 1.24(i), 1.64(i) and 1.38(i) (with $\mathbb{R}^N$ instead of Ω), give an upper bound for the last term in the right-hand side above,

$$\Big|\int_{\mathbb{R}^N} \big(\widetilde{\mathcal{M}}_\varepsilon(\varphi) - \widetilde{\mathcal{Q}}_\varepsilon(\varphi)\big)\big(\psi - \widetilde{\mathcal{M}}_\varepsilon(\psi)\big)\,dx\Big| \leqslant C\varepsilon\|\varphi\|_{L^p(\mathbb{R}^N)}\|\nabla\psi\|_{L^{p'}(\mathbb{R}^N)}. \tag{12.13}$$

The next to the last term, according to equalities (12.9), becomes

$$\begin{aligned}&\int_{\mathbb{R}^N} \Big(\widetilde{\mathcal{M}}_\varepsilon(\varphi) - \widetilde{\mathcal{Q}}_\varepsilon(\varphi)\Big)\widetilde{\mathcal{M}}_\varepsilon(\psi)\,dx\\ &\qquad = \int_{\mathbb{R}^N} \widetilde{\mathcal{M}}_\varepsilon(\breve{\varphi})\Big(\widetilde{\mathcal{M}}_\varepsilon(\breve{\psi}) - \widetilde{\mathcal{Q}}_\varepsilon(\breve{\psi})\Big)\,dx,\end{aligned} \tag{12.14}$$

which, by using Proposition 1.64(iv), is evaluated as

$$\begin{aligned}\Big|\int_{\mathbb{R}^N} \big(\widetilde{\mathcal{M}}_\varepsilon(\varphi) - \widetilde{\mathcal{Q}}_\varepsilon(\varphi)\big)\,\widetilde{\mathcal{M}}_\varepsilon(\psi)\,dx\Big| &\leqslant C\varepsilon\|\breve{\varphi}\|_{L^p(\mathbb{R}^N)}\|\nabla\breve{\psi}\|_{L^{p'}(\mathbb{R}^N)}\\ &\leqslant C\varepsilon\|\varphi\|_{L^p(\mathbb{R}^N)}\|\nabla\psi\|_{L^{p'}(\mathbb{R}^N)}.\end{aligned}$$

This estimate, combined with (12.13) and $(12.7)_1$, proves (12.10).
(ii) Now, taking into account equality (12.12) and (12.14), one writes

$$\int_{\mathbb{R}^N} \frac{\varphi - \widetilde{\mathcal{Q}}_\varepsilon(\varphi)}{\varepsilon}\psi\,dx = \int_{\mathbb{R}^N} \frac{\varphi - \widetilde{\mathcal{M}}_\varepsilon(\varphi)}{\varepsilon}\psi\,dx + \int_{\mathbb{R}^N} \frac{\widetilde{\mathcal{M}}_\varepsilon(\varphi) - \widetilde{\mathcal{Q}}_\varepsilon(\varphi)}{\varepsilon}\psi\,dx$$

$$\begin{aligned}\int_{\mathbb{R}^N} \frac{\widetilde{\mathcal{M}}_\varepsilon(\varphi) - \widetilde{\mathcal{Q}}_\varepsilon(\varphi)}{\varepsilon}\psi\,dx &= \int_{\mathbb{R}^N} \widetilde{\mathcal{M}}_\varepsilon(\breve{\varphi})\frac{\widetilde{\mathcal{M}}_\varepsilon(\breve{\psi}) - \widetilde{\mathcal{Q}}_\varepsilon(\breve{\psi})}{\varepsilon}\,dx\\ &\quad + \int_{\mathbb{R}^N} \big(\widetilde{\mathcal{M}}_\varepsilon(\varphi) - \widetilde{\mathcal{Q}}_\varepsilon(\varphi)\big)\frac{\psi - \widetilde{\mathcal{M}}_\varepsilon(\psi)}{\varepsilon}\,dx.\end{aligned}$$

Then (1.40) (with $\Omega = \mathbb{R}^N$) and (1.76) imply the convergence

$$\frac{1}{\varepsilon}\big(\widetilde{\mathcal{M}}_\varepsilon(\breve{\psi}) - \widetilde{\mathcal{Q}}_\varepsilon(\breve{\psi})\big) \to -\sum_{i=1}^N \mathcal{M}_Y(y_i)\frac{\partial\breve{\psi}}{\partial x_i} \qquad \text{strongly in } L^{p'}(\mathbb{R}^N).$$

Finally, this convergence, $(12.7)_2$ and (1.40) (with $\Omega = \mathbb{R}^N$), give

$$\lim_{\varepsilon\to 0}\int_{\mathbb{R}^N} \frac{\varphi - \widetilde{\mathcal{Q}}_\varepsilon(\varphi)}{\varepsilon}\,\psi\,dx = -\int_{\mathbb{R}^N}\breve{\varphi}\Big(\sum_{i=1}^N \mathcal{M}_Y(y_i)\frac{\partial\breve{\psi}}{\partial x_i}\Big)\,dx,$$

from which the conclusion follows after performing an integration by parts and a change of variable. □

Proposition 12.9 (Compare with Proposition 1.38). *Suppose* $p \in [1, +\infty)$. *Let* φ *be in* $L^p(\mathbb{R}^N)$.
(i) Then,

$$\big\|\widetilde{\mathcal{T}}_\varepsilon \circ \widetilde{\mathcal{Q}}_\varepsilon(\varphi) - \widetilde{\mathcal{M}}_\varepsilon \circ \widetilde{\mathcal{Q}}_\varepsilon(\varphi)\big\|_{L^\infty(Y;W^{-1,p}(\mathbb{R}^N))} \leqslant C\varepsilon\|\varphi\|_{L^p(\mathbb{R}^N)}, \tag{12.15}$$

with the constant C *depending only on* Y.
(ii) For every y *in* Y*, one has the following weak convergence in* $W^{-1,p}(\mathbb{R}^N)$*:*

$$\frac{1}{\varepsilon}\Big(\widetilde{\mathcal{T}}_\varepsilon \circ \widetilde{\mathcal{Q}}_\varepsilon(\varphi)(\cdot\,,y) - \widetilde{\mathcal{M}}_\varepsilon \circ \widetilde{\mathcal{Q}}_\varepsilon(\varphi)\Big) \rightharpoonup y^c\cdot\nabla\varphi. \tag{12.16}$$

Proof. (i) For ψ in $W^{1,p'}(\mathbb{R}^N)$,

$$\begin{aligned}\int_{\mathbb{R}^N}\big(\widetilde{\mathcal{M}}_\varepsilon(\varphi) - \widetilde{\mathcal{M}}_\varepsilon\circ\widetilde{\mathcal{Q}}_\varepsilon(\varphi)\big)\,\psi\,dx &= \int_{\mathbb{R}^N}\big(\varphi - \widetilde{\mathcal{Q}}_\varepsilon(\varphi)\big)\,\widetilde{\mathcal{M}}_\varepsilon(\psi)\,dx\\ &= \int_{\mathbb{R}^N}\big(\varphi - \widetilde{\mathcal{Q}}_\varepsilon(\varphi)\big)\,\psi\,dx + \int_{\mathbb{R}^N}\big(\varphi - \widetilde{\mathcal{Q}}_\varepsilon(\varphi)\big)\,\big(\widetilde{\mathcal{M}}_\varepsilon(\psi) - \psi\big)\,dx.\end{aligned} \tag{12.17}$$

By Proposition 1.64(i), (12.10) and (1.40) (with $\Omega = \mathbb{R}^N$), one has the estimate

$$\|\widetilde{\mathcal{M}}_\varepsilon(\varphi) - \widetilde{\mathcal{M}}_\varepsilon\circ\widetilde{\mathcal{Q}}_\varepsilon(\varphi)\|_{W^{-1,p}(\mathbb{R}^N)} \leqslant C\varepsilon\|\varphi\|_{L^p(\mathbb{R}^N)}. \tag{12.18}$$

Then, proceeding as in the proof of (12.10), we obtain

$$\|\widetilde{\mathcal{T}}_\varepsilon\circ\widetilde{\mathcal{Q}}_\varepsilon(\varphi)(\cdot,y) - \widetilde{\mathcal{M}}_\varepsilon(\varphi)\|_{W^{-1,p}(\mathbb{R}^N)} \leqslant C\varepsilon\|\varphi\|_{L^p(\mathbb{R}^N)}, \qquad \forall y\in Y.$$

This, together with (12.18), gives (12.15).
(ii) Again, from equalities (12.17) with ψ in $W^{1,p'}(\mathbb{R}^N)$,

$$\begin{aligned}&\int_{\mathbb{R}^N}\frac{1}{\varepsilon}\big(\widetilde{\mathcal{M}}_\varepsilon(\varphi) - \widetilde{\mathcal{M}}_\varepsilon\circ\widetilde{\mathcal{Q}}_\varepsilon(\varphi)\big)\,\psi\,dx\\ &\qquad= \int_{\mathbb{R}^N}\frac{1}{\varepsilon}\widetilde{\mathcal{R}}_\varepsilon(\varphi)\,\psi\,dx + \int_{\mathbb{R}^N}\big(\varphi - \widetilde{\mathcal{Q}}_\varepsilon(\varphi)\big)\,\frac{\widetilde{\mathcal{M}}_\varepsilon(\psi) - \psi}{\varepsilon}\,dx.\end{aligned}$$

Then, from convergences (12.11), (1.40) (with $\Omega = \mathbb{R}^N$) and (1.73)(i), one gets

$$\frac{1}{\varepsilon}\big(\widetilde{\mathcal{M}}_\varepsilon(\varphi) - \widetilde{\mathcal{M}}_\varepsilon \circ \widetilde{\mathcal{Q}}_\varepsilon(\varphi)\big) \rightharpoonup -\sum_{i=1}^N \mathcal{M}_Y(y_i)\frac{\partial \varphi}{\partial x_i} \quad \text{weakly in } W^{-1,p}(\mathbb{R}^N). \tag{12.19}$$

Now, (12.9)(i) and (iii) imply that for every y in Y,

$$\int_{\mathbb{R}^N} \frac{\widetilde{\mathcal{T}}_\varepsilon \circ \widetilde{\mathcal{Q}}_\varepsilon(\varphi)(\cdot, y) - \widetilde{\mathcal{M}}_\varepsilon(\varphi)}{\varepsilon}\,\psi\,dx = \int_{\mathbb{R}^N} \breve{\varphi}\,\frac{\widetilde{\mathcal{T}}_\varepsilon \circ \widetilde{\mathcal{Q}}_\varepsilon(\breve{\psi})(\cdot, y) - \widetilde{\mathcal{M}}_\varepsilon(\breve{\psi})}{\varepsilon}\,dx.$$

At this point, recall the strong convergence proved in Corollary 1.68,

$$\frac{1}{\varepsilon}\widetilde{\mathcal{T}}_\varepsilon \circ \big(\widetilde{\mathcal{Q}}_\varepsilon - \widetilde{\mathcal{M}}_\varepsilon\big)(\breve{\psi})(\,\cdot\,, y) \to \sum_{i=1}^N \frac{\partial \breve{\psi}}{\partial x_i} y_i \quad \text{strongly in } L^p(\mathbb{R}^N \times Y).$$

The dependence of the left-hand side is affine with respect to each y_i (it is in $\mathbf{Q}_1(Y)$). Therefore it implies the point-wise convergence for every $y \in Y$,

$$\frac{1}{\varepsilon}\widetilde{\mathcal{T}}_\varepsilon \circ \big(\widetilde{\mathcal{Q}}_\varepsilon - \widetilde{\mathcal{M}}_\varepsilon\big)(\breve{\psi})(\,\cdot\,, y) \to \sum_{i=1}^N \frac{\partial \breve{\psi}}{\partial x_i} y_i \quad \text{strongly in } L^p(\mathbb{R}^N).$$

Recalling that from (1.19), $\widetilde{\mathcal{T}}_\varepsilon \circ \widetilde{\mathcal{M}}_\varepsilon = \widetilde{\mathcal{M}}_\varepsilon$, we finally get for every $y \in Y$

$$\begin{aligned}\lim_{\varepsilon\to 0}\int_{\mathbb{R}^N} \frac{\widetilde{\mathcal{T}}_\varepsilon \circ \widetilde{\mathcal{Q}}_\varepsilon(\varphi)(\cdot, y) - \widetilde{\mathcal{M}}_\varepsilon(\varphi)}{\varepsilon}\psi\,dx &= \int_{\mathbb{R}^N} \breve{\varphi}\Big(\sum_{i=1}^N \frac{\partial \breve{\psi}}{\partial x_i} y_i\Big)dx \\ &= -\int_{\mathbb{R}^N} \varphi\Big(\sum_{i=1}^N \frac{\partial \psi}{\partial x_i} y_i\Big)dx = \Big\langle \sum_{i=1}^N \frac{\partial \varphi}{\partial x_i} y_i\,,\,\psi\Big\rangle_{W^{-1,p}(\mathbb{R}^N), W^{1,p}(\mathbb{R}^N)}.\end{aligned}$$

Together with (12.19), this implies (12.16). □

Proposition 12.10. *Suppose $p \in [1, +\infty)$.*
(i) There exits a constant $C = C(Y)$, such that for every φ in $W^{1,p}(\mathbb{R}^N)$,

$$\big\|\widetilde{\mathcal{T}}_\varepsilon(\nabla\widetilde{\mathcal{Q}}_\varepsilon(\varphi)) - \nabla\widetilde{\mathcal{Q}}_\varepsilon(\varphi)\big\|_{L^\infty(Y; W^{-1,p}(\mathbb{R}^N))} \leqslant C\varepsilon\|\nabla\varphi\|_{L^p(\mathbb{R}^N)}. \tag{12.20}$$

(ii) The following weak convergence in $W^{-1,p}(\mathbb{R}^N)^N$ holds for $y \in Y$:

$$\frac{1}{\varepsilon}\big(\widetilde{\mathcal{T}}_\varepsilon(\nabla\widetilde{\mathcal{Q}}_\varepsilon(\varphi))(\,\cdot\,, y) - \nabla\widetilde{\mathcal{Q}}_\varepsilon(\varphi)(\,\cdot\,)\big) \rightharpoonup \big(\nabla^2\varphi - \mathit{diag}\,(\nabla^2\varphi)\big)\,y^c, \tag{12.21}$$

where $\mathit{diag}\,(\nabla^2\varphi)$ is the N^2 diagonal matrix with the same diagonal as $\nabla^2\varphi$.

(iii) Let Ω be a bounded domain with Lipschitz boundary. There exists a constant C depending on $\partial\Omega$ and $d(Y)$, such that for every φ in $W^{1,p}(\mathcal{V}_{4\varepsilon d(Y)}(\Omega))$ (cf. notation 1.63),

$$
\begin{aligned}
&\big\|\mathcal{T}_\varepsilon\big(\nabla\mathcal{Q}_\varepsilon(\varphi)\big) - \nabla\mathcal{Q}_\varepsilon(\varphi)\big\|_{L^\infty(Y;(W^{1,p'}(\Omega))')}\\
&\qquad\qquad \leqslant C\varepsilon\|\nabla\varphi\|_{L^p(\Omega)} + C\varepsilon^{1/p'}\|\nabla\varphi\|_{L^p(\Omega^b_{4\varepsilon d(Y)})}.
\end{aligned}
\tag{12.22}
$$

Proof. (i) Let Φ be in $W^{1,p'}(\mathbb{R}^N)$. For $1 \leqslant i \neq j \leqslant N$ and $y \in Y$, one has

$$
\frac{\partial}{\partial y_i}\Big(\widetilde{\mathcal{T}}_\varepsilon\Big(\frac{\partial\widetilde{\mathcal{Q}}_\varepsilon(\varphi)}{\partial x_j}\Big)\Big) = \widetilde{\mathcal{T}}_\varepsilon\Big(\frac{\partial^2\widetilde{\mathcal{Q}}_\varepsilon(\varphi)}{\partial x_i\partial x_j}\Big),
$$

the latter being ε-piece-wise constant with respect to x. Consequently,

$$
\begin{aligned}
&\int_{\mathbb{R}^N}\frac{\partial}{\partial y_i}\Big(\widetilde{\mathcal{T}}_\varepsilon\Big(\frac{\partial\widetilde{\mathcal{Q}}_\varepsilon(\varphi)}{\partial x_j}\Big)\Big)(x,y)\,\Phi(x)\,dx\\
&\qquad\qquad = \varepsilon\int_{\mathbb{R}^N}\widetilde{\mathcal{T}}_\varepsilon\Big(\frac{\partial^2\widetilde{\mathcal{Q}}_\varepsilon(\varphi)}{\partial x_i\partial x_j}\Big)(x,y)\,\widetilde{\mathcal{M}}_\varepsilon(\Phi)(x)\,dx.
\end{aligned}
\tag{12.23}
$$

Using the explicit expression (1.65) of $\dfrac{\partial^2\widetilde{\mathcal{Q}}_\varepsilon(\varphi)}{\partial x_i\partial x_j}$ and equality (1.70), a discrete integration by parts yields

$$
\begin{aligned}
&\int_{\mathbb{R}^N}\widetilde{\mathcal{T}}_\varepsilon\Big(\frac{\partial^2\widetilde{\mathcal{Q}}_\varepsilon(\varphi)}{\partial x_i\partial x_j}\Big)(x,y)\widetilde{\mathcal{M}}_\varepsilon(\Phi)(x)\,dx\\
=&\varepsilon^N|Y|\sum_{\xi\in\mathbf{G}}\Big[\sum_{\kappa\in\mathbf{K}_{ij}}\mathbf{q}^{ij}_\kappa(y)\frac{1}{\varepsilon}\Big(\widetilde{\mathcal{M}}_\varepsilon\circ\mathrm{M}_{\varepsilon\mathbf{b}_i}\Big(\frac{\partial\varphi}{\partial x_i}\Big)(\varepsilon\xi+\varepsilon\kappa+\varepsilon\mathbf{b}_j)\\
&\qquad\qquad -\widetilde{\mathcal{M}}_\varepsilon\circ\mathrm{M}_{\varepsilon\mathbf{b}_i}\Big(\frac{\partial\varphi}{\partial x_i}\Big)(\varepsilon\xi+\varepsilon\kappa)\Big]\widetilde{\mathcal{M}}_\varepsilon(\Phi)(\varepsilon\xi)\\
=&\varepsilon^N|Y|\sum_{\xi\in\mathbf{G}}\Big[\sum_{\kappa\in\mathbf{K}_{ij}}\mathbf{q}^{ij}_\kappa(y)\widetilde{\mathcal{M}}_\varepsilon\circ\mathrm{M}_{\varepsilon\mathbf{b}_i}\Big(\frac{\partial\varphi}{\partial x_i}\Big)(\varepsilon\xi+\varepsilon\kappa)\Big]\\
&\qquad\qquad\qquad\qquad\times\widetilde{\mathcal{M}}_\varepsilon\circ\mathrm{M}_{-\varepsilon\mathbf{b}_j}\Big(\frac{\partial\Phi}{\partial x_j}\Big)(\varepsilon\xi)\\
=&\sum_{\kappa\in\mathbf{K}_{ij}}\mathbf{q}^{ij}_\kappa(y)\int_{\mathbb{R}^N}\widetilde{\mathcal{M}}_\varepsilon\circ\mathrm{M}_{\varepsilon\mathbf{b}_i}\Big(\frac{\partial\varphi}{\partial x_i}\Big)(\cdot+\varepsilon\kappa)\,\widetilde{\mathcal{M}}_\varepsilon\circ\mathrm{M}_{-\varepsilon\mathbf{b}_j}\Big(\frac{\partial\Phi}{\partial x_j}\Big)dx.
\end{aligned}
\tag{12.24}
$$

Then, from Proposition 1.24 (i) and Proposition 1.58, we get

$$
\Big|\int_{\mathbb{R}^N}\widetilde{\mathcal{T}}_\varepsilon\Big(\frac{\partial^2\widetilde{\mathcal{Q}}_\varepsilon(\varphi)}{\partial x_i\partial x_j}\Big)(x,y)\Phi(x)\,dx\Big| \leqslant C\Big\|\frac{\partial\varphi}{\partial x_j}\Big\|_{L^p(\mathbb{R}^N)}\Big\|\frac{\partial\Phi}{\partial x_i}\Big\|_{L^{p'}(\mathbb{R}^N)}.
$$

Therefore, for every $y \in Y$ and $i \neq j$,

$$
\Big\|\frac{\partial}{\partial y_i}\Big(\widetilde{\mathcal{T}}_\varepsilon\Big(\frac{\partial\widetilde{\mathcal{Q}}_\varepsilon(\varphi)}{\partial x_j}\Big)\Big)(\,\cdot\,,y)\Big\|_{W^{-1,p}(\mathbb{R}^N)} \leqslant C\varepsilon\|\nabla\varphi\|_{L^p(\mathbb{R}^N)}.
$$

On the other hand, due to (1.58)

$$\frac{\partial}{\partial y_j}\widetilde{\mathcal{T}}_\varepsilon\Big(\frac{\partial \widetilde{\mathcal{Q}}_\varepsilon(\varphi)}{\partial x_j}\Big)(\,\cdot\,,y) = \frac{1}{\varepsilon}\frac{\partial^2}{\partial y_j^2}\Big(\widetilde{\mathcal{T}}_\varepsilon \circ \widetilde{\mathcal{Q}}_\varepsilon(\varphi)\Big)(\,\cdot\,,y) \equiv 0. \tag{12.25}$$

Thus, the last two results imply for every $y \in Y$ and $j = 1, \dots, N$,

$$\Big\|\nabla_y\, \widetilde{\mathcal{T}}_\varepsilon\Big(\frac{\partial \widetilde{\mathcal{Q}}_\varepsilon(\varphi)}{\partial x_j}\Big)(\,\cdot\,,y)\Big\|_{W^{-1,p}(\mathbb{R}^N)} \leqslant C\varepsilon \|\nabla\varphi\|_{L^p(\mathbb{R}^N)}. \tag{12.26}$$

This means that the map $y \mapsto \widetilde{\mathcal{T}}_\varepsilon(\nabla\widetilde{\mathcal{Q}}_\varepsilon(\varphi))(\,\cdot\,,y)$ is Lipschitz with values in the space $W^{-1,p}(\mathbb{R}^N)^N$. Moreover, it satisfies

$$\|\nabla_y \widetilde{\mathcal{T}}_\varepsilon(\nabla\widetilde{\mathcal{Q}}_\varepsilon(\varphi))\|_{L^\infty(Y;W^{-1,p}(\mathbb{R}^N))} \leqslant C\varepsilon \|\nabla\varphi\|_{L^p(\mathbb{R}^N)}.$$

This, by using the Poincaré-Wirtinger inequality and recalling (1.18), yields

$$\big\|\widetilde{\mathcal{T}}_\varepsilon(\nabla\widetilde{\mathcal{Q}}_\varepsilon(\varphi)) - \mathcal{M}_Y \circ \widetilde{\mathcal{T}}_\varepsilon(\nabla\widetilde{\mathcal{Q}}_\varepsilon(\varphi))\big\|_{L^\infty(Y;W^{-1,p}(\mathbb{R}^N))} \leqslant C\varepsilon \big\|\nabla\varphi\big\|_{L^p(\mathbb{R}^N)},$$

which reads,

$$\big\|\widetilde{\mathcal{T}}_\varepsilon(\nabla\widetilde{\mathcal{Q}}_\varepsilon(\varphi)) - \widetilde{\mathcal{M}}_\varepsilon(\nabla\widetilde{\mathcal{Q}}_\varepsilon(\varphi))\big\|_{L^\infty(Y;W^{-1,p}(\mathbb{R}^N))} \leqslant C\varepsilon \big\|\nabla\varphi\big\|_{L^p(\mathbb{R}^N)^N}.$$

Hence,

$$\big\|\widetilde{\mathcal{T}}_\varepsilon(\nabla\widetilde{\mathcal{Q}}_\varepsilon(\varphi)) - \widetilde{\mathcal{M}}_\varepsilon(\nabla\widetilde{\mathcal{Q}}_\varepsilon(\varphi))\big\|_{W^{1,\infty}(Y;W^{-1,p}(\mathbb{R}^N))} \leqslant C\varepsilon \big\|\nabla\varphi\big\|_{L^p(\mathbb{R}^N)}.$$

Then, inequality (12.20) follows using (12.7)(i) (with $\nabla\widetilde{\mathcal{Q}}_\varepsilon(\varphi)$ in place of φ and taking into account $(1.81)_3$).

(ii) From (12.23), proceeding as the proof of the equalities in Lemma 12.7 with the explicit expression (1.65) of $\dfrac{\partial^2\widetilde{\mathcal{Q}}_\varepsilon(\varphi)}{\partial x_i \partial x_j}$, one has for $i \neq j$ and for every $y \in Y$,

$$\begin{aligned}\int_{\mathbb{R}^N} \frac{1}{\varepsilon}\frac{\partial}{\partial y_i}\Big(\widetilde{\mathcal{T}}_\varepsilon\Big(\frac{\partial \widetilde{\mathcal{Q}}_\varepsilon(\varphi)}{\partial x_j}\Big)\Big)(x,y)\,\Phi(x)\,dx& \\ = \int_{\mathbb{R}^N} \widetilde{\mathcal{M}}_\varepsilon(\breve{\varphi})(x)\widetilde{\mathcal{T}}_\varepsilon\Big(\frac{\partial^2 \widetilde{\mathcal{Q}}_\varepsilon(\breve{\Phi})}{\partial x_i \partial x_j}\Big)(x,y)\,dx.&\end{aligned}$$

Passing to the limit for $\Phi \in \mathcal{D}(\mathbb{R}^N)$, and using (1.75) gives for every $y \in Y$

$$\frac{1}{\varepsilon}\frac{\partial}{\partial y_i}\Big(\widetilde{\mathcal{T}}_\varepsilon\Big(\frac{\partial \widetilde{\mathcal{Q}}_\varepsilon(\varphi)}{\partial x_j}\Big)\Big)(\,\cdot\,,y) \to \frac{\partial^2\varphi}{\partial x_i \partial x_j} \quad \text{in } \mathcal{D}'(\mathbb{R}^N).$$

With (12.26) and (12.25), this implies the convergence

$$\frac{1}{\varepsilon}\nabla_y \widetilde{\mathcal{T}}_\varepsilon(\nabla\widetilde{\mathcal{Q}}_\varepsilon(\varphi))(\,\cdot\,,y) \rightharpoonup \nabla^2\varphi - \mathrm{diag}(\nabla^2\varphi) \quad \text{weakly in } W^{-1,p}(\mathbb{R}^N)^N.$$

On the one hand, since the sequence

$$\Big\{\mathcal{I}_\varepsilon \doteq \frac{1}{\varepsilon}\Big(\widetilde{\mathcal{T}}_\varepsilon(\nabla\widetilde{\mathcal{Q}}_\varepsilon(\varphi)) - \widetilde{\mathcal{M}}_\varepsilon(\nabla\widetilde{\mathcal{Q}}_\varepsilon(\varphi))\Big)\Big\}_\varepsilon,$$

is uniformly bounded in $W^{1,\infty}(Y; W^{-1,p}(\mathbb{R}^N))^N$, one can assume (up to a subsequence, which will be irrelevant at the end of the proof due to the uniqueness of the limit) that

$$\mathcal{I}_\varepsilon \rightharpoonup W(\cdot, y) \quad \text{weakly in } W^{-1,p}(\mathbb{R}^N)^N, \ \forall y \in Y.$$

Since $\mathcal{M}_Y(\mathcal{I}_\varepsilon) = 0$ in $W^{-1,p}(\mathbb{R}^N)^N$, one has

$$\mathcal{M}_Y(W) = 0 \quad \text{in } W^{-1,p}(\mathbb{R}^N)^N.$$

On the other hand,

$$\nabla_y \mathcal{I}_\varepsilon(\,\cdot\,, y) \rightharpoonup \nabla^2\varphi - \text{diag}(\nabla^2\varphi) \quad \text{weakly in } W^{-1,p}(\mathbb{R}^N)^N, \ \forall y \in Y.$$

Hence, for every $y \in Y$

$$W = (\nabla^2\varphi - \text{diag}(\nabla^2\varphi))\, y^c \quad \text{in } W^{-1,p}(\mathbb{R}^N)^N, \tag{12.27}$$

and the whole sequence converges.

Finally, by virtue of (12.7)(ii) (written for $\nabla\widetilde{\mathcal{Q}}_\varepsilon(\varphi)$ in place of φ), convergence (12.21) follows.

(iii) Let Φ be in $W^{1,p'}(\mathbb{R}^N)$. Proceeding as in the proof of (12.24), yields

$$\begin{aligned}\int_\Omega \frac{\partial}{\partial y_i}\Big(\mathcal{T}_\varepsilon\Big(\frac{\partial\widetilde{\mathcal{Q}}_\varepsilon(\varphi)}{\partial x_j}\Big)\Big)(x,y)\Phi(x)\,dx \\ = \varepsilon \int_{\widehat{\Omega}_\varepsilon} \mathcal{T}_\varepsilon\Big(\frac{\partial^2\widetilde{\mathcal{Q}}_\varepsilon(\varphi)}{\partial x_i \partial x_j}\Big)(x,y)\widetilde{\mathcal{M}}_\varepsilon(\Phi)(x)\,dx.\end{aligned}$$

Following along the lines of the computation (12.24), one obtains

$$\begin{aligned}&\Big|\int_{\widehat{\Omega}_\varepsilon} \mathcal{T}_\varepsilon\Big(\frac{\partial^2\widetilde{\mathcal{Q}}_\varepsilon(\varphi)}{\partial x_i \partial x_j}\Big)(x,y)\widetilde{\mathcal{M}}_\varepsilon(\Phi)(x)\,dx\Big| \\ &\leqslant \sum_{\kappa\in\mathbf{K}_{ij}} \mathbf{q}^{ij}_\kappa(y)\Big[\Big|\int_{\widehat{\Omega}_\varepsilon} \widetilde{\mathcal{M}}_\varepsilon \circ \mathrm{M}_{\varepsilon\mathbf{b}_i}\Big(\frac{\partial\varphi}{\partial x_i}\Big)(\cdot + \varepsilon\kappa)\, \widetilde{\mathcal{M}}_\varepsilon \circ \mathrm{M}_{-\varepsilon\mathbf{b}_j}\Big(\frac{\partial\Phi}{\partial x_j}\Big)\,dx\Big| \\ &\qquad + \Big|\int_{\Omega^b_{2\varepsilon d(Y)}} \frac{1}{\varepsilon}\Big(\widetilde{\mathcal{M}}_\varepsilon \circ \mathrm{M}_{\varepsilon\mathbf{b}_i}\Big(\frac{\partial\varphi}{\partial x_i}\Big)(\cdot + \varepsilon\kappa + \varepsilon\mathbf{b}_j)|\widetilde{\mathcal{M}}_\varepsilon(\Phi)\,dx\Big|\Big].\end{aligned}$$

Let now evaluate the two terms in the right-hand side of this inequality. For the first term, one has

$$\Big|\int_{\widehat{\Omega}_\varepsilon} \widetilde{\mathcal{M}}_\varepsilon \circ \mathrm{M}_{\varepsilon \mathbf{b}_i}\Big(\frac{\partial \varphi}{\partial x_i}\Big)(\cdot + \varepsilon\kappa)\, \widetilde{\mathcal{M}}_\varepsilon \circ \mathrm{M}_{-\varepsilon \mathbf{b}_j}\Big(\frac{\partial \Phi}{\partial x_j}\Big)\, dx\Big| \leqslant C\Big\|\frac{\partial \varphi}{\partial x_j}\Big\|_{L^p(\mathbb{R}^N)}\Big\|\frac{\partial \Phi}{\partial x_i}\Big\|_{L^{p'}(\mathbb{R}^N)}.$$

For the second term, making use of $(12.5)_2$, one gets

$$\Big|\int_{\Omega^b_{2\varepsilon d(Y)}} \frac{1}{\varepsilon}\Big(\widetilde{\mathcal{M}}_\varepsilon \circ \mathrm{M}_{\varepsilon \mathbf{b}_i}\Big(\frac{\partial \varphi}{\partial x_i}\Big)(\cdot + \varepsilon\kappa + \varepsilon \mathbf{b}_j)\widetilde{\mathcal{M}}_\varepsilon(\Phi)\, dx\Big|$$
$$\leqslant C\Big\|\frac{\partial \varphi}{\partial x_j}\Big\|_{L^p(\Omega^b_{4\varepsilon d(Y)})}\|\Phi\|_{L^{p'}(\Omega^b_{2\varepsilon d(Y)})} \leqslant C\varepsilon^{1/p'}\Big\|\frac{\partial \varphi}{\partial x_j}\Big\|_{L^p(\Omega^b_{4\varepsilon d(Y)})}\|\Phi\|_{W^{1,p}(\mathbb{R}^N)}.$$

Therefore,

$$\Big|\int_\Omega \frac{\partial}{\partial y_i}\Big(\widetilde{\mathcal{T}}_\varepsilon\Big(\frac{\partial \widetilde{\mathcal{Q}}_\varepsilon(\varphi)}{\partial x_j}\Big)\Big)(\cdot, y)\Phi\, dx\Big| \leqslant C\Big(\varepsilon\Big\|\frac{\partial \varphi}{\partial x_j}\Big\|_{L^p(\mathbb{R}^N)} + \varepsilon^{1/p'}\Big\|\frac{\partial \varphi}{\partial x_j}\Big\|_{L^p(\Omega^b_{4\varepsilon d(Y)})}\Big)\|\Phi\|_{W^{1,p}(\mathbb{R}^N)},$$

so that, for every $y \in Y$,

$$\Big\|\nabla_y\Big(\widetilde{\mathcal{T}}_\varepsilon\Big(\frac{\partial \widetilde{\mathcal{Q}}_\varepsilon(\varphi)}{\partial x_j}\Big)\Big)(\cdot, y)\Big\|_{(W^{-1,p}(\Omega))'} \leqslant C\big(\varepsilon\|\nabla\varphi\|_{L^p(\mathbb{R}^N)} + \varepsilon^{1/p'}\|\nabla\varphi\|_{L^p(\Omega^b_{4\varepsilon d(Y)})}\big).$$

Using the Poincaré-Wirtinger inequality and $(12.6)_1$ (with $\dfrac{\partial \widetilde{\mathcal{Q}}_\varepsilon(\varphi)}{\partial x_j}$ in place of φ), this last inequality gives (12.22) written for $\widetilde{\mathcal{Q}}_\varepsilon(\varphi)$. Finally, as $\mathcal{Q}_\varepsilon(\varphi)$ is the restriction to Ω of $\widetilde{\mathcal{Q}}_\varepsilon(\varphi)$, estimate (12.22) is straightforward. □

Proposition 12.11. *Let p be in $[1, +\infty)$.*
(i) For every $\lambda \in L^p(Y)$ with $\mathcal{M}_Y(\lambda) = 0$ and $\varphi \in L^p(\mathbb{R}^N)$, the following inequality holds:

$$\Big\|\lambda\Big(\Big\{\frac{\cdot}{\varepsilon}\Big\}_Y\Big)\widetilde{\mathcal{Q}}_\varepsilon(\varphi)\Big\|_{W^{-1,p}(\mathbb{R}^N)} \leqslant C\varepsilon\|\varphi\|_{L^p(\mathbb{R}^N)}\|\lambda\|_{L^p(Y)}, \tag{12.28}$$

with a constant C independent of ε.
Moreover,

$$\frac{1}{\varepsilon}\lambda\Big(\Big\{\frac{\cdot}{\varepsilon}\Big\}_Y\Big)\widetilde{\mathcal{Q}}_\varepsilon(\varphi) \rightharpoonup 0 \quad \textit{weakly in } W^{-1,p}(\mathbb{R}^N). \tag{12.29}$$

(ii) For every $\Lambda \in L^p(Y)^N$ such that $\mathcal{M}_Y(\Lambda) = 0$, and $\Phi \in W^{1,p}(\mathbb{R}^N)$, the following inequality holds:

$$\Big\|\Lambda\Big(\Big\{\frac{\cdot}{\varepsilon}\Big\}_Y\Big)\cdot\nabla\widetilde{\mathcal{Q}}_\varepsilon(\Phi)\Big\|_{W^{-1,p}(\mathbb{R}^N)} \leqslant C\varepsilon\|\nabla\Phi\|_{L^p(\mathbb{R}^N)}\|\Lambda\|_{L^p(Y)}, \tag{12.30}$$

with a constant C independent of ε.

Moreover, one has the weak convergence in $W^{-1,p}(\mathbb{R}^N)$

$$\frac{1}{\varepsilon}\Lambda\Big(\Big\{\frac{\cdot}{\varepsilon}\Big\}_Y\Big)\cdot\nabla\widetilde{\mathcal{Q}}_\varepsilon(\Phi)\rightharpoonup -\sum_{i=1}^N\Big(\frac{1}{|Y|}\int_Y y_i^c\Lambda(y)\cdot\mathbf{b}_i\,dy\Big)\frac{\partial^2\Phi}{\partial x_i^2}. \tag{12.31}$$

(iii) Let $\{v_\varepsilon\}_\varepsilon$ be a sequence in $W^{1,p'}(\mathbb{R}^N)$, such that there exist v in $W^{1,p'}(\mathbb{R}^N)$ and $\widehat{v}$ in $L^{p'}(\mathbb{R}^N;W^{1,p'}_{per}(Y))$ satisfying the following convergences:

$$v_\varepsilon\rightharpoonup v\quad \textit{weakly in } W^{1,p'}(\mathbb{R}^N),$$

and

$$\widetilde{\mathcal{T}}_\varepsilon(\nabla v_\varepsilon)\rightharpoonup\nabla v+\nabla_y\widehat{v}\quad\textit{weakly in } L^{p'}(\mathbb{R}^N\times Y)^N.$$

Then, for every $\Lambda\in L^p(Y)^N$ such that $\mathcal{M}_Y(\Lambda)=0$,

$$\begin{aligned}\int_{\mathbb{R}^N}\frac{1}{\varepsilon}\Lambda\Big(\Big\{\frac{\cdot}{\varepsilon}\Big\}_Y\Big)\cdot\nabla\widetilde{\mathcal{Q}}_\varepsilon(\Phi)\,v_\varepsilon\,dx\\ \to\frac{1}{|Y|}\int_{\mathbb{R}^N\times Y}\sum_{i=1}^N\frac{\partial\Phi}{\partial x_i}\frac{\partial v}{\partial x_i}y_i^c\,\Lambda(y)\cdot\mathbf{b}_i\,dxdy\\ +\frac{1}{|Y|}\int_{\mathbb{R}^N\times Y}\Lambda(y)\cdot\nabla\Phi(x)\,\widehat{v}(x,y)\,dxdy.\end{aligned} \tag{12.32}$$

Proof. *(i)* For v in $W^{1,p'}(\mathbb{R}^N)$, by unfolding we have

$$\begin{aligned}&\int_{\mathbb{R}^N}\lambda\Big(\Big\{\frac{\cdot}{\varepsilon}\Big\}_Y\Big)\widetilde{\mathcal{Q}}_\varepsilon(\varphi)v\,dx\\ &\quad=\frac{1}{|Y|}\int_{\mathbb{R}^N\times Y}\widetilde{\mathcal{T}}_\varepsilon\circ\widetilde{\mathcal{Q}}_\varepsilon(\varphi)(x,y)\lambda(y)\widetilde{\mathcal{T}}_\varepsilon(v)(x,y)\,dxdy.\end{aligned} \tag{12.33}$$

Recall $(1.39)_2$ (with $\mathbb{R}^N$ instead of Ω):

$$\|\widetilde{\mathcal{T}}_\varepsilon(v)-v\|_{L^{p'}(\mathbb{R}^N\times Y)}\leqslant C\varepsilon\|\nabla v\|_{L^{p'}(\mathbb{R}^N)}$$

Then, using Proposition 1.64(i) and estimate $(12.15)_1$ in (12.33), we obtain

$$\begin{aligned}\Big|\int_{\mathbb{R}^N}\lambda\Big(\Big\{\frac{\cdot}{\varepsilon}\Big\}_Y\Big)\widetilde{\mathcal{Q}}_\varepsilon(\varphi)v\,dx\Big|&\leqslant\Big|\frac{1}{|Y|}\int_{\mathbb{R}^N\times Y}\widetilde{\mathcal{T}}_\varepsilon\circ\widetilde{\mathcal{Q}}_\varepsilon(\varphi)(x,y)\lambda(y)\,v(x)\,dxdy\Big|\\ &\qquad+C\varepsilon\|\varphi\|_{L^p(\mathbb{R}^N)}\|\lambda\|_{L^p(Y)}\|\nabla v\|_{L^{p'}(\mathbb{R}^N)}\\ &\leqslant\Big|\frac{1}{|Y|}\int_{\mathbb{R}^N\times Y}\widetilde{\mathcal{M}}_\varepsilon\circ\widetilde{\mathcal{Q}}_\varepsilon(\varphi)(x,y)\lambda(y)\,v(x)\,dxdy\Big|\\ &\qquad+C\varepsilon\|\varphi\|_{L^p(\mathbb{R}^N)}\|\lambda\|_{L^p(Y)}\|\nabla v\|_{L^{p'}(\mathbb{R}^N)}.\end{aligned}$$

Observe that the first term in the right-hand side vanishes since $\mathcal{M}_Y(\lambda)=0$ by hypothesis. So, (12.28) is proved.

To prove convergence (12.29), take φ in $\mathcal{D}(\mathbb{R}^N)$ and v in $W^{1,p'}(\mathbb{R}^N)$. We have successively,

$$
\begin{aligned}
&\int_{\mathbb{R}^N} \frac{1}{\varepsilon}\lambda\Big(\Big\{\frac{\cdot}{\varepsilon}\Big\}_Y\Big)\widetilde{\mathcal{Q}}_\varepsilon(\varphi)\, v\, dx \\
&= \frac{1}{\varepsilon|Y|}\int_{\mathbb{R}^N\times Y} \lambda(y)\widetilde{\mathcal{T}}_\varepsilon \circ \widetilde{\mathcal{Q}}_\varepsilon(\varphi)(x,y)\widetilde{\mathcal{T}}_\varepsilon(v)(x,y)\, dxdy \\
&= \frac{1}{|Y|}\int_{\mathbb{R}^N\times Y} \lambda(y)\widetilde{\mathcal{T}}_\varepsilon \circ \widetilde{\mathcal{Q}}_\varepsilon(\varphi)(x,y)\frac{\widetilde{\mathcal{T}}_\varepsilon(v)(x,y)-\widetilde{\mathcal{M}}_\varepsilon(v)(x)}{\varepsilon}\, dxdy \qquad (12.34)\\
&+ \frac{1}{|Y|}\int_{\mathbb{R}^N\times Y} \lambda(y)\frac{\widetilde{\mathcal{T}}_\varepsilon \circ \widetilde{\mathcal{Q}}_\varepsilon(\varphi)(x,y)-\widetilde{\mathcal{M}}_\varepsilon \circ \widetilde{\mathcal{Q}}_\varepsilon(\varphi)(x)}{\varepsilon}\widetilde{\mathcal{M}}_\varepsilon(v)(x)dxdy \\
&+ \frac{1}{\varepsilon|Y|}\int_{\mathbb{R}^N\times Y} \lambda(y)\widetilde{\mathcal{M}}_\varepsilon \circ \widetilde{\mathcal{Q}}_\varepsilon(\varphi)(x)\widetilde{\mathcal{M}}_\varepsilon(v)(x)\, dxdy.
\end{aligned}
$$

The last integral vanishes using again the fact that $\mathcal{M}_Y(\lambda)=0$.

Proposition 1.39(ii) (with $\mathbb{R}^N$ instead of Ω) and (1.74) of Corollary 1.66 give the following strong convergences:

$$
\begin{aligned}
&\frac{1}{\varepsilon}\big(\widetilde{\mathcal{T}}_\varepsilon(v)-\widetilde{\mathcal{M}}_\varepsilon(v)\big) \to y^c\cdot\nabla v \quad \text{strongly in } L^{p'}(\mathbb{R}^N\times Y),\\
&\frac{1}{\varepsilon}\big(\widetilde{\mathcal{T}}_\varepsilon \circ \widetilde{\mathcal{Q}}_\varepsilon(\varphi)(\,\cdot\,,y)-\widetilde{\mathcal{M}}_\varepsilon \circ \widetilde{\mathcal{Q}}_\varepsilon(\varphi)(\cdot)\big) \to y^c\cdot\nabla\varphi \quad \text{strongly in } L^{p}(\mathbb{R}^N\times Y).
\end{aligned}
$$

Passing to the limit in the two remaining integrals in the right-hand side of (12.34) and taking into account the above convergences lead to

$$
\begin{aligned}
\int_{\mathbb{R}^N} \frac{1}{\varepsilon}\lambda\Big(\Big\{\frac{\cdot}{\varepsilon}\Big\}_Y\Big)\widetilde{\mathcal{Q}}_\varepsilon(\varphi) v\, dx \to & \frac{1}{|Y|}\int_{\mathbb{R}^N\times Y}\lambda(y)\, y^c\cdot\nabla v(x)\,\varphi(x)\, dxdy\\
&+\frac{1}{|Y|}\int_{\mathbb{R}^N\times Y}\lambda(y)\, y^c\cdot\nabla\varphi(x)\, v(x)\, dxdy = 0,
\end{aligned}
$$

which implies convergence (12.29) for $\varphi\in\mathcal{D}(\mathbb{R}^N)$. Since $\mathcal{D}(\mathbb{R}^N)$ is dense in $L^p(\mathbb{R}^N)$, the above convergence and estimate (12.28), actually give (12.29) for φ in $L^p(\mathbb{R}^N)$.

(ii) Let v be in $W^{1,p'}(\mathbb{R}^N)$. Proceeding as above yields

$$
\begin{aligned}
&\int_{\mathbb{R}^N} \Lambda\Big(\Big\{\frac{\cdot}{\varepsilon}\Big\}_Y\Big)\cdot\nabla\widetilde{\mathcal{Q}}_\varepsilon(\Phi)\, v\, dx\\
&\qquad = \frac{1}{|Y|}\int_{\mathbb{R}^N\times Y}\Lambda(y)\cdot\widetilde{\mathcal{T}}_\varepsilon\big(\nabla\widetilde{\mathcal{Q}}_\varepsilon(\Phi)\big)(x,y)\mathcal{T}_\varepsilon(v)(x,y)\, dxdy.
\end{aligned}
$$

As before, $\widetilde{\mathcal{T}}_\varepsilon(v)$ can be replaced by v in the right-hand side of this equality. Then the result follows by using (12.20) and the assumption $\mathcal{M}_Y(\Lambda)=0$.

For convergence (12.31), first take $\Phi \in \mathcal{D}(\mathbb{R}^N)$ and, as before, write

$$\begin{aligned}
&\int_{\mathbb{R}^N} \frac{1}{\varepsilon}\Lambda\Big(\Big\{\frac{\cdot}{\varepsilon}\Big\}_Y\Big)\cdot \nabla\widetilde{\mathcal{Q}}_\varepsilon(\Phi)\, v\, dx \\
&=\frac{1}{|Y|}\int_{\mathbb{R}^N\times Y} \Lambda(y)\cdot \widetilde{\mathcal{T}}_\varepsilon\big(\nabla\widetilde{\mathcal{Q}}_\varepsilon(\Phi)\big)(x,y)\frac{\widetilde{\mathcal{T}}_\varepsilon(v)(x,y)-\widetilde{\mathcal{M}}_\varepsilon(v)(x)}{\varepsilon}\, dxdy \\
&+\frac{1}{|Y|}\int_{\mathbb{R}^N\times Y} \Lambda(y)\cdot \frac{\widetilde{\mathcal{T}}_\varepsilon\big(\nabla\widetilde{\mathcal{Q}}_\varepsilon(\Phi)\big)(x,y)-\widetilde{\mathcal{M}}_\varepsilon\big(\nabla\widetilde{\mathcal{Q}}_\varepsilon(\Phi)\big)(x)}{\varepsilon}\widetilde{\mathcal{M}}_\varepsilon(v)(x)\, dxdy.
\end{aligned}$$

Taking into account convergence (12.27), yields

$$\begin{aligned}
&\frac{1}{\varepsilon}\Big[\widetilde{\mathcal{T}}_\varepsilon\big(\nabla\widetilde{\mathcal{Q}}_\varepsilon(\Phi)\big)-\widetilde{\mathcal{M}}_\varepsilon\big(\nabla\widetilde{\mathcal{Q}}_\varepsilon(\Phi)\big)\Big] \\
&\qquad\rightharpoonup \big(\nabla^2\Phi-\mathrm{diag}(\nabla^2\Phi)\big)\, y^c \quad \text{weakly in } L^p(\mathbb{R}^N\times Y)^N.
\end{aligned}$$

On the other hand, due to Proposition 1.39(ii) and recalling (1.21)-(1.74), the above convergence implies that

$$\begin{aligned}
&\int_{\mathbb{R}^N} \frac{1}{\varepsilon}\Lambda\Big(\Big\{\frac{\cdot}{\varepsilon}\Big\}_Y\Big)\cdot \nabla\widetilde{\mathcal{Q}}_\varepsilon(\Phi)\, v\, dx \\
&\qquad\to \frac{1}{|Y|}\int_{\mathbb{R}^N\times Y} \Lambda(y)\cdot\nabla\Phi(x)\, y^c\cdot\nabla v(x)\, dxdy \\
&\qquad\qquad + \frac{1}{|Y|}\int_{\mathbb{R}^N\times Y} \Lambda(y)\cdot\big(\nabla^2\Phi-\mathrm{diag}(\nabla^2\Phi)\big)\, y^c\, v(x)\, dxdy \\
&\qquad= -\frac{1}{|Y|}\int_{\mathbb{R}^N\times Y} \Lambda(y)\cdot\mathrm{diag}(\nabla^2\Phi)\, y^c\, v(x)\, dxdy,
\end{aligned}$$

which is (12.31) for $\Phi \in \mathcal{D}(\mathbb{R}^N)$.

Now, take Φ in $W^{1,p}(\mathbb{R}^N)$ and Ψ in $\mathcal{D}(\mathbb{R}^N)$. From (12.30) we have

$$\begin{aligned}
&\Big|\int_{\mathbb{R}^N} \frac{1}{\varepsilon}\Lambda\Big(\Big\{\frac{\cdot}{\varepsilon}\Big\}_Y\Big)\cdot\Big[\nabla\big(\widetilde{\mathcal{Q}}_\varepsilon(\Phi)-\widetilde{\mathcal{Q}}_\varepsilon(\Psi)\big)\Big]\, v\, dx\Big| \\
&\qquad\leqslant C\|\nabla(\Phi-\Psi)\|_{L^p(\mathbb{R}^N)}\|\Lambda\|_{L^p(Y)}\|\nabla v\|_{L^{p'}(\mathbb{R}^N)}.
\end{aligned}$$

Then convergence (12.31) for Ψ gives

$$\begin{aligned}
&\limsup_{\varepsilon\to 0}\Big|\int_{\mathbb{R}^N} \frac{1}{\varepsilon}\Lambda\Big(\Big\{\frac{\cdot}{\varepsilon}\Big\}_Y\Big)\cdot \nabla\widetilde{\mathcal{Q}}_\varepsilon(\Phi)\, v\, dx \\
&\qquad - \frac{1}{|Y|}\int_{\mathbb{R}^N\times Y}\sum_{i=1}^N \frac{\partial\Phi}{\partial x_i}\frac{\partial v}{\partial x_i} y_i^c\Lambda(y)\cdot\mathbf{b}_i\, dxdy\Big| \\
&\leqslant C\|\nabla(\Phi-\Psi)\|_{L^p(\mathbb{R}^N)}\|\Lambda\|_{L^p(Y)}\|\nabla v\|_{L^{p'}(\mathbb{R}^N)},
\end{aligned}$$

from which (12.31) follows by density.

(iii) It remains to prove (12.32). To do so, we proceed as above, replacing convergence (ii) of Proposition 1.39 by the following one (see Theorem 1.41(ii)):

$$\frac{1}{\varepsilon}\big(\widetilde{\mathcal{T}}_\varepsilon(v_\varepsilon) - \widetilde{\mathcal{M}}_\varepsilon(v_\varepsilon)\big) \rightharpoonup y^c \cdot \nabla v + \widehat{v} \quad \text{weakly in } L^{p'}(\mathbb{R}^N \times Y).$$

This completes the proof of Proposition 12.11. □

Chapter 13

* Strongly oscillating nonhomogeneous Dirichlet condition

This chapter is dedicated to the homogenization of the Dirichlet problem with a strongly oscillating nonhomogeneous boundary data, that is

$$\begin{cases} \operatorname{div}(A^\varepsilon \nabla u_\varepsilon) = 0 & \text{in } \Omega, \\ u_\varepsilon = g_\varepsilon & \text{on } \partial\Omega. \end{cases} \tag{13.1}$$

with $A^\varepsilon(x) \equiv A\Big(\Big\{\dfrac{x}{\varepsilon}\Big\}_Y\Big)$, where A belongs to $\mathcal{M}(\alpha, \beta, Y)$ and g_ε is in $H^{1/2}(\partial\Omega)$.

In [125], the case of a strongly oscillating boundary data is investigated under the assumptions

$$g_\varepsilon \rightharpoonup g \quad \text{weakly in } H^{-1/2}(\partial\Omega) \quad \text{and} \quad \varepsilon^{1/2}\|g_\varepsilon\|_{H^{1/2}(\partial\Omega)} \to 0,$$

in which case the limit of the sequence $\{u_\varepsilon\}_\varepsilon$ satisfies problem (13.19) with the matrix field A^{hom}.

In this chapter, we consider the more complicated case for which the Dirichlet data g_ε is of the form,

$$g_\varepsilon = \Psi\Big(\frac{\cdot}{\varepsilon}\Big)\, G_{|\partial\Omega} \tag{13.2}$$

where G belongs to $H^1(\Omega)$, while Ψ is in $H^1_{per,0}(Y) \cap L^\infty(Y)$ and satisfies a condition (see (13.3)) which guarantees that g_ε is in $H^{1/2}(\partial\Omega)$ and $\varepsilon^{1/2}\|g_\varepsilon\|_{H^{1/2}(\partial\Omega)}$ is bounded (see (13.4)). In the limit problem, the Dirichlet data differs from the limit of the sequence $\{g_\varepsilon\}_\varepsilon$, which is an effect of the homogenization inside Ω (see Section 13.7 for such a situation).

D. Cioranescu et al., *The Periodic Unfolding Method*, Series in Contemporary Mathematics 3, https://doi.org/10.1007/978-981-13-3032-2_13

This kind of problem, with a strongly oscillating boundary condition, occurs when dealing with boundary layer corrector terms for the classical Dirichlet problem (3.3). This will be studied in Chapter 14.

Section 13.1 considers the one-dimensional case and shows that one cannot expect convergence of the full sequence of solutions.

The next section 13.2 introduces the setting for the study in higher dimensions of the space with notations and preliminary results.

Section 13.3 is dedicated to sharp estimates of the solution of this problem. In Section 13.6, we characterize the asymptotic behavior (when ε goes to 0) of the sequence $\{u_\varepsilon\}_\varepsilon$ in the general case, while Section 13.7 is concerned with the case of a domain with a boundary containing a flat part.

Several preliminary results, needed in Section 13.3 as well as in Chapter 14 are introduced in Section 13.9, namely Theorem 13.41 and the key Theorems 13.43 and 13.44. These are generalizations of projection results onto spaces of periodic functions in the Hilbert setting (p=2), estimating the distance from an element of $W^{1,p}(Y)$ to $W^{1,p}_{per}(Y)$ via its "periodicity defect", expressed as the $W^{1-1/p,p}$ norm of the differences of its traces on the opposite faces of the unit cell Y (see Definition 13.33; cf. [120,122]).

In this chapter the domain Ω *is bounded with a Lipschitz or* $\mathcal{C}^{1,1}$ *boundary.* We use the notations introduced in Section 12.1.

13.1 The one dimensional case

In this section, we investigate the solution of problem 13.1 in the one dimensional case.

Let A be in $M\big(\alpha,\beta,(0,1)\big)$ and extend it by 1-periodicity in $\mathbb{R}$. Let Ψ be in $H^1_{per}(0,1)^{(1)}$. Problem (13.1) reads: *Find* $u^\varepsilon_\Psi \in H^1(a,b)$ *such that*

$$\begin{cases} \dfrac{d}{dx}\Big(A\Big(\dfrac{\cdot}{\varepsilon}\Big)\dfrac{du^\varepsilon_\Psi}{dx}\Big) = 0 \quad \text{in } (a,b),\\ u^\varepsilon_\Psi(a) = \Psi\Big(\dfrac{a}{\varepsilon}\Big), \quad u^\varepsilon_\Psi(b) = \Psi\Big(\dfrac{b}{\varepsilon}\Big). \end{cases}$$

Its (explicit) solution is

$$u^\varepsilon_\Psi(x) = \Psi\Big(\frac{a}{\varepsilon}\Big) + \Big[\Psi\Big(\frac{b}{\varepsilon}\Big) - \Psi\Big(\frac{a}{\varepsilon}\Big)\Big]\frac{\displaystyle\int_a^x \frac{1}{A(t/\varepsilon)}dt}{\displaystyle\int_a^b \frac{1}{A(t/\varepsilon)}dt}.$$

It is well-known (see [143]) that if 1, a and b are linearly independent over the rational numbers, then for every $(a_0,b_0)\in[0,1]^2$ there exists a sequence of integers $(N_k)_{k\in\mathbb{N}}$, such that

$$\lim_{k\to+\infty}\big(\{N_k a\},\{N_k b\}\big) = (a_0,b_0).$$

$^{(1)}$ Actually, Ψ continuous and 1-periodic is enough for the analysis in this section.

Choosing the subsequence $\{\varepsilon_k\}_{k\in\mathbb{N}} = \{1/N_k\}_{k\in\mathbb{N}}$, one has

$$\lim_{\varepsilon\to 0}\Big(\Big\{\frac{a}{\varepsilon}\Big\},\Big\{\frac{b}{\varepsilon}\Big\}\Big) = (a_0, b_0) \in [0,1]^2.$$

The corresponding sequence $\{u_\Psi^{e_k}\}_{k\in\mathbb{N}}$ weakly converges to $u_\Psi \in H^1(a,b)$ where

$$u_\Psi(x) = \Psi(a_0) + \big(\Psi(b_0) - \Psi(a_0)\big)\frac{x-a}{b-a} \quad \text{for every } x \in [a,b].$$

For this example, the homogenized coefficient is constant, explicitly given by the formula

$$A^{hom} = \Big(\int_0^1 A(t)^{-1} dt\Big)^{-1},$$

and u_Ψ is the solution of

$$\begin{cases} u_\Psi'' = 0 \quad \text{in} \quad (a,b), \\ u_\Psi(a) = \Psi(a_0), \quad u_\Psi(b) = \Psi(b_0). \end{cases}$$

If Ψ is not constant, one obtains infinitely many limit points for the full sequence $\{u_\Psi^\varepsilon\}_\varepsilon$, each corresponding to a boundary condition $(\Psi(a_0), \Psi(b_0))$ with $(a_0, b_0) \in [0,1]^2$.

13.2 Notations and preliminary results

In dimension higher than 1, the Dirichlet data (13.2) is of very specific type. To ensure that it belongs to $H^{1/2}(\partial\Omega)$, we need more information.

First, the following problem is variational and by the Lax-Milgram theorem has a unique solution Ψ in $H^1_{per,0}(Y)$ for given B in $M(\alpha,\beta,Y)$ and F in $L^2(Y)^N$:

$$\begin{cases} \displaystyle\int_Y (B\,\nabla_y\Psi + F)\,\nabla_y\varphi\, dy = 0, \\ \forall\varphi \in H^1_{per,0}(Y). \end{cases} \tag{13.3}$$

Every Ψ in $H^1_{per,0}(Y)$ is the solution of such a problem, but not all are such that the trace of $\Psi(\frac{\cdot}{\varepsilon})$ is a multiplier for $H^{1/2}(\partial\Omega)$.

If F is in $L^\infty(Y)$ (which is the case for the boundary layer correctors in Chapter 14), then, as was shown in Lemma 3.29, the trace of $\Psi(\frac{\cdot}{\varepsilon})$ is a multiplier for $H^{1/2}(\partial\Omega)$. This observation justifies the following definition:

Definition 13.1 (The map $\mathbb{K}_B$ and the space $\mathcal{H}_B$). *For B in $M(\alpha,\beta,Y)$, let $\mathbb{K}_B$ denote the map which, to F in $L^\infty(Y)^N$, assigns the solution Ψ of problem* (13.3)., *i.e.,*

$$\mathbb{K}_B(F) = \Psi.$$

Its kernel is the non trivial set

$\{F \in L^\infty(Y) | \operatorname{div}_y F = 0$ *in the sense of* $H^1_{per}(Y)\}$ *(cf. footnote*[10] *page 123).*

The range of this map is denoted $\mathcal{H}_B$. *The natural norm on* $\mathcal{H}_B$ *is therefore*

$$\|\Psi\|_{\mathcal{H}_B} \doteq \min\left\{\|F\|_{L^\infty(Y)} \mid \Psi = \mathbb{K}_B(F)\right\}.$$

The space $\mathcal{H}_B$ contains $W^{1,\infty}_{per,0}(Y)$ (to see this, choose $F = -B\nabla_y\Psi$ for $\Psi \in W^{1,\infty}_{per,0}(Y)$!), so it is dense in $H^1_{per,0}(Y)$. By Lemma 3.29 the space $\mathcal{H}_B$ is included in $L^\infty(Y)$ and there exists a constant C depending only upon Y, α and β, but independent of ε, such that

$$\|\Psi\|_{H^1(Y)\cap L^\infty(Y)} \leqslant C\|\Psi\|_{\mathcal{H}_B}.$$

Non trivial examples of elements of $\mathcal{H}_A$ are given by the correctors: definition (3.10) implies that $\widehat{\chi}_j = \mathbb{K}_A(A\mathbf{b}_j)$, $j = 1, \ldots, N$.

Actually, the elements of $\mathcal{H}_B$ satisfy some important estimates that are given in the next result.

Lemma 13.2. *For every* $(\Psi, w) \in \mathcal{H}_B \times H^1(\Omega)$ *the product* $\Psi\left(\frac{\cdot}{\varepsilon}\right) w$ *belongs to* $H^1(\Omega)$. *In particular,*

$$\Psi\left(\frac{\cdot}{\varepsilon}\right)_{|\partial\Omega} w_{|\partial\Omega} \in H^{1/2}(\partial\Omega).$$

Furthermore, there is a constant C *depending only on* Y, α *and* β *(and independent of* ε*), such that*

$$\left\|\Psi\left(\frac{\cdot}{\varepsilon}\right)_{|\partial\Omega} w_{|\partial\Omega}\right\|_{H^{1/2}(\partial\Omega)} \leqslant C\|\Psi\|_{\mathcal{H}_B}\left(\varepsilon^{-1/2}\|w\|_{L^2(\partial\Omega)} + \|\nabla w\|_{L^2(\Omega)}\right). \tag{13.4}$$

It also satisfies the following estimate:

$$\left\|\Psi\left(\frac{\cdot}{\varepsilon}\right) w\right\|_{L^2(\partial\Omega)} \leqslant C\|\Psi\|_{\mathcal{H}_B}\left(\|w\|_{L^2(\partial\Omega)} + \varepsilon^{1/2}\|\nabla w\|_{L^2(\Omega)}\right). \tag{13.5}$$

Proof. Let F be in $L^\infty(Y)$ and $\Psi = \mathbb{K}_B(F)$.

Step 1. Proof of $\Psi\left(\frac{\cdot}{\varepsilon}\right) w \in H^1(\Omega)$.

Corollary 3.30 implies that

$$\Psi\left(\frac{\cdot}{\varepsilon}\right) w \in H^1(\Omega), \qquad \forall w \in H^1(\Omega).$$

By the trace theorem,

$$\left[\Psi\left(\frac{\cdot}{\varepsilon}\right) w\right]_{|\partial\Omega} \in H^{1/2}(\partial\Omega) \quad \text{and} \quad \left[\Psi\left(\frac{\cdot}{\varepsilon}\right) w\right]_{|\partial\Omega} \equiv \Psi\left(\frac{\cdot}{\varepsilon}\right)_{|\partial\Omega} w_{|\partial\Omega}.$$

This last identity is obviously satisfied a.e. in $\partial\Omega$ for w in $C^\infty(\overline{\Omega})$, and by a density argument extends to every w in $H^1(\Omega)$.

Step 2. Proof of (13.4).

Estimates $(12.3)_1$, (3.46) and (3.43) give

$$
\begin{aligned}
&\Big\|\Psi\Big(\frac{\cdot}{\varepsilon}\Big)w\Big\|_{L^2(\Omega^{bl}_{\varepsilon d(Y)})} \leqslant C\|w\|_{L^2(\Omega^{bl}_{\varepsilon d(Y)})}\|\Psi\|_{\mathcal{H}_B}\\
&\qquad \leqslant C\varepsilon^{1/2}\Big(\|w\|_{L^2(\partial\Omega)} + \varepsilon^{1/2}\|\nabla w\|_{L^2(\Omega^{bl}_{2\varepsilon d(Y)})}\Big)\|\Psi\|_{\mathcal{H}_B},\\
&\Big\|\nabla\Big[\Psi\Big(\frac{\cdot}{\varepsilon}\Big)w\Big]\Big\|_{L^2(\Omega^{bl}_{\varepsilon d(Y)})}\\
&\qquad \leqslant C\varepsilon^{-1/2}\Big(\|w\|_{L^2(\partial\Omega)} + \varepsilon^{1/2}\|\nabla w\|_{L^2(\Omega^{bl}_{2\varepsilon d(Y)})}\Big)\|\Psi\|_{\mathcal{H}_B}.
\end{aligned}
\tag{13.6}
$$

Consider the function $\rho_\varepsilon(x) = \inf\Big\{1, \dfrac{\rho(x)}{\varepsilon d(Y)}\Big\}$[2], which takes the value 1 if and only if $\rho \geqslant \varepsilon d(Y)$. Then the function

$$
(1-\rho_\varepsilon)\,\Psi\Big(\frac{\cdot}{\varepsilon}\Big)\,w = 0 \quad \text{for } \rho \geqslant \varepsilon d(Y),
$$

and belongs to $H^1(\Omega)$. So, estimates (13.6) and $(12.3)_1$ imply (after some easy computations),

$$
\Big\|(1-\rho_\varepsilon)\,\Psi\Big(\frac{\cdot}{\varepsilon}\Big)\,w\Big\|_{H^1(\Omega)} \leqslant C\big(\varepsilon^{-1/2}\|w\|_{L^2(\partial\Omega)} + \|\nabla w\|_{L^2(\Omega)}\big)\|\Psi\|_{\mathcal{H}_B}, \tag{13.7}
$$

from which estimate (13.4) is straightforward by the trace theorem.

Step 3. Proof of (13.5).

Finally, (13.5) follows by using estimates (13.7) and (13.6) in $(12.3)_2$. □

13.3 The problem and first estimate

The problem considered in this chapter can now be stated precisely. For $A \in \mathcal{M}(\alpha, \beta, Y)$, set

$$
A^\varepsilon(x) = A\Big(\Big\{\frac{x}{\varepsilon}\Big\}_Y\Big),
$$

which belongs to $\mathcal{M}(\alpha,\beta,\Omega)$. For $G \in H^1(\Omega)$ and $\Psi \in \mathcal{H}_B$, consider the following nonhomogeneous Dirichlet problem:

$$
\begin{cases}
\operatorname{div}(A^\varepsilon \nabla u^\varepsilon_{\Psi,G}) = 0 & \text{in } \Omega,\\
u^\varepsilon_{\Psi,G} = \Psi\Big(\dfrac{\cdot}{\varepsilon}\Big)G & \text{on } \partial\Omega.
\end{cases}
\tag{13.8}
$$

This problem is well-posed.

The first estimate for $u^\varepsilon_{\Psi,G}$ is a an exact application of Lemma 13.2 and is formulated as follows:

[2] See 1.47 for the definition of ρ.

Proposition 13.3. *Let Ω be a bounded domain with a Lipschitz boundary. Let Ψ be in $\mathcal{H}_B$, and G be in $H^1(\Omega)$. The solution $u^\varepsilon_{\Psi,G}$ of problem (13.8) satisfies the estimate*

$$\|u^\varepsilon_{\Psi,G}\|_{H^1(\Omega)} \leqslant C\|\Psi\|_{\mathcal{H}_B}\big(\varepsilon^{-1/2}\|G\|_{L^2(\partial\Omega)} + \|\nabla G\|_{L^2(\Omega)}\big), \tag{13.9}$$

with C independent of ε.

As can be easily seen, this estimate is not uniform for ε going to 0, as soon as Ψ is not constant. This means that the usual H^1 theory cannot be applied to go to the limit for problem (13.8). However, (13.5) is a uniform bound in $L^2(\partial\Omega)$ for the Dirichlet data. Therefore we need a theory for solutions in the case of Dirichlet boundary condition in a space containing $L^2(\partial\Omega)$.

In the next section, we briefly recall the case where the trace space is $H^{-1/2}(\partial\Omega)$. This will essentially require that the boundary $\partial\Omega$ be at least $\mathcal{C}^{1,1}$ and the matrix field be Lipschitz. These considerations do not apply directly to problem (13.8) (because of the oscillating A^ε), but will apply with the homogenized matrix A^{hom} as a tool to obtain estimates for the solution of the original problem (13.8).

13.4 Dirichlet problem with data in $H^{-1/2}(\partial\Omega)$

We start by recalling a classical result in the theory of elliptic partial differential equations (see, for instance [114]).

Proposition 13.4. *Assume that $\partial\Omega$ Lipschitz and $\mathcal{A}$ is in $\mathcal{M}(\alpha,\beta,\Omega)$. For g in $H^{1/2}(\partial\Omega)$, there is a unique solution u in $H^1(\Omega)$ for the nonhomogeneous Dirichlet problem*

$$\begin{cases} \operatorname{div}(\mathcal{A}\nabla u) = 0 & \text{in } \Omega, \\ u = g & \text{on } \partial\Omega, \end{cases} \tag{13.10}$$

satisfying

$$c\|g\|_{H^{1/2}(\partial\Omega)} \leqslant \|u\|_{H^1(\Omega)} \leqslant C\|g\|_{H^{1/2}(\partial\Omega)}$$

where c and C depends only upon α,β and Ω.

In the following we shall use the space $H^1_\rho(\Omega)$, defined by (1.48) as

$$H^1_\rho(\Omega) \doteq \{\varphi \in L^2(\Omega) \mid \rho\nabla\varphi \in L^2(\Omega)^N\},$$

with the norm $\|\varphi\|_{H^1_\rho(\Omega)} \equiv \big(\|\varphi\|^2_{L^2(\Omega)} + \|\rho\nabla\varphi\|^2_{L^2(\Omega)}\big)^{1/2}$.

Note that the elements of $H^1_\rho(\Omega)$ do not all have traces on $\partial\Omega$ (such an example is given by $\rho^{-\ell}, \ell \in (0,1/2)$, which belongs to $H^1_\rho(\Omega)$ but cannot have a trace on $\partial\Omega$ in any functional space). We need a subspace of $H^1_\rho(\Omega)$ which admits traces (in a space which includes $L^2(\partial\Omega)$ because of (13.5)).

Definition 13.5. *Let $\mathcal{A}$ be in $\mathcal{M}(\alpha,\beta,\Omega)$. Introduce the subspace $\mathbf{W}_{\mathcal{A}}$ of $L^2(\Omega)$ defined as*

$$\mathbf{W}_{\mathcal{A}} \doteq \big\{ w \in L^2(\Omega) \cap H^1_{loc}(\Omega) \mid \operatorname{div}(\mathcal{A}\nabla w) = 0 \quad in \;\; \mathcal{D}'(\Omega)\big\}.$$

Lemma 13.6. *Assume that $\partial\Omega$ is Lipschitz and $\mathcal{A}$ is in $\mathcal{M}(\alpha,\beta,\Omega)$. Then, the space $\mathbf{W}_{\mathcal{A}}$ is a subspace of $H^1_\rho(\Omega)$ and the injection from $\mathbf{W}_{\mathcal{A}}$ into $H^1_\rho(\Omega)$ is continuous. Actually,*

$$\|w\|_{L^2(\Omega)} \leqslant \|w\|_{H^1_\rho(\Omega)} \leqslant C\|w\|_{L^2(\Omega)} \quad \text{for every } w \in \mathbf{W}_{\mathcal{A}}, \tag{13.11}$$

where C depends only on $\partial\Omega$, α and β.

Furthermore, if the coefficients of $\mathcal{A}$ belong to $W^{1,\infty}(\Omega)$, then

$$\mathbf{W}_{\mathcal{A}} \subset W^{2,p}_{loc}(\Omega) \quad \text{for every } p \in [1,\infty). \tag{13.12}$$

Proof. For $\delta > 0$, let $\widetilde{\rho}_\delta = (\rho-\delta)^+$. For every w in $\mathbf{W}_{\mathcal{A}}$, the test function $(\widetilde{\rho}_\delta)^2 w$ belongs to $H^1_0(\Omega)$ and has a compact support. Then integrating by parts gives

$$0 = \int_\Omega \operatorname{div}(\mathcal{A}\nabla w)\, \widetilde{\rho}_\delta^{\,2} w \, dx = \int_\Omega \widetilde{\rho}_\delta^{\,2}\, \mathcal{A}\nabla w \nabla w \, dx + 2\int_\Omega w\, \widetilde{\rho}_\delta\, \mathcal{A}\nabla w \nabla \widetilde{\rho}_\delta \, dx.$$

Hence,

$$\int_\Omega \widetilde{\rho}_\delta^{\,2}\, \mathcal{A}\nabla w \nabla w \, dx = -2\int_\Omega w\, \widetilde{\rho}_\delta\, \mathcal{A}\nabla w \nabla \widetilde{\rho}_\delta \, dx,$$

which implies

$$\alpha\|\widetilde{\rho}_\delta \nabla w\|_{L^2(\Omega)} \leqslant C\|\widetilde{\rho}_\delta \nabla w\|_{L^2(\Omega)}\|w\|_{L^2(\Omega)},$$

and then

$$\|\widetilde{\rho}_\delta \nabla w\|_{L^2(\Omega)} \leqslant C\|w\|_{L^2(\Omega)}.$$

Letting δ go to 0, yields

$$\|\rho\nabla w\|_{L^2(\Omega)} \leqslant C\|w\|_{L^2(\Omega)}, \tag{13.13}$$

which completes the proof of (13.11).

To end the proof, observe that inclusion (13.12) is a classical consequence of Sobolev embeddings and of the results in [114, Chapter 9]. □

Remark 13.7. *Let Γ denote the support of g (closed in $\partial\Omega$). Set*

$$\begin{aligned} &\rho_\Gamma(x) \doteq dist(x,\Gamma) \quad \text{for every } x \in \mathbb{R}^N, \\ &\widetilde{\rho}_{\Gamma,\delta} = (\rho_\Gamma - \delta)^+. \end{aligned} \tag{13.14}$$

Since $(\widetilde{\rho}_{\Gamma,\delta})^2 u$ belongs to $H^1_0(\Omega)$, one can use it in place of $(\widetilde{\rho}_\delta)^2 u$ in the former proof. Estimate (13.13) *is then improved to*

$$\|\rho_\Gamma \nabla u\|_{L^2(\Omega)} \leqslant C\|u\|_{L^2(\Omega)}.$$

From now on, we assume that the following hypothesis holds:

$$\partial\Omega \in \mathcal{C}^{1,1} \quad \text{and } \mathcal{A} \in \mathcal{M}(\alpha, \beta, \Omega) \text{ with } (a_{ij})_{1\leqslant i,j\leqslant N} \in W^{1,\infty}(\Omega). \tag{13.15}$$

The next result is a particular case of a classical one ([151, Chapter II, Section 6]). In some sense, it extends the result of Proposition 13.4.

Proposition 13.8 (Trace theorem for the space $\mathbf{W}_\mathcal{A}$). *Every function w in $\mathbf{W}_\mathcal{A}$ admits a trace on $\partial\Omega$ which belongs to $H^{-1/2}(\partial\Omega)$. The corresponding trace operator $\mathbb{T}_\mathcal{A}$ is continuous, one to one and onto from $\mathbf{W}_\mathcal{A}$ to $H^{-1/2}(\partial\Omega)$. Consequently, there exists two constants c and C (depending only on Ω and $\mathcal{A}$, α, β and $\|\mathcal{A}\|_{W^{1,\infty}(\Omega)}$), such that*

$$c\|w\|_{L^2(\Omega)} \leqslant \|\mathbb{T}_\mathcal{A}(w)\|_{H^{-1/2}(\partial\Omega)} \leqslant C\|w\|_{L^2(\Omega)} \quad \textit{for every } w \in \mathbf{W}_\mathcal{A}. \tag{13.16}$$

Note that this statement is similar to Proposition 13.4 but in the more general setting $\mathbf{W}_\mathcal{A}$ and $H^{-1/2}(\partial\Omega)$ (replacing $\mathbf{W}_\mathcal{A} \cap H^1(\Omega)$ and $H^{1/2}(\partial\Omega)$).

Proof of Proposition 13.8.

Step 1. Problem (13.10) *admits one solution for data g in $H^{-1/2}(\partial\Omega)$.*

We first show that for (u, g) satisfying (13.10) (i.e., $g = Tr(u)$ in the usual sense of the map $H^1(\Omega) \to H^{1/2}(\partial\Omega)$), the second inequality of (13.16) holds.

By Green's formula, the variational formulation of Problem (13.10) is

$$\begin{cases} \displaystyle\int_\Omega u \operatorname{div}({}^t\mathcal{A}\nabla\phi)\, dx = \int_{\partial\Omega} g\, ({}^t\mathcal{A}\nabla\phi)\cdot n\, d\sigma, \\ \forall \phi \in H^1_0(\Omega) \cap H^2(\Omega). \end{cases} \tag{13.17}$$

Consider the solution $\overline{\phi} \in H^1_0(\Omega)$ of

$$\operatorname{div}({}^t\mathcal{A}\,\nabla\overline{\phi}) = u \quad \text{in } \Omega.$$

By the regularity of $\partial\Omega$ and $\mathcal{A}$ (see (13.15)), $\overline{\phi}$ belongs to $H^1_0(\Omega) \cap H^2(\Omega)$ and satisfies the estimate,

$$\|\overline{\phi}\|_{H^2(\Omega)} \leqslant C\|u\|_{L^2(\Omega)}.$$

Then, taking $\overline{\phi}$ as test in (13.17), one has successively,

$$\begin{aligned} \int_\Omega |u|^2 dx = \int_{\partial\Omega} g\,({}^t\mathcal{A}\nabla\,\overline{\phi}\,)\cdot n\, d\sigma &\leqslant \|g\|_{H^{-1/2}(\partial\Omega)} \|({}^t\mathcal{A}\nabla\,\overline{\phi}\,)\cdot n\|_{H^{1/2}(\partial\Omega)} \\ &\leqslant C\|g\|_{H^{-1/2}(\partial\Omega)}\|\,\overline{\phi}\,\|_{H^2(\Omega)} \\ &\leqslant C\|g\|_{H^{-1/2}(\partial\Omega)}\|u\|_{L^2(\Omega)}, \end{aligned}$$

whence the estimate,

$$\|u\|_{L^2(\Omega)} \leqslant C\|g\|_{H^{-1/2}(\partial\Omega)}. \tag{13.18}$$

By the density of $H^{1/2}(\partial\Omega)$ in $H^{-1/2}(\partial\Omega)$, this defines a map $\mathbb{S}_{\mathcal{A}}$ from the space $H^{-1/2}(\partial\Omega)$ to $\mathbf{W}_{\mathcal{A}}$.

For g in $H^{-1/2}(\partial\Omega)$, $\mathbb{S}_{\mathcal{A}}(g) = u \in \mathbf{W}_{\mathcal{A}}$ is the solution of problem

$$\begin{cases} \operatorname{div}(\mathcal{A}\nabla u) = 0 & \text{in } \Omega, \\ u = g & \text{on } \partial\Omega. \end{cases} \tag{13.19}$$

Its variational formulation is

$$\begin{cases} \displaystyle\int_\Omega u \operatorname{div}({}^t\!\mathcal{A}\,\nabla\varphi)dx = \langle g, {}^t\!\mathcal{A}\,\nabla\varphi\cdot n\rangle_{H^{-1/2}(\partial\Omega),H^{1/2}(\partial\Omega)}, \\ \forall\varphi \in H^1_0(\Omega) \cap H^2(\Omega). \end{cases} \tag{13.20}$$

Estimate (13.18) holds in the form

$$\|\mathbb{S}_{\mathcal{A}}(g)\|_{L^2(\Omega)} \leqslant C\|g\|_{H^{-1/2}(\partial\Omega)}, \tag{13.21}$$

showing that for given g in $H^{-1/2}(\partial\Omega)$, the solution of (13.20) is unique.

Step 2. The solution map $\mathbb{S}_{\mathcal{A}}$ is one-to-one.

By the standard trace theorem for $H^2(\Omega)$ (see [151]) one has

$$\{\nabla\varphi\cdot n \,|\, \varphi \in H^2(\Omega) \cap H^1_0(\Omega)\} \equiv H^{1/2}(\partial\Omega). \tag{13.22}$$

For every $\varphi \in H^2(\Omega) \cap H^1_0(\Omega)$, since $\varphi = 0$ on $\partial\Omega$, one has

$$\nabla\varphi = (\nabla\varphi\cdot n)n \quad \text{a.e. on } \partial\Omega.$$

Hence,

$${}^t\!\mathcal{A}\,\nabla\varphi\cdot n = ({}^t\!\mathcal{A}\,n\cdot n)(\nabla\varphi\cdot n).$$

By hypothesis(13.15), the function $({}^t\!\mathcal{A}\,n\cdot n)^{-1}$ belongs to $W^{1,\infty}(\partial\Omega)$. As a consequence of (13.22) one has the identity,

$$\{{}^t\!\mathcal{A}\,\nabla\varphi\cdot n \,|\, \varphi \in H^2(\Omega) \cap H^1_0(\Omega)\} \equiv H^{1/2}(\partial\Omega). \tag{13.23}$$

If $u = 0$ in (13.20), this implies $g = 0$ which is the claim.

Step 3. The solution map $\mathbb{S}_{\mathcal{A}}$ is onto.

By (13.23), the map

$$\phi \in H^1_0(\Omega) \cap H^2(\Omega) \mapsto {}^t\!\mathcal{A}\,\nabla\varphi\cdot n \in H^{1/2}(\partial\Omega),$$

is onto. Its kernel, $H^2_0(\Omega)$, admits an orthogonal supplement $\mathbf{H}^\perp$. Hence

$$\phi \in \mathbf{H}^\perp \mapsto {}^t\!\mathcal{A}\,\nabla\varphi\cdot n \in H^{1/2}(\partial\Omega),$$

is continuous, one-to-one and onto. It admits a continuous inverse R which is one-to-one and onto from $H^{1/2}(\partial\Omega)$ to $\mathbf{H}^\perp$.

For every h in $H^{1/2}(\partial\Omega)$, the function $R(h)$ belongs to $\mathbf{H}^{\perp}$ and satisfies

$$({}^t\mathcal{A}\nabla R(h))\cdot n = h \quad \text{a.e. in } \partial\Omega \qquad \text{and} \quad \|R(h)\|_{H^2(\Omega)} \leqslant C\|h\|_{H^{1/2}(\partial\Omega)},$$

hence

$$\operatorname{div}\big({}^t\mathcal{A}\,\nabla R(h)\big) \leqslant C\|h\|_{H^{1/2}(\partial\Omega)}.$$

Let u be in $\mathbf{W}_{\mathcal{A}}$. The linear form on $H^{1/2}(\partial\Omega)$,

$$h \in H^{1/2}(\partial\Omega) \mapsto \int_\Omega u \operatorname{div}\big({}^t\mathcal{A}\,\nabla R(h)\big)dx,$$

is continuous, so there exists $g \in H^{-1/2}(\partial\Omega)$ such that

$$\int_\Omega u \operatorname{div}\big({}^t\mathcal{A}\,\nabla R(h)\big)dx = \big\langle g, h\big\rangle_{H^{-1/2}(\partial\Omega),H^{1/2}(\partial\Omega)}.$$

By the definition of $\mathbf{W}_{\mathcal{A}}$, one has

$$\int_\Omega u \operatorname{div}\big({}^t\mathcal{A}\,\nabla\phi\big)dx = 0 \qquad \text{for every} \quad \phi \in H_0^2(\Omega).$$

Combining the last two equalities shows that u is the solution of (13.20) with the boundary data g. This proves that $\mathbb{S}_{\mathcal{A}}$ is onto.

Step 4. The solution map $\mathbb{S}_{\mathcal{A}}$ is an isomorphism.

By the Banach isomorphism theorem, (13.21) implies inequalities (13.16) with $\mathbb{T}_{\mathcal{A}}$ the inverse of $\mathbb{S}_{\mathcal{A}}$. □

The preceding proposition can be reformulated as follows:

Corollary 13.9. *For every g in $H^{-1/2}(\partial\Omega)$, problem* (13.20) *admits a unique solution in $\mathbf{W}_{\mathcal{A}}$ and the solution operator $\mathbb{S}_{\mathcal{A}}$ is an isomorphism, i.e., the following inequalities hold:*

$$c\|g\|_{H^{-1/2}(\partial\Omega)} \leqslant \|\mathbb{S}_{\mathcal{A}}(g)\|_{L^2(\Omega)} \leqslant C\|g\|_{H^{-1/2}(\partial\Omega)}. \tag{13.24}$$

Proposition 13.10. *Under the hypotheses of Proposition 13.8, if g belongs to $L^\infty(\partial\Omega)$, then $\mathbb{S}_{\mathcal{A}}(g)$ is in $\mathbf{W}_{\mathcal{A}} \cap L^\infty(\Omega)$. Conversely, if u is in $\mathbf{W}_{\mathcal{A}} \cap L^\infty(\Omega)$, then $\mathbb{T}_{\mathcal{A}}(u)$ is in $L^\infty(\partial\Omega)$ and*

$$\begin{aligned} &\|\mathbb{S}_{\mathcal{A}}(g)\|_{L^\infty(\Omega)} = \|g\|_{L^\infty(\partial\Omega)}, \quad \forall g \in L^\infty(\partial\Omega),\\ &\|\mathbb{T}_{\mathcal{A}}(u)\|_{L^\infty(\partial\Omega)} = \|u\|_{L^\infty(\Omega)}, \quad \forall u \in \mathbf{W}_{\mathcal{A}} \cap L^\infty(\Omega). \end{aligned} \tag{13.25}$$

Proof. Step 1. Proof of equality $(13.25)_1$.

First, for g in $H^{1/2}(\partial\Omega) \cap L^\infty(\partial\Omega)$, equality $(13.25)_1$ is a consequence of the weak maximum principle (see [114, Chapter 8]).

Now, let g be in $L^\infty(\partial\Omega)$ and $\{g_n\}_n$ be a sequence in $H^{1/2}(\partial\Omega) \cap L^\infty(\partial\Omega)$ satisfying

$$\begin{aligned} &g_n \to g \quad \text{strongly in } L^p(\partial\Omega) \text{ for every finite } p,\\ &g_n \rightharpoonup g \quad \text{weakly-}* \text{ in } L^\infty(\partial\Omega). \end{aligned}$$

Then

$$\mathbb{S}_{\mathcal{A}}(g_n) \to \mathbb{S}_{\mathcal{A}}(g) \quad \text{strongly in} \quad L^2(\Omega).$$

From the equality

$$\|\mathbb{S}_{\mathcal{A}}(g_n)\|_{L^\infty(\Omega)} = \|g_n\|_{L^\infty(\partial\Omega)},$$

the sequence $\{\mathbb{S}_{\mathcal{A}}(g_n)\}_n$ is uniformly bounded in $L^\infty(\Omega)$. Due to the above strong convergence in $L^2(\Omega)$

$$\begin{aligned} &\mathbb{S}_{\mathcal{A}}(g_n) \to \mathbb{S}_{\mathcal{A}}(g) \quad \text{strongly in} \quad L^p(\Omega) \quad \text{for every } p \text{ finite},\\ &\mathbb{S}_{\mathcal{A}}(g_n) \rightharpoonup \mathbb{S}_{\mathcal{A}}(g) \quad \text{weakly-}* \text{ in} \quad L^\infty(\Omega). \end{aligned}$$

As a consequence of the above strong convergences, one obtains

$$\|\mathbb{S}_{\mathcal{A}}(g)\|_{L^\infty(\Omega)} = \lim_{p\to\infty} \|\mathbb{S}_{\mathcal{A}}(g)\|_{L^p(\Omega)} = \lim_{p\to\infty} \|g\|_{L^p(\partial\Omega)} = \|g\|_{L^\infty(\partial\Omega)}.$$

Step 2. Proof of equality $(13.25)_2$.

In view of $(13.25)_1$, it is enough to show that $\mathbb{T}_{\mathcal{A}}(u)$ is in $L^\infty(\partial\Omega)$ as soon as u is in $\mathbf{W}_{\mathcal{A}} \cap L^\infty(\Omega)$.

Let ω be an open set strictly included in Ω with a Lipschitz boundary. By (13.12), u belongs to $W^{2,p}(\omega)$ for $p \in [1, +\infty)$. Therefore, by the regularity of $\partial\omega$, u belongs to $\mathcal{C}^0(\overline{\omega})$. Then, the weak maximum principle implies

$$\|u_{|\partial\omega}\|_{L^\infty(\partial\omega)} = \|u\|_{L^\infty(\omega)} \leqslant \|u\|_{L^\infty(\Omega)}. \tag{13.26}$$

Since $\operatorname{div}(\mathcal{A}\nabla u) = 0$ in ω, Green's formula gives

$$\begin{cases} \displaystyle\int_\omega u \operatorname{div}({}^t\!\mathcal{A}\nabla\varphi)dx = \int_{\partial\omega} u_{|\partial\omega}\, {}^t\!\mathcal{A}\nabla\varphi \cdot n_{\partial\omega}\, d\sigma,\\ \forall \varphi \in H^2(\omega) \cap H^1_0(\omega). \end{cases} \tag{13.27}$$

Now, consider the covering by open sets of the boundary $\partial\Omega$ (given in Section 12.1) and let $\mathcal{O}$ be one of them. There exists an orthonormal coordinate system $(O; \mathbf{e}_1, \dots, \mathbf{e}_N)$, a map $f : [-\boldsymbol{\alpha}, \boldsymbol{\alpha}]^{N-1} \to \mathbb{R}$, of class $\mathcal{C}^{1,1}$, such that

$$\begin{aligned} \mathcal{O} &= \{(x', x_N) \in \mathbb{R}^N \mid f(x') < x_N < f(x') + \boldsymbol{\beta},\ x' \in (-\boldsymbol{\alpha}, \boldsymbol{\alpha})^{N-1}\},\\ \Gamma &= \{(x', x_N) \in \mathbb{R}^N \mid x_N = f(x'),\ x' \in (-\boldsymbol{\alpha}, \boldsymbol{\alpha})^{N-1}\}. \end{aligned}$$

where $x' = (x_1, \dots, x_{N-1})$ and $\boldsymbol{\alpha}$, $\boldsymbol{\beta}$ are strictly positive constants. Here, Γ is the part of the boundary of Ω parametrized by f.

Set $\eta_0 \in (0, \inf(\boldsymbol{\alpha}, \boldsymbol{\beta})/2]$ and let φ be in $H^2(\mathcal{O}) \cap H^1_0(\mathcal{O})$ with support included in

$$\mathcal{O}_0 = \{(x', x_N) \in \mathbb{R}^N \mid f(x') < x_N < f(x') + \boldsymbol{\beta} - \eta_0,\ x' \in (-\boldsymbol{\alpha}, \boldsymbol{\alpha})^{N-1}\},$$

and such that support of $\nabla\varphi \cdot n_{\partial\mathcal{O}}$ is strictly included in Γ.

For every $\eta \in (0, \eta_0)$, set

$$\mathcal{O}_\eta = \{(x', x_N) \in \mathbb{R}^N \mid f(x') + \eta < x_N < f(x') + \boldsymbol{\beta},\ x' \in (-\boldsymbol{\alpha}, \boldsymbol{\alpha})^{N-1}\},$$
$$\Gamma_\eta = \{(x', x_N) \in \mathbb{R}^N \mid x_N = f(x') + \eta,\ x' \in (-\boldsymbol{\alpha}, \boldsymbol{\alpha})^{N-1}\}.$$

Set

$$\varphi_\eta = \varphi(\cdot - \eta \mathbf{e}_N) \quad \text{and note that } \varphi_\eta \in H^2(\mathcal{O}_\eta) \cap H_0^1(\mathcal{O}_\eta).$$

From (13.27) (with $\mathcal{O}_\eta$ in place of ω) one obtains

$$\int_{\mathcal{O}_\eta} u \operatorname{div}({}^t\!\mathcal{A} \nabla \varphi_\eta) dx = \int_{\Gamma_\eta} u_{|\Gamma_\eta} {}^t\!\mathcal{A} \nabla \varphi_\eta \cdot n_{\partial\mathcal{O}_\eta} \, d\sigma. \tag{13.28}$$

Define $g_\eta \in L^\infty(\Gamma)$ by

$$g_\eta(x', f(x')) = u_{|\Gamma_\eta}(x', f(x') + \eta) \quad \text{for a.e. } x' \in (-\boldsymbol{\alpha}, \boldsymbol{\alpha})^{N-1}.$$

From (13.26), the sequence $\{g_\eta\}_\eta$ is bounded in $L^\infty(\Gamma)$ by $\|u\|_{L^\infty(\Omega)}$. Thus, up to a subsequence, there exists $g^* \in L^\infty(\Gamma)$ such that

$$g_\eta \rightharpoonup g^* \quad \text{weakly-}* \text{ in } L^\infty(\Gamma).$$

Equality (13.28) implies

$$\int_{\mathcal{O}} u \, \mathbf{1}_{\mathcal{O}_\eta} \operatorname{div}({}^t\!\mathcal{A} \nabla \varphi_\eta) dx = \int_\Gamma g_\eta \, {}^t\!\mathcal{A}(\cdot - \eta \mathbf{e}_N) \nabla \varphi \cdot n_{\partial\mathcal{O}} \, d\sigma. \tag{13.29}$$

The left-hand side is actually equal to

$$\int_{\mathcal{O}} u \, \mathbf{1}_{\mathcal{O}_\eta} \operatorname{div}({}^t\!\mathcal{A} \nabla \varphi_\eta) dx = \int_{\mathcal{O}} u \, \mathbf{1}_{\mathcal{O}_\eta} \Big(\sum_{i=1,j=1}^N \frac{\partial \mathcal{A}_{ji}}{\partial x_i} \frac{\partial \varphi_\eta}{\partial x_j} + \mathcal{A}_{ji} \frac{\partial^2 \varphi_\eta}{\partial x_i \partial x_j} \Big) dx.$$

Since

$$\begin{aligned} \mathbf{1}_{\mathcal{O}_\eta} \nabla \varphi_\eta \to \mathbf{1}_{\mathcal{O}} \nabla \varphi \quad & \text{strongly in } L^2(\Omega)^N \\ \mathbf{1}_{\mathcal{O}_\eta} D^2 \varphi_\eta \to \mathbf{1}_{\mathcal{O}} D^2 \varphi \quad & \text{strongly in } L^2(\Omega)^{N \times N}, \end{aligned}$$

passing to the limit in (13.29) yields

$$\int_{\mathcal{O}} u \operatorname{div}({}^t\!\mathcal{A} \nabla \varphi) dx = \int_\Gamma g^* \, {}^t\!\mathcal{A} \nabla \varphi \cdot n_{\partial\mathcal{O}} \, d\sigma.$$

This equality is satisfied for every φ in $H^2(\mathcal{O}) \cap H_0^1(\mathcal{O})$ such that support of $\nabla \varphi \cdot n_{\partial\mathcal{O}}$ is strictly included in Γ. By a density argument, it is also satisfied for every φ in the same space with support included in $\overline{\mathcal{O}}$ and support of $\nabla \varphi \cdot n_{\partial\mathcal{O}}$ included in Γ.

Taking into account the different open sets of the covering of $\partial\Omega$ (see Proposition 12.2) and using a partition of unity associated to this covering, one obtains some $g^{**} \in L^\infty(\partial\Omega)$ such that, for every $\varphi \in H^2(\Omega) \cap H^1_0(\Omega)$,

$$\begin{aligned}\int_\Omega u \operatorname{div}({}^t\mathcal{A}\nabla\varphi)dx &= \int_{\partial\Omega} g^{**}\, {}^t\mathcal{A}\nabla\varphi\cdot n\, d\sigma\\ &= \big\langle \mathbb{T}_\mathcal{A}(u), {}^t\mathcal{A}\nabla\varphi\cdot n\big\rangle_{H^{-1/2}(\partial\Omega),H^{1/2}(\partial\Omega)}.\end{aligned}$$

As a consequence $\mathbb{T}_\mathcal{A}(u) = g^{**}$ is in $L^\infty(\partial\Omega)$, and estimate in $(13.25)_2$ follows from $(13.25)_1$.

Finally, $\mathbb{T}_\mathcal{A}$ is one to one and onto from $\mathbf{W}_\mathcal{A} \cap L^\infty(\Omega)$ to $L^\infty(\partial\Omega)$ and $\mathbb{S}_{\mathcal{A}|_{L^\infty(\partial\Omega)}}$ is its inverse. □

13.5 Uniform estimates for solutions

We now use Lemma 13.2 to obtain the following uniform estimate:

Proposition 13.11. *Let Ω be a bounded domain with a boundary of class $C^{1,1}$. Let Ψ be in $\mathcal{H}_B$, and G be in $H^1(\Omega)$. The solution $u^\varepsilon_{\Psi,G}$ of problem* (13.8) *satisfies the estimates*

$$\|u^\varepsilon_{\Psi,G}\|_{H^1_\rho(\Omega)} \leqslant C\|\Psi\|_{\mathcal{H}_B}\big(\|G\|_{L^2(\partial\Omega)} + \varepsilon^{1/2}\|\nabla G\|_{L^2(\Omega)}\big). \tag{13.30}$$

The constant C is independent of ε.

Note that Corollary 13.9 does not apply directly because the coefficients of the matrix field A^ε are not bounded in $W^{1,\infty}(\Omega)$ with respect to ε (unless it is constant). The proof of estimate (13.30) will require several intermediate results. One of them follows from problem (13.40) below for which Corollary 13.9 does apply since A^{hom} is with constant coefficients.

Lemma 13.12. *Under the hypotheses of Proposition 13.11, the following inequality holds for every w in $H^1_0(\Omega) \cap H^2(\Omega)$:*

$$\Big|\int_\Omega A^{hom}\nabla u^\varepsilon_{\Psi,G}\nabla w\, dx\Big| \leqslant C\varepsilon^{1/2}\|u^\varepsilon_{\Psi,G}\|_{H^1(\Omega)}\|w\|_{H^2(\Omega)}. \tag{13.31}$$

Proof. We actually establish the two estimates (13.36) and (13.39) below, which combined give (13.31).

Let w be in $H^1_0(\Omega) \cap H^2(\Omega)$.

Step 1. Proof of estimate (13.36).

From (12.4) we get

$$\|\nabla w\|_{L^2(\Lambda_\varepsilon)} \leqslant C\big(\varepsilon^{1/2}\|\nabla w\|_{L^2(\partial\Omega)} + \varepsilon\|\nabla^2 w\|_{L^2(\Omega)}\big) \leqslant C\varepsilon^{1/2}\|w\|_{H^2(\Omega)}. \tag{13.32}$$

This estimate and $(1.39)_2$ imply

$$\|\mathcal{T}_\varepsilon(\nabla w) - \nabla w\|_{L^2(\Omega\times Y)} \leqslant C\varepsilon^{1/2}\|w\|_{H^2(\Omega)}. \tag{13.33}$$

Using w as test function in (13.8) and unfolding the result implies

$$\frac{1}{|Y|}\int_{\Omega\times Y} A(y)\mathcal{T}_\varepsilon(\nabla u^\varepsilon_{\Psi,G})(x,y)\ \mathcal{T}_\varepsilon(\nabla w)(x,y)\,dxdy = -\int_{\Lambda_\varepsilon} A^\varepsilon \nabla u^\varepsilon_{\Psi,G}\ \nabla w\,dx.$$

Combining with (13.33) and (13.32), this gives

$$\begin{aligned}\Big|\int_{\Omega\times Y} A(y)\mathcal{T}_\varepsilon(\nabla u^\varepsilon_{\Psi,G})(x,y)\,\nabla w(x)dxdy\Big|&\\ \leqslant C\varepsilon^{1/2}\|u^\varepsilon_{\Psi,G}\|_{H^1(\Omega)}\|w\|_{H^2(\Omega)}.&\end{aligned}\tag{13.34}$$

Now Theorem 13.44 furnishes an element $\widehat{u}^\varepsilon_{\Psi,G}$ in $L^2(\Omega; H^1_{per,0}(Y))$ satisfying

$$\begin{aligned}\|\mathcal{T}_\varepsilon(\nabla u^\varepsilon_{\Psi,G}) - \nabla u^\varepsilon_{\Psi,G} - \nabla_y \widehat{u}^\varepsilon_{\Psi,G}\|_{L^2(Y;(H^1(\Omega))')}&\\ \leqslant C\varepsilon^{1/2}\|\nabla u^\varepsilon_{\Psi,G}\|_{L^2(\Omega)}.&\end{aligned}\tag{13.35}$$

This estimate combined with (13.34) yields

$$\Big|\int_{\Omega\times Y} A(\nabla u^\varepsilon_{\Psi,G} + \nabla_y\widehat{u}^\varepsilon_{\Psi,G})\nabla w\,dxdy\Big| \leqslant C\varepsilon^{1/2}\|u^\varepsilon_{\Psi,G}\|_{H^1(\Omega)}\|w\|_{H^2(\Omega)}. \tag{13.36}$$

Step 2. Proof of estimate (13.39).

Let χ_i in $H^1_{per,0}(Y)\cap L^\infty(Y)$ be the corrector for the adjoint problem defined, for $i \in \{1,\dots,N\}$, by

$$\int_Y A(y)\nabla_y\psi(y)(\nabla_y\chi_i(y) + \mathbf{b}_i)dy = 0, \quad \forall\psi \in H^1_{per,0}(Y). \tag{13.37}$$

These functions satisfy the assumptions of Lemma 3.29 with ${}^tA$ in place of A and $F = {}^tA\mathbf{b}_i$.

Introduce the function $\widehat{w}_\varepsilon$ in $H^1_0(\Omega)$ defined as

$$w_\varepsilon(x) = \varepsilon\rho_\varepsilon(x)\frac{\partial w}{\partial x_i}(x)\,\chi_i\Big(\frac{x}{\varepsilon}\Big) \quad \text{for a.e. } x\in\Omega.$$

By a straightforward computation, using also estimate (13.6) (for ${}^tA$ instead of A and for $F = {}^tA\mathbf{b}_i$) and (12.4),

$$\Big\|\nabla w_\varepsilon - \frac{\partial w}{\partial x_i}\nabla_y\chi_i\Big(\frac{\cdot}{\varepsilon}\Big)\Big\|_{L^2(\Omega)} + \|\nabla w_\varepsilon\|_{L^2(\Omega^{bl}_{\varepsilon d(Y)})} \leqslant C\varepsilon^{1/2}\|w\|_{H^2(\Omega)}. \tag{13.38}$$

Now, in the original problem (13.8), use w_ε as test function, combine the corresponding estimate with (13.38) and unfold the result. Since Λ_ε is included in $\Omega^{bl}_{\varepsilon d(Y)}$, (13.38) allows to control the integral term on the boundary layer Λ_ε, namely $\displaystyle\int_{\Lambda_\varepsilon} A^\varepsilon\nabla u^\varepsilon_{\Psi,G}\ \nabla w_\varepsilon\,dx$. As a consequence, we get the estimate

$$\Big|\int_{\Omega\times Y} A(y)\mathcal{T}_\varepsilon(\nabla u^\varepsilon_{\Psi,G})\,\mathcal{T}_\varepsilon\Big(\frac{\partial w}{\partial x_i}\Big)\nabla_y\chi_i(y)\,dxdy\Big| \leqslant C\varepsilon^{1/2}\|u^\varepsilon_{\Psi,G}\|_{H^1(\Omega)}\|w\|_{H^2(\Omega)},\cdot$$

Applying $(1.39)_2$ to $\dfrac{\partial w}{\partial x_i}$ gives

$$\Big\|\mathcal{T}_\varepsilon\Big(\frac{\partial w}{\partial x_i}\Big)-\frac{\partial w}{\partial x_i}\Big\|_{L^2(\Omega\times Y)}\leqslant C\varepsilon^{1/2}\|w\|_{H^2(\Omega)}.$$

Using this inequality and (13.35) yields,

$$\Big|\int_{\Omega\times Y}A(\nabla u^\varepsilon_{\Psi,G}+\nabla_y\widehat{u}^\varepsilon_{\Psi,G})\frac{\partial w}{\partial x_i}\nabla_y\chi_i\,dxdy\Big|\leqslant C\varepsilon^{1/2}\|u^\varepsilon_{\Psi,G}\|_{H^1(\Omega)}\|w\|_{H^2(\Omega)},$$

which, summing up from $i=1$ to $i=N$, implies

$$\Big|\int_{\Omega\times Y}A(\nabla u^\varepsilon_{\Psi,G}+\nabla_y\widehat{u}^\varepsilon_{\Psi,G})\,\nabla_y\Big(\sum_{i=1}^N\frac{\partial w}{\partial x_i}\chi_i\Big)\,dxdy\Big|\leqslant C\varepsilon^{1/2}\|u^\varepsilon_{\Psi,G}\|_{H^1(\Omega)}\|w\|_{H^2(\Omega)}.$$

Now, by the definition of the original correctors $\widehat{\chi}_i$'s,

$$\int_{\Omega\times Y}A\Big(\nabla u^\varepsilon_{\Psi,G}+\sum_{i=1}^N\frac{\partial u^\varepsilon_{\Psi,G}}{\partial x_i}\nabla_y\widehat{\chi}_i\Big)\,\nabla_y\Big(\sum_{j=1}^N\frac{\partial w}{\partial x_j}\chi_j\Big)\,dxdy=0.$$

Thus, by subtraction

$$\Big|\int_{\Omega\times Y}A\nabla_y\Big(\widehat{u}^\varepsilon_{\Psi,G}-\sum_{i=1}^N\frac{\partial u^\varepsilon_{\Psi,G}}{\partial x_i}\widehat{\chi}_i\Big)\,\nabla_y\Big(\sum_{j=1}^N\frac{\partial w}{\partial x_j}\chi_j\Big)\,dxdy\Big|$$
$$\leqslant C\varepsilon^{1/2}\|u^\varepsilon_{\Psi,G}\|_{H^1(\Omega)}\|w\|_{H^2(\Omega)}.$$

Using definition (13.37) of the χ_j's,

$$\int_{\Omega\times Y}A\nabla_y\Big(\widehat{u}^\varepsilon_{\Psi,G}-\sum_{i=1}^N\frac{\partial u^\varepsilon_{\Psi,G}}{\partial x_i}\widehat{\chi}_i\Big)\,\nabla_y\Big(\sum_{j=1}^N\frac{\partial w}{\partial x_j}\chi_j\Big)\,dxdy$$
$$=-\int_{\Omega\times Y}A\nabla_y\Big(\widehat{u}^\varepsilon_{\Psi,G}-\sum_{i=1}^N\frac{\partial u^\varepsilon_{\Psi,G}}{\partial x_i}\,\widehat{\chi}_i\Big)\nabla w\,dxdy.$$

Finally,

$$\Big|\int_{\Omega\times Y}A\nabla_y\Big(\widehat{u}^\varepsilon_{\Psi,G}-\sum_{i=1}^N\frac{\partial u^\varepsilon_{\Psi,G}}{\partial x_i}\,\widehat{\chi}_i\Big)\nabla w\,dxdy\Big|$$
$$\leqslant C\varepsilon^{1/2}\|u^\varepsilon_{\Psi,G}\|_{H^1(\Omega)}\|w\|_{H^2(\Omega)}.\tag{13.39}$$

Step 3. Proof of estimate (13.31).

Combining (13.36) and (13.39) implies

$$\Big|\int_{\Omega\times Y}A\Big(\nabla u^\varepsilon_{\Psi,G}+\sum_{i=1}^N\frac{\partial u^\varepsilon_{\Psi,G}}{\partial x_i}\nabla_y\widehat{\chi}_i\Big)\nabla w\,dxdy\Big|\leqslant C\varepsilon^{1/2}\|u^\varepsilon_{\Psi,G}\|_{H^1(\Omega)}\|w\|_{H^2(\Omega)}.$$

By the definition (3.19) (see also (3.22)) of the homogenized matrix A^{hom}, this is the claimed estimate (13.31). □

Proof of Proposition 13.11. Recall that A^{hom} is constant, so Corollary 13.9 applies for the solution $\overline{u}^{\varepsilon}_{\Psi,G}$ of the problem (of type (13.19))

$$\begin{cases} \operatorname{div}(A^{hom}\nabla\overline{u}^{\varepsilon}_{\Psi,G}) = 0 & \text{in } \Omega, \\ \overline{u}^{\varepsilon}_{\Psi,G} = \Psi\left(\frac{\cdot}{\varepsilon}\right)G & \text{on } \partial\Omega. \end{cases} \tag{13.40}$$

Clearly $\overline{u}^{\varepsilon}_{\Psi,G}$ belongs to $\mathbf{W}_{A^{hom}} \cap H^1(\Omega)$ and, by (13.5) and (13.24), it satisfies

$$\|\overline{u}^{\varepsilon}_{\Psi,G}\|_{H^1_\rho(\Omega)} \leqslant C\big(\|G\|_{L^2(\partial\Omega)} + \varepsilon^{1/2}\|\nabla G\|_{L^2(\Omega)}\big)\|\Psi\|_{\mathcal{H}_B}. \tag{13.41}$$

Since

$$\int_\Omega A^{hom}\nabla\overline{u}^{\varepsilon}_{\Psi,G}\nabla v\,dx = 0, \quad \forall v \in H^1_0(\Omega),$$

from (13.31) we deduce that for every $w \in H^1_0(\Omega) \cap H^2(\Omega)$,

$$\left|\int_\Omega A^{hom}\nabla(u^{\varepsilon}_{\Psi,G} - \overline{u}^{\varepsilon}_{\Psi,G})\nabla w\,dx\right| \leqslant C\varepsilon^{1/2}\|u^{\varepsilon}_{\Psi,G}\|_{H^1(\Omega)}\|w\|_{H^2(\Omega)}. \tag{13.42}$$

Introduce now the solution w_ε in $H^1_0(\Omega)$ of the adjoint variational problem

$$\int_\Omega A^{hom}\nabla v\,\nabla w_\varepsilon\,dx = \int_\Omega v(u^{\varepsilon}_{\Psi,G} - \overline{u}^{\varepsilon}_{\Psi,G})dx, \quad \forall v \in H^1_0(\Omega). \tag{13.43}$$

Since the boundary of Ω is $\mathcal{C}^{1,1}$, according to classical results,

$$w_\varepsilon \in H^1_0(\Omega) \cap H^2(\Omega) \quad \text{and} \quad \|w_\varepsilon\|_{H^2(\Omega)} \leqslant C\|u^{\varepsilon}_{\Psi,G} - \overline{u}^{\varepsilon}_{\Psi,G}\|_{L^2(\Omega)}, \tag{13.44}$$

where the constant does not depend on ε.

As $u^{\varepsilon}_{\Psi,G} - \overline{u}^{\varepsilon}_{\Psi,G}$ belongs to $H^1_0(\Omega)$, it can be used as v in (13.43), to get

$$\|u^{\varepsilon}_{\Psi,G} - \overline{u}^{\varepsilon}_{\Psi,G}\|^2_{L^2(\Omega)} = \int_\Omega A^{hom}\nabla(u^{\varepsilon}_{\Psi,G} - \overline{u}^{\varepsilon}_{\Psi,G})\,\nabla w_\varepsilon\,dx.$$

The last integral is now bounded by choosing w_ε as w in (13.42), so that, because of estimate (13.44),

$$\|u^{\varepsilon}_{\Psi,G} - \overline{u}^{\varepsilon}_{\Psi,G}\|_{L^2(\Omega)} \leqslant C\varepsilon^{1/2}\|u^{\varepsilon}_{\Psi,G}\|_{H^1(\Omega)}.$$

This estimate, together with (13.9), yields

$$\|u^{\varepsilon}_{\Psi,G} - \overline{u}^{\varepsilon}_{\Psi,G}\|_{L^2(\Omega)} \leqslant C\|\Psi\|_{\mathcal{H}_B}\|G\|_{H^1(\Omega)}.$$

The L^2-estimate of $u^{\varepsilon}_{\Psi,G}$ (part of (13.30)) now easily follows from (13.41).

Choosing $v = \rho^2 u^{\varepsilon}_{\Psi,G}$ in (13.8) as in the proof of Corollary 13.9, we get

$$\|\rho\nabla u^{\varepsilon}_{\Psi,G}\|_{L^2(\Omega)} \leqslant C\|u^{\varepsilon}_{\Psi,G}\|_{L^2(\Omega)},$$

which concludes the proof of (13.30). □

Remark 13.13. *Using notation* (13.14) *and the arguments in Remark 13.7, estimate* (13.30) *can be improved as follows:*

$$\|u^{\varepsilon}_{\Psi,G}\|_{L^2(\Omega)} + \|\rho_\Gamma\nabla u^{\varepsilon}_{\Psi,G}\|_{L^2(\Omega)} \leqslant C\|\Psi\|_{\mathcal{H}_B}\big(\|G\|_{L^2(\partial\Omega)} + \varepsilon^{1/2}\|\nabla G\|_{L^2(\Omega)}\big).$$

13.6 Homogenization

Our aim is now to describe the asymptotic behavior as $\varepsilon \to 0$ of the sequence $\{u^\varepsilon_{\Psi,G}\}_\varepsilon$ of solutions of problem (13.8).

We start by considering the case $G \equiv 1$, that is, the problem

$$\begin{cases} \operatorname{div}(A^\varepsilon \nabla u^\varepsilon_{\Psi,1}) = 0 & \text{in } \Omega, \\ u^\varepsilon_{\Psi,1} = \Psi\Big(\dfrac{\cdot}{\varepsilon}\Big) & \text{on } \partial\Omega, \end{cases} \tag{13.45}$$

for Ψ given in $\mathcal{H}_B$.

The direct homogenization of this problem is not possible in its weak variational form (analogous to (13.17))

$$\int_\Omega u^\varepsilon_{\Psi,1} \operatorname{div}({}^t A^\varepsilon \nabla\phi) dx = \int_{\partial\Omega} \Psi\Big(\frac{\cdot}{\varepsilon}\Big)\, {}^t A^\varepsilon \nabla\phi \cdot n \, d\sigma,$$

even if the matrix field A^ε is Lipschitz (since it is not uniformly Lipschitz). Also, the sequence $\{\Psi(\frac{\cdot}{\varepsilon})_{|\partial\Omega}\}_\varepsilon$ is bounded in $L^\infty(\partial\Omega)$ (and not in $H^{1/2}(\partial\Omega)$ unless Ψ is constant). But the estimate of the next proposition does not allow for the convergence of the trace of $u^\varepsilon_{\Psi,1}$ on $\partial\Omega$, the elements of $H^1_\rho(\Omega)$ not having traces in general.

Proposition 13.14 (Homogenization of (13.45)). *The sequence $\{u^\varepsilon_{\Psi,1}\}_\varepsilon$ is bounded in $H^1_\rho(\Omega)$. All its weak limit-points in $H^1_\rho(\Omega)$ actually belongs to $\mathbf{W}_{A^{hom}} \cap L^\infty(\Omega)$.*

Proof. From (13.30)

$$\|u^\varepsilon_{\Psi,1}\|_{H^1_\rho(\Omega)} \leqslant C\|\Psi\|_{\mathcal{H}_B}.$$

Let $u_\Psi \in H^1_\rho(\Omega)$ be one of the limit point of the sequence $\{u^\varepsilon_{\Psi,1}\}_\varepsilon$ so that, up to a subsequence,

$$u^\varepsilon_{\Psi,1} \rightharpoonup u_\Psi \quad \text{weakly in } H^1_\rho(\Omega). \tag{13.46}$$

Step 1: u_Ψ is in $L^\infty(\Omega)$.

By the weak maximum principle

$$\|u^\varepsilon_{\Psi,1}\|_{L^\infty(\Omega)} \leqslant \|\Psi\|_{L^\infty(Y)},$$

so that

$$u^\varepsilon_{\Psi,1} \rightharpoonup u_\Psi \quad \text{weakly-}* \text{ in } L^\infty(\Omega),$$

and

$$\|u_\Psi\|_{L^\infty(\Omega)} \leqslant \|\Psi\|_{L^\infty(Y)} \leqslant C\|\Psi\|_{\mathcal{H}_B}.$$

Step 2: u_Ψ is in $\mathbf{W}_{A^{hom}}$.

In view of Theorem 1.46, up to a subsequence, there exists

$$\widehat{u}_\Psi \in L^2_\rho(\Omega; H^1_{per,0}(Y)),$$

such that

$$\mathcal{T}_\varepsilon(\rho\nabla u^\varepsilon_{\Psi,1}) \rightharpoonup \rho\nabla u_\Psi + \rho\nabla_y\widehat{u}_\Psi \quad \text{weakly in } L^2(\Omega\times Y)^N. \tag{13.47}$$

In the course of the proof, formula (13.51) will show that the whole sequence $\{\mathcal{T}_\varepsilon(\rho\nabla u^\varepsilon_{\Psi,1})\}_\varepsilon$ converges to this limit.

For arbitrary v in $H^1_0(\Omega)$, choose ρv as test function in (13.8), to get

$$\begin{aligned}\int_\Omega \rho A^\varepsilon \nabla u^\varepsilon_{\Psi,1}\nabla v\,dx &= -\int_\Omega vA^\varepsilon\nabla u^\varepsilon_{\Psi,1}\nabla\rho\,dx\\ &= -\int_\Omega \frac{v}{\rho}A^\varepsilon(\rho\nabla u^\varepsilon_{\Psi,1})\,\nabla\rho\;dx.\end{aligned} \tag{13.48}$$

Since v belongs to $H^1_0(\Omega)$, the quotient v/ρ belongs to $L^2(\Omega)$ (see [37]), and from Proposition 1.9(i) it satisfies

$$\mathcal{T}_\varepsilon\Big(\frac{v}{\rho}\Big) \to \frac{v}{\rho} \quad \text{strongly in } L^2(\Omega\times Y).$$

Now, unfold the expression (13.48) and pass to the limit by using convergence (13.47), to obtain

$$\begin{aligned}&\int_{\Omega\times Y}\rho(x)A(y)\big(\nabla u_\Psi(x)+\nabla_y\widehat{u}_\Psi(x,y)\big)\nabla v(x)\,dxdy\\ &\qquad = -\int_{\Omega\times Y} v(x)A(y)\big(\nabla u_\Psi(x)+\nabla_y\widehat{u}_\Psi(x,y)\big)\nabla\rho(x)\,dxdy,\end{aligned}$$

that is,

$$\int_{\Omega\times Y} A(y)\big(\nabla u_\Psi(x)+\nabla_y\widehat{u}_\Psi(x,y)\big)\nabla\big(\rho(x)v(x)\big)\,dxdy = 0. \tag{13.49}$$

Let $\widehat{\xi}$ be in $H^1_{per,0}(Y)$ and η in $\mathcal{D}(\Omega)$. In (13.48), choose

$$\eta_\varepsilon(x) = \varepsilon\,\widehat{\xi}\Big(\frac{x}{\varepsilon}\Big)\eta(x),$$

as test function. Performing a similar computation as above yields

$$\int_{\Omega\times Y}\rho(x)\,A(y)\big(\nabla u_\Psi(x)+\nabla_y\widehat{u}_\Psi(x,y)\big)\nabla_y\big(\widehat{\xi}(y)\eta(x)\big)\,dxdy = 0.$$

By the density of the tensor product $\mathcal{D}(\Omega)\otimes H^1_{per,0}(Y)$ in $L^2(\Omega; H^1_{per,0}(Y))$,

$$\int_{\Omega\times Y} A(\nabla u_\Psi + \nabla_y\widehat{u}_\Psi)\nabla_y(\rho\varphi)\,dxdy = 0, \quad \forall\varphi\in L^2(\Omega; H^1_{per,0}(Y)). \tag{13.50}$$

Since ρ does not vanish in Ω, this last equality allows to express $\nabla_y\widehat{u}_\Psi$ in terms of the partial derivatives of u_Ψ and the correctors $\widehat{\chi}_i$ (just as in (3.20)),

$$\nabla_y\widehat{u}_\Psi = \sum_{i=1}^N \frac{\partial u_\Psi}{\partial x_i}\,\widehat{\chi}_i. \tag{13.51}$$

Using this formula in (13.49), a simple computation shows that

$$\int_{\Omega} A^{hom} \nabla u_{\Psi} \nabla(\rho\eta) = 0, \quad \forall \eta \in H_0^1(\Omega). \tag{13.52}$$

For ϕ in $H_0^2(\Omega)$, by the same argument as above, ϕ/ρ is in $H_0^1(\Omega)$, so that substituting it to η in (13.52), proves that u_Ψ is in $\mathbf{W}_{A^{hom}}$. □

From here on, until the end of this section, for Ψ given in $\mathcal{H}_B$, we restrict the discussion to the subsequence which was selected to define u_Ψ (see convergence (13.46)). For simplicity it is still denoted $\{\varepsilon\}$, and it is from this subsequence that further subsequences may be extracted below.

One is led to the following definition:

Definition 13.15. *For $\Psi \in \mathcal{H}_B$, the trace of u_Ψ on $\partial\Omega$ (which makes sense by Proposition 13.14) is denoted*

$$\theta_\Psi = \mathbb{T}_{A^{hom}}(u_\Psi).$$

Due to Proposition 13.10, θ_Ψ belongs to $L^\infty(\partial\Omega)$ and satisfies

$$\|\theta_\Psi\|_{L^\infty(\partial\Omega)} \leqslant C\|\Psi\|_{\mathcal{H}_B}.$$

Remark 13.16. *As noted above, the sequence of traces $\Big\{\Psi\Big(\frac{\cdot}{\varepsilon}\Big)_{|\partial\Omega}\Big\}_\varepsilon$ is bounded in $L^\infty(\partial\Omega)$, so admits weak-$*$ limit points in that space. One would expect that these limit points coincide with the traces $\mathbb{T}_{A^{hom}}(u_\Psi)$ of the weak limit points in $H^1_\rho(\Omega)$ of the sequence $\{u^\varepsilon_{\Psi,1}\}_\varepsilon$. This is the case in one dimension. In higher dimensions, Remark 13.29 shows that this is not true in general.*

We are now in position to state the main results in this subsection for the solutions of problem (13.8).

Theorem 13.17. *For every $G \in H^1(\Omega)$, there is some $u_{\Psi,G}$ in $H^1_\rho(\Omega)$ such that,*

$$u^\varepsilon_{\Psi,G} \rightharpoonup u_{\Psi,G} \quad \text{weakly in } H^1_\rho(\Omega), \tag{13.53}$$

where the limit function $u_{\Psi,G}$ is the solution of the homogenized problem,

$$\begin{cases} \operatorname{div}\big(A^{hom}\nabla u_{\Psi,G}\big) = 0 & \text{in } \Omega,\\ u_{\Psi,G} = \theta_\Psi\, G_{|\partial\Omega} & \text{on } \partial\Omega, \end{cases} \tag{13.54}$$

in the sense of Corollary 13.9 , i.e., for every $\phi \in H^2(\Omega) \cap H^1_0(\Omega)$,

$$\int_\Omega u_{\Psi,G} \operatorname{div}\big({}^tA^{hom}\,\nabla\phi\big)dx = \big\langle \theta_\Psi\, G_{|\partial\Omega}, {}^tA^{hom}\,\nabla\phi\cdot n\big\rangle_{H^{-1/2}(\partial\Omega),H^{1/2}(\partial\Omega)},$$

where $\theta_\Psi \in H^{-1/2}(\partial\Omega)$ is the trace of u_Ψ on $\partial\Omega$.

Proof. Step 1. Proof of convergence (13.53).

From (13.30), the sequence $\{u^\varepsilon_{\Psi,G}\}_\varepsilon$ is bounded in $H^1_\rho(\Omega)$. By Theorem 1.46, up to a subsequence, there exists

$$(u_{\Psi,G}, \widehat{u}_{\Psi,G}) \in H^1_\rho(\Omega) \times L^2_\rho(\Omega; H^1_{per,0}(Y))$$

such that

$$\mathcal{T}_\varepsilon(\rho \nabla u^\varepsilon_{\Psi,G}) \rightharpoonup \rho \nabla u_{\Psi,G} + \rho \nabla_y \widehat{u}_{\Psi,G} \quad \text{weakly in } L^2(\Omega \times Y)^N.$$

By (13.30) again

$$\|u_{\Psi,G}\|_{H^1_\rho(\Omega)} \leqslant C \|\Psi\|_{\mathcal{H}_B} \|G\|_{L^2(\partial\Omega)}. \tag{13.55}$$

At the end of the proof, a uniqueness argument will show that the whole sequence $\{u^\varepsilon_{\Psi,G}\}_\varepsilon$ converges to $u_{\Psi,G}$.

Step 2. The weak limit $u_{\Psi,G}$ *belongs to* $\mathbf{W}_{A^{hom}}$.

The proof follows the lines of Step 2 in Proposition 13.14.

Step 3. Proof of $(13.54)_2$.

Starting with G in $W^{1,\infty}(\Omega)$, we compare the functions $Gu^\varepsilon_{\Psi,1}$ and $u^\varepsilon_{\Psi,G}$, then the traces of their limits. For $w \in H^1_0(\Omega)$, we have successively,

$$\begin{aligned}
\int_\Omega A^\varepsilon \nabla (G u^\varepsilon_{\Psi,1} - u^\varepsilon_{\Psi,G}) \nabla w \, dx &= \int_\Omega A^\varepsilon \nabla (G u^\varepsilon_{\Psi,1}) \nabla w \, dx \\
&= \int_\Omega u^\varepsilon_{\Psi,1} A^\varepsilon \nabla G \, \nabla w \, dx - \int_\Omega w A^\varepsilon \nabla u^\varepsilon_{\Psi,1} \nabla G \, dx \\
&\qquad + \int_\Omega A^\varepsilon \nabla u^\varepsilon_{\Psi,1} \nabla (G w) \, dx \\
&= \int_\Omega u^\varepsilon_{\Psi,1} A^\varepsilon \nabla G \, \nabla w \, dx - \int_\Omega \frac{w}{\rho} A^\varepsilon (\rho \nabla u^\varepsilon_{\Psi,1}) \nabla G \, dx,
\end{aligned}$$

where we made use of problems (13.8) and (13.45) defining $u^\varepsilon_{\Psi,G}$ and $u^\varepsilon_{\Psi,1}$, respectively.

Now, since w is in $H^1_0(\Omega)$, the function w/ρ is in $L^2(\Omega)$ and satisfies

$$\|w/\rho\|_{L^2(\Omega)} \leqslant C \|\nabla w\|_{L^2(\Omega)}.$$

From this estimate as well as from (13.30), it follows that

$$\Big| \int_\Omega A^\varepsilon \nabla (G u^\varepsilon_{\Psi,1} - u^\varepsilon_{\Psi,G}) \nabla w \, dx \Big| \leqslant C \|G\|_{W^{1,\infty}(\Omega)} \|\nabla w\|_{L^2(\Omega)} \|\Psi\|_{\mathcal{H}_B}.$$

Note that the function $G u^\varepsilon_{\Psi,1} - u^\varepsilon_{\Psi,G}$ belongs to $H^1_0(\Omega)$. Therefore, one can take it as w in the inequality above, to get

$$\|\nabla (G u^\varepsilon_{\Psi,1} - u^\varepsilon_{\Psi,G})\|_{L^2(\Omega)} \leqslant C \|G\|_{W^{1,\infty}(\Omega)} \|\Psi\|_{\mathcal{H}_B}.$$

As a consequence, the limit function $Gu_\Psi - u_{\Psi,G}$ belongs to $H_0^1(\Omega)$. Thus,

$$\begin{aligned}\mathbb{T}_{A^{hom}}(u_{\Psi,G}) = \mathbb{T}_{A^{hom}}(Gu_\Psi) &= \mathbb{T}_{A^{hom}}(u_\Psi)G_{|\partial\Omega}\\ &= \theta_\Psi G_{|\partial\Omega} \quad \text{in} \quad H^{-1/2}(\partial\Omega).\end{aligned}$$

This proves the claim for G in $W^{1,\infty}(\Omega)$.

For G in $H^1(\Omega)$, we show that $(13.54)_2$ holds in the sense of Corollary 13.9. Consider a sequence $\{G_n\}_{n\in\mathbb{N}}$ of functions belonging to $W^{1,\infty}(\Omega)$ such that

$$G_n \to G \quad \text{strongly in} \quad H^1(\Omega).$$

From (13.24), one has

$$\begin{aligned}\|\mathbb{T}_{A^{hom}}(u_{F,G}) - \mathbb{T}_{A^{hom}}(u_{F,G_n})\|_{H^{-1/2}(\partial\Omega)} &\leqslant C\|u_{F,G} - u_{F,G_n}\|_{L^2(\Omega)}\\ &\leqslant C\|G - G_n\|_{L^2(\Omega)}\|\Psi\|_{\mathcal{H}_B}.\end{aligned}$$

Passing to the limit gives $(13.54)_2$.

Step 4. Proof that the whole sequence $u^\varepsilon_{\Psi,G}$ *converges.*

Recall that we only consider the sequence defining θ_Ψ. Now, all the weak limit-points of $\{u^\varepsilon_{\Psi,G}\}_\varepsilon$ satisfy (13.54) which has a unique solution. Therefore the whole sequence converges weakly. □

Remark 13.18. *From* (13.49) *and* (13.50), *one obtains*

$$\begin{cases}\displaystyle\int_{\Omega\times Y} A(\nabla u_{\Psi,G} + \nabla_y \widehat{u}_{\Psi,G})(\nabla\Phi + \nabla_y\widehat{\Phi})\,dxdy = 0,\\ \forall(\Phi,\widehat{\Phi}) \in H_0^1(\Omega)\times L^2(\Omega; H^1_{per,0}(Y)) \text{ such that}\\ \displaystyle\Big(\frac{\Phi}{\rho},\frac{\widehat{\Phi}}{\rho}\Big) \in H_0^1(\Omega)\times L^2(\Omega; H^1_{per,0}(Y)),\end{cases} \tag{13.56}$$

which is the limit unfolded formulation.

Proposition 13.19. *The convergence of* $\{u^\varepsilon_{\Psi,G}\}_\varepsilon$ *to* $u_{\Psi,G}$ *is weak in* $L^2(\Omega)$. *It is strong in* $L^2_{\rho^\gamma}(\Omega)$ *for every* $\gamma > 0$, *that is,*

$$\rho^\gamma u^\varepsilon_{\Psi,G} \to \rho^\gamma u_{\Psi,G} \quad \textit{strongly in} \quad L^2(\Omega). \tag{13.57}$$

Proof. By (13.30), the sequence $\{u^\varepsilon_{\Psi,G}\}_\varepsilon$ is uniformly bounded in $L^2(\Omega)$. So, for every $\omega \Subset \Omega$,

$$\begin{aligned}\|\rho^\gamma(u^\varepsilon_{\Psi,G} - u_{\Psi,G})\|^2_{L^2(\Omega)} &\leqslant \|\rho^\gamma(u^\varepsilon_{\Psi,G} - u_{\Psi,G})\|^2_{L^2(\omega)} + \|\rho^\gamma(u^\varepsilon_{\Psi,G} - u_{\Psi,G})\|^2_{L^2(\Omega\setminus\overline{\omega})}\\ &\leqslant \|\rho^\gamma(u^\varepsilon_{\Psi,G} - u_{\Psi,G})\|^2_{L^2(\omega)} + C\big[\mathrm{dist}(\partial\Omega,\omega)\big]^{2\gamma}.\end{aligned}$$

Observe that $\{u^\varepsilon_{\Psi,G} - u_{\Psi,G}\}_\varepsilon$ is uniformly bounded in $H^1(\omega)$, hence this sequence strongly converges to 0 in $L^2(\omega)$. Thus,

$$\limsup_{\varepsilon\to 0} \|\rho^\gamma(u^\varepsilon_{\Psi,G} - u_{\Psi,G})\|^2_{L^2(\Omega)} \leqslant C\big[\mathrm{dist}(\partial\Omega,\omega)\big]^{2\gamma},$$

from which convergence (13.57) follows. □

The argument of convergence of the energy holds in this situation.

Proposition 13.20 (Convergence of the energy).

$$\lim_{\varepsilon\to 0}\int_{\Omega\times Y} A\mathcal{T}_\varepsilon(\rho^{1+\gamma}\nabla u^\varepsilon_{\Psi,G})\,\mathcal{T}_\varepsilon(\rho^{1+\gamma}\nabla u^\varepsilon_{\Psi,G})\,dxdy$$
$$= \int_{\Omega\times Y} A\rho^{1+\gamma}\big(\nabla u_{\Psi,G}+\nabla_y\widehat{u}_{\Psi,G}\big)\,\rho^{1+\gamma}\big(\nabla u_{\Psi,G}+\nabla_y\widehat{u}_{\Psi,G}\big)\,dxdy, \tag{13.58}$$

and

$$\lim_{\varepsilon\to 0}\int_{\Lambda_\varepsilon}\big|\rho^{1+\gamma}\nabla u^\varepsilon_{\Psi,G}\big|^2\,dx = 0. \tag{13.59}$$

Proof. Choosing $\rho^{2\gamma+2}\,u^\varepsilon_{\Psi,G}$ as test-function in (13.8) implies

$$\int_\Omega A^\varepsilon\,\rho^{1+\gamma}\,\nabla u^\varepsilon_{\Psi,G}\rho^{1+\gamma}\,\nabla u^\varepsilon_{\Psi,G}\,dx$$
$$= -(2\gamma+2)\int_\Omega A^\varepsilon\,\rho^{1+\gamma}\,\nabla u^\varepsilon_{\Psi,G}\nabla\rho\,\rho^\gamma\,u^\varepsilon_{\Psi,G}\,dx. \tag{13.60}$$

Unfolding the two sides of this identity gives

$$\int_\Omega A^\varepsilon\,\rho^{1+\gamma}\,\nabla u^\varepsilon_{\Psi,G}\rho^{1+\gamma}\,\nabla u^\varepsilon_{\Psi,G}\,dx$$
$$= \frac{1}{|Y|}\int_{\Omega\times Y} A\mathcal{T}_\varepsilon(\rho^{1+\gamma}\nabla u^\varepsilon_{\Psi,G})\,\mathcal{T}_\varepsilon(\rho^{1+\gamma}\nabla u^\varepsilon_{\Psi,G})\,dxdy$$
$$+\int_{\Lambda_\varepsilon} A^\varepsilon\rho^{1+\gamma}\nabla u^\varepsilon_{\Psi,G}\;\rho^{1+\gamma}\nabla u^\varepsilon_{\Psi,G}\,dx,$$

respectively,

$$\int_\Omega A^\varepsilon\,\rho^{1+\gamma}\,\nabla u^\varepsilon_{\Psi,G}\nabla\rho\,\rho^\gamma\,u^\varepsilon_{\Psi,G}\,dx$$
$$= \frac{1}{|Y|}\int_{\Omega\times Y} A\mathcal{T}_\varepsilon(\rho^{1+\gamma}\nabla u^\varepsilon_{\Psi,G})\,\mathcal{T}_\varepsilon(\nabla\rho)\mathcal{T}_\varepsilon(\rho^\gamma u^\varepsilon_{\Psi,G})\,dxdy \tag{13.61}$$
$$+\int_{\Lambda_\varepsilon} A^\varepsilon\rho^{1+\gamma}\nabla u^\varepsilon_{\Psi,G}\,\nabla\rho\,\rho^\gamma u^\varepsilon_{\Psi,G}\,dx.$$

The last integral above goes to 0. Indeed,

$$\Big|\int_{\Lambda_\varepsilon} A^\varepsilon\rho^{1+\gamma}\nabla u^\varepsilon_{\Psi,G}\nabla\rho\,\rho^\gamma u^\varepsilon_{\Psi,G}\,dx\Big| \leqslant C(\varepsilon d(Y))^{2\gamma}\int_{\Lambda_\varepsilon}|\rho\nabla u^\varepsilon_{\Psi,G}|\,|\nabla\rho|\,|u^\varepsilon_{\Psi,G}|\,dx$$
$$\leqslant C(\varepsilon d(Y))^{2\gamma}\int_\Omega|\rho\nabla u^\varepsilon_{\Psi,G}|\,|\nabla\rho|\,|u^\varepsilon_{\Psi,G}|\,dx$$
$$\leqslant C(\varepsilon d(Y))^{2\gamma}\to 0.$$

By (13.57) the first integral on the right-hand side of (13.61) converges,

$$\lim_{\varepsilon\to 0}\int_\Omega A^\varepsilon\,\rho^{1+\gamma}\,\nabla u^\varepsilon_{\Psi,G}\nabla\rho\,\rho^\gamma\,u^\varepsilon_{\Psi,G}\,dx = \frac{1}{|Y|}\int_{\Omega\times Y} A\rho^{1+\gamma}\big(\nabla u_{\Psi,G}+\nabla_y\widehat{u}_{\Psi,G}\big)\,\nabla\rho\,\rho^\gamma u_{\Psi,G}\,dxdy.$$

Therefore, from equality (13.60) it follows that

$$\begin{aligned}\lim_{\varepsilon\to 0}\bigg(\int_{\Omega\times Y} A\mathcal{T}_\varepsilon(\rho^{1+\gamma}\nabla u^\varepsilon_{\Psi,G})\,\mathcal{T}_\varepsilon(\rho^{1+\gamma}\nabla u^\varepsilon_{\Psi,G})\,dxdy &\\ +\int_{\Lambda_\varepsilon} A^\varepsilon\rho^{1+\gamma}\nabla u^\varepsilon_{\Psi,G}\;\rho^{1+\gamma}\nabla u^\varepsilon_{\Psi,G}\,dx\bigg)& \\ = -(2\gamma+2)\int_{\Omega\times Y} A\rho^{1+\gamma}\big(\nabla u_{\Psi,G}+\nabla_y\widehat{u}_{\Psi,G}\big)\,\nabla\rho\,\rho^\gamma u_{\Psi,G}\,dxdy.&\end{aligned}\tag{13.62}$$

On the other hand, choosing as test functions $\rho^{2\gamma+2}u_{\Psi,G}$ and $\rho^{2\gamma+2}\widehat{u}_{\Psi,G}$ in problem (13.56) gives

$$\int_{\Omega\times Y} A\big(\nabla u_{\Psi,G}+\nabla_y\widehat{u}_{\Psi,G}\big)\big(\nabla(\rho^{2\gamma+2}u_{\Psi,G})+\nabla_y(\rho^{2\gamma+2}\widehat{u}_{\Psi,G})\big)\,dxdy=0,$$

which reads as

$$\begin{aligned}\frac{1}{|Y|}\int_{\Omega\times Y} A\rho^{1+\gamma}\big(\nabla u_{\Psi,G}+\nabla_y\widehat{u}_{\Psi,G}\big)\,\rho^{1+\gamma}\big(\nabla u_{\Psi,G}+\nabla_y\widehat{u}_{\Psi,G}\big)\,dxdy&\\ =-\frac{2\gamma+2}{|Y|}\int_{\Omega\times Y} A\rho^{1+\gamma}\big(\nabla u_{\Psi,G}+\nabla_y\widehat{u}_{\Psi,G}\big)\,\nabla\rho\,\rho^\gamma u_{\Psi,G}\,dxdy.&\end{aligned}\tag{13.63}$$

Comparing (13.62) and (13.63) we get

$$\begin{aligned}\lim_{\varepsilon\to 0}\bigg(\int_{\Omega\times Y} A\mathcal{T}_\varepsilon(\rho^{1+\gamma}\nabla u^\varepsilon_{\Psi,G})\,\mathcal{T}_\varepsilon(\rho^{1+\gamma}\nabla u^\varepsilon_{\Psi,G})\,dxdy &\\ +\int_{\Lambda_\varepsilon} A^\varepsilon\rho^{1+\gamma}\nabla u^\varepsilon_{\Psi,G}\;\rho^{1+\gamma}\nabla u^\varepsilon_{\Psi,G}\,dx\bigg)& \\ = \int_{\Omega\times Y} A\rho^{1+\gamma}\big(\nabla u_{\Psi,G}+\nabla_y\widehat{u}_{\Psi,G}\big)\,\rho^{1+\gamma}\big(\nabla u_{\Psi,G}+\nabla_y\widehat{u}_{\Psi,G}\big)\,dxdy.&\end{aligned}$$

One concludes by applying Lemma 3.11. □

Corollary 13.21. *The following strong convergences hold:*

$$\mathcal{T}_\varepsilon(\rho^{1+\gamma}\nabla u^\varepsilon_{\Psi,G})\to\rho^{1+\gamma}\nabla u_{\Psi,G}+\rho^{1+\gamma}\nabla_y\Big(\sum_{i=1}^N\frac{\partial u_{\Psi,G}}{\partial x_i}\,\widehat{\chi}_i\Big)\quad\text{strongly in } L^2(\Omega\times Y)^N,\tag{13.64}$$

where $\gamma > 0$, *and also*

$$\rho^{1+\gamma}\Big(u^\varepsilon_{\Psi,G} - u_{\Psi,G} - \varepsilon\sum_{i=1}^N \frac{\partial u_{\Psi,G}}{\partial x_i}\widehat{\chi}_i\Big(\frac{\cdot}{\varepsilon}\Big)\Big) \to 0 \quad \textit{strongly in } H^1(\Omega).$$

Proof. Convergence (13.64) is a consequence of that given in (13.58), combined with Lemma 3.11. The second convergence then follows by using also (13.57), Proposition 1.29(ii) and convergence (13.59), in view of the properties of the averaging operator applied to convergence (13.64). □

13.7 The case of flat parts of the boundary

We consider the case of a connected open part Γ of the boundary of Ω which is included in an affine hyperplane $\mathbf{H}$ orthogonal to a non null vector of $\mathbf{G}$. For simplicity, we choose $\mathbf{G} = \mathbb{Z}^N$. Set $\mathbf{H} = \mathbf{b} + \mathbf{H}_0$ (where $\mathbf{b}$ belongs to $\mathbb{R}^N$ and $\mathbf{H}_0$ is a vector hyperplane).

13.7.1 The general case

Proposition 13.22. *Let* $\theta_\Psi = \mathbb{T}_{A^{hom}}(u_\Psi)$ *be the trace (given by definition 13.15) of a limit point of the sequence* $\{u^\varepsilon_{\Psi,1}\}_\varepsilon$*. Then,*

$$\theta_{\Psi|\Gamma} \quad \textit{is constant on } \Gamma.$$

Proof. Set $D(a;R) = \overline{B(a,R)} \cap \overline{\Omega}$ where $a \in \Gamma$ and $R > 0$. One chooses R small enough so that $D(a,R)$ is a half closed ball. Let η_R be a smooth cut-off function whose support is included in $D(a,R)$ and which is equal to 1 in $D(a,R/2)$.

Consider $\mathbf{h}$ a vector in $\mathbf{H}_0$ such that $D(a,R)+\mathbf{h}$ is included in $\overline{\Omega}$. Denote it $D(a+\mathbf{h},R)$. There exists a sequence $\{\mathbf{h}_\varepsilon\}_\varepsilon$ of vectors belonging to $\mathbf{H}_0 \cap \varepsilon\mathbb{Z}^N$ converging to $\mathbf{h}$ and such that $D(a,R)+\mathbf{h}_\varepsilon$ is also included in $\overline{\Omega}$.

Let

$$v_\varepsilon = u^\varepsilon_{\Psi,1}\eta_R(\cdot + a) \quad \text{and} \quad w_\varepsilon = u^\varepsilon_{\Psi,1}\eta_R(\cdot + a + \mathbf{h}_\varepsilon).$$

Then

$$v^\varepsilon \rightharpoonup u_\Psi\eta_R(\cdot + a) \quad \text{and} \quad w_\varepsilon \rightharpoonup u_\Psi\eta_R(\cdot + a + \mathbf{h}) \quad \text{weakly in } H^1_\rho(\Omega).$$

For every $\varphi \in H^1_0(D(a,R))$ the function v_ε satisfies

$$\begin{aligned}\int_{D(a,R)} A^\varepsilon\nabla v_\varepsilon\,\nabla\varphi\,dx = &\int_{D(a,R)} u^\varepsilon_{\Psi,1}A^\varepsilon\nabla\eta_R(\cdot + a)\,\nabla\varphi\,dx\\ &-\int_{D(a,R)} \varphi A^\varepsilon\nabla u^\varepsilon_{\Psi,1}\,\nabla\eta_R(\cdot + a)\,dx,\end{aligned}$$

while w_ε satisfies

$$\begin{aligned}\int_{D(a+\mathbf{h}_\varepsilon,R)} & A^\varepsilon \nabla w_\varepsilon \, \nabla\varphi(\cdot+\mathbf{h}_\varepsilon)\,dx \\ &= \int_{D(a+\mathbf{h}_\varepsilon,R)} u^\varepsilon_{\Psi,1} A^\varepsilon \nabla\eta_R(\cdot+a+\mathbf{h}_\varepsilon)\,\nabla\varphi(\cdot+\mathbf{h}_\varepsilon)\,dx \\ &\quad - \int_{D(a+\mathbf{h}_\varepsilon,R)} \varphi(\cdot+\mathbf{h}_\varepsilon) A^\varepsilon \nabla u^\varepsilon_{\Psi,1}\,\nabla\eta_R(\cdot+a+\mathbf{h}_\varepsilon)\,dx.\end{aligned}$$

After a change of variables, observing that $A^\varepsilon(\cdot-\mathbf{h}_\varepsilon)=A^\varepsilon$ a.e. in $\mathbb{R}^N$, one obtains, for every $\varphi\in H^1_0(D(a,R))$,

$$\begin{aligned}\int_{D(a,R)} A^\varepsilon \nabla w_\varepsilon(\cdot-\mathbf{h}_\varepsilon)\,\nabla\varphi\,dx &= \int_{D(a,R)} u^\varepsilon_{\Psi,1}(\cdot-\mathbf{h}_\varepsilon) A^\varepsilon \nabla\eta_R(\cdot+a)\,\nabla\varphi\,dx \\ &\quad - \int_{D(a,R)} \varphi A^\varepsilon \nabla u^\varepsilon_{\Psi,1}(\cdot-\mathbf{h}_\varepsilon)\,\nabla\eta_R(\cdot+a)\,dx.\end{aligned}$$

Since

$$\Psi^\varepsilon\Big(\frac{\cdot-\mathbf{h}_\varepsilon}{\varepsilon}\Big)=\Psi^\varepsilon\Big(\frac{\cdot}{\varepsilon}\Big) \quad \text{a.e. in } \mathbb{R}^N,$$

the function $w_\varepsilon(\cdot-\mathbf{h}_\varepsilon)-v_\varepsilon$ belongs to $H^1_0(D(a,R))$. Due to estimate (13.30) and to the fact that φ/ρ belongs to $L^2(D(a,R))$, it follows that the sequence $\{w_\varepsilon(\cdot-\mathbf{h}_\varepsilon)-v_\varepsilon\}_\varepsilon$ is uniformly bounded in $H^1_0(D(a,R))$. Then, as in the proof of Theorem 13.17 (Step 3), one has

$$\mathbb{T}_{A^{hom}}(u_\Psi(\cdot-\mathbf{h})\eta_R(\cdot+a))=\mathbb{T}_{A^{hom}}(u_\Psi\eta_R(\cdot+a)) \quad \text{in } L^\infty(\partial D(a,R)\cap\partial\Omega).$$

Consequently,

$$\theta_\Psi(\cdot-\mathbf{h})=\theta_\Psi \quad \text{in } L^\infty(\partial D(a,R/2)\cap\partial\Omega).$$

A connectedness argument now shows that θ_Ψ is constant on Γ. □

13.7.2 A particular case

In this subsection we assume that Γ is included in the hyperplane

$$a_1x_1+\ldots+a_Nx_N=0, \quad (a_1,\ldots,a_N)\in\mathbb{Z}^N\backslash\{(0,\ldots,0)\}.$$

What is important here is that the hyperplane contains the origin (in Section 13.1, this corresponds to $a=0$).

The intersection $\mathbb{H}\doteq\mathbf{H}\cap\mathbb{Z}^N$ is a subgroup of dimension $N-1$. Hence, from Subsection 2.2.4, there exists a basis of $\mathbb{Z}^N$ such that the first $N-1$ vectors lie in the hyperplane. This new basis of $\mathbb{Z}^N$ (and of $\mathbb{R}^N$!) is still denoted $\mathbf{B}\doteq(\mathbf{b}_1,\ldots,\mathbf{b}_N)$, the periodicity cell Y is again the open parallelotope associated with $\mathbf{B}$. Its face on the hyperplane is denoted Y', this is the open parallelotope generated by the vectors $\mathbf{b}_1,\ldots,\mathbf{b}_{N-1}$.

Therefore, without loss of generality, one can assume that the hyperplane $\mathbf{H}$ is equal to $\mathbb{R}^{N-1} \times \{0\}$. One can also assume that

$$\Omega \subset \mathbb{R}^N_+ \doteq \mathbb{R}^{N-1} \times (0, +\infty).$$

The aim of this subsection is to determine the restriction of the trace on Γ of the limit function $u_{\Psi,G}$ given by Theorem 13.17.

To do so, we introduce a specific unfolding operator which mixes the periodic unfolding operator in $\mathbb{R}^{N-1}$ and a scaling in the direction $\mathbf{b}_N$. This operator can be seen as a partial unfolding operator (after the affinity) where $\mathcal{O} = (0, +\infty)$ and $\Omega = \mathbb{R}^{N-1}$ (see Section 1.5).

Definition 13.23 (The unfolding operator for a flat boundary layer). *For ϕ measurable function on $\mathbb{R}^N_+$, the unfolding operator $\mathcal{T}^\#_\varepsilon$ is defined as follows:*

$$\mathcal{T}^\#_\varepsilon(\phi)(x', y', y_N) = \phi\Big(\varepsilon\Big[\frac{x'}{\varepsilon}\Big]_{Y'} + \varepsilon y', \varepsilon y_N\Big),$$

for a.e. $(x', y', y_N) \in \mathbb{R}^{N-1} \times Y^\#$ where $Y^\# \doteq Y' \times (0, +\infty)$.

Introduce the notation

$$\mathbb{W}^p(\mathbb{R}^N_+) \doteq \{\phi \in W^{1,p}_{loc}(\mathbb{R}^N_+) \mid \phi_{|\mathbb{R}^{N-1}\times(0,\zeta)} \text{ belongs to } W^{1,p}(\mathbb{R}^{N-1} \times (0, \zeta)),\ \forall \zeta > 0\}.$$

The following properties of $\mathcal{T}^\#_\varepsilon$ are simply derived from those of the usual unfolding operator in fixed domains:

Lemma 13.24. *Suppose $p \in [1, +\infty]$. The operator $\mathcal{T}^\#_\varepsilon$ is linear and continuous from $L^p(\mathbb{R}^N_+)$ into $L^p(\mathbb{R}^{N-1} \times Y^\#)$. For every ϕ in $L^1(\mathbb{R}^N_+)$, ψ in $L^p(\mathbb{R}^N_+)$ and $\Phi \in \mathbb{W}^p(\mathbb{R}^N_+)$ one has*

(i) $$\frac{1}{|Y|}\int_{\mathbb{R}^{N-1}\times Y^\#} \mathcal{T}^\#_\varepsilon(\phi)(x', y)\, dx'dy = \frac{1}{\varepsilon}\int_{\mathbb{R}^N_+} \phi(x)\, dx,$$

(ii) $$\|\mathcal{T}^\#_\varepsilon(\psi)\|_{L^p(\mathbb{R}^{N-1}\times Y^\#)} = \frac{|\,Y\,|^{1/p}}{\varepsilon^{1/p}}\, \|\psi\|_{L^p(\mathbb{R}^N_+)},$$

(iii) $\forall\, \zeta > 0,\ \mathcal{T}^\#_\varepsilon(\Phi) \to \Phi_{|\Gamma}$ *strongly in* $L^p(\mathbb{R}^{N-1} \times Y' \times (0, \zeta))$.

This operator maps $L^p(\mathbb{R}^N_+)$ into $L^p(\mathbb{R}^{N-1} \times Y^\#)$. If ϕ belongs to $\mathbb{W}^p(\mathbb{R}^N_+)$ ($p \in [1, +\infty]$), one has

$$\nabla_y \mathcal{T}^\#_\varepsilon(\phi) = \varepsilon \mathcal{T}^\#_\varepsilon(\nabla\phi).$$

We introduce the following spaces:

$$\mathbf{L} \doteq \Big\{\Phi \in H^1_{loc}(Y^\#) \mid \nabla_y \Phi \in L^2(Y^\#)^N,\ \Phi(\cdot, y_N) \text{ is } Y' \text{ periodic for a.e. } y_N \in (0, +\infty)\Big\},$$

$$\mathbf{L}_0 \doteq \Big\{\Phi \in \mathbf{L} \mid \Phi(y', 0) = 0 \text{ for a.e. } y' \in Y'\Big\}.$$

We endow $\mathbf{L}$ with the norm

$$\|\phi\|_{\mathbf{L}} = \Big(\|\Phi(\cdot,0)\|^2_{L^2(Y')} + \|\nabla_y\Phi\|^2_{L^2(Y\#)}\Big)^{1/2}.$$

Denote

$$\mathbf{L}^\zeta = \big\{\Phi \in \mathbf{L} \mid \nabla_y\Phi = 0 \quad \text{in } Y' \times (\zeta,+\infty)\big\}.$$

It is known that $\bigcup_{\zeta>0} \mathbf{L}^\zeta$ is dense in $\mathbf{L}$.

From now on, in problems (13.8) we assume that G is such that

$$G \in W^{1,\infty}(\Omega) \quad \text{and} \quad \text{support}(G_{|\partial\Omega}) \subset \text{interior}(\Gamma). \tag{13.65}$$

Then, we extend $u^\varepsilon_{\Psi,G}$ and G by 0 in $\mathbb{R}^N_+\setminus\overline{\Omega}$. These extensions are still denoted $u^\varepsilon_{\Psi,G}$ and G. Lemma 13.24 together with estimates (13.9) and (13.41) imply

$$\begin{aligned}\|\mathcal{T}^\#_\varepsilon(u^\varepsilon_{\Psi,G})\|_{L^2(\mathbb{R}^{N-1}\times Y\#)} &\leqslant \frac{C}{\sqrt{\varepsilon}}\|u^\varepsilon_{\Psi,G}\|_{L^2(\Omega)}\\ &\leqslant \frac{C}{\varepsilon}\|\Psi\|_{\mathcal{H}_B}\big(\|G_{|\partial\Omega}\|_{L^2(\mathbf{H}_{\partial\Omega})} + \varepsilon^{1/2}\|\nabla G\|_{L^2(\Omega)}\big),\end{aligned}$$

and

$$\begin{aligned}\|\nabla_y\mathcal{T}^\#_\varepsilon(u^\varepsilon_{\Psi,G})\|_{L^2(\mathbb{R}^{N-1}\times Y\#)} &\leqslant C\frac{\varepsilon}{\sqrt{\varepsilon}}\|u^\varepsilon_{\Psi,G}\|_{H^1(\Omega)}\\ &\leqslant C\|\Psi\|_{\mathcal{H}_B}\big(\|G_{|\partial\Omega}\|_{L^2(\mathbf{H}_{\partial\Omega})} + \varepsilon^{1/2}\|\nabla G\|_{L^2(\Omega)}\big).\end{aligned}$$

Theorem 13.25. *Let $u^\varepsilon_{\Psi,G}$ be the solution of problem* (13.8) *with G satisfying* (13.65). *Then, there exists a unique $u^\#_{\Psi,G}$ in $L^2(\Gamma;\mathbf{L})$ satisfying*

$$\nabla_y\mathcal{T}^\#_\varepsilon(u^\varepsilon_{\Psi,G}) \rightharpoonup \nabla_y u^\#_{\Psi,G} \quad \text{weakly in } L^2(\mathbb{R}^{N-1}\times Y\#)^N \tag{13.66}$$

and

$$u^\#_{\Psi,G|Y'} = G_{|\Gamma}\,\Psi_{|Y'}. \tag{13.67}$$

Moreover, if $\overline{\Psi}$ is the unique solution in $\mathbf{L}$ of the variational problem,

$$\begin{cases}\displaystyle\int_{Y\#} A(y)\nabla_y\overline{\Psi}(y)\,\nabla_y\Phi(y)\,dy = 0,\\ \overline{\Psi}_{|Y'} = \Psi_{|Y'}, \quad \forall\Phi \in \mathbf{L}_0,\end{cases} \tag{13.68}$$

then $u^\#_{\Psi,G} = G_{|\Gamma}\,\overline{\Psi}$.

Proof. Let $v \in C^\infty(\overline{\Omega})$, such that $v = 0$ on $\partial\Omega\setminus\Gamma$, and $\Phi \in \mathbf{L}_0 \cap \mathbf{L}^\zeta$. Consider the function

$$v_\varepsilon(x) = \varepsilon v(x)\Phi\Big(\Big\{\frac{x}{\varepsilon}\Big\}_Y\Big).$$

Observe that v_ε belongs to $H_0^1(\Omega)$. So, we can take it as test function in problem (13.8). Then, by unfolding the result with $\mathcal{T}_\varepsilon^{\#}$, we obtain

$$\int_{\mathbb{R}^{N-1}\times Y^{\#}} A(y)\nabla_y \mathcal{T}_\varepsilon^{\#}(u^\varepsilon_{\Psi,G})(x',y)\Big(\varepsilon\mathcal{T}_\varepsilon^{\#}(\nabla v)(x',y)\Phi(y) + \mathcal{T}_\varepsilon^{\#}(v)(x',y)\nabla_y\Phi(y)\Big)\,dx'dy = 0. \tag{13.69}$$

From Lemma 13.24(ii), one has for every $\zeta > 0$,

$$\varepsilon\mathcal{T}_\varepsilon^{\#}(\nabla v) \to 0 \quad \text{strongly in } L^\infty(\mathbb{R}^{N-1}\times Y^{\#})^N,$$
$$\mathcal{T}_\varepsilon^{\#}(v) \to v_{|\Gamma} \quad \text{strongly in } L^\infty(\mathbb{R}^{N-1}\times Y'\times(0,\zeta)).$$

Passing to the limit in (13.69) gives

$$\int_{\mathbb{R}^{N-1}\times Y^{\#}} A(y)\nabla_y u^{\#}_{\Psi,G}(x',y)\, v_{|\Gamma}(x')\,\nabla_y\Phi(y)\,dx'dy = 0, \quad \forall\Phi\in \mathbf{L}_0\cap\mathbf{L}^\zeta.$$

As $\mathbf{L}_0\cap\bigcup_{\zeta>0}\mathbf{L}^\zeta$ is dense in $\mathbf{L}_0$, this identity is still true for Φ in $L^2(\Gamma;\mathbf{L}_0)$,

$$\int_{\mathbb{R}^{N-1}\times Y^{\#}} A(y)\nabla_y u^{\#}_{\Psi,G}(x',y)\,\nabla_y\Phi(x',y)\,dx'dy = 0, \quad \forall\Phi\in L^2(\Gamma;\mathbf{L}_0).$$

Recalling (13.67) and problem (13.68), the last assertion of the theorem is easily obtained. Since the solution of (13.68) is unique, convergence (13.66) holds for the whole sequence. □

We now give some properties of the function $\overline{\Psi}$ which are essential for the proof of Theorem 13.28 below.

Lemma 13.26. *There exist two strictly positive constants c and C independent of ζ such that the solution of problem (13.68) satisfies*

$$\begin{aligned}&\int_{Y'\times(\zeta,+\infty)} |\nabla_y\overline{\Psi}(y)|^2dy \leqslant Ce^{-c\zeta}, \quad \forall\zeta\geqslant 0,\\ &\int_{Y^{\#}} y_N^2|\nabla_y\overline{\Psi}(y)|^2dy \leqslant C.\end{aligned} \tag{13.70}$$

Moreover, $\nabla_y\overline{\Psi}$ belongs to $L^1\big(Y^{\#}\big)^N$, and there exits a real number $\overline{\Psi}(\infty)$, such that, as $\zeta\to+\infty$,

$$\overline{\Psi}(\cdot,\zeta)\to\overline{\Psi}(\infty) \quad \text{strongly in } L^2(Y'). \tag{13.71}$$

Furthermore

$$\|\overline{\Psi}-\overline{\Psi}(\infty)\|_{L^2(Y'\times(\zeta,+\infty))} \leqslant C\,e^{-c\zeta/2} \quad \forall\zeta\geqslant 0. \tag{13.72}$$

Proof. Step 1. Proof of estimates (13.70).

For ϕ in $H^1_{loc}(\mathbb{R}_+) \cap \mathbf{L}_0$, take $\Phi(y) = \phi(y_N)$ as test function in (13.68) to get,

$$\sum_{j=1}^{N} \int_{Y'} \Big(\sum_{j=1}^{N} a_{Nj}(y) \frac{\partial \overline{\Psi}}{\partial y_j}(y) \Big) dy' = 0 \quad \text{for a.e. } y_N \in \mathbb{R}_+. \tag{13.73}$$

For every n in $\mathbb{N}$, set

$$U_n = \int_{Y' \times (n,+\infty)} |\nabla_y \overline{\Psi}(y)|^2 \, dy,$$

and

$$\pi_n(t) = \inf\big(\sup(0, t-n), 1\big), \quad \forall t \in \mathbb{R}^+.$$

Then, choosing now $y \mapsto \pi_n(y_N)\overline{\Psi}(y)$ as test function in (13.68), gives

$$\int_{Y^\#} A(y)\nabla_y \overline{\Psi}(y)\, \pi_n(y_N) \nabla_y \overline{\Psi}(y)\, dy \\ + \int_{Y' \times (n,n+1)} A(y)\nabla_y \overline{\Psi}(y)\, \mathbf{b}_N \overline{\Psi}(y)\, dy = 0.$$

Since

$$\pi_n(y_N) = 1 \quad \text{for } \ y_N \geqslant n+1,$$

according to equality (13.73) this implies

$$\alpha U_{n+1} \leqslant \int_{Y' \times (n,n+1)} A(y)\nabla_y \overline{\Psi}(y)\, \mathbf{b}_N \big(\overline{\Psi}(y) - \mathcal{M}_{Y'}(\overline{\Psi})(y_N)\big)\, dy, \tag{13.74}$$

where

$$\mathcal{M}_{Y'}(\overline{\Psi})(y_N) = \frac{1}{|Y'|} \int_{Y'} \overline{\Psi}(y', y_N)\, dy' \quad \text{for a.e. } y_N \in (0, +\infty).$$

From (13.74) and due to the Poincaré-Wirtinger inequality applied in every cross-sections $Y' \times \{y_N\}$, one obtains

$$U_{n+1} \leqslant C(U_n - U_{n+1}), \quad \forall n \in \mathbb{N}.$$

The constant does not depend on n. This implies $(13.70)_1$.

On the other hand, from $(13.70)_1$ one obtains

$$\int_{Y^\#} y_N^2 \, |\nabla_y \overline{\Psi}(y)|^2 \, dy = \sum_{n \in \mathbb{N}} \int_{Y' \times (n,n+1)} y_N^2 \, |\nabla_y \overline{\Psi}(y)|^2 \, dy \\ \leqslant C \sum_{n \in \mathbb{N}} (n+1)^2 \, e^{-cn} \leqslant C,$$

and this is precisely $(13.70)_2$.

Step 2. Proof of (13.71).

By Cauchy-Schwarz inequality and $(13.70)_1$, it follows that for all $k \in \mathbb{N}$,

$$\begin{aligned}\int_{Y'\times(k,+\infty)} |\nabla_y \overline{\Psi}(y)|dy &= \sum_{n\geqslant k}\int_{Y'\times(n,n+1)} |\nabla_y \overline{\Psi}(y)|dy\\ &\leqslant \sum_{n\geqslant k}\Big(\int_{Y'\times(n,n+1)} |\nabla_y \overline{\Psi}(y)|^2 dy\Big)^{1/2}\\ &\leqslant C\sum_{n\geqslant k} e^{-cn/2} \leqslant C\, e^{-ck/2}.\end{aligned}$$

From the trace theorems, for every $k \in \mathbb{N}$ one has

$$\begin{aligned}&\|\overline{\Psi}(\cdot,k)-\overline{\Psi}(\cdot,k+1)\|_{L^2(Y')} \leqslant C\|\nabla_y \overline{\Psi}\|_{L^2(Y'\times(k,k+1))} \leqslant Ce^{-ck/2},\\ &\|\overline{\Psi}(\cdot,k)-\mathcal{M}_{Y'}(\overline{\Psi})(k)\|_{L^2(Y')} \leqslant C\|\nabla_y \overline{\Psi}\|_{L^2(Y'\times(k,k+1))} \leqslant Ce^{-ck/2}.\end{aligned}$$

Then, for every $l \in \mathbb{N}$,

$$\|\overline{\Psi}(\cdot,k)-\overline{\Psi}(\cdot,k+l)\|_{L^2(Y')} \leqslant C\sum_{j=k}^{k+l-1} e^{-cj/2} \leqslant Ce^{-ck/2}.$$

Taking the mean value over the cross-sections $Y'\times\{k\}$ and $Y'\times\{k+l\}$, gives

$$|\mathcal{M}_{Y'}(\overline{\Psi})(k)-\mathcal{M}_{Y'}(\overline{\Psi})(k+l)| \leqslant C\|\overline{\Psi}(\cdot,k)-\overline{\Psi}(\cdot,k+l)\|_{L^2(Y')} \leqslant Ce^{-ck/2}.$$

As a consequence, $\{\overline{\Psi}(\cdot,k)\}_{k\in\mathbb{N}}$ and $\{\mathcal{M}_{Y'}(\overline{\Psi})(k)\}_{k\in\mathbb{N}}$ are Cauchy sequences in $L^2(Y')$, respectively in $\mathbb{R}$. Both converge to the same limit which is a constant function denoted $\overline{\Psi}(\infty)$.

Step 3. Proof of (13.72).

Proceeding as in the previous step, we first show that

$$\int_{Y'} \big|\overline{\Psi}(y',y_N)-\overline{\Psi}(y',[y_N])\big|^2 dy' \leqslant C\, e^{-c[y_N]} \quad \text{for every } y_N \in \mathbb{R}_+,$$

where $[y_N]$ is the integer part of y_N. This, together with the convergences obtained in the previous step imply (13.72). □

Corollary 13.27. *The following estimates hold:*

$$\begin{aligned}&\Big\|\rho_{\mathbf{H}}\nabla\overline{\Psi}\Big(\frac{x}{\varepsilon}\Big)\Big\|_{L^2(\Omega)} \leqslant C\varepsilon^{1/2},\\ &\Big\|\overline{\Psi}\Big(\frac{\cdot}{\varepsilon}\Big)-\overline{\Psi}(\infty)\Big\|_{L^2(\Omega\cap\{x_N>\zeta\})} \leqslant C\, e^{-c\zeta\varepsilon/2} \quad \textit{for every } \zeta>0\end{aligned} \tag{13.75}$$

where $\rho_{\mathbf{H}}(x) = dist(x,\mathbf{H}) = |x_N|$. *The constant does not depend on* ε.

Proof. From $(13.70)_2$ and Lemma 13.24, one has

$$\Big\|\rho_{\mathbf{H}}\nabla\overline{\Psi}\Big(\frac{\cdot}{\varepsilon}\Big)\Big\|^2_{L^2(\Omega)} = \int_\Omega x_N^2\Big|\nabla\overline{\Psi}\Big(\frac{x}{\varepsilon}\Big)\Big|^2 dx = \frac{\varepsilon}{|Y|}\int_{\mathbb{R}^{N-1}\times Y^\#} y_N^2\big|\nabla_y\overline{\Psi}(y)\big|^2 dx'dy.$$

Then, from (13.72) and again Lemma 13.24, estimate $(13.75)_2$ follows. □

Now, we are in a position to give an explicit formula for the restriction of the function θ_Ψ (see Theorem 13.17) to the flat part Γ of Ω.

Theorem 13.28. *All limit points u_Ψ of the sequence $\{u^\varepsilon_{\Psi,1}\}_\varepsilon$ have the same trace on Γ, namely*

$$\mathbb{T}_{A^{hom}}(u_\Psi)_{|\Gamma} = \theta_{\Psi|\Gamma} = \overline{\Psi}(\infty).$$

Moreover, for G given in $H^1(\Omega)$ such that the restriction $G_{|\partial\Omega}$ has a support included in Γ, the whole sequence $\{u^\varepsilon_{\Psi,G}\}_\varepsilon$ converges to the unique solution of

$$\begin{cases}\operatorname{div}\big(A^{hom}\nabla u_{\Psi,G}\big) = 0 & \text{in } \Omega,\\ u_{\Psi,G} = \overline{\Psi}(\infty)\, G_{|\partial\Omega} & \text{on } \partial\Omega.\end{cases}$$

Proof. Step 1. $\Big\|u^\varepsilon_{\Psi,G} - G\,\overline{\Psi}\Big(\frac{\cdot}{\varepsilon}\Big)\Big\|_{H^1_0(\Omega)}$ *is bounded.*

Let ϕ be in $H^1_0(\Omega)$ and $G \in W^{1,\infty}(\Omega)$ with support included in Γ. The function defined for a.e. $y \in Y^\#$ by

$$y \mapsto \sum_{\xi'\in\mathbb{H}} G(\varepsilon\xi' + \varepsilon y)\phi(\varepsilon\xi' + \varepsilon y),$$

belongs to $\mathbf{L}_0$. Using it in (13.68), gives

$$\int_{Y^\#} A(y)\nabla_y\overline{\Psi}(y)\,\varepsilon\Big(\sum_{\xi'\in\mathbb{H}}\nabla_x(G\phi)(\varepsilon\xi' + \varepsilon y)\Big)\,dy = 0,$$

or equivalently (after multiplication by ε^{N-1})

$$\int_{\mathbb{R}^{N-1}\times Y^\#} A(y)\nabla_y\overline{\Psi}(y)\,\varepsilon\mathcal{T}^\#_\varepsilon\big(\nabla_x(G\phi)\big)(x',y)\,dx'dy = 0.$$

Making an inverse unfolding yields

$$\int_\Omega A^\varepsilon(x)\nabla\overline{\Psi}\Big(\frac{x}{\varepsilon}\Big)\,\nabla(G\phi)(x)\,dx = 0.$$

Hence,

$$\begin{aligned}&\int_\Omega A^\varepsilon\nabla\Big(G\overline{\Psi}\Big(\frac{\cdot}{\varepsilon}\Big)\Big)\,\nabla\phi\,dx\\ &\qquad = -\int_\Omega A^\varepsilon\nabla\overline{\Psi}\Big(\frac{\cdot}{\varepsilon}\Big)\,\phi\nabla G\,dx + \int_\Omega A^\varepsilon\overline{\Psi}\Big(\frac{\cdot}{\varepsilon}\Big)\nabla G\,\nabla\phi\,dx.\end{aligned}\tag{13.76}$$

Since ϕ belongs to $H_0^1(\Omega)$,

$$\left\|\frac{\phi}{\rho_{\mathbf{H}}}\right\|_{L^2(\Omega)} \leqslant \left\|\frac{\phi}{\rho}\right\|_{L^2(\Omega)} \leqslant C\|\nabla\phi\|_{L^2(\Omega)}.$$

This estimate and (13.75) give

$$\left|\int_\Omega A^\varepsilon \nabla\overline{\Psi}\Big(\frac{\cdot}{\varepsilon}\Big)\,\phi\nabla G\,dx\right| \leqslant C\varepsilon^{1/2}\|\nabla\phi\|_{L^2(\Omega)}\|\nabla G\|_{L^\infty(\Omega)}. \tag{13.77}$$

Now, recall that

$$\int_\Omega A^\varepsilon \nabla u^\varepsilon_{\Psi,G}\nabla\phi\,dx = 0, \quad \forall\phi\in H_0^1(\Omega).$$

This equality, together with (13.76) and (13.77), yields

$$\begin{aligned}\left|\int_\Omega A^\varepsilon\nabla\Big(u^\varepsilon_{\Psi,G} - \overline{\Psi}\Big(\frac{\cdot}{\varepsilon}\Big)G\Big)\nabla\phi\,dx + \int_\Omega A^\varepsilon\overline{\Psi}\Big(\frac{\cdot}{\varepsilon}\Big)\nabla G\,\nabla\phi\,dx\right| \\ \leqslant C\varepsilon^{1/2}\|\nabla\phi\|_{L^2(\Omega)}\|\nabla G\|_{L^\infty(\Omega)}.\end{aligned} \tag{13.78}$$

On the one hand,

$$w_\varepsilon \doteq u^\varepsilon_{\Psi,G} - G\overline{\Psi}\Big(\frac{\cdot}{\varepsilon}\Big) \in H_0^1(\Omega).$$

On the other hand, estimates $(13.75)_2$ and (13.78) show that the sequence $\{w_\varepsilon\}_\varepsilon$ is uniformly bounded in $H_0^1(\Omega)$.

Step 2. Asymptotic behavior of the sequence $\{w_\varepsilon\}_\varepsilon$.

Again from $(13.75)_2$ one has, for every $\zeta > 0$, the following convergences:

$$\begin{aligned}&\overline{\Psi}\Big(\frac{\cdot}{\varepsilon}\Big) \rightharpoonup \overline{\Psi}(\infty) \quad \text{weakly in } L^2(\Omega),\\ &\overline{\Psi}\Big(\frac{\cdot}{\varepsilon}\Big) \to \overline{\Psi}(\infty) \quad \text{strongly in } L^2(\Omega\cap\{x_N > \zeta\}).\end{aligned}$$

Now, proceeding as in Section 3.1, we obtain that there exist w and $\widehat{w}$ in $L^2(\Omega; H^1_{per,0}(Y))$ such that

$$\begin{aligned}&w_\varepsilon \rightharpoonup w \quad \text{weakly in } H_0^1(\Omega),\\ &\mathcal{T}_\varepsilon(\nabla w_\varepsilon) \to \nabla w + \nabla_y\widehat{w} \quad \text{strongly in } L^2(\Omega\times Y)^N.\end{aligned}$$

The pair $(w,\widehat{w}) \in H_0^1(\Omega)\times L^2(\Omega; H^1_{per,0}(Y))$ is the unique solution of the following variational problem:

$$\begin{cases}\displaystyle\int_{\Omega\times Y} A(y)\Big(\nabla w(x) + \nabla_y\widehat{w}(x,y)\Big)\big(\nabla\phi(x) + \nabla_y\widehat{\phi}(x,y)\big)\,dxdy\\ \displaystyle\quad + \int_{\Omega\times Y} A(y)\,\overline{\Psi}(\infty)\,\nabla G(x)\big(\nabla\phi(x) + \nabla_y\widehat{\phi}(x,y)\big)\,dxdy = 0,\\ \forall\phi\in H_0^1(\Omega),\ \ \forall\widehat{\phi}\in L^2(\Omega; H^1_{per,0}(Y)).\end{cases} \tag{13.79}$$

As in (3.21), we express $\widehat{w}$ in terms of the $\widehat{\chi}_i$'s, ∇w and ∇G, as follows:

$$\widehat{w}(x,y) = \sum_{i=1}^{N} \Big(\frac{\partial w}{\partial x_i}(x) + \overline{\Psi}(\infty) \frac{\partial G}{\partial x_i}(x) \Big) \widehat{\chi}_i(x,y).$$

Replacing this expression in (13.79), gives

$$\int_\Omega A^{hom} \nabla \big(w + \overline{\Psi}(\infty) G \big) \nabla \phi \, dx = 0, \quad \forall \phi \in H_0^1(\Omega),$$

so that the function $w + \overline{\Psi}(\infty) G$ belongs to $\mathbf{W}_{A^{hom}}$.

Step 3. Asymptotic behavior of the sequence $\Big\{ G\overline{\Psi}\Big(\frac{\cdot}{\varepsilon}\Big) \Big\}_\varepsilon$.

For every $\zeta > 0$, estimates $(13.75)_1$ imply the convergences,

$$\begin{aligned} &\rho_{\mathrm{H}} \nabla \overline{\Psi}\Big(\frac{\cdot}{\varepsilon}\Big) \to 0 \quad \text{strongly in } L^2(\Omega)^N, \\ &\nabla \overline{\Psi}\Big(\frac{\cdot}{\varepsilon}\Big) \to 0 \quad \text{strongly in } L^2(\Omega \cap \{x_N > \zeta\})^N. \end{aligned}$$

Besides, from $(13.75)_2$, one also has

$$G\overline{\Psi}\Big(\frac{\cdot}{\varepsilon}\Big) \rightharpoonup G\overline{\Psi}(\infty) \quad \text{weakly in } H_\rho^1(\Omega).$$

Finally, since the limit function $w = u^{\#}_{\Psi,G} - \overline{\Psi}(\infty)G$ belongs to $H_0^1(\Omega)$, the trace of $\overline{\Psi}(\infty)G$ is equal to the trace of $u^{\#}_{\Psi,G}$ in $H^{-1/2}(\partial\Omega)$. Hence

$$\mathbb{T}_{A^{hom}}(u^{\#}_{\Psi,G}) = G_{|\Gamma} \theta_\Psi = G_{|\Gamma} \overline{\Psi}(\infty) \quad \text{in} \quad H^{-1/2}(\Gamma).$$

Thus $\theta_\Psi = \overline{\Psi}(\infty)$ in $H^{-1/2}(\Gamma)$. □

Remark 13.29. *Assume that the matrix field A satisfies*

$$a_{Nj} = 0 \ \text{ a.e. in } Y \ \text{ for } j = 1, \dots, N-1, \quad \text{and } a_{NN} \text{ depends only on } \ y_N.$$

From (13.73) *it follows that*

$$\mathcal{M}_{Y'}(\Psi) = \mathcal{M}_{Y'}(\overline{\Psi})(y_N) = \overline{\Psi}(\infty), \qquad \forall y_N \in [0, +\infty).$$

Hence, for the sequence $\{u^\varepsilon_{\Psi,1}\}_\varepsilon$, the limit Dirichlet boundary condition which is $\mathcal{M}_{Y'}(\Psi)$ on Γ (for every limit point) coincides with that of u_Ψ on Γ.

In the general case, these two boundary conditions can be distinct, which is an effect of the homogenization process in Ω.

13.8 A variant of problem (13.8)

We briefly mention a variant of the results given in Proposition 13.3 and Theorem 13.17 which could be of interest. It uses a different boundary condition with Ψ more general. For Ψ in $H^1_{per}(Y)$ and G in $H^1(\Omega)$, $\big[\Psi\big(\frac{\cdot}{\varepsilon}\big)\, G\big]_{|\partial\Omega}$ does not necessarily belongs to $H^{1/2}(\partial\Omega)$. Here, this is compensated by replacing G by $\mathcal{Q}_\varepsilon(G)$. Corollary 1.72 implies the equivalent of Lemma 13.2 in this setting:

Lemma 13.30. *For every* $(\Psi, G) \in H^1_{per}(Y) \times H^1(\Omega)$, *one has*

$$\Big\|\Psi\Big(\frac{\cdot}{\varepsilon}\Big)\mathcal{Q}_\varepsilon(G)\Big\|_{L^2(\partial\Omega)} \leqslant C\|\Psi\|_{H^1(Y)}\big\{\|G\|_{L^2(\partial\Omega)} + \varepsilon^{1/2}\|\nabla G\|_{L^2(\Omega)}\big\},$$
$$\Big\|\Psi\Big(\frac{\cdot}{\varepsilon}\Big)\mathcal{Q}_\varepsilon(G)\Big\|_{H^{1/2}(\partial\Omega)} \leqslant C\varepsilon^{-1/2}\|\Psi\|_{H^1(Y)}\big\{\|G\|_{L^2(\partial\Omega)} + \varepsilon^{1/2}\|\nabla G\|_{L^2(\Omega)}\big\}.$$

In particular

$$\Big\|\Psi\Big(\frac{\cdot}{\varepsilon}\Big)\Big\|_{L^2(\partial\Omega)} \leqslant C\|\Psi\|_{H^1(Y)}.$$

The constants do not depend on ε.

Now, problem (13.8) can be replaced by

$$\begin{cases} \operatorname{div}(A^\varepsilon \nabla v^\varepsilon_{\Psi,G}) = 0 & \text{in } \Omega, \\ v^\varepsilon_{\Psi,G} = \Psi\Big(\dfrac{\cdot}{\varepsilon}\Big)\mathcal{Q}_\varepsilon(G) & \text{on } \partial\Omega. \end{cases} \tag{13.80}$$

Proposition 13.31 (Compare with Theorem 13.17)**.** *Problem* (13.80) *admits a unique solution satisfying (see* (13.30)*)*

$$\|v^\varepsilon_{\Psi,G}\|_{H^1_\rho(\Omega)} \leqslant C\|\Psi\|_{H^1(Y)}\big(\|G\|_{L^2(\partial\Omega)} + \varepsilon^{1/2}\|\nabla G\|_{L^2(\Omega)}\big).$$

The constant is independent of ε.

Consider a subsequence (still denoted ε*) such that*

$$v^\varepsilon_{\Psi,1} \rightharpoonup v_\Psi \quad \textit{weakly in } H^1_\rho(\Omega).$$

Then, for every $G \in H^1(\Omega)$, *there exists* $v_{\Psi,G} \in H^1_\rho(\Omega)$ *such that (for the same subsequence)*

$$v^\varepsilon_{\Psi,G} \rightharpoonup v_{\Psi,G} \quad \textit{weakly in } H^1_\rho(\Omega).$$

The limit function $v_{\Psi,G}$ *belongs to the space* $\mathbf{W}_{A^{hom}}$ *and is characterized as the solution of problem*

$$\begin{cases} \operatorname{div}(A^{hom}\nabla v_{\Psi,G}) = 0 & \textit{in } \Omega, \\ v_{\Psi,G} = \theta_\Psi\, G_{|\partial\Omega} & \textit{on } \partial\Omega. \end{cases}$$

Here $\theta_\Psi = \mathbb{T}_{A^{hom}}(v_\Psi)$ *belongs to* $L^2(\partial\Omega)$ *and satisfies*

$$\|\theta_\Psi\|_{L^2(\partial\Omega)} \leqslant C\|\Psi\|_{H^1(Y)}. \tag{13.81}$$

Proof. The proof is similar to that of Theorem 13.17. For the proof of estimate (13.81) use (13.24) for $G \in W^{1,\infty}(\Omega)$, to obtain

$$\|\theta_\Psi G_{|\partial\Omega}\|_{H^{-1/2}(\partial\Omega)} = \|\mathbb{T}_{A^{hom}}(u_{\Psi,G})\|_{H^{-1/2}(\partial\Omega)} \leqslant C\|u_{\Psi,G}\|_{H^1_\rho(\Omega)}.$$

This, together with (13.55), implies that

$$\|\theta_\Psi G_{|\partial\Omega}\|_{H^{-1/2}(\partial\Omega)} \leqslant C\|G\|_{L^2(\partial\Omega)}\|\Psi\|_{H^1(Y)}.$$

Consequently, the (linear) form

$$G_{|\partial\Omega} \quad \mapsto \quad \langle \theta_\Psi, G_{|\partial\Omega}\rangle_{H^{-1/2}(\partial\Omega),H^{1/2}(\partial\Omega)} \equiv \langle \theta_\Psi\, G_{|\partial\Omega}, 1\rangle_{H^{-1/2}(\partial\Omega),H^{1/2}(\partial\Omega)},$$

extends uniquely to a linear form on $L^2(\partial\Omega)$ and prove (13.81). □

Remark 13.32 (see Remark 13.29). *From Lemma 13.30, it is easily seen that the sequence* $\{\Psi(\frac{\cdot}{\varepsilon})_{|\partial\Omega}\, \mathcal{Q}_\varepsilon(G)_{|\partial\Omega}\}_\varepsilon$ *is bounded in* $L^2(\partial\Omega)$. *Its weak limit points are the product of* $G_{|\partial\Omega}$ *with the weak limit points of* $\{\Psi(\frac{\cdot}{\varepsilon})_{|\partial\Omega}\}_\varepsilon$, *the latter being actually in* $L^\infty(\partial\Omega)$. *As in the preceding case, these weak limit points are not necessarily the traces of the weak limit points of the sequence* $\{v^\varepsilon_{\Psi,G}\}_\varepsilon$. *This gap is an effect of the homogenization process.*

13.9 Complements

In this section, p belongs to $(1,+\infty)$ and Ω is a bounded domain with a Lipschitz boundary.

13.9.1 Projection operators on spaces of periodic functions

In the following $\mathbf{X}$ denotes a separable reflexive Banach space. The Banach space $W^{1-1/p,p}(Y_k;\mathbf{X})$ is endowed with the norm

$$\|\Psi\|_{W^{1-1/p,p}(Y_k;\mathbf{X})} = \Big(\int_{Y_k}\|\Psi(y)\|^p_{\mathbf{X}}dy + \int_{Y_k\times Y_k}\frac{\|\Psi(y)-\Psi(y')\|^p_{\mathbf{X}}}{|y-y'|^{p+N-2}}dydy'\Big)^{1/p}.$$

Definition 13.33. *For* ϕ *in* $W^{1,p}(Y;\mathbf{X})$, *we call "periodic defect" of* ϕ *the following quantity:*

$$D_{\mathbf{X}}(\phi) \doteq \sum_{j=1}^N \|\phi_{|\mathbf{b}_j+Y_j} - \phi_{|Y_j}\|_{W^{1-1/p,p}(Y_j;\mathbf{X})},$$

which is a continuous semi-norm on $W^{1,p}(Y;\mathbf{X})$ *whose kernel is* $W^{1,p}_{per}(Y;\mathbf{X})$.

The aim of this section is to prove that the distance between a function f in $W^{1,p}(Y;\mathbf{X})$ and the space of periodic functions $W^{1,p}_{per}(Y;\mathbf{X})$ is bounded above by constant times the periodic defect $D_{\mathbf{X}}(f)$ of f. In Proposition 13.34 we investigate the case $\mathbf{X}=\mathbb{R}$, then in Proposition 13.38 we consider the cases $\mathbf{X}=L^p(\mathbb{R}^N)$ and $\mathbf{X}=W^{-1,p}(\mathbb{R}^N)$.

Proposition 13.34. *There exists a continuous projection*

$$\mathbf{p} : W^{1,p}(Y) \longrightarrow W^{1,p}_{per}(Y),$$

and a constant C (depending only upon p and Y) such that

$$\|\phi - \mathbf{p}(\phi)\|_{W^{1,p}(Y)} \leqslant C\, D_{\mathbb{R}}(\phi),$$

for all $\phi \in W^{1,p}(Y)$.

For $p = 2$, the projection $\mathbf{p}$ can be replaced by an orthogonal projection from $H^1(Y)$ onto $H^1_{per}(Y)$. This case has been investigated in [120].

For the proof of Proposition 13.34 and in the sequel, we need some preliminary notations and results. For $k \in \{0, \dots, N\}$, introduce the spaces

$$\begin{aligned} W_0 &= W^{1,p}(Y),\\ W_k &= \Big\{\phi \in W^{1,p}(Y) \mid \phi(y) = \phi(y + \mathbf{b}_i) \ \text{ for a.e. } \ y \in Y_i, \quad i \in \{1, \dots, k\}\Big\},\\ W_N &= W^{1,p}_{per}(Y). \end{aligned}$$

Let $\{\mathbb{P}_k\}_{1\leqslant k\leqslant N}$ be a family of parallelotopes defined by (see Figures 13.1 and 13.1 bis in the case $N = 3$)

$$\begin{cases} \mathbb{P}_1 = (0, 2) \times (-1, 2)^{N-1},\\ \mathbb{P}_k = (-1, 2)^{k-1} \times (0, 2) \times (-1, 2)^{N-k} \quad 2 \leqslant k \leqslant N - 1,\\ \mathbb{P}_N = (-1, 2)^{N-1} \times (0, 2). \end{cases}$$

The face $y_k = 0$ of $\mathbb{P}_k$, denoted $\mathcal{Y}_k$, plays a special role. It contains at its center the face Y_k of Y, surrounded by $2^{N-1} - 1$ of its copies.

The Banach spaces $W^{1-1/p,p}(\partial\mathbb{P}_k)$ and $W^{1-1/p,p}(Y_k)$ that are used widely in this chapter, are equipped with their standard norms.

Recall that for every $k \in \{1, \dots, N\}$, there exists a continuous lifting operator

$$r_k : W^{1-1/p,p}(\partial\mathbb{P}_k) \mapsto W^{1,p}(\mathbb{P}_k),$$

with

$$\|r_k(\Phi_{|\partial\mathbb{P}_k})\|_{W^{1,p}(\mathbb{P}_k)} \leqslant C\|\Phi_{|\partial\mathbb{P}_k}\|_{W^{1-1/p,p}(\partial\mathbb{P}_k)}.$$

Consider the subspace $Tr(W_{k-1})_{|Y_k}$ of $W^{1-1/p,p}(Y_k)$, consisting of the traces on Y_k of the elements of W_{k-1}, $k \in \{1, \dots, N\}$.

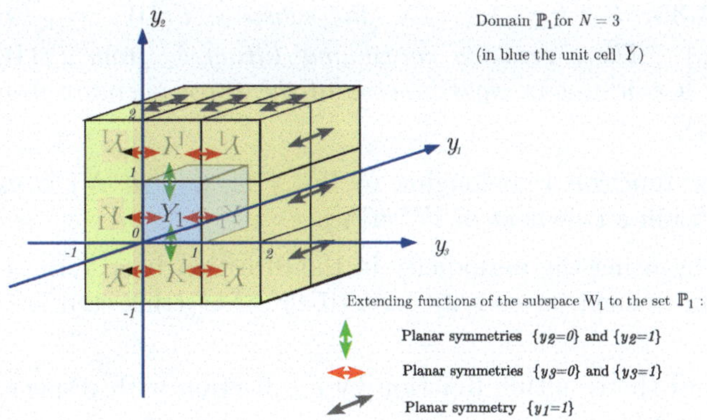

Figure 13.1: The parallelotope $\mathbb{P}_1$ for $N = 3$

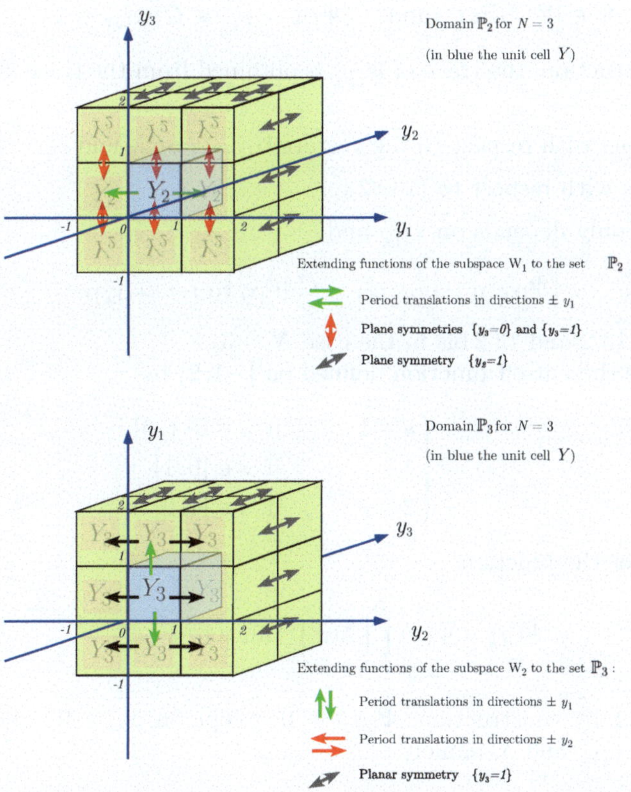

Figure 13.1 bis: The parallelotopes $\mathbb{P}_k$, $k = 2$ and 3, for $N = 3$

Lemma 13.35. *For $k = 1,\dots,N$, the subspace $Tr(W_{k-1})_{|Y_k}$ is closed in $W^{1-1/p,p}(Y_k)$. There exists a continuous lifting $\tilde{r}_k$ from $Tr(W_{k-1})_{|Y_k}$ to W_{k-1} (i.e. a continuous right-inverse of the trace operator from W_{k-1} to $W^{1-1/p,p}(Y_k)$).*

Proof. Every function ψ belonging to W_{k-1} ($1 \leqslant k \leqslant N$) is extended as follows to obtain an element of $W^{1,p}(\mathbb{P}_k)$:

• if $k \geqslant 2$, by using the periodicity in the direction $\mathbf{b}_1,\dots,\mathbf{b}_{k-1}$, extend ψ in a function defined on $(-1,2)^{k-1} \times (0,1)^{N-k+1}$ (this step is omitted for $k = 1$),

• then extend the resulting function by a reflection with respect to the hyperplane $y_k = 1$,

• finally, if $k < N$, extend by reflections with respect to

$$y_{k+1} = 0,\ \ y_{k+1} = 1,\ \dots,\ y_N = 0,\ \ y_N = 1. \tag{13.82}$$

The result is a function Ψ such that

$$\Psi \in W^{1,p}(\mathbb{P}_k) \quad \text{and} \quad \|\Psi\|_{W^{1,p}(\mathbb{P}_k)} \leqslant C\|\psi\|_{W^{1,p}(Y)}.$$

By construction, the trace of $\Psi_{|\mathcal{Y}_k}$ is obtained from the trace $\Psi_{|Y_k} = \psi_{|Y_k}$ by performing

• translations with respect to the vectors $\mathbf{b}_1,\dots,\mathbf{b}_{k-1}$ and

• reflections with respect to (13.82).

Hence $\Psi_{|Y_k}$ only depends on $\psi_{|Y_k}$ and

$$\|\Psi_{|Y_k}\|_{W^{1-1/p,p}(\mathcal{Y}_k)} \leqslant C\,\|\psi_{|Y_k}\|_{W^{1-1/p,p}(Y_k)},$$

(see Figure 13.2 and 13.2 bis in the case $N = 3$).

Let λ be the cut-off function defined on $[-1,2]$ by

$$\lambda(s) = \begin{cases} s+1 & \text{if } s \in [-1,0], \\ 1 & \text{if } s \in [0,1], \\ 2-s & \text{if } s \in [1,2]. \end{cases}$$

Now consider the function

$$\widetilde{\Psi}(y) = \Psi(y)\prod_{i=1}^{N}\lambda(y_i) \quad \text{for a.e. } y \in \mathbb{P}_k.$$

Observe that by construction, $\widetilde{\Psi}_{|\partial\mathbb{P}_k} = 0$ except on $\mathcal{Y}_k = 0$, where it only depends on $\psi_{|Y_k}$ and λ, namely

$$\widetilde{\Psi}_{|\mathcal{Y}_k}(y) = \Psi(y)\prod_{i=1, i\neq k}^{N}\lambda(y_i) \quad \text{for a.e. } y \in \mathcal{Y}_k.$$

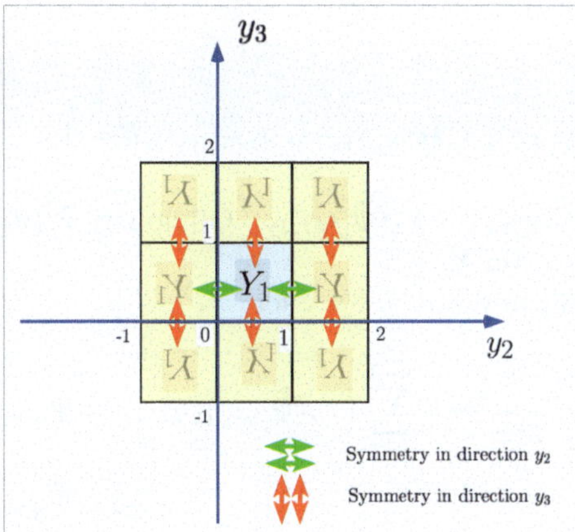

Figure 13.2: The face $\mathcal{Y}_1$ of the parallelotope $\mathbb{P}_1$ for $N = 3$

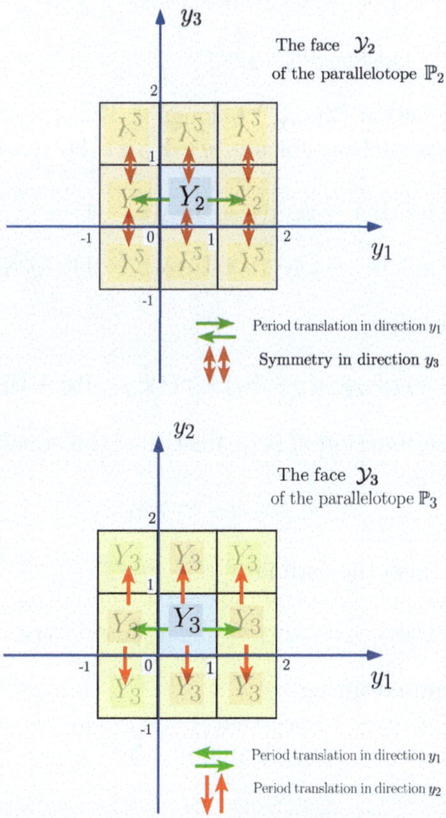

Figure 13.2 bis: The face $\mathcal{Y}_k$ of the parallelotope $\mathbb{P}_k$ ($k = 2, 3$) for $N = 3$

Therefore,

$$\|\widetilde{\Psi}\|_{W^{1-1/p,p}(\partial\mathbb{P}_k)} \leqslant C\|\psi_{|Y_k}\|_{W^{1-1/p,p}(Y_k)}, \quad k = 1,\ldots,N. \tag{13.83}$$

For $k = 1$, set $\widetilde{r}_1(\psi_{|Y_1}) = r_1(\widetilde{\Psi}_{|\partial\mathbb{P}_1})_{|Y}$. Obviously $\widetilde{r}_1(\psi_{|Y_1})$ belongs to W_0 and $\widetilde{r}_1(\psi_{|Y_1})_{|Y_1} = \psi_{|Y_1}$ on Y_1.

For $k \in \{2,\ldots,N\}$ and a.e. $y \in Y$, set

$$\widetilde{r}_k(\psi_{|Y_k})(y) = \frac{1}{2(k-1)}\sum_{i=1}^{k-1}\Big(r_k(\widetilde{\Psi}_{|\partial\mathbb{P}_k})(y) + r_k(\widetilde{\Psi}_{|\partial\mathbb{P}_k})(y-\mathbf{b}_i) + r_k(\widetilde{\Psi}_{|\partial\mathbb{P}_k})(y+\mathbf{b}_i)\Big).$$

By the definition of the map r_k and the properties of $\widetilde{\Psi}$,

$$\begin{aligned} &r_k(\widetilde{\Psi}_{|\partial\mathbb{P}_k})_{|y_\ell=-1} = 0,\\ &r_k(\widetilde{\Psi}_{|\partial\mathbb{P}_k})_{|y_\ell=0} = r_k(\widetilde{\Psi}_{|\partial\mathbb{P}_k})_{|y_\ell=1},\\ &r_k(\widetilde{\Psi}_{|\partial\mathbb{P}_k})_{|y_\ell=2} = 0. \end{aligned}$$

Consequently, the function $\widetilde{r}_k(\psi_{|Y_k})$ belongs to W_{k-1}.

Moreover, by construction, for a.e. $y \in Y_k$, $i \in \{1,\ldots,k-1\}$,

$$\begin{aligned} r_k(\widetilde{\Psi}_{|\partial\mathbb{P}_k})_{|y_k=0}(y-\mathbf{b}_i) &= \psi_{|Y_k}(y-\mathbf{b}_i)\lambda(y_i-1) = \psi_{|Y_k}(y)\lambda(y_i-1),\\ r_k(\widetilde{\Psi}_{|\partial\mathbb{P}_k})_{|y_k=0}(y+\mathbf{b}_i) &= \psi_{|Y_k}(y+\mathbf{b}_i)\lambda(y_i+1) = \psi_{|Y_k}(y)\lambda(y_i+1), \end{aligned}$$

from which it follows that

$$r_k(\widetilde{\Psi}_{|\partial\mathbb{P}_k})(y) + r_k(\widetilde{\Psi}_{|\partial\mathbb{P}_k})(y-\mathbf{b}_i) + r_k(\widetilde{\Psi}_{|\partial\mathbb{P}_k})(y+\mathbf{b}_i) = 2\psi_{|Y_k}(y).$$

This implies that the function $\widetilde{r}_k(\psi_{|Y_k})$ satisfies the equality

$$\widetilde{r}_k(\psi_{|Y_k})_{|Y_k} = \psi_{|Y_k} \tag{13.84}$$

and, due to (13.83), also the estimate

$$\|\widetilde{r}_k(\psi_{|Y_k})\|_{W^{1,p}(Y)} \leqslant C\|\psi_{|Y_k}\|_{W^{1-1/p,p}(Y_k)},$$

so that, it is the required lifting.

At this point, it is easy to see that $Tr(W_{k-1})_{|Y_k}$ is complete (indeed, a Cauchy sequence in $Tr(W_{k-1})_{|Y_k}$ is mapped by $\widetilde{r}_k$ on a Cauchy sequence in the space W_{k-1}!). □

Proposition 13.34 will be obtained by combining the following $N-1$ elementary projection results, which are of interest in their own right.

Lemma 13.36. *For $k \in \{0, \dots, N-1\}$, there is a continuous linear projection $\mathbf{q}_{k+1}$ from W_k to W_{k+1}, and a constant C such that, for every ψ in W_k,*

$$\|\psi - \mathbf{q}_{k+1}(\psi)\|_{W^{1,p}(Y)} \leqslant C\|\psi_{|\mathbf{b}_{k+1}+Y_{k+1}} - \psi_{|Y_{k+1}}\|_{W^{1-1/p,p}(Y_{k+1})}. \quad (13.85)$$

Proof. *Step 1.* For $k = 0, \dots, N$, denote by S_k the isometry on $W^{1,p}(Y)$ which to ϕ assigns its flip $S_k(\phi)$ defined by

$$S_k(\phi)(y) \doteq \phi(y_1, \dots, 1 - y_k, \dots, y_N),$$

and set

$$L_k(\phi)(y) \doteq (1 - y_k)\phi(y) - y_k S_k(\phi)(y).$$

The map L_k is linear and continuous from $W^{1,p}(Y)$ into itself. Clearly, S_k^2 is the identity, and

$$S_k(\phi)_{|Y_k} = \phi_{|\mathbf{b}_k+Y_k} \quad \text{and} \quad S_k(\phi)_{|\mathbf{b}_k+Y_k} = \phi_{|Y_k}. \quad (13.86)$$

Furthermore, S_k maps every $W_j (j = 0, \dots N)$ into itself, while L_k maps every $W_j (j = 0, \dots k-1)$ into itself. Note that L_k does not map any space W_j into itself because the straightforward identity $S_k \circ L_k \equiv -L_k$, implies that $L_k(\phi)$ is anti-periodic with respect to y_k for every ϕ in $W^{1,p}(Y)$.

Note also that

$$L_k(\phi)_{|Y_k} \equiv \phi_{|Y_k}. \quad (13.87)$$

Step 2. For ψ in W_k, set

$$\mathbf{q}_{k+1}(\psi) \doteq \psi + \frac{1}{2} L_{k+1} \circ \tilde{r}_{k+1}\big((S_{k+1}(\psi) - \psi)_{|Y_{k+1}}\big).$$

Obviously, $\mathbf{q}_{k+1}$ is linear continuous from W_k into itself. Moreover, (13.87) together with (13.86) and (13.84), imply estimate (13.85).

It remains to check that the range of $\mathbf{q}_{k+1}$ is in W_{k+1}, that is, to show that for every ψ in W_k, $\mathbf{q}_{k+1}(\psi)$ is periodic in the direction $\mathbf{b}_{k+1}$. Indeed, by (13.84), (13.86) and (13.87) again,

$$\begin{aligned}\mathbf{q}_{k+1}(\psi)_{|Y_{k+1}} &= \psi_{|Y_{k+1}} + \frac{1}{2}(\psi_{|\mathbf{b}_{k+1}+Y_{k+1}} - \psi_{|Y_{k+1}}) \\ &= \frac{1}{2}(\psi_{|\mathbf{b}_{k+1}+Y_{k+1}} + \psi_{|Y_{k+1}}),\end{aligned}$$

while

$$\begin{aligned}\mathbf{q}_{k+1}(\psi)_{|\mathbf{b}_{k+1}+Y_{k+1}} &= \psi_{|\mathbf{b}_{k+1}+Y_{k+1}} - \frac{1}{2}(\psi_{|\mathbf{b}_{k+1}+Y_{k+1}} - \psi_{|Y_{k+1}}) \\ &= \frac{1}{2}(\psi_{|\mathbf{b}_{k+1}+Y_{k+1}} + \psi_{|Y_{k+1}}).\end{aligned}$$

Hence $\mathbf{q}_{k+1}(\psi)$ belongs to W_k. □

Proof of Proposition 13.34. At this point, it is straightforward to check that for $0 \leqslant i < j \leqslant N$, the map $\mathbf{p}_{ij} \doteq \mathbf{q}_j \circ \cdots \circ \mathbf{q}_{i+1}$ is a linear continuous projection from W_i to W_j and satisfies

$$\|\phi - \mathbf{p}_{ij}(\phi)\|_{W^{1,p}(Y)} \leqslant C \sum_{k=i}^{j-1} \|\phi_{|\mathbf{q}_{k+1}+Y_{k+1}} - \phi_{|Y_{k+1}}\|_{W^{1-1/p,p}(Y_{k+1})}.$$

Finally set $\mathbf{p} \doteq \mathbf{p}_{0N}$, this is the required projection. □

Corollary 13.37. *The projection* $\mathbf{p}$ *defined in Proposition 13.34 induces continuous projections from*

$$W^{k,p}(\mathbb{R}^N; W^{1,p}(Y)) \ \ to \ \ W^{k,p}(\mathbb{R}^N; W^{1,p}_{per}(Y)), \quad k \in \mathbb{N}.$$

We still denote these continuous projections $\mathbf{p}$. *We easily obtain from Proposition 13.34, that for every function* f *in* $W^{k,p}(\mathbb{R}^N; W^{1,p}(Y))$

$$\|f - \mathbf{p}(f)\|_{W^{k,p}(\mathbb{R}^N;W^{1,p}(Y))} \leqslant C\, D_{W^{k,p}(\mathbb{R}^N)}(f). \tag{13.88}$$

Moreover,

$$\begin{aligned} &(i)\ \mathbf{p}\Big(\int_{\mathbb{R}^N} \phi(x,\cdot)\,dx\Big) = \int_{\mathbb{R}^N} \mathbf{p}\big(\phi(x,\cdot)\big)\,dx, \qquad \forall \phi \in L^1(\mathbb{R}^N; W^{1,p}(Y)),\\ &(ii)\ \mathbf{p}(\nabla_x \phi) = \nabla_x \mathbf{p}(\phi), \qquad \forall \phi \in W^{1,p}(\mathbb{R}^N; W^{1,p}(Y)). \end{aligned} \tag{13.89}$$

The next proposition plays a crucial role in the error estimates.

Proposition 13.38. *For every function* f *in* $L^p(\mathbb{R}^N; W^{1,p}(Y))$, *its projection* $\mathbf{p}(f)$ *belonging to* $L^p(\mathbb{R}^N; W^{1,p}_{per}(Y))$, *satisfies*

$$\|f - \mathbf{p}(f)\|_{W^{1,p}(Y;W^{-1,p}(\mathbb{R}^N))} \leqslant C\, D_{W^{-1,p}(\mathbb{R}^N)}(f). \tag{13.90}$$

The constant depends only on N *and* p.

Moreover, if ω *is an open subset of* $\mathbb{R}^N$ *such that* $f(x,y) = 0$ *for a.e.* $(x,y) \in \omega \times Y$, *then* $\mathbf{p}(f)(x,y) = 0$ *for a.e.* $(x,y) \in \omega \times Y$.

For the proof, recall the classical result (see [114]): for every function f in $W^{-1,p}(\mathbb{R}^N)$, there exists a unique solution $w \in W^{1,p}(\mathbb{R}^N)$ of the equation

$$-\Delta w + w = f \quad \text{in } \mathcal{D}'(\mathbb{R}^N).$$

It satisfies the estimate,

$$\|w\|_{W^{1,p}(\mathbb{R}^N)} \leqslant C\|f\|_{W^{-1,p}(\mathbb{R}^N)}.$$

Moreover, if $f \in L^p(\mathbb{R}^N)$, then

$$\|w\|_{W^{2,p}(\mathbb{R}^N)} \leqslant C\|f\|_{L^p(\mathbb{R}^N)}.$$

As a consequence, we obtain the following lemma:

Lemma 13.39. *For every $f \in W^{1,p}(Y; W^{-1,p}(\mathbb{R}^N))$, there exists a unique solution $w \in W^{1,p}(Y; W^{1,p}(\mathbb{R}^N))$ of the equation,*

$$-\Delta_x w(y) + w(y) = f(y) \quad \text{in } \mathcal{D}'(\mathbb{R}^N) \quad \text{for a.e. } y \in Y. \tag{13.91}$$

It satisfies for $j = 1, \dots, N$, the a priori estimates

$$\begin{aligned} &\|w\|_{W^{1,p}(Y;W^{1,p}(\mathbb{R}^N))} \leqslant C\|f\|_{W^{1,p}(Y;W^{-1,p}(\mathbb{R}^N))}, \\ &\|w_{|Y_j}\|_{W^{1-1/p,p}(Y_j;W^{1,p}(\mathbb{R}^N))} \leqslant C\|f_{|Y_j}\|_{W^{1-1/p,p}(Y_j;W^{-1,p}(\mathbb{R}^N))}. \end{aligned} \tag{13.92}$$

Moreover, if $f \in L^p(\mathbb{R}^N; W^{1,p}(Y))$, then $w \in W^{2,p}(\mathbb{R}^N; W^{1,p}(Y))$ and

$$\|w\|_{W^{2,p}(\mathbb{R}^N;W^{1,p}(Y))} \leqslant C\|f\|_{L^p(\mathbb{R}^N;W^{1,p}(Y))}.$$

The constants depend only on N and p. □

Proof of Proposition 13.38. Let $w \in W^{2,p}(\mathbb{R}^N; W^{1,p}(Y))$ be the solution of problem (13.91). From (13.92), it follows that w satisfies the estimate,

$$\|w\|_{W^{1,p}(Y;W^{1,p}(\mathbb{R}^N))} \leqslant C\|f\|_{W^{1,p}(Y;W^{-1,p}(\mathbb{R}^N))}.$$

By taking the traces of (13.91) on Y_j and $\mathbf{b}_j + Y_j$ for $j \in \{1, \dots, N\}$, gives

$$\begin{aligned} -\Delta_x\big(w_{|\mathbf{b}_j+Y_j} - w_{|Y_j}\big)(y) - \big(w_{|\mathbf{b}_j+Y_j} - w_{|Y_j}\big)(y) \\ = \big(f_{|\mathbf{b}_j+Y_j} - f_{|Y_j}\big)(y) \quad \text{for a.e. } y \in Y_j. \end{aligned} \tag{13.93}$$

On the other hand, since $w \in W^{2,p}(\mathbb{R}^N; W^{1,p}(Y))$, by (13.89) (ii) it follows that $\mathbf{p}(w)$ is the solution of the equation [(3)]

$$-\Delta_x \mathbf{p}(w)(y) + \mathbf{p}(w)(y) = \mathbf{p}(f)(y) \quad \text{in} \quad L^p(\mathbb{R}^N) \quad \text{for a.e. } y \in Y.$$

Due to estimate (13.88), equality (13.93) and Lemma 13.39, $\mathbf{p}(w)$ satisfies

$$\begin{aligned} \|w - \mathbf{p}(w)\|_{W^{1,p}(Y;W^{1,p}(\mathbb{R}^N))} &\leqslant C\sum_{j=1}^{N} \|w_{|\mathbf{b}_j+Y_j} - w_{|Y_j}\|_{W^{1-1/p,p}(Y_j;W^{1,p}(\mathbb{R}^N))} \\ &\leqslant C\sum_{j=1}^{N} \|f_{|\mathbf{b}_j+Y_j} - f_{|Y_j}\|_{W^{1-1/p,p}(Y_j;W^{-1,p}(\mathbb{R}^N))}. \end{aligned}$$

Since Δ_x is a continuous linear operator from $W^{1,p}(\mathbb{R}^N)$ into $W^{-1,p}(\mathbb{R}^N)$,

$$\|\Delta_x(w - \mathbf{p}(w))\|_{W^{1,p}(Y;W^{-1,p}(\mathbb{R}^N))} \leqslant C\|w - \mathbf{p}(w)\|_{W^{1,p}(Y;W^{1,p}(\mathbb{R}^N))}.$$

Therefore, as $\mathbf{p}(f) = \mathbf{p}(w) - \Delta_x \mathbf{p}(w)$ is in $L^p(\mathbb{R}^N; W^{1,p}_{per}(Y))$ and

$$f - \mathbf{p}(f) = \big(w - \mathbf{p}(w)\big) - \Delta_x\big(w - \mathbf{p}(w)\big),$$

(3) This formula can actually be taken as the definition of $\mathbf{p}$ acting on every element of $W^{1,p}(Y; W^{-1,p}(\mathbb{R}^N))$.

it follows that $f - \mathbf{p}(f)$ satisfies estimates (13.90).

To prove the last statement of the proposition, choose in (13.91) a test function $\psi \in \mathcal{D}(\omega)$, to get for a.e. $y \in Y$,

$$\int_{\mathbb{R}^N} \nabla_x w(y) \nabla \psi \, dx + \int_{\mathbb{R}^N} w(y) \psi \, dx = \int_{\mathbb{R}^N} \psi \, f(y) dx = 0.$$

This together with (13.89)(i)-(ii) implies

$$\begin{aligned}\int_{\mathbb{R}^N} \nabla_x \mathbf{p}(w)(y) \nabla \psi \, dx + \int_{\mathbb{R}^N} \mathbf{p}(w)(y) \psi \, dx = 0 \\ = \int_{\mathbb{R}^N} \mathbf{p}(f)(y) \psi \, dx \quad \text{for a.e. } y \in Y,\end{aligned}$$

which concludes the proof of the proposition. □

13.9.2 Projection theorems for unfolded functions on the space of periodic functions

Theorem 13.40. *For $\phi \in W^{1,p}(\mathbb{R}^N)$, there exists $\widehat{\phi}_\varepsilon \in L^p(\mathbb{R}^N; W^{1,p}_{per}(Y))$, such that*

$$\begin{aligned}&\|\widehat{\phi}_\varepsilon\|_{L^p(\mathbb{R}^N; W^{1,p}(Y))} \leqslant C\big(\|\phi\|_{L^p(\mathbb{R}^N)} + \varepsilon \|\nabla \phi\|_{L^p(\mathbb{R}^N)}\big),\\ &\|\widetilde{\mathcal{T}}_\varepsilon(\phi) - \widehat{\phi}_\varepsilon\|_{W^{1,p}(Y; W^{-1,p}(\mathbb{R}^N))} \leqslant C\varepsilon\big(\|\phi\|_{L^p(\mathbb{R}^N)} + \varepsilon \|\nabla \phi\|_{L^p(\mathbb{R}^N)}\big).\end{aligned}$$

The constant depends only on p and Y.

Proof. First, for $j \in \{1, \dots, N\}$, introduce a "double unfolding operator" for ψ in $L^p(\mathbb{R}^N)$(4),

$$\widetilde{\mathcal{T}}_{\varepsilon j}(\psi)(x, y) = \psi\Big(\varepsilon\Big[\frac{x}{\varepsilon}\Big]_Y + \varepsilon y\Big) \quad \text{for a.e. } (x, y) \in \mathbb{R}^N \times \operatorname{int}\big(\overline{Y \cup (Y + \mathbf{b}_j)}\big).$$

We have, for a.e. $(x, y) \in \mathbb{R}^N \times Y$,

$$\begin{aligned}&\widetilde{\mathcal{T}}_{\varepsilon j}(\psi)(x, y) = \widetilde{\mathcal{T}}_\varepsilon(\psi)(x, y),\\ &\widetilde{\mathcal{T}}_{\varepsilon j}(\psi)(x, y + \mathbf{b}_j) = \widetilde{\mathcal{T}}_\varepsilon(\psi)(x + \varepsilon \mathbf{b}_j, y).\end{aligned}$$

These identities and a simple change of variables give for ψ in $\mathcal{D}(\mathbb{R}^N)$ and Ψ in $W^{1,p'}(\mathbb{R}^N)$,

$$\begin{aligned}&\int_{\mathbb{R}^N} \big(\widetilde{\mathcal{T}}_{\varepsilon j}(\psi)(\cdot, y + \mathbf{b}_j) - \widetilde{\mathcal{T}}_{\varepsilon j}(\psi)(\cdot, y)\big) \Psi \, dx\\ &\qquad = \int_{\mathbb{R}^N} \widetilde{\mathcal{T}}_\varepsilon(\psi)(\cdot, y)\big(\Psi(\cdot - \varepsilon \mathbf{b}_j) - \Psi\big) dx \quad \text{for a.e. } y \in Y.\end{aligned}$$

(4) This is similar to the operator $\mathcal{T}_\varepsilon^{\mathcal{Y}*}$ introduced for the proof of Theorem 4.28.

Since Ψ belongs to $W^{1,p'}(\mathbb{R}^N)$,

$$\|\Psi(\cdot - \varepsilon \mathbf{b}_j) - \Psi\|_{L^{p'}(\mathbb{R}^N)} \leqslant C\varepsilon \|\nabla \Psi\|_{L^{p'}(\mathbb{R}^N)}.$$

Thus, for a.e. $y \in Y$,

$$\begin{aligned} &\Big| \int_{\mathbb{R}^N} \big(\widetilde{\mathcal{T}}_{\varepsilon j}(\psi)(\cdot, y + \mathbf{b}_j) - \widetilde{\mathcal{T}}_{\varepsilon j}(\psi)(\cdot, y)\big) \Psi \, dx \Big| \\ &\qquad \leqslant C\varepsilon \|\nabla \Psi\|_{L^{p'}(\mathbb{R}^N)} \big\|\widetilde{\mathcal{T}}_{\varepsilon}(\psi)(\cdot, y)\big\|_{L^p(\mathbb{R}^N)}, \end{aligned}$$

which implies

$$\big\|\widetilde{\mathcal{T}}_{\varepsilon j}(\psi)(\cdot, y + \mathbf{b}_j) - \widetilde{\mathcal{T}}_{\varepsilon j}(\psi)(\cdot, y)\big\|_{W^{-1,p}(\mathbb{R}^N)} \leqslant C\varepsilon \|\widetilde{\mathcal{T}}_{\varepsilon}(\psi)(\cdot, y)\|_{L^p(\mathbb{R}^N)},$$

so that (recall Proposition 1.8(iii)),

$$\big\|\widetilde{\mathcal{T}}_{\varepsilon j}(\psi)(\cdot, \cdot + \mathbf{b}_j) - \widetilde{\mathcal{T}}_{\varepsilon j}(\psi)\big\|_{L^p(Y; W^{-1,p}(\mathbb{R}^N))} \leqslant C\varepsilon \|\psi\|_{L^p(\mathbb{R}^N)}. \tag{13.94}$$

The space $\mathcal{D}(\mathbb{R}^N)$ being dense in $L^p(\mathbb{R}^N)$, this estimate is valid for every $\psi \in L^p(\mathbb{R}^N)$.

Now, let ϕ be in $W^{1,p}(\mathbb{R}^N)$. Applying (13.94) for ϕ and for $\nabla\phi$, we obtain

$$\begin{aligned} &\big\|\widetilde{\mathcal{T}}_{\varepsilon j}(\phi)(\cdot, \cdot + \mathbf{b}_j) - \widetilde{\mathcal{T}}_{\varepsilon j}(\phi)\big\|_{L^p(Y; W^{-1,p}(\mathbb{R}^N))} \leqslant C\varepsilon \|\phi\|_{L^p(\mathbb{R}^N)}, \\ &\big\|\widetilde{\mathcal{T}}_{\varepsilon j}(\nabla\phi)(\cdot, \cdot + \mathbf{b}_j) - \widetilde{\mathcal{T}}_{\varepsilon j}(\nabla\phi)\big\|_{L^p(Y; W^{-1,p}(\mathbb{R}^N)^N)} \leqslant C\varepsilon \|\nabla\phi\|_{L^p(\mathbb{R}^N)}. \end{aligned}$$

Note that, as for $\widetilde{\mathcal{T}}_{\varepsilon}$, the formula $\nabla_y\big(\widetilde{\mathcal{T}}_{\varepsilon j}(\phi)\big) = \varepsilon \widetilde{\mathcal{T}}_{\varepsilon j}(\nabla\phi)$ also holds for $\widetilde{\mathcal{T}}_{\varepsilon j}$. Therefore, the above estimates give ,

$$\begin{aligned} &\big\|\widetilde{\mathcal{T}}_{\varepsilon j}(\phi)(\cdot, \cdot + \mathbf{b}_j) - \widetilde{\mathcal{T}}_{\varepsilon}(\phi)\big\|_{W^{1,p}(Y; W^{-1,p}(\mathbb{R}^N))} \\ &\qquad \leqslant C\varepsilon\big(\|\phi\|_{L^p(\mathbb{R}^N)} + \varepsilon\|\nabla\phi\|_{L^p(\mathbb{R}^N)}\big). \end{aligned}$$

From this inequality we deduce the estimate of the difference of the traces of $y \mapsto \widetilde{\mathcal{T}}_{\varepsilon}(\phi)(\cdot, y)$ on ∂Y, and consequently, on all the faces Y_j,

$$\begin{aligned} &\big\|\widetilde{\mathcal{T}}_{\varepsilon j}(\phi)(\cdot, \cdot + \mathbf{b}_j) - \widetilde{\mathcal{T}}_{\varepsilon j}(\phi)\big\|_{W^{1-1/p,p}(Y_j; W^{-1,p}(\mathbb{R}^N))} \\ &\qquad \leqslant C\varepsilon\big(\|\phi\|_{L^p(\mathbb{R}^N)} + \varepsilon\|\nabla\phi\|_{L^p(\mathbb{R}^N)}\big). \end{aligned}$$

Finally, the periodic defect of the function $y \mapsto \widetilde{\mathcal{T}}_{\varepsilon}(\phi)(\cdot, y)$ is estimated by

$$\begin{aligned} D_{W^{-1,p}(\mathbb{R}^N)}(\widetilde{\mathcal{T}}_{\varepsilon}(\phi)) &= \sum_{j=1}^{N} \big\|\widetilde{\mathcal{T}}_{\varepsilon}(\phi)(\cdot, \cdot + \mathbf{b}_j) - \widetilde{\mathcal{T}}_{\varepsilon}(\phi)\big\|_{W^{1-1/p,p}(Y_j; W^{-1,p}(\mathbb{R}^N))} \\ &\leqslant C\varepsilon\big(\|\phi\|_{L^p(\mathbb{R}^N)} + \varepsilon\|\nabla\phi\|_{L^p(\mathbb{R}^N)}\big). \end{aligned}$$

Choose $\widehat{\phi}_{\varepsilon} = \mathbf{p} \circ \widetilde{\mathcal{T}}_{\varepsilon}(\phi)$. Because $\mathbf{p}$ is a continuous projection from $L^p(\Omega; W^{1,p}(Y))$ onto $L^p(\Omega; W^{1,p}_{per}(Y))$, we obtain the first estimate of the theorem, the second one follows from (13.90) of Proposition 13.38. $\square$

We will need in the following to work in the sets Ω^b_δ and Ω^{bl}_δ which were defined for $\delta > 0$, by (12.1). that we recall here,

$$\Omega^b_\delta \doteq \{x \in \mathbb{R}^N \mid \rho(x) < \delta\},$$
$$\Omega^{bl}_\delta \doteq \{x \in \Omega \mid \rho(x) < \delta\}.$$

Theorem 13.41. *For every $\phi \in W^{1,p}(\Omega)$, there exists $\widehat{\phi}_\varepsilon \in L^p(\Omega; W^{1,p}_{per}(Y))$, such that*

$$\|\widehat{\phi}_\varepsilon\|_{L^p(\Omega;W^{1,p}(Y))} \leqslant C\big(\|\phi\|_{L^p(\Omega)} + \varepsilon\|\nabla\phi\|_{L^p(\Omega)}\big),$$

and

$$\begin{aligned}\|\mathcal{T}_\varepsilon(\phi) - \widehat{\phi}_\varepsilon\|_{W^{1,p}(Y;(W^{1,p}(\Omega))')} &\leqslant C\varepsilon\big(\|\phi\|_{L^p(\Omega)} + \varepsilon\|\nabla\phi\|_{L^p(\Omega)}\big)\\ &\quad + C\varepsilon^{1/p'}\big(\|\phi\|_{L^p(\Omega^{bl}_{2\varepsilon d(Y)})} + \varepsilon\|\nabla\phi\|_{L^p(\Omega^{bl}_{2\varepsilon d(Y)})}\big).\end{aligned}$$

Furthermore,

$$\|\mathcal{T}_\varepsilon(\phi) - \widehat{\phi}_\varepsilon\|_{W^{1,p}(Y;W^{-1,p}(\Omega))} \leqslant C\varepsilon\big(\|\phi\|_{L^p(\Omega)} + \varepsilon\|\nabla\phi\|_{L^p(\Omega)}\big).$$

The constants do not depend on ε.

Proof. With notation recalled above, set

$$\widehat{\rho}_\varepsilon(x) = \sup\Big\{0, \inf\Big(1, \frac{\rho(x) - \varepsilon d(Y)}{\varepsilon d(Y)}\Big)\Big\} \quad \text{for } x \in \Omega.$$

Let ϕ be in $W^{1,p}(\Omega)$. Consider the function $\phi_\varepsilon = \widehat{\rho}_\varepsilon\phi$ which belongs to $W^{1,p}_0(\Omega)$, and extend it by 0 outside Ω (extension still denoted ϕ_ε). Due to (12.4), it satisfies

$$\begin{aligned}&\|\phi_\varepsilon\|_{L^p(\mathbb{R}^N)} \leqslant \|\phi\|_{L^p(\Omega)},\\ &\|\phi - \phi_\varepsilon\|_{L^p(\Omega)} \leqslant \|\phi\|_{L^p(\Omega^{bl}_{2\varepsilon d(Y)})},\\ &\|\nabla\phi_\varepsilon\|_{L^p(\mathbb{R}^N)} \leqslant \|\nabla\phi\|_{L^p(\Omega)} + \frac{C}{\varepsilon}\|\phi\|_{L^p(\Omega^{bl}_{2\varepsilon d(Y)})},\\ &\|\nabla(\phi - \phi_\varepsilon)\|_{L^p(\Omega)} \leqslant \|\nabla\phi\|_{L^p(\Omega^{bl}_{2\varepsilon d(Y)})} + \frac{C}{\varepsilon}\|\phi\|_{L^p(\Omega^{bl}_{2\varepsilon d(Y)})}.\end{aligned} \tag{13.95}$$

Now, apply Theorem 13.40 to obtain a function $\widehat{\phi}_\varepsilon \in L^p(\mathbb{R}^N; W^{1,p}_{per}(Y))$ such that,

$$\begin{aligned}\|\widehat{\phi}_\varepsilon\|_{W^{1,p}(Y;L^p(\mathbb{R}^N))} &\leqslant C\big(\|\phi_\varepsilon\|_{L^p(\mathbb{R}^N)} + \varepsilon\|\nabla\phi_\varepsilon\|_{L^p(\mathbb{R}^N)}\big)\\ &\leqslant C\big(\|\phi\|_{L^p(\Omega)} + \varepsilon\|\nabla\phi\|_{L^p(\Omega)}\big),\\ \|\widetilde{\mathcal{T}}_\varepsilon(\phi_\varepsilon) - \widehat{\phi}_\varepsilon\|_{W^{1,p}(Y;W^{-1,p}(\mathbb{R}^N))} &\leqslant C\varepsilon\big(\|\phi_\varepsilon\|_{L^p(\mathbb{R}^N)} + \varepsilon\|\nabla\phi_\varepsilon\|_{L^p(\mathbb{R}^N)}\big)\\ &\leqslant C\varepsilon\big(\|\phi\|_{L^p(\Omega)} + \varepsilon\|\nabla\phi\|_{L^p(\Omega)}\big).\end{aligned} \tag{13.96}$$

Observe that, according to Proposition 13.38,

$$\widehat{\phi}_\varepsilon = 0 \quad \text{a.e. in} \quad (\mathbb{R}^N \setminus \overline{\widehat{\Omega}}_\varepsilon)) \times Y.$$

Let Ψ be in $W^{1,p'}(\Omega; L^{p'}(Y))$, extend it in a function, still denoted Ψ, belonging to $W^{1,p'}(\mathbb{R}^N; L^{p'}(Y))$ and satisfying

$$\|\Psi\|_{W^{1,p'}(\mathbb{R}^N;L^{p'}(Y))} \leqslant C\|\Psi\|_{W^{1,p'}(\Omega;L^{p'}(Y))}.$$

Since the support of ϕ_ε is included in $\widehat{\Omega}_\varepsilon$, a simple computation yields

$$\begin{aligned}\big\langle \widetilde{\mathcal{T}}_\varepsilon(\phi_\varepsilon) - \widehat{\phi}_\varepsilon, \Psi\big\rangle_{L^p(Y;W^{-1,p}(\mathbb{R}^N)),L^{p'}(Y;W^{1,p'}(\mathbb{R}^N))} \\ = \int_{\mathbb{R}^N\times Y} (\widetilde{\mathcal{T}}_\varepsilon(\phi_\varepsilon) - \widehat{\phi}_\varepsilon)\Psi\, dxdy = \int_{\Omega\times Y} (\mathcal{T}_\varepsilon(\phi_\varepsilon) - \widehat{\phi}_\varepsilon)\Psi\, dxdy,\end{aligned}$$

Then, thanks to Proposition 1.8 (iii) and (13.95),

$$\|\mathcal{T}_\varepsilon(\phi - \phi_\varepsilon)\|_{L^p(\Omega\times Y)} \leqslant C\|\phi\|_{L^p(\Omega^{bl}_{2\varepsilon d(Y)})}.$$

This inequality and $(12.3)_1$ imply

$$\Big|\int_{\Omega\times Y} \mathcal{T}_\varepsilon(\phi - \phi_\varepsilon)\Psi\, dxdy\Big| \leqslant C\varepsilon^{1/p'}\|\phi\|_{L^p(\Omega^{bl}_{2\varepsilon d(Y)})}\|\Psi\|_{L^{p'}(Y;W^{1,p'}(\Omega))}$$

which, with $(13.96)_2$, gives

$$\begin{aligned}\|\mathcal{T}_\varepsilon(\phi) - \widehat{\phi}_\varepsilon\|_{L^p(Y;(W^{1,p'}(\Omega))')} &\leqslant C\varepsilon\big(\|\phi\|_{L^p(\Omega)} + \varepsilon\|\nabla\phi\|_{L^p(\Omega)}\big) \\ &\quad + C\varepsilon^{1/p'}\|\phi\|_{L^p(\Omega^{bl}_{2\varepsilon d(Y)})}.\end{aligned}$$

If Ψ belongs to $L^{p'}(Y; W^{1,p'}_0(\Omega))$, from $(12.5)_1$, we get

$$\Big|\int_{\Omega\times Y} \mathcal{T}_\varepsilon(\phi - \phi_\varepsilon)\Psi\, dxdy\Big| \leqslant C\varepsilon\|\phi\|_{L^p(\Omega^{bl}_{2\varepsilon d(Y)})}\|\Psi\|_{L^{p'}(Y;W^{1,p'}(\Omega))},$$

that is,

$$\|\mathcal{T}_\varepsilon(\phi) - \widehat{\phi}_\varepsilon\|_{L^p(Y;W^{-1,p}(\Omega))} \leqslant C\varepsilon\big(\|\phi\|_{L^p(\Omega)} + \varepsilon\|\nabla\phi\|_{L^p(\Omega)}\big).$$

In the same way we show that

$$\begin{aligned}\|\nabla_y\mathcal{T}_\varepsilon(\phi) - \nabla_y\widehat{\phi}_\varepsilon\|_{L^p(Y;(W^{1,p'}(\Omega))')} &\leqslant C\varepsilon\big(\|\phi\|_{L^p(\Omega)} + \varepsilon\|\nabla\phi\|_{L^p(\Omega)}\big), \\ &\quad + C\varepsilon^{1/p'}\big(\|\phi\|_{L^p(\Omega^{bl}_{2\varepsilon d(Y)})} + \varepsilon\|\nabla\phi\|_{L^p(\Omega^{bl}_{2\varepsilon d(Y)})}\big), \\ \|\nabla_y\mathcal{T}_\varepsilon(\phi) - \nabla_y\widehat{\phi}_\varepsilon\|_{L^p(Y;W^{-1,p}(\Omega))} &\leqslant C\varepsilon\,\big(\|\phi\|_{L^p(\Omega)} + \varepsilon\|\nabla\phi\|_{L^p(\Omega)}\big),\end{aligned}$$

and this concludes the proof. □

Theorem 13.42. *For every $\phi \in W^{1,p}(\mathbb{R}^N)$, there exists*

$$\widehat{\phi}_\varepsilon \in L^p(\mathbb{R}^N; W^{1,p}_{per}(Y))$$

such that

$$\begin{aligned}&\|\widehat{\phi}_\varepsilon\|_{L^p(\mathbb{R}^N;W^{1,p}(Y))} \leqslant C\|\nabla\phi\|_{L^p(\mathbb{R}^N)},\\ &\|\widetilde{\mathcal{T}}_\varepsilon(\nabla\phi) - \nabla\phi - \nabla_y\widehat{\phi}_\varepsilon\|_{L^p(Y;W^{-1,p}(\mathbb{R}^N))} \leqslant C\varepsilon\|\nabla\phi\|_{L^p(\mathbb{R}^N)}.\end{aligned} \tag{13.97}$$

The constant depends only on p and Y.

Proof. Decompose ϕ as $\phi = \widetilde{\mathcal{Q}}_\varepsilon(\phi) + \widetilde{\mathcal{R}}_\varepsilon(\phi)$. From Corollary 1.68,

$$\|\nabla\widetilde{\mathcal{R}}_\varepsilon(\phi)\|_{W^{-1,p}(\mathbb{R}^N)} \leqslant C\|\widetilde{\mathcal{R}}_\varepsilon(\phi)\|_{L^p(\mathbb{R}^N)} \leqslant C\varepsilon\|\nabla\phi\|_{L^p(\mathbb{R}^N)}.$$

This estimate and those of Proposition 12.10, as well and that derived from Theorem 13.40 applied to $\widetilde{\mathcal{R}}_\varepsilon(\phi)$, easily prove the claim of the theorem. □

The next result is the equivalent of Theorem 13.42 for bounded domains.

Theorem 13.43. *For every $\phi \in W^{1,p}(\Omega)$ there exists $\widehat{\phi}_\varepsilon \in L^p(\Omega; W^{1,p}_{per}(Y))$, such that*

$$\begin{aligned}&\|\widehat{\phi}_\varepsilon\|_{L^p(\Omega;W^{1,p}(Y))} \leqslant C\|\nabla\phi\|_{L^p(\Omega)},\\ &\|\mathcal{T}_\varepsilon(\nabla\phi) - \nabla\phi - \nabla_y\widehat{\phi}_\varepsilon\|_{L^p(Y;W^{-1,p}(\Omega))} \leqslant C\varepsilon\|\nabla\phi\|_{L^p(\Omega)}.\end{aligned} \tag{13.98}$$

The constant depends only on p, Y and $\partial\Omega$.

Proof. Extend ϕ in a function belonging to $W^{1,p}(\mathbb{R}^N)$, still denoted ϕ, and satisfying (1.80). Theorem 13.42 furnishes a function $\widehat{\phi}_\varepsilon \in L^p(\mathbb{R}^N; W^{1,p}_{per}(Y))$, such that

$$\|\widehat{\phi}_\varepsilon\|_{L^p(\Omega;W^{1,p}(Y))} \leqslant \|\widehat{\phi}_\varepsilon\|_{L^p(\mathbb{R}^N;W^{1,p}(Y))} \leqslant C\|\nabla\phi\|_{L^p(\mathbb{R}^N)} \leqslant C\|\nabla\phi\|_{L^p(\Omega)}.$$

Now, for Φ be in $L^{p'}(Y; W^{1,p'}_0(\Omega))$ (extended by 0 in $Y \times (\mathbb{R}^N \backslash \Omega)$),

$$\begin{aligned}\int_{\Omega\times Y} \big(\mathcal{T}_\varepsilon(\nabla\phi) - \nabla\phi - \nabla_y\widehat{\phi}_\varepsilon\big)\Phi\, dxdy = &\int_{\Omega\times Y} \big(\widetilde{\mathcal{T}}_\varepsilon(\nabla\phi) - \nabla\phi - \nabla_y\widehat{\phi}_\varepsilon\big)\Phi\, dxdy\\ &- \int_{\Lambda_\varepsilon\times Y} \widetilde{\mathcal{T}}_\varepsilon(\nabla\phi)\Phi\, dxdy.\end{aligned}$$

Since

$$\int_{\Omega\times Y} \big(\widetilde{\mathcal{T}}_\varepsilon(\nabla\phi) - \nabla\phi - \nabla_y\widehat{\phi}_\varepsilon\big)\Phi\, dxdy = \int_{\mathbb{R}^N\times Y} \big(\widetilde{\mathcal{T}}_\varepsilon(\nabla\phi) - \nabla\phi - \nabla_y\widehat{\phi}_\varepsilon\big)\Phi\, dxdy,$$

estimate $(13.97)_2$ in Theorem 13.42 leads to

$$\Big|\int_{\Omega\times Y} \big(\widetilde{\mathcal{T}}_\varepsilon(\nabla\phi) - \nabla\phi - \nabla_y\widehat{\phi}_\varepsilon\big)\Phi\, dxdy\Big| \leqslant C\varepsilon\|\nabla\phi\|_{L^p(\Omega)}\|\Phi\|_{L^{p'}(Y;W^{1,p'}_0(\Omega))}.$$

Also, by (12.4)

$$\|\Phi\|_{L^{p'}(\Lambda_\varepsilon\times Y)} \leqslant C\varepsilon\|\Phi\|_{L^{p'}(Y;W_0^{1,p'}(\Omega))},$$

hence

$$\begin{aligned}\Big|\int_{\Lambda_\varepsilon\times Y}\widetilde{\mathcal{T}}_\varepsilon(\nabla\phi)\Phi\,dxdy\Big| &\leqslant C\varepsilon\|\widetilde{\mathcal{T}}_\varepsilon(\nabla\phi)\|_{L^p(\mathbb{R}^N\times Y)}\|\Phi\|_{L^{p'}(Y;W_0^{1,p'}(\Omega))}\\ &\leqslant C\varepsilon\|\nabla\phi\|_{L^p(\mathbb{R}^N)}\|\Phi\|_{L^{p'}(Y;W_0^{1,p'}(\Omega))}\\ &\leqslant C\varepsilon\|\nabla\phi\|_{L^p(\Omega)}\|\Phi\|_{L^{p'}(Y;W_0^{1,p'}(\Omega))}\end{aligned}$$

which in turn, taking into account the previous inequality, proves inequality $(13.98)_2$, which concludes the proof of the theorem. □

Theorem 13.44. *For any ϕ in $W^{1,p}(\mathbb{R}^N)$ there exists $\widehat{\phi}_\varepsilon$ in $L^p(\Omega;W^{1,p}_{per}(Y))$, satisfying (with notation (12.1)),*

$$\begin{aligned}&\|\widehat{\phi}_\varepsilon\|_{L^p(\Omega;W^{1,p}(Y))} \leqslant C\|\nabla\phi\|_{L^p(\Omega)},\\ &\|\mathcal{T}_\varepsilon(\nabla\phi)-\nabla\phi-\nabla_y\widehat{\phi}_\varepsilon\|_{L^p(Y;(W^{1,p'}(\Omega))')}\\ &\qquad\qquad \leqslant C\varepsilon\|\nabla\phi\|_{L^p(\Omega)}+C\varepsilon^{1/p'}\|\nabla\phi\|_{L^p(\Omega^b_{5\varepsilon d(Y)})}.\end{aligned}\tag{13.99}$$

The constant depends only on p, Y and $\partial\Omega$[(5)].

Proof. Decompose ϕ as

$$\phi=\widetilde{\mathcal{Q}}_\varepsilon(\phi)+\widetilde{\mathcal{R}}_\varepsilon(\phi),$$

and apply Theorem 13.41 with the function

$$\frac{1}{\varepsilon}\widetilde{\mathcal{R}}_\varepsilon(\phi)_{|\Omega}.$$

We obtain $\widehat{\phi}_\varepsilon$ in $L^p(\Omega;W^{1,p}_{per}(Y))$ such that (see also Corollary 1.68),

$$\begin{aligned}\|\widehat{\phi}_\varepsilon\|_{L^p(\Omega;W^{1,p}(Y))} &\leqslant \frac{C}{\varepsilon}\big(\|\widetilde{\mathcal{R}}_\varepsilon(\phi)\|_{L^p(\mathbb{R}^N)}+\varepsilon\|\nabla\widetilde{\mathcal{R}}_\varepsilon(\phi)\|_{L^p(\mathbb{R}^N)}\big)\\ &\leqslant C\,\|\nabla\phi\|_{L^p(\mathbb{R}^N)},\end{aligned}$$

that is $(13.99)_1$.

To prove $(13.99)_2$, recall that

$$\mathcal{T}_\varepsilon\big(\nabla\widetilde{\mathcal{R}}_\varepsilon(\phi)\big)=\frac{1}{\varepsilon}\nabla_y\big(\mathcal{T}_\varepsilon\circ\widetilde{\mathcal{R}}_\varepsilon(\phi)\big).$$

(5) This theorem can be stated with the function $\widehat{\phi}_\varepsilon$ in $L^p(\Omega;W^{1,p}_{per,0}(Y))$ (by subtracting $\mathcal{M}_Y(\widehat{\phi}_\varepsilon)$ from the original $\widehat{\phi}_\varepsilon$).

Then, from Theorem 13.41, we get successively,

$$\begin{aligned}\|\mathcal{T}_\varepsilon(\nabla\widetilde{\mathcal{R}}_\varepsilon(\phi)) - \nabla_y\widehat{\phi}_\varepsilon\|_{L^p(Y;(W^{1,p'}(\Omega))')} \\ &\leqslant C\varepsilon\big(\|\widetilde{\mathcal{R}}_\varepsilon(\phi)\|_{L^p(\Omega)} + \varepsilon\|\nabla\widetilde{\mathcal{R}}_\varepsilon(\phi)\|_{L^p(\Omega)}\big) \\ &\quad + C\varepsilon^{1/p'}\big(\|\widetilde{\mathcal{R}}_\varepsilon(\phi)\|_{L^p(\Omega^b_{2\varepsilon d(Y)})} + \varepsilon\|\nabla\widetilde{\mathcal{R}}_\varepsilon(\phi)\|_{L^p(\Omega^b_{2\varepsilon d(Y)})}\big) \\ &\leqslant C\varepsilon\|\nabla\phi\|_{L^p(\mathbb{R}^N)} + C\varepsilon^{1/p'}\|\nabla\phi\|_{L^p(\Omega^b_{4\varepsilon d(Y)})}.\end{aligned}$$

Besides, from the following formula (true for all $\Psi \in W^{1,p'}(\Omega)^N$):

$$\int_\Omega \nabla\widetilde{\mathcal{R}}_\varepsilon(\varphi)\cdot\Psi dx = \int_{\partial\Omega}\widetilde{\mathcal{R}}_\varepsilon(\varphi)\Psi\cdot n\, d\sigma - \int_\Omega \widetilde{\mathcal{R}}_\varepsilon(\varphi)\cdot \operatorname{div}(\Psi)dx,$$

and from estimates (1.81) and $(12.4)_2$, it follows that

$$\|\nabla\widetilde{\mathcal{R}}_\varepsilon(\varphi)\|_{(W^{1,p'}(\Omega))'} \leqslant C\varepsilon\|\nabla\varphi\|_{L^p(\mathbb{R}^N)} + C\varepsilon^{1/p'}\|\nabla\varphi\|_{L^p(\Omega^b_{3\varepsilon d(Y)})}.$$

Summarizing the above inequalities, we finally get

$$\begin{aligned}\|\mathcal{T}_\varepsilon(\nabla\widetilde{\mathcal{R}}_\varepsilon(\phi)) - \nabla\widetilde{\mathcal{R}}_\varepsilon(\varphi) - \nabla_y\widehat{\phi}_\varepsilon\|_{L^p(Y;(W^{1,p'}(\Omega))')} \\ \leqslant C\varepsilon\|\nabla\phi\|_{L^p(\mathbb{R}^N)} + C\varepsilon^{1/p'}\|\nabla\phi\|_{L^p(\Omega^b_{4\varepsilon d(Y)})}.\end{aligned} \tag{13.100}$$

On the other hand, by estimate $(12.6)_1$ (with $\nabla\widetilde{\mathcal{Q}}_\varepsilon(\varphi)$ in place of ϕ) and by the second inequality of Proposition 1.64(iii),

$$\begin{aligned}\|\nabla\widetilde{\mathcal{Q}}_\varepsilon(\varphi) - \mathcal{M}_\varepsilon(\nabla\widetilde{\mathcal{Q}}_\varepsilon(\varphi))\|_{(W^{1,p'}(\Omega))'} \\ \leqslant C\varepsilon\|\nabla\varphi\|_{L^p(\mathbb{R}^N)} + C\varepsilon^{1/p'}\|\nabla\varphi\|_{L^p(\Omega^b_{5\varepsilon d(Y)})}.\end{aligned} \tag{13.101}$$

Now, using (12.22), this implies

$$\begin{aligned}\|\widetilde{\mathcal{T}}_\varepsilon(\nabla\widetilde{\mathcal{Q}}_\varepsilon(\varphi)) - \nabla\widetilde{\mathcal{Q}}_\varepsilon(\varphi)\|_{L^\infty(Y;(W^{1,p'}(\Omega))')} \\ \leqslant C\varepsilon\|\nabla\varphi\|_{L^p(\mathbb{R}^N)} + C\varepsilon^{1/p'}\|\nabla\varphi\|_{L^p(\Omega^b_{4\varepsilon d(y)})},\end{aligned}$$

which, together with (13.101), yields

$$\begin{aligned}\|\mathcal{T}_\varepsilon(\nabla\widetilde{\mathcal{Q}}_\varepsilon(\varphi)) - \nabla\widetilde{\mathcal{Q}}_\varepsilon(\varphi)\|_{L^\infty(Y;(W^{1,p'}(\Omega))')} \\ \leqslant C\varepsilon\|\nabla\varphi\|_{L^p(\mathbb{R}^N)} + C\varepsilon^{1/p'}\|\nabla\phi\|_{L^p(\Omega^b_{5\varepsilon d(Y)})}.\end{aligned}$$

This inequality and the former one (13.100), give $(13.99)_2$. □

Chapter 14

Some sharp error estimates

This chapter is dedicated to improvements on the error estimates obtained by Griso in [120,122] and [125] for the Dirichlet problem

$$\begin{cases} -\mathrm{div}\,(A^\varepsilon \nabla u_\varepsilon) = f & \text{in } \Omega, \\ u_\varepsilon = 0 & \text{on } \partial\Omega, \end{cases} \tag{14.1}$$

and for Neumann problem,

$$\begin{cases} -\mathrm{div}\,(A^\varepsilon \nabla u_\varepsilon) = f & \text{in } \Omega, \\ A^\varepsilon \nabla u_\varepsilon \cdot n = 0 & \text{on } \partial\Omega, \\ \displaystyle\int_\Omega u_\varepsilon dx = 0, \end{cases} \tag{14.2}$$

where f is in $L^2(\Omega)$ and

$$A^\varepsilon(x) \equiv A\Big(\Big\{\frac{x}{\varepsilon}\Big\}_Y\Big),$$

with $A \in M(\alpha, \beta, Y)$.

The method we present in this chapter is based on the use of the unfolding operators. Several preliminary results, needed in the proofs and concerning in particular, projection operators on spaces of periodic functions and the notion of periodic defect, were established in Section 13.9 (Theorem 13.41 and the key Theorems 13.43 and 13.44).

Both global and interior error estimates are obtained for the homogeneous Dirichlet problem and for the Neumann problem, studied in Section 14.1 and in Section 14.2, respectively.

In the Dirichlet case we characterize the limit in $H^1_0(\Omega)$ of the sequence

$$\Big\{\frac{1}{\varepsilon}\Big(u_\varepsilon - u_0 - \varepsilon \sum_{j=1}^N \frac{\partial u_0}{\partial x_j}\, \widehat{\chi}_j\Big(\frac{\cdot}{\varepsilon}\Big) - \varepsilon \beta^{\rm Dir}_\varepsilon\Big)\Big\}_\varepsilon,$$

D. Cioranescu et al., *The Periodic Unfolding Method*, Series in Contemporary Mathematics 3, https://doi.org/10.1007/978-981-13-3032-2_14

where the function u_0 is the solution of the homogenized problem and $\beta_\varepsilon^{\mathrm{Dir}}$ is a boundary layer term introduced to take into account the homogeneous Dirichlet condition which must be satisfied by the approximated solution. The asymptotic behavior of this kind of sequence was studied in the previous chapter. We show that each limit point of the sequence $\{\beta_\varepsilon^{\mathrm{Dir}}\}_\varepsilon$ is expressed in terms of the homogenized solution u_0 and a function Θ belonging to $L^\infty(\partial\Omega)^N$. In most cases, there are infinitely many such limit points. This shows that in the general case, it is not possible to give an asymptotic expansion of u_ε beyond the first two classical terms.

In the case where Ω is a convex polyhedral domain, $f_\varepsilon \equiv f$ in $L^2(\Omega)$ and for a well-chosen sequence $\{\varepsilon\}$ (see Subsection 14.1.1 for more details), we show that the solution of the classical Dirichlet homogenization problem (3.3) can be written as

$$\begin{aligned} u_\varepsilon(x) = u_0(x) + \varepsilon \sum_{j=1}^N \frac{\partial u_0}{\partial x_j}(x)\, \widehat{\chi}_j\Big(\frac{x}{\varepsilon}\Big) + \varepsilon\Big(u_1(x) + \varepsilon \widehat{u}_1\Big(x, \Big\{\frac{x}{\varepsilon}\Big\}\Big)\Big) \\ + \varepsilon\Big(\beta^{\mathrm{Dir}}(x) + \varepsilon \widehat{\beta}^{\mathrm{Dir}}\Big(x, \Big\{\frac{x}{\varepsilon}\Big\}\Big)\Big) + \ldots, \end{aligned}$$

where all the functions in the right-hand side are bounded in the L^2-norm, except the term $\widehat{\beta}^{\mathrm{Dir}}(\cdot, \frac{\cdot}{\varepsilon})$ which is bounded in the space $L^2_\rho(\Omega)$.

In the Neumann case, we only characterize the limit in $H^1(\Omega)$ of the sequence

$$\Big\{\frac{1}{\varepsilon}\Big(u_\varepsilon - u_0' - \varepsilon \sum_{j=1}^N \frac{\partial u_0'}{\partial x_j}\, \widehat{\chi}_j\Big(\frac{\cdot}{\varepsilon}\Big) - \varepsilon \beta_\varepsilon^{\mathrm{Ne}}\Big)\Big\}_\varepsilon,$$

where the function u_0' is the solution of the homogenized problem and $\beta_\varepsilon^{\mathrm{Ne}}$ is a kind of corrector for the homogeneous Neumann condition as shown in Section 14.2.

The techniques developed in this chapter can be transposed to a large class of homogenization problems.

Other results concerning error estimates of these problems can be found in [138–140, 171, 172].

In this chapter the domain Ω is bounded with a Lipschitz or $\mathcal{C}^{1,1}$ boundary, and **B** *is an orthonormal basis.*

14.1 Error estimates for the Dirichlet problem

The homogenization result for the Dirichlet problem (14.1) was given in Subsection 3.1.6. It states in particular, that

$$u_\varepsilon \rightharpoonup u_0 \quad \text{weakly in } H_0^1(\Omega),$$

where u_0 in $H_0^1(\Omega)$ is the unique solution of the homogenized problem (3.18),

$$\int_\Omega A^{hom} \nabla u_0 \, \nabla \varphi \, dx = \int_\Omega f \varphi \, dx, \quad \forall \varphi \in H_0^1(\Omega), \tag{14.3}$$

where the standard homogenized matrix A^{hom} is defined by formulas (3.19). Our aim here is to improve upon the corrector result from Theorem 3.27.

For $F \in L^\infty(\Omega)^N$ consider the function

$$\Psi = \mathbb{K}_A(F),$$

given by (13.3) (with $B \equiv A$) and introduce the vector function $\widetilde{d}$ defined as,

$$\widetilde{d}(y) \doteq A(y)\nabla_y \Psi(y) + F(y) - \mathcal{M}_Y\big(A\nabla_y\Psi + F\big) \quad \text{for a.e. } y \in Y, \qquad (14.4)$$

which belongs to $L^2(Y)^N$ and satisfies $\mathcal{M}_Y(\widetilde{d}) = 0$.

Then, the following result holds:

Lemma 14.1. *For every w in $H^1(\Omega)$ and for every $v \in H^1_0(\Omega)$,*

$$\Big|\int_\Omega \widetilde{d}\Big(\frac{\cdot}{\varepsilon}\Big)\cdot \nabla w\, v\, dx\Big| \leqslant C\varepsilon \|\nabla w\|_{L^2(\Omega)}\|\nabla v\|_{L^2(\Omega)}. \qquad (14.5)$$

The constant does not depend on ε.

Furthermore,

$$\lim_{\varepsilon\to 0}\frac{1}{\varepsilon}\int_\Omega \widetilde{d}\Big(\frac{\cdot}{\varepsilon}\Big)\cdot\nabla w\, v\, dx = 0. \qquad (14.6)$$

Let $\{v_\varepsilon\}_\varepsilon$ be a sequence in $H^1_0(\Omega)$ such that there exist v in $H^1_0(\Omega)$ and $\widehat{v}$ in $L^2(\Omega; H^1_{per}(Y))$ satisfying

$$\begin{aligned} &v_\varepsilon \rightharpoonup v \quad \textit{weakly in } H^1(\Omega),\\ &\mathcal{T}_\varepsilon(\nabla v_\varepsilon) \rightharpoonup \nabla v + \nabla_y \widehat{v} \quad \textit{weakly in } L^2(\Omega\times Y)^N. \end{aligned} \qquad (14.7)$$

Then

$$\lim_{\varepsilon\to 0}\frac{1}{\varepsilon}\int_\Omega \widetilde{d}\Big(\frac{\cdot}{\varepsilon}\Big)\cdot\nabla w\, v_\varepsilon\, dx = \frac{1}{|Y|}\int_{\Omega\times Y}\widetilde{d}(y)\cdot\nabla w(x)\,\widehat{v}(x,y)\,dxdy. \qquad (14.8)$$

Proof. Step 1. Proof of (14.5).

Due to Lemma 3.29 the integral $\displaystyle\int_\Omega \widetilde{d}\Big(\frac{\cdot}{\varepsilon}\Big)\cdot\nabla w\, v\, dx$ makes sense since, by (3.44), $\nabla_y\Psi\big(\frac{\cdot}{\varepsilon}\big)v$ belongs to $L^2(\Omega)$. We extend the functions v and w outside of Ω (v by 0 in $\mathbb{R}^N\setminus\overline{\Omega}$ and w as a function belonging to $H^1(\mathbb{R}^N)$), the extensions are still denoted v and w, so that clearly,

$$\int_\Omega \widetilde{d}\Big(\frac{\cdot}{\varepsilon}\Big)\cdot\nabla w\, v\, dx = \int_{\mathbb{R}^N}\widetilde{d}\Big(\frac{\cdot}{\varepsilon}\Big)\cdot\nabla w\, v\, dx. \qquad (14.9)$$

We shall repeatedly use the following estimates (coming from Corollary 1.68, Proposition 1.64(iv) and $(1.39)_1$ with $\Omega = \mathbb{R}^N$, respectively):

$$\begin{aligned} &\|v - \widetilde{\mathcal{Q}}_\varepsilon(v)\|_{L^2(\mathbb{R}^N)} + \varepsilon\|\nabla(v - \widetilde{\mathcal{Q}}_\varepsilon(v))\|_{L^2(\mathbb{R}^N)} \leqslant C\varepsilon\|\nabla v\|_{L^2(\Omega)},\\ &\|\widetilde{\mathcal{Q}}_\varepsilon(v) - \widetilde{\mathcal{M}}_\varepsilon(v)\|_{L^2(\mathbb{R}^N)} \leqslant C\varepsilon\|\nabla v\|_{L^2(\Omega)},\\ &\|\widetilde{\mathcal{M}}_\varepsilon(w) - w\|_{L^2(\mathbb{R}^N)} \leqslant C\varepsilon\|\nabla w\|_{L^2(\mathbb{R}^N)}. \end{aligned} \qquad (14.10)$$

According to estimates (3.44) and $(14.10)_1$ one can replace v by $\widetilde{\mathcal{Q}}_\varepsilon(v)$ in the right-hand side of (14.9). Then one substitutes $\widetilde{\mathcal{M}}_\varepsilon(v)$ by $\widetilde{\mathcal{Q}}_\varepsilon(v)$ in view of the first estimate in Proposition 1.71. The result is

$$\Big|\int_{\mathbb{R}^N} \widetilde{d}\Big(\frac{x}{\varepsilon}\Big)\cdot\nabla w(x)\,v(x)\,dx - \int_{\mathbb{R}^N} \widetilde{d}\Big(\frac{x}{\varepsilon}\Big)\cdot\nabla w(x)\,\widetilde{\mathcal{M}}_\varepsilon(v)(x)\,dx\Big| \leqslant C\varepsilon\|\nabla w\|_{L^2(\Omega)}\|\nabla v\|_{L^2(\Omega)}.$$

Now, applying the unfolding operator to the second integral and using estimate $(14.10)_3$, one obtains

$$\Big|\int_\Omega \widetilde{d}\Big(\frac{x}{\varepsilon}\Big)\cdot\nabla w(x)\,v(x)\,dx - \frac{1}{|Y|}\int_{\mathbb{R}^N\times Y}\widetilde{d}(y)\cdot\widetilde{\mathcal{T}}_\varepsilon(\nabla w)(x,y)\,v(x)\,dxdy\Big| \leqslant C\varepsilon\|\nabla w\|_{L^2(\Omega)}\|\nabla v\|_{L^2(\Omega)}.$$

On the other hand, by Theorem 13.42 in Subsection 13.9.2, there exists some function $\widehat{w}_\varepsilon$ in $L^2(\mathbb{R}^N; H^1_{per}(Y))$ such that

$$\|\widetilde{\mathcal{T}}_\varepsilon(\nabla w) - \nabla w - \nabla_y\widehat{w}_\varepsilon\|_{L^2(Y;H^{-1}(\mathbb{R}^N))} \leqslant C\varepsilon\|\nabla w\|_{L^2(\Omega)},$$

which used in the inequality above, leads to

$$\Big|\int_\Omega \widetilde{d}\Big(\frac{\cdot}{\varepsilon}\Big)\cdot\nabla w\,v\,dx - \frac{1}{|Y|}\int_{\mathbb{R}^N\times Y}\widetilde{d}(y)\cdot\big(\nabla w + \nabla_y\widehat{w}_\varepsilon\big)\,v\,dxdy\Big| \leqslant C\varepsilon\|\nabla w\|_{L^2(\Omega)}\|\nabla v\|_{L^2(\Omega)}.$$

Finally, estimate (14.5) follows easily, by taking into account the fact that

$$\int_Y \widetilde{d}(y)\cdot\big(\nabla w(x) + \nabla_y\widehat{w}_\varepsilon(x,y)\big)\,dy = 0 \qquad \text{for a.e. } x\in\mathbb{R}^N.$$

Step 2. Proof of (14.6).

First, assume w in $\mathcal{D}(\mathbb{R}^N)$. We have successively,

$$\begin{aligned}\frac{1}{\varepsilon}\int_\Omega \widetilde{d}\Big(\frac{x}{\varepsilon}\Big)\cdot\nabla w(x)\,v(x)\,dx &= \frac{1}{\varepsilon|Y|}\int_{\mathbb{R}^N\times Y}\widetilde{d}(y)\cdot\widetilde{\mathcal{T}}_\varepsilon(\nabla w)\,\widetilde{\mathcal{T}}_\varepsilon(v)\,dxdy\\ &= \frac{1}{|Y|}\int_{\mathbb{R}^N\times Y}\widetilde{d}(y)\cdot\frac{\widetilde{\mathcal{T}}_\varepsilon(\nabla w) - \widetilde{\mathcal{M}}_\varepsilon(\nabla w)}{\varepsilon}\,\widetilde{\mathcal{T}}_\varepsilon(v)\,dxdy\\ &\quad+\frac{1}{|Y|}\int_{\mathbb{R}^N\times Y}\widetilde{d}(y)\cdot\widetilde{\mathcal{M}}_\varepsilon(\nabla w)\,\frac{\widetilde{\mathcal{T}}_\varepsilon(v) - \widetilde{\mathcal{M}}_\varepsilon(v)}{\varepsilon}\,dxdy\\ &\quad+\frac{1}{\varepsilon|Y|}\int_{\mathbb{R}^N\times Y}\widetilde{d}(y)\cdot\widetilde{\mathcal{M}}_\varepsilon(\nabla w)\,\widetilde{\mathcal{M}}_\varepsilon(v)\,dxdy.\end{aligned}$$

The last integral is equal to 0 since $\mathcal{M}_Y(\widetilde{d}) = 0$. Proposition 1.25 (with $\Omega = \mathbb{R}^N$) and Proposition 1.39 allow to pass to the limit to get

$$\begin{aligned}
&\lim_{\varepsilon\to 0} \frac{1}{\varepsilon}\int_\Omega \widetilde{d}\Big(\frac{x}{\varepsilon}\Big)\cdot \nabla w(x)\, v(x)\, dx \\
&\quad= \frac{1}{|Y|}\int_{\Omega\times Y} \widetilde{d}(y)\cdot\Big(\sum_{j=1}^N y^c\cdot\nabla\Big(\frac{\partial w}{\partial x_j}\Big)\mathbf{b}_j\Big)\, v\, dxdy \\
&\quad+ \frac{1}{|Y|}\int_{\Omega\times Y} \big(\widetilde{d}(y)\cdot\nabla w\big)\,\big(y^c\cdot\nabla v\big)\, dxdy \\
&\quad= \frac{1}{|Y|}\int_Y\int_\Omega \Big[\sum_{j=1}^N \nabla\Big(\frac{\partial w}{\partial x_j}\Big)\big(\widetilde{d}(y)\cdot\mathbf{b}_j\big)\, v + \big(\widetilde{d}(y)\cdot\nabla w\big)\,\nabla v\Big]\cdot y^c dxdy \\
&\quad= \frac{1}{|Y|}\int_Y\int_\Omega \operatorname{div}\Big[\big(\widetilde{d}(y)\cdot\nabla w\big)\, v\, y^c\Big]\, dxdy.
\end{aligned}$$

The last integral vanishes since v is in $H_0^1(\Omega)$. We conclude by a density argument (of $\mathcal{D}(\mathbb{R}^N)$ in $H^1(\mathbb{R}^N)$), together with estimate (14.5).

Step 3. Proof of (14.8).

Here again, we start with w in $\mathcal{D}(\mathbb{R}^N)$, then proceeding as in Step 2 and thanks to (14.7), Proposition 1.39 and Theorem 1.41, convergence (14.8) follows. This convergence, a density argument and estimate (14.5) allow to conclude in the general case. □

Following the model of (14.4), we introduce the vector functions $\widetilde{d}_j$ (for $j = 1,\dots,N$)

$$\widetilde{d}_j(y) \doteq A(y)\big(\nabla_y\widehat{\chi}_j(y) + \mathbf{b}_j\big) - \mathcal{M}_Y\big(A(\nabla_y\widehat{\chi}_j + \mathbf{b}_j)\big) \text{ for a.e. } y \in Y, \quad (14.11)$$

where $\widehat{\chi}_j$ are the correctors given by (3.10). By definition (3.19) (with $A(y)$ in place of $B(x,y)$), the second term $\mathcal{M}_Y\big(A(\nabla_y\widehat{\chi}_j + \mathbf{b}_j)\big)$ is the j'th column of the matrix A^{hom}.

Proposition 14.2. *Let u_0 be the solution of the homogenized problem* (14.3). *Assume that u_0 belongs to $H^2(\Omega)$. Then the solution u_ε of* (14.1) *satisfies the following estimate:*

$$\begin{aligned}
&\Big|\int_\Omega A\Big(\frac{\cdot}{\varepsilon}\Big)\nabla\Big(u_\varepsilon - u_0 - \varepsilon\sum_{j=1}^N \frac{\partial u_0}{\partial x_j}\,\widehat{\chi}_j\Big(\frac{\cdot}{\varepsilon}\Big)\Big)\,\nabla v dx\Big| \\
&\qquad\qquad \leqslant C\varepsilon\|u_0\|_{H^2(\Omega)}\|\nabla v\|_{L^2(\Omega)}, \quad \forall v \in H_0^1(\Omega).
\end{aligned} \quad (14.12)$$

Proof. Let v be in $\mathcal{D}(\Omega)$. Recall (see Step 1 of Lemma 3.29) that

$$\operatorname{div} d_i\Big(\frac{\cdot}{\varepsilon}\Big) = 0 \text{ in } H_{loc}^{-1}(\mathbb{R}^N) \text{ and in particular in } H^{-1}(\Omega).$$

Multiply (14.11) by $\dfrac{\partial u_0}{\partial x_i}\, v$, integrate over Ω and sum on i, to get

$$\int_\Omega A\Big(\frac{\cdot}{\varepsilon}\Big)\Big[\nabla u_0 + \sum_{i=1}^N \frac{\partial u_0}{\partial x_i}\nabla_y \widehat{\chi}_i\Big(\frac{\cdot}{\varepsilon}\Big)\Big]\nabla v\, dx - \int_\Omega A^{hom}\nabla u_0\,\nabla v dx \\ + \sum_{i=1}^N \int_\Omega \widetilde{d}_i\Big(\frac{\cdot}{\varepsilon}\Big)\,\nabla\Big(\frac{\partial u_0}{\partial x_i}\Big) v\, dx = 0.$$

Since

$$\int_\Omega A^{hom}\nabla u_0\,\nabla v dx = \int_\Omega A^\varepsilon \nabla u_\varepsilon \nabla v\, dx,$$

this rewrites in the form

$$\begin{aligned}\int_\Omega A\Big(\frac{\cdot}{\varepsilon}\Big)\nabla\Big(u_\varepsilon - u_0 - \varepsilon\sum_{j=1}^N \frac{\partial u_0}{\partial x_j}\,\widehat{\chi}_j\Big(\frac{\cdot}{\varepsilon}\Big)\Big)\,\nabla v\, dx \\ = \sum_{i=1}^N \int_\Omega \widetilde{d}_i\Big(\frac{\cdot}{\varepsilon}\Big)\cdot\nabla\Big(\frac{\partial u_0}{\partial x_i}\Big) v\, dx \\ - \varepsilon \sum_{i=1}^N \int_\Omega A\Big(\frac{\cdot}{\varepsilon}\Big)\nabla\Big(\frac{\partial u_0}{\partial x_i}\Big)\,\widehat{\chi}_i\Big(\frac{\cdot}{\varepsilon}\Big)\,\nabla v\, dx.\end{aligned} \tag{14.13}$$

As the $\widehat{\chi}_i$'s belong to $L^\infty(Y)$ (see Lemma 3.29), we first have

$$\Big|\sum_{i=1}^N \int_\Omega A\Big(\frac{\cdot}{\varepsilon}\Big)\nabla\Big(\frac{\partial u_0}{\partial x_i}\Big)\,\widehat{\chi}_i\Big(\frac{\cdot}{\varepsilon}\Big)\,\nabla v\, dx\Big| \leqslant C\|u_0\|_{H^2(\Omega)}\|\nabla v\|_{L^2(\Omega)}$$

and then, from (14.5),

$$\Big|\sum_{i=1}^N \int_\Omega \widetilde{d}_i\Big(\frac{\cdot}{\varepsilon}\Big)\cdot\nabla\Big(\frac{\partial u_0}{\partial x_i}\Big) v\, dx\Big| \leqslant C\varepsilon\|u_0\|_{H^2(\Omega)}\|\nabla v\|_{L^2(\Omega)}.$$

These last two estimates and the density of $\mathcal{D}(\Omega)$ in $H_0^1(\Omega)$, give (14.12). □

We are now in position to state the main theorem of this section.

Theorem 14.3. *Suppose that u_0, the solution of the homogenized problem (14.3) belongs to $H^2(\Omega)$. Then the solution u_ε of (14.1) satisfies the following estimate:*

$$\Big\|u_\varepsilon - u_0 - \varepsilon\sum_{i=1}^N \frac{\partial u_0}{\partial x_i}\widehat{\chi}_i\Big(\frac{\cdot}{\varepsilon}\Big)\Big\|_{H^1(\Omega)} \leqslant C\varepsilon^{1/2}\|u_0\|_{H^2(\Omega)}. \tag{14.14}$$

Let β_ε^{Dir} in $H^1(\Omega)$, be the boundary corrector for the nonhomogeneous Dirichlet condition, i.e., the solution of the following Dirichlet problem: [1]

$$\begin{cases} \operatorname{div}(A^\varepsilon \nabla \beta_\varepsilon^{Dir}) = 0 & \text{in } \Omega, \\ \beta_\varepsilon^{Dir} = -\displaystyle\sum_{i=1}^N \frac{\partial u_0}{\partial x_i} \widehat{\chi}_i\Big(\frac{\cdot}{\varepsilon}\Big) & \text{on } \partial\Omega. \end{cases} \tag{14.15}$$

Then, the following estimate holds:

$$\Big\| u_\varepsilon - u_0 - \varepsilon \sum_{j=1}^N \frac{\partial u_0}{\partial x_j} \widehat{\chi}_j\Big(\frac{\cdot}{\varepsilon}\Big) - \varepsilon \beta_\varepsilon^{Dir} \Big\|_{H^1(\Omega)} \leqslant C\varepsilon \|u_0\|_{H^2(\Omega)}. \tag{14.16}$$

Furthermore, if $\partial\Omega$ is of class $C^{1,1}$, then

$$\begin{aligned} \|u_\varepsilon - u_0\|_{L^2(\Omega)} + \Big\| \rho\Big(\nabla u_\varepsilon - \nabla u_0 - \sum_{i=1}^N \frac{\partial u_0}{\partial x_i} \nabla_y \widehat{\chi}_i\Big(\frac{\cdot}{\varepsilon}\Big)\Big)\Big\|_{L^2(\Omega)^N} \\ \leqslant C\varepsilon \|u_0\|_{H^2(\Omega)}. \end{aligned} \tag{14.17}$$

The constants C are all independent of ε.

Proof. To begin with, note that problem (14.15) is of the form (13.8) from Section 13.3. So, with the notation from (13.8), by linearity, one can write

$$\beta_\varepsilon^{\mathrm{Dir}} = \sum_{i=1}^N u^\varepsilon_{\widehat{\chi}_i, -\partial_i u_0}. \tag{14.18}$$

Observe that by construction,

$$u_0 + \varepsilon \sum_{j=1}^N \frac{\partial u_0}{\partial x_j} \widehat{\chi}_j\Big(\frac{\cdot}{\varepsilon}\Big) + \varepsilon \beta_\varepsilon^{\mathrm{Dir}} \in H_0^1(\Omega).$$

On the other hand, in view of (14.15), $\beta_\varepsilon^{\mathrm{Dir}}$ satisfies

$$\int_\Omega A\Big(\frac{\cdot}{\varepsilon}\Big) \nabla \beta_\varepsilon^{\mathrm{Dir}} \, \nabla v \, dx = 0, \quad \forall v \in H_0^1(\Omega).$$

Using this in (14.12), yields

$$\begin{aligned} \Big| \int_\Omega A\Big(\frac{\cdot}{\varepsilon}\Big) \nabla \Big(u_\varepsilon - u_0 - \varepsilon \sum_{j=1}^N \frac{\partial u_0}{\partial x_j} \widehat{\chi}_j\Big(\frac{\cdot}{\varepsilon}\Big) - \varepsilon \beta_\varepsilon^{\mathrm{Dir}} \Big) \nabla v dx \Big| \\ \leqslant C\varepsilon \|u_0\|_{H^2(\Omega)} \|\nabla v\|_{L^2(\Omega)}. \end{aligned}$$

[1] Due to Lemma 3.29, since $u_0 \in H^2(\Omega)$ and the $\widehat{\chi}_i$'s in $\mathcal{H}_A$, the nonhomogeneous Dirichlet problem (14.15) is well-posed (see also the discussions of Chapter 13).

Choosing in this inequality as test function

$$v = u_\varepsilon - u_0 - \varepsilon \sum_{j=1}^N \frac{\partial u_0}{\partial x_j} \widehat{\chi}_j\Big(\frac{\cdot}{\varepsilon}\Big) - \varepsilon \beta_\varepsilon^{\mathrm{Dir}},$$

easily gives estimate (14.16).

Now, applying Proposition 13.3 to problem (14.15), the corresponding estimate (13.9) takes the form

$$\|\beta_\varepsilon^{\mathrm{Dir}}\|_{H^1(\Omega)} \leqslant \frac{C}{\sqrt{\varepsilon}} \|u_0\|_{H^2(\Omega)}.$$

Using it in (14.16) proves (14.14).

If $\partial\Omega$ is of class $C^{1,1}$, estimate (13.30) in Proposition 13.11 becomes

$$\|\beta_\varepsilon^{\mathrm{Dir}}\|_{H^1_\rho(\Omega)} \leqslant C\|u_0\|_{H^2(\Omega)},$$

which replaced in (14.16), leads to the L^2 and H^1_{loc} estimates of (14.17). □

Estimate (14.16) can be written as $\|U_\varepsilon\|_{H^1(\Omega)} \leqslant C\|u_0\|_{H^2(\Omega)}$, where

$$U_\varepsilon \doteq \frac{1}{\varepsilon}\Big(u_\varepsilon - u_0 - \varepsilon \sum_{j=1}^N \frac{\partial u_0}{\partial x_j} \widehat{\chi}_j\Big(\frac{\cdot}{\varepsilon}\Big) - \varepsilon \beta_\varepsilon^{\mathrm{Dir}}\Big).$$

The limit behavior of U_ε is given in the next result.

Theorem 14.4. *Supposing that u_0, the solution of the homogenized problem (14.3) is in $H^2(\Omega)$, there exist $(u_1, \widehat{u}_1)$ in $H^1_0(\Omega) \times L^2(\Omega; H^1_{per,0}(Y))$, such that*

$$\begin{aligned} &U_\varepsilon \rightharpoonup u_1 \quad \text{weakly in } H^1_0(\Omega),\\ &\mathcal{T}_\varepsilon(U_\varepsilon) \to u_1 \quad \text{strongly in } L^2(\Omega; H^1(Y)),\\ &\mathcal{T}_\varepsilon(\nabla U_\varepsilon) \rightharpoonup \nabla u_1 + \nabla_y \widehat{u}_1 \quad \text{weakly in } L^2(\Omega \times Y)^N. \end{aligned} \tag{14.19}$$

The pair $(u_1, \widehat{u}_1) \in H^1_0(\Omega) \times L^2(\Omega; H^1_{per,0}(Y))$ is the unique solution of problem

$$\begin{cases} \dfrac{1}{|Y|}\displaystyle\int_{\Omega\times Y} A(y)[\nabla u_1(x) + \nabla_y \widehat{u}_1(x,y)]\,\big[\nabla\Psi(x) + \nabla_y\Phi(x,y)\big] dxdy \\ \qquad = \dfrac{1}{|Y|}\displaystyle\int_{\Omega\times Y} F_0(x,y)\Phi(x,y)\, dxdy \\ \qquad\quad - \dfrac{1}{|Y|}\displaystyle\int_{\Omega\times Y} F_1(x,y)\cdot\big[\nabla\Psi(x) + \nabla_y\Phi(x,y)\big] dxdy, \\ \forall \Psi \in H^1_0(\Omega), \ \ \forall\Phi \in L^2(\Omega; H^1_{per,0}(Y)), \end{cases} \tag{14.20}$$

where, for a.e. $(x,y) \in \Omega \times Y$,

$$F_0(x,y) = \sum_{i=1}^N \widetilde{d}_i(y)\cdot\nabla\Big(\frac{\partial u_0}{\partial x_i}\Big)(x), \quad F_1(x,y) = A(y)\Big(\sum_{i=1}^N \widehat{\chi}_i(y)\nabla\Big(\frac{\partial u_0}{\partial x_i}\Big)(x)\Big).$$

Proof. In view of estimate (14.16), the sequence $\{U_\varepsilon\}_\varepsilon$ is uniformly bounded in $H^1_0(\Omega)$. Hence, there exist functions $u_1 \in H^1_0(\Omega)$ and $\widehat{u}_1 \in L^2(\Omega; H^1_{per,0}(Y))$ such that, up to a subsequence, convergences (14.19) hold.

From (14.13), the function U_ε satisfies, for $v \in H^1_0(\Omega)$,

$$\begin{aligned}\int_\Omega A\Big(\frac{\cdot}{\varepsilon}\Big)\nabla U_\varepsilon\,\nabla v\,dx = \sum_{i=1}^N\Big[\frac{1}{\varepsilon}\int_\Omega \widetilde{d}_i\Big(\frac{\cdot}{\varepsilon}\Big)\cdot\nabla\Big(\frac{\partial u_0}{\partial x_i}\Big)v\,dx\\ -\int_\Omega A\Big(\frac{\cdot}{\varepsilon}\Big)\nabla\Big(\frac{\partial u_0}{\partial x_i}\Big)\,\widehat{\chi}_i\Big(\frac{\cdot}{\varepsilon}\Big)\,\nabla v\,dx\Big].\end{aligned} \tag{14.21}$$

We transform this equality by unfolding and then, proceeding as in the proof of Theorem 3.5 , we obtain thanks to convergence (14.8), the equation from (14.20). By Lax-Milgram's theorem, problem (14.20) has a unique solution. Consequently, convergences (14.19) hold for the whole sequence $\{(U_\varepsilon, \mathcal{T}_\varepsilon(\nabla U_\varepsilon)\}_\varepsilon$. □

Introduce the second order correctors $\widehat{\chi}_{ij} \in H^1_{per,0}(Y)$ for the Dirichlet condition, defined for $(i,j) \in \{1,\dots,N\}^2$, as the solutions of the cell problems,

$$\begin{cases}\displaystyle\int_Y A(y)\nabla_y\widehat{\chi}_{ij}(y)\,\nabla_y\Phi(y)\,dy = \int_Y \widetilde{d}_i(y)\cdot\mathbf{b}_j\Phi(y)\,dy\\ \displaystyle\qquad\qquad -\int_Y A(y)\widehat{\chi}_i(y)\mathbf{b}_j\cdot\nabla_y\Phi(y)\,dy,\\ \forall\Phi\in H^1_{per,0}(Y).\end{cases}$$

We now express $\widehat{u}_1$ in terms of the first correctors and the second order correctors for a.e. $(x,y) \in \Omega\times Y$, in the form

$$\widehat{u}_1(x,y) = \sum_{i=1}^N \frac{\partial u_1}{\partial x_i}(x)\,\widehat{\chi}_i(y) + \sum_{i,j=1}^N \frac{\partial^2 u_0}{\partial x_i\partial x_j}\,\widehat{\chi}_{ij}(y).$$

It is easily checked that u_1 is the solution of the homogenized problem,

$$\begin{cases}\displaystyle\int_\Omega A^{hom}\nabla u_1\,\nabla\Psi\,dx = -\int_\Omega\sum_{i,j=1}^N\Big(\int_Y A(y)\nabla_y\widehat{\chi}_{ij}(y)\Big)\frac{\partial^2 u_0}{\partial x_i\partial x_j}\cdot\nabla\Psi\,dx\\ \displaystyle\qquad\qquad -\int_\Omega\sum_{i,j=1}^N\Big(\int_Y A(y)\widehat{\chi}_i(y)\mathbf{b}_j\Big)\frac{\partial^2 u_0}{\partial x_i\partial x_j}\cdot\nabla\Psi\,dx,\\ \forall\Psi\in H^1_0(\Omega).\end{cases}$$

Proposition 14.5. *Under the assumption of Theorem 14.4, the following strong convergences hold:*

$$\mathcal{T}_\varepsilon(\nabla U_\varepsilon) \to \nabla u_1 + \nabla_y\widehat{u}_1 \quad \textit{strongly in} \quad L^2(\Omega\times Y)^N,$$

$$U_\varepsilon - u_1 - \varepsilon\sum_{i=1}^N\frac{\partial u_1}{\partial x_i}\widehat{\chi}_i\Big(\frac{\cdot}{\varepsilon}\Big) - \varepsilon\sum_{i,j=1}^N \mathcal{Q}_\varepsilon\Big(\frac{\partial^2 u_0}{\partial x_i\partial x_j}\Big)\widehat{\chi}_{ij}\Big(\frac{\cdot}{\varepsilon}\Big)\to 0$$

$$\textit{strongly in } H^1(\Omega).$$

Moreover, if u_0 belongs to $H^3(\Omega)$, one has the strong convergence in $H^1(\Omega)$,

$$U_\varepsilon - u_1 - \varepsilon \sum_{i=1}^N \frac{\partial u_1}{\partial x_i}\widehat{\chi}_i\Big(\frac{\cdot}{\varepsilon}\Big) - \varepsilon \sum_{i,j=1}^N \frac{\partial^2 u_0}{\partial x_i \partial x_j}\widehat{\chi}_{ij}\Big(\frac{\cdot}{\varepsilon}\Big) \to 0.$$

Proof. The convergence of the energy, namely

$$\begin{cases} \lim\limits_{\varepsilon\to 0}\displaystyle\int_\Omega A^\varepsilon \nabla U_\varepsilon \nabla U_\varepsilon\, dx \\ \qquad = \dfrac{1}{|Y|}\displaystyle\int_{\Omega\times Y} A\big[\nabla u_1 + \nabla_y \widehat{u}_1\big]\big[\nabla u_1 + \nabla_y \widehat{u}_1\big]\, dxdy, \\ \lim\limits_{\varepsilon\to 0}\displaystyle\int_{\Lambda_\varepsilon} |\nabla U_\varepsilon|^2\, dx = 0, \end{cases} \tag{14.22}$$

follows by reasoning as in the proof of Proposition 3.9. By standard weak lower-semi-continuity and recalling (14.21),

$$\begin{aligned} &\frac{1}{|Y|}\int_{\Omega\times Y} A\big[\nabla u_1 + \nabla_y \widehat{u}_1\big]\big[\nabla u_1 + \nabla_y \widehat{u}_1\big] dxdy \\ &\qquad \leqslant \lim_{\varepsilon\to 0}\sum_{i=1}^N \Big[\frac{1}{\varepsilon}\int_\Omega \widetilde{d}_i\Big(\frac{\cdot}{\varepsilon}\Big)\cdot\nabla\Big(\frac{\partial u_0}{\partial x_i}\Big) U_\varepsilon\, dx \\ &\qquad\qquad - \int_\Omega A\Big(\frac{\cdot}{\varepsilon}\Big)\nabla\Big(\frac{\partial u_0}{\partial x_i}\Big)\,\widehat{\chi}_i\Big(\frac{\cdot}{\varepsilon}\Big)\,\nabla U_\varepsilon\, dx\Big]. \end{aligned}$$

Each term of the above sum is estimated by using (3.45). One has for every $i \in \{1,\dots,N\}$,

$$\begin{aligned} &\Big|\frac{1}{\varepsilon}\int_{\Lambda_\varepsilon} \widetilde{d}_i\Big(\frac{\cdot}{\varepsilon}\Big)\cdot\nabla\Big(\frac{\partial u_0}{\partial x_i}\Big) U_\varepsilon\, dx - \int_{\Lambda_\varepsilon} A^\varepsilon\nabla\Big(\frac{\partial u_0}{\partial x_i}\Big)\,\widehat{\chi}_i\Big(\frac{\cdot}{\varepsilon}\Big)\cdot\nabla U_\varepsilon\, dx\Big| \\ &\qquad \leqslant C\|u_0\|_{H^2(\Omega^b_{\varepsilon d(Y)})}. \end{aligned} \tag{14.23}$$

We now pass to the limit in the left-hand side of this estimate. By (14.8),

$$\begin{aligned} &\lim_{\varepsilon\to 0}\frac{1}{\varepsilon}\int_\Omega \widetilde{d}_i\Big(\frac{\cdot}{\varepsilon}\Big)\cdot\nabla\Big(\frac{\partial u_0}{\partial x_i}\Big) U_\varepsilon\, dx \\ &\qquad = \frac{1}{|Y|}\int_{\Omega\times Y} \widetilde{d}_i(y)\cdot\nabla\Big(\frac{\partial u_0}{\partial x_i}\Big)(x)\,\widehat{u}_1(x,y)\, dxdy. \end{aligned} \tag{14.24}$$

To treat the second term in (14.23), we unfold the right-hand side of equality (14.21) with v replaced by U_ε. Then, by Proposition 1.9(i) and (14.19),

$$\begin{aligned} &\lim_{\varepsilon\to 0}\int_\Omega A^\varepsilon\nabla\Big(\frac{\partial u_0}{\partial x_i}\Big)\,\widehat{\chi}_i\Big(\frac{\cdot}{\varepsilon}\Big)\,\nabla U_\varepsilon\, dx \\ &\qquad = \frac{1}{|Y|}\int_{\Omega\times Y} A\,\nabla\Big(\frac{\partial u_0}{\partial x_i}\Big)\,\widehat{\chi}_i\,\big(\nabla u_1 + \nabla_y \widehat{u}_1\big)\, dxdy, \end{aligned}$$

which, together with (14.24), ends the proof of convergences (14.22).

Proceeding as in the proofs of Corollary 3.10 and Theorem 3.27, we obtain the strong convergences in Proposition 14.5 (the last convergence being a consequence of Lemma 3.29). □

We end this section with a result giving the asymptotic behavior of the sequence $\{\beta_\varepsilon^{\mathrm{Dir}}\}_\varepsilon$.

First, consider a limit point Ψ in $H^1_\rho(\Omega)^N$ of the sequence (see (14.18)),

$$\Big\{\sum_{i=1}^N u^\varepsilon_{\widehat{\chi}_i,\,-\partial_i u_0}\mathbf{b}_i\Big\}_\varepsilon,$$

so that, there exists a subsequence of ε (denoted ε') such that the following convergence holds:

$$\sum_{i=1}^N u^{\varepsilon'}_{\widehat{\chi}_i,\,-\partial_i u_0}\mathbf{b}_i \rightharpoonup \Psi \qquad \text{weakly in } H^1_\rho(\Omega)^N.$$

Let Θ be the trace of Ψ on $\partial\Omega$ in the sense of Proposition 13.8, i.e.,

$$\Theta = \mathbb{T}_{A^{hom}}(\Psi).$$

The function Θ belongs to $L^\infty(\partial\Omega)^N$ by Proposition 13.10. The question is what is the relationship between Θ and the limit of $\{\beta_\varepsilon^{\mathrm{Dir}}\}_\varepsilon$. Here is the answer.

Theorem 14.6. *Let $\beta_{\varepsilon'}^{Dir}$ be the solution of problem* (14.15). *One has the convergence,*

$$\beta_{\varepsilon'}^{Dir} \rightharpoonup \beta^{Dir} \qquad \textit{weakly in } H^1_\rho(\Omega),$$

where the function $\beta^{Dir}\in H^1_\rho(\Omega)$ is the solution of the problem

$$\begin{cases}\operatorname{div}\big(A^{hom}\nabla\beta^{Dir}\big) = 0 & \textit{in} \quad \Omega,\\ \beta^{Dir} = \Theta\cdot\nabla u_0 & \textit{on} \quad \partial\Omega,\end{cases}$$

in the following sense:

$$\begin{cases}\displaystyle\int_\Omega \beta^{Dir}\operatorname{div}\big({}^tA^{hom}\,\nabla\phi\big)dx = \int_{\partial\Omega}\big(\Theta\cdot\nabla u_{0|\partial\Omega}\big)\big({}^tA^{hom}\,\nabla\phi\cdot n\big)\,d\sigma,\\ \forall\phi\in H^2(\Omega)\cap H^1_0(\Omega).\end{cases}$$

Furthermore,

$$\beta_{\varepsilon'}^{Dir} - \beta^{Dir} - \varepsilon'\sum_{i=1}^N \frac{\partial\beta^{Dir}}{\partial x_i}\widehat{\chi}_i\Big(\frac{\cdot}{\varepsilon'}\Big) \to 0 \quad \textit{strongly in } H^1_{loc}(\Omega). \tag{14.25}$$

Proof. By Theorem 13.17, the sequence $\{\beta_{\varepsilon'}^{\mathrm{Dir}}\}_{\varepsilon'}$ converges weakly in $H^1_\rho(\Omega)$ to the function

$$\beta^{\mathrm{Dir}} \doteq \sum_{i=1}^N u_{\widehat{\chi}_i,\,-\partial_i u_0},$$

whose trace in $H^{-1/2}(\partial\Omega)$ is given by

$$\mathbb{T}_{A^{hom}}(\beta^{\mathrm{Dir}}) = -\sum_{i=1}^{N} \theta_{\widehat{\chi}_i} \frac{\partial u_0}{\partial x_i}{}_{|\partial\Omega} \doteq \Theta \cdot \nabla u_{0|\partial\Omega}.$$

Finally, convergence (14.25) is a consequence of Corollary 13.21. □

14.1.1 An example

To conclude this section, we consider a particular type of convex polyhedral domain Ω: we assume that Ω is the convex hull of a finite set of points belonging to $\mathbf{G}$ (see Figure 14.1). We also assume that $\varepsilon = 1/k$, $k \in \mathbb{N}^*$.

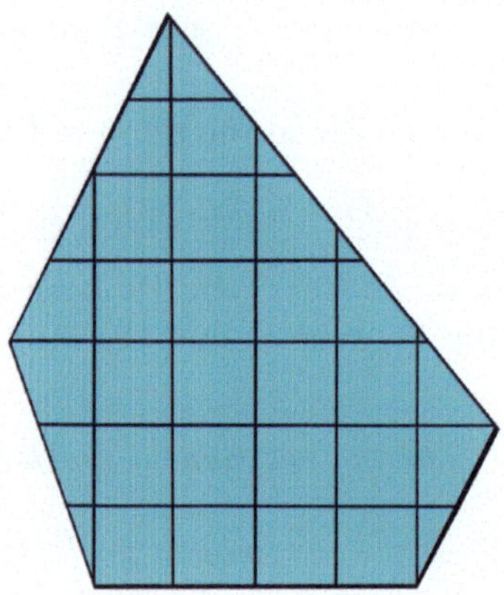

Figure 14.1: A convex polyhedral domain

Under these conditions the whole sequence $\{\beta_\varepsilon^{\mathrm{Dir}}\}_\varepsilon$ converges weakly. Furthermore, the following convergences hold:

$$\frac{1}{\varepsilon}\Big(u_\varepsilon - u_0 - \varepsilon \sum_{j=1}^{N} \frac{\partial u_0}{\partial x_j}\, \widehat{\chi}_j\Big(\frac{\cdot}{\varepsilon}\Big)\Big) \rightharpoonup \beta^{\mathrm{Dir}} + u_1 \quad \text{weakly in } H^1_\rho(\Omega),$$

as well as

$$\frac{1}{\varepsilon}\Big(u_\varepsilon - u_0 - \varepsilon \sum_{j=1}^{N} \frac{\partial u_0}{\partial x_j}\, \widehat{\chi}_j\Big(\frac{\cdot}{\varepsilon}\Big)\Big) - \beta^{\mathrm{Dir}} - \varepsilon\widehat{\beta}^{\mathrm{Dir}}\Big(\cdot, \Big\{\frac{\cdot}{\varepsilon}\Big\}\Big)$$
$$- u_1 - \varepsilon \widehat{u}_1\Big(\cdot, \Big\{\frac{\cdot}{\varepsilon}\Big\}\Big) \to 0 \quad \text{strongly in } H^1_{loc}(\Omega),$$

where β^{Dir} is given by Theorem 14.6 (with Θ determined on each face of $\partial\Omega$ thanks to Theorem 13.28), $\widehat{\beta}^{\mathrm{Dir}}$ is given by (13.51), and $(u_1, \widehat{u}_1)$ is given by Theorem 14.4.

Remark 14.7. *By [132], the solutions of the homogenized problems* (3.18) *and* (3.33) *are in* $H^2(\Omega)$ *for a right-hand side in* $L^2(\Omega)$*, when* Ω *is a convex polyhedron whose edges are orthogonal to vectors in* $\mathbb{Q}^N$ *(as in the example presented above). In this case, in general, there can be infinitely many limit points* Θ *and they are constant on each face of* $\partial\Omega$ *(see Proposition 13.22, Theorem 13.28 and also [111, 112]).*

14.2 Error estimates for the Neumann problem

In this section we consider the Neumann problem (14.2). It was shown in Subsection 3.1.2 that

$$u_\varepsilon \rightharpoonup u_0' \quad \text{weakly in } V(\Omega),$$

where

$$V(\Omega) = \{w \in H^1(\Omega) \mid \mathcal{M}_\Omega(w) = 0\},$$

u_0' is the unique solution of the homogenized problem (3.33), that is,

$$\int_\Omega A^{hom} \nabla u_0' \, \nabla \varphi \, dx = \int_\Omega f \, \varphi \, dx, \quad \forall \varphi \in V(\Omega). \tag{14.26}$$

The homogenized matrix A^{hom} is defined by (3.19).

Our aim is to give a corrector result, the equivalent for this problem, of Theorem 14.3 for the Dirichlet case from the preceding section; the procedure here will be the same.

We start with some properties of the function $\widetilde{d}$ defined by (14.4), completing the ones given in Lemma 14.1.

Lemma 14.8. *For every* $(v, w) \in H^1(\Omega) \times H^1(\Omega)$,

$$\begin{aligned} &\Big| \int_\Omega \widetilde{d}\Big(\frac{\cdot}{\varepsilon}\Big) \cdot \nabla v \, w \, dx \Big| \\ &\qquad \leqslant C \varepsilon^{1/2} \|w\|_{H^1(\Omega)} \big(\varepsilon^{1/2} \|\nabla v\|_{L^2(\Omega)} + \|\nabla v\|_{L^2(\Omega^b_{5\varepsilon d(Y)})} \big). \end{aligned} \tag{14.27}$$

If v *belongs to* $H^1_0(\Omega)$*, then*

$$\Big| \int_\Omega \widetilde{d}\Big(\frac{\cdot}{\varepsilon}\Big) \cdot \nabla v \, w \, dx \Big| \leqslant C \varepsilon \|w\|_{H^1(\Omega)} \|\nabla v\|_{L^2(\Omega)}. \tag{14.28}$$

Moreover, for every $(V, w) \in H^1(\Omega)^N \times H^1(\Omega)$,

$$\Big| \int_\Omega \widetilde{d}\Big(\frac{\cdot}{\varepsilon}\Big) \cdot V \, w \, dx \Big| \leqslant C \varepsilon \|w\|_{H^1(\Omega)} \|V\|_{H^1(\Omega)}. \tag{14.29}$$

The constants C *do not depend on* ε.

Proof. Step 1. Proof of (14.27).

We proceed as in the proof of Lemma 14.1, following the lines of the proof of (14.5). We replace $\mathbb{R}^N$ by $\widehat{\Omega}_\varepsilon$ and the estimates from (14.10) by the following ones:

$$\begin{aligned}
&\|w - \mathcal{Q}_\varepsilon(w)\|_{L^2(\widehat{\Omega}_\varepsilon)} + \varepsilon\|\nabla(w - \mathcal{Q}_\varepsilon(w))\|_{L^2(\widehat{\Omega}_\varepsilon)} \leqslant C\varepsilon\|\nabla w\|_{L^2(\Omega)},\\
&\|\mathcal{Q}_\varepsilon(w) - \mathcal{M}_\varepsilon(w)\|_{L^2(\widehat{\Omega}_\varepsilon)} \leqslant C\varepsilon\|\nabla w\|_{L^2(\Omega)},\\
&\|\mathcal{M}_\varepsilon(w) - w\|_{L^2(\widehat{\Omega}_\varepsilon)} \leqslant C\varepsilon\|\nabla w\|_{L^2(\Omega)}.
\end{aligned}$$

First, one has

$$\begin{aligned}
\Big|\int_{\widehat{\Omega}_\varepsilon} \widetilde{d}\Big(\frac{\cdot}{\varepsilon}\Big)\cdot\nabla v\, w\, dx - \frac{1}{|Y|}\int_{\Omega\times Y} \widetilde{d}(y)\cdot\mathcal{T}_\varepsilon(\nabla v)(x,y)\, w(x)\, dxdy\Big|&\\
\leqslant C\varepsilon\|\nabla w\|_{L^2(\Omega)}\|\nabla v\|_{L^2(\Omega)}.&
\end{aligned}\tag{14.30}$$

Estimates (3.46) and (12.4) imply that (recall the notations (12.1))

$$\Big|\int_{\Lambda_\varepsilon} \widetilde{d}\Big(\frac{\cdot}{\varepsilon}\Big)\cdot\nabla v\, w\, dx\Big| \leqslant C\varepsilon^{1/2}\|w\|_{H^1(\Omega)}\|\nabla v\|_{L^2(\Omega^{bl}_{2\varepsilon d(Y)})}.\tag{14.31}$$

Then, (14.30) and (14.31) lead to

$$\begin{aligned}
&\Big|\int_{\Omega} \widetilde{d}\Big(\frac{\cdot}{\varepsilon}\Big)\cdot\nabla v\, w\, dx - \frac{1}{|Y|}\int_{\Omega\times Y} \widetilde{d}\cdot\mathcal{T}_\varepsilon(\nabla v)\, w\, dxdy\Big|\\
&\qquad\leqslant C\varepsilon\|\nabla w\|_{L^2(\Omega)}\|\nabla v\|_{L^2(\Omega)} + C\varepsilon^{1/2}\|w\|_{H^1(\Omega)}\|\nabla v\|_{L^2(\Omega^{bl}_{2\varepsilon d(Y)})}.
\end{aligned}$$

By Theorem 13.44 there exists $\widehat{v}_\varepsilon \in L^2(\Omega; H^1_{per}(Y))$ such that

$$\begin{aligned}
&\|\mathcal{T}_\varepsilon(\nabla v) - \nabla v - \nabla_y\widehat{v}_\varepsilon\|_{L^2(Y;(H^1(\Omega))')}\\
&\qquad\leqslant C\varepsilon\|\nabla v\|_{L^2(\Omega)} + C\varepsilon^{1/2}\|\nabla v\|_{L^2(\Omega^{b}_{5\varepsilon d(Y)})}.
\end{aligned}$$

Therefore

$$\begin{aligned}
&\Big|\int_{\Omega} \widetilde{d}\Big(\frac{\cdot}{\varepsilon}\Big)\cdot\nabla v\, w\, dx - \frac{1}{|Y|}\int_{\Omega\times Y} \widetilde{d}(y)\cdot(\nabla v + \nabla_y\widehat{v}_\varepsilon)\, w\, dxdy\Big|\\
&\qquad\leqslant C\|w\|_{H^1(\Omega)}\big(\varepsilon\|\nabla v\|_{L^2(\Omega)} + \varepsilon^{1/2}\|\nabla v\|_{L^2(\Omega^{b}_{5\varepsilon d(Y)})}\big),
\end{aligned}$$

which is precisely estimate (14.27), since

$$\int_Y \widetilde{d}(y)\cdot(\nabla v(x) + \nabla_y\widehat{v}_\varepsilon(x,y))\, dy = 0 \quad \text{for a.e. } x\in\Omega.$$

Step 2. Proof of (14.28).

It is similar to that of Step 1 of the proof of Lemma 14.1.

Step 3. Proof of (14.29).

Note first that due to (3.46) and (12.4),

$$\Big|\int_{\Lambda_\varepsilon} \widetilde{d}\Big(\frac{\cdot}{\varepsilon}\Big)\cdot V\, w\, dx\Big| \leqslant C\varepsilon\big(\|w\|_{L^2(\partial\Omega)} + \varepsilon^{1/2}\|\nabla w\|_{L^2(\Omega^{bl}_{2\varepsilon d(Y)})}\big) \times\big(\|V\|_{L^2(\partial\Omega)^N} + \varepsilon^{1/2}\|V\|_{H^1(\Omega^{bl}_{2\varepsilon d(Y)})}\big). \tag{14.32}$$

Then, proceeding as in the proof of (14.27), we get

$$\Big|\int_{\widehat{\Omega}_\varepsilon} \widetilde{d}\Big(\frac{\cdot}{\varepsilon}\Big)\cdot V\, w\, dx - \frac{1}{|Y|}\int_{\widehat{\Omega}_\varepsilon\times Y} \widetilde{d}(y)\cdot \mathcal{T}_\varepsilon(V)(x,y)\, w(x)\, dxdy\Big| \leqslant C\varepsilon\|\nabla w\|_{L^2(\Omega)}\|V\|_{L^2(\Omega)}.$$

In this estimate, one can substitute V by $\mathcal{T}_\varepsilon(V)$ to obtain a similar one since (see $(1.39)_2$ in Proposition 1.38).

$$\|\mathcal{T}_\varepsilon(V) - V\|_{L^2(\widehat{\Omega}_\varepsilon\times Y)} \leqslant C\varepsilon\|V\|_{H^1(\Omega)}. \tag{14.33}$$

On the other hand, as $\mathcal{M}_Y(\widetilde{d}) = 0$,

$$\int_{\widehat{\Omega}_\varepsilon\times Y} \widetilde{d}(y)\cdot V(x)\, w(x)\, dxdy = 0.$$

Thus

$$\Big|\int_{\widehat{\Omega}_\varepsilon} \widetilde{d}\Big(\frac{\cdot}{\varepsilon}\Big)\cdot V\, w\, dx\Big| \leqslant C\varepsilon\|\nabla w\|_{L^2(\Omega)}\|V\|_{L^2(\Omega)}.$$

Inequality (14.29) then follows with (14.32). □

We now introduce the corrector $\beta^{\rm Ne}_\varepsilon \in H^1(\Omega)$ for the homogeneous Neumann condition. It will play an essential role in the estimates for the solution of problem (14.2). It is defined by

$$\begin{cases} \displaystyle\int_\Omega A^\varepsilon\nabla\beta^{\rm Ne}_\varepsilon\,\nabla v\, dx = -\sum_{i=1}^N \frac{1}{\varepsilon}\int_\Omega \widetilde{d}_i\Big(\frac{\cdot}{\varepsilon}\Big)\cdot\nabla v\,\frac{\partial u_0'}{\partial x_i}\, dx,\\ \displaystyle\int_\Omega \beta^{\rm Ne}_\varepsilon\, dx = 0,\\ \forall v\in H^1(\Omega), \end{cases} \tag{14.34}$$

where $\widetilde{d}_i$ is given by (14.11).

Proposition 14.9. *The function β^{Ne}_ε defined by* (14.34)*, satisfies the following estimates:*

$$\begin{aligned} &\|\nabla\beta^{Ne}_\varepsilon\|_{L^2(\Omega)} \leqslant \frac{C}{\sqrt{\varepsilon}}\|u_0'\|_{H^2(\Omega)},\\ &\|\beta^{Ne}_\varepsilon\|_{L^2(\Omega)} \leqslant C\|u_0'\|_{H^2(\Omega)}. \end{aligned} \tag{14.35}$$

Proof. First, estimate $(14.35)_1$ is an immediate consequence of the variational formulation (14.34) and of estimate (14.27).

Now, we prove $(14.35)_2$. Due to (14.29), for every test function $U \in H^2(\Omega)$ in (14.34), one has

$$\Big|\int_\Omega A^\varepsilon \nabla \beta_\varepsilon^{\rm Ne}\, \nabla U\, dx\Big| \leqslant C\|u_0'\|_{H^2(\Omega)}\|U\|_{H^2(\Omega)}. \tag{14.36}$$

We now proceed as in the proof of the L^2-estimate of (13.30). Transforming the integral of (14.36) by unfolding gives

$$\begin{aligned}\Big|\frac{1}{|Y|}\int_{\widehat{\Omega}_\varepsilon\times Y} A(y)\mathcal{T}_\varepsilon(\nabla\beta_\varepsilon^{\rm Ne})\,\mathcal{T}_\varepsilon(\nabla U)\,dxdy + \int_{\Lambda_\varepsilon} A^\varepsilon\nabla\beta_\varepsilon^{\rm Ne}\,\nabla U\,dx\Big|&\\ \leqslant C\|u_0'\|_{H^2(\Omega)}\|U\|_{H^2(\Omega)}.&\end{aligned}$$

On the one hand, from (12.4),

$$\|\nabla U\|_{L^2(\Lambda_\varepsilon)} \leqslant \|\nabla U\|_{L^2(\Omega^{bl}_{\varepsilon d(Y)})} \leqslant C\varepsilon^{1/2}\|U\|_{H^2(\Omega)},$$

therefore,

$$\Big|\int_{\Lambda_\varepsilon} A^\varepsilon\nabla\beta_\varepsilon^{\rm Ne}\,\nabla U\,dx\Big| \leqslant C\varepsilon^{1/2}\|\beta_\varepsilon^{\rm Ne}\|_{H^1(\Omega)}\|U\|_{H^2(\Omega)}.$$

Using (14.33), one replaces $\mathcal{T}_\varepsilon(\nabla U)$ by ∇U. That yields

$$\Big|\int_{\widehat{\Omega}_\varepsilon\times Y} A(y)\mathcal{T}_\varepsilon(\nabla\beta_\varepsilon^{\rm Ne})\,\nabla U\,dxdy\Big| \leqslant C\|u_0'\|_{H^2(\Omega)}\|U\|_{H^2(\Omega)}. \tag{14.37}$$

On the other hand, by Theorem 13.44, there is an element $\widehat{\beta}_\varepsilon^{\rm Ne}$ in the space $L^2(\Omega; H^1_{per}(Y))$, such that (recall estimate $(14.35)_1$)

$$\begin{aligned}\|\mathcal{T}_\varepsilon(\nabla\beta_\varepsilon^{\rm Ne}) - \nabla\beta_\varepsilon^{\rm Ne} - \nabla_y\widehat{\beta}_\varepsilon^{\rm Ne}\|_{L^2(Y;(H^1(\Omega))')} &\leqslant C\varepsilon^{1/2}\|\nabla\beta_\varepsilon^{\rm Ne}\|_{L^2(\Omega)}\\ &\leqslant C\|u_0'\|_{H^2(\Omega)}.\end{aligned} \tag{14.38}$$

To summarize, (14.37)-(14.38) give

$$\Big|\int_{\Omega\times Y} A(\nabla\beta_\varepsilon^{\rm Ne} + \nabla_y\widehat{\beta}_\varepsilon^{\rm Ne})\,\nabla U\,dxdy\Big| \leqslant C\|u_0'\|_{H^2(\Omega)}\|U\|_{H^2(\Omega)}. \tag{14.39}$$

Now we proceed as in Step 2 of the proof of Proposition 13.3. Let $\overline{\chi}_i$ ($i \in \{1,\dots,N\}$) in $H^1_{per}(Y)$ be the adjoint correctors defined by (13.37). Consider the function $\widehat{U}_\varepsilon$ in $H^1(\Omega)$ defined for a.e. $x \in \Omega$ by

$$\widehat{U}_\varepsilon(x) = \varepsilon\frac{\partial U}{\partial x_i}(x)\overline{\chi}_i\Big(\frac{x}{\varepsilon}\Big).$$

By Corollary 1.72 and (12.4),

$$\begin{aligned}
&\Big\|\nabla\widehat{U}_\varepsilon - \frac{\partial U}{\partial x_i}\nabla_y \overline{\chi}\Big(\frac{\cdot}{\varepsilon}\Big)\Big\|_{L^2(\Omega)} \leqslant C\varepsilon\|U\|_{H^2(\Omega)}\|\overline{\chi}_i\|_{L^2(Y)},\\
&\|\nabla\widehat{U}_\varepsilon\|_{L^2(\Omega)} \leqslant C\|U\|_{H^2(\Omega)}\|\overline{\chi}_i\|_{H^1(Y)},\\
&\|\nabla\widehat{U}_\varepsilon\|_{L^2(\Omega^{bl}_{\varepsilon d(Y)})} \leqslant C\varepsilon^{1/2}\|U\|_{H^2(\Omega)}\|\overline{\chi}_i\|_{H^1(Y)}.
\end{aligned} \tag{14.40}$$

Choosing $\widehat{U}_\varepsilon$ as test function in (14.34), estimates (14.40) together with inequality (14.27), lead to

$$\Big|\int_\Omega A^\varepsilon\nabla\beta^{\rm Ne}_\varepsilon\,\nabla\widehat{U}_\varepsilon\,dx\Big| \leqslant C\|U\|_{H^2(\Omega)}\|\overline{\chi}_i\|_{H^1(Y)}.$$

Since Λ_ε is included in $\Omega^{bl}_{\varepsilon d(Y)}$, using estimate $(14.40)_3$ and the H^1-estimate in (14.35) of $\beta^{\rm Ne}_\varepsilon$, we get

$$\Big|\int_{\Lambda_\varepsilon} A^\varepsilon\nabla\beta^{\rm Ne}_\varepsilon\,\nabla\widehat{U}_\varepsilon\,dx\Big| \leqslant C\|U\|_{H^2(\Omega)}\|\overline{\chi}_i\|_{H^1(Y)},$$

hence,

$$\Big|\int_{\widehat{\Omega}_\varepsilon} A^\varepsilon\nabla\beta^{\rm Ne}_\varepsilon\,\nabla\widehat{U}_\varepsilon\,dx\Big| \leqslant C\|U\|_{H^2(\Omega)}\|\overline{\chi}_i\|_{H^1(Y)}.$$

Thanks to $(14.40)_1$ one can replace in the integral $\nabla\widehat{U}_\varepsilon$ by $\dfrac{\partial U}{\partial x_i}\nabla_y\overline{\chi}_i\Big(\dfrac{\cdot}{\varepsilon}\Big)$ and preserve the same type of estimate. Unfolding the result yields

$$\begin{aligned}
\Big|\int_{\widehat{\Omega}_\varepsilon\times Y} A\mathcal{T}_\varepsilon(\nabla\beta^{\rm Ne}_\varepsilon)\,\mathcal{T}_\varepsilon\Big(\frac{\partial U}{\partial x_i}\Big)\nabla_y\overline{\chi}_i\,dxdy\Big|&\\
\leqslant C\|u_0'\|_{H^2(\Omega)}\|U\|_{H^2(\Omega)}\|\overline{\chi}_i\|_{H^1(Y)}&.
\end{aligned}$$

This inequality, estimate $(1.39)_2$ and a summation from $i=1$ to N, give

$$\begin{aligned}
\Big|\int_{\Omega\times Y} A(\nabla\beta^{\rm Ne}_\varepsilon + \nabla_y\widehat{\beta}^{\rm Ne}_\varepsilon)\Big(\sum_{i=1}^N\frac{\partial U}{\partial x_i}\nabla_y\overline{\chi}_i\Big)\,dxdy\Big|&\\
\leqslant C\|u_0'\|_{H^2(\Omega)}\|U\|_{H^2(\Omega)}\|\overline{\chi}_i\|_{H^1(Y)}&.
\end{aligned}$$

A similar computation using the original correctors $\widehat{\chi}_i$'s defined by (3.10), leads to

$$\int_{\Omega\times Y} A\Big(\nabla\beta^{\rm Ne}_\varepsilon + \sum_{i=1}^N\frac{\partial\beta^{\rm Ne}_\varepsilon}{\partial x_i}\nabla_y\widehat{\chi}_i\Big)\nabla_y\Big(\sum_{j=1}^N\frac{\partial U}{\partial x_j}\overline{\chi}_j\Big)\,dxdy = 0.$$

Consequently,

$$\begin{aligned}
\Big|\int_{\Omega\times Y} A\nabla_y\Big(\widehat{\beta}^{\rm Ne}_\varepsilon - \sum_{i=1}^N\frac{\partial\beta^{\rm Ne}_\varepsilon}{\partial x_i}\widehat{\chi}_i\Big)\nabla_y\Big(\sum_{j=1}^N\frac{\partial U}{\partial x_j}\overline{\chi}_j\Big)\,dxdy\Big|&\\
\leqslant C\|u_0'\|_{H^2(\Omega)}\|U\|_{H^2(\Omega)}&.
\end{aligned}$$

Going back to definition (13.37) of the $\overline{\chi}_j$'s, one obtains

$$\Big|\int_{\Omega\times Y} A\Big(\nabla_y \widehat{\beta}_\varepsilon^{\mathrm{Ne}} - \sum_{i=1}^{N} \frac{\partial \beta_\varepsilon^{\mathrm{Ne}}}{\partial x_i} \nabla_y \widehat{\chi}_i\Big) \nabla U \, dx dy\Big| \leqslant C \|u_0'\|_{H^2(\Omega)} \|U\|_{H^2(\Omega)}.$$

This estimate and (14.39) imply

$$\Big|\int_{\Omega\times Y} A\Big(\nabla \beta_\varepsilon^{\mathrm{Ne}} + \sum_{i=1}^{N} \frac{\partial \beta_\varepsilon^{\mathrm{Ne}}}{\partial x_i} \nabla_y \widehat{\chi}_i\Big) \nabla U \, dx dy\Big| \leqslant C \|u_0'\|_{H^2(\Omega)} \|U\|_{H^2(\Omega)}.$$

By definition of the homogenized matrix A^{hom}, this reads

$$\Big|\int_{\Omega} A^{hom} \nabla \beta_\varepsilon^{\mathrm{Ne}} \, \nabla U \, dx\Big| \leqslant C \|u_0'\|_{H^2(\Omega)} \|U\|_{H^2(\Omega)}. \tag{14.41}$$

Now, let V_ε in $H^1(\Omega)$ be the solution of the following variational problem:

$$\begin{cases} \displaystyle\int_{\Omega} A^{hom} \nabla v \, \nabla V_\varepsilon \, dx = \int_{\Omega} v \, \beta_\varepsilon^{\mathrm{Ne}} \, dx, \\ \displaystyle\int_{\Omega} V_\varepsilon \, dx = 0, \\ \forall v \in H^1(\Omega). \end{cases}$$

It is known that V_ε belongs to $H^2(\Omega)$ and satisfies

$$\|V_\varepsilon\|_{H^2(\Omega)} \leqslant C \|\beta_\varepsilon^{\mathrm{Ne}}\|_{L^2(\Omega)},$$

with a constant not depending on ε. As a consequence, taking $U \equiv V_\varepsilon$ in estimate (14.41), gives the required L^2-estimate $(14.35)_2$ of $\beta_\varepsilon^{\mathrm{Ne}}$. □

We are now in a position to give the main theorem of this section.

Theorem 14.10. *Let u_0' be the solution of the homogenized problem* (14.26). *Assume that it belongs to $H^2(\Omega)$. Then the solution u_ε of the Neumann problem* (14.2) *satisfies the following estimate:*

$$\Big\| u_\varepsilon - u_0' - \varepsilon \sum_{i=1}^{N} \frac{\partial u_0'}{\partial x_i} \widehat{\chi}_i\Big(\frac{\cdot}{\varepsilon}\Big)\Big\|_{H^1(\Omega)} \leqslant C \varepsilon^{1/2} \|u_0'\|_{H^2(\Omega)}. \tag{14.42}$$

Furthermore,

$$\Big\| u_\varepsilon - u_0' - \varepsilon \sum_{i=1}^{N} \frac{\partial u_0'}{\partial x_i} \widehat{\chi}_i\Big(\frac{\cdot}{\varepsilon}\Big) - \varepsilon \beta_\varepsilon^{\mathrm{Ne}}\Big\|_{H^1(\Omega)} \leqslant C \varepsilon \|u_0'\|_{H^2(\Omega)}. \tag{14.43}$$

Suppose that $\partial\Omega$ is of class $C^{1,1}$. Then u_ε satisfies

$$\begin{aligned} \|u_\varepsilon - u_0'\|_{L^2(\Omega)} + \Big\| \rho\Big(\nabla u_\varepsilon - \nabla u_0' - \sum_{i=1}^{N} \frac{\partial u_0'}{\partial x_i} \nabla_y \widehat{\chi}_i\Big(\frac{\cdot}{\varepsilon}\Big)\Big)\Big\|_{L^2(\Omega)} \\ \leqslant C \varepsilon \|u_0'\|_{H^2(\Omega)}. \end{aligned} \tag{14.44}$$

The constants C are independent of ε.

Proof. By the definition (3.19) of A^{hom} (in our case $b_{ij}(x,y) \equiv a_{ij}(y)$) and the definition (14.11) of the $\widetilde{d}_i$'s,

$$\sum_{i=1}^{N} \int_\Omega \widetilde{d}_i\Big(\frac{\cdot}{\varepsilon}\Big) \cdot \nabla v \, \frac{\partial u_0'}{\partial x_i}\, dx = \int_\Omega A^\varepsilon \Big[\nabla u_0' + \sum_{i=1}^{N} \frac{\partial u_0'}{\partial x_i} \nabla_y \widehat{\chi}_i\Big(\frac{\cdot}{\varepsilon}\Big)\Big] \nabla v \, dx \\ - \int_\Omega A^{hom} \nabla u_0' \, \nabla v dx.$$

Therefore,

$$\begin{aligned} &\int_\Omega A^\varepsilon \nabla \Big[u_\varepsilon - u_0' - \varepsilon \sum_{i=1}^{N} \frac{\partial u_0'}{\partial x_i} \widehat{\chi}_i\Big(\frac{\cdot}{\varepsilon}\Big) - \varepsilon \beta_\varepsilon^{\mathrm{Ne}}\Big] \nabla v \, dx \\ &\qquad = -\varepsilon \sum_{i=1}^{N} \int_\Omega A^\varepsilon \nabla \Big(\frac{\partial u_0'}{\partial x_i}\Big) \widehat{\chi}_i\Big(\frac{\cdot}{\varepsilon}\Big) \nabla v \, dx. \end{aligned} \tag{14.45}$$

which implies (14.43), as the $\widehat{\chi}_i$'s belong to $L^\infty(Y)$ and u_0' to $H^2(\Omega)$.

Since $\|\beta_\varepsilon^{\mathrm{Ne}}\|_{H^1(\Omega)} \leqslant C\varepsilon^{-1/2}\|u_0'\|_{H^2(\Omega)}$, then (14.42) is a direct consequence of (14.43).

To go further, we shall estimate the function $\rho^2 \nabla \beta_\varepsilon^{\mathrm{Ne}}$. By definition, $\rho^2 \beta_\varepsilon^{\mathrm{Ne}}$ belongs to $H_0^1(\Omega)$. From (14.28), for every $i \in \{1, \dots, N\}$,

$$\Big| \int_\Omega \widetilde{d}_i\Big(\frac{\cdot}{\varepsilon}\Big) \cdot \nabla (\rho^2 \beta_\varepsilon^{\mathrm{Ne}}) \, \frac{\partial u_0'}{\partial x_i} \, dx \Big| \\ \leqslant C \|u_0'\|_{H^2(\Omega)} \Big(\|\rho \nabla \beta_\varepsilon^{\mathrm{Ne}}\|_{L^2(\Omega)} + \|\beta_\varepsilon^{\mathrm{Ne}}\|_{L^2(\Omega)} \Big).$$

Taking $\rho^2 \beta_\varepsilon^{\mathrm{Ne}}$ as test function in problem (14.34) and using the above inequality, one gets

$$\Big| \int_\Omega A^\varepsilon (\rho \nabla \beta_\varepsilon^{\mathrm{Ne}}) \, (\rho \nabla \beta_\varepsilon^{\mathrm{Ne}}) \, dx \Big| \leqslant 2 \Big| \int_\Omega A^\varepsilon (\rho \nabla \beta_\varepsilon^{\mathrm{Ne}}) \, (\beta_\varepsilon^{\mathrm{Ne}} \nabla \rho) \, dx \Big| \\ + C \|u_0'\|_{H^2(\Omega)} \big(\|\rho \nabla \beta_\varepsilon^{\mathrm{Ne}}\|_{L^2(\Omega)} + \|\beta_\varepsilon^{\mathrm{Ne}}\|_{L^2(\Omega)} \big),$$

which implies the estimate

$$\|\rho \nabla \beta_\varepsilon^{\mathrm{Ne}}\|_{L^2(\Omega)} \leqslant C \|u_0'\|_{H^2(\Omega)}.$$

This, together with the L^2-estimate of $\beta_\varepsilon^{\mathrm{Ne}}$, yields (14.44). The proof of Theorem 14.10 is complete. □

As for the Dirichlet problem, we are now interested in the asymptotic behavior of the quantity

$$U_\varepsilon' \doteq \frac{1}{\varepsilon} \Big(u_\varepsilon - u_0' - \varepsilon \sum_{j=1}^{N} \frac{\partial u_0'}{\partial x_j} \widehat{\chi}_j\Big(\frac{\cdot}{\varepsilon}\Big) - \varepsilon \beta_\varepsilon^{\mathrm{Ne}} \Big),$$

which, by (14.43) satisfies,

$$\|U'_\varepsilon\|_{H^1(\Omega)} \leqslant C\|u'_0\|_{H^2(\Omega)}.$$

Theorem 14.11. *Under the hypotheses of Theorem 14.10, there exist u'_1 in $H^1(\Omega)$ and $\widehat{u}'_1$ in $L^2(\Omega; H^1_{per,0}(Y))$, such that*

$$\begin{aligned} &(i) \quad U'_\varepsilon \rightharpoonup u'_1 \quad \text{weakly in } H^1(\Omega),\\ &(ii) \quad \mathcal{T}_\varepsilon(U'_\varepsilon) \to u'_1 \quad \text{strongly in } L^2(\Omega; H^1(Y)),\\ &(iii) \quad \mathcal{T}_\varepsilon(\nabla U'_\varepsilon) \rightharpoonup \nabla u'_1 + \nabla_y \widehat{u}'_1 \quad \text{weakly in } L^2(\Omega\times Y)^N, \end{aligned} \tag{14.46}$$

and the pair $(u'_1, \widehat{u}'_1) \in H^1(\Omega) \times L^2(\Omega;\, H^1_{per,0}(Y))$ is the unique solution of the problem

$$\begin{cases} \dfrac{1}{|Y|}\displaystyle\int_{\Omega\times Y} A(y)\big[\nabla u'_1(x) + \nabla_y \widehat{u}'_1(x,y)\big]\big[\nabla\Psi(x) + \nabla_y\Phi(x,y)\big]dxdy \\ \qquad = -\dfrac{1}{|Y|}\displaystyle\int_{\Omega\times Y} F(x,y)\cdot\big[\nabla\Psi(x) + \nabla_y\Phi(x,y)\big]dxdy, \\ \displaystyle\int_\Omega u'_1(x)\,dx = 0, \\ \forall\Psi \in H^1(\Omega), \quad \forall\Phi \in L^2(\Omega;\, H^1_{per,0}(Y)), \end{cases} \tag{14.47}$$

where for a.e. $(x,y) \in \Omega\times Y$

$$F(x,y) = A(y)\Big(\sum_{i=1}^N \nabla\Big(\frac{\partial u'_0}{\partial x_i}\Big)(x)\widehat{\chi}_i(y)\Big).$$

Proof. By estimate (14.43), $\{U'_\varepsilon\}_\varepsilon$ is uniformly bounded in $H^1(\Omega)$. Hence, there exist functions $u'_1 \in H^1(\Omega)$ and $\widehat{u}'_1 \in L^2(\Omega; H^1_{per,0}(Y))$ such that, up to a subsequence, convergences (14.46) hold.

From (14.45), the function U'_ε satisfies, for all v in $H^1(\Omega)$,

$$\int_\Omega A^\varepsilon \nabla U'_\varepsilon\, \nabla v\, dx = -\sum_{i=1}^N \int_\Omega A^\varepsilon \nabla\Big(\frac{\partial u'_0}{\partial x_i}\Big)\widehat{\chi}_i\Big(\frac{\cdot}{\varepsilon}\Big)\,\nabla v\, dx. \tag{14.48}$$

Transforming this equality by unfolding, proceeding as in the proof of Theorem 3.5, one obtains the equation from (14.47) due to convergences (14.46). This problem has a unique solution, and consequently, convergences (14.46) hold for the whole sequence $\{(U'_\varepsilon, \mathcal{T}_\varepsilon(\nabla U'_\varepsilon)\}_\varepsilon$. □

Introduce the second order correctors for Neumann condition $\widehat{\chi}'_{ij}$ for $(i,j) \in \{1,\dots,N\}^2$. They are solutions of the following cell problems:

$$\begin{cases} \widehat{\chi}'_{ij} \in H^1_{per,0}(Y), \\ \displaystyle\int_Y A(y)\nabla_y\widehat{\chi}'_{ij}(y)\,\nabla_y\Phi(y)\,dy = -\int_Y A(y)\widehat{\chi}_i(y)\mathbf{b}_j\cdot\nabla_y\Phi(y)\,dy, \\ \forall\Phi \in H^1_{per,0}(Y). \end{cases} \tag{14.49}$$

The solution $\widehat{u}_1'$ is expressed in terms of the first correctors $\widehat{\chi}_i$ ($i \in \{1, \ldots, N\}$) defined by (3.10), the second order correctors for Neumann condition defined by (14.49), and the partial derivatives of u_1' and u_0', as follows:

$$\widehat{u}_1'(x,y) = \sum_{i=1}^{N} \frac{\partial u_1'}{\partial x_i}(x)\widehat{\chi}_i(y) + \sum_{i,j=1}^{N} \frac{\partial^2 u_0'}{\partial x_i \partial x_j} \widehat{\chi}_{ij}'(y) \quad \text{for a.e. } (x,y) \in \Omega \times Y,$$

so that u_1' is the solution of the homogenized problem,

$$\begin{cases} \displaystyle\int_\Omega A^{hom} \nabla u_1' \, \nabla \Psi \, dx = -\int_\Omega \sum_{i,j=1}^{N} \Big(\int_Y A(y)\widehat{\chi}_i(y)\mathbf{b}_j \Big) \frac{\partial^2 u_0'}{\partial x_i \partial x_j} \cdot \nabla \Psi \, dx, \\ \displaystyle\int_\Omega u_1'(x)\, dx = 0, \\ \forall \Psi \in H^1(\Omega). \end{cases}$$

Proposition 14.12. *Under the assumption of Theorem 14.11, the following strong convergences hold:*

$$\mathcal{T}_\varepsilon(\nabla U_\varepsilon) \to \nabla u_1' + \nabla_y \widehat{u}_1' \quad \textit{strongly in} \quad L^2(\Omega \times Y)^N,$$

$$U_\varepsilon - u_1' - \varepsilon \sum_{i=1}^{N} \frac{\partial u_1'}{\partial x_i} \widehat{\chi}_i\Big(\frac{\cdot}{\varepsilon}\Big) - \varepsilon \sum_{i,j=1}^{N} \mathcal{Q}_\varepsilon\Big(\frac{\partial^2 u_0'}{\partial x_i \partial x_j}\Big) \widehat{\chi}_{ij}'\Big(\frac{\cdot}{\varepsilon}\Big) \to 0 \quad \textit{strongly in } H^1(\Omega).$$

Moreover, if u_0' belongs to $H^3(\Omega)$ then

$$U_\varepsilon - u_1' - \varepsilon \sum_{i=1}^{N} \frac{\partial u_1'}{\partial x_i} \widehat{\chi}_i\Big(\frac{\cdot}{\varepsilon}\Big) - \varepsilon \sum_{i,j=1}^{N} \frac{\partial^2 u_0'}{\partial x_i \partial x_j} \widehat{\chi}_{ij}'\Big(\frac{\cdot}{\varepsilon}\Big) \to 0 \quad \textit{strongly in } H^1(\Omega).$$

Proof. The convergence of the energy,

$$\lim_{\varepsilon \to 0} \int_\Omega A^\varepsilon \nabla U_\varepsilon' \, \nabla U_\varepsilon' \, dx = \frac{1}{|Y|} \int_{\Omega \times Y} A\big[\nabla u_1' + \nabla_y \widehat{u}_1'\big]\big[\nabla u_1' + \nabla_y \widehat{u}_1'\big] dx dy, \tag{14.50}$$

$$\lim_{\varepsilon \to 0} \int_{\Lambda_\varepsilon} |\nabla U_\varepsilon'|^2 \, dx = 0$$

is proved in the same way as in the Dirichlet case. By weak lower-semi-continuity and (14.48),

$$\frac{1}{|Y|} \int_{\Omega \times Y} A\big[\nabla u_1' + \nabla_y \widehat{u}_1'\big]\big[\nabla u_1' + \nabla_y \widehat{u}_1'\big] dx dy \leqslant -\lim_{\varepsilon \to 0} \sum_{i=1}^{N} \int_\Omega A^\varepsilon \nabla\Big(\frac{\partial u_0'}{\partial x_i}\Big) \widehat{\chi}_i\Big(\frac{\cdot}{\varepsilon}\Big) \nabla U_\varepsilon' \, dx.$$

Since the $\widehat{\chi}_i$'s belong to $L^\infty(Y)$ and u_0' to $H^2(\Omega)$, it follows that

$$\Big|\sum_{i=1}^{N}\int_{\Lambda_\varepsilon} A^\varepsilon\nabla\Big(\frac{\partial u_0'}{\partial x_i}\Big)\,\widehat{\chi}_i\Big(\frac{\cdot}{\varepsilon}\Big)\,\nabla U_\varepsilon'\,dx\Big| \leqslant C\|u_0'\|_{H^2(\Omega^{bl}_{\varepsilon d(Y)})}.$$

Now, we transform by unfolding the right-hand side of the equality (14.48) written for $v = U_\varepsilon'$. From Proposition 1.9(i) and convergences (14.46), we get

$$\lim_{\varepsilon\to 0}\int_\Omega A^\varepsilon\nabla\Big(\frac{\partial u_0'}{\partial x_i}\Big)\,\widehat{\chi}_i\Big(\frac{\cdot}{\varepsilon}\Big)\,\nabla U_\varepsilon'\,dx$$
$$= \frac{1}{|Y|}\int_{\Omega\times Y} A(y)\nabla\Big(\frac{\partial u_0'}{\partial x_i}\Big)\,\widehat{\chi}_i(y)\,(\nabla u_1' + \nabla_y\widehat{u}_1')\,dxdy,$$

which proves convergences (14.50).

Finally, proceeding as in the proof of Corollary 3.10, the strong convergences from Proposition 14.12 are easily obtained. □

Correction to: Homogenization of an elastic thin plate

Correction to:
Chapter 11 in: D. Cioranescu et al., *The Periodic Unfolding Method*, Series in Contemporary Mathematics 3, https://doi.org/10.1007/978-981-13-3032-2_11

The original version of the chapter was inadvertently published without the proof for Lemma 11.11, which has now been included. The correction chapter has been updated with the change.

The updated version of this chapter can be found at
https://doi.org/10.1007/978-981-13-3032-2_11

D. Cioranescu et al., *The Periodic Unfolding Method*, Series in Contemporary Mathematics 3, https://doi.org/10.1007/978-981-13-3032-2_15

List of Figures

D. Cioranescu et al., *The Periodic Unfolding Method*, Series in Contemporary Mathematics 3, https://doi.org/10.1007/978-981-13-3032-2

Bibliography

[1] E. Acerbi, V. Chiadò Piat, G. Dal Maso and D. Percivale, An extension theorem from connected sets, and homogenization in general periodic domains. Nonlinear Anal. 18, 5 (1992), 481–496.

[2] R.A. Adams, J.J.F Founier, *Sobolev Spaces*, Academic Press, 2003.

[3] G. Allaire, Homogenization of the Navier-Stokes equations in open sets perforated with tiny holes I. Abstract framework, a volume distribution of holes. Arch. Rational Mech. Anal., 113 (1984), 209–259. September 1991, Volume 113, Issue 3, pp 209-259 — Cite as

[4] G. Allaire, Homogenization of the Stokes flow in a connected porous medium, Asymptotic Analysis, 3 (1989), 203–222.

[5] G. Allaire, Homogenization and two-scale convergence, SIAM J. Math. Anal., 23 (1992), 1482–1518.

[6] G. Allaire and M. Briane, Multiscale convergence and reiterated homogenization, Proc. Royal Soc. Edinburgh, 126 A (1996), 297–342.

[7] G. Allaire and C. Conca, Bloch wave homogenization and spectral asymptotic analysis, J. Math. Pures Appl., 2, 77 (1998), 153–208.

[8] G. Allaire, C. Conca and M. Vanninathan, Spectral asymptotics of the Helmholtz model in fluid-solid structures, International J. Numer. Methods Engrg., 9, 46 (1999), 1463–1504.

[9] G. Allaire, A. Damlamian and U. Hornung, Two-scale convergence on periodic surfaces and applications, in *Proceedings of the International Conference on Mathematical Modelling of Flow through Porous Media, 1995*, A. Bourgeat and C. Carasso eds., World Scientific Publications, Singapore,1996.

[10] G. Allaire and F. Murat, Homogenization of the homogeneous Neumann problem with nonisolated holes, including Appendix with A.K. Nandakumar, Asymptotic Analysis 7 (1993), 81–95.

D. Cioranescu et al., *The Periodic Unfolding Method*, Series in Contemporary Mathematics 3, https://doi.org/10.1007/978-981-13-3032-2

[11] N. Ansini, The nonlinear sieve problem and applications to thin films, Asymptotic Analysis, 39, 2 (2004), 113–145.

[12] N. Ansini, J.-F. Babadjian and C.-I. Zeppieri, The Neumann sieve problem and dimensional reduction: a multiscale approach, Math. Models and Methods in Appl. Sciences, 17, 5 (2007) 681–735.

[13] N. Ansini and A. Braides, Asymptotic analysis of periodically-perforated nonlinear media, J. Math. Pures Appl., 81, 5, (2002), 439-451. Erratum: J. Math. Pures Appl., 84, 5 (2005) 147–148.

[14] T. Arbogast, J. Douglas and U. Hornung, Derivation of the double porosity model of single phase flow via homogenization theory, SIAM J. Math. Anal., 21 (1990), 823–836.

[15] J.-M. Arrieta and M. Villanueva-Pesqueira, Thin domains with non-smooth periodic oscillatory boundaries. J. Math. Anal., Appl., 446, 1 (2017), 130–164.
doi: 10.1016/j.jmaa.2016.08.039

[16] H. Attouch, Famille d'opérateurs maximaux monotones et mesurabilité, Annali di Matematica pura et applicata, 5, Vol. CXX (1979), 35-111.

[17] H. Attouch, *Variational Convergence for Functions and Operators*, Pitman London 1984.

[18] H. Attouch, G. Buttazzo and G. Michaille, *Variational Analysis in Sobolev and BV Spaces: Applications to PDEs and Optimization*, Series on Optimization, SIAM-MPS, Philadelphia 2006.

[19] I. Babuška, Homogenization and its application. Mathematical and computational problems. Numerical solution of partial differential equations, III in *Proc. Third Symposium (SYNSPADE), Univ. Maryland, College Park (1975)*, Academic Press, New York, 1976, 89–116.

[20] N. Bakhvalov and G. Panasenko, *Homogenization: Averaging Processes in Periodic Media*, translated from the Russian (Nauka, Moscow, 1984) by D. Letes, Kluwer Academic Publishers Group, Dordrecht, 1989.

[21] A. Bensoussan, J.-L. Lions and G. Papanicolaou, *Asymptotic Analysis for Periodic Structures*, Stud. Math. Appl. 5, North Holland 1978.

[22] D. Blanchard, A. Gaudiello and G. Griso, Junction of a periodic family of elastic rods with a $3d$ plate. Part I. J. Math. Pures Appl., 88 (2007) 1–33.

[23] D. Blanchard, A. Gaudiello and G. Griso, Junction of a periodic family of elastic rods with a thin plate. Part II., J. Math. Pures Appl., 88 (2007), 149–190.

[24] D. Blanchard and G. Griso, Microscopic effects in the homogenization of the junction of rods and a thin plate, Asymptotic Analysis, 56, 1 (2008), 1–36.

[25] D. Blanchard and G. Griso, Decomposition of deformations of thin rods: application to nonlinear elasticity, Analysis and Applications, 7, 1 (2009), 21–71.

[26] D. Blanchard and G. Griso, Asymptotic behavior of structures made of straight rods, Journal of Elasticity, 108, 1 (2012), 85–118.

[27] D. Blanchard and G. Griso, A simplified model for elastic thin shells, Asymptotic Analysis, 76, 1 (2012), 1–33.

[28] D. Blanchard and G. Griso, Junction between a plate and a rod of comparable thickness in nonlinear elasticity, Journal of Elasticity, 112, 2 (2013), 79–109.

[29] D. Blanchard and G. Griso, Asymptotic behavior of a structure made by a plate and a straight rod, Chinese Annals of Mathematics, Series B 34, 3 (2013), 399–434.

[30] A. Blasselle and G. Griso, Mechanical modeling of the skin, Asymptotic Analysis, 74, 3-4 (2011), 167–198.

[31] A. Bossavit, G. Griso and B. Miara, Modelling of periodic electromagnetic structures. Bianisotropic materials with memory effects, J. Math. Pures Appl., 84 (2005), 819–850.

[32] G. Bouchitté and I. Fragalà, Homogenization of thin structures by two-scale method with respect to measures, SIAM J. Math. Anal., 32, 6 (2001), 1198-1226.
doi:10.1137/S0036141000370260

[33] A. Bourgeat, S. Luckhaus and A. Mikelic, Convergence of the homogenization process for a double-porosity model of immiscible two-phase flow, SIAM. J. Math. Anal., 27 (1996), 1520–1543.

[34] A. Braides and A. Defranceschi, *Homogenization of Multiple Integrals*, Oxford Lecture Series in Mathematics ans its Applications 12, Oxford University Press, 1998.

[35] A. Braides and D. Lukkassen, Reiterated homogenization of integral functionals, Math. Models and Methods in Appl. Sciences, 10, 1 (2000), 1–25.

[36] H. Brezis, *Opérateurs maximaux monotones et semi-groupes de contractions dans les espaces de Hilbert*, North Holland Mathematics Studies, 5. North Holland Amsterdam-London, American Elsevier Publishing Co. Inc., New York 1973.

[37] H. Brezis, *Functional Analysis, Sobolev Spaces and Partial Differential Equations*, Universitext series, Springer, New York Dordrecht Heidelberg London, 2011.
doi 10.1007/978-0-387-70914-7

[38] M. Briane, A. Damlamian and P. Donato, H-convergence in perforated domains, in Nonlinear Partial Diff. Eqs. and Their Applications, Collège de France Seminar vol. XIII, D. Cioranescu and J.-L. Lions eds., 62–100, Pitman Research Notes in Mathematics Series, Longman, New York 1998.

[39] R. Brizzi and J.-P. Chalot, Boundary homogenization and Neumann boundary value problem, Ricerche Mat., 46 (1997), 341–387.

[40] DAG Bruggeman, Berechnung verschiedener physikalisher konstanten von heterogenen substanzen, Ann. Physik. 24 (1935), 634.

[41] B. Cabarrubias and P. Donato, Homogenization of a quasilinear elliptic problem with nonlinear Robin boundary conditions, Applicable Anal., 91, 6 (2011), 1–17.
https://doi.org/10.1080/00036811.2011.619982

[42] B. Cabarrubias and P. Donato , Homogenization of some evolution problems in domains with small holes, Electron. J. Diff. Equ., 169 (2016), 1–26.

[43] D. Caillerie, Equations aux dérivées partielles à coefficients périodiques dans des domaines cylindriques aplatis, C. R. Acad. Sci. Paris, Série 1, 290 (1980), 143–146.

[44] D. Caillerie, Thin elastic and periodic plates, Math. Models and Methods in Appl. Sciences, 6, 1 (1984), 159-191.

[45] L. Carbone and R. De Arcangelis, *Unbounded Functionals in the Calculus of Variations. Representation, Relaxation, and Homogenization*, Chapman &Hall/CRC 2001.

[46] L. Carbone and C. Sbordone, Some properties of Γ-Limits of integral functionals, Ann. Mat. Pura Appl. 4, 122 (1979), 1–60.

[47] J. Casado-Díaz, Two-scale convergence for nonlinear Dirichlet problems in perforated domains, Proc. Royal Soc. Edinburgh, 130 A (2000), 249–276.

[48] J. Casado-Díaz, M. Luna-Laynez, A multiscale method to the homogenization of elastic thin reticulated structures. Homogenization, 2001 (Naples), 155–168, Gakuto Internat. Ser. Math. Sci. Appl., 18, Gakkōtosho, Tokyo, 2003.

[49] J. Casado-Díaz, M. Luna-Laynez and J. D. Martín, An adaptation of the multi-scale methods for the analysis of very thin reticulated structures, C. R. Acad. Sci. Paris, Série 1, 332 (2001), 223–228.

[50] J. Casado-Díaz, M. Luna-Laynez and J. D. Martín, A new approach to the analysis of thin reticulated structures, in *Homogenization, 2001 (Naples)*, 257–262, Gakuto Internat. Ser. Math. Sci. Appl., 18, Gakkōtosho, Tokyo, 2003.

[51] C. Castaing and M. Valadier, *Convex Analysis and Measurable Multifunctions*, Lecture Notes in Math. 580, Springer, 1977.

[52] G. Chechkin, Homogenization in perforated domains, in *Topics on concentration phenomena and problems with multiple scales*, Lect. Notes Unione Mat. Ital., 2, Springer, Berlin, 2006.

[53] G. Chechkin, D. Cioranescu, A. Damlamian, and Andrey L. Piatnitski, On boundary value problem with singular inhomogeneity concentrated on the boundary, J. Math. Pures Appl., 98 (2012), 115–138.

[54] D. Chenais, On the existence of a solution in a domain identification problem, Journal Math. Anal. and Appl., 52 (1975), 189-219.

[55] V. Chiadò Piat and A. Piatnitski, Γ-convergence approach to variational problems in perforated domains with Fourier boundary conditions, ESAIM: COCV, 16 (2010), 148–175.

[56] P.-G. Ciarlet, *The finite element method for elliptic problems* Studies in Mathematics and its Applications, Vol. 4, North-Holland Publishing Co., Amsterdam-New York-Oxford, 1978.

[57] P. Ciarlet & P.-G. Ciarlet, Another approach to linearized elasticity and a new proof of Korn's inequality, Math. Models and Methods in Appl. Sciences, 15, 2 (2005) 259–271.

[58] P.-G. Ciarlet and P. Destuynder, A justification of the two-dimensional linear plate model, J. Mécanique, 2, 18 (1979), 315–344.

[59] D. Cioranescu, A. Damlamian, P. Donato, G. Griso and R. Zaki, The periodic unfolding method in domains with holes, SIAM J. Math. Anal., 44, 2 (2012), 718–760.

[60] D. Cioranescu, A. Damlamian and G. Griso, Periodic unfolding and homogenization, C. R. Acad. Sci. Paris, Série 1, 335 (2002), 99–104.

[61] D. Cioranescu, A. Damlamian and G. Griso, The Stokes problem in perforated domains by the periodic unfolding method. in *Proceedings Conference New Trends in Continuum Mechanics*, Ed. M. Suliciu, Theta , Bucarest, 2005, 67–80.

[62] D. Cioranescu, A. Damlamian and G. Griso, The periodic unfolding method in homogenization, SIAM J. of Math. Anal., 40, 4 (2008), 1585–1620.

[63] D. Cioranescu, A. Damlamian, G. Griso and D. Onofrei, The periodic unfolding method for perforated domains and Neumann sieve models, J. Math. Pures Appl., 89 (2008), 248–277.

[64] D. Cioranescu, A. Damlamian and R. De Arcangelis, Homogenization of nonlinear integrals via the periodic unfolding method, C. R. Acad. Sci. Paris, Série 1, 339 (2005), 77–82.

[65] D. Cioranescu, A. Damlamian and R. De Arcangelis, Homogenization of integrals with pointwise gradient constraints via the periodic unfolding method, Ricerche di Matematica, 55, 1 (2006), 31–54.

[66] D. Cioranescu, A. Damlamian and R. De Arcangelis, Homogenization of quasiconvex integrals via the periodic unfolding method, SIAM J. of Math. Anal., 37, 5 (2006), 1435–1453.

[67] D. Cioranescu, A. Damlamian and Li Tatsien, The homogenization of the elastic torsion problem and the related conductors problem, in *Partial Differential Equations: Theory, Control and Approximation. In Honor of the Scientific Heritage of Jacques-Louis Lions*, Springer-Verlag Berlin Heidelberg 2014, 718–760.

[68] D. Cioranescu, A. Damlamian and J. Orlik, Homogenization via unfolding in periodic elasticity with contact on closed and open cracks, Asymptotic Analysis, 82 (2013) 201–232.
doi 10.3233/ASY-2012-1141

[69] D. Cioranescu and P. Donato, Homogénéisation du problème de Neumann non homogène dans des ouverts perforés, Asymptotic Analysis, 1 (1988), 115–138.

[70] D. Cioranescu and P. Donato, *An Introduction to Homogenization.* Oxford Lecture Series in Mathematics ans its Applications 17, Oxford University Press, 1999.

[71] D. Cioranescu, P. Donato and H. Ene, Homogenization of the Stokes problem with non homogeneous slip boundary conditions, Mathematical Methods in Applied Sciences, 19 (1996), 857–881.

[72] D. Cioranescu, P. Donato and R. Zaki, The periodic unfolding method in perforated domains, Portugaliae Mathematica, 4 (2006), 467–496.

[73] D. Cioranescu, P. Donato and R. Zaki, Asymptotic behavior of elliptic problems in perforated domains with nonlinear boundary conditions, Asymptotic Analysis, 53, 4 (2007), 209–235.

[74] D. Cioranescu and F. Murat, Un terme étrange venu d'ailleurs, in *Nonlinear partial differential equations and their applications, Collège de France Seminar*, II & III, ed. H. Brezis and J.L. Lions, Research Notes in Math., 60-70, Pitman, Boston 1982, 98–138, 154–178.

[75] D. Cioranescu and F. Murat, A strange term coming from nowhere, *Topics in the Mathematical Modelling of Composite Materials*, A. Cherkaev and R. Kohn editors, Progress in Nonlinear Differential Equations Appl., 31, Birkhäuser-Verlag, Boston 1997, 45–88.

[76] D. Cioranescu and A. Ould Hammouda, Homogenization of elliptic problems in perforated domains with mixed boundary conditions, Rev. Roumaine Math. Pures Appl., 53, 5-6 (2008), 389–406.

[77] D. Cioranescu and J. Saint Jean Paulin, Homogenization in open sets with holes, J. Mathematical Anal. and Appl., 71 (1979), 590–607.

[78] D. Cioranescu and J. Saint Jean Paulin, *Homogenization of Reticulated Structures*, Applied Mathematical Sciences 136, Springer-Verlag New York, 1999.

[79] C. Conca and P. Donato, Non-homogeneous Neumann problems in domains with small holes, RAIRO Model. Math. Anal. Numer., 22, 4 (1988), 561–607.

[80] A. Corbo Esposito, C. D'Apice and A. Gaudiello, Homogenization in a perforated domain with both Dirichlet and Neumann boundary conditions on the holes, Asymptotic Analysis, 2002

[81] A. Corbo Esposito, P. Donato, A. Gaudiello and C. Picard, Homogenization of the p-Laplacian in a domain with oscillating boundary, Commun. Appl. Nonlinear Anal., 4, (1997), 1–23.

[82] G. Dal Maso, *An Introduction to* Γ*-Convergence*, Birkhäuser Verlag, 1993.

[83] G. Dal Maso and F. Murat, Asymptotic behavior and correctors for linear Dirichlet problems with simultaneously varying operators and domains, Ann. Inst. H. Poincaré, Analyse Nonlinéaire, 21 (2004), 445–486.

[84] A. Damlamian, Convergence et dualité des fonctions convexes dans les espaces de Banach réflexifs, unpublished, 1975.

[85] A. Damlamian, Le problème de la passoire de Neumann, Rend. Sem. Mat. Univ. Politec. Torino, 43, 3 (1985), 427–450.

[86] A. Damlamian, An elementary introduction to periodic unfolding, in *Proc. of the Narvik Conference 2004*, A. Damlamian, D. Lukkassen, A. Meidell, A. Piatnitski edts, Gakuto Int. Series, Math. Sci. Appl., vol. 24, Gakkōkotosho (2006), 119–136.

[87] A. Damlamian, From linear to nonlinear by the homogenizations of cracks, in *Current advances in nonlinear analysis and related topics*,T.Aiki, N.Kenmochi, M.Niezgódka, M.Ôtani edts,Gakuto Internat. Ser. Math. Sci. Appl., 32, Gakkotosho, Tokyo, 2010, 25-40.

[88] A. Damlamian and P. Donato, Which sequences of holes are admissible for periodic homogenization with Neumann boundary condition?, ESAIM : Control, Optimization and Calculus of Variations. Special Volume: A Tribute to Jacques-Louis Lions, 8 (2002), 555–585.

[89] A. Damlamian and P. Donato, H^0-convergence and iterated homogenization, Asymptotic Analysis, 39 (1) (2004), 45–60.

[90] A. Damlamian and N. Meunier, The "strange term" in the periodic homogenization for multivalued Leray–Lions operators in perforated domains, Ricerche mat.,59 (2010), 281–312
doi: 10.1007/s11587-010-0087-4.

[91] A. Damlamian, N. Meunier and J. Van Schaftingen, Periodic homogenization for monotone multivalued operators, Nonlinear Analysis, 67 (2007), 3217–3239.

[92] A. Damlamian, N. Meunier and J. Van Schaftingen, Periodic homogenization for convex functionals using Mosco convergence, J. Ricerche Mat., 57 (2008), 209–249.
doi: 10.1007/s11587-008-0038-5

[93] A. Damlamian and K. Pettersson, Homogenization of oscillating boundaries, Discrete and Continuous Dynamical Systems, 23, 1&2 (2009), 197–219.
doi: 10.3934/dcds.2009.23.197

[94] A. Damlamian and M. Vogelius, Homogenization limits of the equations of elasticity in thin domains, SIAM J. Math. Anal., 18, 2 (1987), 435–451.

[95] E. De Giorgi, Sulla convergenza di alcune successioni d'integrali del tipo dell'area, Rend. Mat., 6, 8(1975), 277–294.

[96] E. De Giorgi and S. Spagnolo, Sulla convergenza degli integrali dell'energia per operatori ellittici del secondo ordine, Boll. Un. Mat. Ital., 4, 8 (1973), 391–411.

[97] C.-J. de la Vallée-Poussin, *Intégrales de Lebesgue, Fonctions d'ensemble, Classe de Baire, Paris, 1916.*

[98] T. Del Vecchio, The thick Neumann's sieve, Ann. Mat. Pura. Appl., 4, 147 (1987), 363–402.

[99] P. Donato, A. Gaudiello and L. Sgambati, Homogenization of bounded solutions of elliptic equations with quadratic growth in periodically perforated domains, Asymptotic Analysis, 16 (1998), 223–243.

[100] P. Donato, O. Guibé and A. Oropeza, Homogenization of quasilinear elliptic problems with nonlinear Robin conditions and L^1 data, in press in J. Math. Pures Appl., Available on: https://doi.org/10.1016/j.matpur.2017.10.002

[101] P. Donato and K. H. Le Nguyen, Homogenization of diffusion problems with a nonlinear interfacial resistance, Nonlinear Differ. Equ. Appl. 22 (2015), 1345–1380. DOI 10.1007/s00030-015-0325-2

[102] P. Donato, H. Le Nguyen and R. Tardieu, The periodic unfolding method for a class of imperfect transmission problems, Journal of Math. Sci. , 176, 6 (2011), 891–927.

[103] P. Donato and Z.Yang, The periodic unfolding method for the wave equation in domains with holes, Adv. Math. Sci. Appli., 22, 2 (2012), 521–551.

[104] P. Donato and Z.Yang, The periodic unfolding method for the heat equation in perforated domains, Sci. China Math, 59 (2016), 891–906. doi: 10.1007/s11425-015-5103-4

[105] N. Dunford and J. T. Schwartz, *Linear Operators. Part I: General Theory*, Wiley Classics Lib., Wiley Interscience, New. York 1988.

[106] A. Ene and J. Saint Jean Paulin, On a model of fractured porous media, in *Mathematical modelling of flow through porous media* Proc. Conference St. Etienne, A. Bourgeat, C. Carasso, S. Luckhaus and A. Mikelic eds., 402–409, World Scientific, 1995.

[107] H. Ene and E. Sanchez-Palencia, Equation et phénomènes de surface pour l'écoulement dans un modèle de milieu poreux, Journal Mécanique, 14 (1975), 73–108.

[108] G. Francfort, F. Murat and L. Tartar, Homogenization of monotone operators in divergence form with x-dependent multivalued graphs, Annali di Matematica, 188 (2009), 631–652.
doi: 10.1007/s10231-009-0094-9

[109] A. Gaudiello, O. Guibét and F. Murat, Homogenization of the brush problem with a source term in L^1, Arch. Rational Mech. Anal., 225 (2017), 1–64.
doi: 10.1007/s00205-017-1079-2,

[110] G. Geymonat, F. Krasucki and J.J. Marigo, Sur la commutativité des passages à la limite en théorie asymptotique des poutres composites, C. R. Acad. Sci. Paris, Série 1, 305, 2 (1987), 225–228.

[111] D. Gérard-Varet and N. Masmoudi, Homogenization in polygonal domains, J. Eur. Math. Soc., 13 (2010), 1477–1503.

[112] D. Gérard-Varet and N. Masmoudi, Homogenization and boundary layers, Acta Math., 209, 1 (2012), 133–178.

[113] M. Ghergu, G. Griso, H. Mechkour and B. Miara, Homogenization of thin piezoelectric perforated shells, M2AN Math. Model. Numer. Anal., 41, 5 (2007), 875–895.

[114] D. Gilbarg and N. S. Trudinger, *Elliptic Partial Differential Equations of Second Order*, Springer Verlag, Heldelberg, New York, 1997.

[115] J. Gobert, Une inégalité fondamentale de la théorie de l'Elasticité, Bull. Soc. Roy. Sci. Liège, 31 (1962), 182–191.

[116] G. Griso, Analyse asymptotique de structures réticulées. Thèse Université Pierre et Marie Curie (Paris VI), 1996.

[117] G. Griso, Asymptotic behavior of structures made of plates, Analysis and Applications, 3, 04 (2004), 325–356.

[118] G. Griso, Asymptotic behavior of curved rods by the unfolding method, Math. Methods in the Applied Sciences, 27, 17 (2004), 2081–2110.

[119] G. Griso, Thin reticulated structures, in *Progress in Partial Differential Equations. The Metz Surveys 3,* ed. M. Chipot, J. Saint Jean Paulin and I. Shafrir, Pitman (London), 1994, 161–182.

[120] G. Griso, Error estimate and unfolding for periodic homogenization, Asymptotic Analysis, 40, 3-4 (2004), 269–286.

[121] G. Griso, Les méthodes d'éclatement en homogénéisation périodique et en élasticité linéarisée. Thèse d'Habilitation, Université Pierre et Marie Curie, 2005.

[122] G. Griso, Interior error estimates for periodic homogenization, Analysis and Applications, 4 (2006), 61–79.

[123] G.Griso, Asymptotic behavior of structures made of curved rods, Analysis and Applications, 6, 1 (2008), 11-22.

[124] G. Griso, Decomposition of displacements of thin structures, J. Math. Pures Appl., 9, 89 (2008), 199–233.

[125] G. Griso, Error estimates in periodic homogenization with a non-homogeneous Dirichlet condition, Asymptotic Analysis, 87, 1-2 (2014), 91–121.

[126] G. Griso and E. Rohan, On homogenization of diffusion-deformation problem in strongly heterogeneous media, Ricerche Mat., 56, 2 (2007), 161–188.

[127] G. Griso and M.-T. Sanchez, Homogenization of an elasticity problem for a catalyst support by using the unfolding method, Asymptotic Analysis, 67 (2010), 45–84.

[128] G. Griso and M. Villanueva-Pesqueira, Straight rod with different order of thickness, Asymptotic Analysis, 94, 3-4 (2015), 255–291.

[129] G. Griso and L. Merzougui, Junctions between two plates and a family of beams, Math. Models and Methods in Appl. Sciences, 41, 1 (2018), 58-79.
doi: 10.1002/mma.4594

[130] G. Griso, A. Migunova and J. Orlik, Asymptotic analysis for domains separated by a thin layer made of periodic vertical beams, Journal of Elasticity, 128, 2 (2017), 291–331.

[131] G. Griso and B. Miara, Homogenization of periodically heterogeneous thin beams, Chin. Ann. Math., Series B, 39, 3 (2018), 397–426.
doi: 10.1007/s11401-018-0075-7

[132] P. Grisvard, *Singularities in boundary value problems*, Masson, Paris, 1992.

[133] Z. Hashin and S. Shtrikman, A variational approach to the theory of the elastic behaviour of multiphase materials, J. Mech. Phys. Solids, 11 (1963), 127–140.

[134] W. Jäger and A. Mikelič, On the boundary conditions at the contact interface between a porous medium and a free fluid, Annali Sc. Norm. Sup. Pisa, 3 (1996), 403–465.

[135] W. Jäger and A. Mikelič, Homogenization of the Laplace equation in a partially perforated domain, in *Homogenization*, Ser. Adv. Math. Appl. Sci., 50, World Sci. Publ., 1999.

[136] W. Jäger and A. Mikelič, On the interface boundary condition of Beavers, Joseph and Saffman, SIAM J. Appl. Math., 60, 4 (2000), 1111–1127.

[137] V. Jikov, S. Kozlov and O. Oleinik, *Homogenization of Differential Operators and Integral Functionals*, translated from the Russian by G. A. Yosifian. Springer-Verlag, Berlin, 1994.

Under V. Jikov, see also V. Zhikov [191–193].

[138] C. Kenig and Z. Shen, Homogenization of elliptic boundary value problems in Lipschitz domains, Math. Ann., 350 (2011), 867–917.

[139] C. Kenig and F. Lin and Z. Shen, Convergence rates in L^2 for elliptic homogenization problems, Arch. Rational Mech. Anal., 203, 3 (2012), 1009–1036.

[140] C. Kenig and F. Lin and Z. Shen, Homogenization of elliptic systems with Neumann boundary conditions, J. Amer. Math. Soc., 26 (2013), 901–937.

[141] D. Kinderlehrer and G. Stampacchia, *An Introduction to Variational Inequalities and Their Applications*, Academic Press, 1980.

[142] S. Kaizu, The Poisson equation with semilinear boundary conditions in domains with many tiny holes, J. Faculty of Science, Univ.Tokyo. Sect. 1 A, 36, 1 (1989), 43–64.

[143] L. Kuipers and H. Niederreiter, *Uniform Distribution of Sequences*, Wiley, New York, 1974.

[144] M. Lenczner, Homogénéisation d'un circuit électrique, C. R. Acad. Sci. Paris, Série 2, 324 (1997), 537–542.

[145] M. Lenczner and D. Mercier, Homogenization of periodic electrical networks including voltage to current amplifiers, SIAM Multiscale Model. Simul., 2, 3 (2004), 359–397.

[146] M. Lenczner and G. Senouci-Bereksi, Homogenization of electrical networks including voltage-to-voltage amplifiers, Math. Models and Methods in Appl. Sciences, 9, 6 (1999), 899–932.

[147] M. Lenczner, M. Kader M. and P. Perrier, Modèle à deux échelles de l'équation des ondes à coefficients oscillants, C. R. Acad. Sci. Paris, Série 1, 328 (2000), 335–340.

[148] J. Leray and J.-L. Lions, Quelques résultats de Višik sur les problèmes elliptiques nonlinéaires par les méthodes de Minty-Browder, Bull. Soc. Math. France, 93 (1965), 97–107.

[149] J.-L. Lions, Remarks on Darcy's law, IMA, Journal of Appl. Math., 46 (1991), 29–38.

[150] J.-L. Lions, D. Lukkassen, L. E. Persson and P. Wall, Reiterated homogenization of monotone operators, Chin. Ann. Math., Series B, 22, 1 (2001), 1–12.

[151] J.-L. Lions and E. Magenes, *Problèmes aux limites non homogènes et applications*, vol. I, Dunod, Paris, 1968.

[152] R. Lipton and M. Avellaneda, Darcy's law for slow viscous flow past a stationary array of bubbles, Proc. Royal Soc. Edinburgh, 114 A (1990), 71–79.

[153] D. Lukkassen and G. Milton, On hierarchical structures and reiterated homogenization, in *Proc. Conf. Function Spaces, Interpolation Theory and Related Topics in Honour of J. Peetre on his 65th Birthday 2000*, Walter de Gruyter, Berlin 2002, 311–324.

[154] D. Lukkassen, G. Nguetseng and P. Wall, Two-scale convergence, International J. of Pure Appl. Math., 2, 1 (2002), 35–86.

[155] P. Marcellini, Periodic solutions and homogenization of non linear variational problems, Ann. Mat. Pura Appl., 4, 117 (1978), 139–152.

[156] V. Marchenko and E. Khruslov, *Homogenization of Partial Differential Equations* Translated from the 2005 Russian original by M. Goncharenko and D. Shepelsky, Progress in Mathematical Physics, 46, Birkhääuser Boston, Inc., Boston 2006.

[157] O. Martio, John domains, bi-Lipschitz balls and Poincaréé inequality. Rev. Roumaine Math. Pures Appl., 33, 1-2 (1988), 107–112.

[158] C.C. Mei and B. Vernescu, *Homogenization Methods for Multiscale Mechanics*, World Scientific Publishing Co., NJ, 2010.

[159] G. Minty, Monotone (nonlinear) operators in a Hilbert space, Duke Math. J., 29 (1962), 341–346.

[160] G. Milton, *The Theory of Composites*, Cambridge University Press, Cambridge, 2002.

[161] C. B. Morrey, *Multiple Integrals in the Calculus of Variations*, Grundlehren Math. Wiss., 130, Springer-Verlag, Berlin 1966.

[162] U. Mosco, Convergence of convex sets and of solutions of variational inequalities, Advances in Math., 3 (1969), 510–585.

[163] F. Murat, The Neuman sieve, in *Nonlinear Variational Problems , Isola d'Elba 1983*, Research Notes in Math., 127, Pitman, Boston 1985, 24–32.

[164] F. Murat and L. Tartar, H-Convergence, in *Topics in the Mathematical Modelling of Composite Materials*, A. Cherkaev and R. Kohn editors, Progress in Nonlinear Differential Equations Appl., 31, Birkhäuser-Verlag, Boston 1997, 21–44.

[165] M. Neuss-Radu, Some extensions of two-scale convergence, C. R. Acad. Sci. Paris, Série 1, 322 (1996), 899–904.

[166] G. Nguetseng, Problemes d'écrans perforés pour l'équation de Laplace, RAIRO Model. Math. Anal. Numer., 19, 1 (1985), 33–63.

[167] G. Nguetseng, A general convergence result for a functional related to the theory of homogenization, SIAM J. Math. Anal., 20 (1989), 608–629.

[168] O. Oleinik, A. Shamaev and G.Yosifian, *Mathematical Problems in Elasticity and Homogenization*, Studies in Mathematics and its Applications, 26, North-Holland, Amsterdam, 1992.

[169] D. Onofrei, The unfolding operator near a hyperplane and its application to the Neumann Sieve model, AMSA, 16 (2006), 239–258.

[170] D. Onofrei and B. Vernescu, Asymptotics of a spectral problem associated with the Neumann Sieve, Analysis and Applications, 3, 1 (2005), 69–87.

[171] D. Onofrei and B. Vernescu, Error estimates for periodic homogenization with non-smooth coefficients, Asymptotic Analysis, 94 (3-4) (2015), 255–291. Asymptotic Analysis, 54, 1-2 (2007), 103–123.

[172] D. Onofrei and B. Vernescu, Asymptotic analysis of second-order boundary layer correctors, Applicable Anal., 91, 6 (2012), 1097–1110.

[173] A. Ould Hammouda, Homogenization of a class of Neumann problems in perforated domains, Asymptotic Analysis 71 (1-2), (2011), 33–57.

[174] A. Ould-Hammouda and R. Zaki, Homogenization of a class of elliptic problems with nonlinear boundary conditions in domains with small holes, Carpathian Journal Math., 31 (1), (2015), 77–88

[175] A. Pankov, *G-convergence and homogenization of nonlinear partial differential operators*, Mathematics and its Applications, 422, Kluwer Academic Publishers, Dordrecht, 1997.

[176] C. Picard, Analyse limite d'équations variationelles dans un domaine contenant une grille, RAIRO Model. Math. Anal. Numer., 21, 2 (1987), 293–326.

[177] M. Ptashnyk, Locally periodic unfolding method and two-scale convergence on surfaces of locally periodic microstructures, SIAM Journal on Multiscale Modeling and Simulation (2014).
doi: 10.1137/140978405

[178] P.-A. Raviart and J.-M. Thomas, *Introduction à l'analyse numérique des équations aux dérivées partielles*, Mathématiques Appliqués pour la Maîtrise, Masson, Paris, 1983.

[179] E. Sanchez-Palencia, *Non Homogeneous Media and Vibration Theory*, Lectur Notes in Physics, Springer Verlag, 1980.

[180] E. Sanchez-Palencia, Boundary value problems in domains containing perforated walls, in *Nonlinear partial differential equations and their applications, Collège de France Seminar*, III (Paris, 1980-1981), Res. Notes in Math., 70, Pitman, Boston 1982 , 309–325.

[181] E. Sanchez-Palencia, Perturbation of eigenvalues in thermoelasticity and vibration of systems with concentrated masses, in: *Trends and Appl. of Pure Math. to Mechanics*, Lecture Notes in Physics, 195, Springer Berlin, 1984, 346–368.

[182] E. Sanchez-Palencia and H. Tchatat, Vibration des systèmes élastiques avec masses concentrées, Rendiconti Rend. Sem. Mat. Univ. Politec. Torino, 42, 3 (1984), 43–63.

[183] J. Simon, Representation of distributions with explicit antiderivatives up to the boundary, in *Progress in Partial Differential Equations. The Metz Surveys 2,* (1992), Pitman, Longman Sci. Tech., Harlow, 1993, 201–215.

[184] E. Stein, *Singular Integrals and Differentiability Properties of Functions*, Princeton University Press, 1971.

[185] L. Tartar, Incompressible flow in a porous medium–convergence of the homogenization process, Appendix to [179].

[186] L. Tartar, *The General Theory of Homogenization. A Personalized Introduction*, Lecture Notes of the Unione Matematica Italiana, 7, Springer-Verlag, Berlin 2009.

[187] R. Temam, *Navier-Stokes Equations*, North Holland, 1984.

[188] M. Višik, Boundary value problems for quasilinear strongly elliptic systems of equations having divergence form, (in Russian), Dokl. Akad. Nauk SSSR, 138 (1961), 518–521.

[189] M. Višik, Trudy Moskovskogo matematičeskogo Obščestva, (in Russian), 12 (1963), 125–184.

[190] R. Zaki, Homogenization of a Stokes problem in a porous medium by the periodic unfolding method, Asymptotic Analysis, 79 (2012), 229–250.

[191] V. Zhikov, Connectedness and homogenization, examples of fractal conductivity, Mat. Sbornik, 187, 8 (1996), 3–40.

[192] V. Zhikov, On an extension of the method of two-scale convergence and its applications, Mat. Sbornik, 191, 7 (2000), 31–72 (Russian version); Sbornik Mathematics 191 (7) (2000), 973–1014 (English version). doi: 10.1070/SM2000v191n07ABEH000491

[193] V. Zhikov, On some estimates of homogenization theory, Dokl. Ros. Akad. Nauk, 406, 5 (2006), 597–601 (Russian version); Dokl. Math., 73 (2006), 96–99 (English version).

Under V. Zhikov, see also V. Jikov [137].

Notations and Symbols

Symbol	Description
p'	The conjugate of $p \in [1, +\infty]$, $\frac{1}{p} + \frac{1}{p'} = 1$ (for Lebesgue and Sobolev spaces)
p^*	The Sobolev conjugate of $p \in [1, N)$, $\frac{1}{p^*} = \frac{1}{p} - \frac{1}{N}$ (from the Sobolev embedding theorems)
$L^p(\mathcal{O})$	The real valued Lebesgue space for the measure space $\mathcal{O}$ ($p \in [1, +\infty]$)
$L^p(\mathcal{O}; \mathcal{X})$	The $\mathcal{X}$-valued Lebesgue space for the measure space $\mathcal{O}$ and the Banach space $\mathcal{X}$
$\mathcal{D}(\mathcal{O})$	The space of real-valued infinitely differentiable functions with compact support in the domain $\mathcal{O}$
$W^{s,p}(\mathcal{O})$	The Sobolev space for the domain $\mathcal{O} \subset \mathbb{R}^N$ ($p \in [1, +\infty]$, $s \in (0, +\infty)$; $H^s(\mathcal{O})$ for $p = 2$
$W_0^{1,p}(\mathcal{O}; \Gamma)$	The subspace of $W^{1,p}(\mathcal{O})$ whose elements vanish on the subset Γ of $\overline{\mathcal{O}}$ ($p \in [1, +\infty]$; $H_0^1(\mathcal{O}; \Gamma)$ for $p = 2$
$W^{s,p}(\mathcal{O}; \mathcal{X})$	The vector-valued Sobolev space for the domain $\mathcal{O} \subset \mathbb{R}^N$ and the Banach space $\mathcal{X}$ ($p \in [1, +\infty]$, $s \in (0, +\infty)$; $H^s(\mathcal{O}; \mathcal{X})$ for $p = 2$
$\mathbf{1}_{\mathcal{O}}$	The characteristic function of the set $\mathcal{O}$
$\lvert \mathcal{O} \rvert$	The Lebesgue measure of the measurable set $\mathcal{O}$ in $\mathbb{R}^N$
$\Subset$	$\omega \Subset \mathcal{O}$: ω is a subset of $\mathcal{O}$ such that $\overline{\omega}$ is compact in $\mathcal{O}$

D. Cioranescu et al., *The Periodic Unfolding Method*, Series in Contemporary Mathematics 3, https://doi.org/10.1007/978-981-13-3032-2

$\mathbf{G}$	The group of macroscopic periods	6, 156
$M(\alpha, \beta, \mathcal{O})$	A set of matrix field on the domain $\mathcal{O}$	100
$\mathcal{M}_{\mathcal{O}}$	The average operator over the domain $\mathcal{O}$	81
$\mathbb{M}_{\mathbf{a}}$	The average operator over the line segment $\mathbf{a}$	44
$\mathcal{M}_\varepsilon$	The local average operator in fixed domains	17
$\mathcal{M}_\varepsilon^*$	The local average operator in perforated domains	163
$\mathcal{M}_\varepsilon^b$	The local boundary average operator in perforated domains	184
$\mathcal{M}_\varepsilon^{bl}, \mathcal{M}_\varepsilon^{bl\pm}$	The boundary layer local average operator in domains with small holes and its one-sided versions	292, 294
$\widehat{\Omega}_\varepsilon, \Lambda_\varepsilon$	The largest union of ε-cells included in the domain Ω, its complement	7
Ω_ε^*	The domain Ω periodically perforated by holes of size ε	157
$\widehat{\Omega}_\varepsilon^*, \Lambda_\varepsilon^*$	The largest union of perforated ε-cells included in the domain Ω, its complement in Ω_ε^*	157
$\Omega_{\varepsilon,\delta}^*$	The domain Ω ε-periodically perforated by holes of size $\varepsilon\delta$	289
$\Omega_\delta^{int}, \Omega_\delta^{bl}, \Omega_\delta^b$	Sets derived from the domain Ω	34, 123, 405
$\widetilde{\mathcal{Q}}_\varepsilon, \mathcal{Q}_\varepsilon^o, \mathcal{Q}_\varepsilon$	The $\mathbf{Q}_1$-interpolating operators in fixed domains	45, 56, 57
$\mathcal{Q}_\varepsilon^*$	The $\mathbf{Q}_1$-interpolating operators in perforated domains	175
$\widetilde{\mathcal{R}}_\varepsilon, \mathcal{R}_\varepsilon$	The remainder scale-splitting operators in fixed domains	52, 57
$\mathcal{R}_\varepsilon^*$	The remainder scale-splitting operators in perforated domains	178
$\Sigma_\varepsilon, \Sigma_\varepsilon^\pm$	The ε-boundary layers	291, 294
$\mathcal{T}_\varepsilon$	The unfolding operator in fixed domains	7
$\mathcal{T}_\varepsilon^Y, \mathcal{T}_\delta^Z$	The iterated unfolding operators in fixed domains	38, 40
$\mathcal{T}_\varepsilon^*$	The unfolding operator in perforated domains	160
$\mathcal{T}_\varepsilon^b$	The boundary unfolding operator in perforated domains	183
$\mathcal{T}_{\varepsilon,\delta}$	The unfolding operator in domains with small holes	289
$\mathcal{T}_{\varepsilon,\delta}^{bl}$	The unfolding operator for boundary layers with small holes	292, 345
$\widehat{\mathcal{T}}_\varepsilon$	The partial unfolding operator with artificial periods	260
$\mathcal{T}_\varepsilon^A$	The partial unfolding operator in converging domains	270
$\mathfrak{T}_\delta$	The rescaling operator for thin plates	363
$\mathfrak{T}_{\varepsilon,\delta}$	The rescaling-unfolding operator for thin plates	367
$\mathcal{T}_\varepsilon^\#$	The unfolding operator for a flat boundary layer	448
$\mathcal{U}_\varepsilon$	The averaging operator in fixed domains	20
$\mathcal{U}_\varepsilon^*$	The averaging operator in perforated domains	164
$\mathcal{U}_\varepsilon^b$	The boundary averaging operator in perforated domains	184

Index

D. Cioranescu et al., *The Periodic Unfolding Method*, Series in Contemporary Mathematics 3, https://doi.org/10.1007/978-981-13-3032-2

The manufacturer's authorised representative in the EU is Springer
Nature Customer Service Centre GmbH, Europaplatz 3, 69115 Heidelberg,
Germany. If you have any concerns regarding our products, please
contact ProductSafety@springernature.com

Printed and bound by CPI Group (UK) Ltd, Croydon, CR0 4YY
25/09/2026
02222884-0014